Selected Titles in This Series

Volume

2.2 William H. Kazez, Editor
Geometric Topology
1997

2.1 William H. Kazez, Editor
Geometric Topology
1997

1 B. Greene and S.-T. Yau, Editors
Mirror Symmetry II
1997

Geometric Topology

AMS/IP

Studies in Advanced Mathematics

Volume 2, Part 1

Geometric Topology

1993 Georgia International
Topology Conference
August 2–13, 1993
University of Georgia
Athens, Georgia

William H. Kazez, Editor

American Mathematical Society · International Press

Shing-Tung Yau, Managing Editor

1991 *Mathematics Subject Classification*. Primary 57–02, 57–06.

Library of Congress Cataloging-in-Publication Data

Georgia International Topology Conference (1993 : University of Georgia)
Geometric topology : 1993 Georgia International Topology Conference, August 2–13, 1993, University of Georgia, Athens, Georgia / William H. Kazez, editor.
p. cm.—(AMS/IP studies in advanced mathematics, ISSN 1089-3288; v. 2)
Includes bibliographical references.
ISBN 0-8218-0652-1 (set : acid-free paper). —ISBN 0-8218-0654-8 (pt. 1 : acid-free paper). —ISBN 0-8218-0653-X (pt. 2 : acid-free paper)
1. Topology—Congresses. I. Kazez, William H. (William Hilal), 1954– . II. Title. III. Series.
QA611.A1G456 1993
514—dc20 96-38266
CIP

∞ The paper used in this book is acid-free and falls within the guidelines
established to ensure permanence and durability.

10 9 8 7 6 5 4 3 2 1 02 01 00 99 98 97

TABLE OF CONTENTS

PART 1

PART 2

Preface

The 1993 Georgia International Topology Conference was held at the University of Georgia in Athens, Georgia from August 2–13, 1993. 274 people from 20 countries attended the conference. The proceedings include research articles, expository articles, and problem sets. A wide variety of topics in geometric topology are covered including many which were discussed at the conference.

I would like to thank the authors for the work they have put into their papers and the referees for the fine job they have done. Many of the authors have commented that they found the referees' comments particularly useful.

My colleagues John Hollingsworth, Gordana Matić, and Clint McCrory contributed to every aspect of the conference. Simon Donaldson, Bob Edwards, Mike Freedman, Dave Gabai, and Bill Thurston suggested many of the people who spoke at the conference. Dave Gabai and Rob Kirby put a lot of energy both into organizing problem sessions at the conference and into writing up the very thorough problem sets that appear in this volume. Gail Suggs and Sharon Southwick helped with the day to day running of the conference and in the preparation of many of the manuscripts. I appreciate all of their help.

The conference would not have been possible without the very generous financial support we received from the University of Georgia (in particular, the Office of the Vice President for Academic Affairs, the Office of the Vice President for Research, the Franklin College of Arts and Sciences, and the Department of Mathematics), the National Science Foundation (grant number DMS-9300677), and the International Science Foundation. We gratefully acknowledge their support.

I appreciate the guidance, encouragement and inspiration provided by previous organizers of the conference: M.K. Fort 1961, Jim Cantrell and C.H. Edwards, Jr. 1969, Jim Cantrell 1977, and Clint McCrory and Ted Shifrin 1985.

W.H.K.

Athens, Georgia
February, 1995

Dedication

These 1993 Georgia Topology Conference proceedings are dedicated to Professor James C. Cantrell on the occasion of his retirement from the University of Georgia after 31 years of service. His direction and inspiration are responsible for the continuation of the initial 1961 Georgia Topology Conference as a major international conference held every eight years, with smaller summer topology conferences held in Athens annually for over three decades. It is therefore our pleasure to offer this most appropriate dedication in honor and recogition of Jim Cantrell's career of leadership and friendship with topologists everywhere.

AMS/IP Studies in Advanced Mathematics
Volume **2**, 1997 (Part 1)

POLYHEDRAL MINIMAL SURFACES, HEEGAARD SPLITTINGS AND DECISION PROBLEMS FOR 3-DIMENSIONAL MANIFOLDS

J.H. RUBINSTEIN

1. Introduction.

W. Haken [6], [7] introduced the theory of normal surfaces to solve the triviality problem for knots, i.e to give an algorithm to determine if an embedded circle in the 3-sphere bounds an embedded disk or not. He also gave an algorithm to detect embedded, 2-sided, closed, incompressible surfaces in closed, triangulated, irreducible 3-manifolds, although his algorithm had no bound for termination. An efficient algorithm with such a bound was constructed by Jaco and Oertel [14]. Combining this algorithm with Hemion's solution of the conjugacy problem in the mapping class group of surfaces ([10]), which was also done by Hatcher and Thurston [9], there follows an algorithm to decide if two closed triangulated irreducible 3-manifolds are homeomorphic, assuming one of the 3-manifolds contains an embedded 2-sided incompressible surface.

In this paper, a brief outline is given of a solution of the recognition problem or homeomorphism problem for the 3-sphere.

Theorem 1. *There is an algorithm to decide if a triangulated closed 3-manifold is homeomorphic to the 3-sphere.*

Remarks.

1. A different proof, using thin position, has been given by A. Thompson [1].
2. This result was given in lectures at the Technion in 1992. The author would especially like to thank J. Hass, P. Shalen and A. Thompson for helpful comments and Y. Moriah and B. Wynryb for their hospitality in organising the program.
3. W. Haken has an unpublished program to solve the recognition problem for the 3-sphere.

Combining the methods used for Theorem 1 with the iteration technique of J. Pitts and the author [19], we obtain the following extension;

Theorem 2. *There is an algorithm to find the minimal Heegaard genus of a triangulated closed orientable irreducible 3-manifold. Moreover there is an algorithm to construct all minimal genus Heegaard splittings of such a manifold, up to isotopy and Dehn twisting about embedded incompressible tori.*

Corollary 1. *There is an algorithm to solve the homeomorphism problem for 3-manifolds which are closed, triangulated, orientable, irreducible and have Heegaard genus at most two.*

Remark. This follows from Theorem 2, by using the fact such 3-manifolds are 2-fold branched covering of the 3-sphere over 3-bridge links.

In the next section, we will describe a class of *almost normal surfaces.* These are a natural generalisation of Haken's normal surfaces and can be viewed as polyhedral minimal surfaces with index of instability one. From this viewpoint, normal surfaces should be thought of as having index zero. Since our methods apply in the 3-manifold with neighbourhoods of the vertices removed, normal surfaces are then *stable* minimal surfaces.

Index one minimal surfaces arise from sweepouts associated with Heegaard splittings by the minimax procedure of J. Pitts and the author ([17], [18], [19]). A fundamental result of [19] says roughly that a strongly irreducible Heegaard splitting either can be isotoped to be minimal (with index of instability one for a generic metric) or to the boundary of a regular neighbourhood of a closed embedded one-sided surface, with a single unknotted tube attached, crossing the neighbourhood. If the neighbourhood is expressed as an interval bundle over the non-orientable surface, the tube is a union of interval fibers.

A fundamental step in the proof of Theorem 2 is the following polyhedral analogue of the main result of [19].

Theorem 3. *Suppose M is a triangulated, closed, orientable, irreducible 3-manifold. Then any strongly irreducible Heegaard splitting of M can be isotoped to an almost normal surface.*

Remark. The case where the splitting is isotopic to the boundary of the regular neighbourhood of a closed embedded non-orientable surface, with a single unknotted tube attached, also gives an almost normal surface.

Using other ideas from [19], the following general finiteness result for Heegaard splittings can be obtained.

Theorem 4. *Suppose M is a closed, orientable and irreducible 3-manifold. Then M has finitely many Heegaard splittings of a fixed genus, up to isotopy and Dehn twisting about embedded incompressible tori.*

Remarks.

1. This result was proved by Johannson [13] in case M is Haken, ie. contains an embedded incompressible surface. Hass has obtained a similar result (unpublished).
2. If M has a Riemannian metric of negative sectional curvature, then the result was proved by Pitts and the author [19].
3. This answers a conjecture of Waldhausen [22].

My thanks to Peter Scott for many suggestions and corrections. Note that all the results announced above are accompanied by brief discussions and outlines of proof technique. A more detailed account will be given elsewhere.

2. Normal and almost normal surfaces.

A normal surface in a triangulated 3-manifold M is a union of normal disk types. We will adopt the terminology of [8]. For each tetrahedron, a normal disk is a properly embedded disk which is either triangular or a quadrilateral. All the boundary arcs join two different edges. It is easy to see that up to isotopy, there are four triangular disk types and three quadrilaterals in each tetrahedron. So if there are t tetrahedra, then there are $7t$ normal disk types in M. Each normal surface has coordinates which are written as a vector $\bar{n} = (n_1, n_2, \ldots n_{7t})$. Each n_i counts the number of copies of the i-th normal disk type in the normal surface. For embedded surfaces, it is straightforward to show there is a unique way of constructing the surface, given its normal coordinate vector. The *complexity* of a normal surface is the number of vertices in its cell decomposition as a union of normal disk types.

Haken defines *compatibility equations*, using boundary arcs of the normal disk types. Each triangular face of a tetrahedron has three different arc types, joining different edges of the face. This gives $6t$ arc types and for each, there is a compatibility equation. There are two tetrahedra containing the given face in their boundaries and therefore four disk types with a given arc type in their boundaries, two triangles and two quadrilaterals. Equating the numbers in the two tetrahedra of these disk types gives an equation

$$n_i + n_j = n_k + n_l$$

The solution of these $6t$ equations, intersected with the non-negative vectors of R^{7t}, gives a cone-like region,

$$R = \{(x_1, \ldots.., x_{7t}) : x_i \geq 0, x_i + x_j = x_k + x_l \ldots.\}$$

The integer points in R are then normal surfaces. A *fundamental normal surface* has coordinate vector $\bar{n}$ with the property that if $\bar{n} = \bar{m} + \bar{p}$, for $\bar{m}$, $\bar{p}$ integer points in R, then either $\bar{m} = \bar{o}$ or $\bar{p} = \bar{o}$. Haken shows there are finitely many fundamental normal surfaces and these can be found by a simple algorithm .

We define an *almost normal surface* to be a union of pieces which are all normal disk types except one *exceptional piece.* The latter is either a disk or an annulus properly embedded in a tetrahedron. In the case of an exceptional disk, it must be an octagon with each boundary edge a normal arc type (i.e an arc running between different edges of the tetrahedron). It is easy to check there are three such exceptional octagons in each tetrahedron. An exceptional annulus consists of two normal disk types joined by an unknotted tube. In this case, unknotted refers to the fact that the tube is boundary parallel, i.e the annulus is boundary compressible. There are twenty-five different such exceptional annuli.

For a choice of octagon as exceptional disk type, we define the coordinate vector of an almost normal surface to be a vector of non-negative integers in $\mathbf{R}^{10t}$. Moreover we require there is exactly one non-zero octagon coordinate and this has value one. So we introduce $3t$ new coordinates corresponding to the $3t$ choices of octagon. Define R^a to be the cone of non-negative solutions in $\mathbf{R}^{10t}$ of the $6t$ compatibility equations. Note that each such equation will involve six variables, two triangular disks, two quadrilaterals and two octagons. Since an octagon meets every quadrilateral and we are only interested in embedded surfaces, we never get both octagons and quadrilaterals in the same tetrahedron.

An *almost normal fundamental surface* has coordinate vector $\bar{n}$ in R^a with the property that $\bar{n} = \bar{m} + \bar{p}$, for $\bar{m}, \bar{p}$ integer points in R^a, implies $\bar{m} = \bar{o}$ or $\bar{p} = \bar{o}$. It is easy to check in this case, that for any almost normal surface with coordinate vector $\bar{n}$, if $\bar{n} = \bar{m} + \bar{p}$ with $\bar{m}, \bar{p}$ integer vectors in R^a then either $\bar{m}$ or $\bar{p}$ must correspond to a normal surface, with $\bar{p}$ or $\bar{m}$ respectively an almost normal surface. Moreover, Haken's method then gives a simple algorithm to find almost normal fundamental surfaces, if they exist.

Remarks. In R^a, we noted before that embedded surfaces correspond to vectors with non-zero coordinates for at most one octagon-type or one quadrilateral type in each tetrahedron.

Conversely, such a vector in R^a, with integer coordinates, uniquely specifies an embedded surface. Moreover if $\bar{n}$ is such a vector and $\bar{n} = \bar{m} + \bar{p}$, with $\bar{m}$, $\bar{p}$ in R^a with integer coordinates, then both $\bar{m}$ and $\bar{p}$ correspond to embedded surfaces.

In the following, we are mainly interested in the subcollection of almost normal surfaces with exactly one octagonal disk. The case of exceptional piece being an annulus turns out to be natural, but not crucial in the results.

A curve of intersection of two embedded normal surfaces is called *regular* if it consists of arcs, one for each pair of crossing disk types which are both triangular or one quadrilateral and one triangular. A *regular exchange* along such a curve is the unique cut and paste which gives two new normal surfaces which are disjoint near the curve.

3. The recognition problem for the 3-sphere.

Step 1. Choose a maximal family Σ of disjoint embedded normal 2-spheres in the triangulated 3-manifold M. By results of Kneser and Haken [8], Σ can be explicitly constructed and there is a bound on the number of components of Σ. Cut M open along Σ to form pieces $M_1, \ldots.., M_k$, bounded by normal 2-spheres.

Claim. A component M_i with at least two boundary components is a punctured ball, i.e a 3-ball with disjoint open 3-balls removed from its interior.

The explanation is that there must be a tetrahedron with a connected component of intersection with M_i containing pieces of different boundary 2-spheres, say S_1 and S_2, coming from Σ. Join S_1 and S_2 by an unknotted tube in this tetrahedron, giving an almost normal 2-sphere S. Then S can be normalised (cf [8]) by moves not crossing either S or any sphere in Σ. The result is that S "shrinks" to some points or a disjoint family of embedded normal 2-spheres. In the latter case, by maximality of Σ, these 2-spheres are parallel to members of Σ. By pushing S off itself in either direction, the result of the two normalisation processes is a sweepout of M_i by a 2-sphere or family of embedded 2-spheres. In either case, it is straightforward to deduce that M_i is homeomorphic to a punctured ball and the claim is proved.

Step 2. Next we sketch why the existence of an embedded almost normal 2-sphere S with an octagonal piece in a component M_j with connected boundary implies M_j is homeomorphic to a 3-ball. The argument is similar to Step 1. Push S slightly off itself to either side and then perform normalising moves in the complement of S and the boundary of M_j. This gives a sweepout of M_j by embedded 2-spheres, establishing that M_j is a 3-cell.

Note that normal and almost normal surfaces form a "barrier" in the process of normalising surfaces. This is very similar to the use of the maximum principle when analytic minimal surfaces are used as barriers.

Step 3. The hard part is to show if M_j has a single boundary 2-sphere and is a 3-ball, then M_j contains an embedded almost normal 2-sphere with an octagonal piece.

Clearly M_j has a sweepout S_t by embedded 2-spheres, i.e there is a continous map $f : S^2 \times [0,1] \to M_j$ with $f(S^2 \times \{0\})$ a point, f restricted to $S_2 \times \{t\}$ an embedding to S_t for $0 < t \leq 1$ and S_1 equal to the boundary of M_j. The 2-site argument originating in [16] can be used to show that there is an "efficient" sweepout for which the maximum number of transverse intersections of S_t with the edges of the triangulation of M is as small as possible. This can be arranged to occur for a value $t = t_o$ such that S_{t_0} is an almost normal embedded 2-sphere with exceptional piece an octagon. Note that this efficient sweepout is analogous to a "thin position" projection of D. Gabai [4].

Step 4. Finally the algorithm to construct an embedded almost normal 2-sphere with an octagonal piece in M_j can be viewed in either of two ways. In the first, it is easy to show that if there is such a 2-sphere then there is a fundamental one. Haken's algorithm can then be slightly modified to find all fundamental almost normal surfaces and to check if any is a 2-sphere.

Alternatively, Haken's method can be used to show that if there is a normal disk with the same boundary as some octagonal piece then there is a fundamental normal disk with this property. Note that the normal disk will have interior in the complement of the truncated tetrahedron which contains the octagon. (The tetrahedron must be truncated by slicing off a small tetrahedron near each vertex). Since there are only finitely many octagonal disk types, each can be checked to see if there is a compatible fundamental normal disk with the same boundary.

Note that this can be carried out in every component M_j with a connected boundary to check whether or not M_j is homeomorphic to a 3-ball. This completes the recognition algorithm.

4. Strongly irreducible splittings and almost normal surfaces.

Next we give a sketch of the proof of Theorem 3. In the final section, it is shown how to deduce Theorems 2 and 4, together with Corollary 1, from Theorem 3.

Step 1. Again cut M open along a maximal collection of disjoint embedded normal 2-spheres Σ to give pieces $M_1, ..., M_k$. Any component with several boundary spheres will be a punctured cell, by the argument in Section 3. At most one of the pieces with a connected boundary is not a 3-cell, since M is irreducible. By Theorem 1, we can check which one this is, say M_1, if we wish.

Step 2. Let S be a strongly irreducible Heegaard splitting of M. Clearly S can be used to define a sweepout S_t of M_1. So for $0 < t < 1$, S_t is an embedding and for $t = 0$ or 1, S_0 is a graph and S_1 is a graph attached to the boundary 2-sphere of M_1. The idea is to use the iterated minimax procedure of Pitts and the author [19], to show that "eventually" we can find an efficient sweepout so that S_{t_0} is an almost normal embedding in M_1 or S is isotopic to the boundary of a regular neighourhood $N(K)$ of an embedded non-orientable surface K, with a single unknotted tube attached.

In the latter case, since S is strongly irreducible, it is easy to see that K must be geometrically incompressible, i.e if an embedded disk D in M intersects K in ∂D only then ∂D must be contractible in K, where ∂D denotes the boundary of D. But then, Haken's normalisation procedure applies and K can be isotoped to be a normal surface (in M_1 if we like). Finally $N(K)$ can be chosen thin enough so that its boundary is also normal, with two parallel normal disk types for each one of K. Then S is isotopic to an almost normal surface with special piece an annulus, given by attaching an unknotted tube between any two such normal disk types of the boundary of $N(K)$.

For completeness, a brief description of the iteration technique of [19] is now given, with suitable modifications to deal with polyhedral rather than analytic surfaces. Assume S bounds a handlebody A and a punctured handlebody B in M_1. Run the minimax S_t across M_1. The minimax surface $S^{(1)}$ is obtained by isotopy and some surgeries of S along some disks. By strong irreducibility of S (and a result of [21]), if the essential surgery disks are denoted by $\mathcal{D}$ then all these disks are disjoint meridian disks for one of A or B. Here an inessential surgery disk has a contractible boundary curve on S. Without loss of generality, suppose $\mathcal{D}$ lies in A. $S^{(1)}$ must be almost normal, with an octagon as special disk, since an annular special disk only gives a sweepout decreasing surface complexity on one side. (Complexity here is the number of transverse intersections with the 1-skeleton of the triangulation minus the number of components of the surface.) If $\mathcal{D}$ is empty then $S^{(1)}$ is isotopic to S and the theorem is proved. Let $N(\mathcal{D})$ denote a neighbourhood of $\mathcal{D}$, so we can view $S^{(1)}$ as bounding a region isotopic to $B \cup N(\mathcal{D})$, which will be denoted by X.

We push $S^{(1)}$ off itself into X and normalise by Haken moves. Strong irreducibility of S implies $S^{(1)}$ is incompressible in X and so is isotopic to a normal surface $S^{(2)}$. Next we attach thin tubes to $S^{(2)}$ to obtain a new surface isotopic to S. There is now a new sweepout of the region between $S^{(2)}$ and the boundary of M_1 by S. The minimax procedure gives a new almost normal surface $S^{(3)}$ in this region. Continue on until $S^{(2k+1)}$ is obtained by no nontrivial surgeries on S. In this case we have obtained the requisite almost normal surface isotopic to S. Notice that there is always a nontrivial amount of the triangulation of M_1 between $S^{(i)}$ and $S^{(i+1)}$, so the procedure terminates after finitely many steps. (A more complicated argument is needed in the analytic case).

If the essential surgery disks lie in B, then there is a second possibility, namely the region bounded by some $S^{(2k-1)}$ could be a twisted interval bundle over an embedded non-orientable surface T. In this case, $S^{(2k)}$ may be a double covering of T and the process stops. By strong irreducibility, it is easy to see S is isotopic to the boundary of a small regular neighbourhood $N(T)$, with tubes attached crossing $N(T)$. By the argument of [19], there is in fact a single unknotted tube attached to form S. This completes the sketch of the proof of Theorem 3.

5. Algorithms to find almost normal surfaces.

The key idea in the proofs of the remaining results is the following:

Proposition 1. *Suppose M is a closed, orientable, irreducible 3-manifold. There is a bounded collection of surfaces, which can be explicitly constructed, and contain all strongly irreducible Heegaard splittings of bounded genus, up to isotopy and Dehn twisting about incompressible tori.*

Proof. The approach is to use Theorem 3 and Haken's philosophy, to write the resulting almost normal surfaces as Haken sums of fundamental normal and almost normal surfaces. We will first show all the fundamental surfaces which arise can be assumed to have non-positive Euler characteristic. Next, it is established that there is a bound on the number of surfaces of zero Euler characteristic in such a Haken sum. As Euler characteristic is additive under Haken sums, there is a bounded number of sums of the finitely many fundamental surfaces of negative and zero Euler characteristic which need to be considered. This gives the desired bounded collection of surfaces, containing all strongly irreducible Heegaard splittings, up to isotopy and Dehn twisting about incompressible tori. □

Step 1. *Suppose S is an almost normal surface in M_1 with coordinate vector $\bar{n}$ in R^a. Here M_1 is defined as in Step 1 of the previous section. So M_1 has a normal boundary sphere S_1 and contains no vertices of the triangulation of M or normal 2-spheres not parallel to S_1. Assume $\bar{n} = \bar{m} + \bar{p}$ with neither $\bar{m} = \bar{o}$ nor $\bar{p} = \bar{o}$. Then we can suppose that neither $\bar{n}$ nor $\bar{p}$ has a component surface with positive Euler characteristic.*

The reason is that firstly, if such a component was a normal or almost normal 2-sphere, this would either contradict the assumption that M_1 has no normal 2-spheres except for its boundary, or would imply M_1 was a 3-cell by Theorem 1. In the latter case we would deduce M is homeomorphic to the 3-sphere and so has only a single genus zero strongly irreducible Heegaard splitting [23]. Secondly, if M_1 contains a normal or almost normal projective plane, since M is irreducible we see it is homeomorphic to real projective 3-space. By [2], then M has a unique genus one strongly irreducible Heegaard splitting. So it suffices to assume all the components of the summands of the almost normal Heegaard surface S have non-positive Euler characteristic.

Step 2. *If S has a summand with zero Euler characteristic, then it suffices to assume this is either an incompressible torus or a torus bounding a solid torus.*

In fact there are two other possibilities for such a summand; either a Klein bottle L or a torus bounding a cube with knotted hole in M_1. In the former case, let T be the torus boundary of a small regular neighbourhood $N(L)$. If T compresses in M, then M is a prism manifold and it is known that such manifolds have a unique strongly irreducible Heegaard splitting of genus two (see [21]). If T is incompressible and S has coordinate vector $\bar{n}$, L has coordinate vector $\bar{m}$ and $\bar{n} = a\bar{m} + \bar{p}$, for some integer $a \geq 2$, then we can write $\bar{n} = [\frac{a}{2}]\bar{m}' + \bar{p}'$, where $\bar{m}' = 2\bar{m}$ is the coordinate vector for T. So we are reduced to the case of an incompressible torus summand. Note that $\bar{p}'$ could be represented by a Klein bottle with coefficient one. We can ignore this case as we are only interested in bounding the coefficient of any summand with zero Euler characteristic. Note that for $a \geq 2$, L must be normal rather than almost normal.

In the latter case, there is a normal or almost normal torus T bounding a (possibly punctured) cube with knotted hole in M_1. It is straightforward to find a normal disk D in M_1, filling in the hole of T. Then $T \cup D$ is the image of a piecewise normal 2-sphere. This can be pushed off itself (in the complement of the cube with knotted hole) and acts as a barrier for normalisation. So the sphere shrinks to a point or to the boundary of M_1, depending on whether the boundary of M_1 is inside or outside the cube with knotted hole. In either case, we see that M_1 is a 3-cell and M is homeomorphic to S^3.

Step 3. *Assume S has coordinate vector $\bar{n}$, T has coordinate vector $\bar{m}$ and is the boundary of a solid torus X and $\bar{p}$ is the coordinate vector of a surface U, with $\bar{n} = a\bar{m} + \bar{p}$, where $a \geq 3$. If $T \cap U$ contains a meridian curve for X, then S is isotopic to a smaller complexity almost normal surface.*

Consider a parallel copies of T intersected by U. Regular exchanges are performed at all the curves of intersection to give S. Clearly U is almost normal and T is normal.

By [21], if meridian curves C_1, C_2 on adjacent parallel copies of T are essential in S, we get a contradiction to S being strongly irreducible. For C_1, C_2 bound disks on opposite sides of S. We conclude that at most one copy of T can have a meridian curve which is essential on S. But then two such inessential curves will be parallel on S. The annulus they bound on S can be replaced by a thin annulus of U. This can be achieved by an isotopy of S and decreases complexity.

Step 4. *Suppose $S = aT + U$, where $a \geq 3$ and T bounds a solid torus X, as in Step 3. Assume also that some curve C of intersection of U and T is inessential on U. Then S can be isotoped to an almost normal surface of smaller complexity.*

Without loss of generality, C can be chosen as an innermost inessential curve on U. By Step 3 we may assume that the disk D that C bounds on U is boundary parallel in X or in $M - intX$. Hence C bounds a disk D' in T. Consider the direction of the regular exchange at C. If $D \cup D' = V$ is a normal 2-sphere, then V must be parallel to the boundary of M_1. Hence S cannot cross V and D'. But then clearly U does not cross $intD'$ and V is a component of $aT + U$. This is a contradiction. We conclude that $(T - intD') \cup D$ must be a normal torus.

Let C', C'' be parallel curves to C on parallel copies of T. (We suppose for convenience that $a = 3$). So C, C', C'' are all components of the intersection of U with copies of T and are connected by thin annuli of U. By strong irreducibility, since C, C', C'' bound parallel copies of D, at most one of these curves can be essential on S. So for example, if C, C' are inessential on S, we can replace the disk that C' bounds on S by a copy of D and a thin annulus of U. This reduces the complexity of S by an isotopy.

Step 5. *Assume $S = aT + U$, where $a \geq 3$ and T bounds a solid torus X, as in step 3, and all curves C of intersection of U and T are essential on U. If some curve C is inessential on T then after an isotopy of S, we can write S as a sum of $aT + U'$ or $(a-2)T + U''$, where U' or U'' meets T in only essential curves on T.*

Choose an innermost curve C of $U \cap T$ bounding a disk D' on T. Suppose first C bounds a disk D on S. Note that $T \cap intD = 0$, since $U \cap T$ has no inessential curves on U by Step 4. We can isotope S to $S' = (S - intD) \cup D'$. Then $S' = aT + U'$, with $T \cap U' = T \cap U - \{C\}$, after a slight perturbation. Therefore it can be assumed that every such innermost inessential curve C is essential on S. Define a tube of U as a small annulus meeting T in an inessential curve on T but essential on the annulus. Then, if $U \cap T$ contains only innermost inessential curves on T, it is easy to see that the regular exchanges between the parallel copies of T and U attach the tubes of U only along the first and last copy of T. So if $U'' = U' + 2T$ is the result of adding the first and last copy of T in this way to U', then U'' meets $(a-2)T$ in essential curves only on T.

Next assume $U \cap T$ contains a curve C' bounding a disk D' on T including some curves C of $U \cap T$ which are all innermost inessential curves on T. The regular exchange between U and T at C' and the curves C inside D' may connect the tubes of U to the second or $(a-1)$ copy of T. For this to happen, the tubes at C and C' must be attached on the same side of the copies of T. Without loss of generality, assume the tube at C' is connected to the first copy of T and that at C to the second copy. Now if C' does not bound a disk on S, we contradict the strong irreducibility of S, since C and C' bound disks on opposite sides of S. Assume then that C' does bound a disk D on S and let V be the embedded 2-sphere $D \cup D'$. (Note that the innermost tubes of U attached to D' along curves C cannot lie inside D on S, as C is essential on S). V bounds a 3-cell denoted by Y in M. The strongly irreducible splitting S intersects V in a single curve C', after a small perturbation. This is again a contradiction, since $S \cap Y$ must then be a disk but contains essential curves C on S. The conclusion is that the tube at C' must be attached on the opposite side of the copies of T to the tubes at the curves C. If all the inessential loops of $U \cap T$ are of the type of C' and C, we can again replace U by $U'' = U' + 2T$ and then U'' meets $(a-2)T$ in essential curves only. Finally if there are inessential curves C'' bounding disks D'' with loops of type C' and C inside D'', then the tube of U at C'' must be attached on the same side as that for either C or C'. This is impossible by the above argument and in all cases, the inessential loops of $U \cap T$ can be eliminated.

Step 6. *Suppose $S = aT + U$, where $a \geq 3$ and T bounds a solid torus. By steps 4 and 5, we may assume all curves of $U \cap T$ are essential on both U and T.. For convenience, we also assume M is atoroidal, ie. has no embedded incompressible tori. If $S = U + aT$ for a is sufficiently large, then there is an isotopy taking S to an almost normal surface S' of smaller complexity.*

The isotopy of S is a Dehn twist about T and possibly some other disjoint tori bounding solid tori. In the final step we consider the case where M has incompressible tori, so S is modified by ambient homeomorphisms, rather than just isotopies.

Consider the components of $U \cap X$. Since $S = U + aT$, we can arrange that $S \cap X = U \cap X$. By the solid torus lemma of [15], since S is strongly irreducible, all the components of $S \cap X$ are boundary parallel essential annuli except possibly for one which is either an annulus with a tube attached or two annuli connected by a tube.

We distinguish two cases; $S \cap T$ has components which are non-trivial multiples of a generator of $\pi_1(X)$ or the curves of $S \cap T$ are parallel to a core circle of X.

In both cases, the first step is to enlarge X to a *maximal* solid torus X' with a normal torus T' as boundary. It is easy to see this can be done in M_1. (There may be several choices of X' and T'; we arbitrarily pick one.) Suppose M is not a "small" Seifert fiber space, ie. with at most three exceptional fibers and S^2 as orbit surface. A useful property of X' is that any normal annulus or disk which is properly embedded in $M_1 - intX'$ and has essential boundary on T' in the case of an annulus, is parallel into T' as a normal surface. This means that the annulus or disk can be deformed into T' without changing complexity.

The reason is that otherwise, since M_1 is irreducible, atoroidal and not a small Seifert fiber space, the normal annulus or disk is isotopic into T'. Adding the product region between the normal annulus or disk to X', we obtain a larger solid torus X'' with piecewise normal boundary T''. Then T'' can be normalised in $M_1 - intX''$. If T'' either shrinks to a circle or compresses, it follows that M is either S^3 or a lens space, ie. a small Seifert fiber space. If T'' is isotopic to a normal torus T^*, then X' is not a maximal solid torus with normal boundary. This completes the argument for the asserted property of X'.

Next we observe that T' is a *fundamental* normal torus. Suppose on the contrary that $T' = T_1 + T_2$ is a non-trivial Haken sum. Since T' is in M_1, we cannot have either T_1 or T_2 being a normal 2-sphere. The case where T_1 or T_2 is a projective plane implies M is $\mathcal{R}P^3$, ie. a small Seifert fiber space, and similarly if T_1 or T_2 is a Klein bottle. So we conclude both T_1 and T_2 are tori. First,we deal with the case that $T_1 \cap T_2$ contains curves which are inessential on both T_1 and T_2.

Assume C is such a curve and C bounds D_i in T_i, $i = 1, 2$. If $D_1 \cap D_2$ has curves other than C, choose an innermost one. Replacing C by this curve, we may suppose $D_1 \cap D_2 = C$. Now examine the regular exchange at C. If $D_1 \cup D_2$ is a normal 2-sphere, then we contradict the fact that M_1 has only the boundary normal 2-sphere. If D_1 and D_2 can be switched for T_1 and T_2 to give normal (possibly singular) tori, we can write T' as a Haken sum of surfaces with fewer intersection curves (after if necessary doing regular exchanges on the immersed tori to make them embedded).

Secondly, if all curves of $T_1 \cap T_2$ are essential on each, it is easy to see there are normal annuli in $M_1 - intX'$ which are not normally parallel into T', contradicting the property established earlier. Finally suppose $T_1 \cap T_2$ contains some curve C which is inessential on say T_2 and is a meridian curve for a solid torus bounded by T_1. If C is innermost on T_2, then C bounds a disk D on T_2 with $intD$ disjoint from T_1. Now it can be shown that if X_2 is the solid torus bounded by T_2, then C can be chosen so that C bounds a disk D' on T' with D' contained in X_2. But then D is outside X' and this contradicts the property established earlier. We conclude that T' must be fundamental and can now discuss the two cases where curves of $S \cap T'$ are multiple of cores or cores of X', assuming for the moment that M is not a small Seifert fiber space.

In the case of multiples of cores, S can be isotoped till $S \cap X'$ is a single annulus A which is boundary parallel but does not contain a core circle for X'. Let $T' = A' \cup A''$ where $\partial A' = \partial A'' = \partial A$ and A is parallel to A'. Then A'' is boundary compressible in one of the two handlebodies bounded by S in M. If we isotope S along the disk of this boundary compression, the result is that $S \cap X'$ can be changed into a copy of T' pushed slightly into $intX'$ joined by a tube to the boundary. So X' is a handle of one of the handlebodies bounded by S.

Consequently S can be used to run a minimax across $M_1 - intX'$. We get a new almost normal surface isotopic to S and missing X'. If the new S is written as a sum of fundamental normal and almost normal surfaces, no summands can meet X'. So there are fewer summands with Euler characteristic zero (and large coefficient). The process terminates with a bound on the complexity of S.

Finally, in the second case where curves of $S \cap T'$ are cores of X', S can be isotoped to meet X' in a single boundary parallel annulus which contains a core circle for X'. Separate sweepouts of X' and

$M_1 - intX'$ can be described, using $S \cap X'$ and $S \cap (M_1 - intX')$. For X', we require a bound for the maximum complexity which is independent of the choice of boundary circle of the annulus $S \cap X'$. To do this, sweepout X' by a pair of meridian disks, starting and ending at one disk with multiplicity two. Now these disks can be assumed to cross the annulus $S \cap X'$ in a single arc. The region between two such arcs on $S \cap X'$ is a quadrilateral which can be pushed across the cylinder between the two meridian disks in X' to the quadrilateral on the other side of T'. The procedure is to do this first when the meridian disks are close together and the corresponding quadrilaterals are very thin. Then pull the meridian disks apart, gradually lengthening the quadrilaterals. A bound for the complexity of the sweepout is the maximum over the meridian disk pairs plus the complexity of T'. This is independent of the choice of curves of $S \cap T'$.

To complete the discussion of this case, $S \cap (M_1 - intX')$ can be used to sweep across $M_1 - intX'$. The same argument as in Section 4 gives an almost normal surface properly embedded in $M_1 - intX'$, which we again denote by $S \cap (M_1 - intX')$. This surface can be written as a sum of normal and almost normal fundamental surfaces in $M_1 - intX'$. By our property established earlier, none of these surfaces can be an annulus or disk. A Möbius band can be easily excluded for the same reason. So the only zero or positive Euler characteristic surface in this sum must be a torus or Klein bottle. But now we can iterate the previous discussion if $S \cap (M_1 - intX')$ can be written as a sum, where the coefficient of same fundamental torus is large. Since the number of (disjoint) normal fundamental tori is determined and finite, the process must terminate. We conclude that a bound to the maximum complexity of a sweepout of M_1 by S can be found. Note the sweepouts of X_1 and $M_1 - intX'$ can be "glued" together by using annuli running around T', for each value of the sweepout parameter. Since T' bounds a solid torus, even though the annuli may have to use many copies of T', we can "untwist" the annuli by rotating the solid torus X'. So the "gluing penalty" can be restricted to the complexity of T' plus the increase in the maximum complexity of the sweepout of X' caused by a rotation of X' around a core circle by 2π. This gives a bound on the maximum complexity of a sweepout. Note that the bound on the maximum complexity of a sweepout is *not* the same as a bound on the complexity of the almost normal surface S. The reason is that to find S, we may need to iterate the minimax procedure, as in Section 4. However it is not too difficult to still obtain a bound on the complexity of S, by the following argument.

The sweepout gives a minimax almost normal surface, obtained from S by compressions on one side. Hence this is constructible; from all such almost normal surfaces, pick out those which bound handlebodies on one side and are either incompressible or bound handlebodies on the other side. Both these properties can be checked by Haken's algorithms. Note that the surfaces may not necessarily be connected - in that case, all the components bound disjoint handlebodies.

If the surface is constructed from S by nontrivial compressions, the previous argument can be repeated by sweeping out the complement of the handlebodies in M_1 using S again. As above, there is a new bound on the maximum complexity of such a sweepout. Note that as in the iterated minimax argument, the almost normal surface is pushed off itself and normalised on the incompressible side, before the sweepout is started. This process terminates after a bounded number of steps, by Kneser's method, and a constructible bound for the complexity of S is then found.

To complete this section, we need to discuss the case when M is a small Seifert fiber space. If M is S^3, a lens space, or a prism manifold, then the (strongly) irreducible Heegard splittings are unique. In the other cases where M has three exceptional fibers and S^2 as orbit surface, we argue as follows. If we run the sweepout and obtain an almost normal surface S which can be written as a sum of fundamental surfaces with small coefficients, then S can be found as required. On the other hand, if some torus has a large coefficient, then as above S can be isotoped to meet the solid torus bounded by this torus in a single essential annulus. Moreover if the solid torus is enlarged to a maximal one X' with boundary a normal torus T', the only place where the above argument to bound the maximum complexity of the sweepout fails, is the property of T' that any essential normal annulus which lies in $M_1 - intX'$, is normally parallel into T'. If such an annulus is not homotopic into T', then X' must be a neighbourhood of an exceptional fiber of M. But then S can be isotoped to meet X' in an annulus and so meets T' in two essential circles. The technique of [21] then applies to show S is a standard vertical or horizontal splitting of M. So again we can construct S and this case is completed.

Step 7. *Suppose M has embedded incompressible tori and $S = aT + U$, where T has Euler characteristic zero and a is large. Then S can be altered by Dehn twisting along incompressible tori and isotopy to give a surface with smaller complexity.*

There are various possibilities to discuss. If M is a graph manifold (including the special case of a Seifert fiber space) the (strongly) irreducible Heegard splittings are discussed in [21]. So we may assume M has a characteristic variety V and $M - intV$ contains atoroidal components other than $T^2 \times [0,1]$. Let W be the characteristic graph variety, ie. the union of V with all the components of $M - intV$ which are homeomorphic to $T^2 \times [0,1]$. By [21], S can be isotoped to meet W in a standard splitting. This means S meets each torus component of ∂W in inessential curves only and is a union of horizontal and vertical splittings of the Seifert fibered components of V. Now if any component is vertical, it is easy to see that S is not strongly irreducible. This is because S has disjoint compressing disks on either side, giving vertical incompressible tori as the resulting surface. The only remaining case is where $S \cap V$ is horizontal, ie. a union of incompressible surfaces which are all fibers of a description of V as a surface bundle over S^1. This includes the case V has components of the form $T^2 \times [0,1]$ which we call *isolated tori.*

As in the previous case of a solid torus, we can enlarge all the components of the characteristic graph variety until they are maximal in M_1. So the boundary tori are normal and moreover have the key property that there is no normal essential annulus or disk properly embedded in $M_1 - intW$, which is not normally parallel into ∂W.

Write the minimax almost normal surface S as a sum of U and multiples of normal tori T_i, ie $S = U + a_1T_1 + \ldots\ldots + a_kT_k$, where U is a sum of fundamental normal and almost normal surfaces of negative Euler characteristic and possible some Klein bottles of multiplicity one. There is a finite number of possibilities for U. Also the key property implies that any T_i, which is incompressible, must be in W. Hence if all T_i are incompressible, there is a bound on the number and slope of curves of $U \cap \partial W$, by taking *all* choices for U. Hence there is a bound on the complexity of the horizontal surface $S \cap W$, up to Dehn twists about incompressible tori. We can then sweepout $M_1 - intW$ using $S \cap (M_1 - intW)$, find a minimax surface and a bound on the maximum complexity of the sweepout. Note that when gluing together sweepouts of W and $M_1 - intW$, a Dehn twist about components of ∂W may also be needed to bound the maximum complexity of the total sweepout. The argument then can be completed as in Step 6.

Next assume some T_k with a large coefficient bounds a solid torus. As in Step 6, enlarge the solid torus to a maximal one X' with normal boundary T'. If we can show T' is fundamental, the result follows by the argument in Step 6. Suppose $T' = T + T^*$. As before, it is easy to show T and T^* must be tori or Klein bottles. Moreover curves of $T \cap T^*$ which are inessential on both surfaces can be eliminated. If a component of $T \cap T^*$ is essential on T but inessential on T^*, we get a contradiction to the maximality of X' by finding a normal disk properly embedded in $M_1 - intX'$.

Finally if all curves of $T \cap T^*$ are essential on both surfaces, then a small regular neighbourhood $N(T \cup T^*)$ is Seifert fibered. T' lies in this Seifert fibered space and clearly $N(T \cup T^*) \cup X' = Y$ is also Seifert fibered. If this space is a solid torus, we contradict the maximality of X' as ∂Y is piecewise normal. If Y has an incompressible torus boundary component, then it lies in W. We conclude that either T' is fundamental or X' is contained in W. In the former case, the same approach can be used as in Step 6 to split the sweepout into one for X' and another for $M_1 - intX'$. If there are no such solid tori resulting from T_i with large coefficients, then all such T_i lie in W. In this case, the same method applies as for the situation when all T_i are incompressible. This completes the proof of Proposition 1.

5.1. Proof of Theorem 2. Suppose first that M is a non Haken 3-manifold which satisfies the hypotheses of Theorem 2. We are required to give a procedure to find all the minimal genus Heegard splittings of M, up to isotopy.

Certainly an upper bound for the Heegaard genus of M can be found by taking a neighbourhood of the 1-skeleton of the triangulation of M as a Heegaard splitting. By [3], all minimal Heegaard splittings are strongly irreducible, since M is non Haken. Hence by Proposition 1, there is a finite constructible collection of surfaces containing all such splittings, up to isotopy. It is easy to check using Haken's technique and Theorem 1, which of these constructible surfaces is a Heegaard splitting. So Theorem 2 follows in this case.

Finally suppose M is Haken, ie. contains an orientable incompressible surface. If a minimal genus Heegaard decomposition S is strongly irreducible, the above argument applies to construct a representative in its isotopy class. Otherwise, by [3], there are disjoint compressions of the surface S in both handlebodies and a maximal collection of such compressions gives a separating orientable incompressible surface R. Now by [19], S can be isotoped to meet $N(R)$, a small regular neighbourhood of R, in a copy of R with two standard tubes attached on opposite sides of each component of R. The other ends of these tubes are on the two corresponding components of $\partial N(R)$. Let $N_1, ..., N_m$ be the closures of the components of $M_1 - R$. Note that since R separates M_j no component of R can occur more than once as a boundary surface of some N_j. Let $N(N_j)$ be a small regular neighbourhood of N_j. Take $S \cap N(N_j)$

and pinch out one tube for each component of the boundary. This gives a surface S_j which defines a compression body decomposition of $N(N_j)$.

If S_j is not strongly irreducible, we can find a further orientable incompressible surface R' in $int\, N(N_j)$ and S_j can be further split along R'. There is obtained, after a bounded number of steps, a collection of disjoint orientable incompressible surfaces, again denoted by R, so that S splits along R into *strongly irreducible* compression body decompositions, S_j. Then the argument in Proposition 1 can be easily applied to give an algorithm to find all these (minimal genus) S_j. Moreover, R can be found either using Haken's method [6] or Jaco-Oertel [14]. There are finitely many choices for the negative Euler characteristic components of R (up to Dehn twists about incompressible tori) and for each, we can construct all possible S_j. Note that for the tori and Klein bottle components of R, we can use the previous discussion of the way Heegaard splittings meet characteristic graph varieties. Then S can be reconstituted by "tubing together" the surfaces S_j and "deleting" the extra copy of R. This gives a procedure for finding minimal genus S and completes the proof.

5.2. Proof of Corollary 1. Suppose M, M' are closed orientable irreducible triangulated 3-manifolds and M has Heegaard genus at most two. By Theorem 2, we can decide if M' has the same Heegaard genus as M and can construct all minimal genus Heegaard splittings. If M is S^3 or a lens space, then we can immediately tell if the same is true for M'. Moreover given the genus one splitting, it can be deduced if M' is a lens space of the same homeomorphism type as M. So we are reduced to the case of Heegaard genus 2.

Now it is well known that any 3-manifold of Heegaard genus 2 is a 2-fold branched cover of S^3. Moreover given the Heegaard splitting, such a branched covering can be explicitly constructed. We proceed to do this for all genus 2 splittings of M and of M'. Then M is homeomorphic to M' if and only if the collection of branch sets in S^3 for M coincides with that for M'. In fact it is sufficient to show this for a choice of branched covering for M and for M', but a priori there is no way of telling which choice to make. Now Haken's method together with Gordon-Luecke [5] shows that it can be decided if two knots are equivalent. So if M and M' have branch sets which are knots, the homeomorphism problem can be solved by this approach. For example, if M and M' are Z_2-homology spheres, this is the case.

Finally, if the branch sets for M and M' are links, we have to use Thurston's orbifold theorem (see e.g [12]). It follows that M (and M') is either Seifert fibered, has a non-trivial characteristic variety or is hyperbolic. In the second case M is Haken, so the homeomorphism problem is solvable by Haken's method. In the first and third, it suffices to solve the isomorphism problem for $\pi_1(M)$ and $\pi_1(M')$, unless M is a lens space. The latter case has already been discussed and in any case M has a genus one rather than genus two splitting. In the other cases, there are well-known procedures for solving the isomorphism problem for the fundamental groups and this completes the proof of Corollary 1.

References

1. A. Thompson, *Thin position and an algorithm to recognise the 3-sphere*, Preprint.
2. F. Bonahon and J-P. Otal, *Scindement de Heegaard des espaces lenticulaires*, Ann. Sci. de l'Ecole Norm. Sup. **16** (1983) 451–467.
3. A. Casson and C. Gordon, *Reducing Heegaard splittings*, Topology Appl. **27** (1987) 275–283.
4. D.Gabai, *Foliations and the topology of 3-manifolds, III*, J. Differential Geometry **26** (1987) 479–536.
5. C. Gordon and J. Luecke, *Knots are determined by their complements*, J. Amer. Math. Soc. **2** (1989) 371–415.
6. W. Haken, *Theorie der Normalflächen*, Acta Mathematica **105** (1961) 245–375.
7. W. Haken, *Über das Homöomorphie problem der 3-Mannigfaltigkeiten I*, Math. Zeit. **80** (1962) 89–120.
8. W. Haken, *Some results on surfaces in 3-manifolds*, Studies in Modern Topology, no.5, Math. Assoc. Amer. (P.J.Hilton, ed.) 1968, 39–98.
9. A. Hatcher and W. Thurston, *A presentation of the mapping class group of a closed orientable surface*, Topology **19** (1980) 221–237.
10. G. Hemion, *On the classification of homeomorphisms of 2-manifolds and the classification of 3-manifolds*, Acta Mathematica **142** (1979) 123–155.
11. G. Hemion, *The Classification of knots and 3-Dimensional spaces*, Oxford University Press, Oxford, 1992.
12. C. Hodgson, *Notes on the orbifold theorem*, Preprint.
13. K. Johannson, *Heegaard surfaces in Haken 3-manifolds* , Bull. Amer. Math. Soc. **23** (1990) 91–98.
14. W. Jaco and U. Oertel, *An algorithm to decide if a 3-manifold is a Haken manifold*, Topology **23** (1984) 195–209.
15. Y. Moriah and J.H. Rubinstein, *Heegaard structures of negatively curved 3-manifolds*, Preprint.
16. J.T. Pitts, *Existence and regularity of minimal surfaces on Riemannian manifolds*, Mathematical Notes 27, Princeton University Press, Princeton, New Jersey 1981.
17. J. Pitts and J.H. Rubinstein, *Existence of minimal surfaces of bounded topological type in 3-manifolds*, Proc. of the Centre for Math. Analysis, vol 10, Aust. Nat. Univ., Canberra, 1985, 163–176.

18. J. Pitts and J.H. Rubinstein, *Applications of minimax to minimal surfaces and the topology of 3-manifolds*, Proc. of the Centre for Math. Analysis, vol 12, Aust. Nat. Univ., Canberra, 1987, 137–170.
19. J. Pitts and J.H. Rubinstein, *Minimal surfaces and finiteness of Heegaard splittings of 3-manifolds*, in preparation.
20. J. H. Rubinstein, *An algorithm to recognise the 3-sphere*, Proceedings of ICM, Zurich, 1994, to appear.
21. J.H. Rubinstein and M. Scharlemann, *Characteristic graph varieties and Heegaard splittings of 3-manifolds*, in preparation.
22. F. Waldhausen, *Some problems on 3-manifolds*, Proc. Sympos. Pure Math., vol.32., Amer. Math. Soc., Providence RI, 1987, 313–322.
23. F. Waldhausen, *Heegaard Zerlegungen der 3-sphäre*, Topology **7** (1968) 195–203.

DEPARTMENT OF MATHEMATICS UNIVERSITY OF MELBOURNE, PARKVILLE, VIC 3052, AUSTRALIA
E-mail address: rubin@mundoe.maths.mu.oz.au

AMS/IP Studies in Advanced Mathematics
Volume **2**, 1997 (Part 1)

SURGERY NUMBERS OF 3-MANIFOLDS: A HYPERBOLIC EXAMPLE

DAVID AUCKLY

ABSTRACT. This paper defines the surgery number and the Dehn surgery number of any 3-manifold. Many examples which illustrate the difficulties involved in computing surgery numbers are given, and it is shown that the Dehn surgery number is non-trivial for some hyperbolic manifolds.

In this article, we will define the surgery number and the Dehn surgery number of any 3-manifold. These numbers are topological invariants by their very definition, but they are nearly impossible to compute in general. We will give a series of examples which suggest questions and demonstrate the difficulties involved in computing surgery numbers. The last example in this paper is a hyperbolic manifold with non-trivial Dehn surgery number. In particular, we answer question 3.6(c) from the 1978 Kirby problem list by constructing an irreducible and in fact hyperbolic homology sphere which is not Dehn surgery on any knot [6].

Surgery and Dehn surgery are generic ways of constructing 3-manifolds. A manifold, N, is obtained by surgery on a link in a manifold M, if

$$N - M = \partial[I \times M \bigcup_{n(S^1 \times D^2)} n(D^2 \times D^2)]$$

where the isotopy class of the identification of $n(S^1 \times D^2)$ with $\{1\} \times N(L)$ is called a framing. Likewise, a manifold is Dehn surgery on a link in M if

$$N = (M - \overset{\circ}{N}(L)) \bigcup_{nT^2} n(D^2 \times S^1) \ .$$

Any surgery description of N from M has a corresponding Dehn surgery description.

A Dehn surgery construction corresponds to a surgery construction if and only if the new meridian intersects the old meridian in exactly

1991 *Mathematics Subject Classification.* 57.

one point where the meridian of a knot is the boundary of a fiber in a tubular neighborhood. Every closed orientable 3-manifold may be obtained from surgery on S^3.

Definition. The *surgery metric* is:

$S(M,N) =$ The smallest number of components in a link in M in a surgery description of N.

The *Dehn surgery metric* is:

$D(M,N) =$ The smallest number of components in a link in M in a Dehn surgery description of N.

Let the (Dehn) surgery number of M be $(DS(M) = DS(S^3, M))$, $S(M) = S(S^3, M)$. Let $h(M)$ be the size of a minimal generating set for $H_1(M)$. The weight of a gorup is the size of the smallest set which normally generates the group. Put another way, it is the smallest number of new relations that can be added to the group in order to kill it.

Fact. $h(M) \leq Wt(\pi_1(M)) \leq DS(M) \leq S(M)$.

The first inequality follows because $H_1(M)$ is the abelianization of $\pi_1(M)$. The last is a trivial consequence of the observation that surgery is a special case of Dehn surgery.

If M is Dehn surgery on a link, L, then

$$M = (S^3 - \mathring{N}(L)) \bigcup_{nT^2} n(D^2 \times S^1) \, .$$

Consider,

$$M \bigcup_{nT^2} N(L) = S^3 \bigcup_{nT^2} n(D^2 \times S^1) \, .$$

By Van Kampen's Theorem, $M \bigcup_{nT^2} N(L)$ is simply connected and $\pi_1(M \bigcup_{nT^2} N(L))$ is obtained from $\pi_1(M)$ by adding n relations, thus $\mathrm{Wt}(\pi_1(M)) \leq n$.

The weight is not as useful as it first appears. It is difficult to compute the weight of a group, and the weight is often not equal to the Dehn surgery number. We will prove that the manifold in Figure 1 has Dehn surgery number 2, but the weight of its fundamental group is 1.

It is a straightforward exercise to compute

$$\pi_1(\Sigma\# - \Sigma) = \pi_1(\Sigma)\#\pi_1(\Sigma) \text{ where } \pi_1(\Sigma) = \langle x, y \mid (xy)^2 = x^3 = y^5 \rangle \, .$$

The fundamental group of Σ is normally generated by an element of order 5 and it is also normally generated by an element of order 3, so $\pi_1(\Sigma\# - \Sigma)$ is normally generated by the word which sets the element of

FIGURE 1. $\Sigma \# -\Sigma$

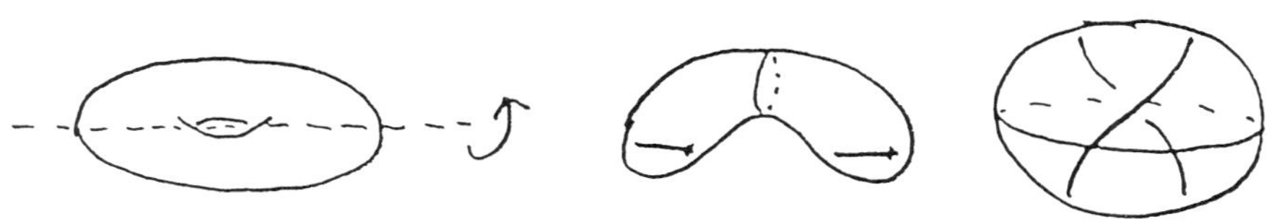

FIGURE 2. 2-fold cover of D^3 branched over 2-arcs.

order 5 in the first factor equal to the element of order 3 in the second factor.

The Dehn surgery number is also related to the unknotting number. To see how, assume that K has unknotting number n. Let M_K be the 2-fold branched cover of S^3 branched over K, and take one small ball around each of the n crossings used to unknot K. The inverse image of one of the small balls is a $D^2 \times S^1$ (see Figure 2).

Cutting out a $S^1 \times D^2$ in the cover and gluing it in differently corresponds to cutting out a ball with two arcs in the base of the covering and gluing the ball in with the opposite crossing. The 2-fold cover of S^3 branched over the unknot is S^3, so M_K is Dehn surgery on a link with n components. We, therefore, have:

Fact. $DS(M_K) \leq u(K)$.

It seems plausible that the number of non-trivial, non-parallel 2-spheres in a manifold should give a lower bound on the Dehn surgery number of the manifold.

The correct conjecture is hard to make because of examples like Figure 4. See [8]. It would also be hard to come up with a correct conjecture as to when incompressible tori imply large surgery number, because of examples like Figure 5.

FIGURE 3. This probably has $DS(M) = 4$.

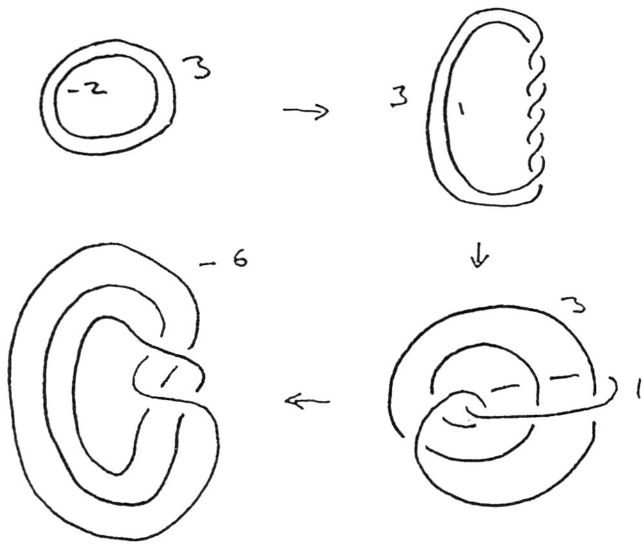

FIGURE 4. The trouble with 2-sphere bounds.

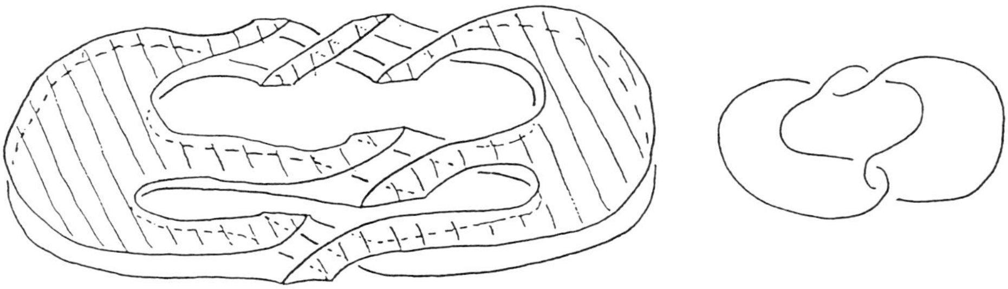

FIGURE 5. An interesting T^2.

Figure 5 shows a $T^2 - 2B^2$ embedded in S^3 with boundary on the boundary of a regular neighborhood of the figure eight knot. This completes to an incompressible torus in -4-framed surgery on the figure eight knot.

It is intuitively clear that it should be possible to construct a homology sphere which is not surgery on any knot by gluing two knot exteriors together. This does not always work. The manifold in Figure 6 is the union of two non-trivial knot exteriors, but it is surgery on a knot.

One indication of how difficult it is to prove anything about surgery numbers is given by the Gordon-Luecke theorem [3].

Theorem. *(Gordon-Luecke) Non-trivial Dehn surgery on a non-trivial not has Dehn surgery number* 1.

If a manifold is non-trivial Dehn surgery on a non-trivial knot, then its Dehn surgery number is less than or equal to 1. The point is that the manifold could be S^3 in which case the Dehn surgery number would be 0. The theorem is that this is not the case. The Gordon-Luecke theorem was around as the P' conjecture for about eighty years.

In light of the above facts, it is natural to ask if there is an irreducible (i.e., no non-trivial 2 spheres) homology 3-sphere ($h(M) = 0$) with Dehn surgery number bigger than one. What about a hyperbolic example? It would also be interesting to have examples of 3-manifolds so that $h(M) \ll \mathrm{Wt}(\pi_1(M))$, $\mathrm{Wt}(\pi_1(M)) \ll DS(M)$ and $DS(M) \ll S(M)$. These conditions are most-likely generic.

Consider lens spaces as examples for the last inequality. John Berge has the following theorem.

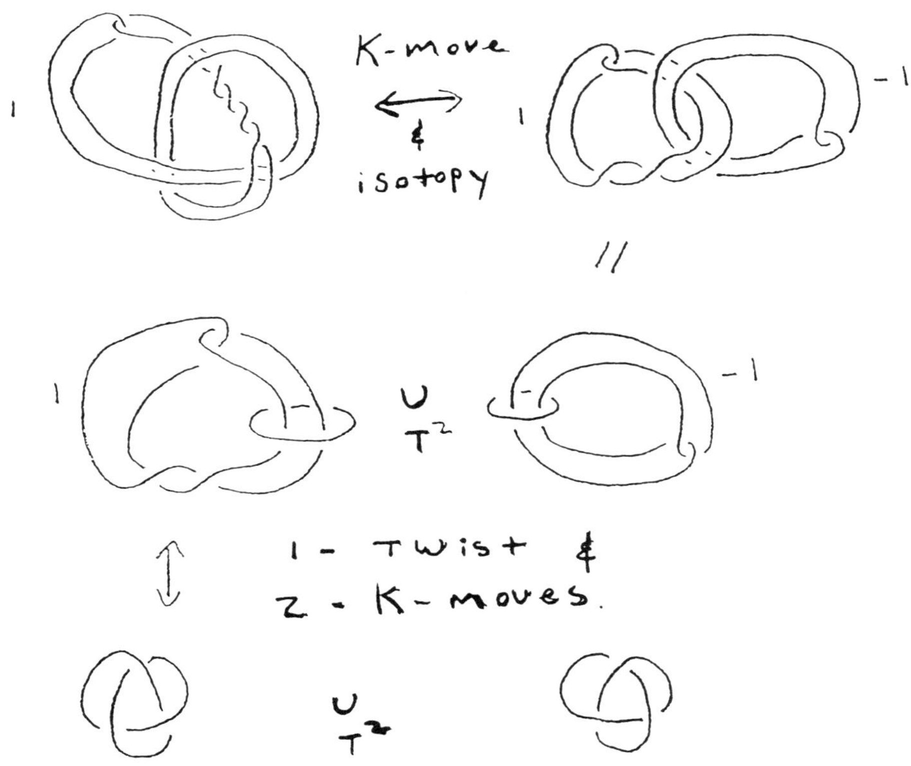

FIGURE 6. Union of a right and a left trefoil.

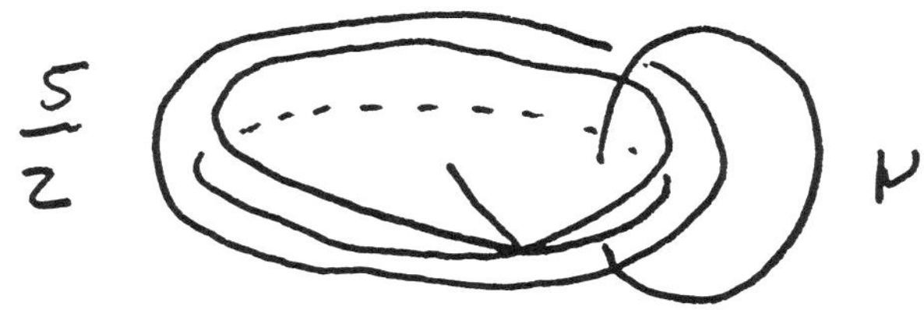

FIGURE 7. Linking pairing for $-L(5,2)$.

Theorem. *(Berge) If K is a simple closed curve on the boundary of a standardly embedded genus two handlebody in S^3 which represents a free generator of both the fundamental group of the handlebody and the fundamental group of its complement, then there is an integral surgery on K which yields a lens space.*

He conjectures that these are the only knots with surgeries which yield lens spaces. From this conjectured form, it is possible to write down a conjectural list of all lens spaces with surgery number one. Since every lens space has Dehn surgery number one, all of the remaining lens spaces would have $S(L) < DS(L)$.

The linking pairing of a rational homology sphere is a slightly obscure tool which can be used to show that the surgery number is bigger than the Dehn surgery number [4]. If M is a rational homology sphere, then $H_1(M)$ is finite, so we may define a pairing on $H_1(M)$ as follows. Let α and β be homology cycles. Some multiple of α, say $n\alpha$, is the boundary of a surface, F. Define the linking pairing of α and β, $\lambda(\alpha,\beta)$, to be the intersection number of F and β divided by n. This is well defined mod $\mathbb{Z}$.

As a concrete example consider $-L(5,2)$. This lens space is $5/2$ Dehn surgery on the unknot in S^3. H_1 is $\mathbb{Z}_5$ generated by the meridian, μ. Five times the meridian is homologous to twice the longitude which bounds a singular disk intersecting the meridian twice, thus $\lambda(\mu,\mu) = 2/5$. See figure 7.

If M is the boundary of a simply connected 4-manifold, W, then the linking pairing of M may be computed from the intersection form of W.

Let

$$Q_W : H^2(W;\partial W) \otimes H^2(W;\partial W) \xrightarrow{\cup} H^4(W;\partial W) \cong \mathbb{Z}$$

be the intersection pairing. This induces a pairing

$$Q'_W : U \times U \to \mathbb{Q}/\mathbb{Z}$$
$$\text{where } U = H^2(W;\partial W) \otimes \mathbb{Q}/H^2(W;\partial W) \ .$$

A straightforward argument with Poincaré duality and the homology exact sequence of the pair (W, M) will show that the above pairing is isomorphic to the linking pairing on M. Looking back at our example, we can see that if $-L(5,2)$ were surgery on a knot then the framing would have to be ± 5. This would imply that $-L(5,2)$ was the boundary of a simply connected 4-manifold with intersection form

$$Q_W : \langle t\rangle \otimes \langle t\rangle \to \mathbb{Z} \ , \qquad Q_W(t,t) = \pm 5 \ .$$

so that the linking pairing would be isomorphic to

$$Q'_W : \langle u \mid 5u = 0\rangle \times \langle u \mid 5u = 0\rangle \to \mathbb{Q}/\mathbb{Z} \ , \text{ where } u = 1/5t \ ,$$
$$\text{so} \quad Q'_W(Ku, Ku) = \pm K^2/5 \ .$$

This contradicts our computation of the linking pairing since 2 is not a square mod 5. We must, therefore, conclude that $-L(5,2)$ is not surgery on any knot. It is, however, surgery on the Hopf link so $DS(-L(5,2)) = 1$ and $S(-L(5,2)) = 2$.

Up to now, we have just used tools from Algebraic topology to address surgery number questions. Gauge theory may also be applied to these questions. The relevant theorem is in [1]. It is really a corollary of a theorem of Taubes [11].

Theorem. *Let K be a compact simply connected 4-manifold with boundary and a non-standard unimodular definite intersection form. If $\Sigma = \partial K$ is a homology sphere and T is a compact oriented 4-manifold such that*

$$\partial T = \Sigma \amalg -\Sigma \amalg M$$
$$H_2(T) = 0$$
$$\textit{and} \quad \pi_1(T) = \textit{Normal Closure}(\pi_1(M)) \ .$$

Then, M is not the boundary of any simply connected definite 4-manifold.

This is related to the surgery number question via the Lemma below.

Lemma. *If M is a homology sphere which is Dehn surgery on a knot, then M is the boundary of a simply connected, definite 4-manifold.*

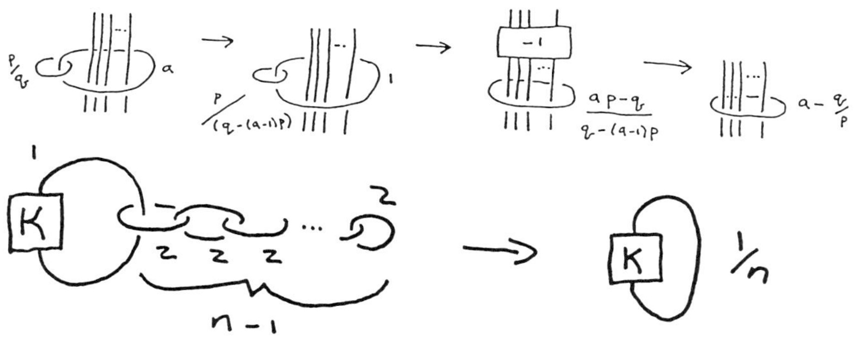

FIGURE 8. Surgery on a chain.

See Figure 8 for a pictorial proof of this.

The Poincaré homology sphere, Σ, is the boundary of the non-standard E_8 manifold. We can construct a T with boundary $\Sigma \perp\!\!\perp -\Sigma \perp\!\!\perp (\Sigma\# - \Sigma)$ by adding a 3-handle to $I \times (\Sigma\# - \Sigma)$. See Figure 9.

It is easy to see that this T satisfies the requirements of the theorem so we conclude that $\Sigma\#-\Sigma$ is not Dehn surgery on any knot. To construct an irreducible or a hyperbolic example, we will add a cobordism onto the bottom of the T.

In our preprint we constructed cobordism to an irreducible example by adding a 1-handle – 2-handle pair to $I \times (\Sigma\# - \Sigma)$ [1]. In [7] Chuck Livingston used link concordance to prove that every homology 3-sphere is homology cobordant to an irreducible homology 3-sphere. Robert Myers showed that Livingston's construction actually produces a homology cobordism to a simple Haken and therefore hyperbolic 3-manifold [9]. We only need to check that the fundamental group of this cobordism is normally generated by the hyperbolic component, and it will follow that the longneck T satisfies the hypothesis of the theorem. Before we describe the Livingston-Myers cobordism, we will describe the example in our preprint. This is interesting because our original construction gives an explicit surgery description of the irreducible example.

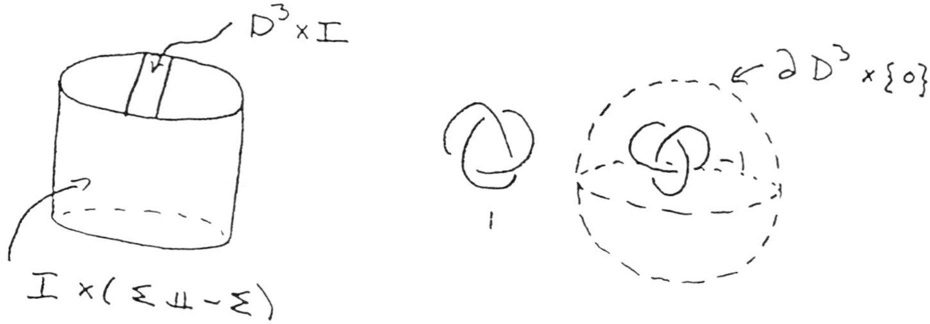

FIGURE 9. How to add a 3-handle.

When a 1-handle – 2-handle pair is added to $I \times (\Sigma \# - \Sigma)$ we get the manifold in figure 11.

Performing a K-move on the unknotted 1-framed component will untie the -1-framed component. Then a K-move on the -1-framed component will generate the following picture of the same manifold.

Figure 12 is an explicit surgery description of an irreducible homology sphere with Dehn surgery number 2.

We will now go back to the Livingston-Myers cobordism. To construct the cobordism, Myers began with $4D^2$'s embedded in a D^4. We will describe this embedding by pictures at $t = 5$, 4, and 3 in $D^4 = [0,5] \times D^3$, where t is the coordinate of the first factor. See Figure 13.

Disks embedded in D^4 in this manner form an object called a tangle concordance. Let $X^4 = D^4 - \mathring{N}(4D^2)$ and let $S_t = (\{t\} \times D^3) \cap X$. Then

$$\partial X = S_0 \bigcup_{F_4 \times \{0\}} (F_4 \times I) \bigcup_{F_4 \times \{1\}} S_5$$

where F_4 is the surface of genus four. Now $\Sigma \# - \Sigma$ has a genus 4 Heegaard splitting. This means that

$$\Sigma \# - \Sigma = (D^2 - 4D^2) \times I \bigcup_f (D^2 - 4D^2) \times I$$

where $f : F_4 \to F_4$ is an orientation reversing homeomorphism [10].

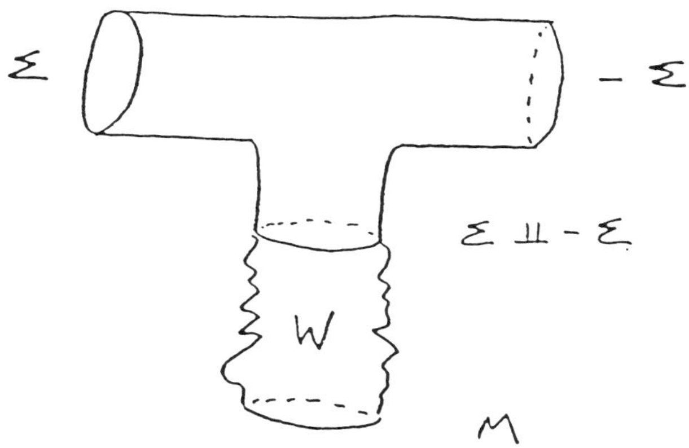

FIGURE 10. Longneck T.

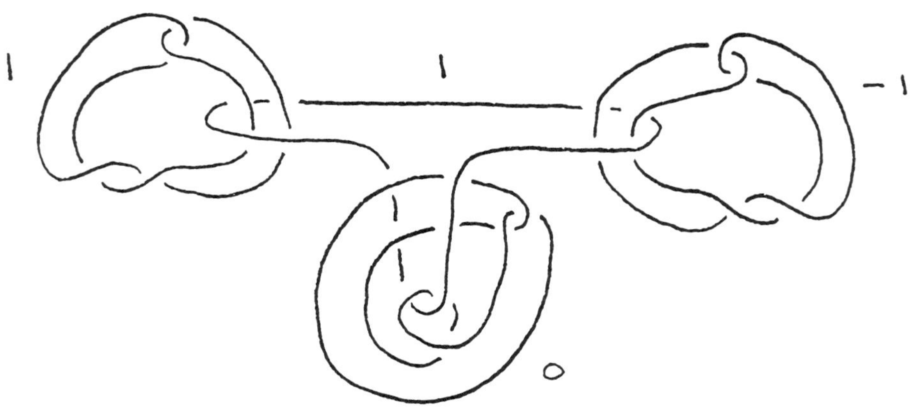

FIGURE 11. The Thing.

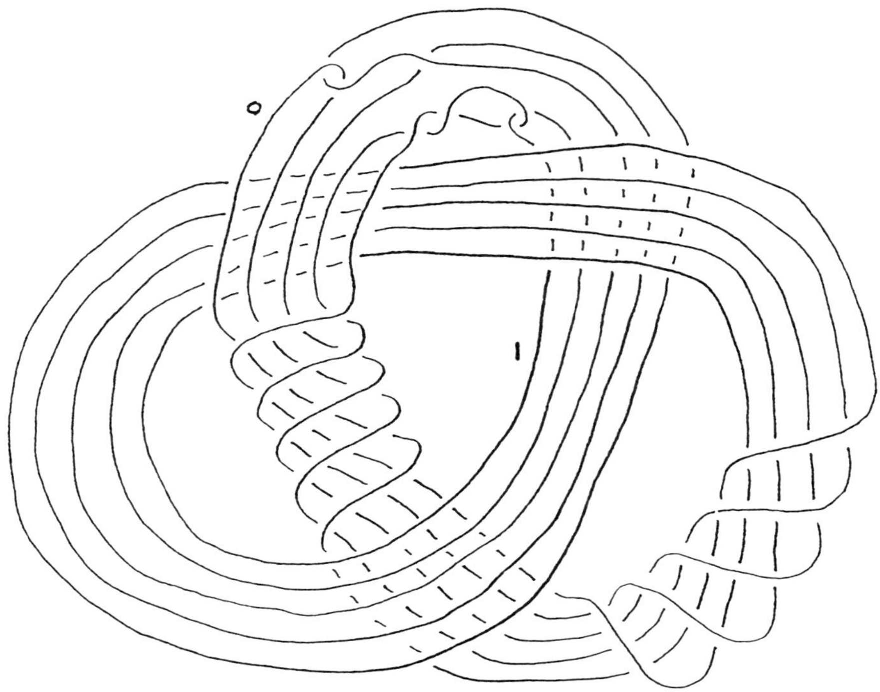

FIGURE 12. The same thing.

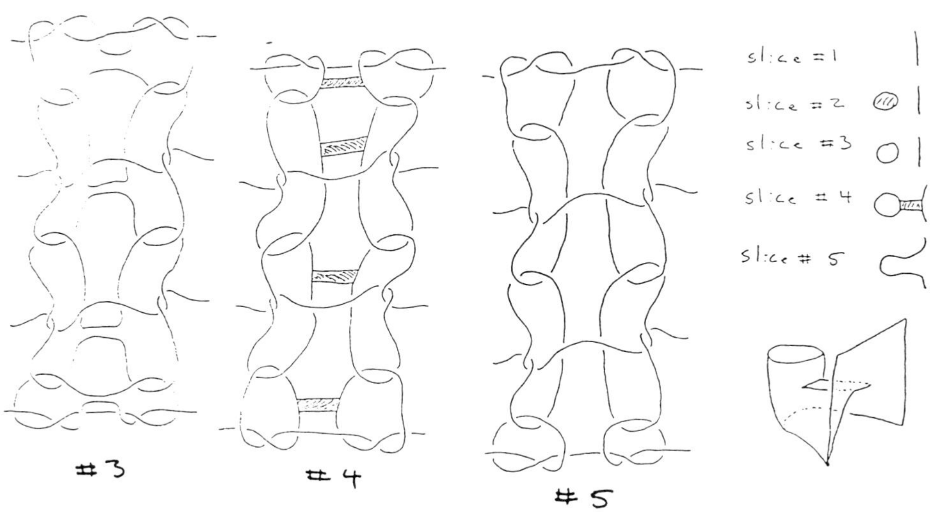

FIGURE 13. The concordance.

This homeomorphism can be extended to a homeomorphism,

$$\bar{f} : F_4 \times I \to F_4 \times I \ .$$

We put all of this together to construct the cobordism, W. Let $W = X \bigcup_{\bar{f}} (-X)$. Myers proved that W is a homology cobordism between $\Sigma \# - \Sigma$ and a hyperbolic homology sphere. It only remains to check that W satisfies the fundamental group condition. The condition on the fundamental group follows from Van Kampen's theorem by the same argument that was used to prove that the weight of the fundamental group is a bound on the Dehn surgery number.

I would like to thank Paul Melvin, Bob Gompf, Karen Uhlenbeck, Dan Freed, John Luecke, Cameron Gordon, Lorenzo Sadun, Oleg Viro, Chuck Livingston, and Geoff Mess for helpful conversations.

References

1. D. Auckly, *An irreducible homology sphere which is not Dehn surgery on a knot*, Preprint.
2. J. Berge, *Some Knots With Surgeries Yielding Lens Spaces*, Preprint.
3. C. McA. Gordon and J. Luecke, *Knots are determined by their complements*, Journal of AMS **22** (1989) 371–415.
4. F. Hirzebruch, W. Neumann and S. Koh, *Differential Manifolds and Quadratic Forms.*, Marcel Dekker, New York (1971).
5. R. Kirby, *A calculus for framed links in* S^3, Invent. Math. **45** (1978) 35–36.
6. R. Kirby, *Problems in low dimensional manifold theory*, Proc. Symposia, Pure Math. **32** (1978).
7. C. Livingston, *Homology cobordisms of 3-manifolds, knot concordances, and prime knots*, Pacific J. Math. **94** (1981) 193–206.
8. L. Moser, *Elementary surgery along a torus knot*, Pacific J. Math. **38** (1971) 737–745.
9. R. Myers, *Homology cobordisms, link concordances, and hyperbolic 3-manifolds*, Trans. Amer. Math. Soc. **278** (1983) 271–288.
10. D. Rolfsen, *Knots and Links*, Publish or Perish, Berkeley (1976).
11. C. Taubes, *Gauge theory on asymptotically periodic 4-manifolds*, J. Diff. Geom. **25** (1987) 363–430.

Department of Mathematics, The University of Texas at Austin, Austin, TX 78712-1082

E-mail address: dav@math.utexas.edu

AMS/IP Studies in Advanced Mathematics
Volume **2**, 1997 (Part 1)

NON-HYPERBOLIC MANIFOLDS OBTAINED BY DEHN SURGERY ON HYPERBOLIC KNOTS

MARIO EUDAVE-MUÑOZ

INTRODUCTION.

Let k be a knot in S^3. Let $M_k = S^3 - int\eta(k)$ the exterior of k. Denote by $M_k(r)$ the 3-manifold obtained by performing r-Dehn surgery on k, where r is a slope on ∂M_k. $\Delta(r,s)$ denotes the minimal geometric intersection number of two slopes r, s on ∂M_k. For an excellent exposition of the main problems on Dehn surgery on knots see the survey paper of C.McA. Gordon [11].

If k is not a satellite and not a torus knot, then by results of W. Thurston [21], k is hyperbolic and $M_k(r)$ is hyperbolic for all but finitely many r. We consider the following problem: Suppose k is an hyperbolic knot; when is it possible that $M_k(r)$ is not a hyperbolic manifold? or, when does $M_k(r)$ contain an incompressible torus, or, when is it a Seifert fibre space?

In [8] it was shown that if k is a non-satellite, strongly invertible knot and $M_k(r)$ contains an incompressible torus then $\Delta(r,\mu) \leq 2$, where μ is a meridian of k; in other words, r is homologous to several meridians and at most two longitudes. It was proved also that if $M_k(r)$ contains incompressible tori and $\Delta(r,\mu) = 2$, then there is an incompressible torus T in $M_k(r)$ which intersects the reglued solid torus in two meridianal disks, or in other words, there is a properly embedded, incompressible, ∂-incompressible, punctured torus in M_k whose boundary consists of two curves on ∂M_k of slope r. Gordon and Luecke [14] have announced a proof of these results for all hyperbolic knots. The last statement can be seen as a kind of generalization of the cabling conjecture, which says that only certain surgery on cable knots yields reducible manifolds, or more explicitly, if $M_k(r)$ is reducible then there is a properly embedded, essential annulus on M_k whose boundary has slope r on ∂M_k.

Here we show an infinite family of knots k, where k is strongly invertible and hyperbolic, and for some r so that $\Delta(r,\mu) = 2$, $M_k(r)$ contains an incompressible torus T which hits the surgered solid torus twice. Such torus divides the manifold $M_k(r)$ into two Seifert fibred spaces with base a disk and 2 exceptional fibres. The simplest example is 37/2-surgery on the (-2,3,7) pretzel knot. This family of knots is quite general, and it could be complete, in the sense that any hyperbolic knot having a half integer surgery producing a toroidal manifold may belong to this family. These knots also admit two integral surgeries which yield a Seifert fibre space with base a sphere and at most 3 exceptional fibres. Then for these knots there are 3 different nontrivial surgeries producing non-hyperbolic manifolds. It is also satisfied that any two of those exceptional surgeries are at distance 1. The trivial surgery on a knot k produces S^3, which does not admit an hyperbolic structure, so from now on, as many authors do, when talking about non-hyperbolic surgeries we will include the trivial surgery. So the knots mentioned above have four non-hyperbolic surgeries. An infinite subfamily of these knots have two more non-hyperbolic surgeries, both are integral, once produces a Seifert fibre space with base a sphere and 3 exceptional fibres and the other produces a manifold containing an incompressible torus which hits the surgered solid torus twice, and divides the manifold into two Seifert fibred spaces with base a disk and 2 exceptional fibres. This family of knots admitting six non-hyperbolic surgeries was first discovered by C. Hodgson and J. Weeks [15]. Other examples of hyperbolic knots having several non-hyperbolic surgeries are given in [2].

We show a family of prime tangles (B, t), with the property of having sums with rational tangles which produce the trivial knot, a doubly composite knot, and two Montesinos knots or links. By taking the double covering of the 3-ball B branched along the strings of the tangle (B, t), we get the exterior of a knot k having the desired properties. $\Delta(r, s)$ also denotes the distance between two rational tangles r and s [8]. In section 1 we describe the tangles (B, t) and prove some of its properties. In section 2 we use the results of section 1 to state the existence and properties of the knots k. In section 3 we make an explicit construction of the knots k, by constructing the double branched covering of the tangles (B, t). In section 4 we discuss some open problems and propose some conjectures. Through the paper we work mainly with figures. By a link we mean a knot or a link of two components.

I would thank John Luecke, Cameron Gordon, Marty Scharlemann and Jeff Weeks for stimulating conversations. I am especially grateful to Steven Bleiler for pointing out to me the examples in [2], there it is shown a family of hyperbolic knots having five non-hyperbolic surgeries, one of them a half integer surgery producing a manifold which has an incompresible torus. After unsuccesfully trying to fit their example into the family of knots I had previously, i.e. the knots denoted by

FIGURE 1

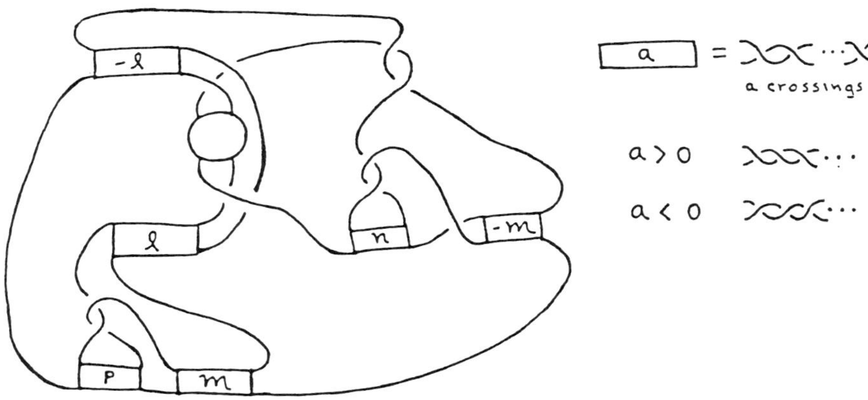

FIGURE 2

$k(l,m,n,0)$, I discover another family of knots with the same properties, which is denoted by $k(l,m,0,p)$. It can be checked that the example which is given in figure 10 of [2] correspond to $k(3,-1,0,p)$ in our notation.

1. TANGLES WITH EXOTIC SUMS.

Consider the tangle $B(l,m,n,p)$ pictured in Figure 2, where l,m,n are integers, and a box indicates the number of crossings of two strands. Assume that always one of n or p is equal to 0. If $p=0$ assume that $m \neq 0$, $l \neq 0, \pm 1$, $(l,m) \neq (2,1)$, $(-2,-1)$, and $(m,n) \neq (1,0)$, $(-1,1)$. If $n=0$ assume that $m \neq 0,1$, $l \neq 0, \pm 1$, $(l,m,p) \neq (-2,-1,0)$, $(2,2,1)$.

Proposition 1.1. *The tangle $B(l,m,n,p)$ has the following properties, where μ, r, s, t, u, v, w, denote the tangles shown in Figure 3,*

(a) $B(l,m,n,p)+\mu$ is the trivial knot (see Figure 4).

To see that just untwist crossings on the boxes with opposite numbers.

FIGURE 3

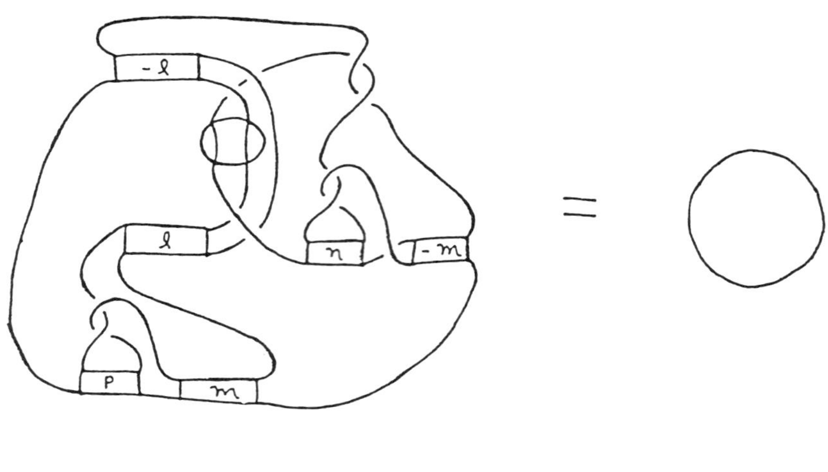

FIGURE 4

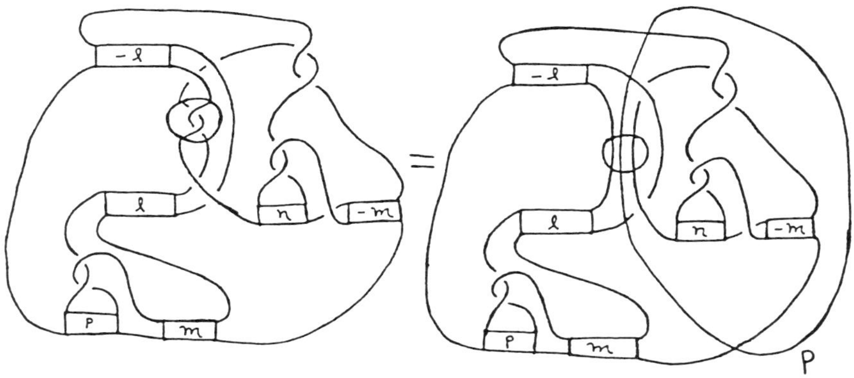

FIGURE 5

(b) $B(l, m, n, p) + r$ *is a sum of two prime tangles, i.e. it is doubly composite. Let* P *be the decomposing sphere pictured above, also called Conway sphere. Note that* P *intersects the tangle* r *in a disk (see Figure 5). In fact, this knot is expressed as a sum of the tangles* (B_1, t_1) *and* (B_2, t_2) *pictured below. Both of these tangles are partial sum of two rational tangles, i.e., they are Montesinos tangles (see Figure 6).*

(c) $B(l, m, n, p) + s$ *is a Montesinos link, formed with at most 3 rational tangles. It is a 2-bridge link only when* (l, m, n, p) *is one of the following:* $(2, m, n, p)$, $(-3, -1, n, 0)$, $(-2, -2, n, 0)$, $(l, 1, \pm 1, 0)$, $(l, 2, 0, p)$, $(-2, -1, 0, \pm 1)$ *(see Figure 7).*

(d) $B(l, m, n, p) + t$ *is a Montesinos link, formed with at most 3 rational tangles. It is a 2-bridge link only when* (l, m, n, p) *is one of the following:* $(-2, m, n, p)$, $(l, -1, 0, p)$, $(l, -1, 2, 0)$, $(l, -2, 1, 0)$, $(2, 2, n, 0)$, $(3, 1, n, 0)$, $(2, 3, 0, 1)$, $(3, 2, 0, 1)$, $(2, 2, 0, 2)$ *(see Figure 8).*

(e) $B(2, m, n, 0) + u$ *is a sum of two prime tangles when* $m \neq \pm 1$, $(m, n) \neq (2, 0)$, *in fact, it is a sum of two Montesinos tangles. Note that the Conway sphere intersects the tangle* u *in a disk. When* $m = \pm 1$ *or* $(m, n) = (2, 0)$ *it is a Montesinos link formed with 3 non-integral rational tangles (see Figure 9).*

(f) $B(l, -1, 0, p) + v$ *is a sum of two prime tangles when* $l \neq -2$, *in fact, it is a sum of two Montesinos tangles. Note that the Conway sphere intersects the tangle* v *in a disk. When* $l = -2$ *it is a Montesinos link formed with 3 non-integral rational tangles (see Figure 10).*

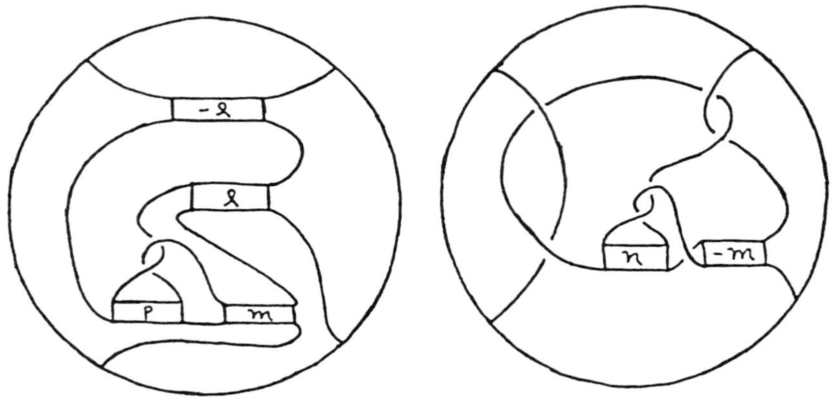

FIGURE 6. (B_1, t_1) (B_2, t_2)

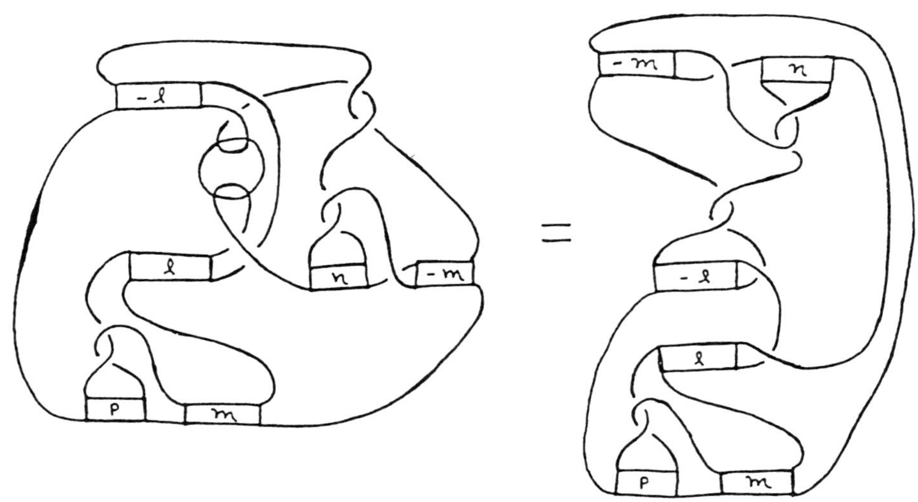

FIGURE 7

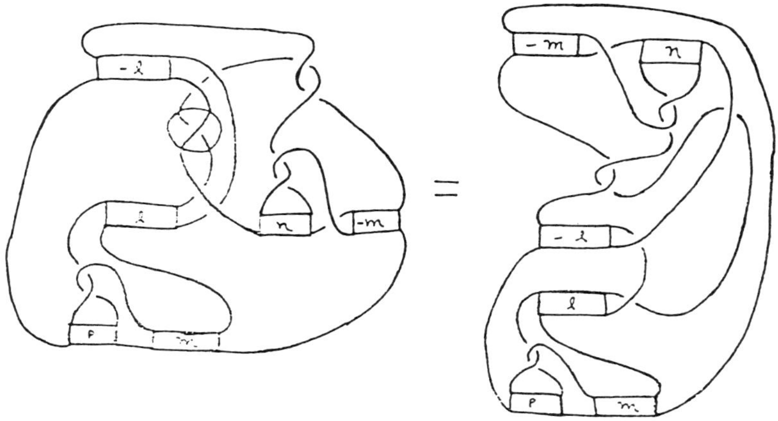

FIGURE 8

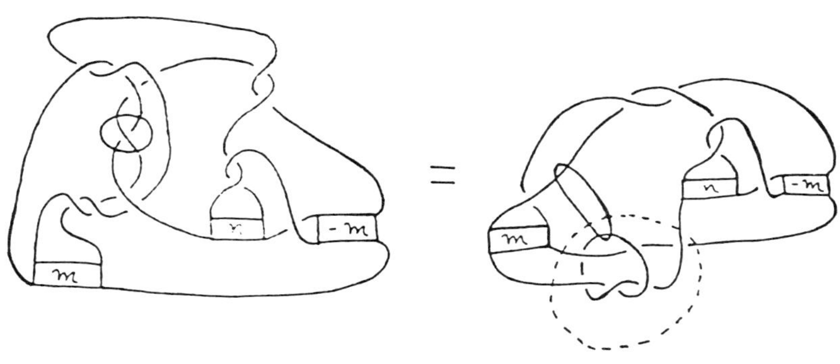

FIGURE 9

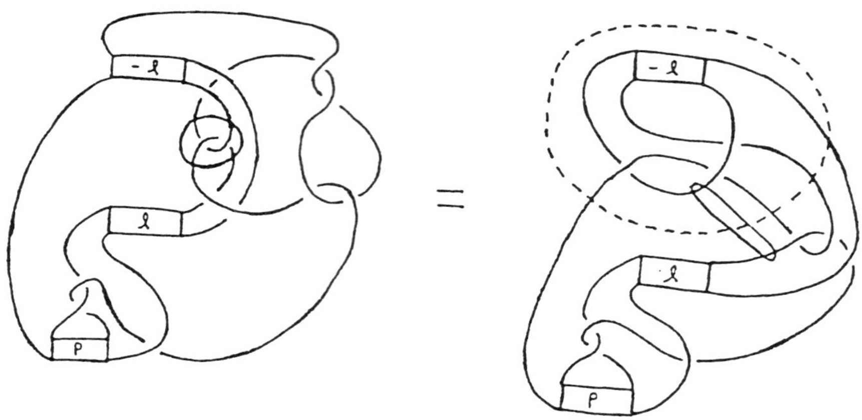

FIGURE 10

(g) $B(2,-1,n,0)+w$ is a sum of two prime tangles, in fact, it is a sum of two Montesinos tangles. Note that the Conway sphere intersects the tangle w in a disk (see Figure 11).
Furthermore $\Delta(\mu,r)=2$, $\Delta(\mu,s)=1$, $\Delta(\mu,t)=1$, $\Delta(\mu,u)=1$, $\Delta(\mu,v)=1$, $\Delta(\mu,w)=1$, $\Delta(r,s)=1$, $\Delta(r,t)=1$, $\Delta(r,u)=3$, $\Delta(r,v)=3$, $\Delta(r,w)=5$, $\Delta(s,t)=1$,$\Delta(s,u)=1$, $\Delta(s,v)=2$, $\Delta(s,w)=2$, $\Delta(t,u)=2$, $\Delta(t,v)=1$,$\Delta(t,w)=3$, $\Delta(u,v)=3$, $\Delta(u,w)=1$, $\Delta(v,w)=4$.

Proof. Most of the proof follows immediately from the figures. In cases (b), (c), (d) it is obvious that after making the sum of the tangles the resulting link has the desired property. In the remaining cases a few moves are needed to put the link in the desired form. For case (b) note that the tangles (B_1,t_1) and (B_2,t_2) are prime tangles, in fact they are rational tangles only when the parameters attain some banned value. In cases (c), (d) it is not difficult to see that one of the rational tangles in the decomposition of the link can be integral only when the parameters attain one of the special values, so by the classification of Montesinos links it follows that a 2-bridge link is obtained only in the special cases.

It may be possible that some other subfamily of the given tangles admit more exotic sums of rational tangles. Note also that $B(2,-1,n,0)$ has six exotic sums with rational tangles, and $B(2,-1,0,0)$ has seven

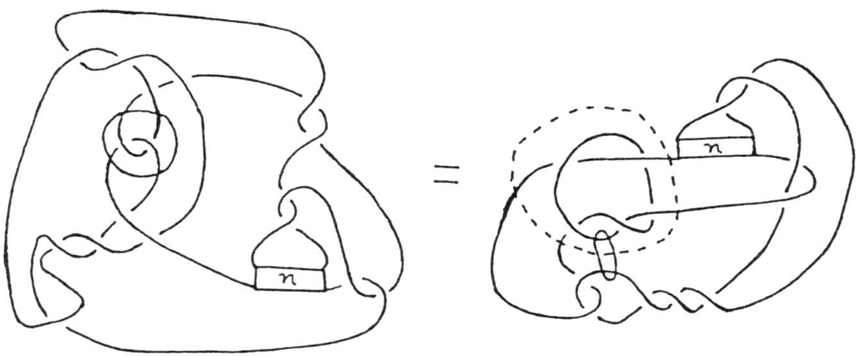

FIGURE 11

exotic sums. □

Lemma 1.2. *$B(l,m,n,p)$ is a prime tangle.*

Proof. This tangle does not contain local knots, for it admits, by 1.1(a), a sum with a rational tangle which produces the trivial knot. It is non-trivial, because by 1.1(b), it can be summed with a rational tangle to get a doubly composite knot; note that 2-bridge knots are doubly prime [1, th.2.2]. □

Let $k = B(l,m,n,p) + r$ as in 1.1(b). Then unknotting number $(k) = 1$. Note that the obvious unknotting crossing intersects the Conway sphere P.

Lemma 1.3. *There is no unknotting crossing for k disjoint from the sphere P.*

Proof. Suppose first that $p = 0$. Suppose there is an unknotting crossing disjoint from P. Then the crossing is performed inside B_1 or B_2, suppose first it is on B_2. Note that $(B_1, t_1)+$ Figure 12 is the trivial knot (see Figure 13), and because (B_1, t_1) is a prime tangle, in fact a Montesinos tangle, this is the only way we can get a trivial knot by summing rational tangles to (B_1, t_1) [3], [9, th.4]; also remember that the sum of two prime tangles is never trivial [17, th.1]. So if there is an unknotting crossing for k which lies on B_2, it has to transform (B_2, t_2) into the tangle shown in Figure 14.

FIGURE 12

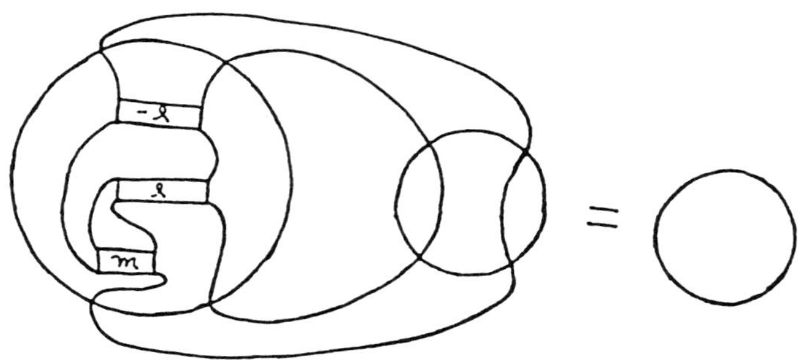

FIGURE 13

FIGURE 14

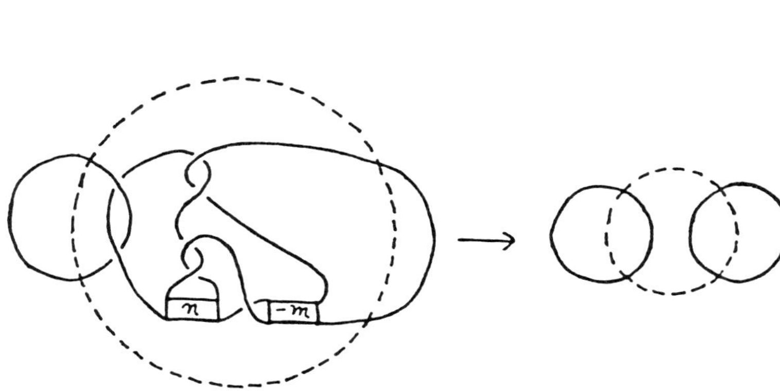

FIGURE 15

FIGURE 16

This implies that doing a crossing change on the link pictured in Figure 15 we get the trivial link. Note that this link is composite, for the right string of (B_2, t_2) is knotted. But this is impossible, for [7, th.2] implies that a composite link of two components cannot be transformed into the trivial link by performing a crossing change (this has also been observed in [13]).

Suppose the unknotting crossing for k lies on (B_1, t_1). Note that the only way to sum a trivial tangle to (B_2, t_2) and get the trivial knot is the pictured below, and this sum gives the trivial knot only if $n = 0, 1$ [9, th.4]. So the tangle (B_1, t_1) has to be transformed by doing a crossing change into the tangle shown in Figure 16. This may be possible only if l is even, or if both l and m are odd. An argument as in the first case yields a contradiction (see Figure 17).

When n=0 a similar argument can be done. □

Note that the knot k as above produce examples which show that case (b) of corollary 1.5 of [8] does happens.

By playing a little bit with the figures of the tangles it is not difficult to settle down the following proposition. By the mirror image of a tangle we mean the mirror image obtained from a given projection. Two tangles are equivalent if there is an homeomorphism between them, which may not fix their boundaries.

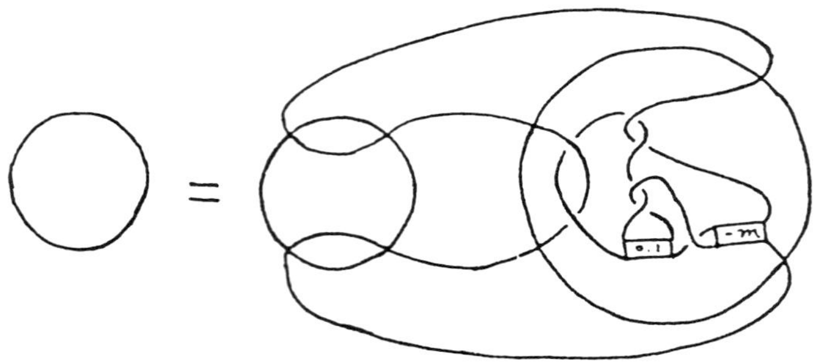

FIGURE 17

Proposition 1.4. *The tangles $B(l,m,n,p)$ have the following properties:*

(a) The mirror image of the tangle $B(l,m,n,0)$ is the tangle $B(-l,-m,1-n,0)$.

(b) The mirror image of the tangle $B(l,m,0,p)$ is the tangle $B(-l,1-m,0,1-p)$.

(c) $B(l,\pm1,n,0) = B(-l\pm1,\pm1,n,0)$.

(d) $B(2,-1,n,0) = B(-3,-1,n,0) = B(2,2,0,n)$.

(e) The knot $k = B(l,m,n,p)+r$ has no mutants, i.e. any of the mutants obtained from the tangle decomposition of k given in 1.1(b) is equivalent to itself.

2. HYPERBOLIC KNOTS WITH EXOTIC SURGERIES.

Let $k(l,m,n,p)$ be the double covering of the 3-ball B branched along the strings of the tangle $B(l,m,n,p)$. $k(l,m,n,p)$ is the exterior of a knot in S^3, which we also denote by $k(l,m,n,p)$; this is because $B(l,m,n,p)$ admits a sum with a rational tangle which produces the trivial knot. The double covering of a 3-ball branched along the strings of a rational tangle is a solid torus; so the double covering of S^3 branched along a link which is a sum of $B(l,m,n,p)$ and a rational tangle is obtained by making Dehn surgery on $k(l,m,n,p)$ [19]. If P is a Conway sphere for a knot or link k, then the preimage of P in the double covering of S^3 branched along that knot is an incompressible torus. Also the double covering of S^3 or a 3-ball branched along a Montesinos link or tangle is a Seifert fibre space [18]. These observations and proposition 1.1 imply the following theorem, where l, m, n, p are integer parameters with the same restrictions as for the tangles $B(l, m, n, p)$. The slopes μ, r, s, t, u, v and w are in correspondence with the tangles with the same name in proposition 1.1, where μ is the meridian. The coordinates of these slopes depend, of course, of the parameters l, m, n, p.

Theorem 2.1. *There is a family of knots $k(l, m, n, p)$ such that*

(b) For some non-integral slope r, with $\Delta(r, \mu) = 2$, the manifold $M_k(r)$ contains an incompressible torus T.

(c) For an integral slopes s, the manifold $M_k(s)$ is a Seifert fibre space with base a disk and at most 3 exceptional fibres. It is a Lens space only when (l, m, n, p) is one of the following: $(2, m, n, p)$, $(-3, -1, n, 0)$, $(-2, -2, n, 0)$, $(l, 1, \pm 1, 0)$, $(l, 2, 0, p)$, $(-2, -1, 0, \pm 1)$.

(d) For an integral slopes t, the manifold $M_k(t)$ is a Seifert fibre space with base a disk and at most 3 exceptional fibres. It is a Lens space only when (l, m, n, p) is one of the following: $(-2, m, n, p)$, $(l, -1, 0, p)$, $(l, -1, 2, 0)$, $(l, -2, 1, 0)$, $(2, 2, n, 0)$, $(3, 1, n, 0)$, $(2, 3, 0, 1)$, $(3, 2, 0, 1)$, $(2, 2, 0, 2)$.

(e) For $k(2, m, n, 0)$ there is an integral slope u, such that when $m \neq \pm 1$ and $(m, n) \neq (2, 0)$, the manifold $M_k(u)$ contains an incompressible torus. When $m = \pm 1$ or $(m, n) = (2, 0)$, $M_k(u)$ is a Seifert fibre space with base a sphere and 3 exceptional fibres.

(f) For $k(l, -1, 0, p)$ there is an integral slope v, such that when $l \neq -2$, the manifold $M_k(v)$ contains an incompressible torus. When $l = -2$ $M_k(v)$ is a Seifert fibre space with base a sphere and 3 exceptional fibres.

(g) For $k(2, -1, n, 0)$ there is an integral slope w, such that the manifold $M_k(w)$ contains an incompressible torus.

All the incompressible tori obtained after performing Dehn surgery intersect the reglued solid torus in two disks, and divide the corresponding manifold into two Seifert fibre spaces with base a disk and two exceptional fibers.

Furthermore $\Delta(r, s) = 1$, $\Delta(r, t) = 1$, $\Delta(r, u) = 3$, $\Delta(r, v) = 3$, $\Delta(r, w) = 5$, $\Delta(s, t) = 1$, $\Delta(s, u) = 1$, $\Delta(s, v) = 2$, $\Delta(s, w) = 2$, $\Delta(t, u) = 2$, $\Delta(t, v) = 1$, $\Delta(t, w) = 3$, $\Delta(u, v) = 3$, $\Delta(u, w) = 1$, $\Delta(v, w) = 4$.

Proposition 2.2. *Any knot $k(l, m, n, p)$ is simple, i.e. there is no incompressible and non ∂-parallel annulus or torus in its exterior. Thus all these knots are hyperbolic.*

Proof. Suppose R is an incompressible and non ∂-parallel torus on the exterior of $k(l, m, n, p)$. As k is a strongly invertible knot, there is an equivariant incompressible torus [16, th.4.3], call it R. There are two cases:

1) R is incompressible in $M_k(r)$.

It is not difficult to show that the torus T is the unique incompressible torus in $M_k(r)$, and T is clearly equivariant. So R and T are isotopic, and there is an equivariant isotopy carrying T to R [4]. Remember that T double branch covers the Conway sphere P, so this implies that P can be isotoped to be disjoint from the rational tangle which was summed to $B(l,m,n,p)$ in 1.1(b), then there is an unknotting crossing for $B(l,m,n,p)+r$ disjoint from P. This is in contradiction with 1.3.

2) R is compressible in $M_k(r)$.

As the slope r is non-integral, [6, th.2.4.4] implies that $k(l,m,n,p)$ is a cable knot with R the cabling torus. Then by [10, § 7] for some integral slope q, at distance 1 from the slope r, $M_k(q)$ is a reducible manifold. This is impossible in our case, because doing surgery on the two integral slopes at distance 1 from r, i.e. s and t, produces Seifert fibre spaces, hence irreducible manifolds.

If $k(l,m,n,p)$ is non-simple and non-satellite then it is a torus knot. But this is not possible, for the only surgery on a torus knot which produces a manifold containing an incompressible torus is the 0-surgery on the trefoil knot. □

Corollary 2.3. *There is an infinite family of hyperbolic knots such that four different Dehn surgeries produce non-hyperbolic manifolds. An infinite subfamily admits six surgeries producing non-hyperbolic manifolds.*

In fact, the families $k(2,m,n,0)$ and $k(l,2,0,p)$ admit five non-hyperbolic surgeries. The family $k(2,-1,n,0)$ admit six of such surgeries, and the knot $k(2,-1,0,0)$ admits seven such surgeries. It follows fron the results in section 3 that the knots $k(2,-1,n,0)$ are the mirror image of the knots obtained by doing 1/n surgeries on the unknotted component of the *Whitehead sister* link (see [2, figure 3]); it also follows that the knot $k(2,-1,0,0)$ is the mirror image of the (-2,3,7) pretzel knot.

Some of the Seifert fibre spaces obtained by surgery on $k(l,m,n,p)$ have finite non-cyclic fundamental group, here we do not establish exactly which knots accept this kind of surgeries (look at [2] and [5] for results and examples about finite surgeries). In theory it is possible to calculate the invariants which determine the Seifert fibre spaces $M_k(s)$, $M_k(t)$, and the pieces of $M_k(r)$, $M_k(u)$, $M_k(v)$ and $M_k(w)$, and calculate the corresponding surgery coefficients. This will be done elsewhere; here we are only concerned with figures.

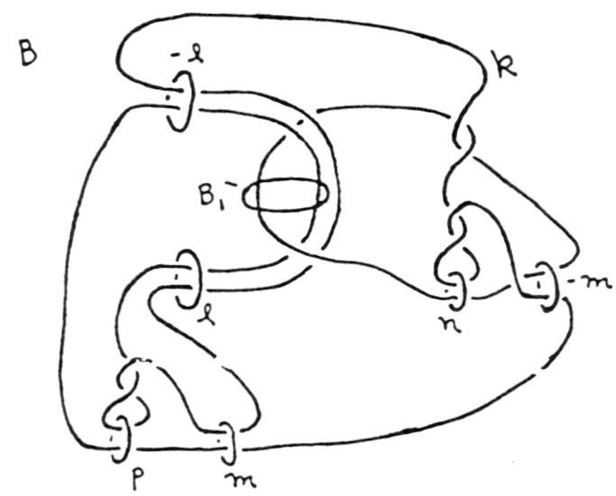

FIGURE 18

Corollary 2.4. *The tangle $B(l,m,n,p)$ is atoroidal, i.e. there is no incompressible torus in the complement of its strings. Every sphere intersecting the strings of the tangle in four points is trivial, i.e. either it is the boundary of a 3-ball which determines a trivial tangle with its intersection with the strings of the tangle, or it is parallel to ∂B relative to the strings of the tangle.*

Proof. If this does not happen, then taking the preimage of the essential torus or sphere on the double branched covering of the tangle $B(l,m,n,p)$, produces an incompressible, non ∂-parallel torus on the exterior of $k(l,m,n,p)$, which contradicts 2.2. □

3. AN EXPLICIT CONSTRUCTION OF THE KNOTS $k(l,m,n,p)$.

We will give now an explicit description of the knots $k(l,m,n,p)$. Consider the Figure 18. The trivial knot k is expressed as the sum of two tangles, each determined by the intersection of k with the balls B and B_1. The circles labeled a_l, a_m, a_n, or simply l,m,n, etc. indicate that we are performing i half twists, $i=l,m,...$, along the disk bounded by the circle a_i; this disk intersects the knot k twice. Note that $B(l,m,n,p)$ is obtained from the tangle $(B, B\cap k)$ by doing these twists.

The knot k is trivial, so it can be isotoped so that it looks like a standard circle, this is made explicit in the following sequence of figures, from Figure19 to Figure 21.

Now from Figure 21 take the double covering of S^3 branched along k, to get Figure 22.

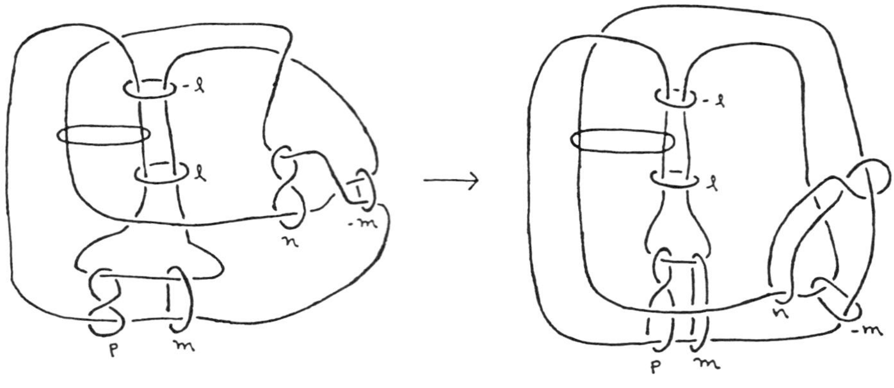

FIGURE 19

Note that the preimage of each of the curves $a_l, a_m, a_n, \ldots$, consists of two curves, which we label J_i and J_i', $i = l, m, \ldots$. Denote the double branched covering of the 3-ball B_1, by B_1 again, this is a solid torus; denote the core of B_1 by k_1, this is a trivial knot. Take a small ball B_i, such that the circle a_i lies on its boundary. Suppose B_i is chosen so that $B_i \bigcap k$ consists of two trivial arcs, i.e. $(B_i, B_i \bigcap k)$ looks like the $1/0$-rational tangle. Making i half twist along the circle a_i is equivalent to remove the tangle $(B_i, B_i \bigcap k)$ and replace it by a i-rational tangle. The preimage of B_i on the double cover of S^3 branched along k is a solid torus B_i'. The preimage of the curve a_i consists of two curves J_i and J_i' contained on $\partial B_i'$; we can suppose that one of them, say J_i, is a core of B_i', and the other one, J_i', is a longitude on $\partial B_i'$, which covers a latitude of the tangle $(B_i, B_i \bigcap k)$. It follows that the double cover of S^3 branched along k, but after making i-half twists along the curve a_i, is obtained by making $(-i)$-Dehn surgery on the knot J_i, with respect to the longitude J_i'. Let λ_i be a standard longitude of J_i; note that $\lambda_i = J_i$ if $i = l, -l, n, p$, and that $J_i = 4\mu_i + \lambda_i$ if $i = m, -m$, where μ_i denotes a meridian of J_i. So we have the following

Proposition 3.1. *The knot $k(l, m, n, p)$ is obtained from k_1 by the following surgery description, where the symbol i^*, $i = l, m, \ldots$, means that we are performing Dehn surgery on J_i with slope $(1, -i)$, if $i = l, -l, n$ or p, and slope $(1 - 4i, -i)$ if $i = m$ or $-m$.*

Note that the curves J_l and J_{-l} bound an annulus L, which intersects the solid torus B_1 in a disk. Note also that the boundary of L on a neighborhood of J_l is precisely the longitude λ_l, similarly for J_{-l}.

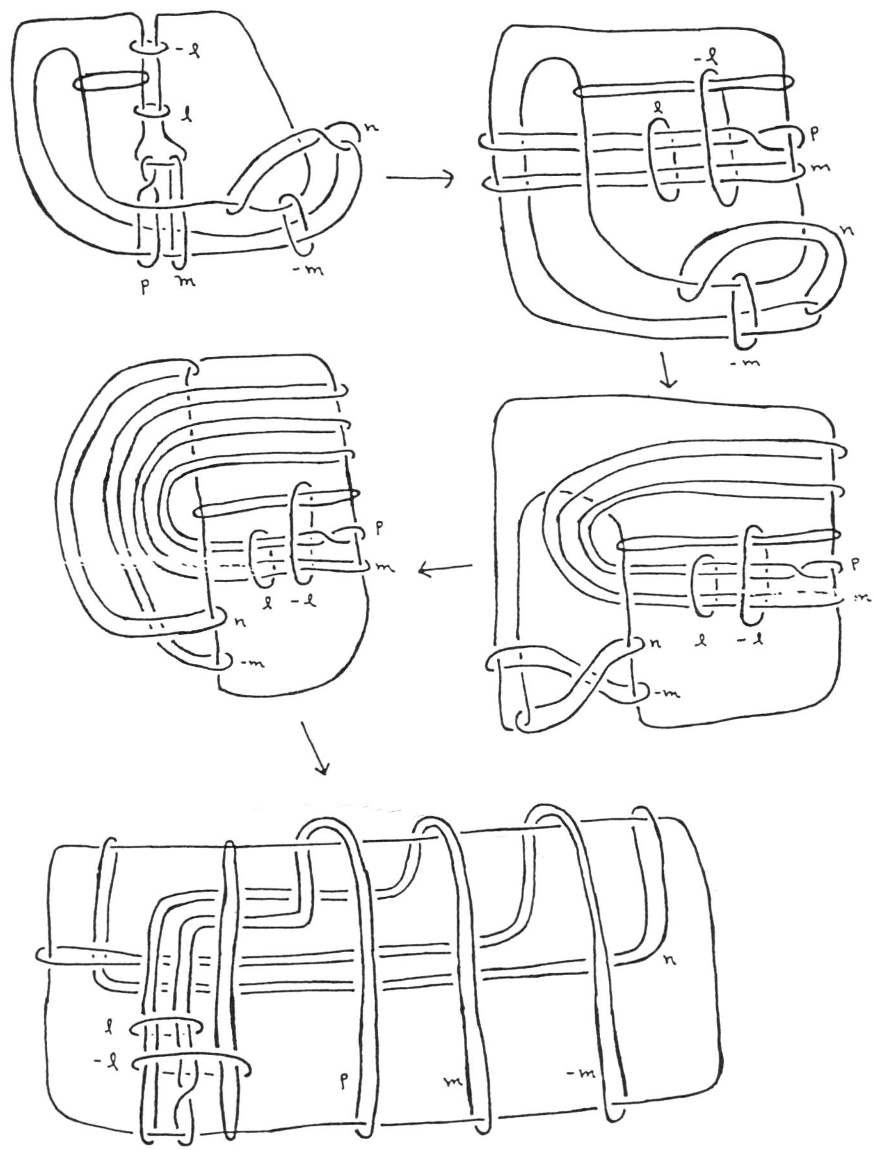

FIGURE 20

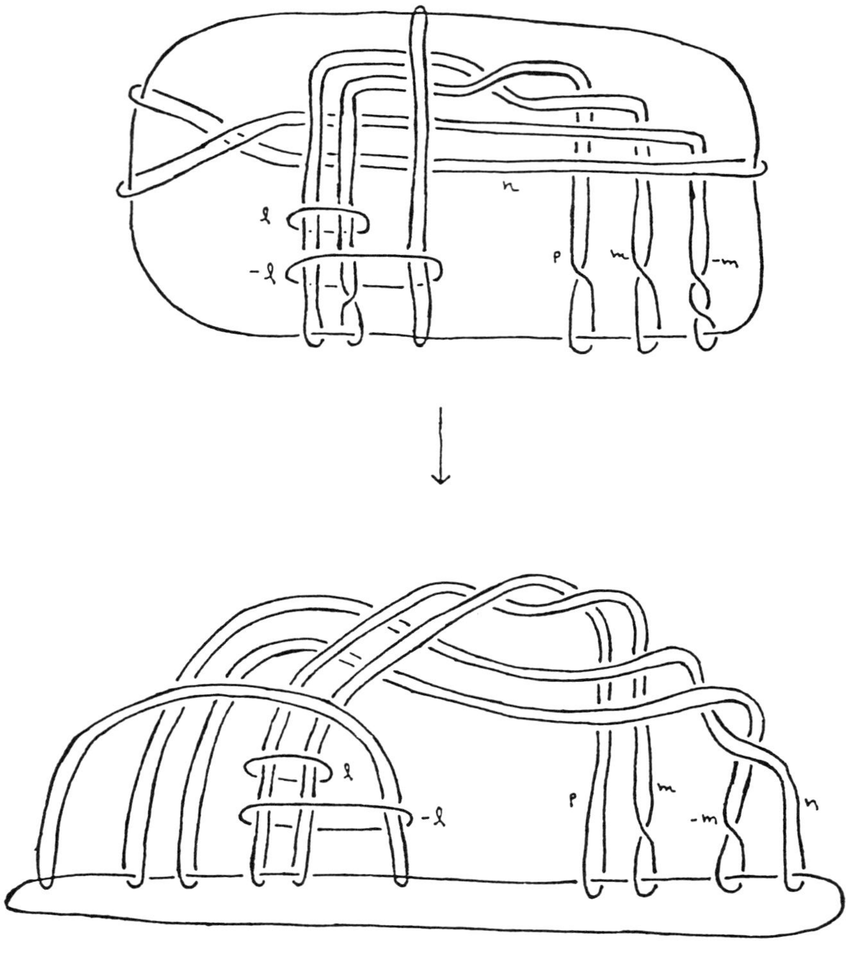

FIGURE 21

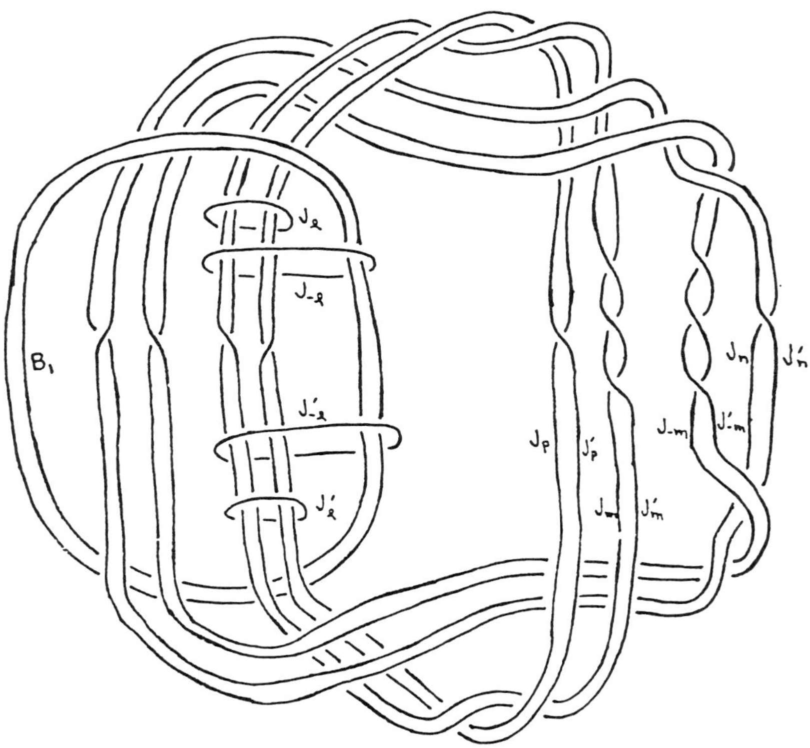

FIGURE 22

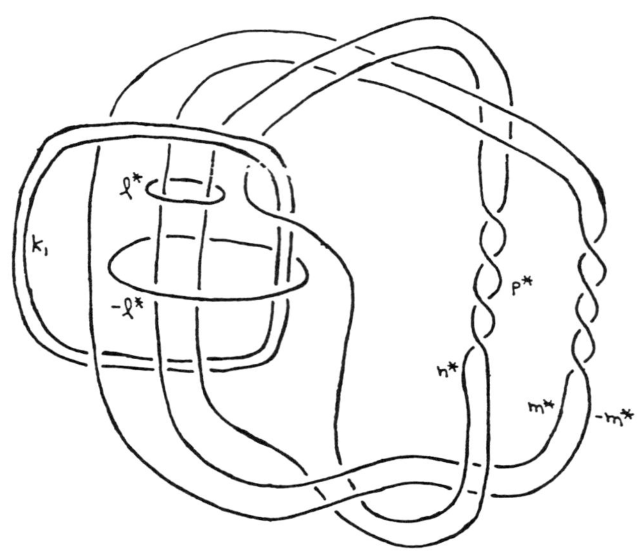

FIGURE 23

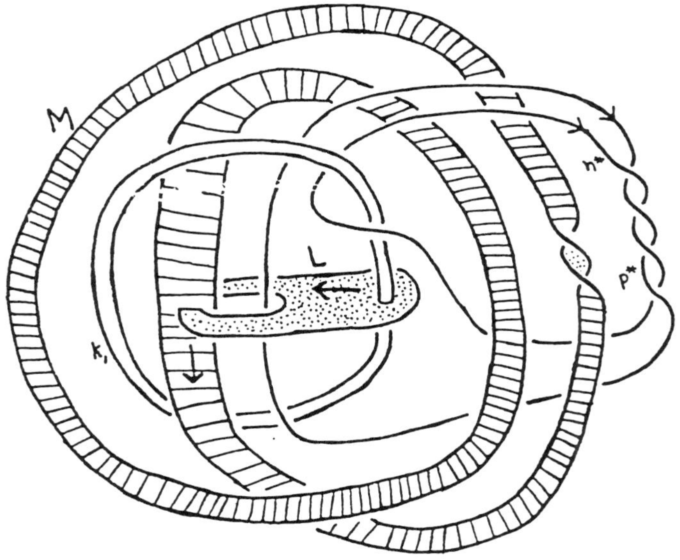

FIGURE 24

Similarly, the curves J_m and J_{-m} bound an annulus M, which is disjoint from B_1, and which intersects L in one arc. Again, ∂M determines the longitudes λ_m and λ_{-m}. Note that the curves J_n and J_p are disjoint from the annuli L, M; note also that the longitudes λ_n and λ_p are prefered longitud in the usual sense of the knots J_n and J_p.

Making l^*-surgery on J_l, and then $-l^*$-surgery on J_{-l}, is equivalent to making l Dehn twists along the annulus L. Similarly for the annulus M. These observations imply the following theorem. The Figure 24 has been simplified.

Theorem 3.2. *The knot $k(l, m, n, p)$ is obtained from the trivial knot k_1, by performing l Dehn twists along the annulus L and m Dehn twists along the annulus M, and then by doing n Dehn twists on the complement of the curve J_n or by doing p Dehn twists on the complement of the curve J_p.*

The arrows in the Figure 24 indicate the directions of the twists in the case that l, m, n and p are positive. This can be determined experimentally, by comparing the knots obtained when the parameters

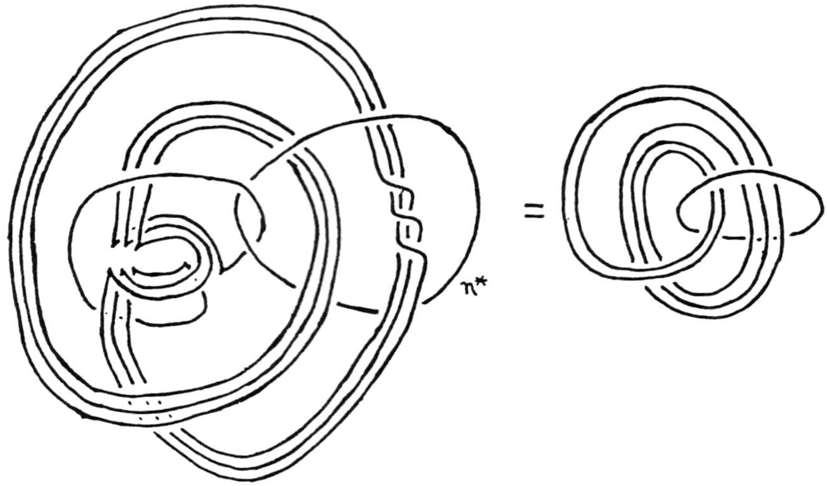

FIGURE 25

get some banned value and some allowed value which differ only in a sign.

Example 1. The knots $k(3,1,n,0)$ are described in Figure 25.

The knot $k(3,1,1,0)$ is shown in Figure 26, it is not difficult to check that it is the (-2,3,7) pretzel knot. Note that the knot $k(3,1,0,0)$, a degenerate case, is the trefoil knot. In fact, the knots $k(l,1,0,0)$ are torus knots, and $k(l,1,n,0)$ is a kind of twisted torus knot.

It follows from proposition 1.4 that the mirror image of the knots $k(l,m,n,0)$ and $k(l,m,0,p)$ are the knots $k(-l,-m,1-n,0)$ and $k(-l,1-m,0,1-p)$ respectively. It also follows that $k(l,\pm 1,n,0) = k(-l \pm 1, \pm 1, n, 0)$, and that $k(2,-1,n,0) = k(-3,-1,n,0) = k(2,2,0,n)$, and of course, there may be some other repetitions among these knots. Therefore the mirror image of the knot $k(2,-1,1-n,0)$ is the knot $k(3,1,n,0)$ pictured above, this is a family having six non-hyperbolic surgeries.

It is not difficult to check that the link $k(3,1,0,0) \cup J_n$ is equivalent to the Whitehead sister link, which is pictured in [2, figure 3]. It was first discovered by Hodgson and Weeks [15] that by doing 1/n surgeries on the unknotted component of this link produces an infinite family of

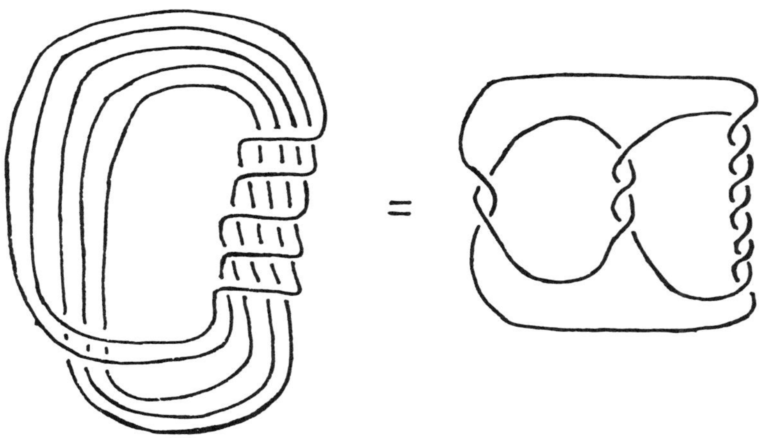

FIGURE 26

knots admitting six non-hyperbolic surgeries.

It follows from the Figure 24 that the knots $k(l, m, n, p)$ are closed braids inside a solid torus, namely the complement of the curve J_n or J_p. The braid index, denoted by br, of these closed braids is as follows:

(a) l and m have the same sign then $br(k(l, m, n, 0)) = 2lm - 1$.

(b) If l and m have different sign then $br(k(l, m, n, 0)) = 2|lm| + 1$.

(c) l and m have the same sign then $br(k(l, m, 0, p)) = 2lm - l - 1$.

(d) If l and m have different sign then $br(k(l, m, 0, p)) = 2|lm| + l + 1$.

Note that the minimal braid index possible is 5; this is attained for $k(2, -1, n, 0)$, $k(-2, 1, n, 0)$, $k(3, 1, n, 0)$, $k(-3, -1, n, 0)$, $k(2, 2, 0, p)$ and $k(-2, -1, 0, p)$, but all these families are equivalent, up to mirror images. As we said before, this is the remarkable family admitting six non-hyperbolic surgeries.

It should be that for all n or all p, except for finitely many, the braid index of the knots $k(l, m, n, p)$ will coincide with the index of the corresponding closed braid, this looks intuitively clear, but I do not know whether is true or not, for I have not found any reference. Note also that the knots $k(l, m, n, 0)$ and $k(l, m, 0, p)$ are both obtained from $k(l, m, 0, 0)$ by twisting on different axis, but these knots should be in general different, for they seem to have different braid indeces.

As we said before, there are some repetitions on the list of knots given here; note however that a result in [20] implies $k(l, m, n_1, 0) \neq$

$k(l, m, n_2, 0)$ if $n_1 \neq n_2$; the same happens for the knots $k(l, m, 0, p)$. It would be interesting to have a classification of these knots.

The Computer program SNAPPEA, developed by Jeff Weeks [22], was used to check that some of the knots $k(l, m, n, p)$ admit non-hyperbolic surgeries, obtaining always the desired results. This program was also used to find repetitions among such knots, and it showed that the knots given by Bleiler and Hodgson in [2, figure 10] coincide with the knots $k(3, -1, 0, p)$.

It follows from theorem 2.1(a) that for each of the knots $k(l, m, n, p)$ there exist a properly embedded surface Q, of genus 1, which is incompressible and ∂-incompressible, and whose boundary consist of two curves of slope r. This surface can in fact be constructed explicitly; this will be done in a sequel of this paper.

4. Some open problems.

The main problem considered in this article is not completely solved, i.e. it is not known exactly on which hyperbolic knots and which surgeries produce non-hyperbolic manifolds. We propose some conjectures which may direct future research, and whose solution may help to understand the hyperbolic knots with bad surgeries. Other papers treating this problem are [2] and [5]. In what follows k denotes an hyperbolic knot in S^3.

It will be interesting to find an universal bound for the number of non-hiperbolic surgeries on a knot k. There is some support to expect that, for example: It is known [2] that $M_k(r)$ has a Riemannian metric of negative curvature for all but at most 24 values of r. The figure eight knot admits 10 surgeries producing non-hyperbolic manifolds, and the (-2,3,7) pretzel knot admits 7 such surgeries, these are 1/0, 20, 19, 18,17, 16 and 37/2 (This knot correspond to $k(3, 1, 1, 0)$ in our list). However these are the known examples with more non-hyperbolic surgeries, so it seems reasonable to conjecture:

1. If k is not the figure eight knot, then there are at most 7 surgeries on k producing non-hyperbolic manifolds.

Infinitely many of the knots of theorem 2.1 have six surgeries producing non-hyperbolic manifolds, however I do not know if infinitely many of these knots have more than six non-hyperbolic surgeries, so we ask:

2. Are there infinitely many hyperbolic knots admitting seven or more non-hyperbolic surgeries?

It will be interesting to know the knots and surgeries which produce manifolds containing an incompressible torus. [8] and [14] give some information. We have the following conjectures:

3. Suppose that for a slope r, with $\Delta(r,\mu) = 1$, $M_k(r)$ has an incompressible torus. Then there is an incompressible torus which hits the surgered solid torus twice.

Perhaps nothing is known about this problem, and I do not know any counterexample. The examples shown in this paper, i.e. the knots $k(2,m,n,0)$, $k(l,2,0,p)$, $k(2,-1,n,0)$ and other examples given in [2] satisfy the conjecture.

4. Suppose that for a slope r, with $\Delta(r,\mu) = 2$, $M_k(r)$ has an incompressible torus. Then k is equivalent to one of the knots $k(l,m,n,p)$ of theorem 2.1.

It is reasonable to expect a characterization of such knots like the proposed in this conjecture, for these knots are in some sense a generalization of cable knots, where a twice punctured torus is used instead of a twice punctured sphere. For such a knot k, there is a twice puctured essential torus T lying on its exterior; Gordon and Luecke [14] have shown that T divides M_k into two handlebodies, and that k must be strongly invertible. This is clearly satisfied by the knots $k(l,m,n,p)$.

If k is strongly invertible, and r is such a surgery, then $M_k(r)$ has an equivariant torus T, which hits the surgered solid torus twice; passing to the quotient under the involution, $M_k(r)$ will be a double cover of S^3 branched along a knot k', which has unknotting number equal to 1. T will cover a Conway sphere P, because [8, 1.6] implies that T cannot cover a torus. It follows that the crossing which unknots k' meets the sphere P once and it cannot be isotoped to miss it, i.e., k' and P are like in lemma 1.3. This implies that the characterization of the knots k admitting such surgeries is equivalent to the following:

Find all the knots k which satisfy:

(a) k admits a Conway sphere P. (b) k has an unknotting crossing which necessarily intersects P once. (c) There is no other Conway sphere or essential torus disjoint from the given unknotting crossing.

It is reasonable to expect that the knots k in lemma 1.3 are all the knots satisfying these conditions.

5. There is at most one non-integral slope r such that $M_k(r)$ contains an incompressible torus. By [12], there are at most 2 such surgeries.

Conjecture 4 should imply conjecture 5, or in other words, the knots of theorem 2.1 should satisfy this conjecture.

It follows from a theorem of Gordon [12] that if two surgeries r, s on k, where k is not the figure eight knot, produce manifolds containing incompressible tori then $\Delta(r,s) \leq 5$, this and [14] imply that there are at most 7 surgeries producing manifolds which contain an incompressible torus. The (-2,3,7) pretzel knot admits 3 such surgeries, these have surgery coefficients 20, 16 and 37/2; the figure eight knot also admits 3 such surgeries, so we have the following:

6. There are at most 3 surgeries on k producing manifolds which contain incompressible tori.

There are many examples of non-torus knots with integral surgeries producing Seifert fibre spaces, but I do not know any example with a non-integral surgery, so we have:

7. Suppose that for a slope r, $M_k(r)$ is a Seifert fibre space. Then $\Delta(r, \mu) = 1$.

References

1. S.A. Bleiler, *Knots prime on many strings*, Trans. Amer. Math. Soc. **282** (1984) 385-401.
2. S.A. Bleiler and C.G. Hodgson, *Spherical space forms and Dehn filling*, Preprint.
3. S.A. Bleiler, M. and Scharlemann, *A projective plane in $\boldsymbol{R}^4$ with three critical points is standard*, Topology **27** (1988) 519-540.
4. F. Bonahon and L. Siebenmann, *Geometric splittings of knots, and Conway's algebraic knots*, Preprint.
5. S. Boyer and X. Zhang, *Finite surgery on knots*, Preprint.
6. M. Culler, C.McA. Gordon, J. Luecke and P. Shalen, *Dehn surgery on knots*, Ann. of Math. **125** (1987) 237-300.
7. M. Eudave-Muñoz, *Primeness and sums of tangles*, Trans. Amer. Math. Soc. **306** (1988) 773-790.
8. M. Eudave-Muñoz, *Essential tori obtained by surgery on a knot*, Pacific J. of Math. **167** (1995) 81-116.
9. J.C. Gómez-Larrañaga, *Graph of tangles*, Trans. Amer. Math. Soc. **286** (1984) 817-830.
10. C.Mc.A. Gordon, *Dehn surgery and satellite knots*, Trans. Amer. Math. Soc. **275** (1983) 687-708.
11. C.McA. Gordon, *Dehn surgery on knots*, Proceedings of the International Congress of Mathematics, Kyoto, Japan 1990, The Japan Mathematical Society (1991) 631-642.
12. C.McA. Gordon, *Boundary slopes of punctured tori in 3-manifolds*, to appear.
13. C.McA. Gordon and J. Luecke, *Links with unlinking number one are prime* Proc. Amer. Math. Soc. **120** (1994) 1271-1274.
14. C.McA. Gordon and J. Luecke, Abstracts AMS **14** (1993), p. 381., and a lecture at the Georgia International Topology Conference, 1993.
15. C. Hodgson and J. Weeks, *A census of closed hyperbolic 3-manifolds*, in preparation.
16. W.H. Holzmann, *An equivariant torus theorem for involutions*, Trans. Amer. Math. Soc. **326** (1991) 887-906.

17. W.B.R. Lickorish, *Prime knots and tangles*, Trans. Amer. Math. Soc. **267** (1981) 321-332.
18. J.M. Montesinos, *Variedades de Seifert que son recubridores cíclicos ramificados de dos hojas*, Bol. Soc. Mat. Mexicana (2) **18** (1973) 1-32.
19. J.M. Montesinos, *Surgery on links and double branched covers of* S^3, Annals of Math. Studies no. 84, Princeton Univ. Press, Princeton, N.J. (1975) 227-260.
20. T. Shibuya, on the cobordism of compound knots which are T-congruent, Kobe J. Math. **2** (1985) 71-74.
21. W. Thurston, *Three dimensional manifolds, Kleininan groups and hyperbolic geometry*, Bull. Amer. Math. Soc. **6** (1982) 357-381.
22. J. Weeks, SNAPPEA, the Hyperbolic Structures Computer Program, 1992. (For a description, see C.Adams, SNAPPEA: The Weeks Hyperbolic 3-manifolds Program, Notices of the A.M.S. **37** (1990) 273-275.

Instituto de Matemáticas, Universidad Nacional Autónoma de México, Circuito Exterior, Ciudad Universitaria, México, D.F. 04510, MEXICO

E-mail address: eudave@serxidor.unam.mx

AMS/IP Studies in Advanced Mathematics
Volume **2**, 1997 (Part 1)

CYCLIC SURGERY AND BOUNDARY SLOPES

S. BOYER* AND X. ZHANG+

1. Introduction.

Let K be a knot in the 3-sphere S^3, $N(K)$ a regular neighborhood of K in S^3 and $M = S^3 - intN(K)$. The manifold obtained by Dehn surgery on K with slope r is denoted by $M(r)$. A surgery is called a *cyclic surgery* if the resultant 3-manifold $M(r)$ has cyclic fundamental group. According to Thurston [28], knots in S^3 can be partitioned into three classes, namely the torus knots, whose exteriors M are Seifert fibred, the satellite knots, whose exteriors M contain an essential torus, and the hyperbolic knots, whose complements admit a complete hyperbolic metric of finite volume. The cyclic surgeries on torus knots and satellite knots have been classified [20], [31], [1], [32] and so we shall be concerned with cyclic surgeries on hyperbolic knots in this paper. It is known that for a hyperbolic knot K in S^3, (i) any non-trivial cyclic surgery slope of K must be integral, (ii) K admits at most two non-trivial cyclic surgeries and if two, they are two succesive integers [8, Corollary 1].

A surface in a compact orientable 3-manifold is called *essential* if it is properly embedded, orientable, incompressible, and non-boundary-parallel. A slope $r \subset \partial M$ is called a *boundary* slope if there is an essential surface F in M such that ∂F is a non-empty set of parallel, simple closed curves on ∂M of slope r. The most frequently occurring boundary slopes for knots in S^3 are the trivial slope $1/0$ and the integral slopes. Non-integral, non-trivial boundary slopes for knots in S^3 are also known to exist [26], [13], [33].

The purpose of this paper is to describe some close connections between cyclic surgeries and boundary slopes. Recall that a link L in S^3 is

---

1991 *Mathematics Subject Classification.* 1980 Mathematics Subject Classification, (1991 Revision). Primary 57M25, 57R65.

*Partially supported by grants: NSERC OGP 0009446 and FCAR EQ 3518.
+Supported by a postdoctoral fellowship from the Centre de Recherches Mathematiques

called *non-sufficiently large* (NSL) if $M = S^3 - intN(L)$ does not contain any closed, essential surfaces. The results described herein suggest and support the following conjecture.

Conjecture. *Let $K \subset S^3$ be an NSL hyperbolic knot. If K admits a non-trivial cyclic surgery, then K possesses at least one non-integral boundary slope.*

Remarks. (1) It follows from [8, Theorem 2.0.3] that the trivial slope $1/0$ is not a boundary slope for an NSL knot K in S^3.
(2) There is no known example of a sufficiently large hyperbolic knot in S^3 which admits a non-trivial cyclic surgery.
(3) The converse of the conjecture is not true, there are NSL hyperbolic knots which have non-integral boundary slopes but do not admit non-trivial cyclic surgery. For example, the knots 10_{128}, 10_{130}, 10_{132}, 10_{133}, 10_{133}, 10_{134}, 10_{136}, 10_{140}, 10_{141}, 10_{143}, 10_{146}, 10_{147} are NSL hyperbolic Montesinos knots [24] and each of them has a non-integral boundary slope [13]. On the other hand, they are also periodic knots [5, Table I], and thus do not admit a non-trivial cyclic surgery [30].

The following theorem gives a sufficient condition for the conjecture to be true. It is a generalization of [33, Theorem 1].

Theorem 1.1. *Let $K \subset S^3$ be an NSL hyperbolic knot. Suppose that K admits a non-trivial cyclic surgery with slope m and suppose that either $M(m+1)$ or $M(m-1)$ is an irreducible manifold containing no essential torus. Then K possesses at least one non-trivial, non-integral boundary slope.*

Remarks. (1) For a knot $K \subset S^3$, if $M(r)$ is reducible then M contains a genus zero essential surface with boundary slope r. It is conjectured that on a hyperbolic knot in S^3, no surgery can yield a reducible manifold.
(2) For a hyperbolic knot $K \subset S^3$, if $M(r)$ contains an essential torus, then M contains a genus one, essential surface with boundary slope r.

Theorem 1.1 is a consequence of the following theorem.

Theorem 1.2. *Let $K \subset S^3$ be a non-torus knot. Suppose that K admits a non-trivial cyclic surgery with slope m where m is not a boundary slope. Then if either $M(m+1)$ or $M(m-1)$ is an atoroidal, Haken manifold, K possesses at least one non-integral, non-trivial boundary slope.*

The proofs of Theorems 1.1 and 1.2, given in §2, are based on the main results and techniques developed in [6], [7] and [8]. The referee has pointed out to us that a version of our argument was known to Culler and Shalen and had in fact been used by them in unpublished work to show that the $(-2,3,7)$-pretzel knot has a non-integral boundary slope.

Applying Theorems 1.1 and 1.2 and results from [18], [19], [12], [13] and [8], we shall prove the following in §3.

Corollary 1.3. *If a non-torus alternating knot has no non-integral, non-trivial boundary slope, then it does not admit a non-trivial cyclic surgery.*

Remarks. (1) For alternating knots, all known boundary slopes are either the trivial slope or even integral slopes.
(2) A non-trivial alternating knot does not admit a non-trivial even order cyclic surgery [23]. We note that there is an elementary homological proof of this fact obtained through consideration of the double branched covering of S^3, branched along the given knot.
(3) Charles Delman and Rachel Roberts have recently announced a proof that any manifold resulting from non-trivial surgery on a non-torus alternating knot admits an essential lamination. In particular, the fundamental groups of these manifolds are not cyclic.

We mention two consequences of Corollary 1.3. The first one yields an alternative proof of the main result of [27].

Corollary 1.4. *([27]) A non-torus 2-bridge knot does not admit a non-trivial cyclic surgery.*

Proof. Any 2-bridge knot is an alternating knot [2] and does not have a non-integral, non-trivial boundary slope [12, Proposition 2]. □

Corollary 1.5. *A non-torus Montesinos knot $K = K(p_1/q_1, ..., p_n/q_n)$, with all p_i/q_i having the same sign, does not admit a non-trivial cyclic surgery.*

Proof. Such a Montesinos knot is an alternating knot (the standard alternating projections of the p_i/q_i tangles of K fit together to form an alternating projection of K) and does not admit a non-integral, non-trivial boundary slope [13, P.462]. □

Finally, in §4 we give two families of knots which satisfy the hypotheses of Theorem 1.1, and we close the paper with some open questions.

2. Proof of Theorems 1.1 and 1.2.

We assume that the reader is familiar with the terminology of [8, Chapter 1]. Let M be a compact, connected, orientable 3-manifold. Let $R(M)$ denote the set of representations of $\pi_1(M)$ in $SL_2(C)$, let $X(M)$ denote the set of characters of representations in $R(M)$, and let $t : R(M) \to X(M)$ denote the natural map which sends each representation $\rho \in R(M)$ to its character χ_ρ. Both $R(M)$ and $X(M)$ are complex, affine, algebraic sets and t is a surjective, regular map between them. For each $\gamma \in \pi_1(M)$, the function $f_\gamma : X(M) \to \mathbf{C}$ defined by $f_\gamma(\chi_\rho) = (\mathrm{trace}(\rho(\gamma)))^2 - 4$ is a regular function.

Now let $M = S^3 - intN(K)$ for some knot $K \subset S^3$. Suppose that $X(M)$ contains an algebraic curve (i.e. a 1-dimensional, irreducible subvariety) X_0 such that X_0 contains the character of an irreducible representation, and suppose further that f_α is a non-constant function on X_0 for each non-trivial element $\alpha \in L = H_1(\partial M; \mathbf{Z}) = \pi_1(\partial M)$. Let $\tilde{X}_0$ be the smooth, projective model of X_0. Note that as $\tilde{X}_0$ is birationally equivalent to X_0, each function f_α corresponds to a meromorphic function on $\tilde{X}_0$. It is shown in chapter 1 of [8] that such a curve X_0 of $X(M)$ determines a norm $\|\cdot\|$ on $V = H_1(\partial M; \mathbf{R})$ with the following properties:

(2.1) For each $\alpha = (m, l) \in L, \alpha \neq (0,0)$, $\|\alpha\| = degree\, f_\alpha$ and so is a positive integer. Note here that every slope m/l corresponds to the pair of primitive elememts $\pm(m, l) \in L$.

(2.2) Let $s = min\{\|\alpha\|; \alpha \in L, \alpha \neq (0,0)\}$ and let B be the ball of radius s in V. Then B is a compact, convex, finite sided polygon which is symmetric about the origin (i.e. $-B = B$). Note that $intB \cap L = (0,0)$.

(2.3) To each vertex of B, there is an associated boundary slope $r = m/n$ such that $\pm(m, n)$ lie on the line passing through the vertex and the origin.

(2.4) Suppose that $r = m/l$ is not a boundary slope and that $M(r)$ has cyclic fundamental group. Then if $\alpha = (m, l)$ and $x \in \tilde{X}_0$, the multiplicity of x as a zero of f_α is minimal amongst the corresponding multiplicities of the functions f_δ, $\delta \in L$, $\delta \neq (0,0)$. Thus $\pm(m, l) \in \partial B$ and by (2.3) they are not vertices of B.

(2.5) If we normalize the area of a parallelogram spanned by any pair of generators of L to be 1, then Area $B \leq 4$.

Proof of Theorem 1.2. Assume all the conditions hypothesized in Theorem 1.2. We shall argue by contradiction, so we shall also suppose that K admits no non-integral, non-trivial boundary slope.

Without loss of generality, we may suppose that $M(m-1)$ is an atoroidal, Haken manifold. According to Thurston [28], $M(m-1)$ is either a Seifert fibred manifold or a hyperbolic manifold. In the former case, the Haken and atoroidal conditions force the Seifert manifold $M(m-1)$ to have base orbifold $S^2(p,q,r)$, the 2-sphere with exactly 3 cone points of orders $2 \leq p \leq q \leq r$. Further, notice that in this case, it follows from [15, VI.13] that $H_1(M(m-1)) \cong \mathbf{Z}_{m-1}$ is infinite, and so $m-1 = \lambda$, the longitudinal slope. We must also have that $1/p + 1/q + 1/r < 1$, for if the sum is greater than 1, then $\pi_1(M(\lambda))$ is finite, and if the sum equals 1, then $M(\lambda)$ contains an essential torus. Thus $S^2(p,q,r)$ is a hyperbolic orbifold and so there is a discrete, faithful representation $\bar{\rho}_1 : \pi_1^{orb}(S^2(p,q,r)) \to PSL_2(\mathbf{R}) \subset PSL_2(\mathbf{C})$. In the case where $M(m-1)$ is hyperbolic, there is also a discrete, faithful representation of $\bar{\rho}_1 : \pi_1(M(m-1)) \to PSL_2(\mathbf{C})$. Note that in either eventuality, $\bar{\rho}_1$ is necessarily irreducible. Further, in the hyperbolic case, the nontrivial elements of the image of $\bar{\rho}_1$ are hyperbolic elements of $PSL_2(\mathbf{C})$, while in the Seifert case, the nontrivial elements of the image of $\bar{\rho}_1$ are either hyperbolic elements of $PSL_2(\mathbf{C})$ or are elements of finite order.

We shall identify $R(M(m-1))$ as a subset of $R(M)$ through composition with the natural surjection $\pi_1(M) \to \pi_1(M(m-1))$. Similarly, $X(M(m-1))$ may be considered a subset of $X(M)$. Note that in particular, we may think of $\bar{\rho}_1$ as a homomorphism $\pi_1(M) \to PSL_2(\mathbf{C})$.

There is a central extension of $\pi_1(M)$ by $\mathbf{Z}_2$, say $\phi : \hat{\pi} \to \pi_1(M)$, and a representation $\hat{\rho}_1 \in R(\hat{\pi})$ such that $\bar{\rho}_1 = \phi \circ \hat{\rho}_1$. Since such extensions are classified by $H^2(\pi_1(M); \mathbf{Z}_2) \cong H^2(M; \mathbf{Z}_2) \cong 0$ ([14, Theorem 10.3]), $\hat{\pi} \cong \pi_1(M) \times \mathbf{Z}_2$. Thus $\bar{\rho}_1$ lifts to a representation $\rho_1 \in R(M)$. Let Y be an irreducible component of $X(M)$ which contains χ_{ρ_1}. Now $\mu = (1,0) \in L = \pi_1(\partial M)$ normally generates $\pi_1(M)$ and so $\rho_1(\mu) \neq \pm I$. Then by [6, Proposition 3.2.1], $\dim(Y)$ is at least 1. Let X_1 be a curve in Y which contains χ_{ρ_1}. Our goal is to show that X_1 provides us with a norm of the type described above (Lemma 2.9).

Lemma 2.6. *Let $\alpha \in L$ be primitive and suppose that f_α is identically zero on X_1. Then there is an $\epsilon \in \{\pm 1\}$ such that $\rho(\alpha) = \epsilon I$ for each $\rho \in R(M)$ such that $\chi_\rho \in X_1$.*

Proof. The condition that f_α is identically zero on X_1 means that $\rho(\alpha)$ is either $\pm I$ or a parabolic element of $SL_2(\mathbf{C})$ for each $\rho \in R(M)$ such that $\chi_\rho \in X_1$. We must show that the latter case never occurs. To that end, we note that by [6, Lemma 1.4.2], the set of characters in X_1 corresponding to reducible representations forms a closed algebraic subset of X_1, and hence there are only finitely many points in X_1, say $x_1, ..., x_n$, which are characters of reducible representations. In particular, $U = X_1 - \{x_1, x_2, ..., x_n\}$ is a Zariski open neighbouhood of $t(\rho_1)$ in

X_1 consisting entirely of characters of irreducible representations. Assume now that $\rho(\alpha)$ is parabolic for some $\rho \in t^{-1}(U)$. Then by [6, Proposition 1.5.4], there is a Zariski open neighborhood U_* of χ_ρ in U such that $\rho'(\alpha) \neq \pm I$ for any $\rho' \in t^{-1}(U_*)$. Now by hypothesis, the function I_α defined on the connected set X_1 by $I_\alpha(\chi_\rho) = \text{trace}(\rho(\alpha))$ is constant and takes on the value 2ϵ at χ_{ρ_1}, for some $\epsilon \in \{\pm 1\}$. Hence it is constantly 2ϵ and as $\rho'(\alpha) \neq \pm I$ for any $\rho' \in t^{-1}(U_*)$, $\rho'(\alpha)$ is necessarily a parabolic element of $SL_2(\mathbf{C})$ for any such ρ'. But now we observe that μ commutes with α in $\pi_1(M)$, and so its image by any $\rho' \in t^{-1}(U_*)$ must either be $\pm I$ or parabolic. In particular, $\text{trace}(\rho'(\mu)) = \pm 2$ for any $\rho' \in t^{-1}(U_*)$, which implies that f_μ is constantly zero on U_*, and thus on X_1. In particular, $f_\mu(\chi_{\rho_1}) = 0$. As ρ_1 has no parabolic elements in its image, we must have that $\rho_1(\mu) = \pm I$, which is clearly impossible as $\rho_1(\mu)$ normally generates $\rho_1(\pi_1(M))$. This contradiction shows us that for each $\rho \in t^{-1}(U)$, we have $\rho(\alpha) = \pm I$. By the continuity of the evaluation map and the connectivity of $t^{-1}(X_1)$, this identity must hold with fixed sign for each $\rho \in R(M)$ such that $\chi_\rho \in X_1$. □

Set $\beta_0 = (m-1, 1) \in L$.

Lemma 2.7. *f_{β_0} is not a constant function on X_1.*

Proof. Suppose, on the contrary, that f_{β_0} is a constant function on X_1. Then $f_{\beta_0} \equiv f_{\beta_0}(\chi_{\rho_1}) = 0$ on X_1 and so by Lemma 2.6, there is an $\epsilon \in \{\pm 1\}$ such that $\rho(\beta_0) = \epsilon I$ for each $\rho \in R(M)$ such that $\chi_\rho \in X_1$. Hence, composition with the natural surjection $SL_2(\mathbf{C}) \to PSL_2(\mathbf{C})$ shows that $\bar{\rho}_1$ is not rigid in $\text{Hom}(\pi_1(M(\beta_0)), PSL_2(\mathbf{C}))$, i.e. there are representations arbitrarily close to $\bar{\rho}_1$ in $\text{Hom}(\pi_1(M(\beta_0)), PSL_2(\mathbf{C}))$ which are not conjugate to $\bar{\rho}_1$. On the other hand, $M(\beta_0)$ is either a closed, hyperbolic manifold or a closed, atoroidal, Haken, Seifert fibred manifold. In either case, $\bar{\rho}_1$ is rigid. This contradiction will complete the proof of Lemma 2.7.

First of all, when $M(\beta_0)$ is hyperbolic, the rigidity of $\bar{\rho}_1$ is a consequence of the Mostow rigidity theorem [21] and the finiteness of the outer automorphism group of a finite-volumed, complete hyperbolic 3-manifold [29, Corollary 5.7.4].

Now consider the Seifert fibred case. We first observe that any irreducible representation $\bar{\rho} \in \text{Hom}(\pi_1(M(\beta_0)), PSL_2(\mathbf{C}))$ must send the class of the fibre (which is central in $\pi_1(M(\beta_0))$) to $\pm I$, and so by continuity, the same holds for every $\bar{\rho}$ which comes from $t^{-1}(X_1)$. Thus every such representation factors through $\pi_1^{orb}(S^2(p,q,r))$ and so we need only show that $\bar{\rho}_1$ is rigid in $\text{Hom}(\pi_1^{orb}(S^2(p,q,r)), PSL_2(\mathbf{C}))$.

To that end, first note that using [15, VI.13], we can construct an irreducible Seifert manifold W with base orbifold $S^2(p,q,r)$, which is also non-Haken. By [6, Propositions 2.2.1 and 2.3.1], W must have a zero-dimensional $SL_2(\mathbf{C})$-character variety, and so by [6, Proposition 1.5.2], any irreducible representation of $\pi_1(W)$ is rigid. In particular, this is the case for the composition of the natural surjection $\pi_1(W) \to \pi_1^{orb}(S^2(p,q,r))$ with $\bar{\rho}_1$, i.e., $\bar{\rho}_1$ is rigid in $\mathrm{Hom}(\pi_1^{orb}(S^2(p,q,r)), PSL_2(\mathbf{C}))$. Thus we are done. □

Lemma 2.8. *If there is a primitive element $\alpha \neq \pm\beta_0$ in L such that $f_\alpha(\chi_{\rho_1}) = 0$, then $M(\beta_0)$ is Seifert fibred and we may assume that $(p,q,r) = (p,q,pq)$ where gcd $(p,q) = 1$.*

Proof. Consider the hyperbolic case first. Now μ and β_0 form a basis of L and so if $f_\alpha(\chi_{\rho_1}) = 0$ for some primitive element $\alpha \in L$ which is not $\pm\beta_0$, we may write $\alpha = p\mu + q\beta_0$ where $p \neq 0$. The condition $f_\alpha(\chi_{\rho_1}) = 0$ implies that trace$(\rho_1(\alpha)) = \pm 2$, which in turn implies, as we are in the hyperbolic case, that $\bar{\rho}_1(\alpha)$ is $\pm I$. But the meridian class normally generates $\pi_1(M)$ and thus $\bar{\rho}_1(\mu)$ is a hyperbolic element of $PSL_2(\mathbf{C})$. In particular, it has infinite order. Then $\bar{\rho}_1(\alpha) = \bar{\rho}_1(\mu)^p \bar{\rho}_1(\beta_0)^q = \bar{\rho}_1(\mu)^p$ also has infinite order. Thus $f_\alpha(\chi_{\rho_1}) \neq 0$, which proves Lemma 2.8 in the hyperbolic case.

Suppose now that $M(\beta_0)$ is Seifert fibred with base orbifold $S^2(p,q,r)$, the 2-sphere with exactly 3 cone points of orders $2 \leq p \leq q \leq r$. The argument for the hyperbolic case works here also as long as $\bar{\rho}_1(\mu)$ is a hyperbolic element of $PSL_2(\mathbf{C})$. Assuming the contrary, the choice of $\bar{\rho}_1$ implies that $\bar{\rho}_1(\mu)$ is an element of finite order in $\bar{\rho}_1(\pi_1(M(\beta_0))) = \bar{\rho}_1(\pi_1^{orb}(S^2(p,q,r)))$, which is isomorphic to the (p,q,r)-triangle group $\Delta(p,q,r) = \{x,y,z : x^p = y^q = z^r = xyz = 1\}$. This implies that $\bar{\rho}_1(\mu)$ is conjugate in $\Delta(p,q,r)$ to some power of x, y or z, say z, and so $\Delta(p,q,r)$ is normally generated by z. Abelianizing and setting $z = 0$ shows that gcd $(p,q) = 1$. Further, we have already observed that in the Seifert case, $H_1(M(\beta_0))$ is infinite, i.e. has order 0. Thus if $(p,a_1),(q,a_2),(r,a_3)$ are the Seifert invariants for the exceptional fibres in $M(\beta_0)$, there is an integer b such that $H_1(M(\beta_0))$ has order $|pqa_3 + pra_2 + qra_1 + bpqr|$ ([15, VI.9]). Then $0 = |pqa_3 + pra_2 + qra_1 + bpqr|$ and so in particular, p divides qr, q divides pr and r divides pq. Since gcd $(p,q) = 1$, we must have $r = pq$. This completes the proof of Lemma 2.8. □

Set $\beta_1 = (m,1) \in L$ and $\beta_2 = (m+1,1) \in L$

Lemma 2.9. *X_1 provides a norm on $H_1(\partial M; \mathbf{R})$ with properties* (2.1)–(2.5) *above.*

Proof. Let $\tilde{X}_1$ be the smooth projective model of X_1. We must show that each function f_α, $\alpha \in L = H_1(\partial M; \mathbf{Z})$, $\alpha \neq (0,0)$, is non-constant on $\tilde{X}_1$. To accomplish this, it is enough to show that f_α is non-constant on $\tilde{X}_1$ for each primitive element α of L. Suppose to the contrary that there is a primitive element $\alpha_0 \in L$ such that f_{α_0} is constant on $\tilde{X}_1$.

According to Lemma 2.7, f_{β_0} is non-constant on $\tilde{X}_1$ and so
(2.10) $\alpha_0 \neq \pm\beta_0$.
Further, by 3 of [4], $\pm\alpha_0$ is the unique slope on M for which f_{α_0} is constant.3 of [4] may also be used to show that
(2.11) $\pm\alpha_0$ is a boundary slope, and so in particular, $\alpha_0 \neq \pm\beta_1$;
(2.12) as β_1 is a cyclic slope which is not a boundary slope, if $x \in \tilde{X}_1$, then the multiplicity of x as a zero of f_{β_1} is minimal amongst the corresponding multiplicities for the functions $f_\alpha, \alpha \in L, \alpha \neq n\alpha_0, n \in \mathbf{Z}$;
(2.13) $\Delta(\alpha_0, \beta_1) = 1$.
We also have
(2.14) $\Delta(\alpha_0, \mu) \leq 1$,
for otherwise, α_0 would be a non-trivial, non-integral boundary slope of K, something which we have assumed does not exist. It follows from these observations that
(2.15) $\alpha_0 = \pm\mu$ or $\pm\beta_2$.

From (2.11), f_{β_1} is non-constant on $\tilde{X}_1$, so let $x \in \tilde{X}_1$ be one of its zeros. From (2.12), x is also a zero of f_α for each $\alpha \in L$. In particular this shows that f_{α_0} is identically zero on $\tilde{X}_1$. According to Lemma 2.6, there is an $\epsilon \in \{\pm 1\}$ such that $\rho(\alpha_0) = \epsilon I$ for each $\rho \in R(M)$ such that $\chi_\rho \in X_1$. Considering $\rho = \rho_1$, it is impossible for $\alpha_0 = \mu$, and so $\alpha_0 = \beta_2$. But then $\rho_1(\mu)^2 = \rho_1(\beta_2)\rho_1(\beta_0^{-1}) = \pm I$, and so $\bar{\rho}_1(\mu^2) = \pm I \in PSL_2(\mathbf{C})$. It follows that the abelianisation of the image of $\bar{\rho}_1$ is a subgroup of $\mathbf{Z}_2$.

Next notice that by applying Lemma 2.8, we see that we are in the Seifert fibred case, and $M(\beta_0)$ has base orbifold $S^2(p,q,pq)$ where gcd $(p,q) = 1$ and $2 \leq p < q \leq r$. But then by choice of $\bar{\rho}_1$, its image is the (p,q,pq)-triangle group whose abelianisation is, under our conditions, $\mathbf{Z}_{pq}$ where $pq \geq 6$. This contradicts the observation in the last line of the previous paragraph. Thus it is impossible for there to be a slope α_0 such that f_{α_0} is constant and so Lemma 2.9 has now been proven. □

We can now complete the proof of Theorem 1.2. For let $\|\cdot\|_1$ be the norm provided by X_1 and let B_1 be the corresponding fundamental polygon. If K has no non-integral, non-trivial boundary slopes, then from the properties (2.1) - (2.5), B_1 is necessarily a parallelogram with vertices $\pm(m-1,1)$ and $\pm(m+1,1)$. Now using property (2.4) we see that f_{β_1} and f_{β_0} have the same zero sets in $\tilde{X}_1$, and their zeros all have the same multiplicities. But χ_{ρ_1} is a zero of f_{β_0} on $\tilde{X}_1$ which is not a zero f_{β_1}, i.e. if $f_{\beta_1}(\chi_{\rho_1}) = 0$, then as $\bar{\rho}_1$ has no parabolics in its image, $\bar{\rho}_1(\beta_1) = \pm I \in PSL_2(\mathbf{C})$. Now $\bar{\rho}_1$ also sends β_0 to the identity, and so $\pm I = \bar{\rho}_1(\beta_1\beta_0^{-1}) = \bar{\rho}_1(\mu)$, which is clearly impossible. The proof of Theorem 1.2 is now finished.

Proof of Theorem 1.1. We will first show, under the hypothesis of Theorem 1.1, that m is not a boundary slope, and next that if K has no non-integral, non-trivial boundary slope, then one of $M(m-1)$ and $M(m+1)$ is an atoroidal, Haken 3-manifold. This contradicts Theorem 1.2 and so K must admit a non-integral, non-trivial boundary slope.

Without loss of generality, we may suppose that $M(m-1)$ is irreducible and contains no essential torus. First note that by [11], the manifold $M(0)$ is not an S^2-bundle over the circle. Since K is NSL, it now follows from [8, Theorem 2.0.3] that if $M(r)$ is not a Haken manifold and is not a connected sum of two non-trivial lens spaces, then r is not a boundary slope. Hence, in particular, the slopes μ and m are not boundary slopes, and if $m-1$ is a boundary slope, then $M(m-1)$ is Haken.

As K is a hyperbolic knot, there is a norm $\|\cdot\|$ on $V = H_1(\partial M; \mathbf{R})$, satisfying the properties (2.1)-(2.5). Hence if M has no non-integral, non-trivial boundary slope, then the fundamental polygon B is necessarily a parallelogram with vertices $\pm(m-1,1)$ and $\pm(m+1,1)$. Appealing to (2.3), we deduce that both $m-1$ and $m+1$ are boundary slopes of K. As noted above, this implies that $M(m-1)$ is a Haken manifold and so the proof of Theorem 1.1 is complete. □

We record the following result, which may be proven using the technique of this section.

Theorem 2.16. *Suppose that $K \subset S^3$ is a hyperbolic knot whose meridian slope is not a boundary slope (e.g. K could be an NSL knot) and let r be a slope on M, the exterior of K, such that $M(r)$ is a hyperbolic 3-manifold. Then there is a norm $\|\cdot\|$ on $H_1(\partial M; \mathbf{R})$ with the properties* (2.1) – (2.5) *such that r is not contained in the associated fundamental polygon.*

3. Proof of Corollary 1.3.

Let $K \subset S^3$ be a non-torus alternating knot having no non-integral, non-trivial boundary slopes. We note that a non-trivial alternating knot in S^3 is either a $(2,p)$-torus knot, or a hyperbolic knot, or a composite knot (see [18, Corollary 1] and [22]). Now it is known that a satellite knot which admits a non-trivial cyclic surgery must be a cabled knot, and thus a composite knot does not admit any non-trivial cyclic slopes. Hence we may assume then that the given knot K is a hyperbolic knot.

Lemma 3.1. *Let K be an alternating knot K. If K has a genus one essential surface in $M = S^3 - intN(K)$ with non-trivial slope r, then r is an integer slope and K is either a 2-bridge knot or a pretzel knot of type $(1/q_1, 1/q_2, 1/q_3)$ (Hence K is NSL by [24]). Further, if m is an integer, then the slopes $m-1$ and $m+1$ cannot both be boundary slopes associated to genus one essential surfaces in M.*

Remark. In independent work [25], Bob Patton has classified the alternating knots which admit surgeries containing essential tori.

Proof. Our strategy is to apply results from [19] to reduce the proof of Lemma 3.1 to that of Lemmas 3.2 and 3.3 below. We refer to [19] for the definitions of *standard alternating diagram* and *twist-crossing number* of an alternating knot.

Suppose that there is a genus one essential surface in M with boundary slope $r \neq 1/0$. Applying [19, Theorem 2], we see that K admits a standard alternating diagram whose twist-crossing number is at most 6. All standard alternating diagrams of twist-crossing number up to 5 are listed in [19, Figure 1.2] and each of these diagrams visibly represents a 2-bridge knot. We list in Figure 3.1 all of the standard alternating diagrams of twist-crossing number 6. Each of the diagrams represents a Montesinos knot of type $(p_1/q_1, p_2/q_2, p_3/q_3)$ with all p_i/q_i having the same sign. Therefore Lemma 3.1 follows from the following two lemmas. □

Lemma 3.2. *Let K be a 2-bridge knot. If r is the boundary slope of a genus one essential surface in the exterior of K, then r is an integer divisible by 4.*

Proof. Essential surfaces in exteriors of 2-bridge knots are classified in [12], where it is shown, in the notation of that paper, that they are all of the form $S_n(n_1, ..., n_{k-1})$ ([12, Theorem 1]). A calculation of the Euler characteristic of the surface $S_n(n_1, ..., n_{k-1})$ implies that if it is a connected, orientable, genus one surface, then $k = 2$ (cf. the proof of [12, Theorem 2 (a)]). But when $k = 2$, the boundary slope of the surface is an integer divisible by 4 ([12, Proposition 2]). We note that the number $n_0^+ - n_0^-$ in the proposition is always an even integer for any 2-bridge knot.

FIGURE 3.1. Stamdard alternating diagrams with twisting-crossing number 6

Lemma 3.3. *Let K be a Montesinos knot of type $(p_1/q_1, p_2/q_2, p_3/q_3)$ with all $q_i \geq 2$ and with all p_i having the same sign. If K has a boundary slope which bounds a genus one essential surface, then $p_1 = p_2 = p_3 = \pm 1$, i.e. K is a pretzel knot, and the slope is an integer divisible by 4. Moreover K has at most one such boundary slope.*

Proof. We shall assume the reader is familiar with the machinery, notation, and terminology of [13]. It is shown there that for the given knot K, one can produce a list of candidate surfaces, properly embedded in its exterior, including all incompressible and ∂-incompressible surfaces (orientable and non-orientable). Therefore Lemma 3.3 follows if we can show that the existence of a connected, orientable, and genus one candidate surface implies all the conclusions of Lemma 3.3.

A candidate surface for K is generated by an edge path system $(\gamma_1, \gamma_2, \gamma_3)$ in the diagram $\mathcal{D}$ of Figure 1.3 of [13], which satisfies properties (E1)-(E4) given on page 457 of that paper. We remark that there is no augmented surface for K as K consists of only three p_i/q_i-tangles. As all p_i/q_i have the same sign, each edge path γ_i starts at the vertex $< p_i/q_i >$ of the diagram $\mathcal{D}$ and cannot stop before reaching the left border of the strip part φ of $\mathcal{D}$. Therefore the system is either of type II or of type III. It follows, in particular, that any boundary slope of a candidate surface of K is an even integer, as is explained in the first example on page 462 of [13].

Suppose that S is a connected, orientable, genus one candidate surface with m boundary components, and which is generated by some edge path system $(\gamma_1, \gamma_2, \gamma_3)$ satisfying properties (E1)-(E4) of [13]. Then $\chi(S) = -m$, where χ denote the Euler characteristic. Note that each γ_i consists of at least one edge of $\mathcal{D}$ since γ_i ends either at a rational point on the left border of φ or at the vertex $< \infty >$. □

Claim 3.4. *Each γ_i consists of exactly one edge of $\mathcal{D}$.*

Proof of the claim. The idea is to show that otherwise S would have the homotopy type of the wedge product of at least $m+2$ circles and thus $\chi(S) \leq -m-1$, contradicting the fact that $\chi(S) = -m$. Suppose then that the claim is false and consider the 1-dimensional spine $C \subset S$ constructed as follows.

Let $S_i = S \cap B_i$, where B_i is the 3-ball containing the p_i/q_i-tangle ([13]). Now S_i is constructed according to the edge path γ_i and is contained in $\partial B_i \times [0,1]$ in such a way that the projection to $[0,1]$ is a Morse function on S_i whose critical points are all saddles. We also note that S_i is m-sheeted since the boundary slope of S is integral. Let C_i be the 'core' of S_i, obtained by the obvious level preserving deformation retraction of S_i (i.e. $\partial B_i \times u$, $u \in [0,1]$). Then C_i is a not necessarily connected 1-complex in S_i. Removing all valence 4 vertices of C_i produces a 1-dimensional manifold, in fact a set of arcs. When S_1, S_2, and S_3 are glued to form the surface S, we may assume that C_1, C_2, and C_3 are identified along their ends in $\partial B_i \times 1$ so that a 1-complex C in S is formed which is a deformation retract of S.

In order to estimate $\chi(S)$ we make the following observations: (i) the set of valence 4 vertices of each C_i corresponds precisely to the set of saddles of S_i, (ii) $C_i \cap \partial B_i \times 1$ consists of $2m$ points, (iii) if the system which generates S is of type III, then all the ends of C_i on $\partial B_i \times 1$, $i = 1,2,3$, are pairwise glued, (iv) if the system is of type II, then these ends are glued in triples (one end from each C_i) so that $2m$ valence 3 vertices are created.

Now we estimate $\chi(C) = \chi(S)$, which amounts to estimating the number of circles in C (i.e. the number of free generators in $\pi_1(C)$). If the system is of type III, then each γ_i consists of at least two edges of $\mathcal{D}$. This implies that each S_i has at least $2m$ saddles, which in turn implies that the number of circles in each C_i is at least $m/2$ if m is even, and $(m+1)/2$ if m is odd. Hence there are a total of at least $m+1$ circles in all of the C_i. In constructing C through the gluing of the C_i, $i = 1,2,3$, at least one extra circle is created. Therefore there are at least $m+2$ circles in C and thus $\chi(C) = 1 - (\#$ of circles of $C)$ $\leq -m-1$, a contradiction.

Now suppose that the system is of type II. For the moment let us think of each C_i as a disjoint union of m arcs with their $2m$ ends on $\partial B_i \times 1$. This is precisely the case when each γ_i consists of exactly one edge of $\mathcal{D}$, ignoring the bottom $2m$ arc ends of C_i. Then C is a graph with $3m$ edges and $2m$ valence 3 vertices. Therefore one can easily see, by calculating the Euler characteristic of the graph, that there are at least $m+1$ circles in C. Now if some γ_i consists of more than one edge of $\mathcal{D}$, then S_i has at least one extra saddle other than its bottom m saddles. The corresponding extra valence 4 vertices of C_i will then produce at least one extra circle in C. Therefore there are at least $m+2$ circles in C and thus $\chi(C) \leq -m-1$. This completes the proof of the claim. □

Now we return to the proof of Lemma 3.3. All p_i/q_i have the same sign and therefore the claim above and property (E3) of [13] force each γ_i to be the edge $< 0, p_i/q_i >$ of $\mathcal{D}$. Then $p_1 = p_2 = p_3 = \pm 1$. What is left to show is that the boundary slope (of S) determined by this unique system is divisible by 4.

To compute this boundary slope we use the method of [13], assuming that all $p_i = 1$. Note that to be a knot, either all q_i are odd numbers or exactly one q_i, say q_3, is an even number. In the former case, S has the same twist number as the Seifert surface of K on the candidate list and thus the boundary slope of S is 0. In the latter case, the Seifert surface of K on the candidate list has twist number $2(3-q_1-q_2)$ and S has twist number 6. Thus the boundary slope of S is $6-2(3-q_1-q_2) = 2(q_1+q_2)$ which is divisible by 4. □

Now there are no genus zero essential surfaces in the exterior of a hyperbolic alternating knot ([19, Theorem 1]), and so applying Theorem 1.1 and Lemma 3.1 we see that if K is NSL, then K does not admit non-trivial cyclic surgery (cf. the remarks following Theorem 1.1). Hence we may assume that K is sufficiently large.

Menasco has proven [18, Theorems 2] that if K is sufficiently large, then for any essential closed surface S in $M = S^3 - intN(K)$, there is an embedded annulus A in M such that $A \cap S$ is one boundary component of A and that $A \cap \partial M$ is the other boundary component, which furthermore has the meridian slope. Hence by [8, Theorem 2.4.3], S is either incompressible in $M(r)$ for any non-trivial slope r, or S is compressible in $M(r)$ exactly for those slopes r which are integer slopes or the meridian slope. To prove Corollary 1.3, we obviously may assume that the latter case always holds, i.e.

3.5 We may assume that a closed essential surface in M becomes compressible in $M(r)$ if and only if r is an integer slope or the meridian slope.

To obtain a contradiction, let us assume that m is a cyclic surgery slope of K.

Claim 3.6. *m is not a boundary slope.*

Otherwise by [8, Theorem 2.0.3], there is a closed essential surface S in M which remains essential in $M(r)$ for any slope r satisfying $\Delta(r, m) > 1$ (note that [11] implies that the fourth possibility of [8, Theorem 2.0.3] cannot arise). This is impossible, given the constraint on the essential surfaces in M described in (3.5). □

As K is a hyperbolic knot, there is norm $\|\cdot\|$ on $H_1(\partial M; R)$ satisfying the conditions (2.1) - (2.5) listed in §2. It follows that the fundamental polygon B of the norm is necessarily a parallelogram with vertices $\pm(m-1, 1)$ and $\pm(m+1, 1)$, as K has no non-integral, non-trivial boundary slope. Hence by property (2.3), each of $m-1$ and $m+1$ is a boundary slope.

Claim 3.7. *$M(m-1)$ and $M(m+1)$ are Haken, atoroidal 3-manifolds.*

By [19], both $M(m-1)$ and $M(m+1)$ are irreducible, and furthermore, applying [8, Theorem 2.0.3] and (3.5), we see that both $M(m-1)$ and $M(m+1)$ are Haken. As we have assumed that K is not NSL, Lemma 3.1 implies that $M(m-1)$ and $M(m+1)$ do not contain any incompressible tori. □

With Claims 3.6 and 3.7 in hand, we may conclude, using Theorem 1.2, that a hyperbolic alternating knot K which admits a non-trivial, cyclic surgery, also admits a non-integral, non-trivial boundary slope. As this eventuality is explicitly excluded by the hypotheses of Corollary 1.3, the corollary holds. □

4. Examples and open questions.

Here we produce two infinite families of knots satisfying all the conditions in Theorem 1.1.

Example 4.1. Let $K_n \subset S^3$ be the knot with the surgery description shown in Figure 4.1 (a). Each K_n, $|n| \geq 2$, satisfies the conditions given in Theorem 1.1 and thus has at least one non-integral, non-trivial boundary slope.

Proof. There is already much known about this class of knots. For instance, if $M_n = S^3 - intN(K_n)$, then each $M_n(9n)$ is a lens space ([9, §4]). Further, when $n \geq 2$, each K_n is a hyperbolic knot and each of $M_n(9n+1)$ and $M_n(9n-1)$ is a Seifert fibred space over S^2 with at most three singular fibres and finite first homology ([3, Example 10.2]). Note

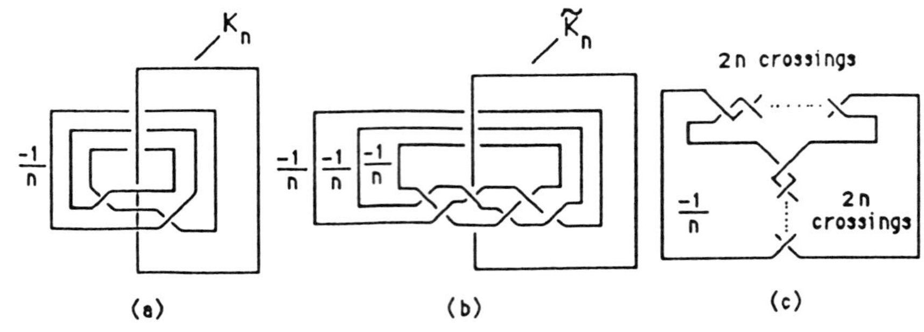

FIGURE 4.1.

that by [15, VI.15], each of $M_n(9n+1)$ and $M_n(9n-1)$ is an irreducible, non-Haken manifold. Thus we only need to show that each K_n is an NSL knot.

Now K_0 is the trivial knot, so we shall assume $n \neq 0$. The three-fold branched cyclic cover of S^3 branched over K_n can be obtained by the Dehn surgery on the Borromean link such that each component of the link has surgery coefficient $-1/n$. This is described in Figure 4.1 (b) where $\widetilde{K}_n$ is the branched knot in the cover. Then by the Kirby-Rolfsen calculus, the same manifold can also be obtained by $-1/n$-surgery on the 2-bridge knot shown in Figure 4.1 (c). By [12], 2-bridge knots are NSL and all their boundary slopes are even integers. Thus the branched cover is irreducible and non-Haken. Now applying [10], we conclude that K_n is NSL. □

Remark. It is observed in [9] that the double branched cover of S^3 branched over K_{2n} is the manifold which can be obtained by $-1/n$ surgery on the figure-eight knot. Hence by [29], it is a hyperbolic manifold as long as $n \geq 2$. It follows that for each $n \geq 2$, K_{2n} is not a Montesinos knot. This family of knots seem to be the first known examples of non-Montesinos knots which admit non-integral, non-trivial boundary slopes (cf. [13]).

Example 4.2. Let $K_n^* \subset S^3$ be the knot with the surgery description shown in Figure 4.2 (a). Each K_n^*, $n \neq 0, 1$, satisfies the conditions given in Theorem 1.1 and thus admits at least one non-integral, non-trivial boundary slope.

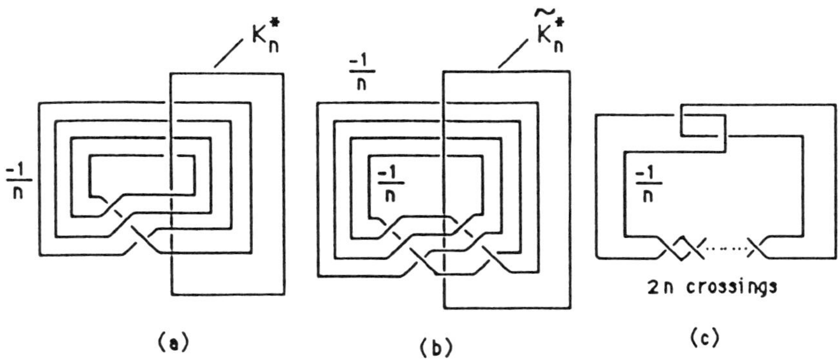

FIGURE 4.2.

Proof. The proof is similar to that of Example 4.1. Let $M_n^* = S^3 - intN(K_n^*)$. It is shown in [17] that each $M_n^*(16n-2)$ is a lens space and in [3, Example 10.3] that each K_n^*, $n \neq 0, 1$, is a hyperbolic knot with $M_n^*(16n-1)$ a Seifert fibred space over S^2 with at most three singular fibres and finite first homology. Hence $M_n^*(16n-1)$ is an irreducible, non-Haken manifold. We must show that each K_n^* is an NSL knot.

K_0^* is the trivial knot and so we shall assume $n \neq 0$. The two-fold branched cover of S^3 branched over K_n^* can be obtained by the Dehn surgery on the Whitehead link with each component having surgery slope $-1/n$. This is described in Figure 4.2 (b), where $\tilde{K}_n^*$ is the branched knot in the cover. Then by the Kirby-Rolfsen calculus, the same manifold can also be obtained by $-1/n$-surgery on the 2-bridge knot (a twisted knot) shown in Figure 4.1 (c). Applying [12] and [10], as in Example 4.1, we conclude that K_n^* is NSL. □

We close with some questions suggested by the results of this paper.

Questions. (1) Is there a sufficiently large hyperbolic knot $K \subset S^3$ which admits a non-trivial cyclic surgery?
(2) Is there a hyperbolic knot $K \subset S^3$ which admits a non-trivial cyclic surgery with slope m, but neither of $M(m-1)$ and $M(m+1)$ is an irreducible non-Haken manifold?
(3) Is there a knot in S^3 which admits a non-trivial cyclic surgery along a boundary slope?

(4) Is there an alternating knot which has a non-integral non-trivial boundary slope?
(5) Is there a non-torus alternating knot which admits a non-trivial cyclic surgery?

We conjecture that the answer to each of these questions is no. Note that the result announced by Delman and Roberts, which is mentioned in the third remark after Corollary 1.3, implies that the answer to question (5) is indeed no.

References

1. S. Bleiler and R. Litherland, *Dehn surgery and satellite knots*, Proc. Amer. Math. Soc. **107** (1990) 1127-1131.
2. C. Bankwitz and H.G. Schumann, *Uber viergeflechte*, Abh. Math. Sem. Univ. Hamburg, **10** (1934) 263-284.
3. S. Boyer and X. Zhang, *Finite Dehn surgery on knots*, Preprint, 1993.
4. —, *The semi-norm and Dehn surgery*, Preprint, 1994.
5. G. Burde and H. Zieschang, *Knots*, de Gruyter 1985.
6. M. Culler and P.B. Shalen, *Varieties of group representations and splittings of 3-manifolds*, Ann. of Math. **117** (1983) 109-146.
7. M. Culler and P.B. Shalen, *Bounded, separating, incompressible surfaces in knot manifolds*, Invent. Math. **75** (1984) 537-545.
8. M. Culler, C. M. Gordon, J. Luecke and P.B. Shalen, *Dehn surgery on knots*, Ann. of Math. **125** (1987) 237-300.
9. R. Fintushel and R. Stern, *Constructing lesns spaces by surgery on knots*, Math. Z. **175** (1980) 33-51.
10. C.M. Gordon and R. Litherland, *Incompressible surfaces in branched coverings*, Smith Conjecture, edited by Morgan and Bass, Academic Press, 1984.
11. D. Gabai, *Foliations and the topology of* 3*-manifolds. III*, J. Diff. Geo. **26** (1987) 479-536.
12. A. Hatcher and W. Thurston, *Incompressible surfaces in* 2*-bridge knot complements*, Invent. Math. **79** (1985) 225-246.
13. A. Hatcher and U. Oertel, *Boundary slopes for Montesinos knots*, Topology **28** (1989) 453-480.
14. P. J. Hilton and U. Stammbach, *A course in homological algebra*, GTM 4, Springer-Verlag, 1970.
15. W. Jaco, *Lectures on three-manifold topology*, Regional conf. ser. in Math. **43** (1977).
16. E. Klassen, *Representations of knot groups in* $SU(2)$, Tran. Amer. Math. Soc. **326** (1991) 795-828.
17. N. Maruyama, *On Dehn surgery along a certain family of knots*, J. of Tsuda College **19** (1987) 261-280.
18. W. Menasco, *Closed incompressible surfaces in alternating knot and link complements*, Topology **23** (1984) 37-44.

19. W. Menasco and M. B. Thistlethwaite, *Surfaces with boundary in alternating knot exteriors*, J. reine angew. Math. **426** (1992) 47-65.
20. L. Moser, *Elementary surgery along a torus knot*, Pacific J. Math. **38** (1971) 737-745.
21. G.D. Mostow, *Quasi-conformal mappings in n-space and the rigidity of hyperbolic space forms*, Publ. Math. I.H.E.S. **34** (1968) 53-104.
22. K. Murasugi, *On the Alexander polynomial of the alternating knot*, Osaka Math J. **10** (1958) 181-189.
23. J.P. Neuzil, *Elementary surgery manifolds and the elementary ideals*, Proc. Amer. Math. Soc. **68** (1978) 225-228.
24. U. Oertel, *Closed incompressible surfaces in complements of star links*, Pacific J. Math. **38** (1981) 737-745.
25. R. Patton, *Incompressible tori in the complement of alternating knots*, to appear in Math. Ann.
26. M. Takahashi, *A negative answer to Hatcher's question*,
27. —, *Two bridge knots have property P*, Memoirs, Amer. Math. Soc. **29** (1981).
28. W. Thurston, *Three manifolds, Kleinian groups and hyperbolic geometry*, Bull. Amer. Math. Soc. **6** (1982) 357-381.
29. —, *The geometry and topology of 3-manifolds*, Princeton Univ. 1979.
30. S. Wang and Q. Zhou, *Symmetry of knots and cyclic surgery*, Tran. Amer. Math. Soc. **330** (1992) 665-676.
31. Y. Wu, *Cyclic surgery and satellite knots*, Topology and its applications **36** (1990) 205-208.
32. X. Zhang, *Cyclic surgery on satellite knots*, Glasgow Math. J. **33** (1991) 125-128.
33. —, *Non-integral boundary slopes exist*, Fund. Math. **138** (1991) 219-224.

DÉPARTEMENT DE MATHÉMATIQUES ET D'INFORMATIQUE, UQAM, P. O. BOX 8888, STATION A, MONTREAL H3C 3P8, CANADA.
E-mail address: zhang@math.uqam.ca

AMS/IP Studies in Advanced Mathematics
Volume **2**, 1997 (Part 1)

RELATIVE EULER NUMBER AND FINITE COVERS OF GRAPH MANIFOLDS

JOHN LUECKE AND YING-QING WU

1. INTRODUCTION.

In this paper we show that every non-trivial graph manifold M has a finite cover that contains a foliation which is very close to a fibration over the circle — the foliation restricted to each vertex manifold of M is a fibration over the circle. On the other hand, we show that this foliation can not always be made to be a fibration over the circle, by exhibiting a large class of graph manifolds that have no finite cover that fibers over the circle. These computations are done using a "relative Euler number" for framed Seifert fibered spaces that are assigned to the vertex manifolds of M. This relative Euler number behaves naturally under finite covers. Using the relative Euler number we define a torsion for each vertex manifold of M, and show that if the torsions are all positive then M is not a surface bundle over S^1. Using the naturality of Euler number with respect to covers, we show that under the same conditions in fact no finite cover of M will fiber over the circle. This situation is analogous to that of Seifert fibered spaces. It follows from [1] that every Seifert fibered space with base orbifold of negative characteristic is covered by a manifold admitting a horizontal foliation. On the other hand a Seifert fibered space is covered by a surface bundle over S^1 if and only if its Euler number or its orbifold characteristic is zero.

We also use this relative Euler number to define a covering invariant $\sigma(M)$ (in the sense of [11] and [5]). In particular it follows that if $\sigma(M)$ is non-zero then M will have the property that homeomorphic covers of

1991 *Mathematics Subject Classification.* Primary 57N10, 57M10.
Key words and phrases. Covering space, graph manifolds, Euler number, fibration .
Partially supported by NSF Grant DMS 9158090, the Sloan Foundation, and the O'Donnell Foundation.
Partially supported by NSF Grant DMS 9102633

M will have the same covering degree. It seems rare that $\sigma(M)$ is zero. For example, we show that if M is a knot complement in the 3-sphere then $\sigma(M)$ is non-zero.

We refer the reader to [4, 3, 7] for definitions and basic facts about Seifert fibered spaces, and to [7] for the theory of 2-dimensional orbifolds, especially their Euler characteristic. If N is a Seifert fibered space, we use $O(N)$ to denote its orbifold, and use $\chi(O(N))$ to denote the Euler characteristic of its orbifold, called the orbifold characteristic of N.

After Waldhausen [10, page 87], an orientable, connected, compact 3-manifold is a *graph manifold* if and only if there is a collection of tori $\mathcal{T} = T_1 \cup \ldots \cup T_k$ in the interior of M such that each component of $M - \text{Int}N(\mathcal{T})$ is a circle bundle. Here $N(\mathcal{T})$ denotes a regular neighborhood of $\mathcal{T}$. Waldhausen then defines a *simple* graph manifold and shows that any graph manifold is a connected sum of simple graph manifolds, lens spaces, and $S^1 \times S^2$'s. A simple graph manifold is characterized by the fact that there is a collection of tori $\mathcal{T}$ such that each component of $M - \text{Int}N(\mathcal{T})$ is a Seifert fibered space with incompressible boundary. When k is minimal, $\mathcal{T}$ is called a *canonical splitting surface*, and each component of $M - \text{Int}N(\mathcal{T})$ is called a *vertex manifold.* We call a simple graph manifold M *trivial* if either $\mathcal{T}$ is empty or it consists of a single torus cutting M into two twisted I-bundles over a Klein bottle. In the first case M is a Seifert fibered space; in the second case M is covered by a torus bundle over S^1, so it admits either Nil or Sol geometry [7].

Suppose M is a nontrivial graph manifold. It is easy to show that M has a double cover $\widetilde{M}$ which has the property that each vertex manifold of $\widetilde{M}$ has negative orbifold characteristic. Since we only discuss properties concerning finite covers, we make the following

Convention. All graph manifolds are assumed simple and nontrivial, and have the property that all of their vertex manifolds have negative orbifold characteristic. We also assume $\mathcal{T}$ is a canonical splitting surface.

Given a graph manifold M, define an associated graph $\Gamma = \Gamma(M)$ as follows. Each vertex manifold M_v of M determines a vertex v of Γ, and each component T_e of $\mathcal{T}$ determines an edge e of Γ, so that if $N(T_e)$ has one boundary component in each of M_v and $M_{v'}$ then e has endpoints on v and v'. Denote by M_V the union of the vertex manifolds. The product structure of $N(\mathcal{T})$ determines an involution ϕ on ∂M_V, and M is homeomorphic to M_V/ϕ.

The paper is organized as follows. In section 2 we define and explore the relative Euler number of a Seifert fibered space. In section 3 we define the torsion of a vertex manifold and use it to give examples of graph manifolds that have no finite cover that fibers over the circle. In section 4, we construct covers of graph manifolds that are needed in sections 3 and 5. In section 5 we show that a graph manifold has a finite cover containing a horizontal foliation. Finally in section 6 we define the covering invariant, $\sigma(M)$, and apply it to the question of when homeomorphic covers of M must have the same covering degree.

2. Relative Euler number of a framed Seifert fibered space.

In this section we will define a relative Euler number for Seifert fibered spaces with a specified set of curves, called a framing, on their boundary. Proposition 2.1 gives a relation between relative Euler numbers corresponding to different framings. Proposition 2.3 shows that the relative Euler number has a natural property under finite covers, which is similar to the naturality of the Euler number of a closed Seifert fibered space.

If α, β are oriented curves on an oriented surface T, we use $\alpha \cdot \beta$ to denote the algebraic intersection number of α with β. Note that $\alpha \cdot \beta = -\beta \cdot \alpha$.

We recall the definition of the Euler number of a Seifert fibered space.

Suppose M is a closed Seifert fibered space. Let $V_1, \dots, V_n$ be a nonempty set of disjoint fibered solid tori in M such that $X = M - \bigcup \mathrm{Int} V_i$ has no singular fibers. Then there is a surface F in X intersecting each fiber exactly once. Let T_i be the torus ∂V_i, let m_i be the curve $F \cap T_i$, and l_i a fiber of M on T_i. Orient m_i, l_i so that $l_i \cdot m_i = 1$ with respect to the orientation of T_i induced from that of X. Now let α_i be a meridian curve of V_i. Then α_i represents some $p_i m_i + q_i l_i$ in $H_1(T_i)$. The *Euler number* of M is defined by $e(M) = \sum q_i/p_i$. This definition is equivalent to the standard one [6, 7]. It turns out that $e(M)$ is independent of the choice of V_i and F, so it is an invariant of the Seifert fibration of M.

A covering $p : \widetilde{M} \to M$ of Seifert fibered spaces is called a (u, v) covering if each regular fiber of M is covered by u regular fibers of $\widetilde{M}$, and the restriction of p on each regular fiber has degree v. Thus the covering degree of $p : \widetilde{M} \to M$ is uv. One of the important properties of $e(M)$ is the naturality: If $p : \widetilde{M} \to M$ is a (u, v) covering, then $e(\widetilde{M}) = e(M)u/v$.

Now consider a Seifert fibered space M with nonempty boundary $\partial M = T_1 \cup \dots \cup T_n$. A *framing* of M is a set of curves $\alpha = \alpha_1 \cup \dots \cup \alpha_n$, such that $\alpha_i = \alpha \cap T_i$ is not isotopic to a fiber of M. Denote by $M(\alpha)$ the manifold obtained from M by Dehn filling along α, that is, $M(\alpha)$ is obtained from M by gluing a solid torus V_i to each T_i so that α_i bounds a disk in V_i. The Seifert fibration of M has a unique extension over $M(\alpha)$.

Definition. The *relative Euler number* of the framed Seifert fibered space (M, α) is defined to be $e(M, \alpha) = e(M(\alpha))$.

Suppose a, b, c are curves on an oriented torus T so that neither of a, b is isotopic to c. Write

$$\delta(a, b, c) = \frac{a \cdot b}{(a \cdot c)(b \cdot c)}.$$

Note that we need to choose an orientation for the curves to define $\delta(a, b, c)$, but since each curve appears twice in the formula, $\delta(a, b, c)$ is independent of such choice.

If α and β are framings of a Seifert fibered space M, we define

$$\Delta(\alpha, \beta) = \sum_i \delta(\alpha_i, \beta_i, l),$$

where α_i and β_i are the curves of α and β on the torus T_i, and l is a fiber of M.

Proposition 2.1. *If α and β are framings of a Seifert fibered space M, then $e(M, \alpha) = e(M, \beta) + \Delta(\alpha, \beta)$.*

Proof. Suppose $\alpha = \alpha_1 \cup \ldots \cup \alpha_n$. Let $V_{n+1}, \ldots, V_m$ be fibered solid tori in M such that $X = M - \bigcup \mathrm{Int} V_j$ has no singular fibers. For $j = n+1, \ldots, m$, let α_j be a curve on $T_j = \partial V_j$ which bounds a meridian disk in V_j. Write $\alpha' = \alpha \cup \alpha_{n+1} \cup \ldots \cup \alpha_m$, and $\beta' = \beta \cup \alpha_{n+1} \cup \ldots \cup \alpha_m$. By definition we have $e(M, \alpha) = e(M(\alpha)) = e(X(\alpha'))$. Similarly $e(M, \beta) = e(X(\beta'))$.

Let F be a surface in X intersecting each fiber of X once. Suppose $\alpha_i = p_i m_i + q_i l_i$, with l_i a fiber of X and m_i a component of ∂F. Choose orientations for m_i, l_i so that $l_i \cdot m_i = 1$ with respect to the orientation of T_i induced from that of X. Suppose $\beta_i = r_i m_i + s_i l_i$. Since the framings of α' and β' are the same on ∂V_j, we have

$$e(X(\alpha')) - e(X(\beta')) = \sum_{i=1}^{n} \frac{q_i}{p_i} - \sum_{i=1}^{n} \frac{s_i}{r_i} = \sum_{i=1}^{n} \frac{q_i r_i - p_i s_i}{p_i r_i}.$$

Since $l_i \cdot m_i = -m_i \cdot l_i = 1$, we have

$$\begin{aligned}\alpha_i \cdot l_i &= (p_i m_i + q_i l_i) \cdot l_i = -p_i \\ \beta_i \cdot l_i &= (r_i m_i + s_i l_i) \cdot l_i = -r_i \\ \alpha_i \cdot \beta_i &= (p_i m_i + q_i l_i)(r_i m_i + s_i l_i) = q_i r_i - p_i s_i\end{aligned}$$

Therefore,

$$e(M, \alpha) = e(M, \beta) + \sum_i \frac{\alpha_i \cdot \beta_i}{(\alpha_i \cdot l_i)(\beta_i \cdot l_i)} = e(M, \beta) + \Delta(\alpha, \beta). \qquad \square$$

A surface F properly embedded in M is called a *transverse surface* if it is transverse to the fibers of M. A framing α is called a *boundary framing* (with respect to F) if there is a transverse surface F so that on each torus T_i of ∂M the curve α_i is a component of $F \cap T_i$. It is well known that a closed Seifert fibered space N has $e(N) = 0$ if and only if there is a closed transverse surface. It follows that β is a boundary framing if and only if $e(M, \beta) = 0$. This fact and Proposition 2.1 can be used to compute $e(M, \alpha)$ without deleting the singular fibers:

Corollary 2.2. *Suppose β is a boundary framing. Then $e(M, \alpha) = \Delta(\alpha, \beta)$.* $\square$

Remark 1. There is another definition of $e(M)$ for closed Seifert fibered space which makes it clearer that $e(M)$ is the obstruction to a cross section of the Seifert fibration: Let V be a regular neighborhood of a regular fiber. Let (α, l) be the meridian–fiber pair of $T = \partial V$. Choose a proper surface in $X = M - \text{Int}V$ transverse to the fibers and let β be the corresponding boundary framing. If $\beta = p\alpha + ql$, then $e(M) = e(X, \alpha) = e(X, \beta) + \Delta(\alpha, \beta) = -q/p$. In other words, M has Euler number $-q/p$ if the boundary framing of X is a (p, q) curve with respect to the meridian–fiber pair of T. From the point of view of Dehn surgery, M has Euler number $-q/p$ if and only if a p/q surgery (with respect to the meridian-fiber coordinate) on a regular fiber produces a manifold with zero Euler number.

Remark 2. Proposition 2.1 and Corollary 2.2 can be thought of as describing a Dehn surgery formula for Euler number. For example, to compute the Euler number for the (r, s) surgery on a (p, q) torus knot, one notices that the longitude l is a boundary framing, and the fiber α of the Seifert fibration of the knot complement is a $(pq, 1)$ curve with respect to the meridian-longitude pair, so one can use Corollary 2.2 to show that the Euler number of the Seifert fibered space after (r, s) surgery is $r/pq(r - pqs)$ up to convention of orientations.

A framed Seifert fibered space $(\widetilde{M}, \widetilde{\alpha})$ is called a cover of (M, α) if there is a covering $\varphi : \widetilde{M} \to M$ so that each $\widetilde{\alpha}_i$ is a component of $\varphi^{-1}(\alpha)$. It is a (u, v) cover if $\varphi : \widetilde{M} \to M$ is such. If φ can be extended to a cover of $\widetilde{M}(\widetilde{\alpha})$ over $M(\alpha)$, then by the naturality of Euler number we have $e(\widetilde{M}, \widetilde{\alpha}) = e(M, \alpha)u/v$. In general there is no such extension of φ, but the naturality still holds.

Proposition 2.3. (Naturality of relative Euler number.) *Suppose $(\widetilde{M}, \widetilde{\alpha})$ is a (u, v) cover of (M, α). Then $e(\widetilde{M}, \widetilde{\alpha}) = e(M, \alpha)u/v$.*

Proof. Let β be a boundary framing of M with respect to a transverse surface F, let $\widetilde{\beta}$ be the boundary framing in $\widetilde{M}$ corresponding to the lifting $\widetilde{F}$ of F. Suppose $\widetilde{T}_j$ is a component of $\partial\widetilde{M}$ covering a torus T_i on ∂M. Let u_j be the number of fibers on $\widetilde{T}_j$ that cover a fiber of T_i. Let $\varphi_*[\widetilde{\alpha}] = n_\alpha[\alpha]$, $\varphi_*[\widetilde{\beta}] = n_\beta[\beta]$. Then by counting the total number of intersections of one component of the preimage of β with all preimages of l we have $(\widetilde{l} \cdot \widetilde{\alpha})u_j = (l \cdot \alpha)n_\alpha$. Similarly, $(\widetilde{l} \cdot \widetilde{\beta})u_j = (l \cdot \beta)n_\beta$. Since the preimage of α has $u_j v/n_\alpha$ components, for the same reason we get $(\widetilde{\alpha} \cdot \widetilde{\beta})(u_j v/n_\alpha) = (\alpha \cdot \beta)n_\beta$. Therefore

$$\delta(\widetilde{\alpha_j}, \widetilde{\beta_j}, \widetilde{l_j}) = \frac{(\widetilde{\alpha} \cdot \widetilde{\beta})}{(\widetilde{l} \cdot \widetilde{\beta})(\widetilde{l} \cdot \widetilde{\alpha})} = \frac{(\alpha \cdot \beta)(n_\alpha n_\beta/u_j v)}{(l \cdot \beta)(l \cdot \alpha)(n_\beta n_\alpha/u_j^2)}$$
$$= \frac{(\alpha \cdot \beta)}{(l \cdot \beta)(l \cdot \alpha)} \frac{u_j}{v} = \delta(\alpha_i, \beta_i, l_j)\frac{u_j}{v}.$$

Summing over all $\widetilde{T}_j$ that cover T_i, we have $\sum u_j = u$. Hence

$$\sum\{\delta(\widetilde{\alpha}_j, \widetilde{\beta}_j, \widetilde{l}_j) \mid \widetilde{T}_j \text{ covers } T_i\} = \delta(\alpha_i, \beta_i, l_i)u/v.$$

Summing both sides of the above formula over i, we get $\Delta(\widetilde{\alpha}, \widetilde{\beta}) = \Delta(\alpha, \beta)u/v$. The result now follows from Corollary 2.2. $\square$

We apply the relative Euler number of framed Seifert fibered spaces to graph manifolds via their vertex manifolds. Let M be a graph manifold and M_i a vertex manifold of M. If T_λ is a torus on ∂M_i, let α_λ^i be a curve on T_λ such that $\phi(\alpha_\lambda^i)$ is a fiber of M_V, where $\phi : \partial M_V \to \partial M_V$ is the involution defined in section 1. In other words, if T_λ is a torus between M_i and M_j, then α_λ^i is a fiber of M_j on T_λ. Since the canonical splitting surface is minimal, none of the α_λ^i is a fiber of M_i. When M_i does not contain boundary components of M, $\alpha^i = \bigcup \alpha_\lambda^i$ is a framing on ∂M_i, called the *natural framing* of M_i. Define the *relative Euler number* of M_i as $E(M_i) = e(M_i, \alpha^i)$. When M_i contains some components of ∂M, $M_i(\alpha^i)$ has zero Euler number, so it is natural to define $E(M_i) = 0$.

3. An obstruction to fibration.

The Euler number of Seifert fibered spaces is the obstruction to the space admitting a horizontal surface, hence usually the obstruction to its fibering over the circle. In fact a Seifert fibered space has a finite cover that fibers over the circle if and only if either its Euler number or its orbifold characteristic is zero [2]. The relative Euler number can be used to define a torsion for each vertex manifold of a graph manifold. Proposition 3.1 shows that if all vertex manifolds of M have positive torsion, then M is not a surface bundle over S^1. The main result is Theorem 3.3, which says that under the same assumption, M is not even covered by any surface bundle over S^1. Some examples will be given at the end of the section.

Definition. Given a framed Seifert fibered space (M, α) with $\partial M = T_1 \cup \ldots \cup T_m$ and l_j a fiber of M on T_j, define the *torsion* of (M, α) to be

$$t(M, \alpha) = |e(M, \alpha)| - \sum_{j=1}^{m} \frac{1}{|\alpha_j \cdot l_j|}.$$

If M is a graph manifold, and M_i is a vertex manifold of M, let α^i be the natural framing on ∂M_i defined in section 2. Define the torsion of M_i to be the torsion with respect to the natural framing, i.e. $t(M_i) = t(M_i, \alpha^i)$.

Proposition 3.1. *If $t(M_i) > 0$ for each vertex manifold M_i of M, then M is not a surface bundle over S^1.*

Proof. Assuming the contrary, write $M = F \times I/h$, where $h : F \times 0 \to F \times 1$ is the monodromy map. By Thurston's classification theorem of surface homeomorphisms [9] we may choose h and a set of curves $\mathcal{C} = C_1 \cup \ldots \cup C_k$ with $h(\mathcal{C}) = \mathcal{C}$, so that h^n restricted to each component of $F_V = F - \text{Int}N(\mathcal{C})$ is either periodic or pseudo-Anosov, and maps each component of F_V to itself. Since M is a graph manifold, no component of $F_V \times I/h$ is hyperbolic, so $h^n|_{F_V}$ has no pseudo-Anosov component [8]. Hence $h^n|_{F_V}$ is periodic, and the I bundle structure of $F_V \times I$ gives rise to a Seifert fibration of $F_V \times I/h$. When $\mathcal{C}$ is minimal, $\mathcal{C} \times I/h$ is the canonical splitting surface of M, so $M_V = F_V \times I/h$. In particular, F is transverse to all fibers of M_V.

Let M_i be a component of M. Let k_i be the number of times a regular fiber of M_i intersects F. Choose i so that $k_i \geq k_j$ for all j. For each T_λ on ∂M_i, let β_λ be a component of $T_\lambda \cap F$. The union of these β_λ forms a framing β of M_i. Since β_λ are boundaries of $F \cap M_i$, β is a boundary framing, so $e(M_i, \beta) = 0$. By Proposition 2.1,

$$e(M_i, \alpha^i) = e(M_i, \beta) + \Delta(\alpha^i, \beta) = \sum_{T_\lambda \subset \partial M_i} \delta(\alpha^i_\lambda, \beta_\lambda, l_\lambda)$$

$$= \sum_{T_\lambda \subset \partial M_i} \frac{\alpha^i_\lambda \cdot \beta_\lambda}{(\alpha^i_\lambda \cdot l_\lambda)(\beta_\lambda \cdot l_\lambda)},$$

where l_λ is a fiber of M_i on T_λ, and α^i is the natural framing on ∂M_i. The sum is taken over all λ such that T_λ is on ∂M_i. Let $r_\lambda = |F \cap T_\lambda|$. Then $k_i = |F \cap l_\lambda| = |F \cap T_\lambda|\,|\beta_\lambda \cap l_\lambda| = r_\lambda\,|\beta_\lambda \cdot l_\lambda|$. Suppose the involution $\phi : \partial M_V \to \partial M_V$ maps T_λ to a torus on ∂M_j. Then similarly we have $k_j = r_\lambda\,|\alpha^i_\lambda \cdot \beta_\lambda|$. Hence,

$$|e(M_i, \alpha^i)| \le \sum_\lambda \left| \frac{\alpha^i_\lambda \cdot \beta_\lambda}{(\alpha^i_\lambda \cdot l_\lambda)(\beta_\lambda \cdot l_\lambda)} \right| = \sum_\lambda \frac{k_j}{|\alpha^i_\lambda \cdot l_\lambda| k_i} \le \sum_\lambda \frac{1}{|\alpha^i_\lambda \cdot l_\lambda|}.$$

This contradicts the assumption that $t(M_i, \alpha^i) > 0$. □

Definition. An s^2-fold covering $f : \widetilde{T} \to T$ of tori is called a *characteristic* covering if there is a basis $(\widetilde{a}, \widetilde{b})$ for $\pi_1(\widetilde{T})$, and a basis (a, b) for $\pi_1(T)$, such that $f_*(\widetilde{a}) = sa$ and $f_*(\widetilde{b}) = sb$.

A covering $p : \widetilde{M} \to M$ of Seifert fibered spaces is called *characteristic* or *s-characteristic* if the restriction of p to each boundary component is an s^2-fold characteristic covering. If $\widetilde{M}$ and M are graph manifolds, then p is a *characteristic* covering if its restriction to each piece of $\widetilde{M}$ is characteristic.

In the next section we will prove the following result.

Proposition 3.2. *Let $p : \widetilde{M} \to M$ be a covering of graph manifolds. Then there is a covering $\widetilde{p} : \widehat{M} \to \widetilde{M}$ so that $p \circ \widetilde{p} : \widehat{M} \to M$ is a characteristic covering.*

Theorem 3.3. *If $t(M_i) > 0$ for each vertex manifold M_i of M, then M is not covered by a surface bundle over S^1.* [1]

Proof. If $p : \widetilde{M} \to M$ is a cover such that $\widetilde{M}$ is a surface bundle over S^1, by the above proposition we can find a further cover $\widetilde{p} : \widehat{M} \to \widetilde{M}$ so that $\widehat{p} = p \circ \widetilde{p} : \widehat{M} \to M$ is an s-characteristic cover for some s. Suppose $\widehat{M}_j$ is a vertex manifold that covers a vertex manifold M_i of M, and suppose a component $T_\lambda \subset \partial M_i$ is covered by k components of $\partial \widehat{M}_j$. Then $\widehat{p}$ is a (ks, s) cover, so by Proposition 2.3, $E(\widehat{M}_j) = E(M_i)k$. Also, since $\widehat{p}$

[1] Added in proof: In "Commensurability and virtual fibration for graph manifolds", W. Neumann has now given an algebraic characterization of when a graph manifold has a finite cover that fibers over the circle.

is a characteristic cover, if $\widehat{T}_\mu \subset \partial\widehat{M}_j$ covers T_λ then $\widehat{\alpha}_\mu \cdot \widehat{l}_\mu = \alpha_\lambda \cdot l_\lambda$, where $\widehat{\alpha}_\mu$ is the natural framing of $\widehat{M}_j$, and $\widehat{l}_\mu$ a fiber. Since each torus in ∂M_i is covered by k tori in $\partial\widehat{M}_j$, we have

$$\sum_{\widehat{T}_\mu \subset \partial\widehat{M}_j} \frac{1}{|\widehat{\alpha}_\mu \cdot \widehat{l}_\mu|} = k \sum_{T_\lambda \subset \partial M_i} \frac{1}{|\alpha_\lambda \cdot l_\lambda|}.$$

Therefore $t(\widehat{M}_j) = t(M_i)k$. It follows that if all vertex manifolds of M have positive torsions, then the vertex manifolds of $\widehat{M}$ have positive torsions also. Since $\widehat{M}$ is a cover of the surface bundle $\widetilde{M}$, it is also a surface bundle, which contradicts Proposition 3.1. □

Examples. (1). Let M_1, M_2 be the product $F \times S^1$, where F is an orientable surface with a single boundary component. On each component of ∂M_i, define $m_i = \partial F \times 1$, and $l_i = \{\text{point}\} \times S^1$. Define a map $\phi : \partial M_1 \to \partial M_2$ so that

$$\phi(l_1, m_1) = (l_2, m_2) \begin{pmatrix} a & ad+1 \\ 1 & d \end{pmatrix}$$

where a, d are arbitrary integers. Since the determinant of the matrix is -1, such ϕ do exist. Let $M = M_1 \cup_\phi M_2$. Using Corollary 2.2 one computes that $E(M_1) = -d$, $E(M_2) = a$, and $|l_1 \cdot l_2| = 1$. Therefore, if both $|a|$ and $|d|$ are greater than 1, then $t(M_i) > 0$, so M is not covered by surface bundles.

(2). Suppose M is a closed graph manifold. From the remarks after Corollary 2.2, we see that if we do p/q surgery (with respect to the meridian-fiber coordinate) on a regular fiber of M_i to get M_i', then $E(M_i') = E(M_i) - q/p$. This operation certainly does not change the second term in the definition of the torsion $t(M_i)$. Therefore, if q is sufficiently large, then $t(M_i') > 0$. It follows that for any closed graph manifold M, we can always do surgery on fibers of vertex manifolds to obtain a new graph manifold M' which is not covered by surface bundles.

(3). Let $\mathcal{D}$ be the plumbing of $\{\mathcal{D}_i\}$, where each $\mathcal{D}_i$ is a 2-disk bundle over an orientable closed surface other than 2-sphere [6]. Let Γ be the plumbing diagram. Then the boundary of $\mathcal{D}$ is a graph manifold M. Γ is also the underlying graph of M. For each vertex v_i of Γ, the relative Euler number of M_i is the Euler number of the corresponding disk bundle $\mathcal{D}_i$. Furthermore, the absolute value of the intersection number of circle fibers at each edge of Γ is 1. Therefore, the torsion of M_i is the absolute value of the Euler number of $\mathcal{D}_i$ minus the valence of v_i.

The set of torsions of the vertex manifolds of M gives an obstruction for M to be covered by surface bundles. But the obstruction disappears when M has boundary. A vertex manifold M_i of M that contains components of ∂M has $E(M_i) = 0$, so $t(M_i)$ is negative.

Question. *Suppose a graph manifold M has nonempty boundary. Is M always covered by some surface bundle over S^1?* [2]

4. Characteristic covering.

In this section we will prove Proposition 4.4, which says that whenever we have a cover of graph manifold, there is a further cover which is characteristic. The result is needed in sections 3 and 5.

Lemma 4.1. *Let S be surface with negative Euler characteristic. To each component d of ∂S assume there is associated an integer $n_d > 1$. Then there is a covering $p : \widetilde{S} \to S$ so that if d is a component of ∂S and $\widetilde{d}$ is a component of $p^{-1}(d)$ then $p : \widetilde{d} \to d$ is the n_d-fold cover.*

Proof. One can prove this by direct construction, but we provide the following proof using known results about 2-dimensional orbifolds. One is referred to [7] for definitions and theorems regarding orbifolds. Recall that an orbifold is called a good orbifold if it is neither a 2-sphere with a single cone point nor a 2-sphere with 2 cone points of different cone angle.

Gluing to each boundary component d of ∂S a cone C_d of angle $2\pi/n_d$, we get an orbifold $\widehat{S}$. Since S has negative Euler characteristic, it either has nonzero genus, or is planar with at least 3 boundary components, so $\widehat{S}$ is a good orbifold. Thus there is an orbifold covering $p : F \to \widehat{S}$, where F has no cone point, i.e. it is a surface [7, Theorem 2.5]. The preimage of each cone C_d is a set of disjoint disks, so the restriction of p to each component of $p^{-1}(d)$ is an n_d fold cover. Define $\widetilde{S} = p^{-1}(S)$. Then the restriction of p to $\widetilde{S}$ is the required cover. □

Lemma 4.2. *Suppose N is a Seifert fibered space with non-empty boundary and negative orbifold characteristic. Then there is a characteristic covering $N' \to N$ so that N' is a product.*

Proof. For each component T_i of ∂N, choose a curve α_i which intersects each fiber of ∂N once. Then $\alpha = \cup \alpha_i$ is a framing on N. As before, let $e(N, \alpha)$ be the relative Euler number of (N, α). It is a rational number.

---

[2] Added in proof: In "Graph manifolds with non-empty boundary are covered by surface bundles", S. Wang and F. Yu have now shown that the answer to this question is yes.

Let S be the orbifold of N. Glue a cone of angle $2\pi/k$ to each component of ∂S, giving an orbifold $\overline{S}$. Since $\chi(S) < 0$, we can choose k big enough so that $\chi(\overline{S}) \leq 0$. (Actually $k = 6$ is enough.) Let $q : \widehat{S} \to \overline{S}$ be a covering such that $\widehat{S}$ is an orientable surface without cone points. Since $\chi(\widehat{S}) \leq 0$, $\widehat{S}$ has positive genus. Thus by going to a further cover if necessary, we may assume that $e(N, \alpha)\deg(q)$ is an integral multiple of k.

Let $\widetilde{S} = q^{-1}(S) \subset \widehat{S}$. Then $q : \widetilde{S} \to S$ is a covering of orbifolds. Note that q restricted to any boundary component of $\widetilde{S}$ has degree k. By pulling back the fibration $N \to S$, we get a fibration $\widetilde{N} \to \widetilde{S}$, and a map $p : \widetilde{N} \to N$ that induces $q : \widetilde{S} \to S$, (see [7, Page 431].) The orbifold covering degree of p is $\deg(q)$, and the fiber covering degree is 1. By Proposition 2.3, $e(\widetilde{N}, \widetilde{\alpha}) = e(N, \alpha)\deg(q)$, so it is a multiple of k, where $\widetilde{\alpha} \subset p^{-1}(\alpha)$ is the pull-back framing. Since α intersects each fiber of ∂N once and since the fiber covering degree is 1, $\widetilde{\alpha}$ also intersects each fiber of $\partial\widetilde{N}$ once. The orbifold map q is a degree k map on each boundary of $\widetilde{S}$, so $\deg(p|_{\widetilde{\alpha}_j}) = k$ for all $\widetilde{\alpha}_j \subset \widetilde{\alpha}$.

Since $\widetilde{S}$ is an orientable surface with boundary, $\widetilde{N}$ can be written as a product $\widetilde{N} = \widetilde{S} \times S^1$. For each torus $\widetilde{T}_j \subset \partial\widetilde{N}$, let $\widetilde{m}_j$ be the component of $\widetilde{S} \times 1$ on $\widetilde{T}_j$, let $\widetilde{l}_j$ be a fiber of $\widetilde{N}$, and let $\widetilde{\alpha}_j$ be the framing curve $\widetilde{\alpha} \cap \widetilde{T}_j$. Since $\widetilde{\alpha}_j$ intersects $\widetilde{l}_j$ once, $\widetilde{\alpha}_j$ can be written as $\widetilde{m}_j + k_j\widetilde{l}_j$ for some integer k_j. Notice that $\cup\widetilde{m}_j$ is a boundary framing, so by Corollary 2.2 we have

$$e(\widetilde{N}, \widetilde{\alpha}) = \Delta(\widetilde{\alpha} \cup \widetilde{m}_j) = \sum \frac{\widetilde{\alpha}_j \cdot \widetilde{m}_j}{(\widetilde{\alpha}_j \cdot \widetilde{l}_j)(\widetilde{m}_j \cdot \widetilde{l}_j)} = \sum k_j.$$

Suppose T_i is the torus in ∂N covered by $\widetilde{T}_j$. Then $p : \widetilde{T}_j \to T_i$ is given by $p(\widetilde{\alpha}_j) = k\alpha_i$, and $p(\widetilde{l}_j) = l_i$. We need to find a covering $p' : N' \to \widetilde{N}$, so that if T'_r covers $\widetilde{T}_j$ and if α'_r, l'_r are components of the preimage of $\widetilde{\alpha}_j$ and $\widetilde{l}_j$ respectively, then $p'(\alpha'_r) = \widetilde{\alpha}_j$, and $p'(l'_r) = k\widetilde{l}_j$. It will then follow that $p \circ p' : N' \to N$ is a k-characteristic covering. Since $\widetilde{N}$ is a product and N' is a cover of $\widetilde{N}$, N' will also be a product.

To find such a covering p', it suffices to find a map $f_* : \pi_1(\widetilde{N}) \to \mathbb{Z}_k$ so that $f_*(\widetilde{\alpha}_j) = 0$, and $f_*(\widetilde{l}_j) = 1$. N' will then be the covering corresponding to the kernel of f_*.

Consider $\widetilde{N} = \widetilde{S} \times S^1$. We have

$$\pi_1(\widetilde{N}) = \langle a_1, b_1, \ldots, a_g, b_g, \widetilde{m}_1, \ldots, \widetilde{m}_r \mid \Pi[a_i, b_i]\Pi\widetilde{m}_i = 1\rangle \times \langle\widetilde{l}\rangle,$$

where a_i, b_i are the standard generators for the corresponding closed surface, $\widetilde{l}$ is represented by a fiber, and $\widetilde{m}_j$ by the boundaries of $\widetilde{S}$.

Define $f_* : \pi_1(\widetilde{N}) \to \mathbb{Z}_k$ by

$$f_*(a_i) = f_*(b_i) = 0$$
$$f_*(\widetilde{l}) = 1$$
$$f_*(\widetilde{m}_j) = -k_j.$$

Since $\sum k_j = e(\widetilde{N}, \widetilde{\alpha})$ is a multiple of k, the map is well defined. Also, $f_*(\widetilde{\alpha}_j) = f_*(\widetilde{m}_j + k_j\widetilde{l}_j) = 0$. Hence this is the required map. □

Lemma 4.3. *Suppose $p : \widetilde{N} \to N$ is a covering of Seifert fibered spaces with nonempty boundary and negative orbifold characteristic. Then there is an integer r such that for any multiple s of r there is a cover $\widetilde{p} : \widehat{N} \to \widetilde{N}$ so that the composition $p \circ \widetilde{p}$ is an s-characteristic cover. Furthermore, $\widehat{N}$ is a product.*

Proof. First assume $N = F \times S^1$ for some compact orientable surface F. Then by passing to a further cover we may assume that $\widetilde{N}$ is also a product $\widetilde{N} = \widetilde{F} \times S^1$, and p can be written as $p = p_1 \times p_2 : \widetilde{F} \times S^1 \to F \times S^1$. For each component d of $\partial\widetilde{F}$, let $m_d = \deg(p_1|_d)$. Let $m_t = \deg(p_2)$ be the circle fiber covering degree. Let $r' = \text{l.c.m.}(\{m_d |\ d \text{ is a component of } \partial\widetilde{F}\ \} \cup \{m_t\})$, and let $r = 2r'$.

Suppose s is a multiple of r. For every component d of $\partial\widetilde{F}$ define $n_d = s/m_d$, and $n_t = s/m_t$. By the choice of r, all n_d are even integers, so in particular $n_d > 1$. By Lemma 4.1, there is a cover $f : \widehat{F} \to \widetilde{F}$ so that if d is a component of $\partial\widetilde{F}$ and $\widehat{d}$ a component of $\partial\widehat{F}$ covering d then $f| : \widehat{d} \to d$ is the n_d-fold cover. Let $g : S^1 \to S^1$ be an n_t-fold cover. Then it is easy to see that $\widetilde{p} = f \times g : \widehat{F} \times S^1 \to \widetilde{F} \times S^1$ has the property that $p \circ \widetilde{p} = (p_1 \circ f) \times (p_2 \circ g)$ is an s-characteristic cover.

Now suppose N is not a product. By Lemma 4.2 we can find an s'-characteristic cover $p' : N' \to N$ so that N' is a product. Let N'' be a common finite cover of N' and $\widetilde{N}$. Consider the cover $p'' : N'' \to N'$. By the above, there is an integer r so that for any multiple s of r we have a cover $\widehat{p} : \widehat{N} \to N''$ so that the composition $p'' \circ \widehat{p} : \widehat{N} \to N'$ is an s-characteristic cover. Choose $\overline{r} = s'r$. Then for any multiple $\overline{s} = s's$ of $\overline{r}$, the composition $p' \circ p'' \circ \widehat{p} : \widehat{N} \to N$ is a $\overline{s}$-characteristic cover (because both p' and $p'' \circ \widehat{p}$ are characteristic), and it factors through $\widetilde{N}$, as required. Since N' is a product, N'' is also a product. □

Proposition 4.4. *Let $p : \widetilde{M} \to M$ be a covering of graph manifolds. Then there is a covering $\widetilde{p} : \widehat{M} \to \widetilde{M}$ so that $p \circ \widetilde{p} : \widehat{M} \to M$ is a characteristic covering. Furthermore each vertex manifold in $\widehat{M}$ is a product.*

Proof. For each vertex manifold $\widetilde{M}_{\bar{v}}$ of $\widetilde{M}$, there is a vertex manifold M_v of M so that the restriction of p is a covering of Seifert fibered spaces $p_{\bar{v}} : \widetilde{M}_{\bar{v}} \to M_v$. By Lemma 4.3 there is an integer $r_{\bar{v}}$ and a cover $\widetilde{p}_{\bar{v}} : \widehat{M}_{\bar{v}} \to \widetilde{M}_{\bar{v}}$ so that $p_{\bar{v}} \circ \widetilde{p}_{\bar{v}} : \widehat{M}_{\bar{v}} \to M_v$ is an s-characteristic cover, where s can be any multiple of $r_{\bar{v}}$. Fix an s which is a common multiple of all $r_{\bar{v}}$. Choose an integer $k_{\bar{v}}$ for each vertex $\bar{v}$ in the graph $\widetilde{\Gamma} = \Gamma(\widetilde{M})$ so that $D = k_{\bar{v}} \deg(\widetilde{p}_{\bar{v}})$ are the same for all $\bar{v}$. Let $N_{\bar{v}}$ be $k_{\bar{v}}$ copies of $\widehat{M}_{\bar{v}}$, and let $q_{\bar{v}} = \cup \widetilde{p}_{\bar{v}} : N_{\bar{v}} \to \widetilde{M}_{\bar{v}}$ be the corresponding covering map. Note that $\deg(q_{\bar{v}}) = D$, and the map $p \circ q_{\bar{v}} : N_{\bar{v}} \to M_v$ is an s-characteristic cover for all $\bar{v}$. We need to show that we can glue these $N_{\bar{v}}$ together to form a cover $\widetilde{p} : \widehat{M} \to \widetilde{M}$.

Suppose $\widetilde{M}_v$ and $\widetilde{M}_{v'}$ are vertex manifolds corresponding to the endpoints of an edge e in $\widetilde{\Gamma}$. Let $\widetilde{T}_e$ and $\widetilde{T}'_e$ be the tori in $\widetilde{M}_v$ and $\widetilde{M}_{v'}$ corresponding to e, and let $\widehat{T}_e, \widehat{T}'_e$ be components of $q_v^{-1}(\widetilde{T}_e)$ and $q_{v'}^{-1}(\widetilde{T}'_e)$, respectively. Let $\widetilde{f} : \widetilde{T}_e \to \widetilde{T}'_e$ be the gluing map. We want to find a lifting $\widehat{f}$ of $\widetilde{f}$ so that the following diagram commutes.

$$\begin{array}{ccc} \widehat{T}_e & \xrightarrow{\widehat{f}} & \widehat{T}'_e \\ {\scriptstyle q_v}\downarrow & & \downarrow{\scriptstyle q_{v'}} \\ \widetilde{T}_e & \xrightarrow{\widetilde{f}} & \widetilde{T}'_e \end{array}$$

To simplify the notation, still use e, v, v' to denote the image of e, v, v' in Γ under the map $\widetilde{\Gamma} \to \Gamma$ induced by p. We can extend the above diagram to

$$\begin{array}{ccc} \widehat{T}_e & & \widehat{T}'_e \\ {\scriptstyle q_v}\downarrow & & \downarrow{\scriptstyle q_{v'}} \\ \widetilde{T}_e & \xrightarrow{\widetilde{f}} & \widetilde{T}'_e \\ {\scriptstyle p_v}\downarrow & & \downarrow{\scriptstyle p_{v'}} \\ T_e & \xrightarrow{f} & T'_e \end{array}$$

Pick consistent base points. Then $p_{v'} \circ \widetilde{f} \circ q_v : \widehat{T}_e \to T'_e$ and $p_{v'} \circ q_{v'} : \widehat{T}'_e \to T'_e$ are both the s^2-fold characteristic covers. Thus $\operatorname{im}(p_{v'} \circ \widetilde{f} \circ q_v)_* = \operatorname{im}(p_{v'} \circ q_{v'})_* =$ the characteristic subgroup of index s^2 in $\pi_1(T'_e)$. Since $(p_{v'})_*$ is injective we must have

$$\operatorname{im}(\widetilde{f} \circ q_v)_* = \operatorname{im}(q_{v'})_*.$$

Thus $\widetilde{f} \circ q_v$ and $q_{v'}$ must be equivalent covers, so there is a homeomorphism $\widehat{f} : \widehat{T}_e \to \widehat{T}'_e$ making the above diagram commute.

The map $p_v \circ q_v : \widehat{T}_e \to \widetilde{T}_e \to T_e$ is an s^2-fold characteristic map, so $\deg(q_v|_{\widehat{T}_e}) = s^2 / \deg(p_v|_{\widetilde{T}_e})$. Thus the number of tori in N_v that cover $\widetilde{T}_e$ is

$$\deg(N_v \to \widetilde{M}_v) / \deg(q_v|_{\widetilde{T}_e}) = D \deg(p_v|_{\widetilde{T}_e})/s^2$$

Similarly for $N_{v'}$. Since $\deg(p_v|_{\widetilde{T}_e}) = \deg(p_{v'}|_{\widetilde{T}'_e})$, it follows that the number of tori in $q_v^{-1}(\widetilde{T}_e)$ is the same as that in $q_{v'}^{-1}(\widetilde{T}'_e)$. Hence we can use the above lifting map $\widehat{f}$ to glue each component of $q_v^{-1}(\widetilde{T}_e)$ to a component of $q_{v'}^{-1}(\widetilde{T}'_e)$. After doing this for all e in $\widetilde{\Gamma}$ we get a manifold $\widehat{M} = (\bigcup N_v)/\{\widehat{f}\}$ and a map $\widetilde{p} : \widehat{M} \to \widetilde{M}$ so that $p \circ \widetilde{p} : \widehat{M} \to M$ is a characteristic map. If $\widehat{M}$ is not connected, just pick a component. □

5. Foliations of finite covers.

In section 3 it was shown that many graph manifolds have no finite cover that fibers over the circle. In this section we will show that all graph manifolds (recall our conventions on graph manifolds), on the other hand, have finite covers that almost fiber over the circle. Specifically, we show that there are finite covers which admit "horizontal foliations"– that is, foliations transverse to the Seifert-fibers of each of the vertex manifolds. These horizontal foliations come from fibering each vertex manifold over the circle so that the fibers match up along the canonical splitting surface. Consequently, although these foliations are not often fibrations, they admit a very strong transverse structure.

This situation is analogous to the case of Seifert-fibered spaces. Here again, many Seifert-fibered spaces do not have finite covers that fiber over the circle but it follows from [1] that Seifert-fibered spaces with non-positive base orbifold will have finite covers that admit horizontal foliations.

The main theorem of this section is Theorem 5.5. Propositions 5.1 and 5.4 will be used in the proof of Theorem 5.5.

Proposition 5.1. *Let M be a graph manifold in which every vertex manifold is a product. Let s be a positive integer. Then there is a characteristic cover $p : \widetilde{M} \to M$ such that*

(1) *the preimage under p of any vertex manifold is connected,*
(2) *if e is an edge of $\Gamma = \Gamma(M)$ and T the corresponding component of the canonical splitting surface, then $|p^{-1}(T)|$ is an integral multiple of s.*

Proof. Let M_v be a vertex manifold of M. By hypothesis we may write $M_v = F_v \times S^1$ for some surface F_v with nonempty boundary and nonpositive Euler characteristic. Attach cones of angle $2\pi/3$ to the components of ∂F_v, giving a good orbifold, $\overline{F}_v$, of negative orbifold characteristic. Let $q_v : \widehat{F}_v \to \overline{F}_v$ be a covering such that $\widehat{F}_v$ is an orientable (connected) surface without cone points. $\overline{F}_v$ will have positive genus, so we may assume, after passing to a further cover, that the degree of q_v is a multiple of $3s$. Let $\widetilde{F}_v = q_v^{-1}(F_v) \subset \widehat{F}_v$. Then $q_v : \widetilde{F}_v \to F_v$ is a connected covering which restricts to a 3-fold cover on each component of $\partial \widetilde{F}_v$. Furthermore, there is a positive integer r_v such that if C is a component of ∂F_v, then $|q_v^{-1}(C)| = r_v s$.

Let $p_v : \widetilde{M}_v \to M_v$ be the cover obtained by taking the cover q_v in the F_v direction and taking the 3-fold cover in the circle direction. Then p_v is 3-characteristic, $\widetilde{M}_v$ is connected, and the preimage of any component of ∂M_v consists of $r_v s$ tori.

We apply the above construction to M_v for every vertex $v \in \Gamma$. We may assume, by first passing to further covers q_v, that $r_{v_1} = r_{v_2} = r$ for any vertices v_1, v_2 in Γ. Fix an orientation on the edges of Γ and let e be one such edge. Let M_{v_1} and M_{v_2} be the vertex manifolds at the head and tail of e. Let T_e^1 be the component of ∂M_{v_1} corresponding to e and T_e^2 the component of ∂M_{v_2} corresponding to e. Let f_e be the gluing homeomorphism identifying T_e^1 and T_e^2. If $\widetilde{T}_e^1, \widetilde{T}_e^2$ are components of $p_{v_1}^{-1}(T_e^1), p_{v_2}^{-1}(T_e^2)$ (resp.) then there is a lift, $\widetilde{f}_e : \widetilde{T}_e^1 \to \widetilde{T}_e^2$, of f_e, since both covers are 3-characteristic. Since $|p_{v_1}^{-1}(T_e^1)| = |p_{v_2}^{-1}(T_e^2)| = rs$, we may use such lifts to identify $p_{v_1}^{-1}(T_e^1)$ and $p_{v_2}^{-1}(T_e^2)$. Doing this for all edges e of Γ gives the desired cover $p : \widetilde{M} \to M$. □

Lemma 5.2. *Let* $\begin{pmatrix} a & c \\ b & d \end{pmatrix}$ *be an integral matrix with* $d = 0$ *and determinant* -1. *Then there is a positive integer,* s, *and a collection of coprime pairs of integers,* $\{(p_i, q_i) \mid i = 1, \dots, s\}$, *such that* $q_i \neq 0$, $bp_i + dq_i \neq 0$ *for every* i, *and*

$$\frac{p_1}{q_1} + \cdots + \frac{p_s}{q_s} = 0;$$
$$\frac{ap_1 + cq_1}{bp_1 + dq_1} + \cdots + \frac{ap_s + cq_s}{bp_s + dq_s} = 0.$$

Proof. Since multiplying the matrix by ± 1 will not change the solutions we may assume that $c = b = 1$. Set $s = |a| + 4$.

If $a \geq 0$ set

$$p_1 = p_1 = 1; \qquad p_3 = p_4 = \cdots = p_{|a|+4} = -1;$$
$$q_1 = q_2 = 2; \qquad q_3 = q_4 = \cdots = q_{|a|+4} = a + 2.$$

If $a < 0$ set

$$p_1 = p_1 = -1; \qquad p_3 = p_4 = \cdots = p_{|a|+4} = 1;$$
$$q_1 = q_2 = 2; \qquad q_3 = q_4 = \cdots = q_{|a|+4} = |a| + 2.$$

These values solve the required problem. □

Lemma 5.3. *Let* $\begin{pmatrix} a & c \\ b & d \end{pmatrix}$ *be an integral matrix with determinant* -1 *and assume* $b \neq 0$, $d \neq 0$. *Then there are coprime pairs of integers* (p, q) *and* (p', q') *such that* $q, q' \neq 0$ *and*

(a) $$\frac{ap + cq}{bp + dq} + \frac{-ap + cq}{-bp + dq} > 0;$$

(b) $$\frac{ap' + cq'}{bp' + dq'} + \frac{-ap' + cq'}{-bp' + dq'} < 0.$$

Proof. Let $f(x) = 2(cd - abx^2)/(d^2 - b^2x^2)$. If $q \neq 0$, then

$$f(\frac{p}{q}) = \frac{ap + cq}{bp + dq} + \frac{-ap + cq}{-bp + dq}$$

So we must find coprime pairs of integers (p, q) and (p', q') so that $f(p/q) > 0$ and $f(p'/q') < 0$.

Case I: $a \neq 0, c \neq 0$

First notice that $cd/ab > 0$ ($ad = bc - 1$ implies that ad and bc have the same sign, hence so will a/d and b/c). Let $\lambda = \sqrt{cd/ab}$. Then λ is a root of f (note that $f(\lambda)$ is well defined since $\lambda^2 \neq d^2/b^2$). Furthermore,

$$f'(\lambda) = \frac{4\lambda bd}{(d^2 - b^2\lambda^2)^2} \neq 0$$

Thus by taking p/q, p'/q' very close to λ and on opposite sides of λ on the real line we will have $f(p/q) > 0$ and $f(p'/q') < 0$.

Case II: $a = 0$

In this case, $b = c = \pm 1$. Thus

$$f(x) = 2\frac{\pm d}{d^2 - x^2}$$

By choosing $|p/q| < |d|$ and $p'/q' > |d|$, or vice versa, we prove the lemma.

Case III: $c = 0$

In this case, $\{a, d\} = \{+1, -1\}$. Thus

$$f(x) = 2\frac{\pm bx^2}{1 - b^2x^2}$$

If $0 < x \ll 1$ and $y \gg 0$, then $f(x)$ and $f(y)$ will be of opposite sign. Cases I,II,III exhaust the possibilities of Lemma 5.3. □

Proposition 5.4. *Let $\begin{pmatrix} a & c \\ b & d \end{pmatrix}$ be an integral matrix with determinant -1 and assume $b \neq 0$. Then there is a positive integer, s, and a collection of coprime pairs of integers, $\{(p_i, q_i) \mid i = 1, \dots, s\}$, such that $q_i \neq 0$, $bp_i + dq_i \neq 0$ for every i, and*

$$\frac{p_1}{q_1} + \cdots + \frac{p_s}{q_s} = 0;$$

$$\frac{ap_1 + cq_1}{bp_1 + dq_1} + \cdots + \frac{ap_s + cq_s}{bp_s + dq_s} = 0.$$

Proof. The special case that $d = 0$ is treated in Lemma 5.2 above. So we assume that $d \neq 0$. In this case we may apply Lemma 5.3 to find coprime integers (p, q) and (p', q') satisfying conditions (a) and (b) of that lemma.

Define integers r, t, u, v as follows

$$\frac{u}{v} = \frac{ap + cq}{bp + dq} + \frac{-ap + cq}{-bp + dq} > 0;$$

$$\frac{r}{t} = \frac{ap' + cq'}{bp' + dq'} + \frac{-ap' + cq'}{-bp' + dq'} < 0.$$

Take $|vr|$ copies of the pair (p, q); $|vr|$ copies of the pair $(-p, q)$; $|tu|$ copies of the pair (p', q'); and $|tu|$ copies of $(-p', q')$ to form a collection of $s = 2|vr| + 2|tu|$ coprime pairs of integers. This collection will satisfy the requirements of the Proposition, for we have

$$\sum_{|vr|} \left(\frac{p}{q} + \frac{-p}{q}\right) + \sum_{|tu|} \left(\frac{p'}{q'} + \frac{-p'}{q'}\right) = 0;$$

$$\sum_{|vr|} \left(\frac{ap + cq}{bp + dq} + \frac{-ap + cq}{-bp + dq}\right) + \sum_{|tu|} \left(\frac{ap' + cq'}{bp' + dq'} + \frac{-ap' + cq'}{-bp' + dq'}\right)$$
$$= |vr|\frac{u}{v} + |tu|\frac{r}{t} = |ru| - |ur| = 0.$$

□

Theorem 5.5. *Let M be a graph manifold. Then there is a finite cover $p : \widetilde{M} \to M$ such that $\widetilde{M}$ admits a foliation that restricts on each vertex manifold of $\widetilde{M}$ to a surface bundle over S^1.*

Proof. By applying Proposition 4.4 to the trivial cover, we may assume that each vertex manifold of M is a product. To each edge of Γ assign an orientation and for each vertex manifold, M_v, pick a product structure $F_v \times S^1$. Let e be an edge of Γ. Let M_v and M'_v be the vertex manifolds corresponding to the tail and head of e, and let T_e and T'_e be the components of ∂M_v and $\partial M'_v$ corresponding to e. Then T_e and T'_e are identified in M by a *gluing homeomorphism* $f_e : T_e \to T'_e$. Now T_e, T'_e are parametrized by the curves $\{l_e, m_e\}, \{l'_e, m'_e\}$ representing $\{$circle fiber, boundary of the surface$\}$ of M_v, M'_v (resp.). In these coordinates f_e is represented by a *gluing matrix* $\begin{pmatrix} a_e & c_e \\ b_e & d_e \end{pmatrix}$ of determinant -1. Note that in these coordinates $b_e \neq 0$ since otherwise this would mean that the Seifert-fibers of M_v and M'_v would be identified under f_e (and this would contradict the minimality of the canonical splitting surface). So we can apply Proposition 5.4 above to this matrix, giving a positive integer s_e and a collection of ordered pairs $\{(p^e_i, q^e_i) \mid i = 1, \dots, s_e\}$.

Let s be the least common multiple of $\{s_e \mid e \in \Gamma\}$. Let $p : \widetilde{M} \to M$ be the cover obtained by applying Proposition 5.1 using this s.

Again let e, M_v, M'_v be as above. Let $\widetilde{M}_v = p^{-1}(M_v)$ and $\widetilde{M}'_v = p^{-1}(M'_v)$. Let $\widetilde{F}_v, \widetilde{F}'_v$ be the preimage of F_v, F'_v in $\widetilde{M}_v, \widetilde{M}'_v$. Take boundary framings on $M_v, M'_v, \widetilde{M}_v, \widetilde{M}'_v$ corresponding to the surfaces $F_v, F'_v, \widetilde{F}_v, \widetilde{F}'_v$ respectively. Let $\widetilde{T}_e, \widetilde{T}'_e$ be components of $p^{-1}(T_e), p^{-1}(T'_e)$ corresponding to the same component of the splitting surface of $\widetilde{M}$. So we have the commuting diagram

$$\begin{array}{ccc} \widetilde{T}_e & \xrightarrow{\widetilde{f}_e} & \widetilde{T}'_e \\ {\scriptstyle p}\big\downarrow & & \big\downarrow{\scriptstyle p} \\ T_e & \xrightarrow{f_e} & T'_e \end{array}$$

The circle fiber and the boundary framing will be denoted l_e, m_e on T_e; $\widetilde{l}_e, \widetilde{m}_e$ on $\widetilde{T}_e$; l'_e, m'_e on T'_e; and $\widetilde{l}'_e, \widetilde{m}'_e$ on $\widetilde{T}'_e$. Note that since p is characteristic, $\widetilde{l}_e$ (resp. $\widetilde{l}'_e$) intersects $\widetilde{m}_e$ (resp. $\widetilde{m}'_e$) once, and the gluing matrix associated to $\widetilde{f}_e$ (in terms of these bases) will be the same as the gluing matrix for f_e. Recall that $s_e, \{(p^e_i, q^e_i) \mid i = 1, \dots, s_e\}$ are associated to f_e through Proposition 5.4. To $\widetilde{T}_e$ we assign the framing $p^e_i \widetilde{l}_e + q^e_i \widetilde{m}_e$ and to $\widetilde{T}'_e$ we assign the framing $(a_e p^e_i + c_e q^e_i)\widetilde{l}'_e + (b_e p^e_i + d_e q^e_i)\widetilde{m}'_e$ for some i. Note that these curves are identified under $\widetilde{f}_e$. By construction, $|p^{-1}(T_e)| = |p^{-1}(T'_e)| = r_e s_e$ for some integer r_e. We make the assignment of framings so that each ordered pair (p^e_i, q^e_i), $i = 1, \dots, s_e$, is assigned to exactly r_e components of $p^{-1}(T_e)$.

Doing this for all edges of Γ assigns a framing α_v to each vertex manifold $\widetilde{M}_v$ in $\widetilde{M}$ (where v denotes the vertex of Γ). Since $\partial\widetilde{F}_v$ gives a boundary framing for $\widetilde{M}_v$, Corollary 2.2 says that

$$\begin{aligned} e(\widetilde{M}_v, \alpha_v) &= \Delta(\alpha, \partial\widetilde{F}_v) \\ &= \sum_{v=\text{tail of e}} r_e \sum_{i=1}^{s_e} \frac{p_i^e}{q_i^e} + \sum_{v=\text{head of e}} r_e \sum_{i=1}^{s_e} \frac{a_e p_i^e + c_e q_i^e}{b_e p_i^e + d_e q_i^e} = 0 \end{aligned}$$

That is, α_v also describes a boundary framing for $\widetilde{M}_v$. A surface corresponding to this boundary framing will be a fiber in a fibration of $\widetilde{M}_v$ over the circle (to see this fibration cut up $\widetilde{M}_v$ along this surface giving an I-bundle over the surface which will be a product bundle). By construction, the fibers in the fibration over the circle that one obtains on $\widetilde{M}_v$ and on an adjacent vertex manifold $\widetilde{M}'_v$ will be isotopic on a splitting torus between them (corresponding to a shared edge $\tilde{e}$ of $\widetilde{\Gamma}$). Therefore the fibrations on the vertex manifolds piece together to give the desired foliation on $\widetilde{M}$. □

Remark. The resulting foliation of M does not always give a fibration over the circle (it could not by the previous section) for the same reason that the fibers of each fibration do not piece together to give a surface. Let $\widetilde{M}_v$ and $\widetilde{M}'_v$ be adjacent along an edge $\tilde{e}$ in $\widetilde{\Gamma}$, let $\widetilde{T}$ be the splitting surface corresponding to $\tilde{e}$, and let $\widetilde{S}_v, \widetilde{S}'_v$ be fibers in the fibrations of $\widetilde{M}_v, \widetilde{M}'_v$. Then $|\widetilde{S}_v \cap \widetilde{T}|$ may differ from $|\widetilde{S}'_v \cap \widetilde{T}|$ (even though each component of $\widetilde{S}_v \cap \widetilde{T}$ is isotopic on $\widetilde{T}$ to any component of $\widetilde{S}'_v \cap \widetilde{T}$). Thus in trying to construct a surface of $\widetilde{M}$ using the fibers of the fibrations of the vertex manifolds one may pick up "holonomy" as one follows a circuit of $\widetilde{\Gamma}$. On the other hand the structure on the foliation one obtains in this construction is very specific and one could capture this phenomenon by saying that the foliation has a transverse affine structure with structure group consisting of rational dilations and contractions.

6. A covering invariant.

In this section we will use the relative Euler number to define an invariant $\sigma(M)$ for a graph manifold M. Theorem 6.1 shows that it has a certain covering property. If $\sigma(M)$ is nonzero, then homeomorphic covers of M have the same covering degree. As an example, we show in Proposition 6.3 that for knot complements this invariant does not vanish.

Let M be a graph manifold. We use χ_i to denote the orbifold characteristic of a vertex manifold M_i.

Definition. The σ invariant of M is defined by

$$\sigma(M) = \sum_{E(M_i)\neq 0} \frac{\chi_i^2}{|E(M_i)|}.$$

When all $E(M_i) = 0$, define $\sigma(M) = 0$.

If M were a Seifert fibered space then this would be a well-known covering invariant. See for example [2, 11].

A question of Thurston [5, Problem 3.16] asks whether there is a real valued 3-manifold invariant C such that if M is a k-fold cover of N then $C(M) = kC(N)$. We call C a *covering invariant* if it satisfies this property. The Gromov norm is a covering invariant of compact 3-manifolds. When M is a graph manifold, however, the Gromov norm vanishes. In [11] Wang and Wu defined a new invariant $\omega(M)$ for graph manifolds, which was shown to be a covering invariant. The following theorem shows that the σ invariant defined above is a covering invariant.

Theorem 6.1. *If $p : \widetilde{M} \to M$ is a k-fold covering of graph manifolds, then $\sigma(\widetilde{M}) = k\sigma(M)$.*

Proof. Suppose $\widetilde{M}_j$ is a component of $p^{-1}(M_i)$, and suppose the restriction of p to $\widetilde{M}_j$ is a (u_j, v_j) cover. The natural framing α^i of M_i lifts to the natural framing $\widetilde{\alpha}^j$ of $\widetilde{M}_j$, so by Proposition 2.3 we have $E(\widetilde{M}_j) = E(M_i)u_j/v_j$. In particular, $E(M_i) \neq 0$ if and only if $E(\widetilde{M}_j) \neq 0$. By definition, u_j is the number of regular fibers in $\widetilde{M}_j$ that cover a regular fiber in M_i, so $u_j = \widetilde{\chi}_j/\chi_i$, where $\widetilde{\chi}_j$ is the orbifold characteristic of $\widetilde{M}_j$. Let $k_j = u_jv_j$ be the covering degree of $\widetilde{M}_j$ over M_i. Then

$$E(\widetilde{M}_j) = \frac{E(M_i)u_j}{v_j} = \frac{E(M_i)u_j^2}{k_j} = \frac{E(M_i)\widetilde{\chi}_j^2}{\chi_i^2 k_j}.$$

Therefore, when $E(M_i) \neq 0$ we have

$$\frac{\widetilde{\chi}_j^2}{E(\widetilde{M}_j)} = \frac{\chi_i^2}{E(M_i)}k_j.$$

Taking absolute value on both sides and summing over all preimages of M_i, we get

$$\sum \left\{ \frac{\widetilde{\chi}_j^2}{|E(\widetilde{M}_j)|} \mid \widetilde{M}_j \text{ covers } M_i \right\} = \frac{\chi_i^2}{|E(M_i)|} \sum k_j = \frac{\chi_i^2}{|E(M_i)|}k.$$

The result now follows by summing both sides over those i with $E(M_i) \neq 0$. $\square$

As is pointed out in [5, Problem 3.16], covering invariants are related to the following problem. What 3-manifolds satisfy

(*): If M_1 and M_2 are homeomorphic finite covers of M, then the covering degrees are equal.

It was conjectured ([11] and [5, Problem 3.16(A)]) that M satisfies (*) if and only if it is not covered by either a (surface)$\times S^1$ or a torus bundle over S^1[3]. Among the category of geometric 3-manifolds, the conjecture is known to be true except for graph manifolds [11]. If some covering invariant, for example $\sigma(M)$, is non-zero, then the covering degree is determined by $\deg(p_i) = \sigma(\widetilde{M})/\sigma(M)$, so M satisfies (*). Hence we have the following

Corollary 6.2. *If some vertex manifold M_i of M has non-zero relative Euler number $E(M_i)$, then M satisfies (*).* □

Remark 1. If (N, α) is a framed Seifert manifold then $e(N, \alpha) = 0$ means that α is a boundary framing. In particular, N fibers over a 1-orbifold so that the fiber intersects ∂N in curves isotopic to components of α. Now let M be a closed graph manifold with canonical splitting surface $\mathcal{T} = T_1 \cup \ldots \cup T_k$. Then $\sigma(M) = 0$ implies that each vertex manifold M_v of M fibers over a 1-orbifold so that if T_i is on ∂M_v then the fiber of this fibration of M_v intersects T_i in curves isotopic to the Seifert fiber of the vertex manifold adjacent to M_v across T_i.

Remark 2. The σ invariant is different from the ω invariant defined in [11]. The ω invariant depends only on the fiber intersection numbers and the orbifold characteristic of vertex manifolds of M, and is insensitive to the relative Euler numbers. So for example if we do some $1/n$ surgeries on regular fibers of vertex manifolds of M, then $\sigma(M)$ will change, but $\omega(M)$ remains the same. Also, if each vertex manifold of M contains some boundary of M, then by definition $\sigma(M)$ vanishes, but $\omega(M)$ may be non-zero.

Example. Suppose a graph manifold M is a closed surface bundle over S^1, so M can be written as $M = F \times I/\varphi$, where $\varphi : F \times 0 \to F \times 1$ is the monodromy map. By passing to a finite cover we may assume that $\varphi = \tau_1^{k_1} \ldots \tau_n^{k_n}$, where the k_i's are non-zero integers, τ_i is a Dehn twist along an essential curve C_i, and $C_1, \ldots, C_n$ are mutually disjoint and nonparallel. A vertex manifold M_i corresponds to a component F_i of

[3]Added in proof: In "Covering degrees are uniquely determined by the graph manifolds involved", F. Yu and S. Wang have now verified this conjecture.

$F - \bigcup \mathrm{Int}N(C_i)$. The relative Euler number of M_i can be computed as follows. If $\partial F_i = C_{j_1} \cup \ldots \cup C_{j_r}$, (some C_i may appear twice), then

$$E(M_i) = \frac{1}{k_{j_1}} + \ldots + \frac{1}{k_{j_r}}.$$

Thus, if some of these sums are non-zero, then $\sigma(M) \neq 0$, and M satisfies (*). In particular, if some F_i has only one boundary curve, then $\sigma(M) \neq 0$.

As an application of the σ invariant, we show that the conjecture above is true for knot complements. Write $X(K) = S^3 - \mathrm{Int}N(K)$.

Proposition 6.3. *Let K be a knot in S^3. Then $X(K)$ satisfies (*) if and only if K is not a torus knot.*

Proof. By Theorem 6.1 and [11, Theorem 2.5], it suffices to show that if $X(K)$ is a graph manifold, then $\sigma(X(K)) \neq 0$. It is well known that $X(K)$ is a graph manifold if and only if K is a connected sum of iterated torus knots but is not a torus knot.

First suppose K is a (p,q) cable of an (r,s) torus knot $K_{r,s}$. In order that K is not just the torus knot $K_{r,s}$, we must have $q > 1$. Let T be a torus decomposing $X(K)$ into $X(K_{r,s})$ and a (p,q) cabling space $C(p,q)$. Let (m,l) be the meridian-longitude pair of $K_{r,s}$ on T. Then l bounds a Seifert surface in $X(K_{r,s})$, which is a transverse surface, so $e(X(K_{r,s}), l) = 0$. Let α be a fiber of $C(p,q)$ on T, let β be a fiber of $X(K_{r,s})$ on T. Then with respect to the (m,l) coordinate α is a (p,q) curve, and β is an $(rs, 1)$ curve. By Proposition 2.1, we have

$$\begin{aligned} E(X(K_{r,s})) &= e(X(K_{r,s}), \alpha) = e(X(K_{r,s}), l) + \Delta(\alpha, l) \\ &= \delta(\alpha, l, \beta) = \frac{\alpha \cdot l}{(\alpha \cdot \beta)(l \cdot \beta)} = \frac{p}{(p - qrs)(rs)} \neq 0. \end{aligned}$$

Note that since p, q are coprime, and $q > 1$, we have $p - qrs \neq 0$, so the above computation makes sense. In particular, by definition of the σ invariant we have $\sigma(X(K)) \neq 0$.

With the same computation one can show that if K is a nontrivial iteration of torus knot, i.e if K is a (p_1, q_1) cable of a (p_2, q_2) cable of ... of a (p_r, q_r) torus knot, and $r > 1$, then the extremal vertex manifold of $X(K)$ has relative Euler number $E = p_r/(p_r - q_r p_{r-1} q_{r-1})(p_{r-1} q_{r-1})$. Hence $\sigma(X(K)) \neq 0$. More generally, if $K = K_1 \# \ldots \# K_r$, and if some K_i is a nontrivial iteration of torus knot, then $X(K_i)$ embeds into $X(K)$, so some vertex manifold of $X(K)$ has non-zero relative Euler number, and $\sigma(X(K)) \neq 0$.

The remaining case is that $K = K_1 \# \ldots \# K_r$, $r > 1$, and all K_i are torus knots. Then $X(K) = X(K_1) \cup \ldots \cup X(K_r) \cup C(r)$, where $C(r)$ is a composition space, i.e, $C(r) = D_r \times S^1$, where D_r is a disk with r holes. Let $T = X(K_1) \cap C(r)$; let (m, l) be a meridian-longitude pair of K_1 on $T = \partial X(K_1)$. Then $X(K_1)$ is glued to $C(r)$ so that m is mapped to some (point)$\times S^1$ on $\partial D_r \times S^1$, and l is mapped to some $C_i \times$(point), where C_i is a boundary component of D_r. Let α be a fiber of $C(r)$, and let β be a fiber of $X(K_1)$. Then with respect to the (m, l) coordinates, α is a $(1, 0)$ curve, β is a $(pq, 1)$ curve if K_1 is a (p, q) torus knot, and the $(0, 1)$ curve l is the boundary of a Seifert surface of K_1. Hence,

$$\begin{aligned} E(X(K_1)) &= e(X(K_1), \alpha) = e(X(K_1), l) + \Delta(\alpha, l) \\ &= \delta(\alpha, l, \beta) = \frac{\alpha \cdot l}{(\alpha \cdot \beta)(l \cdot \beta)} = \frac{1}{1 \cdot (pq)} = \frac{1}{pq}. \end{aligned}$$

Therefore $\sigma(X(K)) \neq 0$, and $X(K)$ satisfies (*). □

References

1. D.Eisenbud, U. Hirsch and W. Neumann, *Transverse foliations on Seifert bundles and self-homeomorphisms of the circle*, Comment. Math. Helv. **56** (1981) 638–660.
2. D. Gabai, *On 3-manifolds finitely covered by surface bundles*, Low-dimensional topology and Kleinian groups, D.B.A. Epstein, London Math. Soc. Lecture notes **112** (1986) 145–156, Cambridge University Press.
3. J. Hempel, *3-manifolds*, Annals of Math Studies **86** (1976) Princeton University press.
4. W. Jaco, *Lectures on Three-manifold Topology*, Regional Conference Series in Mathematics **43** (1981).
5. R. Kirby, *Problems in low dimensional manifold theory*, Algebraic and Geometric topology, R.J. Milgram, Proc. Symposia in Pure Math. **32** (1978) 273–312 Amer. Math. Soc.
6. W.D. Neumann and F. Raymond, *Seifert manifolds, plumbing, μ-invariant and orientation reversing maps*, Lecture notes in math. 664 (Springer, Berlin 1978) 162–195.
7. P. Scott, *The geometry of 3-manifolds*, Bull. London Math. Soc. **15** (1983) 401–478.
8. W. Thurston, *Three-dimensional manifolds, Kleinian groups and hyperbolic geometry*, Bull. Amer. Math. Soc. **6** (1982) 357–381.
9. W. Thurston , *On the geometry and dynamics of diffeomorphisms of surfaces*, Bull. Amer. Math. Soc. **19** (1988) 417–431.
10. F. Waldhausen, *Eine Klasse von 3-dimensionalen Mannifaltigkeiten I*, Invent. Math. **3** (1967) 308–333.

11. S. Wang and Y-Q. Wu, *Covering invariants and cohopficity of 3-manifold groups*, Proc. London Math. Soc. **86** (1994) 203-224.

UNIVERSITY OF TEXAS AT AUSTIN, AUSTIN, TX 78712
E-mail address: wu@math.uiowa.edu

AMS/IP Studies in Advanced Mathematics
Volume **2**, 1997 (Part 1)

GEOMETRIC INVARIANTS FOR 3-MANIFOLDS

MINGQING OUYANG

INTRODUCTION

Recent research has indicated that a new class of geometrically defined invariants reveals many insights about 3-manifolds.

A geometric structure on a manifold is a complete Riemannian metric which is locally homogeneous. The work of Thurston ([26], [27]) implies that there are essentially eight relevant geometries needed for geometric structures on 3-manifolds and most 3-manifolds admit a hyperbolic structure. By the Mostow-Prasad rigidity theorem, a complete hyperbolic 3-manifold of finite volume has a unique hyperbolic structure. It follows that any geometric invariant is actually a topological invariant for such a hyperbolic 3-manifold.

Among the best known geometric invariants for 3-manifolds are the volume, the Chern-Simons invariant and the η-invariant.

The volume of a geometric 3-manifold is the best understood and the most computable among these three invariants.

The Chern-Simons invariant grew out of attempts to develop a combinatorial formula for the first Pontrjagin number of a 4-manifold. It was first defined in [5] for a closed Riemannian 3-manifold by integrating a certain 3-form over the manifold, only well-defined mod $\boldsymbol{Z}$.

The η-invariant of a self-adjoint elliptic operator was originally introduced by Atiyah, Patodi and Singer in [3] for odd dimensional Riemannian manifolds in terms of the spectrum of the operator. The η-invariant we are interested in here is that of the signature operator. Such an invariant has a topological interpretation which allows us to compute it without using analytic tools. Roughly speaking, it measures the extent to which the Hirzebruch signature formula fails for a Riemannian 4-manifold with boundary. In dimension 3, the η-invariant can be thought of as a real-valued generalization of the Chern-Simons invariant as shown in [4].

Thurston in his 1982 expository article [27] made the following comments:

"There is another promising invariant (other than the volume) of oriented hyperbolic 3-manifolds, about which practically nothing has been known until recently. This is the eta invariant, or the Chern-Simons invariant which is obtained by reducing the eta invariant mod a constant. ······"

"Unfortunately, the eta invariant, unlike the volume, has been very difficult to compute, basically because in its definition some very arbitrary choices are involved, and these choices are difficult to make in any reasonable way."

Progress over the last decade has made the η-invariant and the Chern-Simons invariant almost as computable as the volume.

In this paper, we discuss some recent work on the η-invariant as well as on the Chern-Simons invariant and their applications in low-dimensional topology. The rest of this paper is organized as follows:

1. Seifert Fibred Geometric 3-Manifolds
2. Closed Hyperbolic 3-Manifolds
3. Cusped Hyperbolic 3-Manifolds
4. Hyperbolic Links

1. Seifert Fibred Geometric 3-Manifolds.

A Seifert fibred 3-manifold admits one of the six exceptional geometries in Thurston's Geometrization Program. It can be thought of as a circle bundle over a 2-orbifold (An orbifold is a space locally modelled on $\boldsymbol{R}^n$ modulo finite group actions). The relevant geometry is determined by the Euler number e of its fibration and the Euler characteristic χ of its base orbifold as follows:

	$\chi > 0$	$\chi = 0$	$\chi < 0$
$e = 0$	$\mathbf{S^2 \times E^1}$	$\mathbf{E^3}$	$\mathbf{H^2 \times E^1}$
$e \neq 0$	$\mathbf{S^3}$	**Nil**	$\widetilde{\mathbf{PSL}}$

(refer to the article [24] of Scott, for example.)

The volume of a Seifert fibred 3-manifold with the relevant geometry is well understood. In fact, as observed by Neumann in [17], with a natural normalization of the metrics on the geometries, we have $\mathrm{Vol}(M) = 4\pi^2\chi^2/|e|$ if M is such that $e\chi \neq 0$; If M has the ***Nil***-geometry, then $\mathrm{Vol}(M) = l^2|e|$ where l is the fiber length; In all other cases, the volume is indeterminate, depending on the geometric structure.

For the η-invariant as well as the Chern-Simons invariant, our starting point is the following formula due to Atiyah, Patodi and Singer:

Theorem 1.1. *([3]) Let N be a 4-dimensional Riemannian manifold with $\partial N = M$. Assume that N has a product metric near the boundary. Then*

$$\eta(M) = \frac{1}{3}\int_N p_1 - \mathrm{Sign}(N) \tag{1}$$

where p_1 is the first Pontrjagin form of the Riemannian metric on N and Sign(N) is the signature of N.

In order to use (1), it is required that the metric on N is a product near the boundary. But this may not be the case in practice. In [6], Gilkey remedied this by adding a so-called "boundary correction term" to (1). We briefly discuss his formulation here.

Let ω and Ω be the connection form and the curvature form of the Riemannian metric g on N. Denote $C = M \times [0,1]$ and identify $M \times \{0\}$ with $M = \partial N$. Let g_0 be the product metric on C. Extend the metric g on N to a metric g_1 on $C \cup N$ such that g_1 is product near $M \times \{1\}$. Then g_0 and g_1 agree near $M \times 1$. Denote ω_0, ω_1 and Ω_0, Ω_1 the corresponding connection forms and curvature forms of g_0 and g_1 respectively.

Let $\omega = \omega_1 - \omega_0$ and $\omega_t = (1-t)\omega_0 + t\omega_1$. Denote by Ω_t the curvature form of ω_t.

Define

$$Tp_1 = \frac{2}{3}\int_0^1 p_1(\omega, \Omega_t)dt.$$

Then as in [6],

$$\eta(M) = \frac{1}{3}\int_N p_1 - \int_M Tp_1 - \mathrm{Sign}(N). \tag{2}$$

In general, both integral terms in the above formula are difficult to compute. But in certain particular cases, one can come out with more tractable expressions. For example, let $D_r(F) \longrightarrow F$ be a disk bundle of radius r over a Riemann surface F whose fiber metric has a constant curvature, then as Komuro showed in [10], the η-invariant of the corresponding circle bundle $S_r(F)$ can be explicitly computed by

$$\eta(S_r(F)) = \frac{2}{3}e\{\frac{\pi r^2}{\mathrm{Vol}(F)}\chi - (\frac{\pi r^2}{\mathrm{Vol}(F)})^2 e^2\} + \frac{1}{3}e - \mathrm{sign}(e)$$

where e is the Euler number of the fibration, χ is the Euler characteristic of F and Vol(F) is the volume of the Riemann surface F.

A Seifert fibred 3-manifold with the relevant geometry can be thought of as a circle bundle of a certain radius over a Riemannian 2-orbifold.

Thus one hopes that a similar computation can be carried out. However the problem here is that the corresponding disk bundle is no longer a manifold. It is instead a 4-orbifold. Thus we cannot apply the above formula (1) directly. Fortunately, Kawasaki in [8] and [9] provided us with the necessary machinery by generalizing the classical Hirzebruch signature theorem to the orbifold case and subsequently extending (1) to the form we want. Roughly speaking, he showed the following

Theorem 1.2. *([8], [9]) Let N be a 4-dimensional Riemannian orbifold with $\partial N = M$ a Riemannian manifold. Assume that N has a product metric near the boundary. Then*

$$\eta(M) = \frac{1}{3}\int_N p_1 - \mathrm{Sign}(N) + \mathcal{O}(\Sigma N) \tag{3}$$

where $\mathcal{O}(\Sigma N)$ is a topological term associated to ΣN, the singular set of N.

Combining (2) and (3) we have the following formula

$$\eta(M) = \frac{1}{3}\int_N p_1 - \int_M Tp_1 - \mathrm{Sign}(N) + \mathcal{O}(\Sigma N). \tag{4}$$

By applying (4), we generalized Komuro's work to the following

Theorem 1.3. *([19]) Let $E \longrightarrow F$ be a complex line bundle over an oriented, closed Riemannian 2-orbifold F. Suppose that the orbifold charts of E are of the form $(\boldsymbol{C}\times\boldsymbol{C})/\boldsymbol{Z}/\alpha$ where $\boldsymbol{Z}/\alpha$ acts on the first coordinate by multiplication by $e^{2\pi i\beta/\alpha}$ and on the second by $e^{2\pi\gamma i/\alpha}$ for some β and γ prime to α. Choose a fiber metric in E such that it has a constant curvature. Then the η-invariant of the circle bundle of radius r is given by*
(5)

$$\eta(S_rE) = \frac{2}{3}e\{\frac{\pi r^2}{\mathrm{Vol}(F)}\chi - (\frac{\pi r^2}{\mathrm{Vol}(F)})^2e^2\} + \frac{1}{3}e - \mathrm{sign}(e) + \sum_{i=1}^{n} 4s(\beta_i\gamma_i, \alpha_i)$$

where e is the (rational) Euler number of of the fibration, χ is the (rational) Euler characteristic of the base orbifold F and $s(\beta_i\gamma_i, \alpha_i)$ is the classical Dedekind sum.

The main part of the proof is to figure out the term $\mathcal{O}(\Sigma E)$ associated to the singular set of E. The computations for the other terms are generalizations of those in [10].

By applying (5) to the Seifert fibred geometric 3-manifolds, we obtained explicit expressions for the η-invariant of the Seifert fibred geometric 3-manifolds under each geometry. For instance, we derived the following

Corollary 1.4. *([19]) Let $T^1(\boldsymbol{H}^2)$ be the unit tangent bundle of the hyperbolic plane. Equip $PSL(2,\boldsymbol{R})$ with the metric induced by the identification $PSL(2,\boldsymbol{R}) = T^1(\boldsymbol{H}^2)$. Let $\Gamma \subseteq PSL(2,\boldsymbol{R})$ be a co-compact Fuchsian group of signature $\{g;\alpha_1,\cdots,\alpha_n\}$. Then*

$$\eta(PSL(2,\boldsymbol{R})/\Gamma) = \frac{1}{6}(2g+4+7n) - \sum_{i=1}^{n}(\frac{1}{3}\alpha_i + \frac{5}{6\alpha_i}).$$

A similar formula for the the η-invariant of $PSL(2,\boldsymbol{R})/\Gamma$ associated to the Dirac operator was obtained by Seade and Steer in [25].

Surprisingly, the η-invariant of a Seifert fibred geometric 3-manifold only depends on the topology under most of these geometries, even though the metrics on these geometric manifolds are far from unique.

The Chern-Simons invariant and the η-invariant were formulated in completely different manners. However, as shown by Atiyah, Patodi and Singer, they are closely related by the following:

Theorem 1.5. *([4]) Let M be a closed, oriented Riemannian 3-manifold. Then*

$$CS(M) \equiv \frac{3}{2}\eta(M) + \frac{1}{2}\sigma(M) \pmod{\boldsymbol{Z}} \tag{6}$$

where $\sigma(M) = \#$ of the 2-primary summands in $H_1(M;\boldsymbol{Z})$.

Let M be a Seifert fibred 3-manifold with Seifert invariant $(g;(\alpha_1,\beta_1),\cdots,(\alpha_n,\beta_n))$ and let l be the number of even α_i's. Then as computed in [19],

$$\begin{aligned} \sigma(M) &= l-1 \;\; if \;\; l \geq 1 \\ &= 1 \;\; if \;\; l=0 \;\; \text{and} \;\; \textstyle\sum \beta_i \;\; \text{is} \;\; \text{even} \\ &= 0 \;\; if \;\; l=0 \;\; \text{and} \;\; \textstyle\sum \beta_i \;\; \text{is} \;\; \text{odd}. \end{aligned} \tag{7}$$

By using (6) and (7), we derived an elementary formula for the Chern-Simons invariant of each Seifert fibred geometric 3-manifold.

As shown by Hirsch in [7], every compact 3-manifold can be topologically immersed into $\boldsymbol{R}^4$. Chern and Simons in [5] proved that if a closed Riemannian 3-manifold can be conformally immersed into the Euclidean space $\boldsymbol{E}^4$, then its Chern-Simons invariant has to vanish. As an application, we showed that some families of the Seifert fibred geometric

3-manifolds cannot be conformally immersed into $\boldsymbol{E}^4$. For example, we have the following

Corollary 1.6. *([19]) Let F be a closed surface of even genus with a hyperbolic geometry. Then the unit tangent bundle of F with the induced metric cannot be conformally immersed into $\boldsymbol{E}^4$.*

2. Closed Hyperbolic 3-Manifolds.

The computer program "SnapPea", developed by Jeff Weeks and others, allows us to decompose a hyperbolic 3-manifold into ideal tetrahedra and compute the volume of the manifold in great generality.

Thurston in [27] speculated a complex analytic relation between the volume and the Chern-Simons invariant of a hyperbolic 3-manifold. An explicit conjecture was formulated by Neumann and Zagier in [18] and later proved by Yoshida in [29]. Consider the hyperbolic Dehn-surgery space for a cusped hyperbolic 3-manifold. Roughly speaking, there is an analytic function defined on the space whose real part is the volume of the surgered manifold plus the length(s) of the surgered-in geodesic(s) and the imaginary part (mod half) is the Chern-Simons invariant of the surgered manifold plus the torsion of the surgered-in geodesic(s). Recently, Neumann in [16] used such a relation to derive a simplicial formula for the Chern-Simons invariant. We discuss briefly his formulation.

Let M be an oriented complete finite volume hyperbolic 3-manifold with h cusp. Suppose M has an ideal triangulation $M = \Delta(z_1^0) \cup \cdots \cup \Delta(z_n^0)$ where each $\Delta(z_j^0)$ is an ideal tetrahedron described by a complex parameter z_j^o after choosing an edge for each $\Delta(z_j^0)$. Denote

$$\boldsymbol{z^0} = (\log z_1^0, \cdots, \log z_n^0, \log(1 - z_1^0), \cdots, \log(1 - z_n^0)).$$

Then $\boldsymbol{z^0}$ is determined by the consistency and cusp relations in the form

$$\boldsymbol{z^0} U = \pi i \boldsymbol{d}$$

where U is an integer $2n \times (n + 2h)$-matrix and $\boldsymbol{d} = (d_1, \cdots, d_{n+2h})$ is some integer vector. The consistency relation requires that each edge closes up after glueing all tetrahedra together. The cusp condition says that every simple closed curve in a cusp torus has a parabolic holonomy. Let $\boldsymbol{c} = (c_1', \cdots, c_n', c_1'', \cdots, c_n'')$ be a solution to the equation

$$\boldsymbol{c} U = \boldsymbol{d}.$$

Let $\boldsymbol{u} = (z_1, \cdots, z_n)$ and denote $M_{\boldsymbol{u}} = \Delta(z_1) \cup \cdots \cup \Delta(z_n)$. If the metric completion of $M_{\boldsymbol{u}}$ is the result of a hyperbolic Dehn surgery on M, then, as in [16], the above analytic function is computed as

$$(8) \qquad f(\boldsymbol{u}) = \alpha - \frac{i}{\pi^2}\sum_{j=1}^{n}(R(z_j) - \frac{i\pi}{2}(\overline{c_j'}\log(1-z_j) - \overline{c_j''}\log z_j))$$

where α is a constant depending on the choice of the solution $\boldsymbol{C}$ and combinatorics of the triangulation. $R(z)$ is the Rogers dilogarithm function defined by

$$R(z) = \frac{1}{2}\log(z)\log(1-z) - \int_0^z \log(1-t)d\log t$$

with log the standard branch on $\boldsymbol{C} - (-\infty, 0]$. $\overline{c_j'}$ (resp. $\overline{c_j''}$) denotes the complex conjugate of c_j' (resp. c_j'').

As mentioned above, the Chern-Simons invariant is essentially the imaginary part of $f(u)$ up to the torsion of the surgered-in geodesic(s) which is easily computable. Formula (8) plus some work of Hodgson and others has led to the calculations of the Chern-Simons invariant being affixed to SnapPea.

As we mentioned earlier, the η-invariant can be thought of as a real-valued generalization of the Chern-Simons invariant. As such, it contains more information. For example, Meyerhoff and Ruberman in [15] obtained some manifolds with the same Chern-Simons invariant but different η-invariant.

We now describe briefly Yoshida's formulas for the η-invariant.

Suppose that X is a Riemannian 3-manifold and L is a link in X. Let α be a certain framing on X and $\mathcal{F}$ be a framing defined on $X - L$ having a special singularity at L. Then as Yoshida showed in [29]

$$(9) \qquad \eta(X) = \frac{1}{12}\int_{s(X-L)} Q - \frac{1}{6\pi}\tau(L,\alpha) + \frac{2}{3}d(\alpha,\mathcal{F}) + \delta(X,\alpha)$$

where s is the section defined by $\mathcal{F}$ and Q is the Chern-Simons 3-form. $\tau(L,\alpha)$ is the torsion of L with respect to the framing α. $d(\alpha,\mathcal{F}) \in \boldsymbol{Z}$ is the difference between the two framings and $\delta(X,\alpha)$ is the Hirzebruch δ-invariant of the framed manifold (X,α) defined by

$$\delta(X,\alpha) = \frac{1}{3}p_1(N) - \mathrm{Sign}(N)$$

with N a 4-manifold bounded by X and $p_1(N)$ the relative Pontrjagin number.

Formula (9) is seemingly more complicate than formula (1) in Section 1. But as we will see, it contains several terms which can be calculated by some functorial properties. Now, let M be an oriented complete hyperbolic 3-manifold of finite volume with h cusps. Denote by $U(M)$

the hyperbolic Dehn surgery space of M. Let M_u be a hyperbolic 3-manifold represented by some $u \in U(M)$ whose metric completion $\bar{M}_u$ is the result of a Dehn-surgery on the cusps of M. Then from (9) Yoshida derived the following formula:

$$\begin{aligned}\eta(\bar{M}_u) = \frac{1}{3}\operatorname{Im} f(u) - \frac{1}{6\pi}\tau(\gamma, \alpha_u) & + \frac{2}{3}d(\alpha_u, \mathcal{F}) + \delta(\bar{M}_u, \alpha_u) \\ & + \frac{1}{6\pi}(\tau(L', \kappa_u) - \tau(L', \alpha_u))\end{aligned} \tag{10}$$

where $f(u)$ is the analytic function we mentioned above. The γ are the surgered-in geodesic(s). α_u is a certain framing on M_u and κ_u is some framing defined on a neighborhood each component of the link L' with $L' \cup \gamma$ equal to the link L in (9).

However, the arbitrary choices of the framings involved in the above formula make it difficult to apply in practice. For M equal to the figure-eight knot complement in S^3, Yoshida in [29] came out with a simpler and more explicit formula without involving the extra structure on M. It would be much desirable to find an analogous formula valid in general, and we now proceed in steps to such a formula ((13) below).

In [20], the following generalization of Yoshida's figure-eight knot formula is derived.

Theorem 2.1. ([20]) *Let M be a one-cusped hyperbolic 3-manifold. Suppose that the oriented basis $(\boldsymbol{l}, \boldsymbol{m})$ can be chosen at the cusp so that $\boldsymbol{l}$ is a "longitude"(i.e., it represents 0 in $H_1(M, \boldsymbol{Z})$). Then*

$$\eta(M(p,q)) = \frac{1}{3}\operatorname{Im}(f(u(p,q))) - \frac{1}{6\pi p}\operatorname{Im}(v(p,q)) + 4s(q,p) - \frac{q}{3p} \tag{11}$$

where $f(u)$ is, up to a constant, the complex analytic function in (10). $v(p,q)$ is the logarithm of the holonomy of $\boldsymbol{l}$ and $s(q,p)$ is the classical Dedekind sum.

The main idea of the proof goes as follows. There is a degree one map between $M(p,q)$ and the lens space $L(p,q)$ which is covered by a bundle map between their tangent bundles. As in [29], we construct explicitly two framings on $L(p,q)$ and use (9) to obtain a formula for $\eta(L(p,q))$. By comparing it with a computation of Atiyah, Patodi and Singer via the finite-sheeted cover $S^3 \to L(p,q)$ as in [4], we get a formula for the last two terms in (9) associated to the framings. We then use the functorial properties of these terms to pull them back to $M(p,q)$ via the bundle map. Our formula follows from (10) and some explicit computations for the torsion terms.

The above theorem holds, for example, if M is a hyperbolic knot complement in S^3. But due to the homological restriction required in the proof, the method cannot be generalized to the case of several cusps.

In [13], Meyerhoff and Neumann obtained the following

Theorem 2.2. *([13]) Let M be an oriented complete hyperbolic 3-manifold of finite volume with h cusps. Denote $(\boldsymbol{p},\boldsymbol{q}) = (p_1, q_1; ...; p_h, q_h)$ and $M(\boldsymbol{p},\boldsymbol{q})$ the result of (p_k, q_k)-Dehn surgery with respect to a fixed meridian-longitude pair $(\boldsymbol{m}_k, \boldsymbol{l}_k)$ on the k-cusp for all k's. Then for each $p_k^2 + q_k^2 \to \infty$,*

$$\begin{aligned}\eta(M(\boldsymbol{p},\boldsymbol{q})) &= \frac{1}{3}\operatorname{Im}(f(\boldsymbol{u}(\boldsymbol{p},\boldsymbol{q}))) - \sum_{k=1}^{h} \frac{1}{6\pi p_k}\operatorname{Im}(v_k) \\ &\quad + \sum_{k=1}^{h}(4s(q_k,p_k) - \frac{q_k}{3p_k}) - \operatorname{sign}(Y(\boldsymbol{p},\boldsymbol{q}))\end{aligned} \tag{12}$$

where $f(\boldsymbol{u})$ and v_k are as before. $Y(\boldsymbol{p},\boldsymbol{q})$ is the 4-manifold obtained by pasting (the lens space $L(p_k, q_k)$)$\times$ (the unit interval I) to $M(\boldsymbol{p},\boldsymbol{q}) \times I$ along the copy of $(D^2 \times S^1)_k$ for $k = 1, \cdots, h$. Its signature sign$(Y(\boldsymbol{p},\boldsymbol{q}))$ can be computed explicitly from Wall's nonadditivity formula ([28]) and depends only on Ker$(H_1(\text{cusps}; \boldsymbol{Q}) \to H_1(M; \boldsymbol{Q}))$ and the elements $\boldsymbol{l}_1, ..., \boldsymbol{l}_h$, $p_1\boldsymbol{m}_1 + q_1\boldsymbol{l}_1, ..., p_h\boldsymbol{m}_h + q_h\boldsymbol{l}_h$ of H_1(cusps).

The proof given in [13] only works for each p_k or q_k sufficiently large.

By Thurston's hyperbolic Dehn-surgery theorem ([26]), if we exclude finitely many (p_k, q_k)'s for each cusp, then all the resulting manifolds have a hyperbolic structure.

Recently, we proved the following

Theorem 2.3. *([21]) The formula (12) is true for all hyperbolic Dehn surgeries.*

The proof for the 1-cusped case involves passing a hyperbolic (p, q)-surgery to a sufficiently large (P, Q)-surgery for which the formula (12) above is valid. By comparing (10) and (12), we see that the δ-invariant of $M(P, Q)$ can be expressed in terms of the Dedekind sum $S(Q, P)$ and the different degree between two framings used in (10). Construct a 4-manifold whose boundary consists of $M(p, q)$, $M(P, Q)$ and some lens space $L(a, b)$. By choosing a suitable pair of (a, b), we can extend the framing on $M(P, Q)$ to the 4-manifold. This enables us to compute the δ-invariant for $M(p, q)$ by using Wall's nonadditivity formula for the signature and a delicate relation among the Dedekind sums. For the case of several cusps, we can choose a suitable pair (a_k, b_k) for each cusp such that the additional signature term remains the same. Thus a similar argument works.

By combining our formula with Neumann's simplicial expression for $f(u)$ as in (8), we obtained the following

Theorem 2.4. *([21]) For every hyperbolic $(\boldsymbol{p},\boldsymbol{q})$-Dehn surgery,*
(13)

$$\begin{aligned}\eta(M(\boldsymbol{p},\boldsymbol{q})) = \quad & \alpha - \frac{1}{3\pi^2}\operatorname{Im}\sum_{j=1}^{n}(iR(z_j) + \frac{\pi}{2}(\overline{c_j'}\log(1-z_j) - \overline{c_j''}\log z_j)) \\ & - \sum_{k=1}^{h}\frac{1}{6\pi p_k}\operatorname{Im}(v_k) + \sum_{k=1}^{h}(4s(q_k,p_k) - \frac{q_k}{3p_k}) \\ & - \operatorname{sign}(Y(\boldsymbol{p},\boldsymbol{q})).\end{aligned}$$

Remark. The constant α in the above formula can be determined via a "boot-strapping" procedure, using surgery relations discovered between hyperbolic 3-manifolds.

Such a formula is ready to be included in the SnapPea program.

A knot or link in the 3-sphere is said to be hyperbolic if its complement admits a geometric structure locally isometric to hyperbolic 3-space. We finish this section with an interesting application of the η-invariant.

Corollary 2.5. *([21]) If M is a one-cusped hyperbolic 3-manifold, then $M(p,q) \neq M(p,q')$ if $q \neq q'$ and p^2+q^2, $p^2+(q')^2$ are sufficiently large. In particular, for sufficiently large surgery coefficients, two different surgeries along a hyperbolic knot in S^3 never produce the same manifold.*

To prove this, one simply uses formula (11) to show that $\eta(M(p,q)) \neq \eta(M(p,q'))$ for sufficiently large surgery coefficients. Then the Mostow rigidity theorem implies that $M \neq M'$.

3. Cusped Hyperbolic 3-Manifolds.

Both the Chern-Simons invariant and the η-invariant were originally defined for a closed Riemannian manifold.

Meyerhoff in [12] extended the definition of the Chern-Simons invariant to cusped hyperbolic 3-manifolds by using a singular frame field. Such an invariant is only well-defined modulo one-half.

Let M be an oriented, complete hyperbolic 3-manifold of finite volume with n cusps. Denote by $(\boldsymbol{m}_i,\boldsymbol{l}_i)$ a fixed meridian-longitude pair at the i-th cusp of M. Let (p_i,q_i) be a pair of coprime integers for every $1 \leq i \leq n$ and $(\boldsymbol{p},\boldsymbol{q}) = (p_1,q_1;...;p_n,q_n)$. Denote by $M(\boldsymbol{p},\boldsymbol{q})$ the result of the Dehn surgery to kill the class $p_i\boldsymbol{m}_i + q_i\boldsymbol{l}_i$ for every $1 \leq i \leq n$. In joint work with Meyerhoff [14], we defined the η-invariant of M with respect to the chosen meridian-longitude pairs as

$$\begin{aligned}\eta(M) = \ & \lim_{p_i^2+q_i^2\to\infty, i=1,\cdots,n} \tfrac{1}{2}(\eta(M(\boldsymbol{p},\boldsymbol{q})) \\ & + \operatorname{sign}(Y(\boldsymbol{p},\boldsymbol{q})) + \eta(M(\boldsymbol{p},-\boldsymbol{q})) \\ & + \operatorname{sign}(Y(\boldsymbol{p},-\boldsymbol{q})))\end{aligned}$$

where $\operatorname{sign}(Y(\boldsymbol{p},\boldsymbol{q}))$ is the signature of the 4-manifold $Y(\boldsymbol{p},\boldsymbol{q})$ as in Section 2.

Proposition 3.1. *Let M be a cusped hyperbolic 3-manifold. Then the η-invariant completely determines the Chern-Simons invariant as follows:*

$$CS(M) \equiv \frac{3}{2}\eta(M) \pmod{\frac{1}{2}}.$$

Just as for the closed case, the η-invariant of a cusped hyperbolic 3-manifold can be computed from an ideal tetrahedral decomposition. The following combinatorial formula was derived in [14]:

$$\eta(M) = \alpha - \frac{1}{3\pi^2}\operatorname{Im}\Big(\sum_{j=1}^{n}(iR(z_j^0) + \frac{\pi}{2}(\overline{c_j'}\log(1-z_j^0) - \overline{c_j''}\log z_j^0))\Big)$$

where α is the same constant as in (13) of Section 2 and hence can be determined in the same way. The other ingredients are also the same as in (13).

Proposition 3.2. *([14]) Let L be a hyperbolic knot or link in S^3. Denote by L^* its mirror image. Then we have $\eta(S^3 - L^*) = -\eta(S^3 - L)$. In particular, if L is an amphicheiral hyperbolic knot or link (e.g. the figure-eight knot or the Borromean rings), then $\eta(S^3 - L) = 0$.*

Thus the η-invariant detects the chirality of hyperbolic knots and links.

Given a knot or link in S^3. A new knot or link can be obtained by taking out a tangle, flipping it over and gluing it back in. Such an operation is called a mutation. The resulting knot or link tends to differ from the original one. However, many link invariants are preserved by mutation, e.g. the HOMFLY polynomial (hence the Jones polynomial). Recently, the Witten invariant (for $SU(2)$) has also been shown to be preserved by mutation. For a hyperbolic knot or link, we know that a mutation does not change the volume or the Chern-Simons invariant from the work of Meyerhoff and Ruberman in [15] and [23]. In [14], we showed the following

Theorem 3.3. *([22]) The η-invariant of a hyperbolic knot or link complement in S^3 is also preserved by a mutation.*

4. Hyperbolic Links.

The η-invariant for a cusped hyperbolic 3-manifold as defined in [14] depends on the choices of meridian-longitude pairs on the cusps. Given a hyperbolic knot or link L in S^3, we define the η-invariant of L to be $\eta(S^3 - L)$ determined by the complete hyperbolic structure on $S^3 - L$ and the topological meridians and longitudes of L. It is not hard to see that what we get is actually a well-defined link invariant for a hyperbolic knot or link. The work of Thurston indicates that most "randomly chosen" links are hyperbolic. Indeed, as shown by Menasco in [11], every nonsplittable, prime alternating link which is not a torus link is hyperbolic. Let L be a hyperbolic link with a trivial component such that the disk bounded by the component intersects the link in precisely two points. In [23] as well as in [1], it is demonstrated that by cutting $S^3 - L$ open along the disk, twisting a half twist (which permutes the intersections) and reidentifying, we obtain the complement of a new hyperbolic link L' in S^3. In [22], we prove the following:

Theorem 4.1. *([22]) The η-invariant of L and L' differs by $\pm 1/6$ depending upon the direction of the twisting.*

Main ideas of the proof: As shown in [14], there exist a link $\mathcal{L} \subset S^3 - L$, a framing $\mathcal{F}$ on $S^3 - (L \cup \mathcal{L})$ and another framing κ on a neighborhood of each component of $\mathcal{L} \subset S^3 - L$ such that

$$\eta(L) = \frac{1}{12\pi^2} \int_{s(S^3-(L\cup\mathcal{L}))} Q - \frac{1}{6\pi} \sum_{K \subseteq \mathcal{L}} \tau(K, \kappa) \tag{14}$$

where Q is the Chern-Simons 3-form, s is the section defined by $\mathcal{F}$ and $\tau(K, \kappa)$ is the torsion of $K \subseteq \mathcal{L}$ with respect to the framing κ. We construct a framing $\mathcal{F}'$ on $S^3 - (L \cup \mathcal{L})$ such that it is invariant under the twisting. By using $\mathcal{F}'$ in (14), we show that the Chern-Simons integral stays the same under the twisting and the torsion term is changed by $\pm 1/6$ depending upon the direction of the twisting.

Remark. A similar method was used in [14] to prove Theorem 3.3. The general technique was first developed by Meyerhoff and Ruberman in [15] to study the effects of various mutations on the Chern-Simons invariant.

Note that twisting in different directions yields isometric hyperbolic manifolds. In [2], it is proved that if we add some particular trivial components to a hyperbolic link, then the resulting link is still hyperbolic.

Thus the above operations produce a large class of links with isometric complements but different η-invariant. Many of these links even have the same multi-variable Alexander polynomials.

Acknowledgements. I would like to thank the referee for many helpful suggestions for improving the writing.

References

1. C. Adams, *Thrice-punctured spheres in hyperbolic 3-manifolds*, Trans. Amer. Math. Soc. 287 (1985), 645-656.
2. C. Adams, *Augmented alternating link complements are hyperbolic*, London Math. Soc. Lecture Notes 112 (D. B. A. Epstein, ed.), 115-130.
3. M. Atiyah, V. Patodi and I. Singer, *Spectral asymmetry and Riemannian geometry. I*, Math. Proc. Cambridge Philo. Soc. **77** (1975) 43-69.
4. M. Atiyah, V. Patodi and I. Singer, *Spectral asymmetry and Riemannian geometry. II*, Math. Proc. Cambridge Philo. Soc. **78** (1975) 402-433.
5. S. Chern and J. Simons, *Characteristic forms and geometric invariants*, Ann. of Math. **99** (1974) 48-69.
6. P. Gilkey, *The boundary integrand in the formula for the signature and Euler characteristic of a Riemannian manifold with boundary*, Adv. in Math. **15** (1975) 344-360
7. M. Hirsch, *Immersion of manifolds*, Trans. Amer. Math. Soc. **93** (1959) 242-276.
8. T. Kawasaki, *The signature theorem for V-manifolds*, Topology **17** (1978) 75-83.
9. T. Kawasaki, *The index of elliptic operators over V-manifolds*, Nagoya Math. J. **84** (1981) 135-157.
10. M. Komuro, *On the Atiyah-Padoti-Singer η-invariant for S^1-bundles over Riemann surfaces,* J. Fac. Sci. Univ. Tokyo Sect. IA Math. **30** (1984) No.3, 525-548.
11. W. Menasco, *Closed incompressible surfaces in alternating knot and link complements* Topology **23** (1984),37-44.
12. R. Meyerhoff, *Density of the Chern-Simons invariant for hyperbolic 3-manifolds*, Low-dimensional Topology and Kleinian Groups (D. B. A. Epstein, ed.) Cambridge University Press (1986), 217-239.
13. R. Meyerhoff and W. Neumann, *An asymptotic formula for the η-invariant of hyperbolic 3-manifolds*, Comment. Math. Helvetici **67** (1992) 28-46.
14. R. Meyerhoff and M. Ouyang, *The η-invariant of cusped hyperbolic 3-manifolds*, to appear.
15. R. Meyerhoff and D. Ruberman, *Mutation and the η-invariant*, J. Differential Geom. **31** (1990) 101-130.

16. W. Neumann, *Combinatorics of triangulations and the Chern-Simons invariant for hyperbolic 3-manifolds*, in Topology'90, Proceedings of the Research Semester on Low Dimensional Topology at Ohio State University, de Gruyter Verlag, 1992.
17. W. Neumann and M. Jankins, *Seifert manifolds*, lecture notes, Brandeis Univ., 1981, to appear in Cambridge University Press.
18. W. Neumann and D. Zagier, *Volumes of hyperbolic 3-manifolds*, Topology **24** (1985) 307-332.
19. M. Ouyang, *Geometric invariants for Seifert fibred 3-manifolds*, to appear in Trans. Amer. Math. Soc. **346** (1994) 641-659.
20. M. Ouyang, *On the η-invariant of some hyperbolic 3-manifolds*, Topology Appl. **64** (1995) 149-164.
21. M. Ouyang, *A simplicial formula for the η-invariant of hyperbolic 3-manifolds*, Topology, to appear.
22. M. Ouyang, *Mutation and the η-invariant of hyperbolic links*, J. of Knot Theory and its Ramifications, to appear.
23. D. Ruberman, *Mutation and volumes of knots in S^3*, Invent. Math. **90** (1987) 189-215.
24. P. Scott, *The geometries of 3-manifolds*, Bull. London. Math. Soc. **15** (1983) 401-487.
25. J. Seade and B. Steer, *A note on the eta function for quotients of $PSL_2(\boldsymbol{R})$ by co-compact Fuchsian groups*, Topology **26** (1987) 79-91.
26. W. Thurston, *The geometry and topology of 3-manifolds*, lecture notes, Princeton Univ., 1978.
27. W. Thurston, *Three dimensional manifolds, Kleinian groups and hyperbolic geometry*, Bull. Amer. Math. Soc. **6** (1982) 357-381.
28. C. T. C. Wall, *Non-additivity of the signature*, Invent. Math. **7** (1969) 269-274.
29. T. Yoshida, *The η-invariant of hyperbolic 3-manifolds*, Invent. Math. **81** (1985) 473-514.

DEPARTMENT OF MATHEMATICS, OHIO STATE UNIVERSITY, COLUMBUS, OH 43210

E-mail address: mouyang@math.lsa.umich.edu

AMS/IP Studies in Advanced Mathematics
Volume **2**, 1997 (Part 1)

LINK HOMOTOPY IN SIMPLY CONNECTED 3-MANIFOLDS

N. HABEGGER

Let $\mathcal{HL}(k, M)$ denote the set of homotopy classes of k-component links in M^3. If $k = 1$, then $\mathcal{HL}(1, M)$ is naturally in bijection with the set of conjugacy classes of $\pi_1(M)$ and in particular is a homotopy invariant of M. For $k > 1$, however, $\mathcal{HL}(k, M)$ depends on more than just $\pi_1(M)$ as can be seen by the following example:

Let $M = S^1 \times \mathbb{R}^2 \subset S^1 \times S^2 = N$. Consider the links $L = S^1 \times 0 \cup 1 \times \partial D_0^2$ and $L' = S^1 \times 0 \cup 1 \times \partial D_1^2$ where $D_0^2 \subset \mathbb{R}^2$ is a 2-disk containing 0 in its interior, whereas $D_1^2 \subset \mathbb{R}^2$ does not. Then L is homotopic to L' in N (since ∂D_0^2 is contractible to a point in $S^2 \times 0$) but not in M. To see this let $L_0 = S^1 \times 0$. Then (see [M]) Milnor's link group for L_0 in M, denoted $\mathcal{G}(L_0)$ is defined and is equal to $\mathbb{Z} \times \mathbb{Z}$, with $1 \times \partial D_0^2 = \alpha$ representing $(0,1)$ and $1 \times \partial D_1^2 = \beta$ representing $(0,0)$. If $L = L_0 \cup \alpha$ were homotopic to $L' = L_0 \cup \beta$, there would be an isomorphism of $\mathcal{G}(L_0)$ (given by a self-homotopy of L_0 to itself) taking $[\alpha]$ to a conjugate of $[\beta]$. But this is impossible since $[\alpha] \neq 0$ and $[\beta] = 0$. Consequently the inclusion mapping $M \to N$, while an isomorphism on π_1, induces a map $\mathcal{HL}(2, M) \to \mathcal{HL}(2, N)$ which is surjective but not injective.

For simply connected manifolds, the preceding phenomenon does not occur:

Theorem 1. *Let $M \subset N$ be an inclusion of simply connected 3-manifolds. The induced mapping $\mathcal{HL}(k, M) \to \mathcal{HL}(k, N)$ is a bijection.*

Remark. It is enough to show this for $M = B^3$, a ball. Combining the above theorem with the results of [2], we obtain a classification of $\mathcal{HL}(k, M)$, M simply connected. The proof given below makes use of several techniques of that paper. Milnor asked whether for compact M, $\mathcal{HL}(k, M)$ depends only on the homotopy type of the pair $(M, \partial M)$. Theorem 1 shows that any counterexample would have to be non-simply connected.

Partially supported by NSF grants DMS 88-02818 and DMS 91-11663

In particular, link homotopy cannot be used to detect a fake 3-sphere. (A criterion of Bing asserts that if M^3 is closed and simply connected, then $M^3 = S^3$ provided for all k, $\mathcal{L}(k, B^3) \to \mathcal{L}(k, M)$ is surjective, where $\mathcal{L}(k, N)$ denotes the set of *isotopy* classes of k component links in N. See [4].)

Before proving Theorem 1, we recall some ideas from [2]. A k-component string link σ is a collection of k disjoint proper embeddings $\sigma_i : I \to D^2 \times I$ satisfying the boundary conditions $\sigma_i(0) = p_i \times 0$, $\sigma_i(1) = p_i \times 1$, where $p_i \in D^2$ are k fixed distinct points. We will write σ for $\cup \sigma_i(I)$ and will call $\sigma_i(I)$ the ith string.

The inclusions of $D^2 \backslash \{p_i\} \times 0$ and $D^2 \backslash \{p_i\} \times 1$ into $D^2 \times I \backslash \sigma$ induce isomorphisms on the *reduced* fundamental groups and hence the composite

$$RF_k = R\pi_1(D^2 \backslash \{p_i\} \times 0) \overset{\simeq}{\to} R\pi_1(D^2 \times I \backslash \sigma) \overset{\simeq}{\leftarrow} R\pi_i(D^2 \backslash \{p_i\} \times 1) = RF_k$$

is an automorphism, denoted φ_σ, of the reduced free group RF_k. (Here, reduced means we introduce relations which say that each generator commutes with its conjugates, and for link groups that each meridian commutes with its conjugates. See [2] and [3].) One of the main results of [2] is that $\varphi_\sigma = \varphi_{\sigma'}$ if and only if σ is homotopic (rel boundary) to σ'.

Let $B^3_+ = \{(x, y, z) \in \mathbb{R}^3, x^2 + y^2 + z^2 \leq 1, z \geq 0\}$ denote the upper half ball, and $D^2_+ = \{(x, y) \in \mathbb{R}^2, x^2 + y^2 \leq 1, x \geq 0\}$, the half disk. A string link in $D^2_+ \times I$ can be viewed as lying in B^3_+, with lower boundary $\partial_0 B^3_+ = B^3_+ \cap \mathbb{R}^2 \times 0$, via composition with the map

$$(x, y, t) \to (x \cos \pi t, y, x \sin \pi t).$$

The main difficulty in the proof of Theorem 1 is resolved via the following.

Lemma 2. *Suppose $\pi_1(M) = 1$ and let B^3_+ be embedded in M with $\partial_0 B^3_+ = \partial M \cap B^3_+$. Suppose σ and σ' are string links in B^3_+ which are homotopic in M. Then they are homotopic in B^3_+.*

Proof. By Van Kampen's theorem, for $\sigma \subset B^3_+$, the inclusion $B^3_+ \backslash \sigma \mapsto M \backslash \sigma$ induces an isomorphism on fundamental groups, and hence also on reduced fundamental groups. Hence φ_σ is also obtained from the composition

$$RF_k = R\pi_1(D^2_+ \backslash \{p_i\} \times 0) \overset{\simeq}{\to} R\pi_1(M \backslash \sigma) \overset{\simeq}{\leftarrow} R\pi_1(D^2_+ \backslash \{p_i\} \times 1) = RF_k$$

The proof given in [2] that φ_σ depends only on the homotopy class of σ therefore applies also to homotopies in M as well. Thus σ homotopic to σ' in M implies $\varphi_\sigma = \varphi_{\sigma'}$ which implies σ is homotopic to σ' in B^3_+. $\square$

Remark. The above result has an analogue for isotopy: Suppose σ, σ' are braids in B^3_+ which are isotopic in M. Then they are isotopic in B^3_+. For the proof, replace the reduced free group by the free group and apply the Artin representation theorem (see [1]) for the braid group.

We next have the following "connectivity" lemma.

Lemma 3. *Let N^3 be obtained from M^3 via glueing manifolds M_i to M along disjoint surfaces $S_i \subset \partial M_i$, $S_i \subset \partial M$ and suppose the pairs (M_i, S_i) are 1-connected. Then*

$$\mathcal{HL}(k, M) \to \mathcal{HL}(k, N)$$

is surjective.

Proof. Assume inductively the proposition is true for $k-1$ in place of k. Let $L \cup \alpha$ be a k component link in N where L is a $k-1$ component link in N and α is a closed curve in $N\backslash L$. L is homotopic to L' in M and a general position argument shows that up to isotopy of α, α can be kept disjoint from any crossing changes. Thus $L \cup \alpha$ is homotopic to $L' \cup \alpha'$ with L' lying in M. But as the pairs (M_i, S_i) are 1-connected, the part of α' lying in M_i can be homotoped into a neighborhood of S_i and hence into M. □

Remark. Due to possible interference of the other $k-1$ components, the weaker hypothesis of 1-connectivity of the pair (N, M) does not seem to suffice to prove the lemma.

Proof of Theorem 1. It is enough to prove this for $M = B^3 \subset \text{interior}N$. The pair $(N\backslash \text{int}\, B^3, \partial B^3)$ is 1-connected since $N\backslash int\, B^3$ is simply connected by Van Kampen's theorem. Hence lemma 3 implies $\mathcal{HL}(k, B^3) \to \mathcal{HL}(k, N)$ is surjective. It remains to prove injectivity. □

Let L, L' be k-component links in B^3 which are homotopic in N. We will show that L and L' are homotopic in B^3 to the closures $\hat{\sigma}$ and $\hat{\sigma}'$ in B^3 of string links σ, σ' in B^3_+ such that σ, σ' are homotopic as string links in $N' = N\backslash\text{int}B^3_-$. (The closure of a string link in B^3_+ is the link in $B^3 = B^3_+ \cup B^3_-$ which is the union of the given string link in B^3_+ and the "trivial" string link in B^3_-. See [2].) Since $\pi_1(N') = 1$, lemma 2 applies to show that σ is homotopic to σ' in B^3_+. Thus, in B^3, $\hat{\sigma}$ is homotopic to $\hat{\sigma'}$ and hence L and L' are homotopic in B^3.

Up to isotopy, we may replace L by the closure $\hat{\sigma}$ of a string link σ. This is achieved by choosing an embedded disk intersecting each component transversally in a point with $+1$ intersection number and then isotoping this disk (in B^3) into standard position, say $B^2_- = B^3_- \cap 0 \times \mathbb{R}^2$.

Now the disk base of L can be isotoped (rel $\partial_1 B^2_- = \partial B^2_- \cap S^1$) during the deformation of L to L' to a disk base for L'. However, the disk base for L' will in general not be contained entirely in B^3. Nevertheless, we claim that after certain self-homotopies of L', we may suppose that the disk base does lie in B^3. Assuming this can be done, we may isotope the disk base into standard position exhibiting L' as the closure $\hat{\sigma}'$ of a string link σ'. Moreover, σ, σ' are homotopic in N' since they are disk-based homotopic. (See corollary 2.4 of [2]. Since the disk B^2_- retracts to a neighborhood of $\partial_1 B^2_-$,the deformation can be taken rel B^2_- and even rel B^3_-.) This proves theorem 1 modulo the above claim.

In order to isotope the disk base into B^3, note that since the disk retracts to a neighborhood of a 1-complex, the argument in the proof of lemma 3 can be used to homotope the disk base into B^3, where we may assume that each arc of the 1-complex remains disjoint from the other arcs. The idea is then to replace the homotopy obtained above for this 1-complex by isotopies and self-homotopies of the disk-based link.

The 1-complex in question is a collection of segments a_i in B^2_- joining a base point in $\partial_1 B^2_-$ to the individual components L'_i of L' (which we take as base points for these components.

Suppose inductively that a_j lies in B^3 for $j < i$ and consider a homotopy of the parts of a_i lying outside of B^3 into a regular neighborhood of ∂B^3. By general position, we may assume this homotopy is obtained by effecting a sequence of crossing changes on a_i and isotopies both of which are fixed on B^3.

Consider a crossing of a_i with itself. This crossing can also be achieved by an isotopy involving only a_i and L'_i and a self-homotopy (which is an isotopy of the diskbase) of L'_i as follows (compare with 2.6 of [2] where a similar argument is used to show that the trivial link has only one basing, up to homotopy):

First retract the segment a_i towards the basepoint, pulling a small neighborhood of the basepoint of L'_i along a band in a neighborhood of a_i. In the neighborhood of the crossing, this has the effect of replacing the 2 segments involved in the crossing with two bands, the boundary of which are segments of L'_i. Crossing the two bands amounts to 4 crossing changes of L'_i. Pulling back L'_i along the bands obtained after the crossing returns L'_i to its original position, with the crossing change effected on a_i.

Thus after a series of such self homotopies of L_i (fixing a_j for $j < i$), a_i lies in a regular neighborhood of B^3 and hence can be pulled into B^3. This completes the inductive step and hence the proof of theorem 1.

REFERENCES

1. Birman, J. *Braids, Links, and Mapping Class Groups*, Annals of Math Studies, vol 82, Princeton University Press, Princeton, N.J., 1974.
2. Habegger, N. and Lin, X. S., *The classification of Links up to Homotpy*, Journal of the A.M.S. **3** No. 2, (1990) 389–419.
3. J. Milnor, *Link groups*, Annals of Math. **59** (1954) 177-195.
4. Rolfsen. *Knots and Links*, Publish or Perish, Berkeley, Calif., 1976.

E-mail address: habegger@joe.math.uga.edu, habegger@math.univ-nantes.fr

AMS/IP Studies in Advanced Mathematics
Volume **2**, 1997 (Part 1)

HOMOTOPY, ISOTOPY AND GENUINE LAMINATIONS OF 3-MANIFOLDS

DAVID GABAI* AND WILLIAM H. KAZEZ

1. Introduction

The notion of an essential lamination in a 3-manifold was introduced in [8] as a generalization of both incompressible surfaces and Reebless foliations. The hope is that the major results for Haken manifolds will generalize to the much larger class of manifolds containing essential laminations. This sort of study of the properties of 3-manifolds containing essential laminations as well as the question of "how much larger" are two of the outstanding problems in the subject.

Definition 1.1. A *complementary region* of an essential lamination μ is a component of $M - \mu$. A *closed complementary region* of μ is the completion in the path metric of a complementary region. An essential lamination μ in M is a *genuine lamination* if μ contains a non I-bundle closed complementary region.

Remark 1.2.

(1) If each complementary region of the essential lamination μ is an I-bundle, then μ extends to a foliation without Reeb components.

(2) If μ is a genuine lamination in a non-Haken manifold M, then M is necessarily atoroidal, that is $\pi_1(M)$ contains no subgroup isomorphic to $\mathbb{Z} + \mathbb{Z}$; for otherwise by [4, Corollary 8.6] or [2], M is a non-Haken Seifert fibre space. But by [1] or [3] every essential lamination in a non-Haken Seifert fibre space can be isotoped to be transverse to the Seifert fibres and hence the complementary regions are I-bundles.

(3) The point of distinguishing genuine laminations from essential laminations, indeed the point of this paper and its sequel [6], is that a non I-bundle complementary region is a "characteristic seam" of the manifold and hence this class of manifolds is

*This research was partially supported by NSF Grant DMS-9200584.

amenable to new methods of study. In [7] we show the mapping class group of a manifold with genuine lamination is finite.

The main result of this paper is a theorem about a very common type of genuine lamination.

Theorem 1.3. *Let M be a closed 3-manifold containing an essential lamination λ, such that some complementary region of λ is homeomorphic to $\mathbb{R}^2 \times S^1$ and such that the corresponding closed complementary region is not an I-bundle. If $f : M \to M$ is a homeomorphism homotopic to the identity map, then f is isotopic to the identity map.*

Corollary 1.4. *If $f, g : M \to N$ are homotopic homeomorphisms and M contains a genuine lamination λ such that some complementary region of λ is a non-I-bundle solid torus, then f is isotopic to g.*

Remark 1.5. In [9] it is shown that in some sense most manifolds obtained by surgery on hyperbolic knots in S^3 satisfy the hypothesis of the theorem, see [8, Theorem 5.3] for the case of fibred knots.

Theorem 4.6 is a special case of the following generalization of a theorem of Palmeira [11].

Theorem 4.6. *If $\lambda \subset M$ is an essential lamination then $(\tilde{M}, \tilde{\lambda})$ is homeomorphic to $(\mathbb{R}^2, \ell) \times \mathbb{R}$ where $\ell \subset \mathbb{R}^2$ is a lamination.*

2. Outline

Definition 2.1. Let d_n be a disk minus $n \geq 3$ boundary points. Let $a_1, \cdots, a_n$ be the boundary components of d_n in cyclical order. Let $N(\cup_i a_i)$ be the fibred neighborhood drawn in Figure 1 of the lamination $\cup_i a_i$.

Let z be an interior point of d_n and let $s_1, \cdots, s_n$ be properly embedded rays in d_n intersecting only at z. Furthermore, the s_k should intersect $N(\cup_i a_i)$ transversely to the I-bundle fibres, and the I-bundle fibres intersected should be precisely those which connect a_{k-1} and a_k. We assume that the rotation of d_n by $2\pi/n$ preserves $\cup_i a_i, N(\cup_i a_i)$, and $\cup_i s_i$. Let h be the rotation of d_n by $2\pi k/n$, let $C = (d_n \times I)/h$ and let $\gamma = z \times S^1$. Theorem 1.3 applies to manifolds M which contain a closed complementary region homeomorphic to C in the case that $n > 2$. From now on we will view C as a subset of M.

The proof of Theorem 1.3 consists of the following steps.

(1) After an isotopy of f, $f \mid \gamma = id_\gamma$.
(2) If $p : \tilde{M} \to M$ is the universal covering projection, then $(\tilde{M}, p^{-1}(\gamma)) = (\mathbb{R}^2, X) \times \mathbb{R}$, where X is a locally finite set of points in $\mathbb{R}^2$.

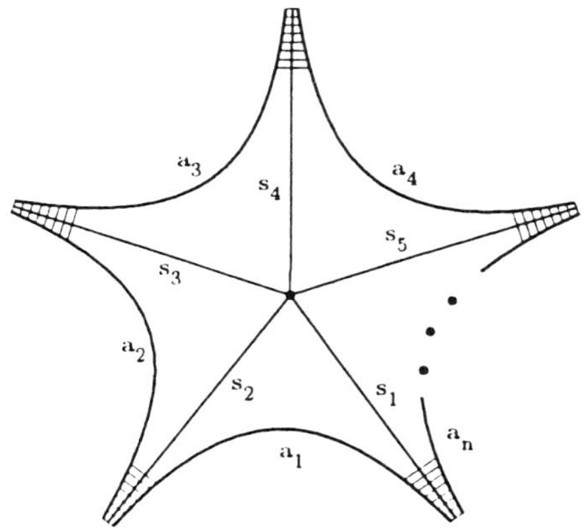

FIGURE 1

(3) f fixes a free link and hence, by a theorem of Hass and Scott [10, Theorem 3.3], f is isotopic to id_M.

3. STEP 1

Step 1. If $f : M \to M$ is homotopic to the identity map then after isotoping f there exists a homotopically non-trivial simple closed curve in M that is mapped to itself by f.

In the proof of the theorem we shall first assume h is the identity map so that $C = d_n \times S^1$. Let $\gamma = z \times S^1$, $A_i = a_i \times S^1$, and $S_i = s_i \times S^1$.

Let $p : \hat{M} \to M$ be the covering space of M corresponding to $\pi_1(C, *) \subset \pi_1(M, *)$. $\hat{M}$ is homeomorphic to $S^1 \times \mathbb{R}^2$ by [5]. $C \subset M$ has a canonical lift $\hat{C}$ to $\hat{M}$. Let $\hat{A}_i$ and $\hat{S}_i$ denote the lifts of A_i and S_i contained in $\hat{C}$. By [5, Lemma 10] each $\hat{A}_i$ is properly embedded in $\hat{M}$ and hence $\hat{C}$ and $\hat{S}_i$ are also properly embedded in $\hat{M}$.

Let $Y = \cup_i S_i$ and $\hat{Y} = \cup_i \hat{S}_i$. Give M a Riemannian metric and let d be the pullback metric on $\hat{M}$. For a sequence x_i contained in a non-compact space, we say $x_i \to \infty$ if the sequence eventually misses any given compact subset. Let $\{\hat{x}_i\} \subset \hat{S}_a$, $\{\hat{y}_i\} \subset \hat{S}_b$ with $\hat{S}_a \neq \hat{S}_b$ and suppose $\hat{x}_i \to \infty$ and $\hat{y}_i \to \infty$. The key observation in this paper is that the ends of $\hat{Y}$ are *separated at infinity*, that is $d(\hat{x}_i, \hat{y}_i) \to \infty$.

Lemma 3.1. *The ends of $\hat{Y}$ are separated at infinity.*

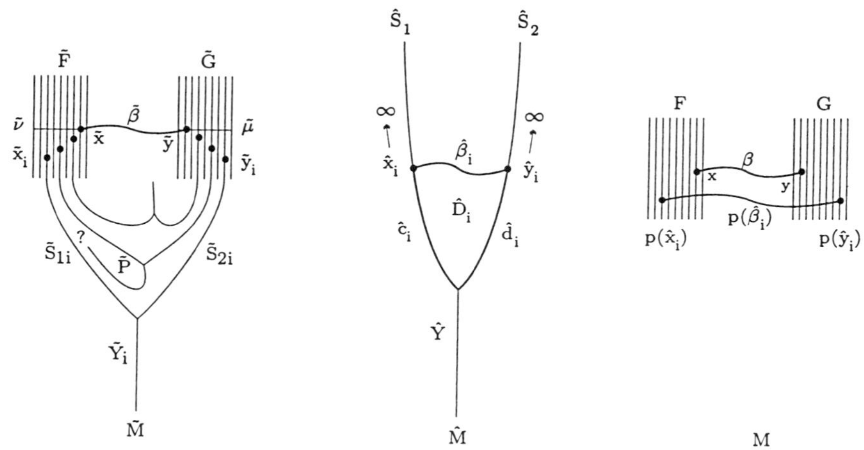
$\tilde{F}$
$\tilde{G}$
$\tilde{\nu}$
$\tilde{\beta}$
$\tilde{\mu}$
$\tilde{x}$
$\tilde{y}$
$\tilde{x}_i$
$\tilde{y}_i$
?
$\tilde{S}_{1i}$
$\tilde{P}$
$\tilde{S}_{2i}$
$\tilde{Y}_i$
$\tilde{M}$
$\hat{S}_1$
$\hat{S}_2$
∞
∞
$\hat{\beta}_i$
$\hat{x}_i$
$\hat{y}_i$
$\hat{D}_i$
$\hat{c}_i$
$\hat{d}_i$
$\hat{Y}$
$\hat{M}$
F
G
β
x
y
$p(\hat{x}_i)$
$p(\hat{\beta}_i)$
$p(\hat{y}_i)$
M

FIGURE 2

Proof. The proof will be by contradiction. It is enough to show that $\hat{S}_1$ and $\hat{S}_2$ must be separated at infinity. Let $\hat{\beta}_i$ be a geodesic from $\hat{x}_i$ to $\hat{y}_i$. Let $\hat{c}_i$ be a path in $\hat{S}_1$ from $\hat{x}_i$ to a point in $\hat{\gamma}$ and let $\hat{d}_i$ be a path in $\hat{S}_2$ from $\hat{y}_i$ to the same point in $\hat{\gamma}$ such that $\hat{\beta}_i \cup \hat{c}_i \cup \hat{d}_i$ is a closed null homotopic curve. Let $\hat{D}_i$ be an immersed disk of null homotopy.

If $d(\hat{x}_i, \hat{y}_i) \not\to \infty$ then by passing to a subsequence we may assume $p(\hat{\beta}_i)$ converges to a geodesic $\beta \subset M$. Let x and y denote the limits of $p(\hat{x}_i)$ and $p(\hat{y}_i)$ respectively. Let F be a flowbox for the lamination contained in $N(\lambda)$ that contains x. Since $p|\hat{S}_1$ is $1-1$ it is not possible that infinitely many $p(\hat{x}_i)$ are contained in one component of $p(\hat{S}_1) \cap F$ and we may assume, after passing to a subsequence, that $p(\hat{x}_i)$ lie on distinct components of $p(\hat{S}_1) \cap F$. Let G be a similar flowbox about y.

Let $\tilde{M}$ be the universal cover of M. Let $\tilde{\beta}$ be a lift of β and choose lifts $\tilde{x}_i, \tilde{y}_i$ and $\tilde{\beta}_i$ such that $\tilde{\beta}_i$ connects $\tilde{x}_i$ and $\tilde{y}_i$ and $\tilde{\beta}_i$ converges to $\tilde{\beta}$. By construction $\hat{\beta}_i$ is homotopic relative to its endpoints to $\hat{c}_i \cup \hat{d}_i \subset \hat{Y}$. By the homotopy lifting property each $\tilde{\beta}_i$ is homotopic relative to its endpoints into some component $\tilde{Y}_i$ of the preimage of Y. In particular $\tilde{x}_i, \tilde{y}_i \in \tilde{Y}_i$. Let $\tilde{S}_{1i}, \tilde{S}_{2i}, \tilde{S}_{3i}, \cdots$ denote the components of the preimage of $S_1, S_2, S_3, \cdots$ that are contained in $\tilde{Y}_i$. Let $\tilde{\nu}$ be an I-bundle fibre for the component of $p^{-1}(F)$ containing $\tilde{x}$ and let $\tilde{\mu}$ be a corresponding fibre for $\tilde{y} \in p^{-1}(G)$.

Claim. There exists j, k such that $\tilde{S}_{kj} \cup \tilde{S}_{k+1j}$ intersects either $\tilde{\nu}$ or $\tilde{\mu}$ at least twice.

Proof of claim. We may start by assuming that for all i the plane $\tilde{P}_i = \tilde{S}_{1i} \cup \tilde{S}_{2i}$ intersects both $\tilde{\nu}$ and $\tilde{\mu}$ exactly once. If $\tilde{P}_j = \tilde{S}_{1j} \cup \tilde{S}_{2j}$ is another such plane then since $\tilde{P}_i$, $\tilde{P}_j$ are separating properly embedded planes, there is a component of $\tilde{M} - (\tilde{P}_i \cup \tilde{P}_j)$ that contains no endpoints of either $\tilde{\nu}$ or $\tilde{\mu}$. This uses the fact that a proper plane in $\tilde{M}$ cannot have exactly one transverse intersection with a closed curve.

By considering three $\tilde{P}_k$'s we can choose i and j so that $\tilde{S}_{3j}$ is between $\tilde{P}_i$ and $\tilde{P}_j$. It follows that $\tilde{S}_{kj}$ is also between $\tilde{P}_i$ and $\tilde{P}_j$ for $k > 2$. If $\tilde{S}_{kj}$ intersects neither $\tilde{\nu}$ or $\tilde{\mu}$ then $\tilde{S}_{1j} \cup \tilde{S}_{kj}$ is a properly embedded plane and $\tilde{\nu} \cup \tilde{P}_i \cup \tilde{\mu} \cup \tilde{P}_j$ contains a curve intersecting $\tilde{S}_{1j} \cup \tilde{S}_{kj}$ in exactly one point. This is impossible and it follows that for all $k > 2$ $\tilde{S}_{kj}$ intersects either $\tilde{\nu}$ or $\tilde{\mu}$. It follows that we can choose k as stated in the claim. □

Without loss of generality suppose $\tilde{P} = \tilde{S}_{kj} \cup \tilde{S}_{k+1j}$ intersects $\tilde{\nu}$ at least twice. $N(\lambda) \cap C$ has the property that each I-bundle fibre that intersects S_k or S_{k+1} also intersects the leaf A_k of λ. It follows that $\tilde{A}_k$ intersects $\tilde{\nu}$ at least twice. This contradicts the fact that an efficient arc cannot be homotoped relative to its endpoints into a leaf [8, Proposition 6.3]. This completes the proof of Lemma 3.1. □

An embedded curve in an open solid torus, $\mathbb{R}^2 \times S^1$, is called a *core* if it represents a generator of $\pi_1(\mathbb{R}^2 \times S^1)$ and is the boundary component of a properly embedded $[0, \infty) \times S^1$. It is well known that cores of open solid tori are unique up to isotopy. Moreover, if the two cores are contained in the interior of a (necessarily unknotted) compact solid torus $\hat{T}$, then the isotopy may be chosen with support contained in $\hat{T}$.

We are given a homeomorphism $f : M \to M$ that is homotopic to id_M. Lift this homotopy to $\hat{M}$ by starting at $id_{\hat{M}}$ and let $\hat{f}$ be the corresponding lift of f. The strategy for proving Step 1 is to isotop f until $\hat{f}(\hat{\gamma}) \subset \hat{C}$. Let $\hat{T} \subset \hat{C}$ be a solid torus containing $\hat{f}(\hat{\gamma})$ and $\hat{\gamma}$. $\hat{f}(\hat{\gamma})$ and $\hat{\gamma}$ are cores of $\hat{M}$ so there exists an isotopy supported in $\hat{T}$ that moves $\hat{f}(\hat{\gamma})$ to $\hat{\gamma}$. Since p maps $\hat{C}$ bijectively to C it follows that f can be isotoped so that $f(\gamma) = \gamma$. Since f induces the identity map on $\pi_1(M)$, f preserves an orientation on γ so after a further isotopy we have $f \mid \gamma = id_\gamma$ as required in Step 1.

By a small isotopy of f we may assume $\hat{f}(\hat{Y})$ is transverse to $\cup_i \hat{A}_i$. By the construction of $\hat{f}$ and the compactness of M there exists a number K such that $d(\hat{x}, \hat{f}(x)) \leq K$ for all $\hat{x} \in \hat{M}$. Consider $\hat{f}(\hat{S}_1) \cap \hat{A}_2$. This must be compact, for otherwise there would exist a sequence of points in the intersection limiting on ∞, and since $\hat{f}(\hat{S}_1)$ is within K of $\hat{S}_1$ and $\hat{A}_2$ is within bounded distance of $\hat{S}_2 \cup \hat{S}_3$, this would contradict the Lemma 3.1.

$\hat{f}(\hat{S}_1) \cap \hat{A}_2$ consists of a finite number of arcs and circles. If there is a circle $\hat{\sigma} \subset \hat{f}(\hat{S}_1) \cap \hat{A}_2$ which is essential in $\hat{f}(\hat{S}_1)$ then it follows by the uniqueness of cores in annuli that $\hat{f}(\hat{\gamma})$ can be isotoped along $\hat{f}(\hat{S}_1)$ into $\hat{C}$.

The next possibility is that $\hat{f}(\hat{S}_1) \cap \hat{A}_2$ contains an inessential circle $\hat{\sigma}$. Suppose $\hat{\sigma}$ bounds a disk $\hat{D}$ in $\hat{f}(\hat{S}_1)$ and a disk $\hat{E}$ in $\hat{A}_2$. $p(\hat{D})$ and $p(\hat{E})$ are embedded disks in M and we may assume they intersect transversely.

Claim. There exists a homeomorphism $g : M \to M$ such that g is isotopic to id_M such that if $\hat{g}$ is the lift of g to $\hat{M}$ then $\hat{g}\hat{f}(\hat{S}_1) \cap \hat{A}_2$ contains 1 less inessential circle.

Proof. If $p(\hat{D})$ and $p(\hat{E})$ intersect only along their boundaries then such a g exists by the irreducibility of M. In general g may be found using an innermost disk argument. □

This procedure may be repeated until $\hat{f}(\hat{S}_1) \cap \hat{A}_2$ contains only arc intersections. Let $\hat{\tau} \subset \hat{f}(\hat{S}_1) \cap \hat{A}_2$ be an arc innermost in $\hat{f}(\hat{S}_1)$.

Claim. There exists a homeomorphism $h : M \to M$ such that h is isotopic to id_M and such that if $\hat{h}$ is the lift of g to $\hat{M}$ then $\hat{h}\hat{f}(\hat{S}_1) \cap \hat{A}_2$ contains 1 less arc of intersection.

Proof. Let $\hat{\tau} \subset \hat{f}(\hat{S}_1) \cap \hat{A}_2$ be an innermost arc in $\hat{f}(\hat{S}_1)$. Then $\hat{\tau}$ together with an arc in $\hat{\gamma}$ bounds an embedded disk $\hat{E}$ in $\hat{f}(\hat{S}_1)$. $p(\hat{E})$ is embedded in M and hence the desired homeomorphism h can be constructed with support contained in a regular neighborhood of $p(\hat{E})$. □

We may now assume $\hat{f}(\hat{S}_1) \cap \hat{A}_2 = \emptyset$. This argument may be repeated until $\hat{f}(\hat{S}_1) \cap \hat{A}_i = \emptyset$ for $i \neq 0, 1$ (This can be done by repeating the above arguments with $\hat{A}_3, \hat{A}_4 \cdots$ or by applying them to $\cup_{i \neq 0,1} \hat{A}_i$ instead of $\hat{A}_2$.) $\hat{A}_2$ separates $\hat{M}$ into two components one of which contains $\hat{C}$. Of course $\hat{A}_1$ is on the $\hat{C}$ side of $\hat{A}_2$ and since $\hat{f}(\hat{S}_1)$ limits on $\hat{A}_1$ so is $\hat{f}(\hat{S}_1)$. In particular $\hat{f}(\hat{\gamma})$ is on the $\hat{C}$ side of $\hat{A}_2$. Similarly $\hat{f}(\hat{\gamma})$ is on the $\hat{C}$ side of $\hat{A}_i$ for $i \neq 0, 1$.

Now consider $\hat{f}(\hat{S}_2) \cap \cup_{i \neq 1,2} \hat{A}_i$. As before if this intersection contains an essential circle then we may complete Step 1. Inessential circles may also be removed as above. This may change $\hat{f}(\hat{S}_1) \cap (\cup_i \hat{A}_i)$ but $\hat{f}(\hat{S}_1) \cap \hat{A}_i$ can be kept empty for $i \neq 0, 1$. Similarly arcs of intersection in $\hat{f}(\hat{S}_2) \cap \cup_{i \neq 1,2} \hat{A}_i$ may be removed while $\hat{f}(\hat{S}_1) \cap \hat{A}_i$ is kept empty for $i \neq 0, 1$.

After repeating this procedure we may assume that if $\hat{f}(\hat{S}_j) \cap \hat{A}_i \neq \emptyset$ then $i = j$ or $i = j - 1$. It follows that $\hat{f}(\hat{\gamma})$ is on the $\hat{C}$ side of $\hat{A}_i$ for all i and therefore $\hat{f}(\hat{\gamma}) \subset \hat{C}$. This completes the proof of Step 1 in the case that C is a product.

For non-trivial d_n bundles C, let $M' \to \hat{M}$ be the cyclic cover of lowest degree such that the inverse image C' of C is the product bundle. Let f' be the lift of f to M' and let the various subsets of C' be denoted A'_i, S'_i, and Y'. As before either we can isotop f until $f'(S'_j) \cap A'_i$ contains no arcs or inessential circles unless $i = j$ or $i = j - 1$. (This does not require $p \mid C'$ to be injective, it is enough that p is a covering map.)

Consequently $f'(\gamma') \cap (\cup_i A'_i) = \emptyset$. Since C' is the only component of $M' - \cup_i A'_i$ that contains a generator of $\pi_1(M')$ we see that $f'(\gamma') \subset C'$. This implies the desired conclusion, $\hat{f}(\hat{\gamma}) \subset \hat{C}$. This completes the proof of Step 1.

4. A PRODUCT THEOREM

We start by giving a brief overview of Gabai and Oertel's proof of $\tilde{M} \cong \mathbb{R}^3$ [8, Construction 6.11]. Complete details are contained in the original article.

Gabai and Oertel define an *order tree* T associated to $\tilde{\lambda} \subset \tilde{M}$. The *points* of T are closed complementary regions of $\tilde{\lambda}$ and leaves of $\tilde{\lambda}$ that are not contained in any closed complementary region (*non-boundary leaves*). The most important example of a *segment* of T is the set of points of T that intersect an I-bundle fibre of a fibred neighborhood of $\tilde{\lambda}$. In a general order tree segments are required to have certain cancellation and conjuction properties. This is used to show that any finite connected 1-complex K contained in $\tilde{M}$ is contained in a ball.

Their construction proceeds as follows. First assume K intersects a fibred neighborhood $N(\tilde{\lambda})$ in I-bundle fibres. The union of these fibres defines a (possibly non-Hausdorff) subtree Γ of T. Γ is finite in the sense that Γ is a finite disjoint union of *vertices* and *intervals*. The vertices are chosen to include all complementary regions of $N(\tilde{\lambda})$ that intersect K. By subdividing edges we can further assume that at least one end of each interval limits on a unique vertex. By thickening a finite number of leaves of $\tilde{\lambda}$ and deleting the interiors of these thickened leaves we can assume that each vertex corresponds to a complementary region of $N(\tilde{\lambda})$.

Remark. We leave it to the reader to check that the operation of [8] discussed in the previous sentence has no effect on the proofs of Theorems 4.4 or 4.6 and that with some minor additional technical work it can be dispensed with entirely. We may also assume, by thickening leaves, that λ has no isolated leaves.

Let $W_1, \dots, W_s$ be the complementary regions of $N(\tilde{\lambda})$ that intersect K and let $V_1, \dots, V_s$ be the closed complementary regions of $\tilde{\lambda}$ such that $W_j \subset V_j$. By $\tilde{\lambda}$ splitting, that is removing a portion of $N(\tilde{\lambda})$, we may assume $K \cap W_j = K \cap V_j$ for all j. In particular, each V_j contains a unique W_j.

Let Δ be the abstract graph with vertices $W_1, \dots, W_s$ such that two vertices are connected by a unique edge if and only if there is a component of $K \cap N(\tilde{\lambda})$ that connects them. In [8] the Digon Lemma is used to show that for each pair W_i, W_j connected by an edge, there exists a 1-handle, i.e., a $D^2 \times I$, in $\tilde{M}$ containing all components of $K \cap N(\tilde{\lambda})$ connecting V_i and V_j. Δ corresponds to the subset $\Delta(\tilde{M})$ which is defined to be the union of $\cup_i W_i$ and all of the 1-handles. Δ is a simplicial tree and $\Delta(\tilde{M})$ is a B^3 missing boundary manifold containing K. It will be more convenient in this paper to replace W_i by V_i in the definition of $\Delta(\tilde{M})$. Since V_i are also B^3 missing boundary manifolds [8, Corollary 6.4], [5, Lemma 5], $\Delta(\tilde{M})$ is still a B^3 missing boundary manifold.

To summarize, the main properties of Δ and the corresponding subset $\Delta(\tilde{M})$ of $\tilde{M}$ are:

(1) Each vertex V of Δ corresponds to a closed complementary region of $\tilde{\lambda}$.
(2) Each edge E of Δ connecting a pair of vertices corresponds to a 1-handle that contains all components of $K \cap N(\tilde{\lambda})$ connecting the corresponding complementary regions of $\tilde{\lambda}$.
(3) The subset $E(\tilde{M})$ of $\Delta(\tilde{M})$ corresponding to an edge E satisfies $(E(\tilde{M}), E(\tilde{M}) \cap \lambda) \cong (I, c) \times \mathbb{R}^2$ where c is a closed subset of I containing 0 and 1.
(4) For each edge E, the intersection of $E(\tilde{M})$ and any complementary region of $\tilde{\lambda}$ is either empty or connected and unknotted in the complementary region [5, Corollary 6].

We expand on this slightly to prove

Lemma 4.1. *Let K be a connected, locally finite, 1-dimensional complex contained in $\tilde{M}$ that intersects $N(\tilde{\lambda})$ in a union of I-bundle fibres. There is a sequence of simplicial trees $\Delta_1, \Delta_2, \ldots$ and a corresponding sequence $\Delta_1(\tilde{M}) \subset \Delta_2(\tilde{M}) \subset \cdots$ of subsets of $\tilde{M}$ such that $K \subset \cup_i \Delta_i(\tilde{M})$ and Δ_{i+1} is obtained from Δ_i in one of three ways.*

(a) *Δ_{i+1} is the union of Δ_i a new edge E and a new vertex V such that E connects V to a vertex of Δ_i.*
(b) *Δ_{i+1} is obtained by adding a new vertex to the interior of an edge of Δ_i thereby replacing the original edge by two new edges.*
(c) *Δ_{i+1} is created by replacing a collection of two or more edges E_j connecting a common vertex V to vertices V_j in Δ_i by a single edge E connecting V and a new vertex V'. In addition the original edges E_j are replaced by new edges E'_j connecting V' and V_j.*

Proof. Let $K_1 \subset K_2 \subset \cdots$ be connected 1-complexes such that $K_{i+1} - K_i$ intersects $N(\tilde{\lambda})$ in a single I-bundle fibre and $\cup_i K_i = K$. Let Δ_i be the tree corresponding to K_i and let $\Delta_i(\tilde{M})$ be the corresponding subset of $\tilde{M}$. To prove Lemma 4.1 it suffices to first show that Δ_{i+1} is obtained from Δ_i by a sequence of Type (a)–(c) moves and then reindex the Δ_i's.

Let $\Gamma_i \subset T$ be the subtree from which Δ_i is constructed. Let I_{i+1} be the closure of $(K_{i+1} - K_i) \cap N(\tilde{\lambda})$. Denote by $I_{i+1}(T)$ the corresponding interval of T. It follows that $I_{i+1}(T) \cap \Gamma_i \neq \emptyset$ and $\Gamma_{i+1} = \Gamma_i \cup I_{i+1}(T)$.

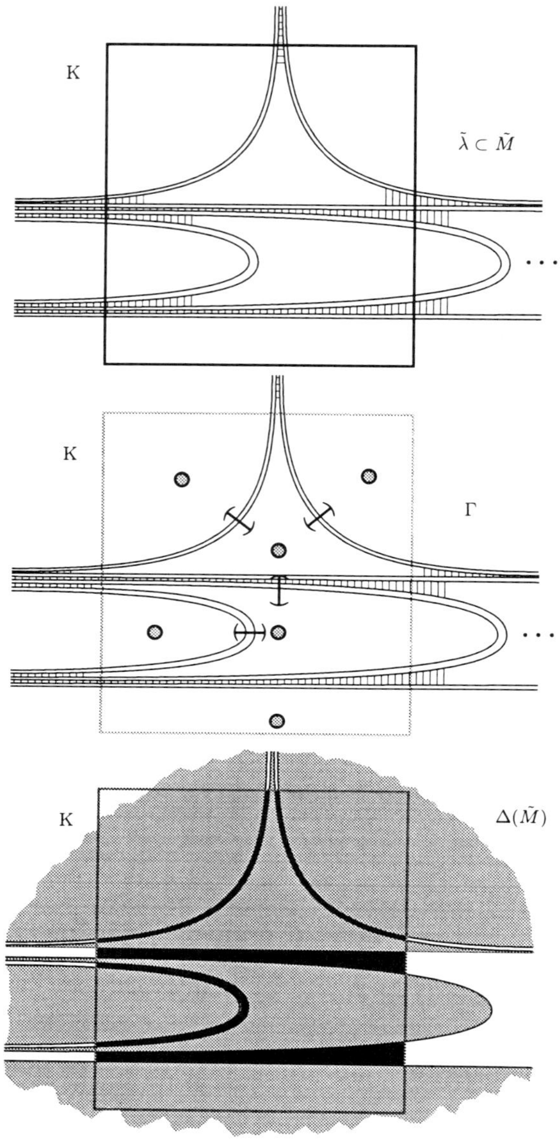

This figure shows a 2-dimensional version of $\tilde{\lambda} \subset \tilde{M}$ and a portion of the interstitial bundle of $N(\tilde{\lambda})$. Notice that a portion of $\tilde{\lambda}$ is the universal cover of a Reeb lamination of an annulus. The leaves shown are exactly the boundary leaves touching the complementary regions of $N(\tilde{\lambda})$ seen in the figure.

This figure shows the change in $N(\tilde{\lambda})$ after splitting. Γ is shown decomposed into a union of vertices and disjoint intervals. The intervals should intersect only non-boundary leaves and therefore should be shorter than drawn. Notice that the top end of the vertical interval limits on the vertex above it and the lowest vertex in the picture. This example does not show the case that both ends of an interval limit on multiple vertices in which case it is necessary to thicken a leaf and delete the interior of the thickened region.

$\Delta(\tilde{M})$ is the realization in $\tilde{M}$ of the tree Δ.

FIGURE 3

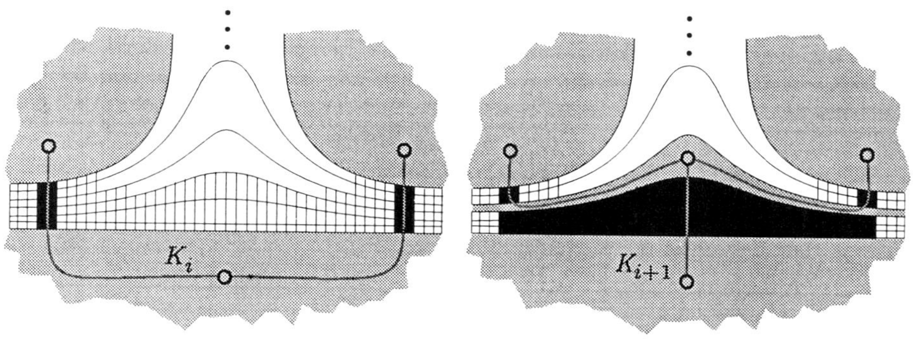

Type (c) move

Now consider the possible relationships between Γ_{i+1} and a decomposition of Γ_i into a disjoint collection of vertices and intervals.

Since each leaf of $\tilde{\lambda}$ is separating and can intersect I_{i+1} at most once, it follows that $J_{i+1} = I_{i+1} - \Gamma_i$ is a possibly empty subinterval of I_{i+1}.

If $J_{i+1} = \emptyset$ then $\Gamma_{i+1} = \Gamma_i$. If the decompositions of Γ_{i+1} and Γ_i are the same then $\Delta_{i+1} = \Delta_i$. If the decompostions are different than an endpoint of I_{i+1} which corresponds to a vertex V in the decomposition of Γ_{i+1} must be an interior point of an interval of the decomposition of Γ_i. Let k be the number of edges of Δ_i corresponding to this interval. The effect of declaring V to be a vertex in Γ_{i+1} is a Type (b) or Type (c) move if $k = 1$ or $k > 1$ respectively.

If J_{i+1} is a single point V, then V is a vertex of Γ_{i+1}. If I is the interval in the decomposition of Γ_i which limits on V, and I when viewed in Γ_{i+1} does not have an end limiting on a unique vertex then first perform the following operation. Thicken a leaf in the interior of I, remove the interior leaves, and the n declare the resulting complementary region to be a vertex. Depending on the number of edges of Δ_i corresponding to I this is a Type (b) or Type (c) move applied to Δ_i. Adding V to the possibly modified Γ_i is a Type (a) move.

The last case is J_{i+1} is a half open interval. Let V be the point of Γ_i that J_{i+1} limits on. If V is already a vertex of Δ_i then Δ_{i+1} is obtained from Δ_i by a Type (a) move. If V is not a vertex of Δ_i then Δ_{i+1} is obtained from Δ_i by first creating a new vertex (by thickening the leaf corresponding to V as in the previous paragraph) and then adding a new vertex and a new edge using a Type (a) move. □

The following result follows from the proof of Lemma 4.1 and the list summarizing the properties of $\Delta(\tilde{M})$.

Lemma 4.2. *Corresponding to the three cases of Lemma 4.1, $\Delta_{i+1}(\tilde{M})$ is obtained from $\Delta_i(\tilde{M})$ in the following ways:*

(a) *$\Delta_{i+1}(\tilde{M})$ is the union of $\Delta_i(\tilde{M})$, a new 1-handle $E(\tilde{M})$ that connects a complementary region of $\Delta_i(\tilde{M})$, and a new complementary region $V(\tilde{M})$.*

(b) *$\Delta_{i+1}(\tilde{M})$ is obtained by replacing a $D^2 \times I$ in the interior of an $E(\tilde{M})$ in $\Delta_i(\tilde{M})$ by a B^3 missing boundary manifold $V(\tilde{M})$. $E(\tilde{M}) \cap V(\tilde{M})$ is unknotted in $V(\tilde{M})$.*

(c) *$\Delta_{i+1}(\tilde{M})$ is created by replacing a collection of two or more 1-handles $E_j(\tilde{M})$ connecting a common complementary region $V(\tilde{M})$ to complementary regions $V_j(\tilde{M})$ in $\Delta_i(\tilde{M})$ by the union of a single 1-handle $E(\tilde{M})$ connecting $V(\tilde{M})$ to a new complementary region $V'(\tilde{M})$, and 1-handles $E_j'(\tilde{M})$ connecting $V'(\tilde{M})$ and $V_j(\tilde{M})$. The restriction of the $E_j(\tilde{M})$'s to each of $E(\tilde{M})$, $V(\tilde{M})$, and $E_j'(\tilde{M})$ is a union of parallel and unknotted $D^2 \times I$'s.*

Lemma 4.3. *$\Delta_i(\tilde{M})$ may be constructed so that $\cup_i \Delta_i(\tilde{M}) = \tilde{M}$.*

Proof. Let I_j be a locally finite collection of I-bundle fibres of $N(\tilde{\lambda})$ and let $N(I_j)$ be I-bundle regular neighborhoods of I_j such that $N(\tilde{\lambda}) = \cup_j N(I_j)$. Choose K satisfying the hypothesis of Lemma 4.1 such that K contains $\cup_j I_j$. Use $N(I_j)$ in the construction of the $\Delta_i(\tilde{M})$. Since $\tilde{M}$ is the union of $N(\tilde{\lambda})$ and the closed complementary regions of $\tilde{M}$ it follows that $\cup_j \Delta_j(\tilde{M}) = \tilde{M}$. □

Step 2 follows from the following more general result.

Theorem 4.4 (Product Theorem). *If $\lambda \subset M$ is a genuine lamination and γ is the core of a solid torus complementary region C (notation as in Definition 2.1) then $(\tilde{M}, \tilde{\lambda}, p^{-1}(\gamma))$ is homeomorphic to $(\mathbb{R}^2, \ell, X) \times \mathbb{R}$ where $\ell \subset \mathbb{R}^2$ is a lamination and $X \subset \mathbb{R}^2$ is a locally finite set of points.*

*Idea of proof.*If V_j (respectively E_j) is a vertex (respectively an edge) of Δ_j, then $V_j(\tilde{M})$ (respectively $E_j(\tilde{M})$) has a product structure. Also each $E_j(\tilde{M})$ standardly intersects each $V_{j+1}(M)$ and $E_{j+1}(M)$. Therefore a product structure on $\Delta_j(\tilde{M})$ extends to a product structure on $\Delta_{j+1}(\tilde{M})$ and the desired product structure on $\tilde{M}$ is obtained by induction.

Proof. To show that $(\tilde{M}, \tilde{\lambda})$ is a product we first examine some choices of product structure on various subsets of $\tilde{M}$. If V is a vertex or E is an edge of $\amalg_i \Delta_i$ denote by $V(\tilde{M})$ and $E(\tilde{M})$ the corresponding subsets of $\tilde{M}$. Each $V(\tilde{M})$ is a B^3 missing boundary manifold with $\partial V(\tilde{M}) \neq S^2$ and therefore there is a homeomorphism $h : (V(\tilde{M}), \partial V(\tilde{M})) \to (v, \partial v) \times \mathbb{R}$ for some B^2 missing boundary manifold v.

Before choosing h we must study the edges of $\amalg_i \Delta_i$. Let P be a component of $\partial V(\tilde{M})$. $h(P) = J \times \mathbb{R}$ where J is a component of ∂v. Identify P with $\mathbb{R} \times \mathbb{R}$ by choosing a homeomorphism between J and $\mathbb{R}$. Let $\{E^i_j\}$ be the subset of edges of $\amalg_i \Delta_i$ such that $E^i_j(\tilde{M}) \cap P \neq \emptyset$ and $E^i_j \in \Delta_i$. $E^i_j(\tilde{M}) \cap P$ must be a disk and will be denoted D^i_j.

By Lemma 4.1 each Δ_{n+1} has one more vertex that Δ_n. Label the vertices of $\amalg_i \Delta_i$ such the vertices of each Δ_n are $V_1, \ldots, V_n$. Let $\{E^n_1, \ldots, E^n_k\}$ be the intersection of $\{E^i_j\}$ and the edges of Δ_n and let $\{D^n_1, \ldots, D^n_k\}$ be the corresponding disks in P. Similarily let $\{E^{n+1}_1, \ldots, E^{n+1}_\ell\}$ be the intersection of $\{E^i_j\}$ and the edges of Δ_{n+1} and let $\{D^{n+1}_1, \ldots, D^{n+1}_\ell\}$ be the corresponding disks in P.

Lemma 4.1 describes the possible relationships between the sets $\{E^n_1, \ldots, E^n_k\}$ and $\{E^{n+1}_1, \ldots, E^{n+1}_\ell\}$. It follows that for each E^{n+1}_i, D^{n+1}_i contains a subset of $\{D^n_1, \ldots, D^n_k\}$ and is disjoint from the other elements of the set. In fact a bit more is true, D^{n+1}_i is either disjoint from $D^n_1 \cup \cdots \cup D^n_k$ or contains $D^n_1 \cup \cdots \cup D^n_k$. To see this, consider the subsets of the order tree T corresponding to the interiors of $E^n_1, \ldots, E^n_k$. By construction these subsets are either equal or disjoint. By assumption no leaf of $\tilde{\lambda}$, in particular P, is isolated and therefore the subsets of T are not disjoint. It follows that D^{n+1}_i can not contain a proper subset of $\{D^n_1, \ldots, D^n_k\}$.

Slightly enlarge each $E^{n+1}_i(\tilde{M})$ so that D^{n+1}_i contains a subset of $\{D^n_1, \ldots, D^n_k\}$ in its interior and is disjoint from the other elements of the set. By Lemma 4.3 $P = \cup_j \text{ interior}(D^n_j)$ so the properties of $\{D^n_j\}$ may be summarized as follows:

(1) If $(i, n) \neq (j, m)$, $\partial D^n_i \cap \partial D^n_j = \emptyset$
(2) $P = \cup^n_j \text{ interior}(D^n_j)$.
(3) $\{\partial D^n_j\}$ is a locally finite set of subsets of P.
(4) $D^n_i \subset D^m_j$ implies $n \leq m$

Fact. Given a collection $\{D^n_j\}$ satisfying 1, 2, 3, 4, there exists a homeomorphism $g : P \to \mathbb{R}^2$ such that for all D^n_j there exists $a_j, b_j \in \mathbb{R}$ with $g(D^n_j) = [a_j, b_j] \times [-n, n]$.

For each vertex V_i in Δ_i identify $V_i(\tilde{M})$ with $(v_i, \partial v_i) \times \mathbb{R}$ by a homeomorphism whose restriction to every boundary component is given by this fact. Such an identification is called a *normal product structure.*

Choose the normal product structure inductivily by considering edges of $\amalg_i \Delta_i$. Suppose that we have chosen normal product structure on $V_1(\tilde{M}), \dots, V_n(\tilde{M})$ and product structures for each edge E_i^n in Δ_n such that

(1) $(E_i^n(\tilde{M}), \tilde{\lambda} \cap E_i^n(\tilde{M})) \cong ([a_i, b_i] \times [-n, n] \times I_i, [a_i, b_i] \times [-n, n] \times C_i)$ where C_i is a closed subset of a closed interval I_i). Alternatively, let $e_i^n = [a_i, b_i] \times I_i$ and write $E_i^n(\tilde{M}) \cong \underset{\sim}{e}_i^n \times I_i$

(2) (Edge compatibility condition) If $E_i^n(\tilde{M})$ connects boundary components P' and P'' of $V'(\tilde{M})$ and $V''(\tilde{M})$ then the normal product structures on $V'(\tilde{M})$ and $V''(\tilde{M})$ map $E_i^n(\tilde{M}) \cap V'(\tilde{M})$ and $E_i^n(\tilde{M}) \cap V''(\tilde{M})$ to $[a_i, b_i] \times [-n, n]$.

Consider the possible ways that V_{n+1} is added to Δ_n in Lemma 4.1.

(a) Let E_{j+1}^{n+1} be the new edge of Δ_{n+1}. Let $E_{j+1}^{n+1}(\tilde{M})$ start at a boundary component P of $V(\tilde{M})$ for some vertex $V \in \Delta_n$. Say the normal product structure on P maps $P \cap E_{j+1}^{n+1}(\tilde{M})$ to $[a, b] \times [-(n+1), (n+1)]$. We guarantee edge compatibility by choosing a homeomorphism

$$(E_{j+1}^{n+1}(\tilde{M}), \tilde{\lambda} \cap E_{j+1}^{n+1}(\tilde{M})) \cong [a, b] \times [-(n+1), (n+1)] \times (I_{j+1}, C_{j+1})$$

and by choosing a normal product structure on $V_{n+1}(\tilde{M})$ that sends $E_{j+1}^{n+1}(\tilde{M}) \cap V_{n+1}(\tilde{M})$ to $[a, b] \times [-(n+1), (n+1)]$.

(b) If V_{n+1} subdivides an edge E_i^n of Δ_n, denote the new edges of Δ_{n+1} by E' and E''. $E'(\tilde{M})$ and $E''(\tilde{M})$ are thickenings of the two components of $\underset{\sim}{E}_i^n(\tilde{M})-$ interior$((V_{n+1}(\tilde{M}))$. Thus the product structure on $E'(\tilde{M})$ is obtained from $E_i^n \cong [a_i, b_i] \times [-n, n] \times I_i$ by restricting the third coordinate and enlarging the first two to $[a_i', b_i'] \times [-(n+1), (n+1)] \times I_i'$. $E''(\tilde{M})$ is treated similarily. Choose a normal product structure on $V_{n+1}(\tilde{M})$ such that $E'(\tilde{M}) \cap \partial V_{n+1}(\tilde{M})$ is mapped to $[a_i', b_i'] \times [-(n+1), (n+1)]$ and similarily for $E''(\tilde{M})$.

(c) Suppose edges $E_1^n, \dots, E_k^n$ each connecting V to V_{i_k} in Δ_n are replaced by an edge E^{n+1} from V to V_{n+1} and edges $E_1^{n+1}, \dots, E_k^{n+1}$ from V_{n+1} to V_{i_k}. Each $E_j^{n+1}(\tilde{M})$ inherits a product structure from $E_j^n(\tilde{M}) \cong [a_j^n, b_j^n] \times [-n, n] \times I_j^n$ by restricting the interval I_j^n and extending to $E_j^{n+1}(\tilde{M}) \cong [a_j^{n+1}, b_j^{n+1}] \times [-(n+1), (n+1)] \times I_j^{n+1}$. The normal product structure on $V(\tilde{M})$ sends $V(\tilde{M}) \cap E^{n+1}(\tilde{M})$ to some $[a, b] \times [-(n+1), (n+1)]$ where $[a, b]$ contains $\cup_j [a_j, b_j]$. For $j = 1, \dots, k$ let I_j' be the portion of I_j^n corresponding to the subset $E_j'(\tilde{M})$ of $E_j(\tilde{M})$ connecting $V(\tilde{M})$ to

$V_{n+1}(\tilde{M})$ Choose a product structure $E^{n+1}(\tilde{M}) \cong [a^{n+1}, b^{n+1}] \times [-(n+1), (n+1)] \times I^{n+1}$ that extends, after parametrizing the I'_j, the product structures $E'_j(\tilde{M}) \cong [a^n_j, b^n_j] \times [-n, n] \times I'_j$. Choose a normal product structure on $V_{n+1}(\tilde{M})$ that satisfies the edge compatibility on $V_{n+1}(\tilde{M}) \cap (E^{n+1}(\tilde{M}) \cup \cup_j E^{n+1}_j(\tilde{M}))$.

Let E^n be an edge of Δ_n that is not modified as in 1, 2, or 3, and let E^{n+1} be the corresponding edge of Δ_{n+1}. $E^{n+1}(\tilde{M})$ is obtained by thickening $E^n(\tilde{M})$ so the product structure of $E^n(\tilde{M})$ may be extended to $E^{n+1}(\tilde{M})$. The definition of normal product structure guarantees that the edge compatibility condition will be satisfied.

We have now chosen product structures $V \cong v \times \mathbb{R}$ and $E \cong e \times [-n, n]$ for vertices and edges of Δ_n. Let $\delta_n = \cup_{V \in \Delta_n} \text{interior}(v) \cup_{E \in \Delta_i} e$ with the natural identifications between the v's and e's. Let $P_n(\tilde{M}) \subset \Delta_n(\tilde{M})$ be the subset corresponding to $\delta_n \times [-n, n]$. $P_1(\tilde{M}) \subset P_2(\tilde{M}) \subset \cdots$ is an exhaustion of $\tilde{M}$ since $\Delta_1(\tilde{M}) \subset \Delta_2(\tilde{M}) \subset \cdots$ is. $\cup_n \delta_n$ is a monotone union of laminated B^2 missing boundary manifolds and therefore is a laminated B^2 missing boundary manifold. The product structure on $P_{n+1}(\tilde{M})$ restricts to the product on $P_n(\tilde{M})$, so it follows that there is a lamination ℓ of $\mathbb{R}^2$ such that $(\tilde{M}, \tilde{\lambda}) \cong (\mathbb{R}^2, \ell) \times \mathbb{R}$.

It remains only to show that this homeomorphism can be chosen to map each component $\tilde{\lambda}$ of $p^{-1}(\gamma)$ to a vertical line. By assumption, $\gamma \subset M - N(\lambda)$ and $\tilde{\gamma}$ is contained in a region $\tilde{C} \cong d_n \times \mathbb{R}$ with $n \geq 3$. Choose i such that $\tilde{\gamma} \subset V_i(\tilde{M})$. If $\tilde{\gamma} \cap \Delta_{i-1}(\tilde{M}) \neq \emptyset$ then $\tilde{\gamma}$ must intersect some $E(\tilde{M})$ where E is an edge Δ_{i-1}. $E(\tilde{M})$ is a neighborhood of I-bundle fibres and digons. $\tilde{\gamma} \subset \tilde{M} - N(\tilde{\lambda})$ and digons can not intersect the core of $\tilde{C} \cong d_n \times \mathbb{R}$ since $n \geq 3$. It follows that $\tilde{\gamma} \cap \Delta_{i-1}(\tilde{M}) = \emptyset$. Consequently the normal product structure on $V_i(\tilde{M})$ may be chosen to map $\tilde{\gamma}$ to a vertical line. This procedure should be performed on all components of $p^{-1}(\gamma)$ thereby completing the proof of Theorem 4.4. □

Corollary 4.5. *$\pi_1(\tilde{M} - p^{-1}(\gamma), *)$ is a free group.*

A simple closed curve $\gamma \subset M$ satisfying Corollary 4.5 is what Hass and Scott [10] call *free*.

Proof of Theorem 1.3. By Step 1 we see that $f : M \to M$ may be isotoped so that $f|\gamma = id_\gamma$. Since γ is a free link in M [10, Theorem 3.3] implies that f is isotopic to id_M. □

Theorem 4.4 may be viewed as a generalization of Palmeira's Theorem [11]. Palmeira shows that any foliation of $\mathbb{R}^3$ by planes is necessarily the product of a foliation of $\mathbb{R}^2$ with $\mathbb{R}$. The genuine lamination hypothesis of Theorem 4.4 is easily dispensed with [7]. Thus one obtains

Theorem 4.6. *If $\lambda \subset M$ is an essential lamination then $(\tilde{M}, \tilde{\lambda})$ is homeomorphic to $(\mathbb{R}^2, \ell) \times \mathbb{R}$ where $\ell \subset \mathbb{R}^2$ is a lamination.*

REFERENCES

1. M. Brittenham, *Essential laminations in Seifert fibred spaces*, Topology **32** (1993) 61-85.
2. A. Casson and D. Jungreis, *Seifert fibered spaces and convergence groups*, Preprint.
3. E. Claus, 1990 U. Texas, Austin thesis.
4. D. Gabai , *Convergence groups are Fuchsian groups*, Ann. Math. **136** (1992) 447-510.
5. D. Gabai and W. H. Kazez, *3-Manifolds with essential laminations are covered by solid tori*, J. London Math. Soc. **47** (1993) 557-576.
6. _____, *Finiteness of the mapping class group for genuine laminations*, in preparation.
7. _____, *Order trees and laminations of the plane*, in preparation.
8. D. Gabai and U. Oertel, *Essential laminations in 3-Manifolds* , Ann. Math. **(2) 130** (1989) 41–73.
9. D. Gabai and L. Mosher, in preparation.
10. J. Hass and P. Scott , *Homotopy and isotopy in 3-manifolds* , Comment. Math. Helvetici. **68** (1993) 341–364.
11. C. F. B. Palmeira, *Open manifolds foliated by planes*, Ann. Math. **107** (1978) 109-131.
12. P. Scott , *Homotopy implies isotopy for some Seifert fibre spaces* , Topology **24** (1985) 341-351.
13. F. Waldhausen, *On Irreducible 3-Manifolds which are Sufficiently Large*, Ann. Math. **87** (1968) 56–88.

253-37, CALIFORNIA INSTITUTE OF TECHNOLOGY, PASADENA, CA 91125
E-mail address: gabai@math.caltech.edu

UNIVERSITY OF GEORGIA, ATHENS, GA 30602
E-mail address: will@alpha.math.uga.edu

AMS/IP Studies in Advanced Mathematics
Volume **2**, 1997 (Part 1)

NON-ASSOCIATIVE TANGLES

DROR BAR-NATAN

ABSTRACT. Following Drinfel'd, Kontsevich and Piunikhin, we study the iterated integral expression for the holonomy of the formal Knizhnik-Zamolodchikov connection, finding that by introducing *non-associative tangles*, tangles whose strands are grouped in some particular way, the computation of these integrals can be reduced to the computation of just two holonomies R and Φ. A category **NAT** of non-associative tangles is thus introduced, and the 'kernel' of the natural forgetful functor **NAT** $\to$ {tangles} is analyzed and is shown to be generated by relations reminiscent of the relations defining a quasitriangular quasi-Hopf algebra. It follows that any solution of these relations can be used to produce tangle (and link) invariants. An iterative combinatorial/algebraic procedure for finding such solutions is described, and thus we have a first completely combinatorial construction of a universal Vassiliev invariant.

1. INTRODUCTION

1.1. Vassiliev invariants and the Kontsevich integrals. A Vassiliev invariant of type $m \in \mathbf{N}$ (see [27, 28, 6]) can be viewed [2] as a knot (or link) invariant whose $m+1$st derivative (in some reasonable sense) vanishes, and thus whose mth derivative is (in some sense) a constant. Upon further thought, this constant turns out to actually be a system of constants indexed by some kind of diagrams, subject to some relations (the "Birman-Lin relations"; the $4T$ and framing independence relations). Such a system of constants is called a *weight system*, and for some time after the introduction of these notions it was not clear whether or not these $4T$ and framing independence relations are the *only* relations satisfied by an mth order derivative of a type m invariant.

This work was supported by NSF grant DMS-92-03382.

Then in the winter of 1991/2 Kontsevich [14] found that if one introduces a certain formal generalization Ω of the Knizhnik-Zamolodchikov connection, writes a certain integral formula for the holonomy of Ω, generalizes a bit and makes a small correction, one finds an integral formula ("the Kontsevich integral") for a type m Vassiliev invariant in terms of a general weight system, thus resolving the above mentioned difficulty.

Somewhat more specifically, using the notation and definitions of Section 4 of [2], the Kontsevich integral formula[1] for a universal $\mathcal{A}^r$-valued Vassiliev invariant $\tilde{Z}(K)$ has the following form[2]:

$$(1) \quad Z(K) = \sum_{m=0}^{\infty} \frac{1}{(2\pi i)^m} \int_{t_1<\ldots<t_m} \sum_{\substack{\text{applicable}\\ \text{pairings}\\ P=\{(z_i,z_i')\}}} (-1)^{\#P_\downarrow} D_P \bigwedge_{i=1}^{m} \frac{dz_i - dz_i'}{z_i - z_i'} \in \mathcal{A}^r,$$

$$\tilde{Z}(K) = Z(K)/(Z(\infty))^{\frac{c}{2}}.$$

The standard mnemonic for (1) is the figure

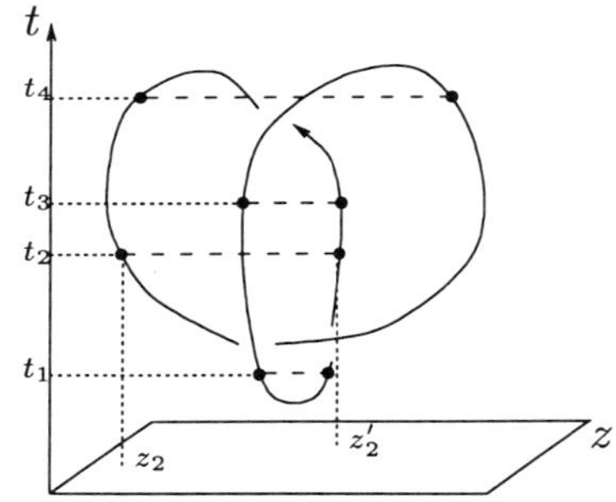

- Sum over all possible choices for the horizontal chords,
- pull the form $d\log(z - z')$ back to $\mathbf{R}_t$ using the ends of each chord,
- wedge together the resulting 1-forms,
- and integrate over $t_1, \ldots, t_m$,

[1]A detailed understanding of the Kontsevich formula is not necessary for understanding this paper. We advise readers unfamiliar with the formula to read the rest of the introduction without too much worry about the details of the formula; these will not be used later in the paper.

[2]Our normalization convention is different than in [2]; with c the number of critical points in a specific embedding of K, we replace $\frac{c}{2} - 1$ by $\frac{c}{2}$. See [2, problem 4.9].

- taking the chord diagram D_P defined by the horizontal chords and the knot as a coefficient.

1.2. A better knot projection. The problem with this lovely integral expression is that, as is, nobody seems to know how to compute it for any knot K. However, as it is known that (1) is independent of the particular embedding chosen for K, it might be that (1) becomes more manageable when the embedding of K is chosen wisely. For example, if ϵ is some very small number, a wiser choice of embedding for the knot displayed in the previous figure is:

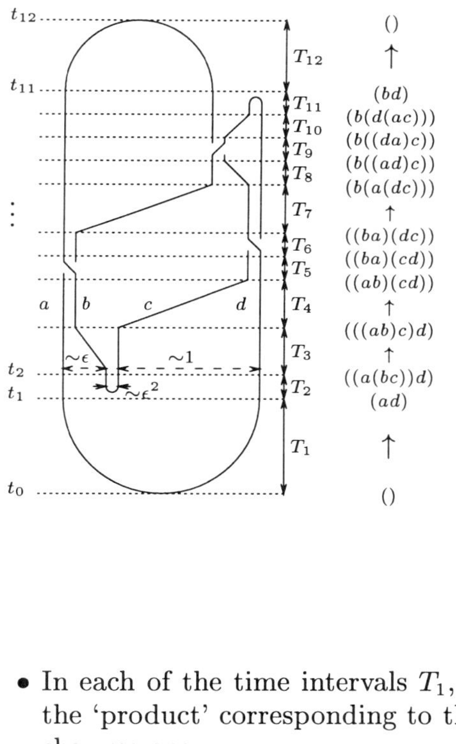

- In all marked time slices, $t_0, \ldots, t_{12}$, all distances between various strands of the knot are approximately equal to some power of ϵ. (At time t_1, say, the distance between the two strands a and d is $\sim 1 + \epsilon + \epsilon^2 \sim 1$).
- Furthermore, pretending that strands are elements in some non-commutative non-associative algebra, in each of the marked time slices the order and distance between the strands gives rise to a complete choice of how to multiply the strands. At time t_2, say, the corresponding 'product' is $((a(bc))d)$, as marked in the right most column of the figure.
- In each of the time intervals $T_1, \ldots, T_{12}$ only one change occurs to the 'product' corresponding to the strands, and only three types of changes occur:
  (i) *Pair creation (annihilation)*, in which a pair of *neighboring strands* is created (or annihilated). *Neighboring strands* are strands for which the distance between them is smaller than the distance between them and any other strand. (intervals T_1, T_2, T_{11}, and T_{12}).

(ii) *Braiding morphism*, in which two neighboring strands are braided. (intervals T_5, T_6, and T_9).
(iii) *Associativity morphism*, in which the associative law is applied once. (intervals T_3, T_4, T_7, T_8, and T_{10}).

1.3. Why is this a better choice? Because the Kontsevich-KZ integrals are multiplicative in a sense explained in [2], and thus it is sufficient to understand the Kontsevich-KZ integrals associated with any of the time intervals $T_1, \dots, T_{12}$. Each of these is easy (or at least *easier*), depending on the interval's type:

(i) *Pair creation (annihilation)*: In this case diagrams that have a chord connecting two strands other than the two newly created do not contribute because on parallel strands $(dz - dz')/(z - z')$ vanishes. The contribution from diagrams in which there is a chord connecting a newly created strand with an 'old' strand is negligible ($\sim O(\epsilon)$) because the domain of integration here would be very small relative to the size of the integrand, and diagrams that only have chords connecting the two strands in the newly created pair vanish because of the framing independence relation. So the only contribution from a pair creation (or annihilation) is the *skeleton* diagram, the diagram having no chords at all.
(ii) *Braiding morphism*: Here we may ignore all strands other than the two being braided, as they are all parallel and too far to interfere with the important two. And the contribution R from diagrams in which all chords connect the two strands being braided is easily computable — it is simply the holonomy of the 2 variable KZ connection, which is an Abelian connection.
(iii) *Associativity morphism*: Here something slightly more complicated happens. The basic holonomy to compute here is that of

(2)  .

Compute this holonomy once and for all and call the result Φ. (the computation might be hard — the connection here is not Abelian. But we need to do it only once[3]). More typically, we will meet an

[3]The magnanimous reader will be forgiving to the fact that this holonomy diverges as $\epsilon \to 0$. There is an easy way to fix it but it would take us too far aside to describe it. Drinfel'd's Φ, as described in [8], is precisely the result of this fix.

associativity morphism like

(3) 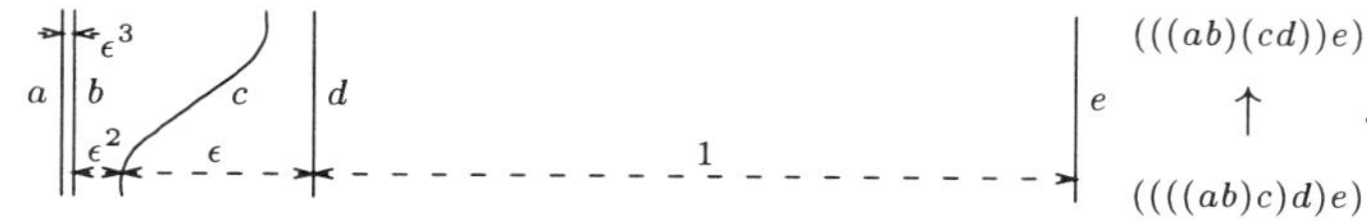

.

Luckily, the above holonomy is easily computable from Φ — clearly, as all other strands are parallel, the only non-vanishing contributions are from diagrams all of whose chords begin (or end) on the strand labeled c. Clearly, chords whose other end is on e give a negligible contribution, as e is too far to matter. Also, *in as much as c is concerned*, strands a and b are equivalent — c is always too far from them to tell them apart. Therefore, chords ending on a give (almost) the same contribution as chords ending on b, and (almost) the same contribution as in (2). Thus the integral corresponding to (3) is computable from Φ.

With the above considerations in mind, it seems natural to replace tangles by *non-associative tangles*, tangles that come equipped with further 'associativity' information, as in our 'wiser' representation of the trefoil knot. The rest of this paper is organized as follows: In Section 2 we realize this idea by introducing the category **NAT** of non-associative tangles and by listing the relations generating the 'kernel' of the obvious forgetful functor **For** : **NAT** $\to$ **PT**, where **PT** is a minor variation of the usual category of tangles. In Section 3 we use an arbitrary solution (R, Φ) to the relations generating the 'kernel' of **For** to construct a functor on **NAT** with values in some category of chord diagrams, and use that functor to construct a universal Vassiliev tangle invariant. The following two sections are devoted to the construction of such a solution, with Section 4 giving some necessary cohomological preliminaries and Section 5 describing the actual solution method. Section 6 contains some remarks, and in Section 7 we show how to use mathematica to compute R and Φ and how to compute the corresponding knot invariant using the methods of the previous sections.

The idea for writing this paper came from reading Piunikhin's [24], in which a universal Vassiliev link invariant is constructed in terms of an appropriate pair (R, Φ), constructed using the Knizhnik-Zamolodchikov equation. Our additions are that we make the construction somewhat clearer and more general (as it applies to tangles and not just to links), and that we give an explicit combinatorial construction of the necessary pair (R, Φ).

During the preparation of this paper and after the research for it was completed, I received other papers dealing with closely related subjects:

T. Q. T. Le and J. Murakami wrote four papers [15, 16, 17, 18] in which they cover essentially the same grounds as Sections 2, 3, and 4 of this paper, and then proceed to a different direction and discuss the rationality of the Kontsevich integral and its relation with the HOMFLY and Kauffman polynomials and with the theory of multiple ζ-numbers. S. Shnider and S. Sternberg [26] wrote a book with an extensive discussion of quasi-Hopf algebras, mostly as deformations of universal enveloping algebras of Lie algebras. P. Cartier [7] formalized the same ideas as in our Sections 2 and 3 in the language of monoidal categories. C. Kassel wrote an account (with many additional results) of Cartier's paper in a chapter of his forthcoming book [12]. N. Bergeron [5] wrote a short paper announcing our joint work on Section 4 here. S. Piunikhin [25] made some low order computations for some specific knots.

One comment has to be made, though. Most of the results in this paper, as well as in most other papers on this subject, are contained, either explicitly or implicitly, in the two seminal papers by Drinfel'd [8, 9].

1.4. Acknowledgement. I would like to thank J. Birman, R. Bott, S. Garoufalidis, D. Kazhdan, S. Piunikhin, B. Sanderson, J. D. Stasheff, S. Sternberg, D. P. Thurston, S. Willerton and H. L. Wolfgang for their many useful comments. Especially I would like to thank V. G. Drinfel'd for making a crucial remark in his short visit to Boston in September 1993, N. Bergeron for for jointly working with me on the results of Section 4, and A. Stoimenow for finding a gap in an earlier version of this paper. This paper is essentially an expanded version of talks I gave in in conferences in Georgia and in Warwick in the summer of 1993. I would like to thank the organizers of these conferences for inviting me to attend and giving me the possibility to speak.

2. The category of non-associative tangles

2.1. The category.

Definition 2.1. A *parenthesized string* is a string (of letters in some alphabet) together with a *balanced* and *complete* choice pairs of parentheses enclosing parts of it. The word *balanced* means that the number of 'open parenthesis' symbols "(" is always bigger than or equal to the number of 'close parenthesis' symbols ")" in any initial segment of the string, with equality if and only if the initial segment is the whole string. The word *complete* means that every pair of parentheses encloses either a pair of letters, or a letter and a substring enclosed in parentheses, or a pair of substrings enclosed in parentheses, and that the full string is enclosed in a pair of parentheses. As an exception, we will also consider the empty and single-character strings enclosed in a pair of parentheses as a parenthesized strings, and so the following is a complete list of all possible parenthesized strings of length at most 3 on an alphabet with a single letter $\uparrow$: $\{(), (\uparrow), (\uparrow\uparrow), ((\uparrow\uparrow)\ \uparrow), (\uparrow\ (\uparrow\uparrow))\}$.

Definition 2.2. If A and B are parenthesized strings and $\star$ is a letter appearing exactly once in A, there is an obvious *substitution operation* which we will denote by $A/\{\star \to B\}$. For example, $((ab)\star)/\{\star \to ((cd)(ef))\} = ((ab)((cd)(ef)))$. Similarly, $(ab)/\{a \to (cd), b \to (ef)\} = ((cd)(ef))$.

Definition 2.3. If A and B are parenthesized strings, the substitution $(ab)/\{a \to A, b \to B\}$ is called *the product* of A and B, and is simply denoted by (AB).

Definition 2.4. A *parenthesized $\updownarrow$-string* is a parenthesized string on the alphabet $\{\uparrow, \downarrow\}$. Often we will represent such a string by a collection of vertical arrows along a horizontal line, whose distances from each other reflect the parenthesization in the obvious way:

$$((\uparrow\downarrow)(\downarrow\uparrow)) \qquad \longleftrightarrow \qquad \uparrow\downarrow\quad\downarrow\uparrow$$

A *parenthesized $\updownarrow\star$-string* is a parenthesized string on the alphabet $\{\uparrow, \downarrow, \star\}$, in which the letter $\star$ appears exactly once.

Definition 2.5. (See also [1, 7, 18, 12]) Let **NAT** (**N**on-**A**ssociative **T**angles) be the category whose objects are parenthesized $\updownarrow$-strings, and whose morphisms are freely generated by the following morphisms: (for a precise definition of "the free category generated by a graph", consult e.g. Mac Lane [21])

(G1) *Associativity morphisms:* A morphism $S^{|\!/\!|}_{A,B,C}$ and a morphism $S^{|\backslash|}_{A,B,C}$ for each parenthesized $\updownarrow\star$-string S, and each triple (A, B, C) of parenthesized $\updownarrow$-strings. The domains of these morphisms are $S/\{\star \to ((AB)C)\}$ and $S/\{\star \to ((A(BC))\}$ and their targets are $S/\{\star \to ((A(BC))\}$ and $S/\{\star \to ((AB)C)\}$ respectively. Graphically, we represent these morphisms as follows:

$(\star)^{|\!/\!|}_{(\uparrow\downarrow),(\downarrow\uparrow),(\uparrow)} \quad \longleftrightarrow$  ;

$(\star)^{|\backslash|}_{(\uparrow\downarrow),(\uparrow\downarrow),(\uparrow)} \quad \longleftrightarrow$ 

(G2) *Braiding morphisms:* A morphism $S^{\diagup\!\!\!\diagdown}_{A,B}$ and a morphism $S^{\diagdown\!\!\!\diagup}_{A,B}$ for each parenthesized $\updownarrow\!\star$-string S, and each pair A, B of parenthesized $\updownarrow$-strings. The domain of these morphisms is $S/\{\star \to (AB)\}$, and the target is $S/\{\star \to (BA)\}$. Graphically, we represent these morphisms as follows:

$((\uparrow \star) \downarrow)^{\diagup\!\!\!\diagdown}_{(\downarrow),(\uparrow\uparrow)} \quad \longleftrightarrow$ 

$((\downarrow \star) \uparrow)^{\diagdown\!\!\!\diagup}_{(\downarrow),(\uparrow\uparrow)} \quad \longleftrightarrow$ 

(G3) *Pair creations (annihilations):* A morphism $S^{\cup\!\!\uparrow}$ for each parenthesized $\updownarrow\!\star$-string S, whose domain is $S/\{\star \to ()\}$ and whose target is $S/\{\star \to (\downarrow\uparrow)\}$. Graphically we will represent such a morphism by a diagram as follows:

$((\uparrow\downarrow)\star)^{\cup\!\!\uparrow} \quad \longleftrightarrow$ 

Similarly, we define morphisms $S^{\uparrow\!\!\cup}$, $S^{\uparrow\!\!\cap}$, and $S^{\cap\!\!\downarrow}$, and regard them as generators in **NAT**. $S^{\uparrow\!\!\cap}$, for example, is a morphism whose domain is $S/\{\star \to (\uparrow\downarrow)\}$ and whose target is $S/\{\star \to ()\}$.

Graphically, we will represent the composition $S_1 \cdot S_2$ of any two composable morphisms S_1 and S_2 by stacking the graphical representation of S_2 on top of that of S_1, as shown on the right.

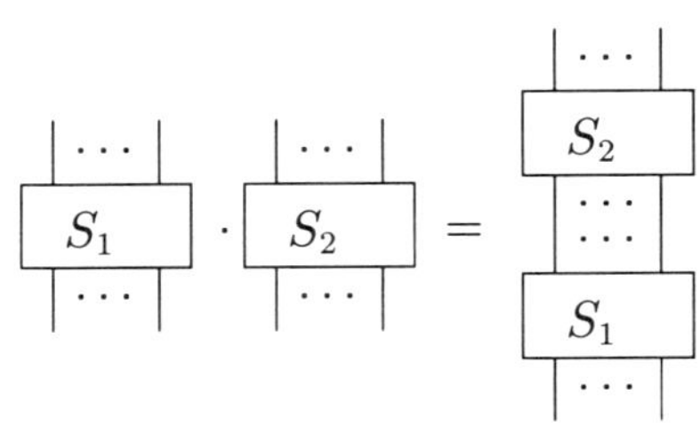

Definition 2.6. Let the categories **NAP** (**N**on-**A**ssociative **P**roducts) and **NAB** (**N**on-**A**ssociative **B**raids) have the same objects as **NAT**. The morphisms of **NAP** are freely generated by the associativity morphisms (G1). The morphisms of **NAB** are freely generated by the associativity morphisms (G1) and the braiding morphisms (G2).

There are a few operations that take an arbitrary morphism M in **NAB** (or **NAP**) into other morphisms in **NAB** (or **NAP**). Let us mention two of these:

(1) Any of the strands of M can be *doubled* — get replaced by a pair of neighboring parallel strands. If M has n strands, we will denote the operation of doubling the kth (counting at the bottom) strand by $1 \otimes \cdots \otimes \Delta \otimes \cdots \otimes 1$, where the Δ is in the kth slot:

$(1 \otimes \Delta)$ 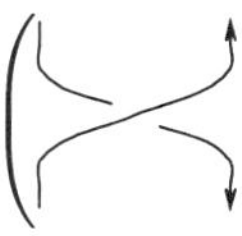 $=$ 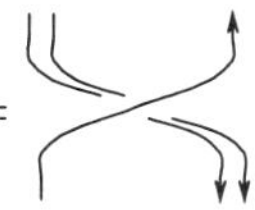 .

There are also some more general *cabling* operations in which more than one of the strands is affected, and the affected strands are replaced by an arbitrary number of ("parenthesized") strands rather than just doubled. For example, if M has two strands and A and B are parenthesized $\updownarrow$-strings, $(\Delta^A \otimes \Delta^B)(M)$ is obtained from M by replacing its first strand by a "bundle" of strands specified by A and replacing its second strand by the bundle specified by B:

$(\Delta^{((\uparrow\downarrow)\uparrow)} \otimes \Delta^{(\downarrow\downarrow)})$ 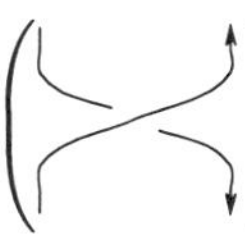 $=$ 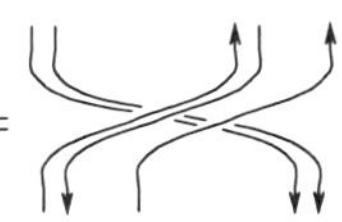 .

(2) Vertical strands can be added to the left and to the right of M. In general, if S is a parenthesized $\updownarrow\star$-string, $S/\{\star \to M\}$ ("M *extended* by S") will be the morphism in **NAB** obtained by replacing the symbol $\star$ in S by M:

$((\downarrow\uparrow)\star) \Big/ \Big\{\star \to$ 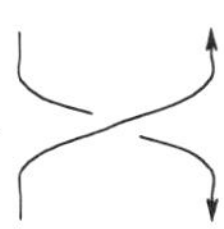 $\Big\} = \Big((\downarrow\uparrow)$ 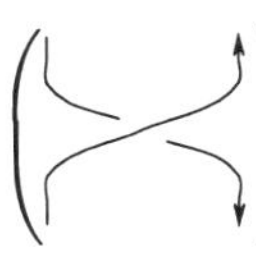 $\Big)$

$=$  .

Notice that the generating morphisms of **NAB** are all obtained from the four 'basic' morphisms |⤢|, |⤡|, ⤫ and ⤬ by the two operations above. For example,

$S^{⤫}_{A,B} = S \Big/ \Big\{\star \to (\Delta^A \otimes \Delta^B)$  $\Big\}$.

2.2. The forgetful functor NAT → PT. Recall that the category **T** of tangles (see e.g. [23]) is the category whose objects are (non-parenthesized) strings of ↑ and ↓ signs, and whose morphisms are *tangles* — disjoint unions of knotted oriented strands (regarded up to ambient isotopy) whose "ends" are the domain and target strings. For a typical example, see figure 1. Similarly, **FT** is the category of *framed tangles* — disjoint unions of knotted oriented strands regarded only up to regular isotopy [13]. The categories **PT** and **PFT** are defined in exactly the same way, only with parenthesized ↕-strings as their objects.

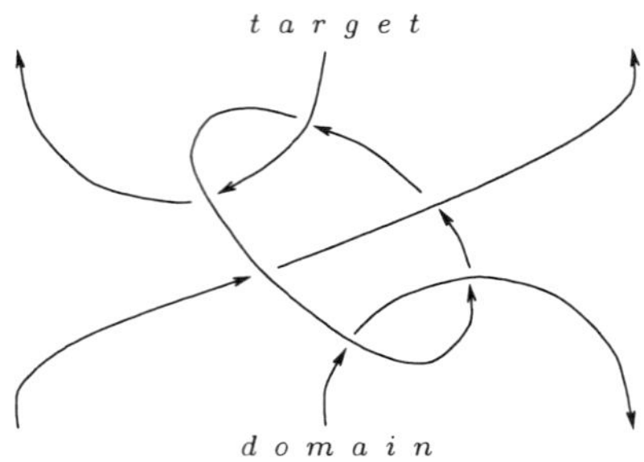

Figure 1. A tangle whose domain is (↑↑↓) and whose target is (↑↓↑).

There are obvious forgetful functors **For** : **NAT** → **PFT** and $\widetilde{\mathbf{For}}$: **NAT** → **PT** whose action on a parenthesized string is simply to do nothing, and whose action on morphisms of **NAT** is evident from their graphical representations in definition 2.5. We will shortly prove that the "kernel" of **For** is generated by the relations (R1)–(R10) below, and that the kernel of $\widetilde{\mathbf{For}}$ is generated by the same relations, with the addition of a single relation (R11).

- *Associativity relations:*

(R1) $S^{|/|}_{A,B,C}$ is equal to an identity morphism if any of A, B, or C is the empty string. Similarly for $S^{|\backslash|}_{A,B,C}$.

(R2) $S^{|\backslash|}_{A,B,C}$ is a two sided inverse of $S^{|/|}_{A,B,C}$:

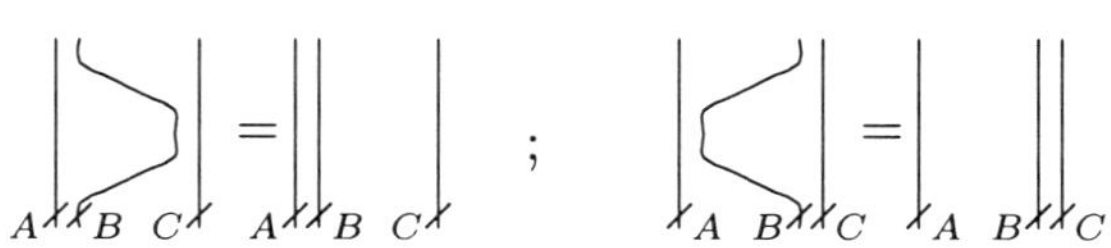

Notice that we've drawn only the "active" part of each relation; strands in the above figure are actually bundles of strands (determined by A, B, or C), and we don't bother to indicate the other "far away" strands in $S^{\uparrow\searrow}$ and $S^{\uparrow\nearrow}$.

(R3) *The pentagon:*(named after the pentagon relation of category theory)

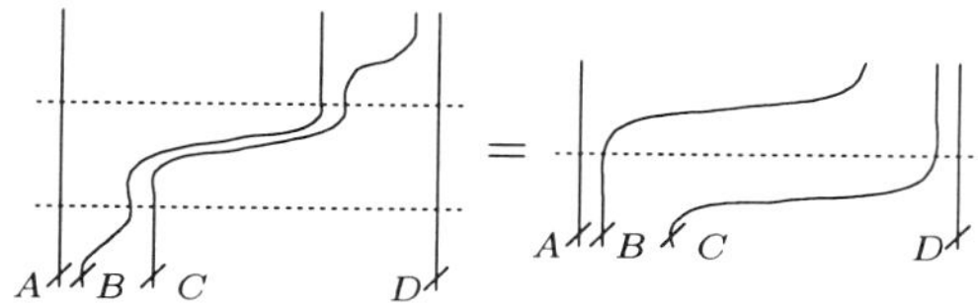

- *Locality relations:*

(R4) *Locality in space:* (for any pair S_A and S_B of morphisms)

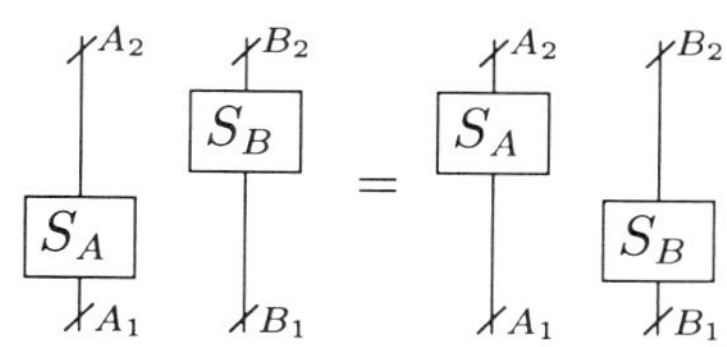

(R5) *Locality in scale:* for any morphisms S_A, S_B, and S_C,

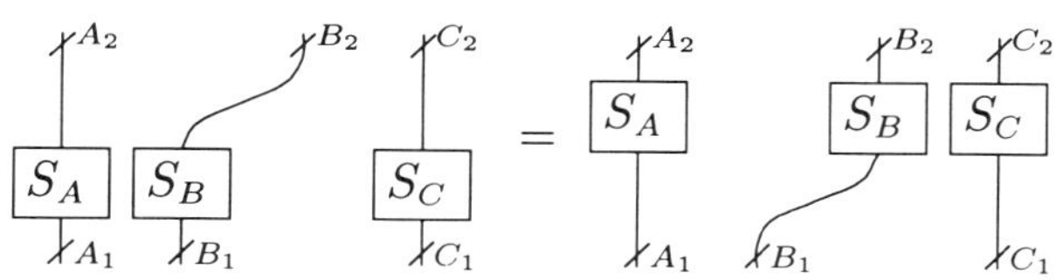

,

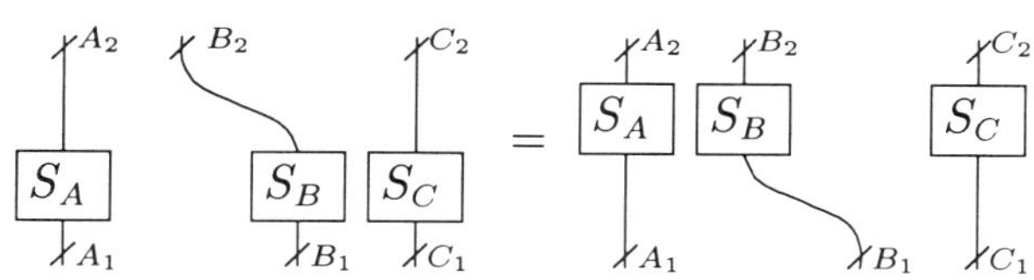

,

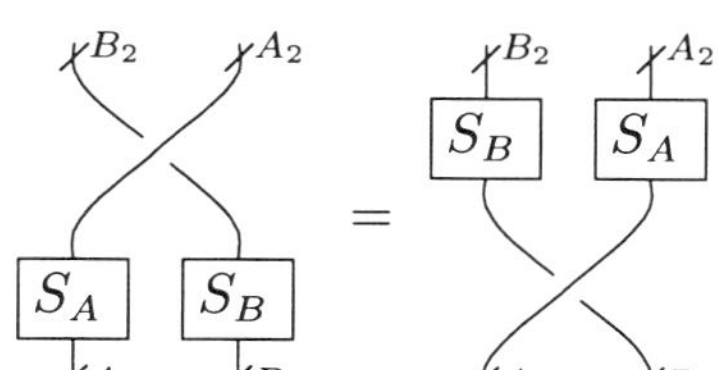

and

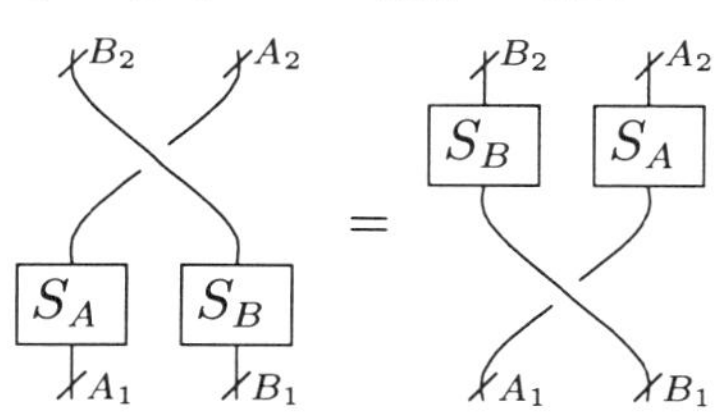

A locality relation (of either kind) will be called *elementary* if one of S_A and S_B (and S_C, when relevant) is one of the generators of **NAB**, while the other one (or two) are the identity morphisms.

- *Braiding relations:*

(R6) $S^{\diagup}_{A,B}$ is equal to an identity morphism if A or B is the empty string. Similarly for $S^{\diagdown}_{A,B}$.

(R7) $S^{\diagdown}_{B,A}$ is a two sided inverse of $S^{\diagup}_{A,B}$:

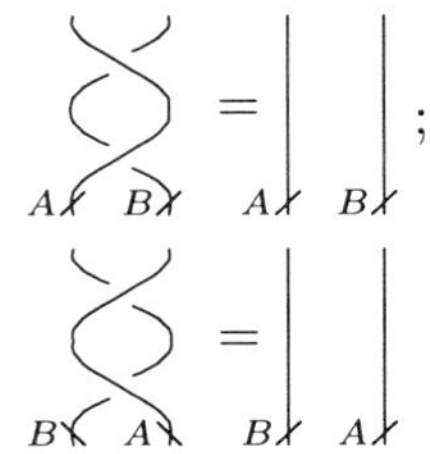

(R8) *The hexagons:* (named after the hexagon relations of category theory)

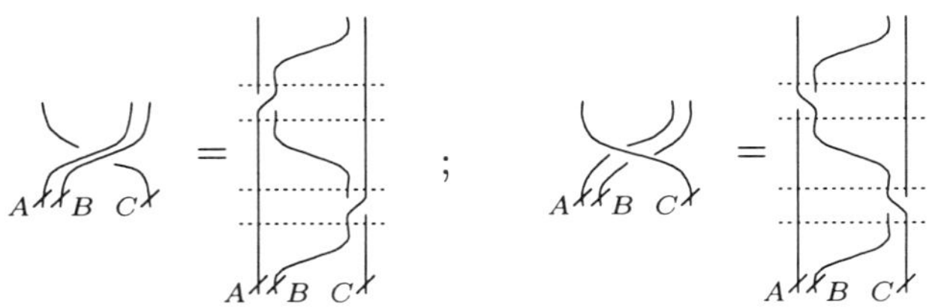

- *Creation/annihilation relations:*

(R9)

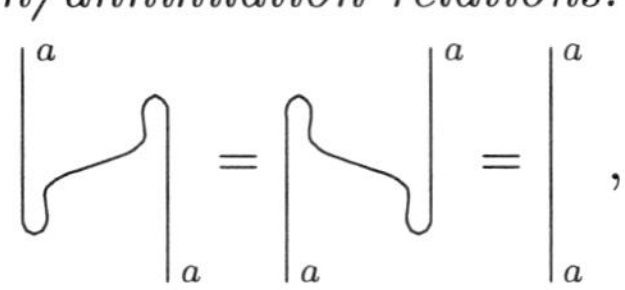

(R10)

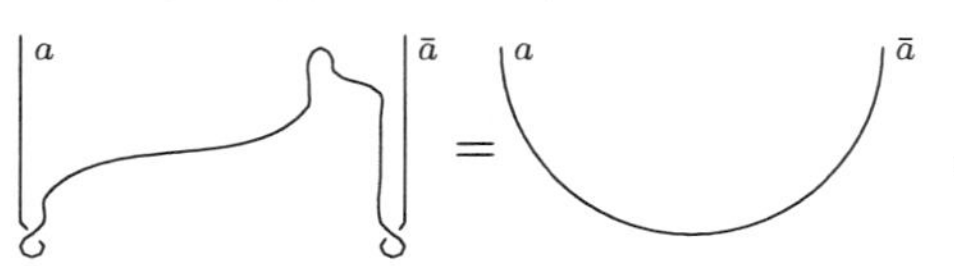

, where $a \in \{\uparrow, \downarrow\}$ is an arrow and $\bar{a}$ is its opposite.

(R11)

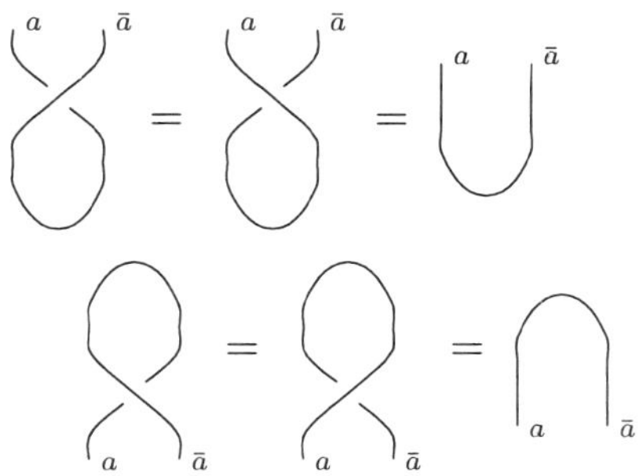

Theorem 1. (*See also* [1, 7, 18, 12]) *The kernel of* **For** *is indeed generated by relations* ($R1$) − −($R10$), *while the kernel of* $\widetilde{\mathbf{For}}$ *is generated by* ($R1$) − −($R11$).

Proof. The proof is basically an amalgamation of the Mac Lane coherence Theorem [21] and the standard facts about "Reidemeister" moves for tangles as in [23]. Roughly speaking, the Mac Lane coherence theorem allows us to ignore 'associativity information' in non-associative tangles and in the relations (R1)–(R11), and when associativity information is ignored in (R1)–(R11), what remains is the usual "Reidemeister" moves for tangles. For more details see e.g. [1]. Let us just comment that the third Reidemeister move is a simple consequence of our locality-in-scale axiom:

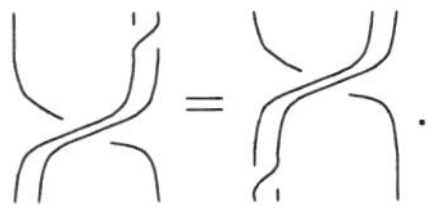

□

Let **PB** (**P**arenthesized **B**raids) be the image via **For** of **NAB**. An additional easy result is that the kernel of **For** : **NAB** → **PB** is generated by (R1)–(R8).

3. A universal Vassiliev invariant

Theorem 1 suggests the following method for constructing invariants of parenthesized framed tangles:

- Construct a functor $\mathbf{Z^{PFT}} : \mathbf{NAT} \to \mathbf{C}$, where **C** is some arbitrary 'target category'. This can be done by simply stating the values of $\mathbf{Z^{PFT}}$ on the generators of **NAT**.
- Check that $\mathbf{Z^{PFT}}$ respects relations (R1)–(R10).

- If it does, it descends to a functor (denoted by the same symbol) $\mathbf{Z}^{\mathbf{PFT}} : \mathbf{PFT} \to \mathbf{C}$, which is the required invariant of parenthesized framed tangles.

Checking in addition the relation (R11), we get a parenthesized tangle invariant $\mathbf{Z}^{\mathbf{PT}} : \mathbf{PT} \to \mathbf{C}$.

In this section, we will carry out this procedure in the case where $\mathbf{C} = \mathbf{AT}$ is a certain category of trivalent graphs, and the idea for the construction of $\mathbf{Z}^{\mathbf{PFT}}$ is borrowed from the considerations of section 1.3.

Recall that in [2, section 1] we've defined $\mathcal{D}^a$ to be the collection of all diagrams made of directed solid lines ("strands"), undirected dashed lines, trivalent vertices in which a dashed line ends on a strand, oriented trivalent vertices in which three dashed lines end, and univalent vertices in which a line (dashed or solid) begins or ends. After picking some characteristic 0 ground field $\mathbf{F}$ we've then set

$$\mathcal{A}^a = \operatorname{span}(\mathcal{D}^a) \,/\, \{\text{anti-symmetric vertices, } STU \text{ and } IHX \text{ relations}\}\ ,$$

where the anti-symmetry, STU, and IHX relations were also discussed in [2]. Notice that these three relations do not involve univalent vertices at all, and so $\mathcal{A}^a$ is the direct sum of its "end-homogeneous" components $\mathcal{A}^a_{ij} = \operatorname{span}\left(\mathcal{D}^a_{ij}\right)/\{\text{same relations}\}$, where $\mathcal{D}^a_{ij}$ is the subset of $\mathcal{D}^a$ containing all diagrams with i univalent vertices in which a strand (solid line) begins (and thus i univalent vertices in which a strand ends) and j univalent vertices in which a dashed line begins (or ends). Notice also that when $j = 0$, the number of trivalent vertices in a diagram has to be even, and so $\mathcal{A}^a_{i0}$ can be graded by half the number of trivalent vertices in a diagram. Let $\bar{\mathcal{A}}^a_{i0}$ be the graded completion of $\mathcal{A}^a_{i0}$. Finally, let the *skeleton* of a diagram $D \in \mathcal{D}^a_{*0}$ be the diagram obtained from D by removing all the dashed lines in it. It is always a diagram of degree 0. We will say that a diagram is a skeleton if it is equal to its own skeleton.

Definition 3.1. Let $\mathbf{AT}$ be the graded category[4] whose objects are (non parenthesized) strings of $\uparrow$ and $\downarrow$ arrows (same as in $\mathbf{T}$). The morphisms of $\mathbf{AT}$ are pairs (S, L) where S is a skeleton diagram whose univalent vertices are in a bijective direction-preserving correspondence with the $\uparrow$ and $\downarrow$ arrows in the domain and target objects, and L is an element of $\bar{\mathcal{A}}^a_{*0}$ that can be presented as a series of diagrams whose skeleton is S. For a typical example, see figure 2.

[4]A *graded category* is a category for which $\operatorname{mor}(A,B)$ is a graded Abelian group for every objects A and B, and for which the composition maps $\operatorname{mor}(A,B) \times \operatorname{mor}(B,C) \to \operatorname{mor}(A,C)$ are graded bilinear.

Let **AB** be the category whose objects are the same objects as those of **AT** and whose morphisms are those morphisms of **AT** in which all the strands connect an arrow (↑ or ↓) in the domain object to an arrow in the target (pointing to the same direction).

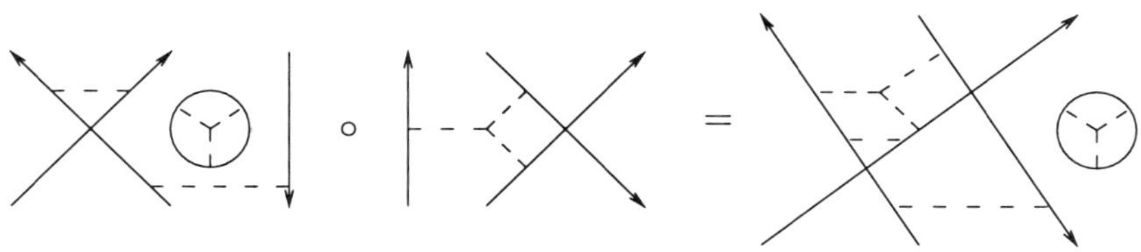

Figure 2. The composition of a degree 4 morphism in $\text{mor}(\uparrow\uparrow\downarrow, \uparrow\uparrow\downarrow)$ with a degree 2 morphism in $\text{mor}(\uparrow\uparrow\downarrow, \uparrow\downarrow\uparrow)$ is a degree 6 morphism in $\text{mor}(\uparrow\uparrow\downarrow, \uparrow\downarrow\uparrow)$. The general composition law of **AT** is obtained from the one indicated here by bilinear extension. Notice that we compose from the bottom to the top, and that in these diagrams apparent quadrivalent vertices are to be ignored, and trivalent vertices are always oriented counterclockwise.

On **AB** it is possible to define operations similar to those defined on **NAB**. Let D be some diagram representing a morphism of **AB**.

(1) Any of the strands of D can be doubled. If l is the kth such line and it is connected to n dashed lines, $(1 \otimes \cdots \otimes \overset{k}{\Delta} \otimes \cdots \otimes 1)(D)$ is the sum of diagrams obtained by doubling the kth strand in the skeleton of D and summing over the 2^n possible ways of 'lifting' D to the new skeleton thus obtained. (Compare with the definition of the Adams operations ψ^q in [2]). In general, it is clear how to define the cabling operation $\Delta^A \otimes \Delta^B \otimes \cdots$ on skeletons. If D_0 is the skeleton of D, let $(\Delta^A \otimes \Delta^B \otimes \cdots)(D)$ be the sum of all possible liftings of D to $(\Delta^A \otimes \Delta^B \cdots)(D_0)$, multiplying each lifting by -1 raised to the number of vertices that change their orientation when lifted:

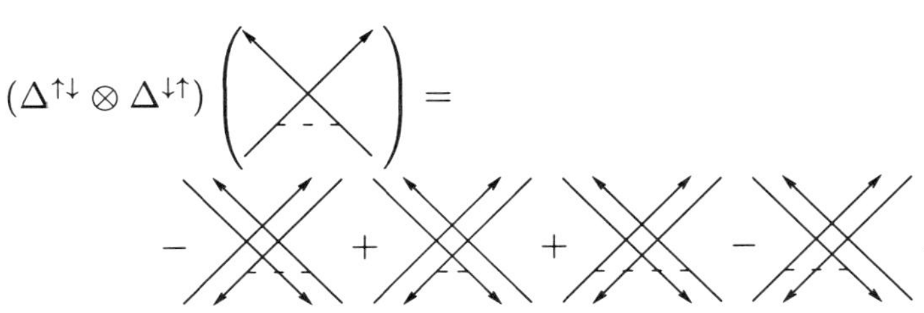

We will use the notation 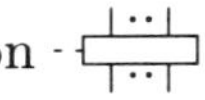 as shorthand for summations as in the above equation. With this notation, we have:

$$(\Delta^A \otimes \Delta^B)$$ 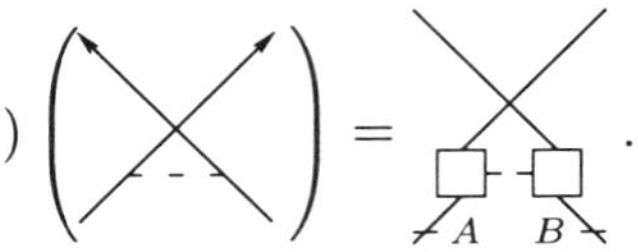

(2) There is no problem with defining the extension operation $S/\{\star \to D\}$ in the same way as in the case of **NAB**.

(3) 'Strand removal' operations are defined as follows: If there are no dashed lines ending on the ith strand of D, $s_i(D)$ is just D with its ith strand removed. Otherwise $s_i(D) = 0$.

Exercise 3.2. Let **AP** be the subcategory of **AB** in which all strands are vertical. Show that the above three operations restrict to **AP**, and their restrictions are self-functors **AP** → **AP**. Formulate and prove the corresponding multiplicativity property of these operations regarded as self-maps (not functors!) **AB** → **AB**.

The following definition will be used in the next section:

Definition 3.3. A morphism D in **AP** is called a *perturbation of the identity* if its degree 0 piece is the identity 1. We say that D is *non-degenerate* if $D_{>0}$ is in the kernels of all the s_i's, where $D_{>0}$ is D with its degree 0 piece removed. That is, D is non-degenerate if every strand in $D_{>0}$ touches some dashed line in D.

3.1. An ansatz for $\mathbf{Z}^{\mathbf{PB}}$. Let us now construct a functor $\mathbf{Z}^{\mathbf{PB}}$: **NAB** → **AB**. Pick an invertible non-degenerate automorphism R of the object $(\uparrow\uparrow)$ of **AP** and an invertible non-degenerate automorphism Φ of the object $(\uparrow\uparrow\uparrow)$, and assume that both R and Φ are perturbations of the identity:

$R = \uparrow\uparrow + \sum(\text{coeff})$ 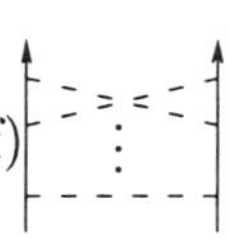 , $\Phi = \uparrow\uparrow\uparrow + \sum(\text{coeff})$ 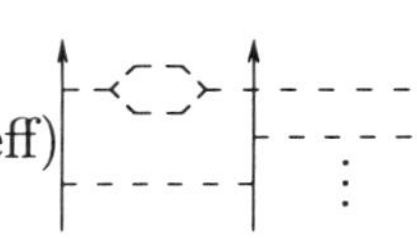 .

On objects $\mathbf{Z}^{\mathbf{PB}}$ acts by forgetting all pairs of parenthesis. On morphisms define

$\mathbf{Z}^{\mathbf{PB}}($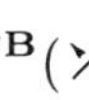$) = R \cdot$  $\stackrel{\text{def}}{=} \tilde{R}$, $\mathbf{Z}^{\mathbf{PB}}($$) =$ $\cdot R^{-1} = \tilde{R}^{-1}$,

$\mathbf{Z}^{\mathbf{PB}}($$) = \Phi$, $\mathbf{Z}^{\mathbf{PB}}($$) = \Phi^{-1}$, and extend this definition in the only way compatible with the cabling and extension operations of **NAB** and **AB**. For example, set

$\mathbf{Z}^{\mathbf{PB}}$ 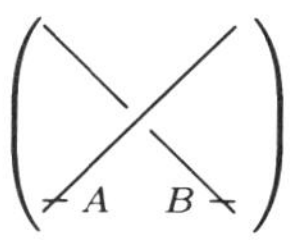

$= (\Delta^A \otimes \Delta^B)(\tilde{R}).$

We wish now to check whether $\mathbf{Z}^{\mathbf{PB}}$ respects the relations (R1)–(R8) generating the kernel of the forgetful functor $\mathbf{NAB} \to \mathbf{PB}$. From the definition of $\mathbf{Z}^{\mathbf{PB}}$ and from exercise 3.2 it readily follows that $\mathbf{Z}^{\mathbf{PB}}$ respects (R2) and (R7) (triviality of 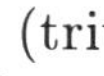 and of ). (R1) and (R6) follow from the non-degeneracy of R and Φ and the fact that they are perturbations of the identity. Locality in space (R4) is obvious, and locality in scale (R5) easily follows from the following lemma:

Lemma 3.4. *When comparable, cablings and extensions commute. In figures, if D is a diagram representing an automorphism of some string A in $\boldsymbol{AB}$, then*

(4) 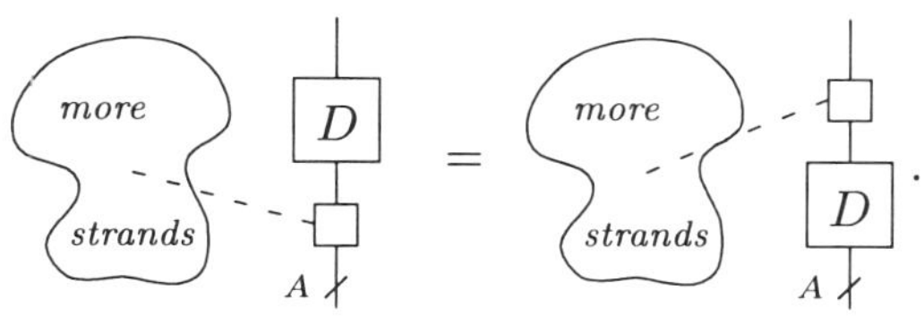

Proof. The proof is essentially the same as the proof of lemma 3.1 in [2]. Namely, add three little hooks near each trivalent vertex of the 'D' part of the diagram

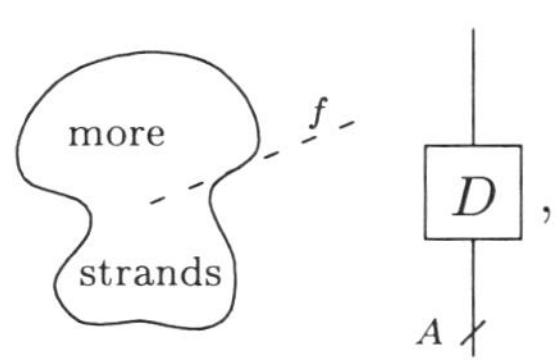

and consider the (properly signed) sum of all the possible ways of connecting f to these hooks. Grouping this sum by vertices we get zero by the IHX and STU relations, and regrouping by arcs we get the difference between the left-hand-side and right-hand-side of (4). □

We see that the only conditions imposed on $\tilde{R}$ and Φ by the relations (R1)–(R8) are the pentagon (R3) and the hexagons (R8). In our notation, these relations read:

$$(\Phi \uparrow) \cdot (1 \otimes \Delta \otimes 1)(\Phi) \cdot (\uparrow \Phi) = (\Delta \otimes 1 \otimes 1)(\Phi) \cdot (1 \otimes 1 \otimes \Delta)(\Phi),$$

$$(\Delta \otimes 1)(\tilde{R}) = \Phi \cdot (\uparrow \tilde{R}) \cdot \Phi^{-1} \cdot (\tilde{R} \uparrow) \cdot \Phi,$$

$$(\Delta\otimes 1)(\tilde{R}^{-1}) = \Phi\cdot(\uparrow \tilde{R}^{-1})\cdot\Phi^{-1}\cdot(\tilde{R}^{-1}\uparrow)\cdot\Phi.$$

Let us rewrite these three relations in terms of R, rather than $\tilde{R}$. First, we need a definition:

Definition 3.5. For a natural number n, let $\mathbf{AP}_n$ be the algebra of all automorphisms (in $\mathbf{AP}$) of the object $\underbrace{\uparrow\uparrow\cdots\uparrow}_{n}$. The unit of $\mathbf{AP}_n$ is the morphism $\uparrow\uparrow\cdots\uparrow$ (no dashed lines), and we will simply denote it by 1. On $\mathbf{AP}_n$ there is a natural action of the symmetric group on n letters $\mathcal{S}_n$, acting by "permuting the strands":

$\Psi =$ 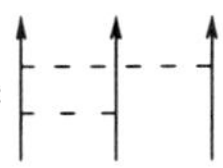, $\Psi^{231} =$ 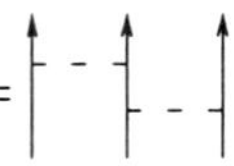.

Furthermore, for each n there are $n!$ injections of $\mathbf{AP}_{n-1}$ into $\mathbf{AP}_n$, each determined by an ordered subset of size $n-1$ of the set $\{1,\dots,n\}$:

$\Phi =$ 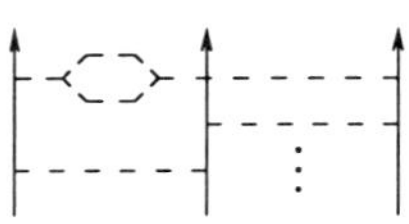, $\Phi^{243} =$ 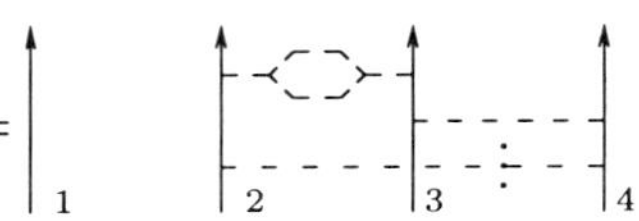.

(The minor ambiguity in the above notation can be resolved by identifying $\mathbf{AP}_{n-1}$ with its image in $\mathbf{AP}_n$ by the map that adds a "free standing" strand on the right, and will never cause any problem.)

The pentagon and the hexagons can now be written as equations in $\mathbf{AP}_4$ and $\mathbf{AP}_3$ respectively; setting $R^+ = R$ and $R^- = (R^{21})^{-1}$ they read:

(⬠) $\Phi^{123}\cdot(1\otimes\Delta\otimes 1)(\Phi)\cdot\Phi^{234} = (\Delta\otimes 1\otimes 1)(\Phi)\cdot(1\otimes 1\otimes\Delta)(\Phi)$,

(⬡$_\pm$) $\quad (\Delta\otimes 1)(R^\pm) = \Phi^{123}\cdot(R^\pm)^{23}\cdot(\Phi^{-1})^{132}\cdot(R^\pm)^{13}\cdot\Phi^{312}.$

Notice that (apart from our somewhat different conventions) these are precisely the pentagon and hexagons of Drinfel'd's [8, 9]. In section 5 we will present the details of a perturbative approach, hinted at in [9], to solving these equations.

3.2. The norm of Φ. The following proposition, also due to Drinfel'd [8, 9], shows that under certain circumstances it is enough to consider just one of the two hexagon equations $\hexagon_{\pm}$:

Definition 3.6. For $\Phi \in \mathbf{AP}_n$ define the *transpose* of Φ to be $\Phi^T = \Phi^{n\dots 21}$. For $\Phi \in \mathbf{AP}_3$ define the *norm squared* of Φ to be $||\Phi||^2 = \Phi \cdot \Phi^T = \Phi \cdot \Phi^{321} \in \mathbf{AP}_3$. We say that Φ is *normalized* if it is of unit norm; i.e., if $||\Phi||^2 = 1$.

Proposition 3.7. *If $R \in \boldsymbol{AP}_2$ and $\Phi \in \boldsymbol{AP}_3$ are perturbations of the identity and R is symmetric (namely, $R = R^{21}$), then any two of the equations $\hexagon_+$, $\hexagon_-$, and $||\Phi||^2 = 1$ implies the third.*

Proof. Assume $\hexagon_-$ and that Φ is normalized, i.e. that $\Phi^{-1} = \Phi^{321}$. Using the latter equality, $\hexagon_-$ can be rewritten as

$$(\Delta \otimes 1)(R^{-1}) = (\Phi^{-1})^{321} \cdot (R^{-1})^{23} \cdot \Phi^{231} \cdot (R^{-1})^{13} \cdot (\Phi^{-1})^{213}.$$

Inverting both sides of this equation and applying the permutation 213 to the result, we get $\hexagon_+$. A similar computation shows that $\hexagon_+$ and $\Phi^{-1} = \Phi^{321}$ imply $\hexagon_-$.

Now assume $\hexagon_{\pm}$. Defining $\mathbf{Z^{PB}}$ as above and noticing that no pentagons can be formed with just a 3-strand non-associative braid, we see that $\mathbf{Z^{PB}}$ descends to a well defined invariant of parenthesized 3-strand braids. In particular, consider the following two braid equalities:

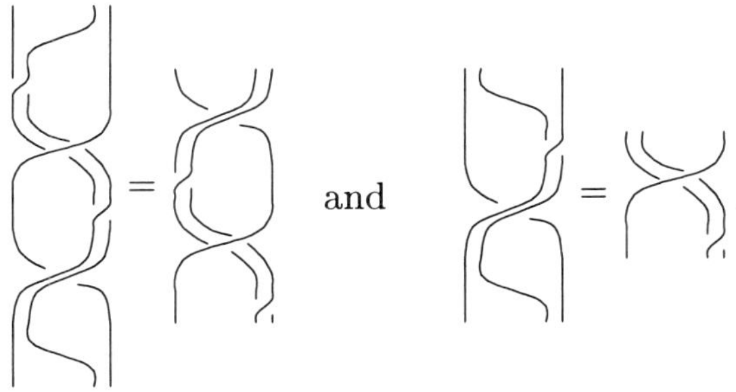

Applying $\mathbf{Z^{PB}}$, we get the equalities (see also [8, proposition 3.5])

$$\Phi^{-1} \cdot \left((\Delta \otimes 1)(R) \cdot R^{12}\right)^2 \cdot \Phi = \left(R^{23} \cdot (1 \otimes \Delta)(R)\right)^2,$$

$$\Phi^{-1} \cdot (\Delta \otimes 1)(R) \cdot R^{12} \cdot (\Phi^{-1})^{321} = R^{23} \cdot (1 \otimes \Delta)(R).$$

Taking the square root of the first equality (possible using power series because both sides are perturbations of the identity) and comparing with the second, we get $\Phi = (\Phi^{-1})^{321}$, as required. $\square$

3.3. An ansatz for $\mathbf{Z}^{\mathbf{PFT}}$. Assume that a solution (R, Φ) to the pentagon and hexagons is given, that $R = R^{21}$ is a non-degenerate perturbation of the identity, and that Φ is a normalized non-degenerate perturbation of the identity. Define an element $Z(\infty) \in \mathbf{AP}_1$ by reversing the middle strand of Φ, regarding the result $(1 \otimes \Delta^{\downarrow} \otimes 1)\Phi$ as a morphism in $\mathbf{AT}$, pre-composing it with the degree 0 morphism , composing the result with the degree 0 morphism 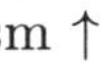, and finally regarding the result as an element of $\mathbf{AP}_1$, as in Figure 3.

$Z(\infty) =$ 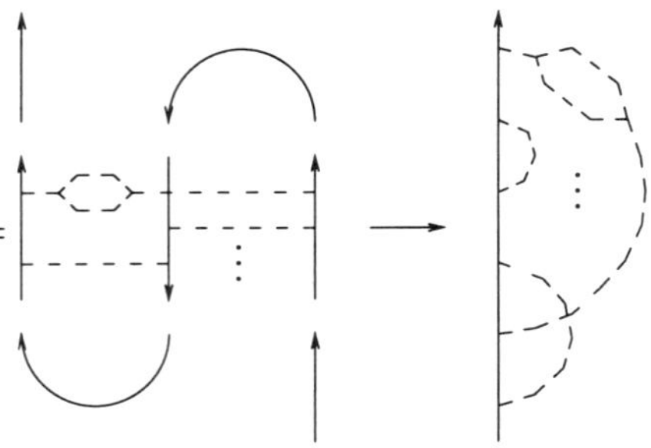

Figure 3. The definition of $Z(\infty)$.

Exercise 3.8. $Z(\infty)$ is the linear combination of chord diagrams corresponding to the non-associative tangle  by some obvious extension $\mathbf{Z}^{\text{naive}}$ of $\mathbf{Z}^{\mathbf{PB}}$. Show that $Z(\infty)$ is also the image of 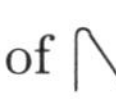, , and  by $\mathbf{Z}^{\text{naive}}$.

Hint 3.9. $\mathbf{Z}^{\text{naive}}$ 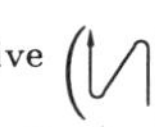 $= \mathbf{Z}^{\text{naive}}$ 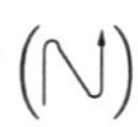 follows from $\Phi^{-1} = \Phi^T$, while $\mathbf{Z}^{\text{naive}}$ 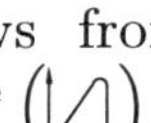 $= \mathbf{Z}^{\text{naive}}$ 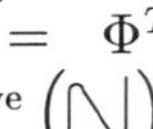 follows by 'closing' the pentagon as on the right, and then using the fact that Φ is a non-degenerate perturbation of the identity.

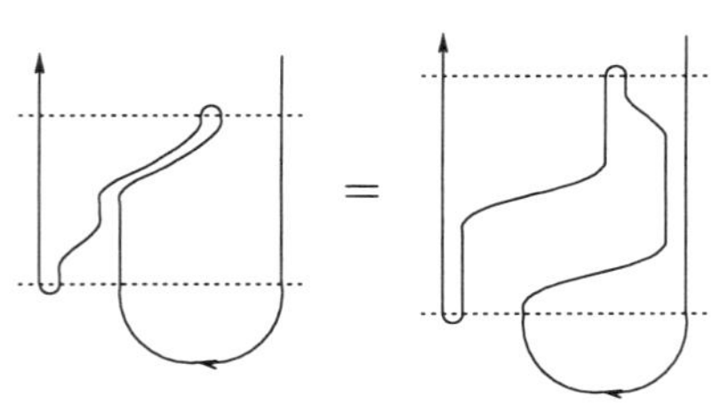

Notice that $\mathbf{AP}_1$ is a commutative algebra; indeed it is simply the algebra $\mathcal{A}$ of [2]. Furthermore, $Z(\infty)$ is also a perturbation of the identity and thus its square root can be computed using a power series expansion. Now extend $\mathbf{Z}^{\mathbf{PB}}$ to a functor $\mathbf{Z}^{\mathbf{PFT}} : \mathbf{NAT} \to \mathbf{AT}$ by defining[5]

$$\mathbf{Z}^{\mathbf{PFT}}(\cup^{\uparrow}) = \mathbf{Z}^{\mathbf{PFT}}(\cap_{\downarrow}) = \mathbf{Z}^{\mathbf{PFT}}({}^{\uparrow}\!\cup) = \mathbf{Z}^{\mathbf{PFT}}({}_{\downarrow}\!\cap) = Z(\infty)^{-1/2},$$

and extending $\mathbf{Z}^{\mathbf{PFT}}$ to more general creation/annihilation morphisms in the natural way.

Theorem 2. (*See also* [7, 18, 12]) $\mathbf{Z}^{\mathbf{PFT}}$ *respects* $(R1) - -(R10)$, *and thus it is descends to an* $\boldsymbol{AT}$*-valued parenthesized framed tangle invariant.*

Proof. (R1), (R2) and (R4)–(R7) hold because they held for $\mathbf{Z}^{\mathbf{PB}}$ (there is no problem with generalizing lemma 3.4 to the current situation). (R3) and (R8) hold by our choice of R and Φ, and the definition of $\mathbf{Z}^{\mathbf{PFT}}$ on creation/annihilation morphisms was clearly cooked up so that (R9) would hold (I wish to thank D. P. Thurston for encouraging me to include exercise 3.8). The remaining relation, (R10), follows by sliding the two factors $\tilde{R}$ and $\tilde{R}^{-1}$ on the left hand side of (R10) until they are adjacent to each other (this is possible using the commutativity of $\mathcal{A}$ proven in [2] and essentially re-proven here as lemma 3.4). When they are adjacent, they can be cancelled using $R = R^{21}$, and the remaining factors can be dealt with using the previous relations. □

3.4. Invariants of unframed tangles. There is a standard way to "renormalize" a regular isotopy invariant Z satisfying $Z\left(\text{ᛕ}\right) = cZ(\uparrow)$ for some scalar c, to become a complete isotopy invariant satisfying $Z\left(\text{ᛕ}\right) = Z(\uparrow)$. Simply replace Z by $Zc^{-writhe}$. We can apply a variant of the same procedure to our $\mathbf{Z}^{\mathbf{PFT}}$, getting an invariant $\mathbf{Z}^{\mathbf{PT}}$ of unframed parenthesized tangles:

Definition 3.10. Let C be the result of composing $\cup^{\uparrow}$ with $(\Delta^{\downarrow} \otimes 1)(R^{-1})$, and regarding the result as an element of $\mathcal{A}$:

(7) 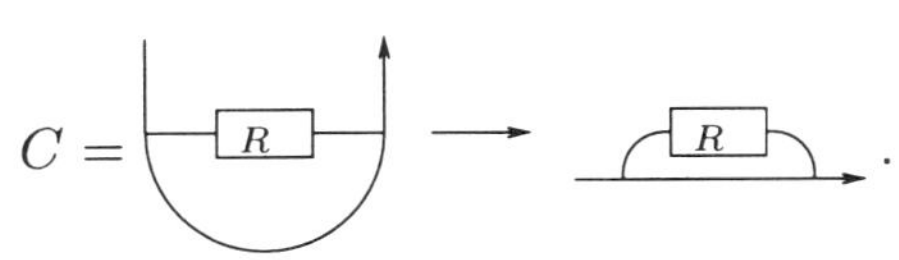

Definition 3.11. The *writhe* of a component of a tangle is half the number of positive crossings it goes through, minus half the number of negative crossings it goes through (counting self-crossings of the same component twice).

[5]In this equality $Z(\infty)^{-1/2}$ is interpreted as a member of five different spaces!

Finally, to compute $\mathbf{Z^{PT}}(M)$ for a given $M \in \mathbf{NAB}$, attach $C^{-writhe}$ to each of the components (i.e., strands) in $\mathbf{Z^{PFT}}(M)$. The usual commutativity property of $\mathcal{A}$ shows that it doesn't matter where the attachement is made. Remembering that the writhes are regular isotopy invariants, the commutativity property also shows that $M \mapsto \mathbf{Z^{PT}}(M)$ is a functor. Clearly, relations (R1)–(R10) are not affected, and meanwhile, we've fixed (R11) to hold for $\mathbf{Z^{PT}}$.

Theorem 3. (*See also* [7, 18, 12]) $\mathbf{Z^{PT}}$ *respects* $(R1) - -(R11)$, *and thus it descends to an* $\boldsymbol{AT}$*-valued parenthesized framed tangle invariant.* □

3.5. Invariants of knots and links. The restriction $\mathbf{Z^K}$ of $\mathbf{Z^{PT}}$ to knots has values in the algebra $\mathcal{A}$ of chord diagrams [2]. Assume now that $R = 1 + \uparrow\!\!-\!\!\uparrow/2$+higher degree terms.

Theorem 4. (*See also* [7, 18, 12]) $\mathbf{Z^K}$ *is a universal Vassiliev invariant.*

In other words, if D is a chord diagram and the singular knot K_D is an embedding (as defined in [2, section 2.2]) of D in $\mathbf{R}^3$, then

$$\mathbf{Z^K}(K_D) = D + \text{higher degree terms.}$$

(This of course means that if W is a degree m weight system, then the numerical knot invariant $W \circ \mathbf{Z^K}$ is Vassiliev of type m and its underlying weight system is W, meaning that we can reconstruct *any* Vassiliev invariant out of $\mathbf{Z^K}$.)

Proof. The proof is essentially identical to the proofs that all standard knot polynomials are series of Vassiliev invariants, as in [2]. The main point is that $R - R^{-1} = \uparrow\!\!-\!\!\uparrow$+higher degree terms, and so every double point in K_D forces a chord in $\mathbf{Z^K}(K_D)$, in just the right place. □

Similarly, one can use $\mathbf{Z^{PT}}$ to define a link invariant $\mathbf{Z^L}$, and $\mathbf{Z^L}$ will have a parallel universality property.

Remark 3.12. Following Drinfel'd's analysis of the non-uniqueness of (R, Φ) Le and Murakami [18] had proven that if R is central, the resulting link invariant $\mathbf{Z^L}$ is independent of the choice of Φ and if in addition $R = \exp \uparrow\!\!-\!\!\uparrow/2$, then $\mathbf{Z^K}$ is equal to the Kontsevich integral $\tilde{Z}$ as in (1).

4. Cohomological preliminaries

Before we can present our algorithm for computing a pair (R, Φ) which solves the equations of the previous section, we need to discuss some cohomological preliminaries. Most of the homological algebra that we will use can be found in [20, section 1.6] (though we need the cohomological versions of the homological results there).

4.1. Defining the complexes.

Definition 4.1. Let $C^n = C^n(\mathbf{AP}) = \mathbf{AP}_n$ and define coface maps $d_i^n : C^n \to C^{n+1}$ $(0 \leq i \leq n+1)$ by

$$d_i^n(\xi) = \begin{cases} \xi^{234\ldots(n+1)} & i = 0 \\ (1 \otimes \cdots \otimes \underset{i}{\Delta} \otimes \cdots \otimes 1)(\xi) & 1 \leq i \leq n \\ \xi^{123\ldots n} & i = n+1 \end{cases}$$

(Namely, d_0^n adds one strand on the left, d_{n+1}^n adds one strand on the right, and otherwise d_i^n doubles the ith strand.) Define a differential $d^n : C^n \to C^{n+1}$ by

$$d^n = \sum_{i=0}^{n+1} (-1)^i d_i^n,$$

and let the *nth formal Hochschild cohomology of* $\boldsymbol{AP}$, $H^n(\mathbf{AP})$, be the cohomology of the resulting complex; namely, set $Z^n = \ker d^n$, $B^n = \operatorname{im} d^{n-1}$, and $H^n = Z^n/B^n$.

Similarly, the *twisted formal Hochschild cohomology of* $\boldsymbol{AP}$, $H^n(\mathbf{AP}_\star \otimes \mathbf{AP}_1)$, will be defined by "leaving the rightmost strand alone":

Definition 4.2. Let $C^n_{\star\otimes 1} = C^n(\mathbf{AP}_\star \otimes \mathbf{AP}_1) = \mathbf{AP}_{n+1}$, define $d_i^n \otimes 1 = d_i^{n+1}$ for $i \leq n$, and set

$$(d_{n+1}^n \otimes 1)(\xi) = \xi^{12\ldots n(n+2)}.$$

Let the differential $d^n \otimes 1$ be the alternating sum of the $d_i^n \otimes 1$'s, and set $Z^n_{\star\otimes 1} = \ker d^n \otimes 1$, $B^n_{\star\otimes 1} = \operatorname{im} d^{n-1} \otimes 1$, and $H^n_{\star\otimes 1} = Z^n_{\star\otimes 1}/B^n_{\star\otimes 1}$. In the context of $C_{\star\otimes 1}$, we will denote the transpose operation (parallel to the transpose of Definition 3.6) by $T \otimes 1$. In other words, for $\xi \in C^n_{\star\otimes 1}$ define $\xi^{T\otimes 1} = \xi^{n\ldots 21(n+1)}$. Similar definitions can be made for $C^n_{1\otimes\star}$, $1 \otimes d$, etc.

Definition 4.3. Let the *symmetric subcomplex* of $C^\star$ be given by

$$C^n_{sym} = C^n_{sym}(\mathbf{AP}) = \left\{ \xi \in C^n : \xi + (-1)^{n(n+1)/2} \xi^T = 0 \right\}.$$

It is easy to check that $C^\star_{sym}$ is indeed a subcomplex of $C^\star$, and so one can define Z^n_{sym}, B^n_{sym}, and H^n_{sym}. Define the groups $C^n_{sym\otimes 1}$, $Z^n_{sym\otimes 1}$,

$B^n_{sym\otimes 1}$, and $H^n_{sym\otimes 1}$ by replacing C and T by $C_{\star\otimes 1}$ and $T\otimes 1$ in the previous sentence.

Proposition 4.4. (*Proof on page 165, see also* [8, Proposition 3.11]) *If n is even, $H^n_{sym}(\boldsymbol{AP})=0$.*

Proposition 4.5. (*Proof on page 166*) *If n is even, $H^n_{sym\otimes 1}(\boldsymbol{AP})=0$.*

One further complex is of interest for us. To define it, we first need the notion of a *shuffle:*

Definition 4.6. Let p and q be non-negative integers. A permutation σ of $\{1,\dots,p+q\}$ is called a *(pq)-shuffle* if σ^{-1} preserves the order of $\{1,\dots,p\}$ and the order of $\{p+1,\dots,p+q\}$. Define $\omega_{pq}:\mathbf{AP}_{p+q}\to\mathbf{AP}_{p+q}$ by

$$\omega_{pq}(\xi)=\sum_{\text{all }(pq)\text{-shuffles }\sigma}(-1)^\sigma\xi^\sigma.$$

Definition 4.7. The Harrison subcomplex [10] of $C^\star$ is defined by

$$C^n_{Harr}=C^n_{Harr}(\mathbf{AP})=\{\xi\in C^n:\omega_{pq}(\xi)=0\text{ whenever }p+q=n\text{ and }p,q\geq 1\}.$$

Make the usual definitions for Z^n_{Harr}, B^n_{Harr}, and H^n_{Harr}.

Exercise 4.8. Show that $C^\star_{Harr}$ is indeed a subcomplex of $C^\star$.

Theorem 5. (*Proof on page 165; the case $n=3$ is in* [18]) *If $n\geq 2$, $H^n_{Harr}(\boldsymbol{AP})=0$.*

Remark 4.9. The cofaces d^n_i together with the strand removal operations s_i of page 154 make all of the complexes discussed here into cosimplicial sets. Thus by the normalization theorem for simplicial cohomology the non-degenerate subcomplex of each of these complexes, defined by

$$\tilde{C}^n_\bullet=\bigcap_i\ker s_i|_{C^n_\bullet},$$

has the same cohomology as the original complex.

Finally, we will briefly need the following definition and proposition:

Definition 4.10. Similarly to Definition 4.2, define $d^n\otimes 1\otimes 1$ by "leaving the two rightmost strands alone", and $1\otimes 1\otimes d^n$ by "leaving the two leftmost strands alone". Finally define $d^p\otimes d^q:\mathbf{AP}_{p+q}\to\mathbf{AP}_{p+q+2}$ by having d^p act on the first p strands and d^q act on the last q strands. For example, $d^1\otimes d^1=(d\otimes 1\otimes 1)(1\otimes d)=(1\otimes 1\otimes d)(d\otimes 1)$.

Proposition 4.11. (*Proof on page 166*) *The kernel Z^{11} of the map $d \otimes d : \boldsymbol{AP}_2 \to \boldsymbol{AP}_4$ is the (non direct) sum of the kernels of $d \otimes 1 : \boldsymbol{AP}_2 \to \boldsymbol{AP}_3$ and $1 \otimes d : \boldsymbol{AP}_2 \to \boldsymbol{AP}_3$. Furthermore, any antisymmetric element of Z^{11} is the antisymmetrization of an element of* $\ker(d \otimes 1)$. *I.e., if $\rho \in Z^{11}$ satisfies $\rho + \rho^T = 0$, then $\rho = \bar{\rho} - \bar{\rho}^T$ for some non-degenerate $\bar{\rho} \in \ker(d \otimes 1)$.*

4.2. Computing some of their cohomologies. Let us start by computing the cohomology of a much simpler related complex.

Definition 4.12. Let $C^n = C^n(\mathbf{F})$ be $\mathbf{F}^n$, the vector space of dimension n over the field $\mathbf{F}$. Let $\{e_j\}_{j=1}^n$ be the standard basis of $\mathbf{F}^n$. Define coface maps $d_i^n : C^n \to C^{n+1}$ $(0 \le i \le n+1)$ by

$$d_i^n(e_j) = \begin{cases} e_{j+1} & i < j \\ e_j + e_{j+1} & i = j \\ e_j & i > j \end{cases},$$

and set $d^n = \sum_{i=0}^{n+1}(-1)^i d_i^n$. Let $H^n(\mathbf{F})$ be the cohomology of the resulting complex.

Remark 4.13. Notice that if we draw the basis vector e_j of $\mathbf{F}^n$ as a sequence of n vertical strands with the j'th one marked by a bead ($\bullet$),

$e_4 =$  $\in \mathbf{F}^5$,

then the d_i^n of the above definition correspond to the d_i^n of Definition 4.1; d_i^n can be viewed as adding one strand on the left, d_{n+1}^n as adding a strand on the right, and otherwise d_i^n can be viewed as doubling the i'th strand, summing over the possible ways of marking the resulting two children strands in case the parent strand was marked.

Lemma 4.14. $\dim H^1(\mathbf{F}) = 1$, *while* $\dim H^n(\mathbf{F}) = 0$ *for* $n > 1$.

Proof. First notice that $C^\star$ is a cosimplicial vector space — one may define codegeneracies $s_i^n : C^n \to C^{n-1}$ $(1 \le i \le n)$ by

$$s_i^n(e_j) = \begin{cases} e_{j-1} & i < j \\ 0 & i = j \\ e_j & i > j \end{cases}.$$

Therefore by the normalization theorem for simplicial cohomology the complex $C^\star$ has the same cohomology as the complex $\tilde{C}^\star$ defined by $\tilde{C}^n = \bigcap_i \ker s_i^n$. But it is clear that $\tilde{C}^n = 0$ unless $n = 1$, and that $\tilde{C}^1 = \mathbf{F}$. $\square$

Definition 4.15. For any natural number k, set $C^n(\mathbf{F}, k) = (\mathbf{F}^n)^{\otimes k}$. The elements of $C^n(\mathbf{F}, k)$ can be pictured as linear combinations of diagrams made of n vertical strands marked by k beads set on k different time levels:

$e_3 \otimes e_2 \otimes e_5 =$ 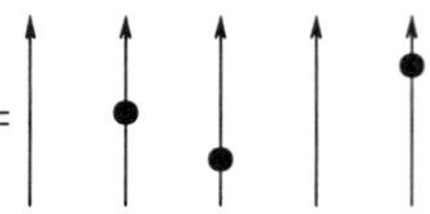 $\in C^5(\mathbf{F}, 3)$.

$C^\star(\mathbf{F}, k)$ inherits the structure of a cosimplicial vector space from $C^n(\mathbf{F})$, and again, the coface maps d_i^n have pictorial interpretations as strand addition and strand doubling as in Definition 4.1. From the Eilenberg-Zilber theorem and the Künneth formula it follows that the only non-vanishing cohomology of $C^\star(\mathbf{F}, k)$ is in degree k, and that $\dim H^k(\mathbf{F}, k) = 1$. Finally, a careful study of the proof of the Eilenberg-Zilber theorem and the Künneth formula reveals that $H^k(\mathbf{F}, k)$ is generated by the class of the "determinant"

$$\delta_k = \sum_{\sigma \in S_k} (-1)^\sigma e_{\sigma 1} \otimes e_{\sigma 2} \otimes \cdots \otimes e_{\sigma k}.$$

(Or else, it is easy to verify directly that $d\delta_k = 0$ and that δ_k is orthogonal to the image of d relative to the Euclidean scalar product on $C^k(\mathbf{F}, k)$.)

Notice that S_n acts on $C^n(\mathbf{F}, k)$ by permuting the e_j's, and so we can define $H^n_{sym}(\mathbf{F}, k)$ and $H^n_{Harr}(\mathbf{F}, k)$ as in the previous section. The total anti-symmetry of δ_k implies that $\delta_k^T = (-1)^{n(n-1)/2}\delta_k$, and therefore $H^n_{sym}(\mathbf{F}, k) = 0$ for any even n.

Definition 4.16. Let k be a natural number and let R be a representation of S_k. Let $C^n(k, R) = (R \otimes (\mathbf{F}^n)^{\otimes k})^{S_k}$ be the S_k-invariant part of $R \otimes (\mathbf{F}^n)^{\otimes k}$, using the given action of S_k on R and the obvious action[6] of S_k on $(\mathbf{F}^n)^{\otimes k}$. $C^\star(k, R)$ inherits a structure of a cosimplicial vector space from $C^\star(\mathbf{F}, k)$, and for each n, $C^n(k, R)$ inherits an S_n action from $C^n(\mathbf{F}, k)$. Thus we can define the cohomologies $H^n(k, R)$, $H^n_{sym}(k, R)$, and $H^n_{Harr}(k, R)$.

Notice that the assignment $R \mapsto H^n(k, R)$ is additive: $H^n(k, R_1 \oplus R_2) = H^n(k, R_1) \oplus H^n(k, R_2)$ for any two representations $R_{1,2}$ of S_k.

Lemma 4.17. *$H^n(k, R)$ vanishes unless $n = k$, and the dimension of $H^k(k, R)$ is equal to the number of times the alternating representation*

[6]This action commutes with the action of S_n discussed in the previous paragraph.

Alt *appears as a summand in the decomposition of* R *into irreducible representations.*

Proof. Let us first check the case of $R = \mathbf{F}S_k$, the regular representation of S_k. In this case $(R \otimes (\mathbf{F}^n)^{\otimes k})^{S_k}$ is isomorphic to $(\mathbf{F}^n)^{\otimes k}$, and the lemma follows from the discussion following Definition 4.15. Furthermore, it is easy to verify that the above isomorphism carries δ_k to $(\sum_\sigma (-1)^\sigma \sigma) \otimes \delta_k$, meaning that already $\dim H^k(\mathrm{Alt}) = 1$. The general case of the lemma now follows from the additivity of $R \to H^n(k, R)$ and from the fact that the regular representation contains every irreducible representation of S_k. □

Corollary 4.18. $H^n_{sym}(k, R)$ *vanishes if* n *is even.* □

Corollary 4.19. $H^n_{Harr}(k, R)$ *vanishes if* $n \geq 2$.

Proof. Using the Eulerian idempotents $e^n_{(l)}$ one gets a decomposition (see [20, pp. 144] and [19, pp. 222])

$$H^n(k, R) = H^n_{(1)}(k, R) \oplus \cdots \oplus H^n_{(n)}(k, R),$$

with $H^n_{(1)}(k, R) = H^n_{Harr}(k, R)$ and $H^n_{(n)}(k, R)$ being the cohomology of the complex obtained by taking the total anti-symmetrization of $C^n(k, R)$ for each n. Lemma 4.17 implies that all of the cohomology $H^n(k, R)$ lies in $H^n_{(n)}(k, R)$, and so there is nothing left for $H^n_{(1)}(k, R) = H^n_{Harr}(k, R)$. □

Proof of Proposition 4.4 *and of Theorem* 5. Given Corollaries 4.18 and 4.19, all we have to do is to show that the complex $C^\star(\mathbf{AP})$ is isomorphic to the complex $C^\star(k, R)$ for some R, or to a direct sum of such complexes, via an isomorphism that respects the S_n actions on $C^n(\mathbf{AP})$ and on $C^n(k, R)$. This is easy. Indeed, let R_k be the space spanned by all Chinese characters[7] having exactly k univalent vertices labeled by the numbers 1 to k, modulo the usual IHX and AS relations. By permuting the labels of the univalent ends, R_k becomes a representation of S_k. $R_k \otimes (\mathbf{F}^n)^{\otimes k}$ can be viewed as the space spanned by Chinese characters whose k univalent ends are ordered and are colored by integers between 1 and n, corresponding to the n basis vectors $\{e_j\}$ of $\mathbf{F}^n$. The S_k invariant subspace $(R_k \otimes (\mathbf{F}^n)^{\otimes k})^{S_k}$ is the space spanned by Chinese characters whose k univalent ends are *unordered* and are colored by integers between 1 and n. Taking the direct sum over k, we get exactly the space $\mathcal{B}^{sl}$ of [3] (with $\Upsilon = \{1, \ldots, n\}$) which is isomorphic

[7] as defined in [2].

to the space $\mathcal{A}^{sl}$ of the same paper. $\mathcal{A}^{sl}$ with $\Upsilon = \{1, \dots, n\}$ is just a different name for our $\mathbf{AP}_n$. Summarizing, we have that

$$C^n(\mathbf{AP}) = \bigoplus_k C^n(k, R_k). \tag{8}$$

Rereading Remark 4.13, we see that the above isomorphisms respect the coface maps, the differentials, and the S_n actions on the two sides of (8). □ □

Proof of Proposition 4.5. Simply repeat the constructions and considerations leading to the proof of Proposition 4.4, only always ignoring the $(n+1)$st strand/basis vector/color. □

Proof of Proposition 4.11. Use the same techniques as in the proof of (8) to reduce the problem to a problem about the kernel of $d \otimes d$ acting on $(\mathbf{F}^2)^{\otimes k}$. The vector space $(\mathbf{F}^2)^{\otimes k}$ is naturally isomorphic to the space $\langle x, y\rangle_k$ of degree k polynomials in two non-commuting variables x and y. Introduce new non-commuting variables $x_{1,2}$ and $y_{1,2}$, and for any polynomial p in the variables x, $x_{1,2}$, y, and $y_{1,2}$ define

$$\begin{aligned} d_x p &= p|_{x \to x_2} - p|_{x \to x_1 + x_2} + p|_{x \to x_1}, \\ d_y p &= p|_{y \to y_2} - p|_{y \to y_1 + y_2} + p|_{y \to y_1}, \end{aligned} \qquad \text{and} \quad p^T = p|_{x \leftrightarrow y}.$$

It is easy to check that using this language, the operators $d \otimes 1$, $1 \otimes d$, $d \otimes d$ and $\rho \mapsto \rho^T$ become d_x, d_y, $d_x \circ d_y$, and $p \mapsto p^T$ respectively, all restricted to $\langle x, y\rangle_k$, and that the condition "ρ is non-degenerate" becomes "$p \in \langle x, y\rangle_k$ vanishes if either x or y is set to 0". To finish the proof of the proposition, simply notice that $\ker d_x$ ($\ker d_y$) is the linear span of the monomials of degree exactly 1 in x (y), and that $\ker d_x \circ d_y$ is the linear span of the monomials of degree exactly 1 in either x or y. □

5. Solving the pentagon and the hexagons

In this section we will present our inductive prescription for solving the pentagon and the hexagons. A similar (and about as complicated) prescription can be read from Drinfel'd's [9, Proposition 3.1]. Our prescription (suggested by [9, Remark 2 following Proposition 5.8]) uses the vanishing of a 'smaller' cohomology, H^4_{Harr}, whereas Drinfel'd's uses the vanishing of the bigger H^4_{sym}. In our case, both groups indeed vanish as proven in the previous section. But it may be that in some similar but different circumstances, (see e.g. section 6.1), the inclusion $H^4_{Harr} \subset H^4_{sym}$ is proper and only the smaller of the two vanishes.

The algebras in which the pentagon ⬠ and the hexagons ⬡$_\pm$ are written are graded, and thus we may hope to solve ⬠ and ⬡$_\pm$ degree by degree. Let us start with a definition:

Definition 5.1. Let $\mathbf{M}$ be a graded module. Denote by $\mathcal{G}_m\mathbf{M}$ the degree m piece of $\mathbf{M}$ and set $\mathcal{F}_m\mathbf{M} = \bigoplus_{i=0}^m \mathcal{G}_i\mathbf{M}$ and $\mathcal{R}_m\mathbf{M} = \bigoplus_{i>m} \mathcal{G}_i\mathbf{M}$.

Start the induction with $R_1 = 1 + \vdash\!\dashv/2$ and $\Phi_1 = 1$. Then for an arbitrary $m > 1$ assume that a non-degenerate $\Phi_{m-1} \in \mathcal{F}_m\mathbf{AP}_3$ and a symmetric non-degenerate $R_{m-1} \in \mathcal{F}_m\mathbf{AP}_2$ solve ⬠ and ⬡$_\pm$ up to and including degree $m-1$, that $R_{m-1} = 1 + \vdash\!\dashv/2 \mod \mathcal{R}_m\mathbf{AP}_2$, and that

$$||\Phi_{m-1}||^2 = 1 \qquad \mod \mathcal{R}_m\mathbf{AP}_3. \tag{9}$$

Set $R_{m,0} = R_{m-1}$ and $\Phi_{m,0} = \Phi_{m-1}$. We will attempt to 'improve' $(R_{m,0}, \Phi_{m,0})$ via a five-step procedure, each time replacing $(R_{m,i-1}, \Phi_{m,i-1})$ $(i = 1, \dots, 5)$ by $(R_{m,i}, \Phi_{m,i})$, with each step bringing us closer to a solution of ⬠ and ⬡$_\pm$ in all degrees up to and including degree m. The result of the last step, $(R_{m,5}, \Phi_{m,5})$, will actually solve ⬠ and ⬡$_\pm$ to the required degree, and will serve as the seed (R_m, Φ_m) for the solution in the following degree.

Let $\mu_i \in \mathbf{AP}_4$ and $\psi_{\pm,i} \in \mathbf{AP}_3$ be the degree m 'mistakes' in ⬠ and ⬡$_\pm$ when using $R_{m,i}$ and $\Phi_{m,i}$:

$$\begin{aligned} 1 + \mu_i = \Phi_{m,i}^{123} \cdot (1 \otimes \Delta \otimes 1)(\Phi_{m,i}) \cdot \Phi_{m,i}^{234} &\cdot (1 \otimes 1 \otimes \Delta)(\Phi_{m,i}^{-1}) \\ &\cdot (\Delta \otimes 1 \otimes 1)(\Phi_{m,i}^{-1}) \end{aligned} \tag{10}$$

$$1 + \psi_{\pm,i} = (\Delta \otimes 1)(R_{m,i}^{\mp 1}) \cdot \Phi_{m,i}^{123} \cdot (R_{m,i}^{\pm 1})^{23} \cdot (\Phi_{m,i}^{-1})^{132} \cdot (R_{m,i}^{\pm 1})^{13} \cdot \Phi_{m,i}^{312} \tag{11}$$

(both equations taken modulo $\mathcal{R}_m\mathbf{AP}$).

Setting $R_{m,i} = R_{m,i-1} + r_i$ and $\Phi_{m,i} = \Phi_{m,i-1} + \varphi_i$, we will search for $r_i \in \mathcal{G}_m\mathbf{AP}_2$ and for $\varphi_i \in \mathcal{G}_m\mathbf{AP}_3$ that will make the mistakes μ_i and $\psi_{\pm,i}$ be 'simpler' than their predecessors μ_{i-1} and $\psi_{\pm,i-1}$. From (10) and (11) it is easy to read the following formulas for the 'new' mistakes in terms of the 'old' mistakes and the correction terms φ_i and r_i:

$$\begin{aligned} \mu_i = \mu_{i-1} + \varphi_i^{234} - (\Delta \otimes 1 \otimes 1)(\varphi_i) + (1 \otimes \Delta \otimes 1)(\varphi_i) - (1 \otimes 1 \otimes \Delta)(\varphi_i) + \varphi_i^{123} \\ = \mu_{i-1} + d\varphi_i, \end{aligned} \tag{12}$$

$$\psi_{\pm,i} = \psi_{\pm,i-1} + \left[\varphi_i^{123} - \varphi_i^{132} + \varphi_i^{312} \pm (r_i^{23} - (\Delta \otimes 1)(r_i) + r_i^{13})\right]. \tag{13}$$

Setting $\psi_i = (\psi_{+,i} + \psi_{-,i})/2$ and $\psi_{diff,i} = (\psi_{+,i} - \psi_{-,i})/2$, we see that (13) is equivalent to the following two equations:

$$\psi_i = \psi_{i-1} + \omega_{21}(\varphi_i); \qquad \psi_{diff,i} = \psi_{diff,i-1} + (d \otimes 1)(r_i). \tag{14}$$

We aim to find φ_i and r_i that will make the left hand sides of the above equations as simple as possible, hopefully even vanishing. The first step towards doing so is to find relations between the determined parts of the right hand sides of the above equations, namely between μ_{i-1} and $\psi_{\pm,i-1}$.

5.1. Finding relations between μ_i and $\psi_{\pm,i}$. Rearranging (10) and then cabling and extending it in some arbitrary way, (and similarly for (11)), we see that it can be regarded as a way of writing the identity of $\mathbf{AP}_n$ (for some n) as a product of five 'variants' of Φ, at the cost of some error proportional to μ. Each variant of Φ, say

,

appears in many different pentagons (and hexagons), and thus the product we just obtained can be expanded further at the cost of some more error terms. Continuing in this manner, if at some point our product becomes a product of pentagons and hexagons, we can simply cross them all out, again at the cost of some more error terms proportional to μ and ψ. What we have left at the end is an expression for the identity in terms of the identity and some error terms — in other words, we have a relation between those error terms, which is precisely what we now seek.

As a first example, start from the inner most pentagon in figure 4. Reading its edges counterclockwise beginning from the vertex labeled $((ab)c)d$, we get equation (10), and we've indicated the error term of that equation in the center of the pentagon. Now each of the R and Φ terms in (10) can be expanded further, by either using a locality relation (a square) or by using a hexagon (or the image of a hexagon equation by some cabling or permutation operation). Locality relations always hold and cost nothing, but each hexagon contributes some error term, as indicated in its center. Continuing to expand in this manner until we reach the outermost hexagon and then replacing that hexagon by its corresponding error term, we get the equation[8]

$$\mu_i^{1234} - \mu_i^{1243} + \mu_i^{1423} - \mu_i^{4123} = \psi_{+,i}^{234} \quad -(\Delta\otimes 1\otimes 1)\psi_{+,i} + (1\otimes\Delta\otimes 1)\psi_{+,i} - \psi_{+,i}^{124}, \tag{15}$$

or

$$\omega_{31}(\mu_i) = (d\otimes 1)(\psi_{+,i}). \tag{16}$$

Similarly, figure 5 proves equation (17), figure 6 proves equation (18), and figure 7 proves equation (19). For a similar proof of (19), see [8, pp. 1449] and [22].

$$\omega_{22}(\mu_i) = (1\otimes 1\otimes d)\psi_{+,i} + (d\otimes 1\otimes 1)\psi_{-,i}^{231} \tag{17}$$

$$\psi_{+,i}^{123} - \psi_{+,i}^{213} + \psi_{-,i}^{321} - \psi_{-,i}^{231} = 0 \tag{18}$$

$$d\mu_i = 0 \tag{19}$$

Proposition 5.2. *If $R_{m,i} = R_{m,i}^{21}$ and $||\Phi_{m,i}||^2 = 1 \mod \mathcal{R}_m \boldsymbol{AP}_3$, then $\psi_{+,i} = -\psi_{-,i}^{213}$ $(= -\psi_{-,i}^{T\otimes 1})$ and $\mu_i = -\mu_i^{4321}$ $(= -\mu_i^T)$.*

Proof. The first assertion follows from a trivial modification of the first few lines of the proof of Proposition 3.7. To prove the second, invert both sides of (10), take the transpose and use $\Phi_{m,i}^{-1} = \Phi_{m,i}^T$ whenever possible, getting the equation

$$1 - \mu_i^T = (1\otimes 1\otimes \Delta)\Phi_{m,i}^{-1}\cdot(\Delta\otimes 1\otimes 1)\Phi_{m,i}^{-1} \quad \cdot\Phi_{m,i}^{123}\cdot(1\otimes\Delta\otimes 1)\Phi_{m,i} \cdot\Phi_{m,i}^{234} \mod \mathcal{R}_m\mathbf{AP}_3.$$

But the right hand side here is again the pentagon, and so by (10)

$$1 - \mu_i^T = 1 + \mu_i,$$

as required. □

---

[8]Actually, the above process doesn't quite give (15), but rather it gives a similar equation in which each of the terms is multiplied by some further factors proportional to $R_{m,i}$ and $\Phi_{m,i}$. But remembering that $R_{m,i}$ and $\Phi_{m,i}$ are perturbations of the identity, that μ_i and $\psi_{\pm,i}$ are of degree m, and taking only the degree m piece, we get equation (15).

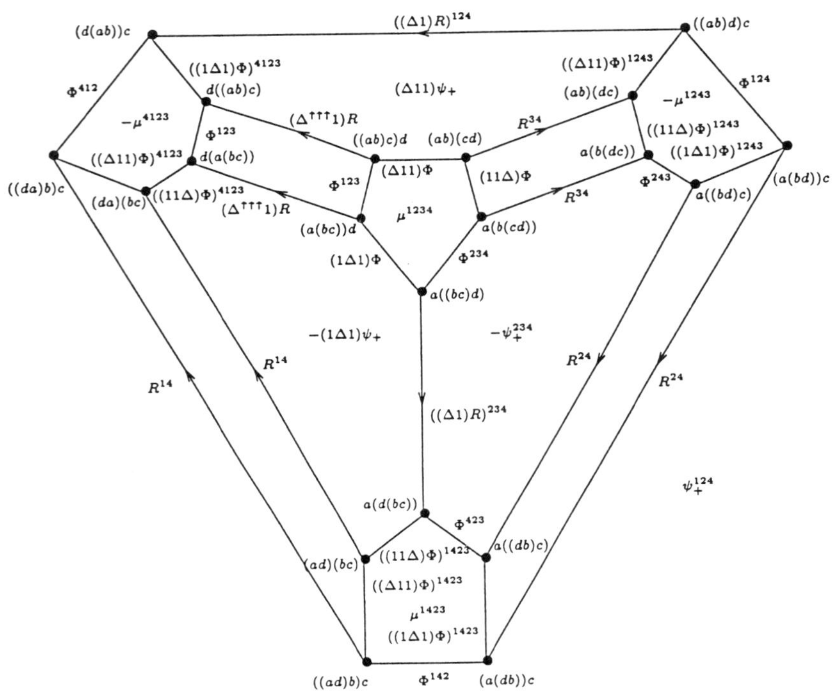

Figure 4. Proof of equation (16). To save space, we've suppressed all the $\otimes$ symbols, and the subscripts m and i. Notice that the orientations of the edges labeled by a variant of Φ can be read from the parenthesized strings labeling their ends, and so we also suppressed these orientations.

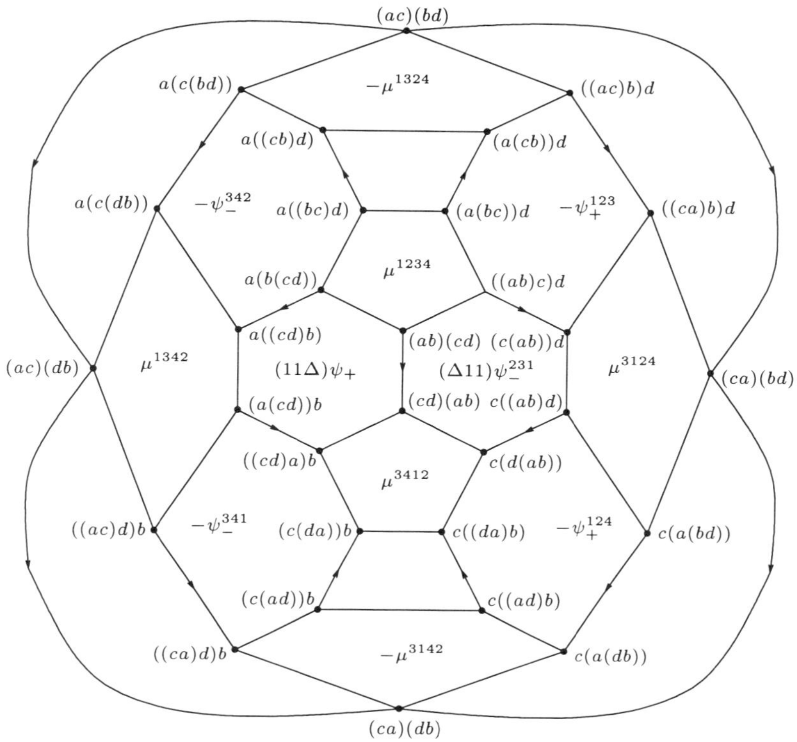

Figure 5. The proof of equation (17)

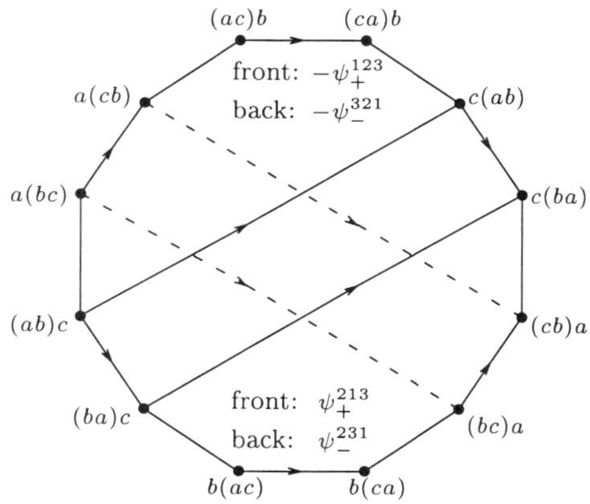

Figure 6. The proof of equation (18). The polyhedron in this figure is a sphere whose equator is a 12-gon, and each of whose hemispheres is made of two hexagons and a square. It is shown from above, with the lower hemisphere ('back') drawn in dashed lines.

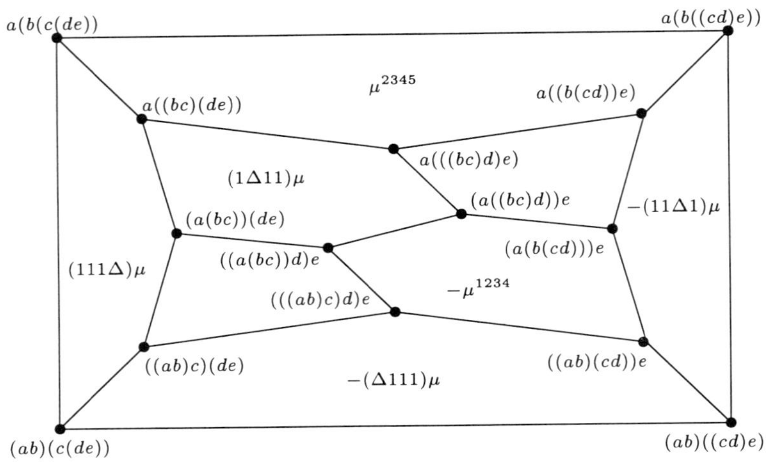

Figure 7. The proof of equation (19)

5.2. First step ($i = 1$), renormalizing Φ. The first step is to guarantee (9) for $\Phi_{m,1}$. This is easy: simply set

$$\Phi_{m,1} = \Phi_{m,0} - \frac{||\Phi_{m,0}||^2 - 1}{2}, \tag{20}$$

and leave R unchanged, namely, set $R_{m,1} = R_{m,0}$. Notice that $\Phi_{m,1}$ is still non-degenerate.

5.3. Second step ($i = 2$), fixing R to eliminate ψ_{diff}.

Lemma 5.3. $\psi_{\mathit{diff},1} \in Z^2_{sym\otimes 1}(\boldsymbol{AP})$.

Proof. Having guaranteed (9), we can use Proposition 5.2, get $\psi_{\mathit{diff},1} = \frac{1}{2}\left(\psi_{+,1} + \psi_{+,1}^{T\otimes 1}\right)$, and thus $\psi_{\mathit{diff},1} \in C^2_{sym\otimes 1}$. Furthermore,

$$\begin{aligned}(d\otimes 1)\psi_{\mathit{diff},1} &= \frac{1}{2}\left((d\otimes 1)\psi_{+,1} - ((d\otimes 1)\psi_{+,1})^{T\otimes 1}\right)\\ &= \frac{1}{2}\left(\mu_1^{1234} - \mu_1^{1243} + \mu_1^{1423} - \mu_1^{4123} - \mu_1^{3214} + \mu_1^{3241} - \mu_1^{3421} + \mu_1^{4321}\right)\end{aligned}$$

by (16). Using the second part of Proposition 5.2 on the last four terms of the above sum, we see that they cancel against the first four. □

Now use Proposition 4.5 to find a non-degenerate r_2 for which $\psi_{\mathit{diff},1} + (d\otimes 1)r_2 = 0$. (It is easy to check that $\psi_{\mathit{diff},1}$ is non-degenerate and thus by Remark 4.9 r_2 can be chosen to be non-degenerate). Setting $R_{m,2} = R_{m,1} + r_2$ and $\Phi_{m,2} = \Phi_{m,1}$, we get from (14) that $\psi_{\mathit{diff},2} = 0$.

5.4. Third step ($i = 3$), symmetrizing R. Unfortunately, our choice of r_2 may have ruined the symmetry of R_m; we need to have some control of the antisymmetric part of r_2:

Lemma 5.4. $(d \otimes d)(r_2^{21} - r_2) = 0$.

Proof.

$$\begin{aligned}(d \otimes d)(r_2^{21} - r_2) &= (d \otimes 1 \otimes 1)(1 \otimes d)r_2^{21} \\ &\quad - (1 \otimes 1 \otimes d)(d \otimes 1)r_2 \\ &= (d \otimes 1 \otimes 1)\psi_{\mathit{diff},1}^{321} - (1 \otimes 1 \otimes d)\psi_{\mathit{diff},1} \\ &= \frac{1}{2}(d \otimes 1 \otimes 1)(\psi_{-,1}^{321} + \psi_{-,1}^{231}) \\ &\quad + \frac{1}{2}(1 \otimes 1 \otimes d)(\psi_{+,1} + \psi_{+,1}^{213}) \\ &= \frac{1}{2}\,\omega_{22}\,(\mu_1) + \frac{1}{2}\,(\omega_{22}(\mu_1))^{2143} \text{ by (17)} \\ &= 0 \qquad\qquad \text{by Proposition 5.2.} \quad \square\end{aligned}$$

Now use Proposition 4.11 on $\rho = r_2^{21} - r_2$ to find a non-degenerate $r_3 \in \ker d \otimes 1 \subset \mathbf{AP}_2$ for which $r_2^{21} - r_2 = r_3 - r_3^{21}$. Setting $R_{m,3} = R_{m,2} + r_3$ and $\Phi_{m,3} = \Phi_{m,2}$ we find that $R_{m,3}$ is symmetric, and $r_3 \in \ker d \otimes 1$ together with (14) and $(d \otimes 1)^2 = 0$ show that we didn't spoil the vanishing of ψ_{diff}.

5.5. Fourth step ($i = 4$), solving the hexagons.

Lemma 5.5. *ψ_3 is totally antisymmetric: if σ is a permutation of $\{1, 2, 3\}$, then $\psi_3^\sigma = (-1)^\sigma \psi_3$.*

Proof. The vanishing of $\psi_{\mathit{diff},3}$ implies that $\psi_3 = \psi_{+,3} = \psi_{-,3}$ and thus Proposition 5.2 implies that $\psi_3^{213} = -\psi_3$, and it only remains to show that $\psi_3^\sigma = \psi_3$ for some 3-cycle σ. Rewrite equation (18) in terms of ψ_3, and use $\psi_3^{213} = -\psi_3$ on the middle two terms. The resulting equation is $2\psi_3 - 2\psi_3^{231} = 0$. $\square$

Now set $r_4 = 0$ and $\varphi_4 = -\psi_3/3$; namely, $R_{m,4} = R_{m,3}$ and $\Phi_{m,4} = \Phi_{m,3} - \psi_3/3$. Equation (14) and the total antisymmetry of ψ_3 imply that $\psi_4 = 0$. We did not touch ψ_{diff}, and so $\psi_{\pm,4} = 0$; namely, the hexagons $⬡_\pm$ hold for $(R_{m,4}, \Phi_{m,4})$ (modulo $\mathcal{R}_m\mathbf{AP}_3$). The antisymmetry of ψ_3 also implies $||\Phi_{m,4}||^2 = ||\Phi_{m,3}||^2 = 1$ (modulo $\mathcal{R}_m\mathbf{AP}_3$). Notice that $\Phi_{m,1}$ is still non-degenerate.

5.6. Fifth step ($i = 5$), solving the pentagon.

Lemma 5.6. $\mu_4 \in Z^4_{Harr}$.

Proof. Now that $\psi_{\pm,4}$ vanish, equations (16) and (17) say that $\omega_{31}(\mu_4) = \omega_{22}(\mu_4) = 0$. $\omega_{31}(\mu_4) = 0$ and $\mu_4 + \mu_4^T = 0$ together imply that $\omega_{13}(\mu_4)$ is also 0, and so $\mu_4 \in C^4_{Harr}$. It only remains to recall equation (19): $d\mu_4 = 0$. $\square$

Now use Theorem 5 and Remark 4.9 to find a non-degenerate $\varphi_5 \in C^3_{Harr}$ with $d\varphi_5 = -\mu_4$, and set $R_{m,5} = R_{m,4}$ and $\Phi_{m,5} = \Phi_{m,4} + \varphi_5$. Equation (12) implies that the pentagon ⬠ now holds, and equation (14) together with $\omega_{21}\varphi_5 = 0$ imply that the hexagons still hold. Finally, as both hexagons now hold up to and including degree m, Proposition 3.7 implies that $||\Phi_{m,5}||^2 = 1 \mod \mathcal{R}_m\mathbf{AP}_3$, and we are ready to go back to step 1 with m replaced by $m + 1$.

6. Odds and ends

6.1. Horizontal chords. The original R and Φ, constructed from the KZ equation as indicated in section 1, contain only horizontal chords. One may hope to have an inductive procedure, similar to the one in section 5, to combinatorially construct such horizontal-chords-only R and Φ. The first step is clear:

Definition 6.1. Let $\mathbf{AP}^{hor}_n$ be the associative algebra generated by formal symbols t^{ij}, $1 \le i, j \le n$, modulo the relations $t^{ij} = t^{ji}$, $[t^{ij}, t^{kl}] = 0$ when $|\{i, j, k, l\}| = 4$ and $[t^{ij} + t^{ik}, t^{jk}] = 0$ when $|\{i, j, k\}| = 3$.

Thinking of t^{ij} as a horizontal chord connecting the ith and the jth vertical strands,

$t^{ij} =$ 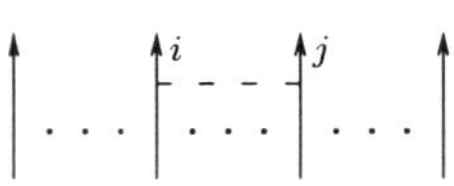,

the relations between the t^{ij} become the $4T$ relations of [2], and so the natural maps $\mathbf{AP}^{hor}_n \to \mathbf{AP}_n$ are well defined. Both locality in space and in scale still hold in $\mathbf{AP}^{hor}$: the former follows from the relation $[t^{ij}, t^{kl}] = 0$ (for appropriate i–l), and the latter from the same argument as in the proof of Lemma 3.4, only this time using the $4T$ relation instead of IHX and STU. The algebras $\mathbf{AP}^{hor}_n$ are graded, and so we may hope that the same inductive procedure as in section 5 would work for $\mathbf{AP}^{hor}$ as well. It does, with the only exception being that we don't know how to prove Theorem 5 in this case.

Conjecture 1. (*Suggested by Drinfel'd's* [9, Remark 2 following Proposition 5.8]) *With the obvious definition,* $\mathcal{G}_m H^4_{Harr}(\boldsymbol{AP}^{hor}) = 0$ *for all* m.

Another variation is to define $\mathbf{AP}_n^{lie}$ in the same way as $\mathbf{AP}_n^{hor}$, only replacing the words "associative algebra" by "lie algebra", and to define $\mathbf{AP}_n^{Lie}$ to be the group generated by the exponentials of elements of $\mathbf{AP}_n^{lie}$. It is easy to check that everything carries through in section 5, and so if we only knew that $\mathcal{G}_m H^4_{Harr}(\mathbf{AP}^{lie}) = 0$ for all m, we could combinatorially construct a solution (R, Φ) in $\mathbf{AP}^{Lie}$. The KZ pair (R, Φ) has that property.

I was able to check conjecture 1 for $m \leq 5$ on a computer, and (as section 7 shows) compute a pair (R_7, Φ_7) in $\mathbf{AP}^{Lie}$ up to degree 7. Using transcendental methods (i.e., the KZ equation), Drinfel'd [9] shows that it should be possible to extend my (or any other partial solution) (R_7, Φ_7) to a solution that works in all degrees.

Notice that if R is in $\mathbf{AP}_2^{hor}$, life is made considerably simpler; step 3 in section 5 becomes unnecessary, and if one reduces modulo diagrams having an isolated chord (the additional relation in $\mathcal{A}^r$), C (of equation (7)) becomes the identity 1, and the reduced $\mathbf{Z}^{\mathbf{PFT}}$ satisfies (R9) automatically.

6.2. The Commuto-Associahedrons. The polyhedra displayed in Figures 4–7 make it tempting to make the following definition:

Definition 6.2. The nth Commuto-Associahedron CA_n is the two dimensional CW complex made of the following cells:

0-cells: All possible products of n elements $a_1, \dots, a_n$ in a non associative non commutative algebra. In other words, all parenthesized strings on the alphabet $\{1, \dots, n\}$ in which every letter appears exactly once.

1-cells: All basic associativities and commutativities. These correspond to generating morphisms of **NAB** in which the orientations of strands and the signs of crossings are disregarded.

2-cells:
- A pentagon sealing every pentagon of the kind appearing in relation (R2).
- A hexagon sealing every hexagon of the kind appearing in relation (R6), just disregarding the difference between overcrossings and undercrossings.
- A square sealing every elementary locality relation of the type considered in (R3) and (R4) (again disregarding the difference between overcrossings and undercrossings). Notice that every elementary locality relation can be written as a product of four morphisms, and so it corresponds to a square in the 1-skeleton of CA_n.

For an example, see Figure 8.

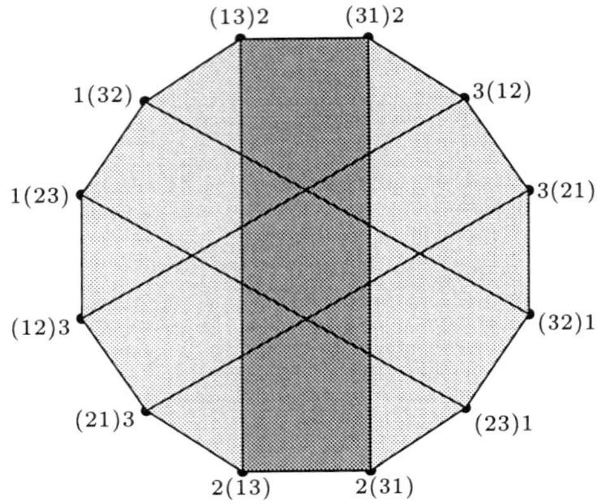

Figure 8. The third commuto-associahedron CA_3. It is made by gluing three sets of a square and two hexagons each into a 12-gon. Only one of these sets is shaded in the figure; the other two are obtained from it by rotations by 60° and 120° respectively. Topologically, the result is a circle with three disks glued in, and has the homotopy type of a wedge of two spheres.

By the Mac Lane coherence theorem, the commuto-associahedrons are all connected and simply connected, and thus their 2nd homology can be computed by simply counting cells and computing the Euler characteristic. This done, one finds that that CA_{3-5} have no 2nd homology other than that displayed in Figures 4–7 (with $a, b, c, \dots$ an arbitrary permutation of $1, 2, 3, \dots$) and other than classes that look like the shell of a prism with basis a square, a pentagon, or a hexagon. The latter classes can also be used to obtain relations between μ and ψ, but the relations thus obtained turn out to be not interesting — they all turn out to be consequences of Lemma 3.4. In other words, in section 5 we've used *all* the relations between μ and ψ that we could generate using our technique.

Notice, however, that Figures 4–7 contain a little more information than just a class in $H_2(CA_n)$ for some n; they also contain information about the the orientations of the crossings involved (disguised as the orientations of the 1-cells marked by a variant of R). It is therefore not a-priori clear that every class in $H_2(CA_n)$ corresponds to a relation between μ and ψ; even the distinction between ψ_+ and ψ_- is lost in the commuto-associahedrons.

Notice also that our definition is somewhat different than the 2-skeletons of Kapranov's permutoassociahedrons [11]. For example, his KP_3 is just a planar 12-gon, as opposed to our CA_3 which is a wedge of two spheres. I do not know if there is a natural way to add higher dimensional cells to our CA_n in a way similar to [11].

7. Some computations

Starting up mathematica [29], loading a definitions file (available at [4]) and testing the $4T$ relation:

```
Mathematica 2.2 for SPARC
Copyright 1988-93 Wolfram Research, Inc.
 -- Open Look graphics initialized --

In[1]:= << NAT.m

In[2]:= t[1,3]**t[1,2] + t[1,3]**t[2,3] - t[2,3]**t[1,3]
```

Out[2]= $t^{12}t^{13}$

Basing the induction and setting $\Phi_{m,1}$ to be as in (20):

```
In[3]:= {R[1]=1+t[1,2]/2, Phi[1]=1}
```

Out[3]= $\{1+\frac{t^{12}}{2},1\}$

```
In[4]:= R[m_,0] := R[m-1]; Phi[m_,0] := Phi[m-1]

In[5]:= R[m_,1] := R[m,0]; Phi[m_,1] := Phi[m,1] =
  ModDegree[m+1,Phi[m,0]-Expand[(-1+NormSquared[Phi[m,0]])/2]]
```

Next, define R^{-1}, Φ^{-1}, $\psi_\pm$, ψ, ψ_{diff} as in section 5. Notice that `Act[{1,2},3][R]` means $(\Delta\otimes 1)(R)$ ("the first strand becomes the pair `{1,2}` and the second becomes number 3"), and that `Act[3,1,2][Phi]` means Φ^{312} ("first becomes number 3, second becomes number 1, third becomes number 2"):

```
In[6]:= RInverse[m_,i___] := RInverse[m,i]=
  ModDegree[m+1,Invert[R[m,i]]]

In[7]:= PhiInverse[m_,i___] := PhiInverse[m,i]=
  ModDegree[m+1,Invert[Phi[m,i]]]

In[8]:= psip[m_,i_] := ModDegree[m+1, -1 +
  Act[{1,2},3][RInverse[m,i]]**
  Phi[m,i]**Act[2,3][R[m,i]]**Act[1,3,2][PhiInverse[m,i]]**
  Act[1,3][R[m,i]]**Act[3,1,2][Phi[m,i]]]

In[9]:= psim[m_,i_] := ModDegree[m+1, -1 +
  Act[{1,2},3][R[m,i]]**Phi[m,i]**
  Act[2,3][RInverse[m,i]]**Act[1,3,2][PhiInverse[m,i]]**
  Act[1,3][RInverse[m,i]]**Act[3,1,2][Phi[m,i]]]

In[10]:= psi[m_,i_]:=Expand[(psip[m,i]+psim[m,i])/2]

In[11]:= psidiff[m_,i_]:=Expand[(psip[m,i]-psim[m,i])/2]
```

Setting $R_{m,2}$ as in section 5.3, and computing $R_{2,2}$ and $\Phi_{2,2}$: (the `OX` here means $\otimes$)

In[12]:=
```
R[m_,2] := R[m,2]=ModDegree[m+1, R[m,1] -
  {APPower[t[1,2],m]}.
  (psidiff[m,1]~InTermsOf~{(d[1]~OX~1)[APPower[t[1,2],m]]})]
```

In[13]:=
```
Phi[m_,2] := Phi[m,1]
```

In[14]:=
```
{R[2,2],Phi[2,2]}
```

Out[14]= $\{1+\frac{t^{12}t^{12}}{8}+\frac{t^{12}}{2},1\}$

As we are dealing with diagrams that have only horizontal chords, R is always symmetric and step 3 is superfluous. Here we follow step 4 as in section 5.5 and compute $R_{2,4}$ and $\Phi_{2,4}$:

In[15]:=
```
R[m_,3] := R[m,2]; Phi[m_,3] := Phi[m,2]
```

In[16]:=
```
R[m_,4] := R[m,3]; Phi[m_,4] :=
  Phi[m,4]=Phi[m,3]-Expand[psi[m,3]/3]
```

In[17]:=
```
{R[2,4],Phi[2,4]}
```

Out[17]= $\{1+\frac{t^{12}t^{12}}{8}+\frac{t^{12}}{2},1+\frac{t^{13}t^{23}}{24}-\frac{t^{23}t^{13}}{24}\}$

Defining μ as in (10), $\Phi_{m,5}$ as in section 5.6, and computing (R_3, Φ_3):

In[18]:=
```
mu[m_,i_] := mu[m,i]=ModDegree[m+1, -1 + Phi[m,i]**
  Act[1,{2,3},4][Phi[m,i]]**Act[2,3,4][Phi[m,i]]**
  Act[1,2,{3,4}][PhiInverse[m,i]]**
  Act[{1,2},3,4][PhiInverse[m,i]]]
```

In[19]:=
```
R[m_,5] := R[m,4]; Phi[m_,5] := Phi[m,5]=Phi[m,4] +
  Expand[HarrisonBasis[3,m].((-mu[m,4])~InTermsOf~
  (d[3] /@ HarrisonBasis[3,m]))]
```

In[20]:=
```
R[m_] := R[m,5]; Phi[m_] := Phi[m,5]
```

In[21]:=
```
{R[3],Phi[3]}
```

Out[21]= $\{1+\frac{t^{12}t^{12}}{8}+\frac{t^{12}t^{12}t^{12}}{48}+\frac{t^{12}}{2},1+\frac{t^{13}t^{23}}{24}-\frac{t^{23}t^{13}}{24}\}$

Just for fun, checking that $\mu_4 \neq 0$ but $d^4\mu_4 = 0$ in degree 4:

In[22]:=
```
{mu[4,4]===0, d[4][mu[4,4]]===0}
```

Out[22]= $\{False, True\}$

Note that $R = \exp$ $/2 = \exp t^{12}/2$, at least up to degree 4:

In[23]:= `ModDegree[5,FormalLog[R[4]]]`

Out[23]= $\frac{t^{12}}{2}$

$\log \Phi_4$ is nothing as simple. To avoid ugly denominators, we need to multiply it by 5760:

In[24]:= `Expand[5760(LogPhi4=ModDegree[5,FormalLog[Phi[4]]])]`

Out[24]= $240\,t^{13}t^{23} - 240\,t^{23}t^{13} - 4\,t^{13}t^{13}t^{13}t^{23} + 12\,t^{13}t^{13}t^{23}t^{13} + 7\,t^{13}t^{13}t^{23}t^{23} - 12\,t^{13}t^{23}t^{13}t^{13} - 14\,t^{13}t^{23}t^{13}t^{23} - 7\,t^{13}t^{23}t^{23}t^{23} + 4\,t^{23}t^{13}t^{13}t^{13} + 14\,t^{23}t^{13}t^{23}t^{13} + 21\,t^{23}t^{13}t^{23}t^{23} - 7\,t^{23}t^{23}t^{13}t^{13} - 21\,t^{23}t^{23}t^{13}t^{23} + 7\,t^{23}t^{23}t^{23}t^{13}$

But it belongs to the Lie algebra generated by t^{13} and t^{23}, and has a reasonable formula in terms of commutators:

In[25]:= `basis=EvenLieBasis[4,{t[1,3],t[2,3]}]`

Out[25]= $\{[t^{13},t^{23}],[t^{13},[t^{13},[t^{13},t^{23}]]],[t^{13},[t^{23},[t^{13},t^{23}]]],[t^{23},[t^{23},[t^{13},t^{23}]]]\}$

In[26]:= `(LogPhi4~InTermsOf~LieExpand[basis]).basis`

Out[26]= $\frac{-[t^{13},[t^{13},[t^{13},t^{23}]]]}{1440} - \frac{7\,[t^{13},[t^{23},[t^{13},t^{23}]]]}{5760} + \frac{[t^{13},t^{23}]}{24} - \frac{7\,[t^{23},[t^{23},[t^{13},t^{23}]]]}{5760}$

Notice that we did not make any effort to find a *non-degenerate* Φ, but luckily, it just came out that way.

Define the correction term $Z(\infty)$ as in section 3.3, define its inverse, and compute the invariant of the unknot (in $\mathcal{A}^r$):

In[27]:= `?AP2CD`

AP2CD[ord,expr] takes an element of the algebra AP_n and closes it to a linear combination of chord diagrams in A^r by traveling around the strands of AP_n in the order specified by ord, going up at first and then down, up,down, ...

In[27]:= `ZInfinity[m_] := ZInfinity[m] = AP2CD[{3,2,1},Phi[m]]`

In[28]:= `ZInfinityInverse[m_] := ZInfinityInverse[m]=ModDegree[m+1,Invert[ZInfinity[m]]]`

In[29]:= `ZInfinityInverse[4]`

Out[29]= $1 - \frac{[1212]}{24} + \frac{[12123434]}{5760} + \frac{[12132434]}{960} - \frac{[12314234]}{5760}$

The notation [12132434] means: write the sequence $1,2,1,3,2,4,3,4$ counterclockwise around a circle and connect by chords any pair of equal numbers, as shown on the right.

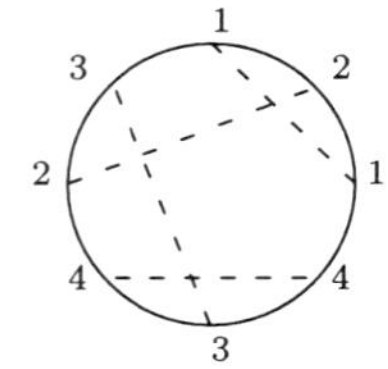

Just for fun, let's evaluate the result for the unknot using the HOMFLY weight system (which, using standard notation, depends on the N of $SL(N)$ and on $h = \log q$), and let us compare it with the well known value of the HOMFLY polynomial on the unknot:

In[30]:= `Expand[WHOMFLY[N,h][ZInfinityInverse[4]]]`

Out[30]= $N - \frac{h^2 N}{24} + \frac{7 h^4 N}{5760} + \frac{h^2 N^3}{24} - \frac{h^4 N^3}{576} + \frac{h^4 N^5}{1920}$

In[31]:= `z=(Exp[h N/2]-Exp[-h N/2])/(Exp[h/2]-Exp[-h/2])`

Out[31]= $\dfrac{-e^{\frac{-(h\,N)}{2}} + e^{\frac{h\,N}{2}}}{-e^{\frac{-h}{2}} + e^{\frac{h}{2}}}$

In[32]:= `Series[z,{h,0,4}]`

Out[32]= $N + \left(\frac{-N}{24} + \frac{N^3}{24}\right) h^2 + \left(\frac{7N}{5760} - \frac{N^3}{576} + \frac{N^5}{1920}\right) h^4 + O(h)^5$

Computing the invariant of the trefoil:

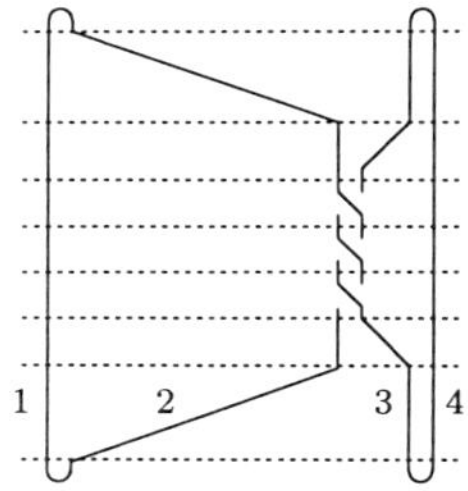

In[33]:= `ModDegree[5,AP2CD[{1,3,4,2},`
`Act[1,2,{3,4}][Phi[4]]**`
`Act[2,3,4][PhiInverse[4]]**`
`Act[2,3][RInverse[4]]**`
`Act[2,3][RInverse[4]]**`
`Act[2,3][RInverse[4]]**`
`Act[3,2,4][Phi[4]]**`
`Act[1,3,{2,4}]`
`[PhiInverse[4]]]**`
`ZInfinityInverse[4]**`
`ZInfinityInverse[4]]`

Notice the conflicting conventions — tangles are composed from bottom to top, while mathematica multiplies from top to bottom.

Out[33]= $1 + \frac{23\,[1212]}{24} - [121323] - \frac{1199\,[12123434]}{5760} + \frac{241\,[12132434]}{960} + \frac{1199\,[12314234]}{5760}$

Let us now observe that this paper does not prove the inconsistency of mathematics. Recomputing the invariant of the trefoil, this time using the presentation in page 141, we get the same answer as before:

In[34]:=
```
ModDegree[5,AP2CD[{1,3,2,4}, PhiInverse[4]**
  Act[{1,2},3,4][Phi[4]]**
  RInverse[4]**Act[3,4][RInverse[4]]**Act[2,1,{3,4}][Phi[4]]**
  Act[1,4,3][PhiInverse[4]]**Act[1,4][R[4]]**
  Act[4,1,3][Phi[4]]]**
  ZInfinityInverse[4]**ZInfinityInverse[4]]
```

Out[34]= $1+\frac{23\,[1212]}{24}-[121323]-\frac{1199\,[12123434]}{5760}+\frac{241\,[12132434]}{960}+\frac{1199\,[12314234]}{5760}$

Finally, typing `ModDegree[8,{FormalLog[R[7]],FormalLog[Phi[7]]}]`, waiting a long time and making some effort to typeset the result in a reasonable form, we find that $\log R=\frac{t^{12}}{2}$ up to degree 7, and that up to the same degree,

$$\log\Phi=\left(\frac{[ab]}{48}-\frac{8[aaab]+[abab]}{11520}+\frac{96[aaaaab]+4[aaabab]+65[aabbab]+68[abaaab]+4[ababab]}{5806080}\right)-(\text{interchange } a\leftrightarrow b),$$

where $[a_1\dots a_m]$ is a short for the iterated bracket $[a_1,[a_2,\dots,[a_{m-1},a_m]\dots]]$, and

$a=t^{12}=$ 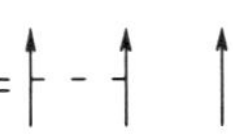 ; $b=t^{23}=$ 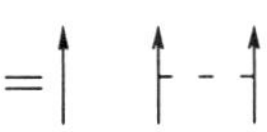 .

References

1. D. Altschuler and A. Coste, *Quasi-quantum groups, knots, three-manifolds, and topological field theory*, Commun. Math. Phys. **150** (1992) 83–107.
2. D. Bar-Natan, *On the Vassiliev knot invariants*, Topology **34** (1995) 423–472.
3. ______, *Vassiliev homotopy string link invariants*, Jour. of Knot Theory and its ramifications **4** (1995) 13–32.
4. ______, *Computer data files*, available via anonymous file transfer from `ftp.math.harvard.edu`, user name `ftp`, subdirectory `dror`. Read the file `README` first. For easier access, point your WWW browser at `http://www.math.harvard.edu/HTML/Individuals/Dror_Bar-Natan.html`.
5. N. Bergeron, *What is new about Hochschild homology*, Harvard Univ. preprint, November 1993.

6. J. S. Birman and X-S. Lin, *Knot polynomials and Vassiliev's invariants*, Invent. Math. **111** (1993) 225–270.
7. P. Cartier, *Construction combinatoire des invariants de Vassiliev-Kontsevich des nœuds*, C. R. Acad. Sci. Paris **316** Série I (1993) 1205–1210.
8. V. G. Drinfel'd, *Quasi-Hopf algebras*, Leningrad Math. J. **1** (1990) 1419–1457.
9. ———, *On quasitriangular Quasi-Hopf algebras and a group closely connected with Gal*($\bar{\mathbb{Q}}/\mathbb{Q}$), Leningrad Math. J. **2** (1991) 829–860.
10. D. K. Harrison, *Commutative algebras and cohomology*, Trans. Amer. Math. Soc. **104** (1962) 191–204.
11. M. M. Kapranov, *The permutoassociahedron, Mac Lane's coherence theorem and asymptotic zones for the KZ equation*, J. Pure and Appl. Alg. **85** (1993) 119–142.
12. C Kassel, *Quantum groups*, Springer GTM, **155**, New York 1994.
13. L. H. Kauffman, *On knots*, Princeton Univ. Press, Princeton, 1987.
14. M. Kontsevich, *Vassiliev's knot invariants*, Adv. in Sov. Math. **16(2)** (1993) 137–150.
15. T. Q. T. Le and J. Murakami, *On Kontsevich's integral for the HOMFLY polynomial and relations of multiple ζ-numbers*, Top. and its Appl., to appear.
16. ———, *Kontsevich integral for Kauffman polynomial*, Max-Planck-Institut Bonn preprint 93-33, March 1993.
17. ———, *Representation of tangles and Kontsevich's integral*, Max-Planck-Institut Bonn preprint, April 1993.
18. ———, *The universal Vassiliev-Kontsevich invariant for framed oriented links*, Max-Planck-Institut Bonn preprint, December 1993.
19. J. L. Loday, *Opérations sur l'homologie cyclique des algèbres commutatives* Inv. Math. **96** (1989) 205–230.
20. ———, *Cyclic Homology*, Springer, New-York, 1992.
21. S. Mac Lane, *Categories for the working mathematician*, Springer GTM **5**, New-York 1971.
22. M. Markl and J. D. Stasheff, *Deformation theory via deviations*, Preprint.
23. N. Yu. Reshetikhin, *Quasitriangle Hopf algebras and invariants of tangles*, Leningrad Math. J. **1** (1990) 491–513.
24. S. Piunikhin, *Combinatorial expression for universal Vassiliev link invariant*, Commun. Math. Phys. **168-1** 1–22.
25. ———, *Addendum to the paper "Combinatorial expression for universal Vassiliev link invariant"*, November 1993, Preprint.
26. S. Shnider and S. Sternberg, *Quantum groups — from coalgebras to Drinfel'd algebras*, International Press, Cambridge MA 1994.
27. V. A. Vassiliev, *Cohomology of knot spaces*, Theory of Singularities and its Applications (Providence) (V. I. Arnold, ed.), Amer. Math. Soc., Providence, 1990.
28. ———, *Complements of discriminants of smooth maps: topology and ap-*

plications, Trans. of Math. Mono. **98**, Amer. Math. Soc., Providence, 1992.
29. S. Wolfram, *Mathematica — a system for doing mathematics by computer*, Addison-Wesley, 1989.

Department of Mathematics, Harvard University, Cambridge, MA 02138

E-mail address: dror@math.harvard.edu

AMS/IP Studies in Advanced Mathematics
Volume **2**, 1997 (Part 1)

POWER SERIES EXPANSIONS AND INVARIANTS OF LINKS

XIAO-SONG LIN

One may think of Vassiliev invariants (see [2], [4], [5], [17] and [25, 26]) as link invariants with certain nilpotency. From this point of view, it is not surprising that Milnor's link invariants ([21, 22]) are of the same nature as Vassiliev invariants ([3], [19]). We give a conceptually clearer proof of this fact here by synthesizing Kontsevich's construction of Vassiliev invariants and Milnor invariants under the framework of power series expansions. The synthesis carried out in the first four sections of this paper motivates the discussion in Section 5 of some questions about infinitesimal structures of discrete groups and the possibility of constructing 3-manifold invariants under the same framework.

1. Expansions of free groups.

Let F_n be the free group generated by $x_1, \cdots, x_n$. Let $\mathcal{P}_n$ be the algebra of formal power series in non-commutative variables $X_1, \cdots, X_n$ over $\mathbb{C}$, i.e. $\mathcal{P}_n = \mathbb{C}[[X_1, \cdots, X_n]]$. The classical *Magnus expansion*

$$M: \ F_n \longrightarrow \mathcal{P}_n$$

is a homomorphism given by

$$M(x_i) = 1 + X_i, \tag{1.1}$$

$$M(x_i^{-1}) = 1 - X_i + X_i^2 - X_i^3 + \cdots . \tag{1.2}$$

If $g \in F_n$, we write

$$M(g) = 1 + \sum_{1 \le i_1, \cdots, i_k \le n} M(i_1, \cdots i_k) \, X_{i_1} \cdots X_{i_k}.$$

This paper is based partly on a talk given at the 1993 Georgia International Topology Conference. The research is supported in part by a Sloan Research Fellowship, NSF Grant DMS–9201091 and NSF Grant DMS–9304580 through the Institute for Advanced Study.

Then

$$M(i_1 \cdots i_k) = \left. \frac{\partial^k g}{\partial x_{i_1} \cdots \partial x_{i_k}} \right|_{x_1 = \cdots = x_n = 1} \tag{1.1}$$

where $\frac{\partial}{\partial x_i}$'s are Fox's free derivatives. See [20]. It can be shown that the Magnus expansion of F_n into $\mathcal{P}_n$ is an injection.

In general, we may define an *expansion* of F_n into $\mathcal{P}_n$ to be a homomorphism $J : F_n \longrightarrow \mathcal{P}_n$ such that

$$J(x_i) = 1 + X_i + \text{ higher order terms.}$$

The proof of the injectivity of the Magnus expansion carries over to any expansion J without difficulties. The nice things about the Magnus expansion are (1) its coefficients are all integers; and (2) they can be expressed in close forms (1.1).

The following is another example of expansions which seems to be quite natural at the first glance. This expansion $E : F_n \longrightarrow \mathcal{P}_n$ is given by

$$E(x_i) = \exp(X_i) = 1 + X_i + \frac{1}{2!}X_i^2 + \frac{1}{3!}X_i^3 + \cdots . \tag{1.2}$$

We are wondering whether the coefficients of $E(g)$, $g \in F_n$ could be expressed by some nice formulae.

We now use K.T. Chen's theory of formal power series connections and their induced monodromy representations (see [8] or [11] for an introduction to Chen's work) to describe one more example of expansions of F_n.

Think of

$$F_n = \pi_1(\mathbb{C} \setminus \{1, \cdots, n\})$$

as the fundamental group of the n-punctured complex plane. Let

$$\omega_i = \frac{1}{2\pi\sqrt{-1}} \frac{dz}{z - i}, \qquad i = 1, \cdots, n$$

be closed 1-forms on $\mathbb{C} \setminus \{1, \cdots, n\}$. Consider a trivial bundle over $\mathbb{C} \setminus \{1, \cdots, n\}$ with $\mathcal{P}_n$ as the fibre and the connection

$$\Omega = \sum_{i=1}^{n} \omega_i X_i$$

with values in $\mathcal{P}_n$. Since Ω is flat (as $d\omega_i = 0$ and $\omega_i \wedge \omega_j = 0$), we get a monodromy representation $I : F_n \longrightarrow \mathcal{P}_n$. If $g \in F_n$ is represented by

a loop $\gamma = \gamma(t)$,
$0 \le t \le 1$ in $\mathbb{C}\setminus\{1,\cdots,n\}$, we have

$$I(g) = 1 + \sum_{1\le i_1,\cdots,i_k\le n} I(i_1\cdots i_k)\; X_{i_1}\cdots X_{i_k}$$

where

$$I(i_1\cdots i_k) = \left(\frac{1}{2\pi\sqrt{-1}}\right)^k \int_{0\le t_1\le\cdots\le t_k\le 1} \bigwedge_{r=1}^{k} \frac{d\gamma(t_r)}{\gamma(t_r)-i_r}. \tag{1.3}$$

These iterated integrals are illustrated in Figure 1 where we may think of the punctures $\{1,\cdots,n\}\subset\mathbb{C}$ as constant paths. One may calculate the first order integrals to get

$$I(x_i) = 1 + X_i + \text{ higher order terms.}$$

So I is an expansion of F_n.

The following is a nice property of the expansion I.

Lemma 1.1. *(Chen, see [7]) $I(g) = \exp$ (Lie element of $\mathcal{P}_n$) for every $g\in F_n$.*

Chen proved this statement in [7] for general monodromy representations derived from formal power series connections. In particular, we have
(1.4)
$$I(x_i) = \exp\;(X_i + \text{multiple commutators in } X_1,\ldots,X_n \text{ involving } X_i)\,.$$

Here if we drop off the i-th puncture, the loop represents x_i in $\mathbb{C}\setminus\{1,\cdots,n\}$ will be homotopic to a constant loop. On the other hand, if we let $X_i = 1$ in $I(x_i)$, we get the expansion of $x_i\in\pi_1\left(\mathbb{C}\setminus\{1,\cdots,\hat{i},\cdots,n\}\right)$. That is why all multiple commutators in $\log I(x_i)$ must involve X_i.

Notice that Lemma 1.1 is also true for the expansion E according to the *Campbell-Hausdorff formula* which says that $\log\,(\exp(X)\exp(Y))$ is a Lie element in X and Y. But the expansion I is better than E in that it can be extended to an expansion of the pure braid group. See Section 3.

All expansions of F_n into $\mathcal{P}_n$ are essentially the same. This is stated precisely in the following lemma.

Lemma 1.2. *If $J_1, J_2 : F_n \longrightarrow \mathcal{P}_n$ are two expansions, then there is an unique automorphism $\Phi : \mathcal{P}_n \longrightarrow \mathcal{P}_n$ such that $J_2 = \Phi\cdot J_1$.*

Proof. We may assume $J_1 = M$, the Magnus expansion. An endomorphism $\Phi: \mathcal{P}_n \longrightarrow \mathcal{P}_n$ can be defined by setting

$$\Phi(X_i) = J_2(x_i) - 1.$$

Obviously, we have $J_2 = \Phi \cdot J_1$. Such a Φ is certainly unique. So, we only need to check that Φ has an inverse. As

$$J_2(x_i) = 1 + X_i + \text{ higher order terms in } X_1, \cdots, X_n,$$

we may solve

$$\Phi(X_i) = X_i + \text{ higher order terms in } X_1, \cdots, X_n \tag{1.5}$$

for X_i as formal power series in $\Phi(X_1), \cdots, \Phi(X_n)$. The endomorphism Ψ of $\mathcal{P}_n$ defined by sending X_i to the solution of (1.5) is easily seen to be a right inverse of Φ, i.e. $\Psi \cdot \Phi = \text{id}$. So $\Psi: \mathcal{P}_n \longrightarrow \mathcal{P}_n$ is onto. On the other hand, since

$$\Psi(X_i) = X_i + \text{ higher order terms},$$

Ψ is injective. This shows Ψ (and therefore Φ) is an automorphism. □

Here is an useful consequence of Lemma 1.2. Let J be an expansion of F_n, $\phi: F_n \longrightarrow F_n$ be an automorphism which sends each x_i to one of its conjugations. Then $J \cdot \phi$ is also an expansion of F_n. Therefore, there is an unique automorphism $\Phi: \mathcal{P}_n \longrightarrow \mathcal{P}_n$ such that

$$J \cdot \phi = \Phi \cdot J.$$

Let $\text{Aut}_0(F_n)$ (resp. $\text{Aut}_0(\mathcal{P}_n)$) be the group of automorphisms of F_n (resp. $\mathcal{P}_n$) which send each x_i to one of its conjugations (resp. send X_i to $X_i +$ higher order terms). Then, Lemma 1.2 implies that

$$\begin{aligned} \text{Aut}_0(F_n) &\longrightarrow \text{Aut}_0(\mathcal{P}_n) \\ \phi &\longmapsto \Phi \end{aligned}$$

is a homomorphism.

Notice that one may think of $\mathcal{P}_n$ as the "tangent space" of F_n at 1 as suggested by (1.1). In this sense, the map $\phi \mapsto \Phi$ may be thought of as going from a diffeomorphism to its tangent map.

2. Link-homotopy and Milnor's μ–invariants.

The *reduced free group* $RF_n = F_n/\text{relations}$ was introduced by Milnor in his study of link-homotopy [21]. Here the relations in RF_n are

$$[x_i,\ g x_i g^{-1}] = 1, \quad \forall g \in F_n$$

where the bracket [,] is the group commutator. The group RF_n has the following geometric meaning.

Link-homotopy is a deformation of links in the 3–space allowing strands from the same component of a link to pass over each other. Consider a trivial link L with n component in $\mathbb{R}^3$. We may identify $\pi_1(\mathbb{R}^3 \backslash L)$ with F_n. Let γ be a simple closed curve in $\mathbb{R}^3 \backslash L$ representing $g \in F_n$. Then the link $L \cup \gamma$ is link-homotopically trivial iff $g = 1$ in RF_n. See [21] or [12].

The set of relations in RF_n can be described in a different way:

$$RF_n \cong F_n \Big/ \begin{array}{l}\text{multiple group commutators in } x_1, \cdots, x_n \\ \text{where some } x_i \text{ appears at least twice} = 1.\end{array}$$

From this description, it is easy to see that RF_n is a nilpotent group of class n (all commutators of length at least $n+1$ are equal to 1).

Let $\mathcal{L}[X_1, \cdots, X_n]$ be the completion of the free Lie algebra generated by
$X_1, \cdots, X_n$. Define

$$\mathcal{L}\,RF_n = \mathcal{L}[X_1, \cdots, X_n] \Big/ \begin{array}{l}\text{multiple algebra commutators in} \\ X_1, \cdots, X_n \text{ where some } X_i \text{ appears} \\ \text{at least twice} = 0.\end{array}$$

This is a nilpotent Lie algebra of class n. By using the Campbell-Hausdorff formula, we may conclude:

Lemma 2.1. *Both expansions $E, I : F_n \longrightarrow \mathcal{P}_n$ reduce to expansions $E_R, I_R : RF_n \longrightarrow \exp(\mathcal{L}\,RF_n)$. Both E_R and I_R are injection.*

For example, since

$$\begin{aligned} E\left([x_1, x_2\, x_1 x_2^{-1}]\right) &= \exp\big([X_1,\ [X_1, X_2]] \\ &\quad + \begin{array}{l}\text{commutators involving either} \\ [X_1,\ [X_1, X_2]] \text{ or } [X_2,\ [X_1, X_2]]\end{array}\big) \\ E\left([x_2, x_1\, x_2 x_1^{-1}]\right) &= \exp\big([X_2,\ [X_2, X_1]] \\ &\quad + \begin{array}{l}\text{commutators involving either} \\ [X_1,\ [X_1, X_2]] \text{ or } [X_2,\ [X_1, X_2]]\end{array}\big), \end{aligned}$$

we have

$$[x_1,\, x_2x_1x_2^{-1}] = 1 \quad \text{and} \quad [x_2,\, x_1x_2x_1^{-1}] = 1$$

iff

$$[X_1,\, [X_1, X_2]] = 0 \quad \text{and} \quad [X_2,\, [X_1, X_2]] = 0.$$

Similar argument works for the expansion I because of (1.4).

We can think of relations in $\mathcal{L}RF_n$ as the infinitesimal version of relations in RF_n. So, the relations in $\mathcal{L}RF_n$ will be referred to as *infinitesimal link-homotopy relations.*

Link-homotopy classification of links in the 3-space was obtained in [12]. We will restrict ourself to the so-called string links in this paper as their classification up to link-homotopy has a particularly nice structure which we want to explore further. All the statements made below without explanation are referred to [12].

A *string link* is a proper imbedding

$$\sigma : \coprod_{i=1}^{n} I_i \longrightarrow \mathbb{C} \times I$$

where $I_i = I = [0,1]$, such that $\sigma(\partial_0 I_i) = i \times 0$ and $\sigma(\partial_1 I_i) = i \times 1$. See Figure 2.

Denote by SL_n the set of ambient isotopy classes of string links with n strings. The multiplication of two string links $\sigma_1, \sigma_2 \in SL_n$ is a string link $\sigma_1\sigma_2$ obtained by stacking σ_1 on the top of σ_2. Under this multiplication, SL_n becomes a semigroup.

Denote by H_n the set of link-homotopy classes of string links with n strings. It turns out that H_n is actually a torsion free nilpotent group. This is best understood in terms of induced automorphisms of RF_n of string links.

Given a string link $\sigma \in SL_n$. The inclusions

$$\begin{array}{c} \pi_1\left(\mathbb{C}\setminus\{1,\cdots,n\}\times 1\right) \\ \downarrow \\ \pi_1\left(\mathbb{C}\times I\setminus\sigma\left(\coprod_{i=1}^{n} I_i\right)\right) \\ \uparrow \\ \pi_1\left(\mathbb{C}\setminus\{1,\cdots,n\}\times 0\right) \end{array}$$

reduce to an automorphism

$$\phi_\sigma : RF_n \longrightarrow RF_n.$$

Intuitively, this amounts to say that every loop in $\mathbb{C}\setminus\{1,\cdots,n\}\times 0$ can be pushed upward to a loop in $\mathbb{C}\setminus\{1,\cdots,n\}\times 1$ modulo link-homotopy relations. Actually, it can be proved that every string link is link-homotopic to a pure braid. Recall that a theorem of Artin represents faithfully the *pure braid group* PB_n into a certain subgroup of $\mathrm{Aut}(F_n)$. We have an analogue of Artin's theorem.

Theorem 2.2. *(Habegger-Lin) The association $\sigma \longmapsto \phi_\sigma$ gives us a faithful representation $H_n \longrightarrow Aut(RF_n)$ whose image is the collection of all $\phi \in Aut(RF_n)$ characterized by the following two properties:*

- *$\phi(x_i)$ is a conjugate of x_i; and*
- *$\phi(x_1\cdots x_n) = x_1\cdots x_n$.*

By this theorem, a string link $\sigma \in H_n$ is determined by n elements

$$\phi_\sigma(x_i)\,x_i^{-1} \in RF_n, \quad i = 1,\cdots,n.$$

Since RF_n is nilpotent, we may choose a basis for RF_n. The coefficients of $\phi_\sigma(x_i)\,x_i^{-1}$ under a chosen basis give us invariants of σ.

For example, we may choose a Hall basis for F_n relative to $\{x_1,\cdots,x_n\}$ and drop off those commutators in this basis where some x_i appears at least twice. The remaining finitely many elements gives us a *reduce Hall basis* for RF_n. The coefficients of $\phi_\sigma(x_i)\,x_i^{-1}$ under a reduced Hall basis are Milnor's *μ-invariants* (with distinct indices).

Let us consider in some detail the case $n = 3$. Then, the following elements may be chosen as a reduced Hall basis:

$$x_1, x_2, x_3,\ [x_1,x_2],\ [x_1,x_3],\ [x_2,x_3],\ [x_1,\ [x_2,x_3]].$$

For a string link $\sigma \in H_3$, we have

$$\begin{aligned}
\phi_\sigma(x_1)\,x_1^{-1} &= [x_1,x_2]^{\mu(12)}\,[x_1,x_3]^{\mu(13)}\,[x_1,\ [x_2,x_3]]^{\mu(123)},\\
\phi_\sigma(x_2)\,x_2^{-1} &= [x_1,x_2]^{\mu(12)}\,[x_2,x_3]^{\mu(23)}\,[x_1,\ [x_1,x_2]]^{\mu(123)},\\
\phi_\sigma(x_3)\,x_3^{-1} &= [x_1,x_3]^{\mu(13)}\,[x_2,x_3]^{\mu(23)}\,[x_1,\ [x_2,x_3]]^{\mu(123)}.
\end{aligned}$$

So, elements in H_3 are classified by the integers

$$\mu(12),\ \mu(13),\ \mu(23),\ \mu(123).$$

The first three integers are linking numbers of corresponding strings whereas the last integer is a "higher order linking number".

Using the expansion $I_R : RF_n \longrightarrow \exp(\mathcal{L}\,RF_n)$, we may transform an automorphism $\phi \in \mathrm{Aut}_0(RF_n)$ to an automorphism $\Phi \in \mathrm{Aut}_0(\mathcal{U}_n)$ by the construction following Lemma 1.2. Here $\mathcal{U}_n$ is the universal enveloping algebra of $\mathcal{L}\,RF_n$.

Lemma 2.3. $\Phi(X_i) \in \mathcal{L}\, RF_n$.

Proof. This follows from the fact that $I_R \cdot \phi = \Phi \cdot I_R$. □

Thus, we may transform an automorphism $\phi_\sigma \in \mathrm{Aut}\,(RF_n)$ corresponding to a string link $\sigma \in H_n$ to an automorphism $\Phi_\sigma \in \mathrm{Aut}\,(\mathcal{L}\, RF_n)$. We have $\Phi_{\sigma_1\sigma_2} = \Phi_{\sigma_1} \cdot \Phi_{\sigma_2}$ for $\sigma_1, \sigma_2 \in H_n$.

A string link $\sigma \in H_n$ now is determined by n elements

$$\Phi_\sigma(X_i) - X_i \in \mathcal{L}\, RF_n, \quad i = 1, \cdots, n.$$

If we choose a Hall basis relative to $\{X_1, \cdots, X_n\}$ for $\mathcal{L}[X_1, \cdots, X_n]$ and drop off those commutators in this basis where some X_i appears at least twice, we get a basis for $\mathcal{L}\, RF_n$. The transition between the coefficients of $\Phi_\sigma(X_i) - X_i$ and μ-invariants can be thought of as resulting from a change of bases for the complex completion of the torsion free nilpotent group RF_n (see, for example, [10]). Therefore, it is clear that coefficients of $\Phi_\sigma(X_i) - X_i$ of length k are linear combinations of μ-invariants of length $\leq k$, and vice versa. Here the *length* of a coefficient is the length of the corresponding commutator.

Finally, the group H_n can be decomposed into semi-direct products of reduced free groups RF_1, $RF_2, \cdots, RF_{n-1}$. As we mentioned before, every string link is link-homotopic to a pure braid. Recall that the pure braid group PB_n can be decomposed as

$$PB_n = F_1 \rtimes F_2 \rtimes \cdots \rtimes F_{n-1}.$$

The free groups F_{j-1} is generated by $\{g_{ij};\quad 1 \leq i < j\}$ where g_{ij} is the pure braid shown in Figure 3. The group H_n can be similarly decomposed as

$$H_n = RF_1 \rtimes RF_2 \rtimes \cdots \rtimes RF_{n-1}.$$

3. Formal power series connections and their monodromies.

This section is a short exposition of Chen's theory of formal power series connections and their monodromies in a special case which was worked out in great detail by Kohno. We first follow the exposition of [15] and then discuss the pure braid group as an example.

Let $\mathcal{M}$ be a smooth manifold. Suppose that we have closed 1-forms $\omega_1, \cdots, \omega_n$ represents a basis for the de Rahm cohomology group $H^1(\mathcal{M})$. We write

$$\omega_i \wedge \omega_j = \sum_{k=1}^{m} c_{ij}^k\, \alpha_k$$

where $\alpha_1, \cdots, \alpha_k$ are linearly independent 2-forms. Consider the connection

$$\Omega = \sum_{i=1}^{n} \omega_i \, X_i$$

on $\mathcal{M}$ with values in $\mathcal{P}_n$. We have

$$\begin{aligned} d\Omega - \Omega \wedge \Omega &= - \left(\sum_{i=1}^{n} \omega_i X_i \right) \wedge \left(\sum_{i=1}^{n} \omega_i X_i \right) \\ &= - \sum_{i<j} \omega_i \wedge \omega_j \, [X_i, X_j] \\ &= - \sum_{i<j} \left(\sum_k C_{ij}^k \, \alpha_k \right) [X_i, X_j] \\ &= - \sum_k \alpha_k \left(\sum_{i<j} c_{ij}^k \, [X_i, X_j] \right). \end{aligned}$$

So, if we think of Ω as a connection with values in

$$\overline{\mathcal{P}}_n = \mathcal{P}_n \Big/ \sum_k c_{ij}^k \, [X_i, X_j] = 0,$$

it is flat. Thus, we may get a homomorphism

$$I : \ \pi_1(\mathcal{M}) \longrightarrow \overline{\mathcal{P}}_n$$

as the monodromy of that flat connection Ω. This homomorphism can be expressed in terms of iterated integrals: If $\gamma = \gamma(t), \ 0 \le t \le 1$ is a loop representing an elements $g \in \pi_1(\mathcal{M})$, then

$$\begin{aligned} I(g) &= 1 + \sum_{k=1}^{\infty} \int_{0 \le t_1 \le \cdots \le t_k \le 1} \Omega(\dot\gamma(t_k)) \wedge \cdots \wedge \Omega(\dot\gamma(t_1)) \\ &= 1 + \sum_{k=1}^{\infty} \int_0^1 \Omega(\dot\gamma(t_k)) \int_0^{t_k} \Omega(\dot\gamma(t_{k-1})) \cdots \int_0^{t_2} \Omega(\dot\gamma(t_1)). \end{aligned}$$

For example, if $x_i \in \pi_1(\mathcal{M})$ represents a cycle dual to ω_i, then

$$I(x_i) = 1 + X_i + \text{ higher order terms.}$$

Thus, we will call I an *expansion* of $\pi_1(\mathcal{M})$ into $\overline{\mathcal{P}}_n$.

We can define a completed Lie algebra

$$\mathcal{L}(\mathcal{M}) = \mathcal{L}[X_1, \cdots X_n] \Big/ \sum_k c_{ij}^k \, [X_i, X_j] = 0.$$

This is the Lie algebra dual to Sullivan's 1-minimal model of $\mathcal{M}$. See the discussion in Section 5. We notice that similar to Lemma 1.1,

$$I(g) \in \exp\left(\mathcal{L}(\mathcal{M})\right)$$

for every $g \in \pi_1(\mathcal{M})$.

An important example of the previous discussion is given by

$$\mathcal{M} = \mathbb{C}^n \setminus \bigcup_{1\le i<j\le n} \{z_i = z_j\},$$

the configuration space of n distinct ordered points in $\mathbb{C}$. We have $\pi_1(\mathcal{M}) = PB_n$. Let

$$\omega_{ij} = \omega_{ji} = \frac{1}{2\pi\sqrt{-1}} d \log(z_i - z_j).$$

Then, $\{\omega_{ij}; \quad 1 \le i < j \le n\}$ represents a basis for $H^1(\mathcal{M})$ and the only kind of relations among wedges of these closed 1-forms are

$$\omega_{ij} \wedge \omega_{j\ell} + \omega_{j\ell} \wedge \omega_{\ell i} + \omega_{\ell i} \wedge \omega_{ij} = 0.$$

See [1].

Let $\{C_{ij} = C_{ji}; \quad 1 \le i < j \le n\}$ be a set of formal non-commutative variables. It is not hard to check that in order to make the connection

$$\Omega = \sum_{1\le i<j\le n} \omega_{ij}\, C_{ij} \tag{3.1}$$

flat, we should assume

$$\begin{cases} [C_{ij},\ C_{k\ell}] = 0, & i, j, k, \ell \text{ are distinct} \\ [C_{ij},\ C_{i\ell} + C_{\ell j}] = 0. \end{cases} \tag{3.2}$$

These relations are called by Kohno *infinitesimal pure braid relations*. Notice that the connection (3.1) is the formal version of the so-called *Knizhnik–Zamolodchikov connection* in conformal field theory [14].

We now define $\mathcal{C}_n$ to be the algebra of formal power series in C_{ij}'s which are subject to the relations (3.2). The monodromy of the connection Ω gives use to an expansion

$$I : PB_n \longrightarrow \mathcal{C}_n.$$

This extends the expansion I for the free group F_n discussed in Section 1 as we may think of F_n as a subgroup of PB_{n+1} generated by $\{g_{i,n+1};\ i = 1, \ldots, n\}$.

We may have a diagrammatic description of the algebra $\mathcal{C}_n$. The object in Figure 4 is defined to be the *chord diagram* of C_{ij}, where the dashed line segment is called a *chord*. Let $F = C_{i_1 j_1} \cdots C_{i_k j_k}$ be a monomial in C_{ij}'s. Then the *chord diagram* of F is obtained by stacking the chord diagrams of $C_{i_1 j_1}, \cdots, C_{i_k j_k}$ together in the given order with $C_{i_1 j_1}$, on the top and $C_{i_k j_k}$ in the bottom.

Now the first relation in (3.2) can be interpreted as allowing the order of two chords in a chord diagram being switched as long as the end points of these two chords are all distinct. The second relation in (3.2) can be depicted as in Figure 5 which is clearly the horizontal version of the *4-term relation* in the theory of Vassilierv link invariants.

Finally, it is easy to calculate that for the pure braid $g_{ij} \in PB_n$ depicted in Figure 3, we have

$$I(g_{ij}) = 1 + C_{ij} + \text{ higher order terms.}$$

4. Kontsevich expansion of string links

The Kontsevich expansion of SL_n generalizes the expansion of PB_n discussed in the previous section. To describe it, we first need to generalize the algebra $\mathcal{C}_n$.

A *chord diagram* with k chords ($k \geq 0$) on n straight strings is a combinatorial pattern of pairing $2k$ points given on these strings. Figure 6 is an example of a chord diagram with 5 chords.

If D_1 and D_2 are two chord diagrams, the multiplication $D_1 D_2$ is a chord diagram obtained by stacking D_1 on the top of D_2. Then $\mathcal{D}_n$ is the completion of the graded algebra generated by chord diagrams (where the grading comes from the number of chords) subject to the following relations:

- if there is a chord whose end points are adjacent to each other on a single string, the corresponding chord diagram is set to be zero;
- 4-term relations as depicted in Figure 7.

Notice that the difference between Figure 5 and Figure 7 is that the chords in Figure 7 may not be all horizontal when the underlying trivial string link is straightened up.

For a string link $\sigma \in SL_n$, we assume that

$$\pi \cdot \sigma : \coprod_{i=1}^{n} I_i \xrightarrow{\sigma} \mathbb{C} \times I \xrightarrow{\pi} I$$

is a Morse function. We write

(4.1)
$$I(\sigma) = 1 + \sum_{k=1}^{\infty} \left(\frac{1}{2\pi\sqrt{-1}}\right)^k \sum_{P_1,\cdots,P_k} \int_{0 \leq t_1 \leq \cdots \leq t_k \leq 1} \bigwedge_{i=1}^{m} \varepsilon_i \frac{dz_i - dz_i'}{z_i - z_i'} D_{P_k \cdots P_1}$$

as an invertible element in $\mathcal{D}_n$. Here:

- 1 is the identity of $\mathcal{D}_n$: the chord diagram with no chords;
- P_i is a horizontal chord on σ with end points

$$(t_i,\ z_i(t_i)) \quad \text{and} \quad (t_i,\ z_i'(t_i));$$

- $\varepsilon_i = -1$ if the orientations of the strings at the end points of P_i are opposite and $\varepsilon_i = 1$ otherwise;
- $D_{P_k \cdots P_1} \in \mathcal{D}_n$ is a chord diagram obtained by straightening σ but keeping the chords $P_1, \cdots, P_k$ there (see Figure 8).

We have to normalize $I(\sigma)$ in order to get a well-defined expansion

$$K : SL_n \longrightarrow \mathcal{D}_n.$$

It is defined as

$$K(\sigma) = I(\sigma) \cdot I(\infty_1)^{-u_1} \cdot \ \cdots \ \cdot I(\infty_n)^{-u_n} \tag{4.2}$$

where u_i is the number of maxima on the i-th string and ∞_i is the string link shown in Figure 9.

For the proof of the invariance of K under isotopy, see [2].

Here are several properties of the Kontsevich expansion K:

- If σ is a pure braid, then $K(\sigma) = I(\sigma)$;
- K is a homomorphism of the semigroup SL_n into the group of units of $\mathcal{D}_n$.

Another important property of the expansion K is related with the notion of *Vassiliev invariants* or *invariants of finite type* (see [2], [4], [5], [17] and [25, 26]). We may extend the notion of *singular links* to *singular string links*. They are immersions

$$\coprod_{i=1}^{n} I_i \longrightarrow \mathbb{C} \times I$$

with transverse double points in the interior. We may define inductively

$$K(\sigma^{(k)}) = K(\sigma_+^{(k-1)}) - K(\sigma_-^{(k-1)})$$

where $\sigma^{(k)}$ is a string link with k double points and $\sigma_+^{(k-1)}$ or $\sigma_-^{(k-1)}$ is obtained from $\sigma^{(k)}$ by resolving one of the k double points into either a positive crossing or a negative crossing. Let $\mathcal{D}_n^{(k)}$ be the completed ideal of $\mathcal{D}_n$ generated by chord diagrams with at least k chords. Then we have

$$K(\sigma^{(k)}) \in \mathcal{D}_n^{(k)}. \tag{4.3}$$

See [2].

We want to study the behavior of K under the reduction $SL_n \longrightarrow H_n$.

Observation I. (Bar-Natan, see [3]). Let $R\mathcal{D}_n = \mathcal{D}_n$ modulo those chord diagrams where there is a chord with two end points on the same string. Then, $K : SL_n \longrightarrow \mathcal{D}_n$ reduces to $K_R : H_n \longrightarrow R\mathcal{D}_n$ and we have a commutative diagram

$$\begin{array}{ccc} SL_n & \xrightarrow{K} & \mathcal{D}_n \\ \downarrow & & \downarrow \\ H_n & \xrightarrow{K_R} & R\mathcal{D}_n. \end{array}$$

It turns out that $R\mathcal{D}_n$ is generated by horizontal diagram $\{C_{ij};\ 1 \leq i < j \leq n\}$ just like H_n is generated by pure braids! This can be seen by playing with 4-term relations together with the reduction made in defining $R\mathcal{D}_n$.

Observation II. We define $\mathcal{L}H_n$ to be the free Lie algebra $\mathcal{L}[C_{ij};\ 1 \leq i < j \leq n]$ modulo two set of relations:

- infinitesimal pure braid relations;
- infinitesimal link-homotopy relations for each set

$$\{C_{ij},\ 1 \leq i < j\}, \quad j = 2, \cdots, n.$$

Then

$$K_R :\ H_n \longrightarrow \exp(\mathcal{L}H_n) \subset R\mathcal{D}_n$$

and

$$K_R(g_{ij}) = 1 + C_{ij} + \text{ higher order terms.}$$

This can be seen from the semi-direct product decomposition

$$H_n = RF_1 \rtimes \cdots \rtimes RF_{n-1}.$$

As pointed out by Bar-Natan, the universal enveloping algebra of $\mathcal{L}H_n$ is $R\mathcal{D}_n$.

Observation III. Using the semi-direct decomposition of H_n and the fact that expansions of RF_j are injective, it is easy to see that K_R is injective also. So we may simply write

$$g_{ij} = 1 + C_{ij} + \text{ higher order terms.}$$

Then, we may solve for C_{ij} and get

$$C_{ij} = (g_{ij} - 1) + \text{ products of two or more } (g_{k\ell} - 1)'\text{s}$$

in the completion with respect to a certain grading . Thus, if we denote by V the underlying vector space of $\mathcal{L}RF_n$, we get a representation

$$(4.4) \qquad R\mathcal{D}_n \longrightarrow \mathrm{End}(V)$$

which makes the following diagram commutative:

$$\begin{array}{ccc} H_n & \xrightarrow{K_R} & R\mathcal{D}_n \\ \downarrow & & \downarrow \\ \mathrm{Aut}(\mathcal{L}RF_n) & \longrightarrow & \mathrm{End}(V). \end{array}$$

Denote by $V^{(r)}$ the subspace of V spanned by commutators of length not smaller than r.

Proposition 4.1. *C_{ij} sends $V^{(r)}$ to $V^{(r+1)}$. In particular, the representation (4.4) is unipotent.*

Proof. Since $g_{ij}(X_k) = X_k +$ commutators and $g_{ij} \in \mathrm{Aut}(\mathcal{L}RF_n)$, $g_{ij}-1$ sends every commutator in $X_1, \cdots, X_n$ of length r to a linear combination of commutators in $X_1, \cdots, X_n\}$ of length $> r$. This proves the first assertion. As $\mathcal{L}RF_n$ is nilpotent of class n, the image of a monomial $C_{i_1 j_1} \cdots C_{i_k j_k}$ in $\mathrm{End}(V)$ is zero when $k \geq n$. □

An immediate corollary of this proposition is that Milnor's μ-invariants for H_n are all of type $\leq n-1$. This fact was originally proved in [3] and [19]. The proof given here is conceptually clear (we hope) than the original ones, thanks to the Kontsevich expansion and the appropriate rearrangement of Milnor's μ-invariants.

Recall that an invariant of (string) links v is of *type* $< k$ if

$$v(\sigma^{(k)}) = 0$$

for all (string) links $\sigma^{(k)}$ with k double points.

Corollary 4.2. *Milnor's μ-invariants for H_n are all of type $\leq n-1$.*

Proof. We consider the coefficients of

$$v_i(\sigma) = \Phi_\sigma(X_i) - X_i \in \mathcal{L}RF_n$$

first. We can identify $\Phi_\sigma \in \mathrm{End}(V)$ with $K_R(\sigma)$ for $\sigma \in H_n$ and then

$$v_i(\sigma^{(k)}) = K_R(\sigma^{(k)})\ (X_i) \in V^{(k+1)}$$

since $K_R(\sigma^{(k)}) \in \mathcal{D}_n^{(k)}$. Thus, coefficients of $\Phi_\sigma(X_i) - X_i$ of length $\leq k$ are of type $\leq k-1$. As length k μ-invariants of σ are linear combinations of coefficients of $\Phi_\sigma(X_i) - X_i$ of length $\leq k$, this implies that Milnor's μ-invariants for H_n are all of type $\leq n-1$. □

Remarks. **1.** Milnor μ-invariants for H_n are μ-invariants with distinct indices. We can capture all μ-invariants with repeated indices by studying link-homotopy after splitting each string into some parallel copies [22]. Analogues of the Habegger-Lin theorem in this setting were established in this author's unpublished thesis (1988, UCSD). Thus, we actually have a whole family of unipotent representations of the algebra $\mathcal{D}_n$ exhausting all μ-invariants.

2. Associated with every finite-dimensional representation of a semi-simple Lie algebra, there is a finite-dimensional representation of the algebra $\mathcal{D}_n$. See [2] and [18]. It will be interesting to know whether this family of representations of $\mathcal{D}_n$ is complementary in some sense to the family of unipotent representations of $\mathcal{D}_n$. Here is a hint about the difference of these two families of representations. Kohno [16] (see also [9]) showed that for the "semi-simple" family, the induced representations of PB_n are all conjugate to representations constructed with respect to certain R-matrices. One may think of this property as an indication of the locality of representations in the "semi-simple" family. On the contrary, representations in the unipotent family are in a sense global: To decide whether two strands at a crossing belong to the same component, one has to take a look at the whole picture of the (string) link in question. This probably explains why we don't have any Jones-type link invariant for link-homotopy (see [13]).

5. Infinitesimal relations of discrete groups.

Let G be a discrete group generated by $\{x_\alpha\}$. Then $G \cong F/R$ where F is the free group generated by $\{x_\alpha\}$ and R is a normal subgroup of F. Elements in R are *relations* of G.

Suppose $\mathcal{P}$ is the algebra of formal power series in $\{X_\alpha\}$. For $r \in R$, let

$$\delta(r) = \begin{array}{l}\text{the homogeneous part of the smallest}\\ \text{degree of } M(r) - 1,\end{array}$$

where $M : F \longrightarrow \mathcal{P}$ is the Magnus expansion. Notice that $\delta(r)$ is a Lie element of $\mathcal{P}$. We call $\delta(r)$ an *infinitesimal relation* of G.

We define a Lie algebra $\mathcal{L}G$ as the completed free Lie algebra $\mathcal{L}[X_\alpha]$ modulo *all* infinitesimal relations of G, i.e.

$$\mathcal{L}G = \mathcal{L}[X_\alpha] \,/\, \{\delta(r);\ r \in R\}.$$

For the groups $G = F_n$, RF_n, PB_n, H_n, the Lie algebras we studied in previous sections are exactly the same as $\mathcal{L}G$. In these cases, the Lie algebra $\mathcal{L}G$ turns out to be the completion of the dual Lie algebra of Sullivan's 1-minimal model of $K(G,1)$ [24]. Thus, in these cases, $\exp(\mathcal{L}G)$ (over the field of rationals) is isomorphic to the Malcev completion of G [6]. In these cases, Chen's method provides in a quite concrete form an expansion $G \longrightarrow \exp(\mathcal{L}G)$ which induces an isomorphism between the Malcev completion of G and $\exp(\mathcal{L}G)$ over the field of real or complex numbers. We therefore ask the following question (without specifying the ground field):

Question 1. Let G be a discrete group.

(1) When is $\exp(\mathcal{L}G)$ isomorphic to the Malcev completion of G?
(2) When there is a homomorphism $I : G \longrightarrow \exp(\mathcal{L}G)$ such that $I(x_\alpha) = 1 + X_\alpha +$ higher order terms?

Question 5.1 (1) must sounds very familiar to the experts. But as it seems that Malcev completion of a group has not been discussed from such a point of view before in the literature, it is probably worthwhile to pose it here for the record. The answer to (2) of Question 5.1 is certainly much more difficult.

Again, although the following example is well-known to the experts, we still want to explain it here briefly in the hope that it will further clarify Question 5.1.

Example 1. Consider the surface group

$$G = \left\langle a_i, b_i,\ i = 1, \cdots, g \,\middle|\, \prod_{i=1}^{g} [a_i, b_i] = 1 \right\rangle.$$

Then

$$\mathcal{L}G = \mathcal{L}\left[A_i, B_i,\ i = 1, \cdots, g\right] \Big/ \sum_{i=1}^{g} [A_i, B_i] = 0.$$

It happens that $\mathcal{L}G$ is the dual Lie algebra of the 1-minimal model of the surface S_g of genus g.

To get an expansion

$$I : G \longrightarrow \exp(\mathcal{L}G),$$

by using Chen's method, all we need is to construct a flat formal power series connection Ω in the variables $A_i, B_i,\ i = 1, \cdots, g$ subject to

$$\sum_{i=1}^{g} [A_i, B_i] = 0 \tag{5.1}$$

on the surface S_g.

Let $\alpha_i, \beta_i \in \Lambda^1(S_g)$ be closed 1-forms dual to a_i, b_i respectively. We have

$$\begin{aligned}\alpha_i \wedge \beta_i &= \omega + d\tau_i, \quad i = 1, \cdots, g,\\ \alpha_i \wedge \alpha_j &= d\xi_{ij}, \quad i < j,\\ \alpha_i \wedge \beta_j &= d\zeta_{ij}, \quad i \neq j,\\ \beta_i \wedge \beta_j &= d\eta_{ij}, \quad i < j.\end{aligned}$$

Here ω is an area form of S_g and $\tau_i,\ \xi_{ij},\ \zeta_{ij},\ \eta_{ij}$ are 1-forms on S_g.

Let

$$\Omega = \Omega_1 + \Omega_2 + \cdots + \Omega_n + \cdots$$

where Ω_n is a homogeneous polynomial in A_i, B_i of degree n with 1-forms on S_g as coefficients.

In order to have $d\Omega - \Omega \wedge \Omega = 0$, we should have

$$\begin{aligned}d\Omega_1 &= 0,\\ d\Omega_n &= \sum_{i+j=n} \Omega_i \wedge \Omega_j, \quad n = 2, 3, \cdots .\end{aligned}$$

Let

$$\Omega_1 = \sum_{i=1}^{g} \alpha_i A_i + \sum_{i=1}^{g} \beta_i B_i,$$

then

$$\begin{aligned}\Omega_1 \wedge \Omega_1 &= \sum_{i=1}^{g} \alpha_i \wedge \beta_i\ [A_i, B_i]\\ &+ \sum_{i<j} \alpha_i \wedge \alpha_j\ [A_i, A_j]\\ &+ \sum_{i\neq j} \alpha_i \wedge \beta_j\ [A_i, B_j]\\ &+ \sum_{i<j} \beta_i \wedge \beta_j\ [B_i, B_j].\end{aligned}$$

Modulo the relation (5.1), we see that there exists Ω_2 such that $d\Omega_2 = \Omega_1 \wedge \Omega_1$.

To get Ω_3, notice that adding closed 1-forms to $\tau_i,\ \xi_{ij},\ \zeta_{ij},\ \eta_{ij}$ will not effect the equality $d\Omega_2 = \Omega_1 \wedge \Omega_1$. So, we may choose the 1-forms $\gamma_i,\ \xi_{ij},\ \zeta_{ij},\ \eta_{ij}$ such that their wedge products with $\alpha_k,\ \beta_k$ are all exact. It is now quite easy to get Ω_3 and this argument can be generalized to construct Ω_4 and so forth.

Thus, the expansion $I : G \longrightarrow \mathcal{L}G$ in this case is obtained by iterated integral:

$$I(g) = 1 + \sum_{k=1}^{\infty} \int_{0\leq t_1 \leq \cdots \leq t_k \leq 1} \Omega(\dot\gamma(t_k)) \wedge \cdots \wedge \Omega(\dot\gamma(t_1))$$

where γ is a loop on S_g representing $g \in \pi_1(S_g) = G$.

See [15] for more examples where the answers to both (1) and (2) of Question 5.1 are affirmative.

It will be interesting to see whether one can extract 3-manifold invariants from expansions of surface groups by considering the action of mapping class groups. This would be analogous to the construction of Milnor invariants. Deeper invariants of 3-manifolds might be constructed if one could understand the infinitesimal relations of Torelli groups and find out a reasonable expansion for Torelli groups. The answer to Question 5.1 for Torelli groups would probably help us to gain more insight in quantum 3-manifold invariants. See [23] and the references there for related ideas along this line.

Acknowledgement. We would like to thank the following persons for helpful discussions, correspondences and comments: D. Bar-Natan, J. Birman, N. Habegger, R. Hain, S. Hurder, V. Jones, N. Yu. Reshetikhin, J. Stasheff and P. Teichner. The referee's effort to make this paper more readable is gratefully acknowledged.

References

1. V.I. Arnold, *The cohomology ring of the colored braid group*, Math. Notes of Acad. of Sci. of USSR **5** (1969) 138–140.
2. D. Bar-Natan, *On the Vassiliev knot invariants*, Preprint, Harvard University, (1992).
3. D. Bar-Natan, *Vassiliev homotopy string link invariants*, Preprint, Harvard University, (1992).
4. J. Birman, *New points of view in knot theory*, Bull. of AMS. **28** (1993) 253–287.
5. J. Birman and X.-S. Lin, *Knot polynomials and Vassiliev's invariants*, Invent. Math. **111** (1993) 225–270.
6. B. Cenkl and R. Porter, *Malcev's completion of a group and differential forms*, Jour. of Diff. Geom. **15** (1980) 531–542.
7. K.-T. Chen, *Integration of paths, geometric invariants and a generalized Baker–Hausdorff formula*, Ann. of Math. **65** (1957) 163–178.
8. K.-T. Chen, *Iterated path integrals*, Bull. of AMS **83** (1977) 831–879.
9. V.G. Drinfeld, *Quasi-Hopf algebras*, Leningard Math. Jour. **1** (1990) 1419–1457.
10. F. Grunewald and D. Segal, *Reflections on classification of torsion free nilpotent groups*, In Group Theory: Essays for Philip Hall, ed. K.W. Gruenberg and J.E. Roseblade, Academic Press, 1984, pp. 121–158.
11. R. Hain and P. Tondeur, *The life and work of Kuo-Tsai Chen*, Ill. Jour. of Math. **34** (1990) 175–188.

12. N. Habegger and X.-S. Lin, *Classification of links up to link-homotopy*, Jour. of AMS **3** (1990) 389–419.
13. V. Jones, *Milnor's work and knot polynomials*, in Topological method in modern mathematics: Proceedings of Milnor symposium, Publish or Perish, (1993).
14. V.G. Knizhnik and A.B. Zamolodchikov, *Current algebra and Wess-Zumino model in two dimension*, Nucl. Phys. **B247** (1984) 83–103.
15. T. Kohno, *On the holonomy Lie algebra and the nilpotent completion of the fundament group of the complement of hypersurfaces*, Nagoya Math. Jour. **92** (1983) 21–37.
16. T. Kohno, *Linear representations of braid groups and classical Yang-Baxter equations*, Contemp. Math. **78** (1988) 339–363.
17. M. Kontsevich, *Vassiliev's knot invariants*, Adv. Sov. Math. **16** (1993) 135–170.
18. X.-S. Lin, *Vertex models, quantum groups and Vassiliev's knot invariants*, Preprint, Columbia University, (1991).
19. X.-S. Lin, *Milnor's link invariants are all of finite type*, Preprint, Columbia University, (1992).
20. W. Magnus, A. Karrass, and D. Solitar, *Combinatorial Group Theory*, Pure and Appl. Math., Vol. XIII, Interscience, N.Y., (1966).
21. J. Milnor, *Link groups*, Ann. of Math. **59** (1954) 177–195.
22. J. Milnor, *Isotopy of links*, In Algebraic Geometry and Topology: A Symposium in Honor of S. Lefschetz, ed. R.H. Fox, D.C. Spencer and A.W. Tucker, Princeton University Press, 1957, pp 280–306.
23. S. Morita, *Mapping class groups of surfaces and three-dimensional manifolds*, Proceedings of ICM 1990, Kyoto, Math. Soc. of Japan, (1991).
24. D. Sullivan, *Infinitesimal computations in topology*, Publ. Math. IHES **47** (1977) 269–331.
25. V.A. Vassiliev, *Cohomology of knot space*, Adv. in Sov. Math. **1** (1990) 23–69.
26. V.A. Vassiliev, *Complements of Discriminants of Smooth Maps: Topology and Application*, Amer. Math. Soc., 1992.

School of Mathematics, Institute for Advanced Study, Princeton, NJ 08540 and Department of Mathematics, Columbia University, New York, NY 10027

E-mail address: xl@math.ias.edu

AMS/IP Studies in Advanced Mathematics
Volume **2**, 1997 (Part 1)

INTRODUCTION TO SPIN TQFT

G. MASBAUM

INTRODUCTION.

This paper is an introduction to joint work with C. Blanchet on topological quantum field theories for surfaces with spin structure [7]. It closely follows the author's talk at the Georgia International Topology Conference at Athens, Georgia, in August 1993. The main results of [7] are described, and some motivation is given for the language in which they are formulated. Details are given only insofar as necessary to state the results, and proofs are omitted altogether. (They can be found in [7].)

The abbreviation 'TQFT' is short for *Topological Quantum Field Theory.* This terminology is used to describe certain properties of certain invariants of 3–manifolds. It was introduced by Atiyah [1] following Witten's [21] interpretation, in terms of quantum field theory, of the Jones polynomial invariant of links in the 3–sphere.

Here, a TQFT is viewed as a functor from a $2+1$-dimensional cobordism category to a category of modules, satisfying certain axioms. The axioms imply that the functor is determined by its values on closed bordisms. These lie in the ground ring, and are 3–manifold invariants. They are sometimes called *quantum invariants.* Existence of quantum invariants was first proven by Reshetikhin and Turaev [17]. Other constructions of invariants were given e.g. in [5, 10, 11, 12, 16, 20]. A construction of TQFT-functors based on the invariants of [5] was given in [6].

The word 'Spin' in the title is to indicate that the invariants in question depend non-trivially on a choice of spin structure on the manifold. Such refined quantum invariants were constructed independently by Kirby-Melvin [10], Turaev [18] (using quantum groups), and Blanchet [4] (using the Kauffman bracket [9].) As a special case, these invariants include (a version of) the well-known classical *Rochlin invariant.*

A formula of [4, 10, 18] asserts that the sum of the spin invariants of a closed 3–manifold is equal to the 'unspun' invariant. The question arises of how this generalises on the level of TQFT's. A partial answer was given already in [6] where it was shown that the V_{8k}–module of a surface is naturally decomposed into a direct sum of submodules associated to spin structures on the surface.

In [7], the methods of [6] are applied to Blanchet's construction of invariants for 3−manifolds with spin structure [4]. A series of Spin TQFT's, denoted by V_{8k}^s, $k \geq 1$, is constructed and computed. The two important properties of the functors V_{8k}^s are that the V_{8k}^s−modules associated to surfaces are free of finite rank, and satisfy a tensor product formula for disjoint unions. However, the latter holds only in a $\mathbf{Z}/2$−graded sense, which is why we speak of $\mathbf{Z}/2$−*graded* TQFT. Rather than viewing this as a flaw of the theory, we consider this an interesting new feature.

The relationship between the Spin TQFT's V_{8k}^s and the 'unspun' TQFT's V_{8k} of [6] is given by a *transfer map* from the unspun theory to the spin theory. This identifies the spin submodules of [6] with the 'zero graded parts' of the V_{8k}^s−modules. In particular, this means that the spin theory is strictly richer than the unspun theory.

This talk is organised as follows. In section 1, we describe the approach to TQFT's adopted in [7], following [6]. The next section contains the statement of the main theorem of [7], and a description of the $\mathbf{Z}/2$−graded tensor product formula. In section 3, we discuss the definition of the cobordism category $\mathcal{C}_{8k}^s$ adopted in [7]. The main point here is that the notion of spin structure is refined to something called w_2−structure, and we try to explain why this is useful. For example, it is used to obtain a Decomposition Theorem for the V_{8k}^s−modules associated to disconnected surfaces (this theorem is closely related to the $\mathbf{Z}/2$−graded tensor product formula.) Section 5 contains a description of the transfer map from the Spin TQFT functors to the unspun TQFT-functors of [6]. Finally, in the last section we give explicit dimension formulae depending on the genus of the surface and the Arf invariant of the spin structure.

1. TQFT-functors and Invariants.

A *cobordism category*, usually denoted by $\mathcal{C}$, is a category together with an empty object $\emptyset$ and a notion of *disjoint union* (denoted by $\amalg$), *orientation reversal* (denoted by a minus sign), and *boundary* (denoted by ∂). These have to satisfy certain axioms which are abstracted from the following basic example. The objects of $\mathcal{C}$ are the oriented closed 2−manifolds. The morphisms $Hom_{\mathcal{C}}(\Sigma_1, \Sigma_2)$ are the compact oriented cobordisms from Σ_1 to Σ_2, i.e. compact oriented 3−manifolds M with boundary $\partial M = -\Sigma_1 \amalg \Sigma_2$. The composition of morphisms is the usual composition of cobordisms.

In many interesting examples of cobordism categories, the objects and morphisms are manifolds equipped with some kind of *structure.* For example, the structure may be a link, a p_1-structure (a weak form of framing, see [6]), a spin structure . . .

We set up some notation. Let Λ be a commutative ring with 1. Consider a covariant functor V from the cobordism category $\mathcal{C}$ to the category of $\Lambda-$modules. Thus, in the basic example, the functor associates to every closed surface Σ a $\Lambda-$module $V(\Sigma)$, and to every cobordism from Σ_1 to Σ_2 a homomorphism $V(\Sigma_1) \to V(\Sigma_2)$.

We shall always assume that $V(\emptyset)$ is the ground ring Λ. Hence, to $M \in Hom_{\mathcal{C}}(\emptyset, \Sigma)$ (in our basic example, this means M is a compact oriented 3−manifold with boundary $\partial M = \Sigma$) the functor associates a homomorphism from Λ to $V(\Sigma)$. Such a homomorphism is naturally identified with an element of $V(\Sigma)$. This element will be denoted by $Z(M)$. In the special case where M is a *closed bordism*, i.e. $M \in Hom_{\mathcal{C}}(\emptyset, \emptyset)$, $Z(M)$ lies in $V(\emptyset) = \Lambda$. In this case, $Z(M)$ is denoted by $< M >$.

The first important requirement on the functor V is the following

1.1. Invariance Property. *$Z(M)$ depends only on the isomorphism class (rel. boundary) of M.*

In particular, $< >: Hom_{\mathcal{C}}(\emptyset, \emptyset) \to \Lambda$ is an invariant of closed bordisms.

Here is the characteristic property of a TQFT-functor.

1.2. TQFT Property. *Assume $\partial M_1 = \Sigma = \partial M_2$, and let*

$$M = M_1 \cup_\Sigma -M_2$$

be the result of gluing together M_1 and M_2 along Σ. Then

$$< M >=< Z(M_1), Z(M_2) >_\Sigma$$

where $< , >_\Sigma$ is a non-degenerate hermitian sesquilinear form on $V(\Sigma)$, functorially associated to Σ.

(We assume the coefficient ring Λ has an involution $\lambda \mapsto \overline{\lambda}$. For example, Λ could be the complex numbers **C**.)

Note. An immediate consequence of the TQFT property is that the natural map

$$V(-\Sigma) \to V(\Sigma)^\star, \quad M \mapsto < \ , M >$$

is injective.

The following key remark is at the base of the construction of TQFT's in [6].

1.3. Key Remark. *Under mild assumptions, every invariant of closed bordisms* $< >: Hom_{\mathcal{C}}(\emptyset, \emptyset) \to \Lambda$ *extends uniquely to a functor* V *with the above properties.*

Here are the assumptions: the invariant is required to be *involutive* (i.e. $< -M > = \overline{< M >}$), and *multiplicative* under disjoint unions. The proof is by an easy *universal construction* (see [6]), which will not be repeated here. [1]

Definition 1.4. A functor V as above is a TQFT-functor if it satisfies in addition the following two properties:
Finiteness. For all objects Σ, the module $V(\Sigma)$ is free of finite rank, and the canonical map $V(-\Sigma) \to V(\Sigma)^\star$ is an isomorphism.
Tensor Product Formula. For all Σ_i, $i = 1, 2$, the canonical map

$$V(\Sigma_1) \otimes V(\Sigma_2) \to V(\Sigma_1 \amalg \Sigma_2)$$

$$Z(M_1) \otimes Z(M_2) \ \mapsto Z(M_1 \amalg M_2)$$

is an isomorphism.

Note. This definition of TQFT is essentially equivalent to the one by Atiyah and Segal (see [1, 3]).

Remark. It should be emphasized that a functor constructed from an invariant of closed bordisms (using the key remark above) is in general *not* a TQFT-functor. First of all, if we start with an invariant of closed bordisms, the universal construction will most of the time produce modules which are not finitely generated. Second, the surjectivity of the natural map $V(\Sigma_1) \otimes V(\Sigma_2) \to V(\Sigma_1 \amalg \Sigma_2)$ means that manifolds with boundary $\Sigma_1 \amalg \Sigma_2$ 'can be disconnected'; i.e. for a connected manifold M with boundary $\Sigma_1 \amalg \Sigma_2$, we may write $Z(M) = \sum_\nu \lambda_\nu Z(M_1^{(\nu)} \amalg M_2^{(\nu)})$ where $\lambda_\nu \in \Lambda$ and $\partial M_i^{(\nu)} = \Sigma_i$. In general, there is no reason why this should be the case. The spin theories will in fact provide examples where the tensor product formula holds only in a $\mathbf{Z}/2-$graded sense.

Note. A construction of TQFT-functors based on (a version of) the 3−manifold invariants of [5] has been given in [6] (see also [13].)

[1] In order to avoid set-theoretical difficulties with the universal construction, we are assuming that isomorphism classes (rel boundary) of morphisms in $\mathcal{C}$ are sets (which is true in all interesting examples.) See also [14].

2. Spin (Generalised) TQFT.

Let $k \geq 1$ be an integer. We shall now describe the (generalised) TQFT-functor V^s_{8k} constructed in [7].[2] It is defined on a cobordism category $\mathcal{C}^s_{8k}$ whose precise definition is somewhat technical (it will be given in section 3.) For now, suffice it to say that $\mathcal{C}^s_{8k}$ is an extension of the basic example of a cobordism category described above (2−dimensional surfaces and 3−dimensional cobordisms), but such that all manifolds are equipped with additional structure. This additional structure includes a spin structure, and a colored banded link or a colored banded trivalent graph (the link or graph may be empty.)

The coefficient ring is

$$\mathbf{Z}[\frac{1}{2k}, A, \kappa]/(\varphi_{16k}(A), \kappa^6 - A^{-6-4k})$$

where $\varphi_d(A)$ is the d-th cyclotomic polynomial.

Thus, A is a primitive $16k$−th root of unity, and κ is a sixth root of a certain power of A. The involution is given by $\overline{A} = A^{-1}$ and $\overline{\kappa} = \kappa^{-1}$.

Theorem 2.1 ([7]). *For $k \geq 1$, there is a generalised TQFT-functor V^s_{8k} defined on the cobordism category $\mathcal{C}^s_{8k}$.*

(We shall explain shortly what is meant by a *generalised* TQFT.)

Convention. The integer k will be fixed throughout, and most of the time suppressed from the notation.

The functor V^s is constructed by first constructing an invariant of closed bordisms in $\mathcal{C}^s$, and then applying the key remark. The invariant is a version of Blanchet's invariant for 3−manifolds with spin structure [4]. His construction used the Kauffman bracket [9], and the element A in the coefficient ring is Kauffman's skein variable, evaluated at a primitive $16k$−th root of unity. [3]

Remark 2.2. If $k = 1$, the invariant is a version of the classical Rochlin invariant. In fact, in this case its value on a closed 3−manifold with spin structure is essentially a power of A (which for $k = 1$ is a primitive

[2]The superscript s stands for *spin*. The indexing by $8k$ is for notational coherence with [6], see section 5.

[3]Blanchet has constructed Spin invariants from the Kauffman bracket at all roots of unity of order not congruent to 8 modulo 16, but those will not be considered in this talk.

16–th root of unity), and the exponent of that power is a version of the Rochlin invariant.[4]

The functor V^s is not a TQFT in the sense of the definition given in the preceding section, since it does not satisfy the Tensor Product Formula: There is an object Σ of $\mathcal{C}^s$ such that $V^s(\Sigma)$ and $V^s(-\Sigma)$ are both zero, but $V^s(-\Sigma \amalg \Sigma)$ is free of rank one, generated by $Z^s(\Sigma \times \mathbf{I})$. (For example, if $k = 1$, there is such an object Σ whose underlying manifold is a torus $S^1 \times S^1$ with a spin structure having non-trivial Arf invariant.)

The functor V^s is, however, a *generalised* TQFT, in the sense that the finiteness property holds, as well as the following generalisation of the Tensor Product Formula.

2.3. Generalised Tensor Product Formula. *There is an object $\widehat{S}$ of the cobordism category $\mathcal{C}^s$ with the following property. For all objects Σ_1 and Σ_2, the natural glueing map*

$$\begin{array}{lll} V^s(\Sigma_1) \otimes V^s(\Sigma_2) & \oplus\ V^s(\Sigma_1 \amalg \widehat{S}) \otimes V^s(-\widehat{S} \amalg \Sigma_2) & \stackrel{\approx}{\longrightarrow} V^s(\Sigma_1 \amalg \Sigma_2) \\ Z^s(M_1) \otimes Z^s(M_2) & & \mapsto Z^s(M_1 \amalg M_2) \\ & Z^s(M_3) \otimes Z^s(M_4) & \mapsto Z^s(M_3 \underset{\widehat{S}}{\cup} M_4) \end{array}$$

is an isomorphism.

Remark 2.4. In the language of bimodules ever algebroids (see [6]), this can be stated in the following way. Consider $\Lambda \oplus \Lambda$ as a Λ–algebroid on two objects, 0 and 1. Define a $\Lambda \oplus \Lambda$–bimodule ${}_-V^s(\Sigma)_-$ by

$$\begin{array}{lll} {}_0V^s(\Sigma)_0 & = & V^s(\Sigma) \\ {}_0V^s(\Sigma)_1 & = & V^s(\Sigma \amalg \widehat{S}) \\ {}_1V^s(\Sigma)_0 & = & V^s(-\widehat{S} \amalg \Sigma) \\ {}_1V^s(\Sigma)_1 & = & V^s(-\widehat{S} \amalg \Sigma \amalg \widehat{S}) \end{array}$$

Then the obvious glueing induces an isomorphism

$${}_-V^s(\Sigma_1)_- \underset{\Lambda \oplus \Lambda}{\otimes} {}_-V^s(\Sigma_2)_- \stackrel{\approx}{\longrightarrow} {}_-V^s(\Sigma_1 \amalg \Sigma_2)_-$$

The following result indicates that the spin theories may be considered as $\mathbf{Z}/2$–graded TQFT's.

---

[4]Recall that the Rochlin invariant takes values in $\mathbf{Z}/16$: it associates to a closed oriented 3–manifold with spin structure, M, the signature mod 16 of an oriented spin 4–manifold with boundary equal to M.

Proposition 2.5. *For all objects Σ of $\mathcal{C}^s$, $V^s(\Sigma) = {}_0V^s(\Sigma)_0$ is isomorphic to ${}_1V^s(\Sigma)_1$, and ${}_0V^s(\Sigma)_1$ is isomorphic to ${}_1V^s(\Sigma)_0$.*

Notation. Motivated by the proposition, we denote ${}_0V^s(\Sigma)_1$ by $\widehat{V}^s(\Sigma)$, and set

$$\mathbf{V}^s(\Sigma) = V^s(\Sigma) \oplus \widehat{V}^s(\Sigma)$$

Then one has the following

2.6. Z/2–graded Tensor Product Formula. *For all objects Σ_1 and Σ_2 in $\mathcal{C}^s$, one has*

$$\mathbf{V}^s(\Sigma_1) \otimes \mathbf{V}^s(\Sigma_2) \approx \mathbf{V}^s(\Sigma_1 \amalg \Sigma_2)$$

3. Structures.

We now discuss the notion of *structures* used in [7] to define the cobordism category $\mathcal{C}^s$. In the next section, this will be applied to explain how the decomposition (implicit in theorem 2.6) of $V^s(\Sigma)$ into tensor products of elementary building blocks is induced geometrically.

First, recall the (or, rather, one possible) definition of a spin structure on a manifold. The stable tangent bundle of an oriented manifold M has a classifying map $\tau_M : M \to BSO$. A *spin structure* on M is a fiber homotopy class of lifts of τ_M to the space $BSpin$.

Remark 3.1. Recall that $BSpin$ can be defined as the homotopy fiber of the map $w_2 : BSO \to K(\mathbf{Z}/2, 2)$ realising the universal second Stiefel-Whitney class $w_2 \in H^2(BSO; \mathbf{Z}/2)$. From this point of view, it is clear that the obstruction to existence of a spin structure is $\tau_M^\star(w_2)$ = the second Stiefel-Whitney class of M, and that the set of spin structures on M, if non-empty, is affinely isomorphic to $H^1(M; \mathbf{Z}/2)$ [15].

In order to define a cobordism category, one needs the concept of a *relative* spin structure, i.e. a spin structure on a cobordism relative to a given spin structure on the two ends. Here, it is necessary to distinguish between an actual lift and its homotopy class, for the following reason.

Let Σ be a closed surface, and let σ_1, σ_2 be two lifts of the stable tangent bundle τ_Σ to $BSpin$ which are fiber homotopic. The question is whether it is possible to identify canonically the modules $V^s(\Sigma, \sigma_1)$ and $V^s(\Sigma, \sigma_2)$. We claim that it is not. Observe that every homotopy between σ_1 and σ_2 gives us a lift σ of the stable tangent bundle of the cylinder $\Sigma \times \mathbf{I}$, and hence (assuming existence of the functor V^s) an isomorphism $Z^s(\Sigma \times \mathbf{I}, \sigma)$ from $V^s(\Sigma, \sigma_1)$ to $V^s(\Sigma, \sigma_2)$. In particular, every self-homotopy of σ_1 gives an automorphism of $V^s(\Sigma, \sigma_1)$. If $V^s(\Sigma, \sigma_1)$ and $V^s(\Sigma, \sigma_2)$ are canonically identified, then in particular every automorphism of $V^s(\Sigma, \sigma_1)$ induced by a selfhomotopy of σ_1 has to be the identity. But, as it turns out, this is not the case for the spin theories V^s_{8k}. (This fact plays an important role in the spin theory, see Lemma 4.1 and Decomposition Theorem 4.2.)

Hence, we must distinguish between the lifts σ_1 and σ_2, even though they are homotopic, and define the same spin structure. We propose the following

Definition 3.2. A w_2-structure on a manifold M is a lift of the stable tangent bundle of M to $BSpin$.

Thus, a spin structure is a homotopy class of w_2-structures. Similarly, a *relative spin structure* on a cobordism M between surfaces Σ_1 and Σ_2 which are equipped with w_2-structures σ_1, σ_2, is defined to be a homotopy class (rel. boundary) of w_2-structures on M extending σ_1 and σ_2.

Replacing the second Stiefel-Whitney class w_2 by the first Pontryagin class p_1, there is a similar definition of a p_1-*structure* on a manifold. It was used in [6] to construct a series of TQFT-functors which are defined on a cobordism category where all manifolds are equipped with a p_1-structure. (The introduction of p_1-structures was used to resolve a similar problem[5] as the one described above.)

We can now give the complete definition of the cobordism category $\mathcal{C}^s$ on which the functor V^s is defined. An object of $\mathcal{C}^s$ is a quadruple $\Sigma = (\Sigma, \alpha, l, \sigma)$ where

— Σ is a closed oriented surface

— α is a p_1-structure on Σ

— l is a colored banded link in Σ (A *banded link* in a surface is an embedding of a finite collection of oriented intervals. A coloring of l is a function from the set of components of l to the set $\{0, 1, \ldots, 4k-2\}$.)

— σ is a w_2-structure on $\Sigma - l$ with the property that it extends across a component of l if and only if that component is colored by an even number.

Similarly, if $\Sigma_i = (\Sigma_i, \alpha_i, l_i, \sigma_i)$, $(i = 1, 2)$, are objects of $\mathcal{C}^s$, a morphism in $Hom_{\mathcal{C}^s}(\Sigma_1, \Sigma_2)$ is a quadruple $M = (M, \widetilde{\alpha}, L, \widetilde{\sigma})$ where

— M is a cobordism from Σ_1 to Σ_2

— $\widetilde{\alpha}$ is a relative p_1-structure on M extending α_1 and α_2

— L is a colored banded trivalent graph in M realising an embedded cobordism from l_1 to l_2 (the definition of a colored banded trivalent graph is as in [6])

[5] customarily called the *framing anomaly*

— $\widetilde{\sigma}$ is a relative spin structure on $M - L$ with the property that it extends across an edge of L if and only if that edge is colored by an even number.

Remark 3.3. Let $\Sigma = (\Sigma, \alpha, l, \sigma)$ be an object of $\mathcal{C}^s$. Up to a non-canonical isomorphism, the module $V^s(\Sigma)$ depends only on the underlying surface, the colored link l, and the spin structure $[\sigma]$ induced by the w_2–structure σ.

Remark 3.4. In [7], a pair (l, σ) as above is called a *singular w_2-structure* on the surface Σ, and homotopy classes of such are called *singular spin structures.* We also express this by saying that the spin structure is *complementary* to the link. There is an obstruction theory for complementary spin structures which nicely parallels the usual obstruction theory for spin structures. This is described in detail in [7], as it is an important technical ingredient in the construction of the Spin TQFT-functors V^s.

4. The Decomposition Theorem.

Let $\Sigma = (\Sigma, \alpha, l, \sigma)$ be an object of $\mathcal{C}^s$. Any relative singular spin structure $\widetilde{\sigma}$ on $\Sigma \times \mathbf{I}$ induces the automorphism $Z^s(\Sigma \times \mathbf{I}, \widetilde{\sigma})$ of $V^s(\Sigma)$. The set of relative singular spin structures on $\Sigma \times \mathbf{I}$ is isomorphic to $H^0(\Sigma; \mathbf{Z}/2)$.

Lemma 4.1. *In this way, the cohomology group $H^0(\Sigma; \mathbf{Z}/2)$ acts on the module $V^s(\Sigma)$.*

Note. In fact, the element $1 \in H^0(\Sigma; \mathbf{Z}/2)$ can be seen to act trivially, hence the action factors through the reduced group $\widetilde{H}^0(\Sigma; \mathbf{Z}/2)$.

This can be used to decompose the module $V^s(\Sigma)$ into isotypic components for this action. For $a \in \widetilde{H}_0(\Sigma; \mathbf{Z}/2)$, put

$$V^s(\Sigma, a) = \{x \in V^s(\Sigma) \ : \ c \cdot x = (-1)^{<c,a>} x \text{ for all } c \in \widetilde{H}^0(\Sigma; \mathbf{Z}/2)\}$$

(where $x \mapsto c \cdot x$ is the action of $\widetilde{H}^0(\Sigma; \mathbf{Z}/2)$ on $V^s(\Sigma)$.) Then we have the following

4.2. Decomposition Theorem [7]. *There is a natural decomposition*

$$V^s(\Sigma) = \bigoplus_{a \in \widetilde{H}_0(\Sigma; \mathbf{Z}/2)} V^s(\Sigma, a)$$

Here is a description of the submodules $V^s(\Sigma, a)$. Let

$$\Sigma = \Sigma_1 \amalg \ldots \amalg \Sigma_n$$

with all Σ_i connected. The reduced homology group $\widetilde{H}_0(\Sigma; \mathbf{Z}/2)$ is isomorphic to $(\mathbf{Z}/2)^{n-1}$. Every element a in $\widetilde{H}_0(\Sigma; \mathbf{Z}/2)$ may be uniquely written as

$$a = \sum_{i=1}^{n-1} a_i([\Sigma_{i+1}] - [\Sigma_i])$$

with $a_i \in \{0, 1\}$ (where $[\Sigma_i]$ is the class of Σ_i in $H_0(\Sigma; \mathbf{Z}/2)$.) There is an obvious gluing map

$$\varphi_a : {}_0V^s(\Sigma_1)_{a_1} \otimes {}_{a_1}V^s(\Sigma_2)_{a_2} \otimes \ldots \otimes {}_{a_{n-1}}V^s(\Sigma_n)_0 \longrightarrow V^s(\Sigma)$$

(the notation ${}_{a_1}V^s(\Sigma_2)_{a_2}$ was defined in remark 2.4.)

4.3. Addendum to the Decomposition Theorem [7]. *The map φ_a is injective, and its image is precisely the submodule $V^s(\Sigma, a)$ of $V^s(\Sigma)$.*

Example. Consider the case $n = 2$, i.e. $\Sigma = \Sigma_1 \amalg \Sigma_2$ and $\widetilde{H}_0(\Sigma; \mathbf{Z}/2) = \mathbf{Z}/2$. Then

$$V^s(\Sigma, a) \approx \begin{cases} V^s(\Sigma_1) \otimes V^s(\Sigma_2) & \text{if } a = 0 \\ V^s(\Sigma_1 \amalg \widehat{S}) \otimes V^s(-\widehat{S} \amalg \Sigma_2) & \text{if } a \neq 0 \end{cases}$$

5. The Transfer Theorem.

In this section, we describe a natural transfer map from the Spin TQFT functor V^s_{8k} (called the *spin theory*) to the TQFT functor V_{8k} (here called the *unspun theory*) constructed in [6].

The functor V_{8k} of [6] has the same coefficient ring as the functor V^s_{8k}, and is defined on a cobordism category $\mathcal{C}_{8k}$ which is obtained from $\mathcal{C}^s_{8k}$ by omitting the w_2–structures.

As before, we fix $k \geq 1$ and write V for V_{8k}, V^s for V^s_{8k}, etc.

In this section, it will be convenient to use the notation Σ for an object of the cobordism category $\mathcal{C}$ (rather than $\mathcal{C}^s$), and to denote an object of $\mathcal{C}^s$ by (Σ, σ) (where σ is a w_2–structure complementary to the link in Σ.)

5.1. Transfer theorem [7]. *There is a natural homomorphism*

$$\Phi_\sigma : V(\Sigma) \to V^s(\Sigma, \sigma)$$

The image of Φ_σ is the submodule $V^s((\Sigma, \sigma), 0)$ of $V^s(\Sigma, \sigma)$.

The homomorphism Φ_σ is called *transfer*. It can be described as follows. Given $\Sigma = (\Sigma, \alpha, l)$, let an element of $V(\Sigma)$ be of the form $Z(M)$, where $M = (M, \widetilde{\alpha}, L)$ and $\partial(M, \widetilde{\alpha}, L) = (\Sigma, \alpha, l)$. Let $Spin(M|L, \sigma)$ denote the set of relative spin structures $\widetilde{\sigma}$ on $M - L$ which extend σ and are complementary to L. Then

$$\Phi_\sigma(Z(M)) = 2^{-b_0(M,\Sigma)} \sum_{\widetilde{\sigma} \in Spin(M|L,\sigma)} Z^s(M, \widetilde{\sigma})$$

Note. It requires proof that Φ_σ is well defined on $V(\Sigma)$.

In order to describe the kernel of Φ_σ, we need to recall the following result of [6].

Theorem 5.2 ([6]). *Let $\Sigma = (\Sigma, \alpha, l)$ be an object of the cobordism category $\mathcal{C}$. There is a natural decomposition*

$$V(\Sigma) = \bigoplus_q V(\Sigma, q)$$

where q runs through the quadratic forms on $H_1(\Sigma - l; \mathbf{Z}/2)$ inducing the intersection form.

(Here, $V(\Sigma, q)$ stands for $V_{8k}(\Sigma, q)$ as defined in [6].)

Recall that a quadratic form on $H_1(\Sigma - l; \mathbf{Z}/2)$ inducing the intersection form is simply a function $q: H_1(\Sigma - l; \mathbf{Z}/2) \to \mathbf{Z}/2$ satisfying $q(a + b) = q(a) + q(b) + a \cdot b$. There is a well known natural bijective correspondence between such quadratic forms and spin structures on $\Sigma - l$, see [2, 8].

Note. This decomposition of $V(\Sigma)$ is induced by a natural geometrically defined action of a Heisenberg type group associated to $H_1(\Sigma - l; \mathbf{Z}/2)$. See [6].

The following result describes the kernel of the transfer map. Let (Σ, σ) be an object of the cobordism category $\mathcal{C}^s$. Let q_σ denote the quadratic form on $H_1(\Sigma - l; \mathbf{Z}/2)$ corresponding to the spin structure $[\sigma]$ on $\Sigma - l$ induced by the w_2−structure σ.

5.3. Addendum to the Transfer Theorem [7]. *The kernel of Φ_σ is the orthogonal complement*[6] *of $V(\Sigma, q_\sigma)$ in $V(\Sigma)$. Hence Φ_σ induces an isomorphism*

$$V(\Sigma, q_\sigma) \xrightarrow{\approx} V^s((\Sigma, \sigma), 0)$$

[6] with respect to the canonical hermitian form $<\ ,\ >_\Sigma$ on $V(\Sigma)$

It is instructive to observe the following corollary, which shows that the unspun theory is essentially the sum of the 'zero graded parts' of the spin theories.

It was observed in [6] that $V(\Sigma, q)$ is non-zero only for those q corresponding to a spin structure on $\Sigma - l$ which is complementary to l. Let us denote the set of complementary spin structures by $Spin(\Sigma|l)$.

Corollary 5.4. *There is a canonical isomorphism*

$$V(\Sigma) \xrightarrow{\approx} \bigoplus_{[\sigma] \in Spin(\Sigma|l)} V^s((\Sigma, \sigma), 0)$$

Note. The right hand side makes sense, because, by its very definition, the submodule $V^s((\Sigma,\sigma),0)$ of $V^s(\Sigma,\sigma)$ is canonically associated not just to the w_2-structure σ, but to the spin structure $[\sigma]$.

6. Dimension formulae.

Let $\Sigma_{g,\varepsilon}$ denote an object of $\mathcal{C}^s$ whose underlying manifold is a closed connected surface of genus g (without link), equipped with a spin structure with Arf invariant $\varepsilon \in \{0,1\}$. Let us denote the ranks of $V^s_{8k}(\Sigma_{g,\varepsilon})$ and $\widehat{V}^s_{8k}(\Sigma_{g,\varepsilon})$ by $d_g^{(\varepsilon)}(8k)$ and $\widehat{d}_g^{(\varepsilon)}(8k)$, respectively.

Theorem 6.1 ([7]). *Let $k \geq 1$. Then*

$$\begin{aligned} d_g^{(\varepsilon)}(8k) &= \frac{1}{4^g}\left(d_g(8k) + (2k)^{g-1}\left((-1)^\varepsilon 2^g - 1\right)\right) \\ \widehat{d}_g^{(\varepsilon)}(8k) &= \frac{1}{4^g}\left(\widehat{d}_g(8k) + (2k)^{g-1}\left(1 - (-1)^\varepsilon 2^g\right)\right) \end{aligned}$$

where $d_g(8k)$ (resp. $\widehat{d}_g(8k)$) are given by the Verlinde Formula (resp. the Twisted Verlinde Formula) in the form given in [6].

Here are the relevant version of the Verlinde Formula and its twisted counterpart:

$$\begin{aligned} d_g(8k) &= (2k)^{g-1} \sum_{j=1}^{4k-1} \left(\sin \frac{\pi j}{4k}\right)^{2-2g} \\ \widehat{d}_g(8k) &= (2k)^{g-1} \sum_{j=1}^{4k-1} (-1)^{j+1} \left(\sin \frac{\pi j}{4k}\right)^{2-2g} \end{aligned}$$

Note. The formula for $d_g^{(\varepsilon)}(8k)$ explicitly appears already in [6].

Remark 6.2. In view of the $\mathbf{Z}/2$-graded tensor product formula, the knowledge of the numbers $d_g^{(\varepsilon)}(8k)$ and $\widehat{d}_g^{(\varepsilon)}(8k)$ completely determines the ranks of the modules $V^s_{8k}(\Sigma)$ for all surfaces with empty link.

Example. Consider the case $k = 1$. Then $d_g^{(0)}(8) = \widehat{d}_g^{(1)}(8) = 1$, and $d_g^{(1)}(8) = \widehat{d}_g^{(0)}(8) = 0$. The number $d_g(8) = 2^{g-1}(2^g + 1)$ is the number of spin structures with Arf invariant zero on a closed connected surface of genus g, and the number $\widehat{d}_g(8) = 2^{g-1}(2^g - 1)$ is the number of spin structures with Arf invariant one.

References

1. M. F. Atiyah, *Topological quantum field theories,* Publ. Math. IHES **68** (1989) 175-186.
2. M. F. Atiyah, *Riemann surfaces and spin structures*, Ann. Scien. Ec. Norm. Sup. 4e série, **4** (1971) 47-62.
3. M. Atiyah, N. Hitchin, R. Lawrence, and G. Segal, *Oxford seminar on Jones-Witten theory* (1988).
4. C. Blanchet, *Invariants of Three-manifolds with spin structure*, Comm. Math. Helv. **67** (1992) 406-427.
5. C. Blanchet, N. Habegger, G. Masbaum, P. Vogel. *Three-manifold invariants derived from the Kauffman bracket*, Topology **31** (1992) 685-699.
6. C. Blanchet, N. Habegger, G. Masbaum, P. Vogel. *Topological quantum field theories derived from the Kauffman bracket*, to appear in Topology (Nantes preprint May 1992).
7. C. Blanchet, G. Masbaum. *Topological quantum field theories for surfaces with spin structure*, Duke Math. J. (to appear).
8. D. Johnson, *Spin structures and quadratic forms on surfaces,* J. London Math. Soc. (2) **22** (1980) 365-377.
9. L. H. Kauffman. *State models and the Jones polynomial*, Topology **26** (1987) 395-401.
10. R. C. Kirby, P. Melvin, *The 3-manifold invariants of Witten and Reshetikhin-Turaev for $sl(2,C)$*, Inv. Math. **105** (1991) 473-545.
11. W. B. R. Lickorish. *Three-manifolds and the Temperley-Lieb algebra*, Math. Ann. **290** (1990) 657-670.
12. W. B. R. Lickorish. *Calculations with the Temperley-Lieb algebra*, Comm. Math. Helv. **67** (1992) 571-591.
13. G. Masbaum, P. Vogel. *Verlinde formulae for surfaces with spin structure,* Contemporary Mathematics **164** (1994) 119-137, Proceedings of the Joint US-Israel workshop on geometric topology, Haifa, June 1992.
14. G. Masbaum, C. Rourke. *Model categories and topological quantum field theories*, Proceedings of the Second Gökova Conference on Geometry and Topology, Gökova, Turkey, 1993.
15. J. Milnor, *Spin structures on manifolds*, l'Ens. Math. **9** (1963) 198-203.
16. H. Morton, P. Strickland. *Satellites and surgery invariants,* in: Knots 90, ed. A. Kawauchi, de Gruyter (1992).
17. N. Yu. Reshetikhin, V. G. Turaev. *Invariants of 3-manifolds via link polynomials and quantum groups*, Invent. Math. **103** (1991) 547-597.

18. V. G. Turaev, *State sum models in low-dimensional topology*, Proc. ICM Kyoto 1990, vol I, 689-698.
19. E. Verlinde. *Fusion rules and modular transformations in 2d conformal field theory,* Nucl. Phys. **B 300** (1988) 360-376.
20. H. Wenzl. *Braids and invariants of 3-manifolds,* Inv. Math. **114** (1993) 235-275.
21. E. Witten. *Quantum field theory and the Jones polynomial*, Comm. Math. Phys. **121** (1989) 351-399.

CNRS, UNIVERSITÉ PARIS VII, U.F.R. DE MATHÉMATIQUES, CASE 7012, 75251 PARIS CEDEX 05, FRANCE

E-mail address: masbaum@mathp7.jussieu.fr

AMS/IP Studies in Advanced Mathematics
Volume **2**, 1997 (Part 1)

REFINED STATE-SUM INVARIANTS OF 3- AND 4-MANIFOLDS

JUSTIN ROBERTS

ABSTRACT. In this paper, Blanchet's techniques for constructing invariants of 3-manifolds with spin structures are combined with the model for the Turaev-Viro invariant introduced in [14], to produce refined state-sum invariants defined by means of triangulations. Generalisations of the analogous 4-dimensional invariant, involving homology classes on the 4-manifold, are also studied, and they are expressed in terms of classical invariants and as state-sums on triangulations. In [14] this equivalence was used to show that (a normalisation of) the Crane-Yetter state sum actually gave the signature of the 4-manifold: here it is used to give state-sum formulae for the second Stiefel-Whitney class, cup products, and Pontrjagin squares of second cohomology classes. Rochlin's theorem follows as a corollary.

1. INTRODUCTION

The paper [14] presented a new skein-theoretic definition ('chain-mail') of an invariant which equalled the square of the modulus of Lickorish's version [11] of the Witten-Reshetikhin-Turaev invariant [13] and which could also be expressed as Kauffman and Lins' version [6] of the Turaev-Viro state-sum [18].

Blanchet refined Lickorish's model in [1] (as Turaev [16] and Kirby and Melvin [9] had refined the model of Reshetikhin and Turaev) to produce invariants of 3-manifolds with a spin structure or first $\mathbb{Z}_2$-cohomology class (these will be referred to by the general name of 'spin' invariants from now on).

The material in [14] was the basis for my talk at the Georgia conference, and I am very grateful to the SERC and to Pembroke College for the funding with which to attend.

In the first part of the present paper, Blanchet's methods are combined with the chain-mail technique to prove existence of a refined 'spin' chain-mail invariant depending on two (co)homology classes on a 3-manifold. The corresponding state-sums and the splitting theorems relating these refinements to the original chain-mail invariant of [14] are discussed, and the refinements are identified in terms of Blanchet's invariants. These generalisations extend the refined state-sums (depending on one cohomology class) originally introduced in [18] and the splitting theorem announced in [17] (see also [15]).

The second part of [14] contained proofs that the 4-manifold invariant defined by Broda [2] was simply an exponential of the signature of the 4-manifold and that it was also given by a state-sum which was a differently-normalised version of that proposed by Crane and Yetter [4].

In the second part of the present paper it is shown that Broda's definition admits a natural generalisation, producing an invariant of a 4-manifold with a second $\mathbb{Z}_2$-homology class. (This has also been noted by Yetter in [19]. His slightly different version is related to this one by a Fourier transform formula.) The refined invariant is once again expressed as a state-sum and in terms of classical invariants. The consequence is that, as well as the signature, the second Stiefel-Whitney class, Pontrjagin squares and cup products of $\mathbb{Z}_2$-cohomology classes may be computed from state-sum formulae. The computation of the Stiefel-Whitney class leads to a new proof of Rochlin's theorem.

This paper mirrors [14] closely in structure, and uses the same skein-theoretic framework. Consequently, not much in the way of background will be given, and some of the material which is very similar will be explained without much detail. For further explanation of the basic skein theory techniques and the handleslide identities, consult the papers of Lickorish [10, 11]: for more on recombination rules and the graphical calculus, see the book of Kauffman and Lins [6], and for information about 4-manifolds and spin structures, see Kirby's book [8].

Section 2 gives a brief outline of the results of Blanchet which will be needed and explains notation; section 3 deals with generalised 3-dimensional (Turaev-Viro) invariants, and section 4 deals with the 4-dimensional case.

2. Blanchet's skein elements

Let $r \geq 3$ be a fixed integer, and let $A = e^{2\pi i/4r}$. Let $\mathcal{S}T$ denote the Kauffman skein space of the solid torus (or annulus) at this particular value A, as defined in [11], for example. Let ϕ_j denote the skein class in $\mathcal{S}T$ obtained by wiring the j-th Jones-Wenzl-Lickorish idempotent $f^{(j)}$

into a segment of the annulus and connecting its opposite sides using j parallel strands. Let Δ_j be the value obtained by wiring ϕ_j into a 0-framed unknot in S^3, thus producing an element of $\mathcal{S}S^3 = \mathbb{C}$.

Lickorish's construction [10, 11] of 3-manifold invariants relies on a special element $\Omega \in \mathcal{S}T$ defined by $\Omega = \eta \sum_{j=0}^{r-2} \Delta_j \phi_j$, where $\eta = (A^2 - A^{-2})/(i\sqrt{2r}) = \sqrt{2/r}\sin(\pi/r)$. With this normalisation, the $\mathcal{S}S^3$-value of a 0-framed unknot with Ω attached is η^{-1}, while ±1-framed unknots with Ω attached give unit-modulus complex numbers $\kappa^{\pm1}$ respectively, where $\kappa = A^{-3-r^2} e^{-i\pi/4}$.

The most important property of Ω is the *handleslide identity*, which states that the complex-number value of a link whose components have 'Ω' inserted along them is invariant under the original Kirby handleslide moves [7].

Blanchet [1] introduced a $\mathbb{Z}_2$-grading $\Omega = \Omega_0 + \Omega_1$ corresponding to even and odd idempotents ϕ_j. This gives rise to a handleslide identity (proved in this form in [11]) in which, for $c, d \in \mathbb{Z}_2 \equiv \{0, 1\}$, Ω_c may be slid over Ω_d, provided Ω_d is then replaced by Ω_{c+d}. In pictures, a black square on a curve with c next to it will be used to denote Ω_c wired onto that curve (see figure 1).

The 'calculus of 'Ω's' is now a little more complicated than in the original case, in particular there is a strong dependence on r mod 4. The proofs in the next section will rely heavily on this calculus, so for convenience certain identities are shown in figures 1 and 2. The table lists the values, denoted ψ_{cd} and $\tilde{\psi}_{cd}$, obtained by wiring Ω_c and Ω_d into $(0,0)$- and $(0,+1)$-framed Hopf links respectively, and the values κ_c obtained by putting Ω_c into a $+1$-framed unknot. Putting Ω_c on the -1-framed unknot gives conjugate values $\overline{\kappa}_c$, whilst Ω_c on a 0-framed unknot gives the value $\frac{1}{2}\eta^{-1}$, regardless of the value of r or c. The value ξ_{r-2} is a 'twist coefficient' defined in figure 1.

3. Spin Turaev-Viro invariants

Because of the strong dependence on r mod 4, it is easiest to study first the simplest case, in which $r \equiv 2(4)$ and the invariants produced correspond to cohomology classes. The spin invariants (the case $r \equiv 0(4)$) and the case of r odd will be mentioned afterwards.

Homology groups will have $\mathbb{Z}_2$-coefficients throughout this paper.

Existence of invariants

As in [14], let M be a closed, connected, oriented 3-manifold with a handle decomposition with d_0, d_1, d_2, d_3 handles of the corresponding indices. Let H be the union of the 0- and 1-handles. Draw the attaching

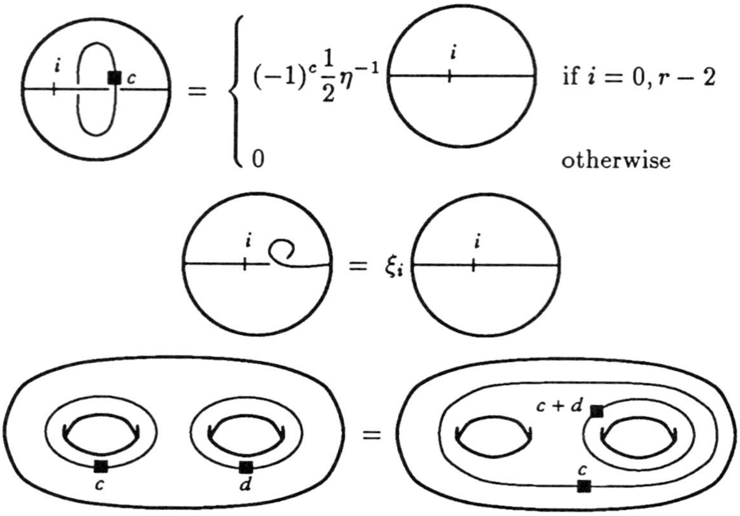

FIGURE 1. Graded Ω-identities and twist coefficient

r mod 4	ψ_{00}	ψ_{01}	ψ_{10}	ψ_{11}	κ_0	A^{r^2}	Δ_{r-2}	ξ_{r-2}
	$\tilde{\psi}_{00}$	$\tilde{\psi}_{01}$	$\tilde{\psi}_{10}$	$\tilde{\psi}_{11}$	κ_1			
2	1	0	0	0	κ	-1	1	1
	1	0	0	0	0			
0	1	0	0	0	0	1	1	-1
	0	0	1	0	κ			
1	$\frac{1}{2}$	$\frac{1}{2}$	$\frac{1}{2}$	$-\frac{1}{2}$	$\frac{1}{\sqrt{2}}e^{-2\pi i/8}.\kappa$	i	-1	i
	$\frac{1}{2}$	$\frac{1}{2}i$	$\frac{1}{2}$	$-\frac{1}{2}i$	$\frac{1}{\sqrt{2}}e^{2\pi i/8}.\kappa$			
3	$\frac{1}{2}$	$\frac{1}{2}$	$\frac{1}{2}$	$-\frac{1}{2}$	$\frac{1}{\sqrt{2}}e^{2\pi i/8}.\kappa$	-1	-1	$-i$
	$\frac{1}{2}$	$-\frac{1}{2}i$	$\frac{1}{2}$	$\frac{1}{2}i$	$\frac{1}{\sqrt{2}}e^{-2\pi i/8}.\kappa$			

FIGURE 2. Table of r-dependent values

curves of the 2-handles slightly inside ∂H, then add meridians of the 1-handles just outside ∂H, and frame them all by thickening them to bands parallel to the surface ∂H. Choose an orientation-preserving embedding of the handlebody H in S^3, and consider the image of the link, called a *chain-mail link* $L \subseteq S^3$.

Definition 3.1. Let $x \in H^1(M), y \in H_2(M)$ be represented by subsets u of 1-handles (those assigned the value 1) and v of 2-handles respectively. Attach Ω_1 to all components of L corresponding to u and v and attach Ω_0 to the rest of L to obtain $\Omega_{uv}L \in \mathcal{S}S^3 = \mathbb{C}$. Define $CH(M,u,v) = (2\eta)^{d_3+d_0}\Omega_{uv}L$.

Theorem 3.2. *The quantity $CH(M,u,v)$ does not depend on the embedding, handle decomposition or representatives for x or y, so is indeed an invariant of the triple (M,x,y).*

Proof. This proof is just a refinement of those in proposition 3.3 and theorem 3.4 of [14]. Any two embeddings of H may be related by allowing 1-handles to 'pass through one another' by means of the standard unknotting move, and by re-framing. Because y is a 2-cycle, an even number of components of v must be slid over a meridian (which is labelled with Ω_0 or Ω_1) in performing an unknotting move. By the $\mathbb{Z}_2$-handleslide identity, the labelling of the meridian is preserved after this slide, and the value of the skein element in $\mathcal{S}S^3$ does not change.

Re-framing (twisting embedded 1-handles across meridian discs) may also be performed by suitable handleslides. Introduce a ±1-framed unknot labelled with Ω_0 and slide all the 2-handle curves in the handle over it, twisting them. Once again, since an even number of odd strands lie in the handle, the unknot is still labelled Ω_0 afterwards. Now slide it over the meridian of the 1-handle and remove it again. (Note: this process requires $\kappa_0 \neq 0$, so it is important that $r \equiv 2(4)$.)

Independence of handle decomposition is proved, as in [14], by considering births or deaths of 0-1-, 1-2-, and 2-3-handle pairs, and handleslides of 1-1- or 2-2-pairs. For handleslides, recall that the chosen homology classes are represented differently after such a move and compare this with the $\mathbb{Z}_2$-handleslide identity. Birth of a 0-1- or 2-3-pair introduces a 0-framed unknot which (to avoid changing the homology classes) is labelled by Ω_0 in S^3. It may be slid over some 'parallel' Ω-curves as in [14], not changing their labels, until it bounds a small disc, then it cancels to a factor which is absorbed by the normalising $(2\eta)^{d_0+d_3}$ factor. Birth of a 1-2-pair adds a new two-component link with both components labelled by Ω_0 and at least one of them a 0-framed unknot in S^3. Since $r \equiv 2(4)$, any such skein element has value 1 in $\mathcal{S}S^3$.

Representatives of (x or) y differ by changing all labels in the (co-)boundary of some (0- or) 3-handles; this may be realised by introducing a small Ω_1 curve at one pole of the boundary sphere, sliding it over all the (2- or) 1-handle circles (changing all of their labels), and cancelling it again. □

As in [14], an alternative proof of invariance of $CH(M, x, y)$ may be given by using triangulations and the classification theorem of Pachner [12]. Independence of choice of representatives follows, as in theorem 3.2, by sliding a small Ω_1-labelled unknot over the whole boundary of a 3-simplex (or, dually, the boundary of a 0-simplex).

Theorem 3.3. *The original invariant* $CH(M) = \eta^{d_3+d_0}\Omega L$ *splits into a sum over all classes* $x \in H^1(M), y \in H_2(M)$*:*

$$CH(M) = \frac{1}{4}\sum_{x,y} CH(M, x, y).$$

Proof. Compute $CH(M)$ via a Heegaard splitting so that there is one 0-handle and one 3-handle. Expand $\Omega = \Omega_0 + \Omega_1$ on all components of the chain-mail link used. Since there are no boundaries or coboundaries in this presentation, $CH(M)$ is given by the above sum plus a sum over other labellings which do not correspond to homology or cohomology classes. Each of these can be transformed using Kirby calculus (just as in Blanchet's paper [1]) into a link which contains a 0-framed unknot linking an algebraically odd number of 'Ω_1's. Expanding these linking strands using $\Omega = \eta \sum \Delta_i \phi_i$ and fusing them together (changing basis using $6j$-symbols so that they are replaced by a single linking strand) produces only odd-numbered idempotents, because fusion preserves the total parity of idempotents. The linking unknot then kills this (as in figure 1) so the term does not appear. □

Relation with Blanchet's invariant

Blanchet's extension [1] of Lickorish's construction goes as follows. Let M be a 3-manifold and pick $z \in H^1(M)$. Represent M by surgery on a framed link L in S^3, and z by a sublink c of L (recall that sublinks c which are in the kernel of the linking matrix (mod 2) correspond to elements of $H^1(M)$). Define $I(M, z)$ to be the value obtained by labelling all components of c with Ω_1, the rest of L by Ω_0, evaluating in $\mathcal{S}S^3 = \mathbb{C}$ and multiplying by $\eta{\kappa_0}^{-\sigma(L)}$. (The factor η was not included in Blanchet's normalisation.) Then $I(M, z)$ is invariant under both types of Kirby moves so is an invariant of (M, z).

In [14] it was shown that any chain-mail link constructed from a Heegaard splitting of M was a surgery presentation for $M\#\overline{M}$. This led to a simple proof of the result, first proved by Turaev and by Walker, that the Turaev-Viro invariant equalled the square of the modulus of the Witten-Reshetikhin-Turaev invariant. The 'spin' analogue is given below, where D denotes Poincaré duality.

Theorem 3.4.

$$CH(M,x,y) = I(M,x).I(\overline{M},x+D(y)) = I(\overline{M},x).I(M,x+D(y)).$$

Proof. This proof will be sketched at the end of the (more complicated) proof of the spin structure version, which is given fully in Theorem 3.6. □

State-sums from triangulations

In [14] it was shown that taking the handle decomposition dual to a given triangulation of M and applying some fusion identities derived from recombination theory (as in Kauffman and Lins [6]) reduced $CH(M)$ to a state-sum which was exactly the normalisation of the Turaev-Viro invariant used in [6]. The 'spin' analogue may also be expressed as a state-sum, although a rather ugly one, so a detailed formula will be omitted.

The basic procedure is exactly as in [14]. In the chain-mail, each meridian links three 2-handle curves, and all these curves are labelled Ω_0 or Ω_1 according to the chosen representatives u, v. Expanding the 2-handle 'Ω's as sums over strands with idempotents attached and applying fusion at each 1-handle reduces $CH(M,x,y)$ to a sum, over certain states, of various simpler skein elements, multiplied by some Δ- and θ-coefficients. Fusion identities for three strands (i,j,k) and an encircling unknot labelled Ω_0 or Ω_1, are shown in figure 3. Two possible terms arise; the first appears only if (i,j,k) is admissible, and the second only if $(i,j,\hat{k})$ is admissible, where $\hat{k}$ denotes $r-2-k$. The framed graph occurring in the second term may be drawn in a variety of ways: it is necessary to choose a particular form at each fusion site, and this is where the state-sum becomes ugly.

The remaining skein elements are graphs consisting of tetrahedron graphs in each 0-handle, possibly joined across 1-handles by strands labelled $r-2$. The quotient skein space of 'maps of the outside' for a 3-ball with 4 boundary points, each labelled 0 or $r-2$ is very simple. If there is an odd number of '$r-2$'s it is empty; if there are 0, 2 or 4 points labelled $r-2$ then it is one-dimensional, with basis vector the empty

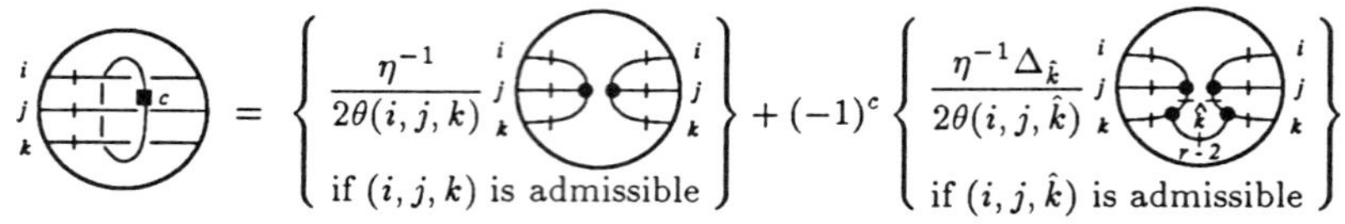

FIGURE 3. Graded fusion identities

skein, one arc, or two arcs labelled $r-2$ respectively. Each tetrahedron network in a 0-handle may be therefore be replaced by a certain multiple (essentially a tetrahedron coefficient itself) of the basis vector. Finally there remains a link in S^3, all of whose components are labelled $r-2$. If $r \equiv 2(4)$ then this contributes the value 1, regardless of knottedness and number of components, because $\Delta_{r-2} = \xi_{r-2} = 1$.

Remark 3.5. The refined state-sum of Turaev and Viro [18] depending on a cohomology class in $H^1(M)$ is obtained in this picture by labelling all 1-handle curves by Ω, and is therefore related to this version by $TV(M, D(y)) = \sum_x CH(M, x, y)$.

The case $r \equiv 0$ mod 4

In this case, the definition of the invariant must be altered, because the crucial r-dependent properties mentioned earlier are no longer true. The correct version is an invariant $CH(M, s, y)$, where s is a spin structure on M and $y \in H_2(M)$.

In order to define the invariant, take a Heegaard splitting for M and embed its 1-skeleton H (with framing) in S^3. There are two spin structures $s_0, s|_H$ induced on H by restriction from S^3 and from (M, s): label each 1-handle meridian of the chain-mail link by Ω_1 if these differ around the handle and by Ω_0 if they agree. Complete the definition of $CH(M, s, y)$ as before by labelling the 2-handle curves using the class y.

The proof of invariance involves consideration of how the spin structures change under the various moves. The principal modifications are that re-framing is now performed by an unknot labelled Ω_1, and that 1-2-birth adds either a $(0,0)$-framed (Ω_0, Ω_0)-labelled Hopf link or a $(0,+1)$-framed (Ω_1, Ω_0)-labelled one.

The splitting Theorem 3.3 in this case is analogous, and the expression in terms of Blanchet invariants is given below. The corresponding state-sum expression is modified at the final stage because of the different values of ξ_{r-2} and Δ_{r-2}.

Theorem 3.6.

$$CH(M,s,y) = I(M,s).I(\overline{M}, s + D(y)) = I(\overline{M}, s).I(M, s + D(y)).$$

Proof. First note that the set of spin structures is affinely isomorphic to $H^1(M)$, so $s + D(y)$ makes sense. Recall that the chain-mail link is a handle picture (attach 1- and 2-handles) for $W_1 = M^{(2)} \times I$, and thus a surgery presentation for the boundary $M\#\overline{M}$. The invariant $CH(M,s,y)$ is given by attaching Ω_0 or Ω_1 to each component of the chain-mail link L. The definition of Blanchet's invariant implies that $CH(M,s,y) = \eta^2.\eta^{-1}.I(M\#\overline{M}, t)$, where $t \in \mathrm{Spin}(M\#\overline{M})$ is the spin structure represented by the sublink of 'Ω_1's under the sequence

$$0 \to \mathrm{Spin}(M\#\overline{M}) \overset{w_2}{\to} H^2(W_2, M\#\overline{M}) \overset{D}{\cong} H_2(W_2),$$

in which W_2 is W_1 with the 1-handles traded to 2-handles, and elements of $H_2(W_2)$ are identified with sublinks. In fact $t = (s, s + D(y)) \in \mathrm{Spin}(M\#\overline{M}) \cong \mathrm{Spin}(M) \oplus \mathrm{Spin}(\overline{M})$, as will be shown below. Recalling the connect-sum rule for I and the fact that $I(M,s) = I(\overline{M}, s)$ then gives the result.

Take H to be a standard unknotted handlebody in S^3. Form W_1 by taking the 4-ball, *adding* 2-handles along the 2-handles curves in H, and *subtracting* thickened slice discs from under the 1-handle curves in $S^3 - H$. It is possible to visualise $M^{(2)}$ in the boundary of W_1 as the handlebody H union half of each 2-handle's boundary ($B^2 \times S^1$ attached along the curve in the boundary of H). Likewise, $\overline{M}^{(2)}$ consists of the complementary handlebody $S^3 - H$, surgered along the meridian curves, union the other halves of the 2-handle boundaries.

Regarding $(M\#\overline{M}, (s, s + D(y)))$ as the spin boundary of W_1 gives rise to an obstruction to extension over W_1 via the sequence

$$\begin{aligned}\mathrm{Spin}(M) \cong \mathrm{Spin}(W_1) &\to \mathrm{Spin}(M\#\overline{M}) \overset{w_2}{\to} H^2(W_1, M\#\overline{M}) \\ &\overset{D}{\cong} H_2(W_1) \cong H_2(M).\end{aligned}$$

The obstruction is in fact y, and gives rise to a characteristic surface K in W_1 made up of the cores of the 2-handles representing y, union a (possibly non-orientable) Seifert surface for their (null-homologous) attaching link in H. By further consideration of this sequence, there is a unique complementary spin structure on $W_1 - K$ which restricts to s on M.

Now glue copies of $S^2 \times B^2$ onto W_1 (with its complementary structure) to obtain W_2 (the 4-ball union 2-handles along *all* curves), in other words fill up each slice disc 'hole' in B^4 and add a genuine 2-handle on top. This gluing can be regarded as a cobordism from spun $M\#\overline{M}$ to itself, and again the relative w_2 map gives an obstruction to extension across the cobordism. If the spin structures $s_0, s|_H$ on H agree then the obstruction vanishes: a difference around a 1-handle of H means that the spin structure will not extend across the core sphere of the $S^2 \times B^2$ associated to that 1-handle's meridian curve. Take a complementary spin structure over this cobordism and glue it onto the original one on W_1 to give one on W_2. The conclusion is that the total obstruction to extension of $(s, s + D(y))$ over W_2 is represented by the sublink of 2-handle curves corresponding to y, union the subset of meridians of 1-handles of H around which $s_0, s|_H$ differ. This is precisely the sublink labelled Ω_1 in the definition of $CH(M, s, y)$, hence the result.

The homology case ($r \equiv 2(4)$) is completely analogous, replacing 'Spin(M)' with '$H^1(M)$' and w_2 by the usual coboundary map δ throughout the proof, and thinking of honest 1-cocycles on the handles of H as representing elements of $H^1(M)$. □

The case of r odd

If r is odd then invariants depending on either cohomology classes or spin structures exist, provided the definition of CH is multiplied by an additional factor 2^{d_1}, which is needed to counteract the values of $\frac{1}{2}$ which occur when 1-2-handle pairs are born (see figure 1). Blanchet [1] proves a theorem showing that the dependence on cohomology class or spin structure of his invariants is very simple. The corresponding result here makes the invariants $CH(M, x, y)$, $CH(M, s, y)$ independent of x and s, and their dependence on y uninteresting.

4. The four-dimensional case

In [2], Broda defined an invariant of 4-manifolds by using the link in S^3 formed by the attaching circles of 1-handles and 2-handles [8]. He attached Ω to all components of the link, evaluated in $\mathcal{S}S^3$ and normalised appropriately: this does indeed give a quantity invariant under the Morse theoretic handle-moves of [3]. In [14] this invariant was shown to be actually just $\kappa^{\sigma(W)}$, and it was reduced, using the method of fusion, to a skein-theoretic analogue of the Crane-Yetter state-sum invariant [4] defined on triangulations of 4-manifolds. That the Crane-Yetter invariant is classical is disappointing, but the proof has the interesting consequence of giving a state-sum formula for computing the signature of a triangulated 4-manifold.

Broda then revised his definition, labelling the 1-handles by Ω and the 2-handles by Ω_0. This version is in fact reducible to a classical value in a similar way, but it is also just a special case of what is suggested by Blanchet's calculus, namely labelling the 2-handles by Ω_1 or Ω_0 according to whether they belong to a subset v of 2-handles representing a chosen class $x \in H_2(W)$. This generalised invariant of a 4-manifold with a second homology class is defined below and then evaluated in terms of classical invariants. Its corresponding triangulation state-sums give formulae for the second Stiefel-Whitney class, the Pontrjagin squares of cohomology classes, and hence for ordinary cup products of classes; furthermore, the Stiefel-Whitney class formula leads to another proof of Rochlin's theorem, which should be compared with the version given by Kirby and Melvin in [9].

Definition 4.1. Let W be a closed, connected, oriented, smooth 4-manifold, so that it has a handle decomposition with one 0-handle and d_i i-handles. Draw the 1- and 2-handles as a link L in $S^3 = \partial B^4$, as in [8], regarding the 1-handles as 0-framed unknots. Let $x \in H_2(W)$ be represented by a subset v of 2-handles. Let $\Omega_{*\cup v}L$ denote the element of $\mathcal{S}S^3 = \mathbb{C}$ obtained by attaching Ω_1 to all 2-handles in v, Ω_0 to the other 2-handles, and the original Ω to all the 1-handles (this is the significance of the '*'). Define $I(W,x) = (2\eta)^{d_3-d_4+1}\Omega_{*\cup v}L$.

Theorem 4.2. *$I(W,x)$ is an invariant of the pair (W,x).*

Proof. The value $\Omega_{*\cup v}$ is invariant under handleslides of 1- over 1- and 2-over 1-handles (by the usual handleslide identity) and under 2-2-slides (by the graded version). Birth of a 1-2-pair introduces a Hopf link with value 1, birth of a 2-3-pair introduces a 0-framed unknot labelled Ω_0, whose value $\frac{1}{2}\eta^{-1}$ is absorbed by the normalising factor, and birth of a 3-4-pair obviously has no effect. These moves relate arbitrary handle decompositions of homeomorphic 4-manifolds [3], so $I(W,x)$ is an invariant of W. Independence of the choice of representative v follows most easily as part of the proof of the following evaluation theorem (it is hard to visualise the attaching 3-handle maps well enough to mimic the proof of Theorem 3.2). □

Theorem 4.3. *Let σ, χ, b_1 and w_2 be the signature, Euler characteristic, first $\mathbb{Z}_2$-coefficient Betti number and second Stiefel-Whitney class of the 4-manifold W, let $x \cdot x$ denote the 'Pontrjagin square' of x (its self-intersection, lifted to $\mathbb{Z}_4$-coefficients), and let D denote Poincaré* *duality. Then*

$$I(W,x)=\begin{cases}(\sqrt{2})^{2-\chi}.(\kappa e^{-2\pi i/8})^{\sigma}.i^{x\cdot x} & r\equiv 1(4)\\ (\sqrt{2})^{2-\chi}.(\kappa e^{2\pi i/8})^{\sigma}.i^{-x\cdot x} & r\equiv 3(4)\\ 2^{b_1}.\kappa^{\sigma} & r\equiv 2(4), x=0\\ 0 & r\equiv 2(4), x\neq 0\\ 2^{b_1}.\kappa^{\sigma} & r\equiv 0(4), x=D(w_2)\\ 0 & r\equiv 0(4), x\neq D(w_2).\end{cases}$$

Proof. The basic method is as follows. Consider the union W_1 of the 0-, 1- and 2-handles of W and trade its 1-handles for 2-handles to make a 4-manifold W_2, which consists of 2-handles attached to B^4 along L and has boundary M, a connect-sum of d_3-d_4+1 copies of $S^1\times S^2$. Copies of $\mathbb{C}P^2$ or its reverse may be added to W_2 to make its signature zero, making it cobordant (rel the boundary M) to a connect-sum of '$S^1\times B^3$'s. By the proof of Kirby's theorem [7], this cobordism may be realised by connect-summing from both sides with copies of $S^2\times S^2$ or its twisted counterpart, to reach a common 4-manifold W_3, whose two handle decompositions are related by 2-2-handleslides alone. In terms of framed link presentations, this means that it is possible first to add some ±1-framed unknots and Hopf links with either $(0,0)$- or $(0,1)$-framings, then to perform original handleslide moves (*not* blow-ups as well), to reach a 'standard link' consisting of d_3-d_4+1 0-framed unknots (presenting the standard $\#S^1\times B^3$) and some more Hopf links.

To evaluate $\Omega_{*\cup v}L$, first add trivial ±1-framed unknots and Hopf links as above. The factors they introduce may be divided out at the end, provided they are labelled so that their values are not zero. (These good labellings depends on r mod 4.) Substitute $\Omega=\Omega_0+\Omega_1$ on the components originally corresponding to 1-handles, obtaining a sum over terms $\Omega_{c\cup v}L$, where all of the components are labelled 0 or 1, and where c is the subset of 1-handle curves labelled Ω_1. Slide each of these labelled links to the standard link, keeping track carefully of labellings (that is, elements of $H_2(W_3)$), and evaluate the (simple) result using the identities in figure 2: the different cases are treated below. Note that the boundary of any 3-chain will be represented after the slide by a subset of the 0-framed unknots, because adding the 3-handles of W to the standard $\#S^1\times B^3$ results in a collection of cancelling 2-3-pairs, and the cobordism from $\#S^1\times B^3$ to W_3 does not affect this. Each 0-framed unknot has the value $\frac{1}{2}\eta^{-1}$, independent of its labelling, and hence $I(W,x)$ is independent of the representative v of $x\in H_2(W)$, completing the proof of Theorem 4.2.

Case 1: r odd. Since v represents an element of $H_2(W)$, an even number of 2-handles labelled Ω_1 passes through each 1-handle unknot. Fusing these always gives an even number (as in the proofs of Theorem 3.3), never $r-2$. So, by the encircling identity of figure 1, Ω may be replaced by $2\Omega_0$, and hence $\Omega_{*\cup v}L = 2^{d_1}\Omega_{\emptyset\cup v}L$. Add σ Ω_0-labelled, -1-framed unknots (it is assumed, without loss of generality, that $\sigma \geq 0$) and enough (say k) (Ω_0, Ω_0)-labelled Hopf links of the two kinds to get a link which is equivalent, by handleslides, to a standard one. Slide handles to this standard form, which has say l Hopf links: the value $\Omega_{\emptyset\cup v}L$ is replaced by $(\frac{1}{2}\eta^{-1})^{d_3-d_4+1}.(\frac{1}{2})^l.i^{\pm v\cdot v}$, with $v\cdot v$ being the total self-linking number of v modulo 4, computed using the linking matrix of the whole standard link. (This fact is verified by observation of the values of labelled Hopf links in figure 2; the sign is positive when $r\equiv 1(4)$, negative when $r\equiv 3(4)$.)

Self-intersection numbers of elements of $H_2(W)$ are computed by choosing representatives and applying the same linking-matrix form, so $v\cdot v = x\cdot x$. Combining this with the definition of $I(W,x)$ and the factor $\overline{\kappa_0}^{-\sigma}.(\frac{1}{2})^{-k}$ correcting for the added unknots and Hopf links,

$$\begin{aligned} &I(W,x)\\ &= \left[(2\eta)^{d_3-d_4+1}\right]\left[2^{d_1}(\tfrac{1}{\sqrt{2}}e^{\pm 2\pi i/8}\kappa^{-1})^{-\sigma}(\tfrac{1}{2})^{-k}\right]\left[(\tfrac{1}{2}\eta^{-1})^{d_3-d_4+1}(\tfrac{1}{2})^l i^{\pm x\cdot x}\right].\end{aligned}$$

Counting components gives $d_3-d_4+1+2l = d_1+d_2+\sigma+2k$: substituting this into the formula gives the result.

Case 2: $r\equiv 0(4)$. Add σ Ω_1-labelled, -1-framed unknots and k Hopf links, either $(0,0)$-framed and (Ω_0,Ω_0)-labelled or $(0,+1)$-framed and (Ω_1,Ω_0)-labelled. Then expand the 'Ω's on the 1-handles to get the sum (over all subsets c of the 1-handle curves) of terms $\Omega_{c\cup v}L$. Sliding these to the standard position and observing the values of Hopf links and 0-framed unknots in figure 2 shows that $\Omega_{c\cup v}L = 0$ unless the sublink b corresponding to $c\cup v$ after the slide is *characteristic*, that is satisfies $Q.b = f$, where Q is the linking matrix and f the vector of diagonal entries of Q. For such a characteristic sublink, $\Omega_{c\cup v}L = (\frac{1}{2})^{d_3-d_4+1}$. The property of being characteristic is preserved under handlesliding and adding the types of labelled Hopf link given above, so $\Omega_{c\cup v} = 0$ unless $c\cup v$ is itself characteristic with respect to the original linking matrix of L, in which case $\Omega_{c\cup v}L = (\frac{1}{2})^{d_3-d_4+1}$.

Characteristic sublinks of L correspond to the set Spin(M) of spin structures on M via the relative Stiefel-Whitney class map

$$0\to \mathrm{Spin}(M) \overset{w_2}{\to} H^2(W_2,M) \overset{D}{\cong} H_2(W_2).$$

Since $H_2(W_2)$ is freely generated by the 2-handles of W_2, this map assigns to any spin structure on M the subset of 2-handles of W_2 over which it does not extend. Identifying 2-handles with components of L, the image of w_2 is the set of characteristic links in L.

Suppose s is a spin structure on M represented by a characteristic link $c \cup v$, with c, v subsets of 1- and 2-handles respectively. Trading handles back, turning W_2 into W_1, the obstruction corresponding to c disappears, and only v remains, representing the image of s under the map w_2 in the sequence

$$0 \to \mathrm{Spin}(W_1) \to \mathrm{Spin}(M) \stackrel{w_2}{\to} H^2(W_1, M) \to H^2(W_1).$$

Now the image of this map equals the preimage of $i^* w_2(W)$ (where $i : W_1 \hookrightarrow W$ is inclusion), because any spin structure on $M = \# S^1 \times S^2$ extends over the 3- and 4-handles of W.

Thus v must represent an element of $H_2(W_1)$ which maps to the dual of $w_2(W)$ in $H_2(W)$. As v was originally chosen as a representative of $x \in H_2(W)$, this means that if $x \neq D(w_2)$ then the invariant vanishes. On the other hand, if $x = D(w_2)$ there are $|\mathrm{Spin}(W)| = |H^1(W)| = 2^{b_1}$ different choices of c such that $c \cup v$ is characteristic, completing the proof.

Case 3: $r \equiv 2(4)$. This is very similar. After handlesliding, only the links which are solutions of $Q.(c \cup v) = 0$ contribute. These are the ones representing preimages of the pullback of the zero class in $H_2(W)$. The whole procedure is the same, with $\mathrm{Spin}(M), \mathrm{Spin}(W)$ replaced by $H^1(M), H^1(W)$. □

Theorem 4.4. *The original invariant* $I(W) = \eta^{d_3 - d_4 + 1} \Omega L$ *splits as*

$$I(W) = 2^{-b_1} \sum_{x \in H_2(W)} I(W, x).$$

Proof. This may be deduced from the evaluations given above and the fact that $I(W) = \kappa^{\sigma}$: in the odd case, it uses an instance of the formula for the Brown invariant of a $\mathbb{Z}_2$-quadratic form

$$\sum_{x \in H_2(W)} i^{x \cdot x} = (\sqrt{2})^{b_2} . e^{2\pi i \sigma / 8},$$

(which may be found, for example, in Gilmer's paper [5]), together with the observation that $\chi = 2 - 2b_1 + b_2$, where b_2 is the second $\mathbb{Z}_2$-Betti number.

Alternatively, the result follows directly, expanding $\Omega = \Omega_0 + \Omega_1$ on all of the 2-handle curves in the skein element ΩL and observing that labellings of the 2-handles not representing elements of $H_2(W)$

contribute 0 (exactly as in Theorems 3.3 and 4.3), whilst $2^{d_3-d_4+1-b_3}$ cycles represent each homology class (this is the number of boundaries of 3-chains). The above formula is then a consequence. □

The state-sum version

Crane and Yetter defined in [4] a 4-dimensional counterpart to the Turaev-Viro invariant, using $15j$-symbols. In [14], this state-sum was shown to arise (in a skein-theoretic normalisation) when evaluating Broda's invariant on a link picture coming from the handle decomposition dual to a triangulation. There is an analogous version for the homology-class invariant $I(W,x)$ above. Choose a subset of 2-simplexes representing the dual cohomology class $\xi = D(x)$ and simply restrict the states in the state-sum to ones where the faces in v are labelled with odd integers, the rest with even. (Because $I(W,x)$ is computed using a *dual* handle decomposition, it should be thought of as giving an invariant of a triangulated 4-manifold with a *cohomology* class.)

Definition 4.5. (Compare [14].) Fix a triangulation T of W and a subset of 2-simplexes v representing $\xi \in H^2(W)$. Let a *state* s be an assignment of integers $0 \leq s(f_j), s(t_k) \leq r-2$ to the 2-simplexes f_j (these labels being odd or even according to whether f_j belongs to v or not) and to the 3-simplexes t_k of a triangulation of W. Call s *admissible* if the pairs of triples occurring in each tetrahedron are admissible. Let $\theta_1(s,t_k)$ and $\theta_2(s,t_k)$ be the two trihedron coefficients arising from these triples and let $\phi(s,h_l)$ be the $15j$-coefficient occurring at the 4-simplex h_l. Let $\Delta(s,f_j) = \Delta_{s(f_j)}$, similarly for $\Delta(s,t_k)$. Define

$$\Sigma(W,T,v) = \sum_s \{\prod_{f_j} \Delta(s,f_j) \prod_{t_k} \Delta(s,t_k)\theta_1(s,t_k)^{-1}\theta_2(s,t_k)^{-1} \prod_{h_l} \phi(s,h_l)\},$$

where the sum is over all admissible states s, and the products are over 2-, 3- and 4-simplexes respectively. Let n_i denote the number of i-simplexes.

Theorem 4.6. *If v represents ξ then*
$2^{n_1-n_0+1}\eta^{-n_0+n_1+n_2-n_3+n_4}\Sigma(W,T,v)$ *equals the invariant* $I(W,D(\xi))$.

Proof. This parallels the version in [14] so closely that it is not worth repeating: it relies on drawing a 'chain-mail' link L in S^3 representing the handle decomposition dual to T, then applying 4-strand fusion (shown in figure 4) at each 1-handle. The only slight difference is that all but one of the 0-handles of W must first be removed by connect-summing with copies of $S^1 \times S^3$, whose invariant is now 2: this gives the formula

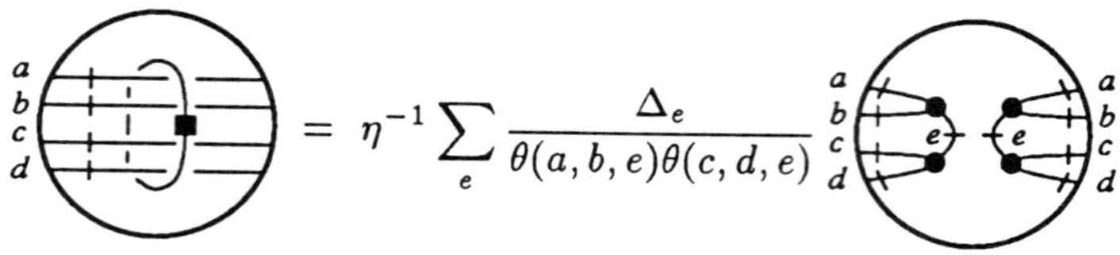

FIGURE 4. Four-strand fusion

$I(W,v) = (2\eta)^{d_3 - d_4 + 1}\eta^{d_0 - 1}\Omega_{*\cup v}L$ for a manifold W with more than one 0-handle. □

The original version in [14] produced a state-sum formula which, setting $r = 3n_2$, determined the signature of a 4-manifold (this choice makes the order of κ greater than twice the absolute value of the signature). The modified version gives state-sums for the Stiefel-Whitney class (leading to a proof of Rochlin's theorem) and for the Pontrjagin squares and cup products of $\mathbb{Z}_2$-cohomology classes. Write $\Sigma_r(W,T,v)$ for the state-sum above, emphasising the choice $A = e^{2\pi i/4r}$.

Corollary 4.7. *The Stiefel-Whitney class $w_2(W)$ may be computed at $r = 4$, using the fact that $\Sigma_4(W,T,v) = 0$ unless v represents w_2. The Pontrjagin square $\xi \cdot \xi \in \mathbb{Z}_4$ of $\xi \in H^2(W)$ may be computed at $r = 3$ using*

$$i^{-\xi\cdot\xi} = \Sigma_3(W,T,v),$$

where v represents ξ. If in addition u represents ζ then the cup product $\xi \cup \zeta \in \mathbb{Z}_2$ is given by

$$(-1)^{\xi\cup\zeta} = \frac{\Sigma_3(W,T,u).\Sigma_3(W,T,v)}{\Sigma_3(W,T,u+v)}.$$

Proof. Simply combine the formula for $I(W,x)$ as a state-sum with its evaluation in terms of classical invariants, and choose the smallest appropriate value of r in each case. The first 'spin' case occurs when $r = 4$. The first 'odd' case is at $r = 3$, when $\kappa = e^{-2\pi i/8}$ (so the signature term is eliminated) and $\eta = 1/\sqrt{2}$. Finally, $(\xi + \zeta)\cdot(\xi + \zeta) = \xi\cdot\xi + \zeta\cdot\zeta + 2(\xi\cup\zeta) \in \mathbb{Z}_4$ gives the cup product formula. □

Corollary 4.8. (Rochlin's theorem) *If W is a smooth closed oriented spin 4-manifold then its signature is divisible by 16.*

Proof. Consider reducing $I_4(W,0) = (\frac{1}{2})^{n_4-1}\Omega_{*\cup\emptyset}L$ to a state-sum by choosing a link L, expanding the 'Ω_0's on the 2-handle curves and applying fusion, as in Theorem 4.6. Observe that at $r=4$, $\Delta_0 = \Delta_2 = 1$, and the only admissible triples are $(0,0,0)$ and $(0,2,2)$, each of whose associated θ-coefficients is 1. Consideration of the four-strand fusion rule in this case shows that the admissible states are precisely labellings of 2-handle curves of L by 0 or 2, such that the sublink s labelled 2 represents a ($\mathbb{Z}_2$-coefficient) 2-cycle. In other words, either 0,2 or 4 strands labelled 2 must pass through every 1-handle 'Ω', and the effect of fusion is to cut and rejoin these in pairs, leaving a product of $15j$-symbol skein elements. The state-sum $\Sigma(W,T,\emptyset)$ therefore reduces to a sum (over the $2^{n_2-n_3+n_4-1+b_1}$ possible 2-cycles) of products (over all 4-simplexes) of $15j$-coefficients alone.

Each $15j$-coefficient is a small link whose components are all labelled 2. Since $\Delta_2 = 1$ and $\xi_2 = -1$, its value is -1 to the power of its total number of crossings. Fixing a state s and taking the product over all 4-simplexes therefore gives -1 to the power of the total number of crossings of the sublink s of L, because the number of crossings is not affected by the way four-strand fusion reconnects the strands. Now the sublink s represents a homology class $x \in H_2(W)$ and the linking matrix represents the intersection form on W, so the total writhe of s (mod 2) equals $D(x) \cup D(x)$. Since W is spin, any class $\xi \in H^2(W)$ satisfies $\xi \cup \xi = w_2 \cup \xi = 0$. Therefore each state s contributes $+1$, $\Sigma_4(W,T,\emptyset) = 2^{n_2-n_3+n_4-1+b_1}$, and $I(W,0) = 2^{b_1}$. On the other hand, Theorem 4.3 gives $I(W,0) = 2^{b_1}e^{2\pi i(-5/16)\sigma(W)}$, and so 16 divides $\sigma(W)$. □

References

1. C. Blanchet, *Invariants on 3-manifolds with spin structure*, Comm. Math. Helv. **67** (1992) 406-427.
2. B. Broda, *Surgical invariants of 4-manifolds*, Preprint and revised preprint, 1993.
3. E. César de Sá, *A link calculus for 4-manifolds*, Topology of low-dimensional manifolds, (ed. R. Fenn) Springer Lecture Notes in Math. Vol. 722, 1979.
4. L. Crane and D. N. Yetter, *A categorical construction of 4d topological quantum field theories*, Quantum Topology, Baadhio and Kauffman, World Scientific, to appear.
5. P. M. Gilmer, *Link cobordism in rational homology spheres*, J. Knot Th. Ram. **2** (1993) 285-320.
6. L. H. Kauffman and S. L. Lins, *Temperley-Lieb recoupling theory and invariants of 3-manifolds*, Princeton University Press, to appear.
7. R. C. Kirby, *A calculus for framed links in S^3*, Invent. Math. **45** (1978) 35-56.

8. _____, *Topology of 4-manifolds*, Springer Lecture Notes in Math. Vol. 1374, 1989.
9. R. C. Kirby and P. Melvin, *The 3-manifold invariant of Witten and Reshetikhin-Turaev for* $sl(2,\mathbb{C})$, Invent. Math. **105** (1991) 473-545.
10. W. B. R. Lickorish, *Calculations with the Temperley-Lieb algebra*, Comm. Math. Helv. **67** (1992) 571-591.
11. _____, *The skein method for 3-manifold invariants*, J. Knot Th. Ram. **2** (1993) 171-194.
12. U. Pachner, *PL homeomorphic manifolds are equivalent by elementary shelling*, Europ. J. Combinatorics **12** (1991) 129-145.
13. N. Y. Reshetikhin and V. G. Turaev, *Invariants of 3-manifolds via link polynomials and quantum groups*, Invent. Math. **103** (1991) 547-597.
14. J. D. Roberts, *Skein theory and Turaev-Viro invariants*, Topology (to appear).
15. V. G. Turaev, *Topology of shadows*, Preprint, 1992.
16. _____, *State-sum models in low-dimensional topology*, Proc. ICM Kyoto, 1991, 689-698.
17. _____, *Quantum invariants of 3-manifolds and a glimpse of shadow topology*, C. R. Acad. Sci. Paris **313** (1991) 395-398.
18. V. G. Turaev and O. Y. Viro, *State-sum invariants of 3-manifolds and quantum 6j-symbols*, Topology **31** (1992) 865-902.
19. D. N. Yetter, *Homologically twisted invariants related to (2+1)- and (3+1)-dimensional state-sum topological quantum field theories*, Preprint, 1993.

Pembroke College, Cambridge, CB2 1RF, England
E-mail address: J.D.Roberts@pmms.cam.ac.uk

AMS/IP Studies in Advanced Mathematics
Volume **2**, 1997 (Part 1)

A FAMILY OF MÖBIUS INVARIANT 2-KNOT ENERGIES

DAVID AUCKLY AND LORENZO SADUN

ABSTRACT. After reviewing energy functionals for 1-dimensional knots and links, we define a family of Möbius invariant energy functionals E_s for surfaces embedded in $\mathbb{R}^n$. These functionals are all finite for smoothly embedded compact surfaces and infinite for self-intersecting immersed surfaces. They treat disconnected surfaces and connected sums of surfaces in a natural way. For sufficiently negative s, E_s is not bounded from below. For sufficiently positive s, the evidence to date suggests that E_s *is* bounded from below, but we have not yet found a proof. We also discuss alternate methods of defining surface energy.

Several years ago Jun O'Hara [6, 7, 8] defined a functional for embeddings of circles in $\mathbb{R}^n$ that he called the energy of a knot. Bryson, Freedman, He and Wang [1, 3] proved that this energy is Möbius invariant, and bounded the complexity of knots by their energy. Since then others, notably Doyle and Schramm [2, 9], have modified and reformulated O'Hara's energy. The formulations are all essentially equivalent, but each exhibits a different important property of the energy. In this paper we generalize some of these ideas to 2 dimensions and define a family of energy functionals for embedded surfaces. These functionals are different from, and postdate, the surface energy defined by Rob Kusner and John Sullivan, which is discussed elsewhere in these proceedings [5].

Section 1 is a review of the 1-dimensional energy. We begin with the unregularized energy formula and discuss its properties. We then list several ideas for regularization and show how they are essentially

This work is partially supported by an NSF Mathematical Sciences Postdoctoral Fellowship and by the Texas Advanced Research Program equivalent. Each of these methods suggests a regularization in 2 dimensions, most of which turn out *not* to be equivalent. The material in this section is not original, but rather reflects past work by many authors [6, 7, 8, 1, 3, 2, 9].

In section 2 we generalize O'Hara's original regularization to 2 dimensions. There is some freedom in the choice of regularization, yielding a 1-parameter family of Möbius invariant surface energies E_s that differ by multiples of the Willmore functional [10]. We compute the energies of several examples, and establish that these energy functionals have all but one of the desirable properties shared by the 1-dimensional energy. The one uncertain property is positivity, which is discussed further in section 4.

In section 3 we consider generalizations of the other 1-dimensional regularizations to 2 dimensions. Some approaches fail, due to a log divergence that has no analog in 1 dimension. One successful approach gives the Kusner-Sullivan energy. Another gives an energy similar to ours. We are presently exploring, with Kusner and Sullivan, this last approach.

In section 4 we consider boundedness. The primary difficulty with our energy functionals E_s is that they are not manifestly bounded from below. Indeed, for s sufficiently negative, E_s is *not* bounded from below. However, for s sufficiently positive, it appears likely that E_s is non-negative, although we have not yet found a proof.

We thank Peter Doyle, Rob Kusner, Dennis Roseman, Oded Schramm and John Sullivan for freely sharing their ideas with us and updating us on work in progress. We thank Rob Kusner, Karen Uhlenbeck and Michael Freedman for useful face-to-face discussions, and thank Boris Apanasov for suggesting this problem to us. Lorenzo Sadun and David Auckly are each partially supported by the Texas Advanced Research Program. Dr. Sadun is also partially supported by an NSF Postdoctoral Fellowship.

1. The 1-dimensional energy.

Let γ_1 and γ_2 be 1-manifolds in $\mathbb{R}^n$, and consider the functional

$$\int\int_{\gamma_1\times\gamma_2} \frac{dx\,dy}{|x-y|^2}, \tag{1.1}$$

where dx and dy are the arclength measures on γ_1 and γ_2. This functional is finite if γ_1 and γ_2 are compact and do not intersect, it diverges if they do intersect, and it is invariant under Möbius transformations.

Möbius invariance is seen in figure 1. Recall that the Möbius group is generated by inversions in spheres. Let $\tilde{x}$ and $\tilde{y}$ be the images of x and

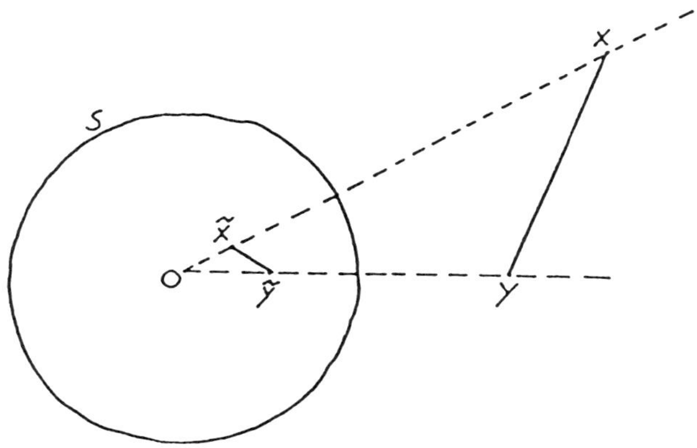

FIGURE 1. Möbius invariance of $dx\,dy/|x-y|^2$.

y under inversion in the sphere S. Since $|x|/|y| = |\tilde{y}|/|\tilde{x}|$, the triangles Oxy and $O\tilde{y}\tilde{x}$ are similar, so $|\tilde{x}-\tilde{y}|^2/|\tilde{x}||\tilde{y}| = |x-y|^2/|x||y|$. Since $|d\tilde{x}| = (|\tilde{x}|/|x|)|dx|$, the invariance of the integral follows.

Next we consider the case where $K = \gamma_1 = \gamma_2$ is a smooth knot or link, with one component possibly extending to the point at infinity. The unregularized "energy of the knot (link)" is

$$E_0(K) = \int\int_{K\times K} |x-y|^{-2} dx\, dy. \tag{1.2}$$

This obviously diverges for any nonempty K. The problem is to find a regularization of the functional, finite for smoothly embedded K, that preserves the desirable formal properties of the unregularized energy. In particular, the regularized energy should, if possible:

1. Be finite for sufficiently smooth knots and links, but infinte for self-intersecting immersions.
2. Be Möbius invariant.
3. Be bounded from below.
4. Satisfy a *connected sum rule.* If K_1 and K_2 are open knots, then, in the limit of large separation, the energy of $K_1 \# K_2$ should approach the sum of the energies of K_1 and K_2.
5. Have the *additive link property* (ALP). The regularization should not affect the force between non-intersecting neighborhoods. The energy of a link should equal the sum of the energy of the components plus the *unregularized* cross terms, and the force between two neighborhoods should be independent of whether they are

in the same link component or not.

Möbius invariance, boundedness, and the connected sum rule together imply that the energy is minimized by straight lines, or equivalently by circles. By scale invariance, a line must have energy zero. If any open knot had energy less than a line, the energy of the connected sum of that knot with itself N times could be made arbitrarily negative, violating boundedness. Therefore all knots must have non-negative energy.

The 1-dimensional regularizations of O'Hara, Schramm, and Doyle and Schramm have all the above properties and are all equivalent, up to unimportant additive constants. However, in generalizing to two dimensions it is easy to lose one or more of these properties. The Kusner-Sullivan energy has the first four properties, but not the ALP. Our energy, described in section 2 below, has the ALP, but we have not yet managed to prove that it is bounded from below.

O'Hara's counterterm regularization

O'Hara [6] was the first to define a sensible regularized energy. Let

$$V(x) = \lim_{\epsilon \to 0} \left(\int_{|x-y|>\epsilon} \frac{dy}{|x-y|^2} \qquad - \frac{2}{\epsilon} \right). \tag{1.3}$$

The *counterterm* $-2/\epsilon$ exactly cancels the divergence of the integral, so the limiting potential is well-defined and finite. We then define the energy of a knot to be

$$E^{(c)}(K) = \int V(x) dx. \tag{1.4}$$

Although O'Hara only defined this energy for closed knots, the same definition works for open knots and for links. Since the counterterm is local, the ALP is immediate. It is also an easy calculation to see that the connected sum rule holds, with

$$E^{(c)}(K_1 \# K_2) = E^{(c)}(K_1) + E^{(c)}(K_2) + O(D^{-2}), \tag{1.5}$$

where D is the separation between K_1 and K_2 in the connected sum. Möbius invariance is, however, more subtle, and was first proven by Bryson et al [1] using a different regularization.

The arclength regularization

O'Hara [7, 8] also defined an energy for knots by subtracting a term involving arclength,

$$E^{(a)}(K) = \int \int_{K \times K} \left(\|x-y\|^{-2} - D(x,y)^{-2} \right) dxdy, \tag{1.6}$$

where $D(x,y)$ is the arclength between x and y. The beauty of this regularization is that it is manifestly non-negative, and equals zero only for K being a straight line.

For closed knots, this arclength-regularized energy turns out to be 4 greater than the counterterm energy, while for open knots it equals the counterterm energy. To see this, hold x fixed and integrate $dy/D(x,y)^2$ over the region $|x-y|>\epsilon$.

$$\int_{|x-y|>\epsilon} D(x,y)^{-2}dy = \frac{2}{\epsilon} - \frac{4}{l} + O(\epsilon), \tag{1.7}$$

where l is the length of the knot. The $O(\epsilon)$ term comes from the difference between cutting off the integral at Euclidean distance ϵ and arclength ϵ. Taking the $\epsilon \to 0$ limit we get that

$$V(x) = -4/l + \int \left(\|x-y\|^{-2} - D(x,y)^{-2}\right) dy. \tag{1.8}$$

Integrating over x we find that $-4/l$ integrates to -4 if l is finite and zero if l is infinite.

Bryson et al [1] proved that $E^{(a)}$ is Möbius invariant up to this difference of 4. If two knots, K_1 and K_2, are related by a Möbius transformation, then $E^{(a)}(K_1) = E^{(a)}(K_2)$ if both knots are open or both are closed, while $E^{(a)}(K_1) = E^{(a)}(K_2) - 4$ if K_1 is open and K_2 is closed. Or to put it another way, $E^{(c)}(K_1) = E^{(c)}(K_2)$ in *all* cases.

For links, Freedman, He and Wang [3] define the link energy to be the sum of the energies of the components plus the unregularized cross-terms. For a link L with n_c closed components, $E^{(a)}(L) = E^{(c)}(L) + 4n_c$. So even though the $E^{(a)}$ energy formula makes a distinction between points on the same or different link components, $E^{(a)}$ does have the ALP.

Schramm's wasted length regularization

Oded Schramm [9] discovered an interesting geometrical interpretation of O'Hara's local potential $V(x)$. Fix x, let S_x be the unit sphere centered at x, let I_x be the inversion in S_x, and let $\tilde{y} = I_x(y)$ be the image of y. Then $|dy|/|x-y|^2 = |I_x^* d\tilde{y}|$, so $\int dy/|x-y|^2$ is just the length of the inverted curve.

This is shown in figure 2. The points on K a distance ϵ from x are p_1 and p_2. The coordinates of $\tilde{p}_{1,2}$ are $x \pm (1/\epsilon)\vec{t} + k\vec{n}/2 + O(\epsilon)$, where $\vec{t}$, $\vec{n}$, and k are the unit tangent, principal normal and curvature of the knot at x. The integral in (1.3) gives the length of the inverted curve from $\tilde{p}_1$ to $\tilde{p}_2$, while $2/\epsilon$ is, to within $O(\epsilon)$, the straight-line distance from $\tilde{p}_1$ to $\tilde{p}_2$. So, to within $O(\epsilon)$, $V_\epsilon(x)$ gives the slack, or "wasted length", in

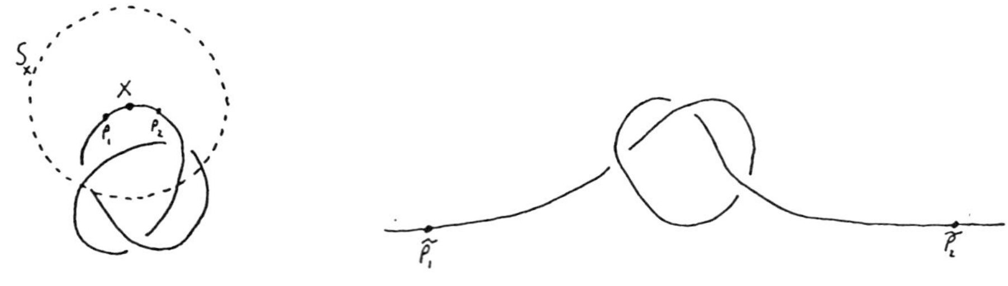

FIGURE 2. Inversion of a knot in S_x.

the inverted knot between $\tilde{p}_1$ and $\tilde{p}_2$. $V(x)$ gives the total slack in the inverted knot.

Doyle and Schramm's Möbius invariant approach

Suppose for simplicity that the knot K is tangent to the x_1 axis at x. Then the slack in the inverted knot $\tilde{y}$ can be written as $\int |d\tilde{y}| - d\tilde{y}_1$, or equivalently as $\int |d\tilde{y}|(1 - \cos(\alpha))$, where α is the angle between the tangent vector at $\tilde{y}$ and the x_1 axis. Since $|d\tilde{y}|$ can be expressed in terms of $|dy|$, we can rewrite the entire energy functional as

$$E^{(m)}(K) = \int\int \frac{(1 - \cos(\alpha))}{|x-y|^2} dxdy. \tag{1.9}$$

The angle α can be understood in terms of the original knot K, without recourse to inversions. Let S_1 be the unique circle, tangent to K at x, that passes through y. $I_x(S_1)$ is the line through $\tilde{y}$ parallel to the x_1 axis, which intersects $I_x(K)$ at angle α. Since Möbius transformations are conformal, the angle between S_1 and K at y is also α. Alternatively, let S_2 be the circle, tangent to K at x that passes through y. The angle between S_1 and S_2, measured at either x or y, is again α.

The functional $E^{(m)}$ is manifestly Möbius invariant, non-negative, and minimized by circles and lines. The ALP is less obvious, but it follows from the equality of $E^{(m)}$ and $E^{(c)}$.

There are a few special cases where the equivalence of $E^{(m)}$ and $E^{(c)}$ breaks down. The two functionals are finite and equal when the knot (or link) is embedded. They are both infinite when there are self-intersections at isolated points. However, they disagree on curves that

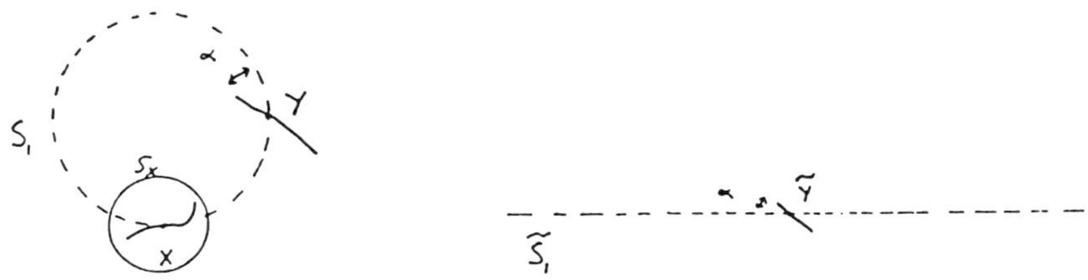

FIGURE 3. Two interpretations of the angle α.

wrap around themselves several times. For example, let L be two superimposed copies of the knot K. If the two copies are given the same orientation, $E^{(m)}(L) = 4E^{(m)}(K)$, while if the orientations disagree, $E^{(m)}(L) = \infty$. By constrast, $E^{(c)}(L) = \infty$ in both cases. (When L contains components that agree on sets of positive measure but then separate, then both $E^{(m)}$ and $E^{(c)}$ are infinite. This is because of the divergent contributions of nearby strands before they join together, or after they pull apart.)

The reason for the discrepancy between $E^{(m)}$ and $E^{(c)}$ is that, at points of self-intersection, the effect of the $\cos(\alpha)$ term is to subtract $2(1+\cos(\beta))/\epsilon$, rather than $2/\epsilon$, where β is the angle between the two intersecting strands. When self-intersection points are absent, or form a set of measure zero, this does not matter, but for multiple copies of the same curve it makes a great difference.

Doyle and Schramm [2, 9] have proposed a family of Möbius invariant knot energies. Given any function $f(\alpha)$ that goes to zero sufficiently fast at $\alpha = 0$, they define the functional

$$E^{(f)}(K) = \int\int f(\alpha)|x-y|^{-2}dxdy. \tag{1.10}$$

This is manifestly Möbius invariant, and is non-negative if f is. For $f \neq 1-\cos(\alpha)$ this is not equivalent to the other energies, and typically does not have the ALP.

2. Energy of surfaces — counterterm approach.

We are now ready to generalize the definition of energy to surfaces embedded in $\mathbb{R}^n$. In this section we present our generalization of O'Hara's counterterm approach, and define a 1-parameter family of surface energies. Generalizations of the other 1-dimensional regularizations will be considered in section 3.

The naive unregularized energy of a k-manifold F is

$$\int_{F\times F} |x-y|^{-2k} d^k x d^k y, \tag{2.1}$$

where $d^k x$ is the k-dimensional volume form on F. Since Möbius transformations are conformal, $d^k x d^k y$ transforms like the k-th power of $dxdy$, so (2.1) is formally Möbius invariant. However, it diverges like ϵ^{-k} at short distance. To get a sensible definition we must remove this divergence.

In 1 dimension the divergence was $2/\epsilon$, regardless of the local geometry, and was easy to subtract off. In 2 dimensions there is a π/ϵ^2 divergence that doesn't depend on local geometry, plus a $\ln(\epsilon)$ divergence that does depend on local geometry. Natural generalizations of some 1-dimensional approaches fail because they do not cancel the log divergence. Kusner and Sullivan [5] have defined an energy that avoids these problems, but which sacrifices the ALP. The challenge is to remove both divergences while preserving the ALP.

We begin by computing the local divergence more precisely. Let

$$V(\epsilon,x) = \int_{|x-y|>\epsilon} |x-y|^{-4} d^2 y. \tag{2.2}$$

For simplicity, take x to be at the origin, and let the tangent plane $T_x F$ be the x_1-x_2 plane. Let

$$L = \begin{pmatrix} \vec{e} & \vec{f} \\ \vec{f} & \vec{g} \end{pmatrix} \tag{2.3}$$

be the second fundamental form at x, where $\vec{e}$, $\vec{f}$, and $\vec{g}$ are vectors normal to the x_1-x_2 plane. Defining

$$\begin{aligned} \vec{A}(\theta) &= \vec{e}\cos^2(\theta)/2 + \vec{f}\cos(\theta)\sin(\theta) + \vec{g}\sin^2(\theta)/2 \\ B(\theta) &= |\vec{e}|^2\cos^2(\theta) + |\vec{f}|^2 + |\vec{g}|^2\sin^2(\theta) + 2\vec{f}\cdot(\vec{e}+\vec{g})\sin(\theta)\cos(\theta) \end{aligned} \tag{2.4}$$

we see that, near x, the surface F takes the form

$$\vec{z} = \vec{A}(\theta) r^2 + O(r^3), \tag{2.5}$$

where $\vec{z} = (y_3, y_4, \dots, y_n)$ and (r, θ) are polar coordinates for T_xF. The area form is

$$(2.6) \qquad d^2y = \sqrt{1 + |\nabla \vec{z}|^2}\; r dr d\theta = r + B(\theta) r^3/2 + O(r^4)\; dr d\theta,$$

while the distance from the origin is

$$(2.7) \qquad \sqrt{r^2 + |\vec{z}|^2} = r + |\vec{A}(\theta)|^2 r^3/2 + O(r^4)$$

and the distance ϵ cutoff is at

$$(2.8) \qquad r = r_\epsilon(\theta) = \epsilon - |\vec{A}(\theta)|^2 \epsilon^3/2 + O(\epsilon^4).$$

Combining (2.4) through (2.8) we get that

$$\begin{aligned} V(\epsilon, x) &= \int_{r > r_\epsilon} [r^2 + |\vec{A}(\theta)|^2 r^4 + O(r^5)]^{-2} [r + B(\theta) r^3/2 + O(r^4)] dr d\theta \\ &= \int_{r > r_\epsilon} [r^{-3} + (B(\theta)/2 - 2|\vec{A}(\theta)|^2)/r + O(1)] dr d\theta \\ &= \int_0^{2\pi} \frac{1}{2} r_\epsilon^{-2} - (B(\theta)/2 - 2|\vec{A}(\theta)|^2) \ln(r_\epsilon) + O(1) d\theta \end{aligned}$$

$$(2.9) \qquad = \pi \epsilon^{-2} - (\pi/8) \Delta(x) \ln(\epsilon) + O(1),$$

where

$$(2.10) \quad \Delta(x) = (\vec{e} + \vec{g})^2 - 4(\vec{e} \cdot \vec{g} - |\vec{f}|^2) = |\vec{e} - \vec{g}|^2 + 4|\vec{f}|^2 = 4(|\vec{H}|^2 - K),$$

where $\vec{H} = (\vec{e} + \vec{g})/2$ is the mean curvature and $K = \vec{e} \cdot \vec{g} - |\vec{f}|^2$ is the Gauss curvature. Δ is a measure of how far a point is from being umbilical. It is the discriminant of the characteristic polynomial of the second fundamental form. If the surface is in $\mathbb{R}^3$, with principal curvatures k_1 and k_2, then $\Delta = (k_1 - k_2)^2$.

The counterterm regularization

To get a suitable regularization, we must find a local expression that has the form (2.9) and transforms correctly under Möbius transformations, and subtract it from $V(\epsilon, x)$. Note that $\ln(\epsilon)$ does *not* transform correctly, since ϵ is a dimensionful quantity. To get a dimensionless quantity, we must divide ϵ by a locally determined length scale. The natural candidate is $\Delta^{-1/2}$, though other choices (such as $2\Delta^{-1/2}$) are equally valid.

We define the local potential energy

$$(2.11) \qquad V_0(x) = \lim_{\epsilon \to 0} V(\epsilon, x) - \pi \epsilon^{-2} + \frac{\pi \Delta(x)}{16} \ln(\Delta(x) \epsilon^2) + \frac{\pi K(x)}{4},$$

where $K(x)$ is the Gauss curvature at x, and define the energy of a surface to be

$$E_0(F) = \int_F V_0(x)d^2x. \tag{2.12}$$

This is not the only reasonable choice. Had we taken the length scale to be $2\Delta^{-1/2}$ rather than $\Delta^{-1/2}$, we would have gotten a potential that differs from V by $\pi\ln(2)\Delta/8$. Other length scales give potentials that differ by other factors of Δ. It therefore makes sense to consider the entire 1-parameter family of potentials $V_s(x) = V_0(x) + s\Delta(x)$ and the corresponding energies

$$E_s(F) = \int V_s(x)d^2x = E_0(F) + s\int \Delta(x)d^2x \tag{2.13}$$

This gives an interplay between the repulsive Möbius potential and the (generally) attractive $\int\Delta$ functional, as first suggested by Rob Kusner in his lecture at this conference.

The integral of Δ is closely related to the Willmore functional $W(F)$, a well-known conformal invariant [10]. Specifically,

$$W(F) = \int_F |\vec{H}|^2 d^2x = \int_F \left(\frac{1}{4}\Delta + K\right) d^2x, \tag{2.14}$$

where $\vec{H} = (\vec{e}+\vec{g})/2$ is the mean curvature. The 2-form $\Delta\ d^2x$ is in fact invariant under all conformal changes of the ambient metric, not just under Möbius transformations, while $\int Kd^2x$ is a topological invariant. $W(F)$ is invariant under Möbius transformations that do not change the Euler characteristic, but decreases by 4π when a point is sent to infinity.

The $\Delta\ln(\Delta)$ term in the regularization is an object better known to measure theorists than to topologists. Minus $\int\Delta\ln(\Delta)d^2x$ is the *entropy* of the Möbius invariant measure Δd^2x relative to the area measure d^2x.

The integral (2.2), while well-defined, can be difficult to compute. To aid computation, we define an equivalent energy using a smooth cutoff function. Let $\rho(r)$ be a function that approaches 1 as $r\to\infty$ and approaches zero rapidly as $r\to 0$. Define the regularized potential

(2.15)
$$V^\rho(x) = \lim_{\epsilon\to 0}\left(\int \frac{\rho(|x-y|/\epsilon)d^2y}{|x-y|^4} - c(\rho)\pi\epsilon^{-2} + \frac{\pi\Delta}{16}\ln(\Delta\epsilon^2) + \frac{\pi}{4}K\right),$$

where $c(\rho) = \int_0^\infty r^{-2}\rho'(r)dr$. When ρ is the characteristic function of $[1,\infty]$, V^ρ is the same as V_0, while if $\rho = \chi_{[R,\infty]}$, V^ρ differs from V_0 by $\pi\Delta\ln(R)/8$. For a general ρ, $V^\rho = V_s$, where $s = -\frac{\pi}{8}\int_0^\infty \rho'(r)\ln(r)dr$.

Varying the cutoff function ρ is equivalent to varying s, and we get exactly the same family of energies as before.

A particularly useful cutoff function is $\rho(r) = r^4/(1+r^2)^2$, which reduces the integral in (2.15) to $\int(\epsilon^2+|x-y|^2)^{-2}d^2y$. This particular cutoff gives V_s with $s=-\pi/16$.

Examples

The plane.

Before exploring the general properties of the energies E_s, we compute a few examples. Our first and simplest example is the plane. Here $K=\Delta=0$, and

$$V(\epsilon,x)-\pi/\epsilon^2=-\pi/\epsilon^2+\int_0^{2\pi}\int_\epsilon^\infty r^{-4}rdrd\theta=0. \tag{2.16}$$

Our energy is zero, as it must be by scale invariance.

The sphere.

The next example is the unit sphere, where $\Delta=0$ and $K=1$. The potential $V(x)$ is constant by symmetry, and we compute it at the north pole.

$$V(\epsilon,N)=2\pi\int_{2\sin^{-1}(\epsilon)}^{\pi}\frac{\sin(\theta)d\theta}{(2\sin(\theta/2))^4}=\pi/\epsilon^2-\pi/4. \tag{2.17}$$

Subtracting the counterterms we get $V_s(N)=0$, and hence $E_s(S^2)=0$, reflecting the Möbius equivalence of the sphere to the plane. Note the effect of the topological $\pi K/4$ counterterm in the definition of V_s. Had this term not been included, the sphere would have had energy $-\pi^2$.

The cylinder.

Our third example is the right circular cylinder of radius 1. In this example the two principal curvatures are unequal, and $\Delta=1$. All points have the same potential, and this is what we compute. Multiplying this potential by 2π gives the energy per unit length of the cylinder.

The natural coordinates on the cylinder are (α,z), where α is the angle around the circle and z is height, and we take our reference point x to be $(0,0)$. It is also useful to convert from α to $t=2\sin(\alpha/2)$, the straight-line distance across the circle, as shown in figure 4. By symmetry, it is sufficient to integrate over the shaded regions in figure 4, and then to multiply the total by 4.

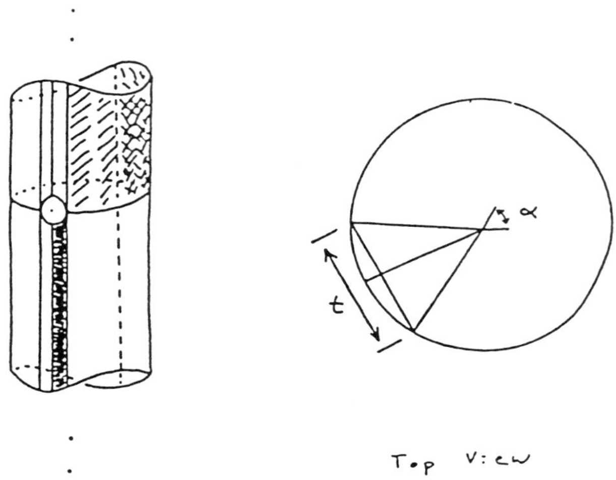

FIGURE 4. The cylinder.

The integral over the lightly shaded region is

$$\begin{aligned}(2.18)\quad \int_{\epsilon}^{2}\int_{0}^{\infty}\frac{2dzdt}{\sqrt{4-t^2}(t^2+z^2)^2} &= \frac{\pi}{2}\int_{\epsilon}^{2}\frac{dt}{t^3\sqrt{4-t^2}}\\ &= \frac{\pi}{32}\left(\frac{2\sqrt{4-\epsilon^2}}{\epsilon^2}+\ln\left(\frac{2+\sqrt{4-\epsilon^2}}{\epsilon}\right)\right)\\ &= \frac{\pi}{32}\left(4\epsilon^{-2}+2\ln(2)-\frac{1}{2}-\ln(\epsilon)+O(\epsilon^2)\right).\end{aligned}$$

To integrate over the heavily shaded region we expand $2/\sqrt{4-t^2} = 1 + t^2/8 + O(t^4)$. We also switch to polar coordinates $r^2 = t^2 + z^2$, $\tan(\theta) = z/t$. Our integral is then, to within $O(\epsilon^2)$,

$$\begin{aligned}&\int_{-\pi/2}^{0}\int_{\epsilon}^{\epsilon\sec(\theta)} r^{-4}(1+r^2\cos^2(\theta)/8)rdrd\theta\\ &\qquad = \int_{-\pi/2}^{0}\epsilon^{-2}\sin^2(\theta)/2-\cos^2(\theta)\ln(\cos(\theta))/8d\theta\\ (2.19)\quad &\qquad = \pi\epsilon^{-2}/8+\pi(2\ln(2)-1)/64.\end{aligned}$$

Adding these two expressions, multiplying by 4, subtracting the counterterms and taking the $\epsilon \to 0$ limit gives

$$(2.20)\qquad V_s(x) = s + \frac{\pi}{8}(3\ln(2)-1).$$

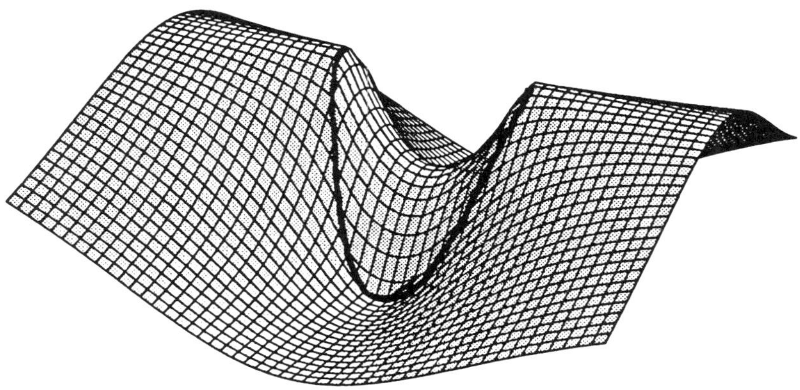

FIGURE 5. The dimple.

Note that, for s sufficiently negative, the cylinder has negative energy per unit length. By taking a very long cylinder with smooth caps at the ends, we can create a smooth surface with arbitrarily negative energy. As a result, for $s < -(3\ln(2) - 1)\pi/8$, the energy functional E_s is unbounded from below.

The "dimple".

Our final example shows that, for any s, there exist surfaces with points where V_s is negative. Let d be a small parameter. Consider the surface in $\mathbb{C}^2$ given by

$$z_1 = r\exp(i\theta), \qquad z_2(r,\theta) = \begin{cases} r^2\exp(2i\theta) & r \leq d \\ d^2\exp(2i\theta) & r \geq d. \end{cases} \tag{2.21}$$

See figure 5. This, of course, is not a smooth surface, but it can be arbitrarily well approximated by smooth surfaces. We compute the potential $V_s(x)$ at $x = 0$ for $d << 1$. The second fundamental form at 0 is $\begin{pmatrix} 2 & 2i \\ 2i & -2 \end{pmatrix}$ and $\Delta(0) = 32$.

The integral for $V(\epsilon, x)$ breaks into two pieces, one for $r > d$ and one for $r < d$. For $r > d$ we have

$$\int_0^{2\pi}\int_d^{\infty} \sqrt{1 + 4d^4/r^2}(r^2 + d^4)^{-2} r\,dr\,d\theta = \pi/d^2 + O(d^2). \tag{2.22}$$

For $r < d$ we have

$$\int_0^{2\pi}\int_{r_\epsilon}^d \sqrt{1+8r^2}(r^2+r^4)^{-2} r\,dr\,d\theta = \int_0^{2\pi}\int_{r_\epsilon}^d (r^{-3}+2r^{-1}+O(r))\,dr\,d\theta$$
$$= \pi + \pi/\epsilon^2 - \pi/d^2 + 4\pi\ln(d/\epsilon) + O(d^2), \tag{2.23}$$

where r_ϵ is the solution to $r_\epsilon^2 + r_\epsilon^4 = \epsilon^2$. Summing these contributions, subtracting the counterterms and taking the $\epsilon \to 0$ limit we get

$$V_s(0) = 32s + \pi + 10\pi\ln(2) + 4\pi\ln(d) + O(d^2). \tag{2.24}$$

No matter how large s is, we can get $V_s(0)$ to be negative by choosing d small enough.

Properties of the surface energy

Having defined our family of energies, we now check that it has the properties 1–5 described in section 1. Since adding a multiple of $\int \Delta d^2x$ leaves properties 1, 2, 4 and 5 unchanged, it suffices to check these properties for E_0.

It is easy to see that E_0 is well-defined and finite for smooth embeddings, and varies continuously with the embedding in the C^k norm, for any $k \geq 3$. When F intersects itself, the potential $V_0(x)$ diverges as r^{-2} as x approaches the self-intersecting set. If the self-intersecting set is discrete the integral $E_0 = \int V_0 d^2x$ diverges logarithmically, while if it has positive dimension the energy diverges even more rapidly.

E_0 is also Möbius invariant. Indeed, the 2-form $V_0(x)d^2x$ is Möbius invariant. This can be proved directly, and we sketch such a proof in the next paragraph. However, Möbius invariance is more easily understood in section 3, where we relate $V_0(x)$ to a quantity that is manifestly Möbius invariant.

Since $V_0 d^2x$ is manifestly invariant under scale changes and Euclidean transformations, we need only check invariance under Möbius transformations I that leave x fixed and for which I_* is the identity on $T_x(\mathbb{R}^n)$. Such a transformation preserves $e - g$ and f, and hence Δ. Since $|x-y|^{-4}d^2y$ is an invariant form, as is d^2x, the change in $V(x)$ can be reduced to the $\epsilon \to 0$ limit of

$$\left(\int_{|I(y)-x|>\epsilon>|y-x|} - \int_{|y-x|>\epsilon>|I(y)-x|}\right) |x-y|^{-4} d^2y \tag{2.25}$$

plus $\pi/4$ times the change in K. By a power-series expansion, this is seen to equal zero in all cases.

Boundedness is problematic. The example of the cylinder shows that E_s is *not* bounded from below for $s < -(3\ln(2)-1)\pi/8$. Furthermore, the example of the dimple shows that V_s cannot be bounded by local information, so a bound on E_s cannot be achieved by integrating a bound on V_s. However, experimentation suggests that, for large s, E_s *is* non-negative and is minimized by spheres. This conjecture is discussed further in section 4.

The connected sum rule follows from Möbius invariance. First suppose that K_1 and K_2 each contain neighborhoods that are exactly spherical. By inverting about a sphere centered in this neighborhood we can assume that K_1 and K_2 are exactly planar outside of compact regions M_1 and M_2, respectively, with finite areas A_1 and A_2. We take $K_1\#K_2$ to be a plane outside of two regions, isometric to M_1 and M_2, that are separated by a distance D. An easy calculation then shows that

$$|E_s(K_1\#K_2) - E_s(K_1) - E_s(K_2)| \leq A_1A_2/D^4. \tag{2.26}$$

Taking $D \to \infty$ gives the connected sum rule for surfaces with spherical neighborhoods. But since *any* surface can be given a small spherical neighborhood at an arbitrarily small cost in energy, the connected sum rule holds for all surfaces. (This is essentially the same argument as found in [4], where a connected sum rule for the Willmore functional is proven).

Finally, E_s manifestly has the ALP, for the same reason that O'Hara's $E^{(c)}$ does. Namely, the regularization at x is independent of whether additional link components are present or not.

3. Alternative approaches to surface energy.

In this section we consider generalizations of other 1-dimensional regularizations. Several natural ideas simply do not work, thanks to the $\ln(\epsilon)$ divergence in $V(\epsilon, x)$. One approach gives functionals similar to the counterterm energies of section 2. Another approach gives the Kusner-Sullivan "first-order" energy, which is discussed elsewhere in these proceedings [5].

Arclength regularization

There are several possible ways to generalize the 1-dimensional arclength formula (1.7). The most obvious is to define

$$E(F) = \int_{F\times F} (|x-y|^{-4} - D(x,y)^{-4})d^2x d^2y, \tag{3.1}$$

but this expression is divergent. At short distance $|x-y|^{-4}$ equals $r^{-4} - 2|\vec{A}(\theta)|^2 r^{-2} + O(r^{-1})$, while $D(x,y)^{-4} = r^{-4} - 8|\vec{A}(\theta)|^2 r^{-2}/3 + O(r^{-1})$. Unless the second fundamental form is identically zero, the difference between these two expressions is $O(r^{-2})$, and is not integrable.

A different generalization is

$$E(F) = \int_{F\times F} (|x-y|^{-2} - D(x,y)^{-2})^2 d^2x d^2y. \tag{3.2}$$

This is finite, as $|x-y|^{-2} - D(x,y)^{-2} = O(1)$, but it is not Möbius invariant. Also, as a practical matter, it is nearly impossible to compute, since just finding the shortest path between two distant points is a difficult endeavor.

In one dimension, arclength was an easy quantity to work with, in that there were only two paths between any two points to choose from, and in that arclength served as a canonical parametrization for any curve. In two dimensions, arclength is neither calculable nor canonical, and does not appear to provide a useful regularization.

The closest thing to a canonical parametrization in 2 dimensions is a map that preserves conformal structure. Kusner has suggested an energy based on this structure. If $f : \mathbb{R}^k \to \mathbb{R}^n$ is a conformal embedding, he defines

$$E(f) = \int_{\mathbb{R}^k\times\mathbb{R}^k} \frac{|f'(z)|^k|f'(w)|^k}{|f(z)-f(w)|^{2k}} - \frac{1}{|z-w|^{2k}} d^kz d^kw. \tag{3.3}$$

This energy is an invariant of the surface, since if $g : \mathbb{R}^k \to \mathbb{R}^n$ is a conformal map with the same image as f, then $E(g) = E(f)$.

In one dimension this expression is well-defined, and gives yet another description of O'Hara's energy $E^{(c)}$. Kusner and Sullivan have applied this approach to graphs, giving a relative energy for distinct embeddings of the same graph. Unfortunately, in 2 dimensions this energy has a log divergence at non-umbilical points. For example, consider the map $f : \mathbb{C} \to \mathbb{C}^2$, $f(z) = (z, z^2)$. Holding $w = 0$ fixed and integrating over z, we see that $|f'(z)|^2|f(z)-f(w)|^{-4} = |z|^{-4} + 2|z|^{-2} + O(1)$. The second term in the integrand cancels the $|z|^{-4}$, but the remaining $2|z|^{-2}$ term is not integrable.

Wasted area regularization

As in the 1-dimensional case, let S_x be the unit sphere centered at x, let I_x be inversion in this sphere, and let $\tilde{y} = I_x(y)$. Then the 2-form $|x-y|^{-4}d^2y$ is the pullback, under I_x, of the area form $d^2\tilde{y}$. As before, the area of $I_x(F)$ is infinite, as $I_x(F)$ extends to infinity. We try to regularize by taking the excess area of $I_x(F)$ relative to some standard surface. The question is, which surface?

Before considering reference surfaces, we investigate what $I_x(F)$ looks like near infinity. Since F takes the form (2.5) near zero, $I_x(F)$ has the form

(3.4)

$$\vec{z} = \vec{A}(\theta)+O(1/r), \qquad \partial\vec{z}/\partial r = O(1/r^2), \qquad \partial\vec{z}/\partial\theta = d\vec{A}/d\theta+O(1/r).$$

The area form is

$$\sqrt{1+|\nabla\vec{z}|^2}\, rdrd\theta = (r+|\vec{A}'(\theta)|^2/2r+O(r^{-2}))drd\theta, \tag{3.5}$$

where $\vec{A}'$ denotes $d\vec{A}(\theta)/d\theta$. Integrating out to $r = R$ gives $\pi R^2 + \pi\Delta\ln(R)/8+O(1)$. These are precisely the divergences we have already seen in (2.9), with $R = 1/\epsilon$.

Our reference surface must have the same asymptotic area as $I_x(F)$. This rules out the natural choice of the plane, whose area is merely πR^2. Another possibility is to take the curve $\vec{z} = \vec{A}(\theta), r = R$, fill it in with a minimal surface, and take the area of that. Unfortunately, the area of such a minimal surface is $\pi R^2 + O(1)$, with no $\ln(R)$ term, so this approach also fails.

A reference surface that *does* work is the fan shape $\vec{z}(r,\theta) = \vec{A}(\theta)$, which we denote M_x. For $r > d$, this is the same as the "dimple" of figure 5. The area form for M_x is exactly $\sqrt{1+|\vec{A}'|^2/r^2}\, rdrd\theta$, which has the form (3.5) for large r. We can therefore define

$$V^{(w)}(x) = \text{Excess area of } I_x(F) \text{ relative to } M_x$$

$$E^{(w)} = \int_F V^{(w)}(x)d^2x \tag{3.6}$$

This definition is the product, and the subject, of a continuing collaboration with Kusner and Sullivan.

Since M_x is not a minimal surface, it is possible for $I_x(F)$ to have less area than M_x, and hence for $V^{(w)}(x)$ to be negative. Indeed, if $I_x(F)$ is a minimal surface for $r < D$ and has the form (3.4) for $r > D$, then $V^{(w)}(x) = -\pi\Delta\ln(D)/8+O(1)$. (The "dimple" of section 2 is a surface of precisely this type, with $D = 1/d$). This shows that $V^{(w)}$, like V_s, cannot be pointwise bounded from below.

$E^{(w)}$ is finite for smooth embeddings and infinite for immersions with self-intersections. It is Möbius invariant, obeys a connected sum rule and has the ALP. We will prove Möbius invariance directly, and then show that $E^{(w)} - E_s$ is the integral of a local and Möbius invariant quantity. This then establishes the other properties for $E^{(w)}$, and shows that E_s is Möbius invariant.

We examine what a Möbius transformation does to $I_x(F)$ and M_x. Here $\tilde{x} = \infty$ is fixed, so the transformation can only act on $I_x(F)$ by rotation, reflection, translation or scale change. We show that in all cases $V^{(w)}(x)d^2x$ is unchanged.

A rotation (or reflection) of $I_x(F)$ corresponds to a rotation (reflection) of F around x, hence a rotation (reflection) of the second fundamental form at x, hence a rotation (reflection) of M_x. $I_x(F)$ and M_x still match asymptotically, and the difference in their areas is unchanged.

A translation of $I_x(F)$ corresponds to a Möbius transformation whose derivative is the identity at x. Such a transformation can change the mean curvature $\vec{H} = (\vec{e} + \vec{g})/2$, but cannot change $\vec{e} - \vec{g}$ or $\vec{f}$. A translation of $I_x(F)$ normal to the x_1-x_2 plane corresponds to a change in $\vec{H}$, which in turn translates M_x the same amount as $I_x(F)$, leaving $V^{(w)}$ unchanged. A translation in the x_1-x_2 plane does not change $\vec{H}$, and so is *not* matched by a translation of M_x. However, the difference in area form between M_x and its translate is absolutely integrable, and integrates to zero, so the shift of $I_x(F)$ relative to M_x does not change $V^{(w)}(x)$.

Finally, rescaling $I_x(F)$ by a factor of λ corresponds to rescaling F by λ^{-1}, rescaling the second fundamental form at x by λ, and so rescaling M_x by λ. $V^{(w)}(x)$ gets rescaled by λ^2, while d^2x gets rescaled by λ^{-1}, leaving $V^{(w)}(x)d^2x$ unchanged. This completes the proof that $V^{(w)}d^2x$, and therefore $E^{(w)}$, is Möbius invariant.

Now we compare $E^{(w)}$ to E_s. We can compute $V^{(w)}$ by taking the area of that part of $I_x(F)$ within distance R of the origin, subtracting the area of the corresponding part of M_x, and taking the limit as $R \to \infty$. The points on $I_x(F)$ of distance less than R from the origin correspond to points on F of distance greater than $\epsilon = 1/R$ from x. This procedure is therefore equivalent to a counterterm regularization, where the counterterm is the area of M_x within distance $R = 1/\epsilon$ of the origin. The distance cutoff is at $r = r_R = R - |\vec{A}(\theta)|^2/(2R) + O(R^{-2})$.

The area is

$$\begin{aligned}
\int_0^{2\pi}\int_0^{r_R} & \sqrt{1+|\vec{A}'(\theta)|^2/r^2}\,r\,dr\,d\theta \\
&= \int_0^{2\pi}\frac{1}{2}\Big(r_R\sqrt{r_R^2+|\vec{A}'|^2}+|\vec{A}'|^2\ln(r_R+\sqrt{r_R^2+|\vec{A}'|^2}) \\
&\qquad\qquad -\frac{1}{2}|\vec{A}'|^2\ln(|\vec{A}'|^2)\Big)\,d\theta \\
&= \int_0^{2\pi}\frac{1}{2}\Big(r_R^2+|\vec{A}'|^2/2+|\vec{A}'|^2\ln(2R)-\frac{1}{2}|\vec{A}'|^2\ln(|\vec{A}'|^2) \\
&\qquad\qquad +O(R^{-2})\big)\,d\theta \\
&= \int_0^{2\pi} R^2/2+|\vec{A}'|^2/4-|\vec{A}(\theta)|^2/2-|\vec{A}'|^2\ln(|\vec{A}'|^2/4R^2)/4 \\
&\qquad\qquad +O(R^{-2})d\theta \\
&= \pi R^2-\pi\Delta\ln(|\Delta/R^2)/16-\pi K/4+\pi\Delta(10\ln(2)-1)/32 \\
&\quad -\frac{\Delta}{32}\int_0^{2\pi}\frac{8|\vec{A}'|^2}{\Delta}\ln\left(\frac{8|\vec{A}'|^2}{\Delta}\right)d\theta+O(R^{-2})
\end{aligned} \tag{3.7}$$

The first 3 terms match the V_0 counterterms. The fourth term corresponds to a change in s. The last term is $\Delta/32$ times the entropy of the S^1 measure $|\vec{A}'(\theta)|^2 d\theta$ relative to its average, $(\Delta/8)d\theta$. This entropy is a locally defined Möbius invariant quantity that depends on the relative size, and the angle between, $\vec{e}-\vec{g}$ and $\vec{f}$. In $\mathbb{R}^3$ we can always choose coordinates such that $\vec{f}=0$, and the integral is $2\pi(1-\ln(2))$. In 4 or more dimensions, the integral can take on any value from 0 to $2\pi(1-\ln(2))$. Thus, although there is no single value of s for which V_s always equals $V^{(w)}$, we always have the inequalities $V_{s_1}\le V^{(w)}\le V_{s_2}$, where $s_1=\pi(1-10\ln(2))/32$ and $s_2=3\pi(1-4\ln(2))/32$.

The fan shape M_x is not the only reference surface that works. Given any cutoff function $\rho(r)$, we can take the surface

$$\vec{z}(r,\theta)=\vec{A}(\theta)\rho(r^2/|\vec{A}_0(\theta)|^2), \tag{3.8}$$

where $\vec{A}_0(\theta)=(e-g)\cos(2\theta)/4+\vec{f}\sin(2\theta)/2$ is the oscillatory part of $\vec{A}(\theta)$. This surface works just as well as M_x, and gives an energy functional that differs from $E^{(w)}$ by local terms such as $\int\Delta d^2x$ and the integral of the entropy of $8|\vec{A}'|^2/\Delta$.

The manifestly Möbius invariant approach

In 1993, Kusner and Sullivan [5] generalized Doyle and Schramms's explicitly Möbius invariant approach to higher dimensions. Let S_x be the sphere, tangent to F at x, that passes through y, and let S_y be the sphere tangent to F at y that passes through x. Let $\alpha(x, y)$ be the angle between S_x and S_y. More precisely, $\cos(\alpha)$ is the inner product of the area form of S_x and the area form of F at y. Then, given any reasonable function $f(\alpha)$ with $f(0) = 0$ they write

$$E^{(f)}(F) = \int_{F \times F} f(\alpha)|x - y|^{-4}. \tag{3.9}$$

This is manifestly Möbius invariant, and is finite for embedded surfaces as long as $f \to 0$ sufficiently quickly as $\alpha \to 0$. Kusner and Sullivan concentrate in particular on $f(\alpha) = (1 - \cos(\alpha))^2$ (or, in k dimensions, $(1 - \cos(\alpha))^k$).

This approach has several advantages over the local regularizations E_s, and one major disadvantage. The advantages are that the energy is manifestly bounded from below, that it requires knowledge only of the tangent planes at x and y, not of the curvature, and that it easily generalizes to higher dimensions.

The disadvantage is that it does not have the additive link property. The force between distinct neighborhoods depends not only on their separation but also on their orientation in space. As such, it does not preserve the long-distance behavior of the unregularized functional (2.1).

To preserve the additive link property, one would have to use $f(\alpha) = 1 - \cos(\alpha)$. This is equivalent to computing the wasted area relative to a plane, or to using the local counterterm π/ϵ^2, and is log divergent at all non-umbilical points. To avoid log divergences, one must either incorporate information about curvature into the regularization (as with E_s) or sacrifice the additive link property (as in [5]).

It is possible to incorporate information about the second derivative into this approach. The integral

$$\int_{F \times F} |x - y|^{-4}\left(1 - \cos(\alpha)\sqrt{1 + |\vec{A}'(\theta)|^2(r^2 + |\vec{z}|^2)^2/r^2}\right) d^2x d^2y \tag{3.10}$$

is equivalent to $E^{(w)}$. Here (r, θ) are coordinates in the tangent plane T_xF and $\vec{z}$ is the component of $y - x$ normal to this plane, and $\vec{A}'$ is related to the second fundamental form at x as before. The complicated second term is just the pullback, under I_x, of the area form of M_x. There is an apparant singularity at points where $r = 0$, $\vec{z} \neq 0$, but this singularity is integrable. Of course, by introducing the second derivative we lose manifest boundedness, as the integrand may be negative.

4. The question of boundedness.

In this section we discuss the boundedness properties of the surface energy functionals E_s defined in section 2. We show that the question may be reduced to evaluating a single number, which we call s_c. We then tell what little is known about this number, and argue that s_c is most likely finite.

The critical parameter

Since the energy of a sphere is zero, for any given s the functional E_s is either

a) Unbounded from below,
b) Bounded from below, but not uniquely minimized by round spheres, or
c) Non-negative, and uniquely minimized by round spheres.

We will prove that there is a (possibly infinite) critical value of s, which we call s_c, such that condition (a) holds for all $s < s_c$ and condition (c) holds for all $s > s_c$. When $s = s_c$, the energy is non-negative, but it is not known whether $E_{s_c}(F) = 0$ implies that F is a round sphere (or a plane, which we consider a sphere of infinite radius).

First consider $W(F) = \int_F \Delta(x) d^2x$. Since $\Delta = |e-g|^2 + 4|\vec{f}|^2 \geq 0$, $W(F)$ is never negative, and is zero only when F is umbilical everywhere, i.e., a sphere. As a result, if $s_2 > s_1$, then $E_{s_2}(F) \geq E_{s_1}(F)$, with equality if and only if F is a sphere.

Define

$$s_c = \sup\{s \mid E_s \text{ is unbounded from below.}\} \tag{4.1}$$

If $s < s_c$, then there exists an $s_- \in (s, s_c]$ such that E_{s_-} is unbounded from below, by the very definition of s_c. Since $E_s \leq E_{s_-}$, E_s is also unbounded from below.

Now suppose that $s > s_c$. This implies that E_s is bounded from below. By the connected sum rule, if $E_s(F)$ were negative for some F, we could get $E_s(F\#F\#\ldots\#F)$ arbitrarily negative, and E_s would be unbounded. Thus E_s must be non-negative. Similarly, $E_{(s+s_c)/2}$ is non-negative. If F is not a round sphere, $E_s(F) > E_{(s+s_c)/2}(F) \geq 0 = E_s(S^2)$, so E_s is uniquely minimized by round spheres.

Finally, we consider E_{s_c}. For any smooth F, $W(F)$ is finite, so $E_{s_c}(F) = \lim_{s\to s_c^+} E_s(F) \geq 0$. This shows that E_{s_c} is non-negative. However, it is not clear whether $E_{s_c}(F) = 0$ implies that F is a sphere.

The remaining task is to determine whether s_c is finite or infinite, and if finite, to estimate its value.

It is easy to see that $s_c > -\infty$. Let F be any non-spherical surface. Then $W(F) > 0$ and, for any $s < -E_0(F)/W(F)$, $E_s(F) < 0$. This shows that $s_c \geq -E_0(F)/W(F)$. In particular, by looking at the cylinder we see that $s_c \geq (1 - 3\ln(2))\pi/8$.

It is not known whether s_c is finite or positive infinite. The evidence collected to date suggests that s_c is finite. However, no proof has been found, and the evidence is far from conclusive.

Evidence for Boundedness

Suppose that E_s is unbounded for all s. Then it should be possible to find examples whose energies are extremely negative even for large values of s. However, all the wild examples we have constructed have had extremely *positive* energies, as might be expected from a well-behaved and bounded functional. While this is hardly proof of boundedness for large s, it is certainly suggestive.

We present some of these examples here. We do so not so much to convince to reader of boundedness as to build intuition. We hope that these examples will suggest methods by which boundedness might be proved, or by which true counterexamples might be constructed.

Our first example is the ellipsoid $x^2/a^2 + y^2/b^2 + z^2/c^2 = 1$ in $\mathbb{R}^3$. If a/b and a/c are both close to 1, then both W and E_0 will be second order in the eccentricity, and for reasonable values of s our energy will be positive. To get candidates for negative energy we must take extreme values of (a, b, c).

Suppose $a >> b, c$, so that our ellipsoid looks like a cigar. In that case most points will lie along the length of the cigar, which is roughly cylindrical. But we have already calculated the energy of the cylinder, and found it to be positive for s positive. Increasing a only increases the length of the cigar and increases the energy.

Now suppose $a << b, c$, so that our ellipsoid looks like a pancake. Most points lie along a flat face of the pancake, with very high energy due to the vicinity of the other face. Even along the rim, one curvature greatly exceeds the other, making the rim cylinder-like, again with positive energy. Finally, if $a >> b >> c$, we have a cylinder whose cross-section is a highly eccentric ellipse, again with extremely high energy.

A similar analysis applies to the Clifford torus $S^1 \times S^1 \in \mathbb{C}^2$, with two different radii. If the two radii are close, then W and E_0 are comparable, and even for reasonably small s, E_s is positive. If one of the radii is much greater than the other, the neighborhood of each point resembles part of a long cylinder, and E_0 is very large and positive.

Next we consider perturbations of the plane. If such a perturbation has a large natural length scale, it can be rescaled to have length scale 1

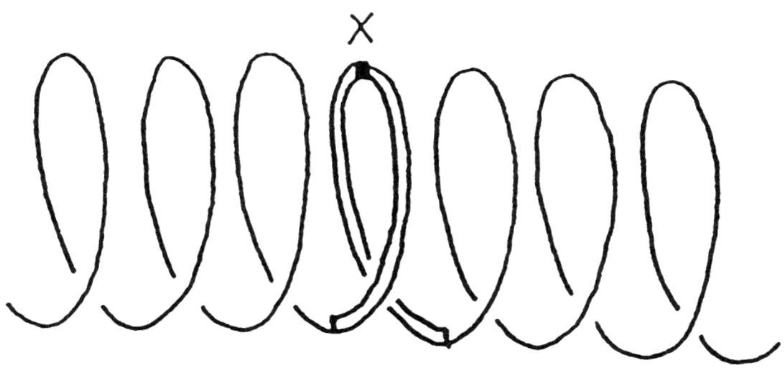

FIGURE 6. The helix.

and a very small amplitude. Both W and E_0 would be quadratic in this amplitude, and so E_s would be positive for s bigger than the ratio of the leading-order coefficients. To get E_s negative for large s one has to take perturbations with very small length scales, or equivalently large amplitudes.

One such example is the plane wave $(u, v, \cos(ku), \sin(ku)) \in \mathbb{R}^4$ for large k. This is the helix $(u, \cos(ku), \sin(ku)) \in \mathbb{R}^3$ crossed with $\mathbb{R}$. See figure 6. As k increases, the loops of the helix get closer and closer. In the $k \to \infty$ limit, the contributions to $V(p)$ from the single loop highlighted in figure 6, minus the counterterms, approach the (finite) energy of the cylinder, while the other loops give more and more positive energy. As a result, $\lim_{k\to\infty} V_s(p) = +\infty$ for all s.

Perturbations with point peaks do no better. Consider the surface $z = c\exp(-(x^2 + y^2)) \in \mathbb{R}^3$. For large c, the sides of the "mountain" get very steep and resemble the sides of a cylinder, whose potential we know to be positive. The base of the mountain (the region of slope ~ 1) may have negative potential, but its area grows only as $\ln(c)$, while the area of the side grows as c. So again, as $c \to \infty$, the energy of the surface goes to $+\infty$.

This by no means exhausts the supply of potential counterexamples. However, it shows that extreme behavior tends to generate extreme positive energy, not negative energy. In particular, when features get stretched they tend to locally resemble planes or cylinders. Planes get positive energy from the presence of other nearly planes (as in the pancake example), while cylinders have positive energy for $s >> (1 - 3\ln(2))\pi/8$. This limiting behavior suggests that s_c is not only finite, but is fairly small.

There are two more classes of examples that should be noted, which we have not yet computed. One is where the perturbation is peaked at small rings, rather than points or lines. An example of this is the surface $z_2 = cz_1^2 \exp(-|z_1|^2) \in \mathbb{C}^2$. The other class is surfaces with many length scales.

5. Conclusions.

We have defined a family of Möbius invariant energy functionals E_s for surfaces embedded in $\mathbb{R}^n$. These functionals are all finite for smoothly embedded compact surfaces and infinite for self-intersecting immersed surfaces. They treat disconnected surfaces and connected sums of surfaces correctly.

For a functional to be useful, it must be bounded from below. For $s < (1 - 3\ln(2))\pi/8$, E_s is not bounded from below. However, for s sufficiently large and positive, the evidence to date suggests that E_s *is* bounded from below, although we have not yet found a proof.

References

1. S. Bryson, M. Freedman, Z.-X. He and Z. Wang, *Möbius Invariance of Knot Energy*, Bull. Amer. Math. Soc **28** (1993) 99–103.
2. P. Doyle, private communication .
3. M. Freedman, Z.-X. He and Z. Wang, *Möbius Energy of Knots and Unknots*, Ann. Math. **139** (1994) 1–50 .
4. R. Kusner, *Comparison Surfaces for the Willmore Problem*, Pacific J. Math. **138** (1989) 317–345.
5. R. Kusner and J.M. Sullivan, *Möbius Energies for Knots and Links, Surfaces and Submanifolds*, in these proceedings.
6. J. O'Hara, *Energy of a Knot*, Topology **30** (1991) 241-247.
7. J. O'Hara, *Family of Energy Functionals of Knots*, Topology Appl. **48** (1992) 147-161.
8. J. O'Hara, *Energy Functionals of Knots*, Topology-Hawaii, World Scientific, K.H. Dovermann, 201–214, 1992.
9. O. Schramm, private communication.
10. T.J. Willmore, *Total Curvature in Riemannian Geometry*, Halstead Press, 1982.

Department of Mathematics, University of Texas, Austin, Texas 78712

E-mail address: sadun@math.utexas.edu

AMS/IP Studies in Advanced Mathematics
Volume **2**, 1997 (Part 1)

INVARIANTS OF FIBRED KNOTS FROM MODULI

HANS U. BODEN

In this paper, invariants of fibred knots are described in terms of unitary representation spaces of fundamental groups of Seifert surfaces. Because these representation spaces admit an interpretation as moduli spaces of parabolic bundles [10], the results of [3, 5, 6] can be used to calculate the invariants. The main result here is Theorem 3. The proof of this theorem involves the theory of parabolic bundles in an essential way, but for the sake of clarity we have suppressed their role. These issues will be addressed in [4].

The knot invariants, denoted $\mu_\alpha(\mathbf{K})$ or simply μ_α when the knot is understood, are defined for fibred knots $\mathbf{K}$ in a rational homology sphere $\mathbf{M}$ and for $\alpha \in \mathbf{SU}_n$ a regular value of the map $\widetilde{\Psi}$ described in Definition 1. They are a generalization of Frohman's invariants, μ_ω, which are invariants of fibred knots defined for ω a regular value of $\widetilde{\Psi}$ in the center of $\mathbf{SU}_n$ [7]. These have already been extended to arbitrary knots in rational homology spheres in the cases

(1) where ω is a regular value of $\widetilde{\Psi}$ in the center of $\mathbf{SU}_n$ in [8], and
(2) where ω is any element in the center of $\mathbf{SU}_n$ in [9].

The invariants presented in this paper form a continuous family of invariants parameterized by conjugacy classes of regular values α of $\widetilde{\Psi}$. The most interesting aspect of this approach is the behavior of μ_α as α is allowed to vary in the set

$$W \stackrel{\text{def}}{=} \mathbf{SU}_n/\text{Ad}.$$

The critical values of Ψ form a union of hyperplanes, giving W a natural chamber structure. Theorem 3 describes the behavior of these invariants in terms of this chamber structure. Part (1) of Theorem 3 states that $\mu_\alpha = \mu'_\alpha$ for α and α' in the same chamber, part (2) compares μ_α and μ_β for α and β in adjacent chambers, and part (3) relates μ_α and μ_ω for α in the interior and ω on the boundary of W.

This work was partially supported by a Rackham Grant from the University of Michigan

The interest in these invariants lies in their ability to detect certain irreducible representations of the knot group: if $\mu_\alpha \neq 0$ then there exists an irreducible unitary representation of the fundamental group of the knot complement with holonomy along the longitude conjugate to α.

To start, we fix some notation. Given an arbitrary group π and a compact Lie group $\mathbf{G}$, let

$$\widetilde{\mathbf{R}}(\pi, \mathbf{G}) = \operatorname{Hom}(\pi, \mathbf{G})$$

denote the space of homomorphisms of π into $\mathbf{G}$. The group $\mathbf{G}$ acts on $\widetilde{\mathbf{R}}(\pi, \mathbf{G})$ by conjugation (denoted by Ad) and we get the quotient

$$\mathbf{R}(\pi, \mathbf{G}) = \widetilde{\mathbf{R}}(\pi, \mathbf{G})/\mathrm{Ad}$$

which is called the space of representations of π in $\mathbf{G}$.

To simplify notation, set $\pi = \pi_1 F$, the fundamental group of a closed Riemann surface F of genus g and $\pi^* = \pi_1 F^*$, the fundamental group of $F^* = F \setminus D^2(p)$, where $D^2(p)$ is a small disk centered at $p \in F$. Then π admits the presentation

$$\pi = \langle a_1, b_1, \dots, a_g, b_g \mid \prod_{i=1}^{g} [a_i, b_i] \rangle, \tag{1}$$

and π^* is simply the free group on the $2g$ generators $a_1, b_1, \dots, a_g, b_g$. This last fact gives an identification

$$\begin{aligned} \widetilde{\mathbf{R}}(\pi^*, \mathbf{G}) &\cong \overbrace{\mathbf{G} \times \cdots \times \mathbf{G}}^{2g}, \\ \rho &\mapsto (A_1, B_1, \dots, A_g, B_g) \end{aligned}$$

where $A_i = \rho(a_i)$ and $B_i = \rho(b_i)$. Setting $\partial = \prod_{i=1}^{g} [a_i, b_i] \in \pi^*$ we define maps $\widetilde{\Psi}$ and Ψ on $\widetilde{\mathbf{R}}$ and $\mathbf{R}$ by evaluation on ∂.

Definition 1. Let $\widetilde{\Psi} : \widetilde{\mathbf{R}}(\pi^*, \mathbf{G}) \longrightarrow \mathbf{G}$ be defined using the above identification and setting $\widetilde{\Psi}(A_1, B_1, \dots, A_g, B_g) = \prod_{i=1}^{g} [A_i, B_i]$. Since $\widetilde{\Psi}(g \cdot \rho) = g \widetilde{\Psi}(\rho) g^{-1}$, we can define $\Psi : \mathbf{R}(\pi^*, \mathbf{G}) \longrightarrow \mathbf{G}/\mathrm{Ad}$ and the following square commutes:

$$\begin{array}{ccc} \widetilde{\mathbf{R}}(\pi^*, \mathbf{G}) & \xrightarrow{\widetilde{\Psi}} & \mathbf{G} \\ {\scriptstyle \mathrm{Ad}}\downarrow & & \downarrow{\scriptstyle \mathrm{Ad}} \\ \mathbf{R}(\pi^*, \mathbf{G}) & \xrightarrow{\Psi} & \mathbf{G}/\mathrm{Ad}. \end{array} \tag{2}$$

By definition, $\mathrm{im}(\widetilde{\Psi}) \subset [\mathbf{G}, \mathbf{G}]$, the commutator subgroup of $\mathbf{G}$. The surjection $\pi^* \to \pi$ induces an inclusion $\mathbf{R}(\pi, \mathbf{G}) \hookrightarrow \mathbf{R}(\pi^*, \mathbf{G})$ defined by pullback, and it is clear that $\rho \in \mathrm{im}(i) \iff \Psi(\rho) = I$.

For the remainder of the paper, we set $\mathbf{G} = \mathbf{U}_n$, the group of unitary $n \times n$ complex matrices. Recall that $\mathbf{SU}_n = [\mathbf{U}_n, \mathbf{U}_n]$, and denote by $\mathbf{Z}_n$ the center of $\mathbf{SU}_n$ and by $\mathbf{PU}_n$ the quotient $\mathbf{SU}_n/\mathbf{Z}_n$.

Suppose now that $\mathbf{K}$ is a fibred knot in a homology sphere $\mathbf{M}$ with spanning surface F^*. Let

$$\varphi : F^* \xrightarrow{\approx} F^*$$

be the monodromy of the fibration, so the knot complement is given by

$$\mathbf{M_K} = F^* \times I / \sim,$$

where $(x, 0) \sim (\varphi(x), 1)$.

In [7], Frohman defines invariants of fibred knots by considering the Lefschetz number of the monodromy action on certain smooth submanifolds of $\mathbf{R}(\pi^*, \mathbf{U}_n)$, namely $\mathbf{R}_\omega$ (defined below), where $\omega \in \mathbf{Z}_n$ is a regular value of Ψ. This is the case if and only if $\omega = e^{2\pi i k/n} I$, where k and n are relatively prime.

We now describe an extension of these invariants. For any $\alpha \in \mathbf{SU}_n$, denote by $C(\alpha)$ the orbit $\mathrm{Ad} \cdot \alpha$ of α under conjugation and set $\widetilde{\mathbf{R}}_\alpha = \widetilde{\Psi}^{-1}(C(\alpha))$. A standard result shows that the critical points of $\widetilde{\Psi}$ are precisely the reducible representations (see [1] or [8]). In particular, if α is a regular value of $\widetilde{\Psi}$, then $\widetilde{\Psi}^{-1}(C(\alpha))$ is smooth and consists entirely of irreducible representations.[1]

Define

$$\mathbf{R}_\alpha = \widetilde{\mathbf{R}}_\alpha / \mathrm{Ad} = \Psi^{-1}(C(\alpha)).$$

Since the adjoint action is a free $\mathbf{PU}_n$ action on the irreducible representations, it follows that for α a regular value, $\mathbf{R}_\alpha$ is a compact oriented manifold of dimension

$$\dim \mathbf{R}_\alpha = (2g-2)n^2 + 2 + \dim C(\alpha).$$

The invariants μ_α are defined as follows.

Definition 2. (1) If P is a compact oriented manifold and $f : P \to P$, then the Lefschetz number of f, denoted $\Lambda(f, P)$, is the algebraic intersection number of the graph of f with the diagonal

[1]It is enough to assume that $\widetilde{\Psi}$ is transverse to $C(\alpha)$, but a simple calculation shows that if $\widetilde{\Psi}$ is transverse to $C(\alpha)$, then α is indeed a regular value.

$\triangle(P)$ in $P \times P$. Further, the Lefschetz polynomial is defined to be

$$L(f, P)(t) = \sum_{0 \leq n} (-1)^n \mathrm{Tr}(f^* : \mathrm{H}^n(P, \mathbb{Q}) \to \mathrm{H}^n(P, \mathbb{Q}))t^n.$$

(2) Let $\mu_\alpha(\mathbf{K})$ denote the Lefschetz fixed point number of the map on the representation variety induced by the monodromy of the knot, i.e.

$$\mu_\alpha(\mathbf{K}) = \Lambda(\varphi^*, \mathbf{R}_\alpha).$$

(3) Let $M_\alpha(\mathbf{K}, t)$ denote the Lefschetz polynomial of the map on the cohomology of the representation variety induced by the monodromy of the knot, i.e.

$$M_\alpha(\mathbf{K}, t) = L(\varphi^*, \mathbf{R}_\alpha).$$

The Lefschetz fixed point theorem implies that the Lefschetz number equals the Lefschetz polynomial evaluated at $t = 1$, thus $\mu_\alpha = M_\alpha(1)$. Of course, if α is conjugate to α', then $\mu_\alpha = \mu_{\alpha'}$. In the other cases, we would like to compare the invariants μ_α and μ_β. One technique for doing this is to let α_t be a path connecting $C(\alpha)$ and $C(\beta)$ in W and to consider the map Ψ restricted to the preimage of this path. By choosing the path carefully (e.g. so it is transverse to the critical values of Ψ), Morse theory for Ψ, in the sense of Bott, constructs a cobordism between $\mathbf{R}_\alpha$ to $\mathbf{R}_\beta$. The invariants μ_α and μ_β can then be studied in terms of the change in the cohomology under this cobordism. This is the rough idea behind our approach.

To begin, identify the critical values of Ψ with a union of hyperplanes in W as follows. Notice that W is an $(n-1)$-simplex. (Choose a positive Weyl chamber in $\mathbf{su}_n$). Use the following natural (but discontinuous!) coordinates on W. Since any $\alpha \in \mathbf{SU}_n$ is conjugate to a matrix of the form

$$\exp(diag(\alpha_1, \dots, \alpha_n)) = \begin{pmatrix} e^{2\pi i \alpha_1} & & 0 \\ & \ddots & \\ 0 & & e^{2\pi i \alpha_n} \end{pmatrix}$$

where $0 \leq \alpha_1 \leq \cdots \leq \alpha_n < 1$ and $\sum_{i=1}^n \alpha_i$ is an integer, it follows that we can use $(\alpha_1, \dots, \alpha_n)$ to give Ad-invariant coordinates to $\mathbf{SU}_n$. The resulting coordinates on W are discontinuous precisely when $\alpha_1 = 0$, which can be seen by considering the coordinates assigned to the path $\alpha_t = \exp(diag(t, \alpha_2 - \frac{t}{n-1}, \dots, \alpha_n - \frac{t}{n-1}))$ for $t \in (-\epsilon, \epsilon)$.

To deal with these discontinuities, decompose $W = \bigcup_{k=0}^{n-1} W_k$, where

$$W_k = \{(\alpha_1, \dots \alpha_n) \in W \mid \sum_{i=1}^{n} \alpha_i = k\},$$

and notice that the coordinates are continuous along each W_k. The hyperplanes $\alpha_1 = 0$ lie in ∂W_k and are called *bad* hyperplanes. Of course, W can be reassembled by identifying these bad hyperplanes to the hyperplanes $\alpha_n = 1$, which lie in $\partial \overline{W_{k+1}}$, in the more or less obvious manner.

Suppose now that $\rho : \pi_1(F^*) \to \mathbf{U}_n$ is a reducible representation. Then up to conjugacy, we have $\mathrm{im}(\rho) \subset \mathbf{U}_{n_1} \times \mathbf{U}_{n_2}$. Because $\gamma = \widetilde{\Psi}(\rho)$ is contained in the commutator subgroup of $\mathrm{im}(\rho)$, which is just $\mathbf{SU}_{n_1} \times \mathbf{SU}_{n_2}$, γ is conjugate to a matrix in block form

$$\begin{pmatrix} \gamma^1 & 0 \\ 0 & \gamma^2 \end{pmatrix}$$

where $\gamma^i \in \mathbf{SU}_{n_i}$ for $i = 1, 2$. Writing $\gamma = \exp(diag(\gamma_1, \dots, \gamma_n))$, we see that $\gamma = \widetilde{\Psi}(\rho)$ for a reducible ρ if and only if there is a proper subcollection $0 \leq \gamma_{\sigma(1)} \leq \cdots \leq \gamma_{\sigma(n_1)} < 1$ with $\sum_{j=1}^{n_1} \gamma_{\sigma(j)}$ an integer.

This shows that the collection of critical values of Ψ (which are the projection under Ad of the critical values of $\widetilde{\Psi}$) are given by a union of hyperplanes $\bigcup_\xi H_\xi$ in W, some of which are *good* (i.e. not bad), others of which are bad. We call the connected components of $W \setminus \bigcup_\xi H_\xi$ *chambers*.

Suppose $\alpha, \beta \in W_k$ are regular values of Ψ in adjacent chambers and are separated by a good hyperplane H. Choose $\gamma \in H$, a generic point lying on the hyperplane separating α and β (genericity means that γ lies on no other hyperplane). Denote by Σ_γ the reducible representations of $\mathbf{R}_\gamma$. Then with a choice of a complex structure J on F, the theorem of Mehta and Seshadri [10] provides an identification of each of these representation spaces with a corresponding moduli space of semistable parabolic bundles. Roughly speaking, the parabolic structure is determined by the eigenspaces and eigenvalues of the matrices (α, β or γ). In this way, each of the representation spaces $\mathbf{R}_\alpha, \mathbf{R}_\beta$, and $\mathbf{R}_\gamma$ inherits the structure of a normal, projective variety. Furthermore, since γ lies on only one hyperplane, there is only one way for a representation in $\mathbf{R}_\gamma$ to reduce, implying that Σ_γ is smooth and is in fact the product of the lower dimensional representation spaces $\mathbf{R}_{\gamma^1} \times \mathbf{R}_{\gamma^2}$. The next theorem follows from Theorem 3.1 of [5] (cf. Theorem 5.3 of [6] in the case of a bad hyperplane).

Theorem 1. *Choose $\alpha, \beta, \gamma \in W$ as above and J a complex structure on F. Then the representation spaces $\mathbf{R}_\alpha$ and $\mathbf{R}_\beta$ are related by a special birational transformation (like a flip in Mori theory), i.e. there are projective maps Φ_α and Φ_β*

$$\begin{array}{ccc} \mathbf{R}_\alpha & & \mathbf{R}_\beta \\ & \phi_\alpha\searrow \swarrow\phi_\beta & \\ & \mathbf{R}_\gamma & \end{array}$$

(depending a priori on J) with the properties that

(1) *along $\mathbf{R}_\gamma \setminus \Sigma_\gamma$, ϕ_α and ϕ_β are isomorphisms,*
(2) *along Σ_γ, ϕ_α and ϕ_β are $\mathbb{CP}^a$ and $\mathbb{CP}^b$ bundles, respectively, and*
(3) $a + b + 1 = \mathrm{codim}_{\mathbb{C}} \Sigma_\gamma$.

We can also compare the spaces $\mathbf{R}_\alpha$ and $\mathbf{R}_\omega$, where α and ω are regular values of Ψ in the interior and on the boundary of W, respectively. By the previous theorem, we can assume that α and ω are not separated by any hyperplanes. The following is a restatement of Proposition 3.4 of [5].

Proposition 2. *There is a natural projection $\psi : \mathbf{R}_\alpha \longrightarrow \mathbf{R}_\omega$ which is a fiber bundle with fiber a flag variety $\mathcal{F}$.*

Theorem 1 and Proposition 2 can now be used to describe the behavior of the invariants μ_α as α is allowed to vary within W.

Theorem 3. *Suppose $\boldsymbol{K}$ is a knot in a homology sphere. The invariants $\mu_\alpha(\mathbf{K})$ depend on $\alpha \in W$ in the following way.*

(1) *If α and α' are regular values of Ψ in the interior of W contained in the same chamber, then*

$$\mu_\alpha = \mu_{\alpha'}.$$

(2) *If α, β and γ are chosen as in Theorem 1, then*

$$\mu_\alpha - \mu_\beta = (\chi(\mathbb{CP}^a) - \chi(\mathbb{CP}^b))\mu_{\gamma^1}\mu_{\gamma^2}.$$

(3) *If α and ω are chosen as in Proposition 2, then*

$$\mu_\alpha = \chi(\mathcal{F})\mu_\omega.$$

Sketch of proof. To prove part (1), notice that $\mathbf{R}_\alpha$ is diffeomorphic to $\mathbf{R}_{\alpha'}$. This follows because α and α' are connected by a path α_t which misses the hyperplanes, hence $\Psi^{-1}(\alpha_t)$ is a product. The map φ^* acts on this product and preserves each fiber, hence the family of maps φ_t^* on $\mathbf{R}_{\alpha_t}$ describes, after pullback to $\mathbf{R}_\alpha$, an isotopy from φ_0^* to the pullback of φ_1^*. Part (1) follows since Lefschetz numbers are invariants of the isotopy class of a map.

For part (2), notice that a direct consequence of Theorem 1 is the identity

$$(*) \qquad \mathrm{P}_t(\mathbf{R}_\alpha) = \mathrm{P}_t(\mathbf{R}_\beta) + (\mathrm{P}_t(\mathbb{CP}^a) - \mathrm{P}_t(\mathbb{CP}^b))\mathrm{P}_t(\Sigma_\gamma),$$

where P_t denotes the Poincaré polynomials taken with $\mathbb{Z}$ coefficients (cf. Corollary 3.2 of [5]). We would like to see that this formula also holds on the level of the Lefschetz polynomials, i.e. that

$$(**) \qquad M_\alpha(t) - M_\beta(t) = (\mathrm{P}_t(\mathbb{CP}^a) - \mathrm{P}_t(\mathbb{CP}^b))M_{\gamma^1}(t)M_{\gamma^2}(t).$$

The direct route is to give a general algorithm for the computation of $M_\alpha(t)$ from which one can conclude $(**)$.

To explain this algorithm, we describe first the Atiyah-Bott-Nitsure method to calculate $\mathrm{P}_t(\mathbf{R}_\alpha)$. Consider the case $\omega \in \mathbf{Z}_n$. By [2], we see that $\mathrm{P}_t(\mathbf{R}_\omega)$ is described as a difference of two terms, both of which are infinite series in t, the first being the cohomology of the classifying space of the gauge group (which is determined by the rank and genus, cf. Theorem 2.15 of [2]) and the second admitting an expression as a power series whose coefficients are polynomial functions in $(1+t)^{2g}$, the Poincaré polynomial of the Jacobian of the surface (see formulas (11.1) – (11.3) of [2]). This formula results from an equivariantly perfect stratification on the space of holomorphic structures on a fixed topological bundle over the Riemann surface (F, J).

Frohman adapted these ideas in order to compute μ_ω in [7]. He expressed the Lefschetz polynomial $M_\omega(t)$ as a difference of two Lefschetz traces (which are formal power series in t), the first of which is determined by the rank and genus and the second of which admits an expression as a power series in t whose coefficients are polynomial functions in $c(t) = L(F^*, \varphi^*)$, the Lefschetz polynomial of the monodromy acting on the Jacobian. Of course, $c(t)$ is just the Alexander polynomial and evaluating $M_\omega(t)$ at $t = 1$ gives a formula for the Lefschetz number μ_ω in terms of the derivatives of the Alexander polynomial of the knot evaluated at $t = 1$ (see Theorems 1.6 and 3.14 of [7]).

The Atiyah-Bott procedure was extended to parabolic bundles in [11] and used to give explicit computations of $\mathrm{P}_t(\mathbf{R}_\alpha)$ for low rank [3]. In general, the procedure expresses the Poincaré polynomial of $\mathbf{R}_\alpha$ as the difference of two infinite series, the first being determined by the rank, genus, and conjugacy class of α (cf. formulas (13-15) of [3]), and the second being a power series whose coefficients are polynomial functions in $(1+t)^{2g}$. As in the non-parabolic case, this formula results from an equivariantly perfect stratification on the space of holomorphic structures on a fixed topological parabolic bundle over (F, J).

This stratification leads to an algorithm for calculating $M_\alpha(t)$. As mentioned before, such a formula exists on the level of cohomology, and the crucial point is to see that the map induced by φ preserves the stratification. This is only true after taking into account the effect of φ on the choice of complex structure J on F.

Having established such a formula, it is relatively easy to see which unstable strata are created and destroyed as the monodromy condition is allowed to vary. (This is another manifestation of equation $(*)$.) Consequently, equation $(**)$ is seen to follow and this implies part (2).

To prove part (3), consider the case $\omega \in \mathbf{Z}_n$. Define $\widehat{\mathbf{R}}_\omega$ to be the quotient of $\widetilde{\mathbf{R}}_\omega$ by conjugation by the maximal torus of $\mathbf{SU}_n$. It is elementary to see that $\widehat{\mathbf{R}}_\omega$ is diffeomorphic to $\mathbf{R}_\alpha$ in this case, and the proof of part (1) shows that μ_α is equal to the Lefschetz number of φ^* on $\widehat{\mathbf{R}}_\omega$. The natural projection $\psi : \widehat{\mathbf{R}}_\omega \longrightarrow \mathbf{R}_\omega$ is a fiber bundle with fiber $\mathcal{F}$, the flag variety of full flags in $\mathbb{C}^n$, and part (3) now follows by observing that the following diagram commutes

$$\begin{array}{ccc} \widehat{\mathbf{R}}_\omega & \xrightarrow{\varphi^*} & \widehat{\mathbf{R}}_\omega \\ {\scriptstyle\psi}\big\downarrow & & \big\downarrow{\scriptstyle\psi} \\ \mathbf{R}_\omega & \xrightarrow{\varphi^*} & \mathbf{R}_\omega. \end{array}$$

An alternative approach to proving part (3) is to use the identity

$$\mathrm{P}_t(\mathbf{R}_\alpha) = \mathrm{P}_t(\mathcal{F})\mathrm{P}_t(\mathbf{R}_\omega)$$

together with an argument similar to that given for part (2).

Remarks. It seems likely that one could define invariants μ_α for arbitrary (non-fibred) knots using either the approach of [8] or that of [9]. The results of [5] are promising for the computation of the latter invariants. A direct consequence of Theorem 1 is that one of the maps ϕ_α, ϕ_β is a small resolution, hence the intersection homology of $\mathbf{R}_\gamma$ is given by the homology of either $\mathbf{R}_\alpha$ or $\mathbf{R}_\beta$. More generally, if γ lies on more than one hyperplane, then $\mathbf{R}_\gamma$ may have a very complicated singular locus Σ_γ, but Conjecture 4.8 of [5] predicts the existence of a small resolution from a nearby non-singular $\mathbf{R}_\alpha$. This conjecture is true if, for example, the hyperplanes containing γ are in general position.

Aknowledgements. I would like to thank Charles Frohman, Arthur

Greenspoon, Andrew Nicas, Richard Wentworth, and Kôji Yokogawa for their advice and correspondence.

References

1. S. Akbulut and J. McCarthy, *Casson's invariant for oriented homology 3-spheres, an exposition* Princeton University Press, Mathematical Notes no. 36 1990.
2. M. Atiyah and R. Bott, *The Yang-Mills equations on a Riemann surface*, Phil. Trans. Roy. Soc. Lond. A. **308** (1982) 524–615.
3. H. U. Boden, *Unitary Representations of Brieskorn spheres*, Duke Math J. **75** (1994) 193–220.
4. H. U. Boden, *Knot invariants and representation varieties*, 1995 in preparation.
5. H. U. Boden and Y. Hu, *Variations of moduli of parabolic bundles*, 1994, Math. Ann. **301** (1995) 539–559.
6. H. U. Boden and K. Yokogawa, *Rationality of moduli spaces of parabolic bundles*, 1995 MPI preprint 95–27.
7. C. Frohman, *Unitary representations of knot groups*, Topology **32** (1993) 121–144.
8. C. Frohman and D. Long, *Casson's invariant and surgery on knots*, Proc. Edin. Math. Soc. **35** (1992) 383–395.
9. C. Frohman and A. Nicas, *An intersection homology invariant for knots in a rational homology 3-sphere*, Topology **33** (1994) 123–158.
10. V. Mehta and C. Seshadri, *Moduli of vector bundles on curves with parabolic structures*, Math. Ann. **248** (1980) 205–239.
11. N. Nitsure, *Cohomology of the moduli of parabolic bundles*, Proc. Ind. Math. Soc. **95** (1986) 61–77.

Max-Planck-Institut für Mathematik, Gottfried-Claren-Str., 26, 53225 Bonn, Germany

E-mail address: hboden@mpim-bonn.mpg.de

AMS/IP Studies in Advanced Mathematics
Volume **2**, 1997 (Part 1)

SYMPLECTIC FORMS ON MODULI SPACES OF FLAT CONNECTIONS ON 2-MANIFOLDS

LISA C. JEFFREY

ABSTRACT. Let G be a compact connected semisimple Lie group. We extend the techniques of Weinstein [17] to give a construction in group cohomology of symplectic forms ω on 'twisted' moduli spaces of representations of the fundamental group π of a 2-manifold Σ (the smooth analogues of $\mathrm{Hom}(\pi_1(\Sigma), G)/G$) and on relative character varieties of fundamental groups of 2-manifolds. We extend this construction to exhibit a symplectic form on the extended moduli space [5] (a Hamiltonian G-space from which these moduli spaces may be obtained by symplectic reduction), and compute the moment map for the action of G on the extended moduli space.

1. INTRODUCTION.

Let Σ be a closed oriented 2-manifold of genus $g \geq 2$; the fundamental group of Σ will be denoted π. Let G be a compact connected Lie group with Lie algebra $\mathfrak{g}$. This paper concerns the moduli space $\mathcal{M} = \mathrm{Hom}(\pi, G)/G$ of conjugacy classes of representations of π into G, and certain more general analogues of $\mathcal{M}$. The space $\mathcal{M}$ has an open dense set on which the structure of a smooth symplectic manifold is defined. In addition to the definition we have given in terms of representations of π, the space $\mathcal{M}$ has two alternative descriptions. The first of these is the gauge theory description: via the holonomy map, $\mathcal{M}$ is identified with the space of gauge equivalence classes of flat connections on a trivial principal G bundle over Σ. The second alternative description appears once one fixes a complex structure on the 2-manifold Σ, so that Σ becomes a Riemann surface; $\mathcal{M}$ is then identified with the space of equivalence classes of semistable holomorphic $G^{\mathbb{C}}$ bundles over Σ.

This material is based on work supported by the National Science Foundation under Grant No. DMS-9306029.

The purpose of this paper is to extend work of Karshon [11] and Weinstein [17] to a more general setting; this paper follows [17] closely and should be read in conjunction with it. These two papers complement the work of Goldman [3]. Goldman originally gave a construction in group cohomology of the symplectic form ω on the space $\mathcal{M}$; in order to prove ω is closed, Goldman used the gauge theory description of the symplectic form. Karshon [11] gave a proof of the closedness of the symplectic form using group cohomology; Weinstein [17] reinterpreted Karshon's construction in the setting of the de Rham-bar bicomplex [2, 15]. In the present work we extend Weinstein's work to construct symplectic forms on relative character varieties of surface groups, and on 'twisted' moduli spaces $\mathcal{M}_\beta$ of bundles on Riemann surfaces (see Section 6 of [1]) associated to an element $\beta \in Z(G)$. The spaces $\mathcal{M}_\beta$ share many properties with $\mathcal{M}$ but in general have less singularities. Indeed, $\mathcal{M}_\beta$ is smooth when $G = SU(n)$ and β is a generator of the center of $SU(n)$: in contrast, even when $G = SU(2)$, the space $\mathcal{M}$ is smooth only in the very special case $g = 2$.

We also give a group cohomology construction of symplectic forms on the extended moduli spaces X and X_β [5]: these are finite dimensional symplectic G-spaces from which $\mathcal{M}$ and $\mathcal{M}_\beta$ may be obtained by symplectic reduction. In [5], symplectic structures on X_β and X (which are in fact the same as the symplectic forms we recover below) were specified using gauge theoretic techniques. Here we compute the moment maps for the G action on X and X_β; up to a normalization factor, these coincide with the moment maps found in [5].

The purpose of the construction of X_β in [5] was to exhibit $\mathcal{M}_\beta$ as the result of finite dimensional symplectic reduction (in contrast to the infinite dimensional quotient construction given in [1]). We make further use of this finite dimensional quotient construction in [8], where we extend the techniques of [7] (which gives a formula for intersection pairings in the cohomology ring of the symplectic quotient $\mathcal{M}_X$ of a finite dimensional Hamiltonian G-space X, in terms of the G-equivariant cohomology $H^*_G(X)$) to treat the intersection pairings in $\mathcal{M}_\beta$, starting from the equivariant cohomology of X_β. By this means we give proofs of formulas (found originally by Witten [18] using physical methods) for the intersection pairings in $H^*(\mathcal{M}_\beta)$.

In this paper, the symplectic forms on X and X_β are constructed explicitly in terms of the Maurer-Cartan form on G and the chain homotopy operator that occurs in the standard proof of the Poincaré lemma. This explicit description of the symplectic form will be important in [8], where we make use of explicit equivariantly closed differential forms representing the relevant classes in de Rham cohomology. In [6] we extend the methods of this paper to give explicit representatives in de Rham cohomology for all the generators of the cohomology ring of $\mathcal{M}_\beta$ (one of which is the cohomology class of the symplectic form); for the applications in [8], we shall need the de Rham representatives for all the generators.

After this work was completed, we received the paper of Huebschmann [4], in which he has obtained similar results independently.

Acknowledgement: We thank A. Weinstein for helpful conversations.

2. Group cohomology.

Let $\mathbb{F} = \mathbb{F}_{2g}$ be the free group on $2g$ generators $x_1, \ldots, x_{2g}$. We introduce a relation $R = \prod_{i=1}^{g}[x_{2i-1}, x_{2i}]$, where $[a, b]$ denotes the commutator $aba^{-1}b^{-1}$. The fundamental group π of a closed 2-manifold of genus g is then given by

$$\pi = \mathbb{F}/\mathcal{R}$$

where $\mathcal{R}$ is the normal subgroup generated by R.

We shall work with Eilenberg-Mac Lane group chains (see [3] section 3.8), and shall denote the differential on the group chain complex by ∂. The p-chains $C_p(\Gamma)$ on a group Γ are $\mathbb{Z}$-linear combinations of elements of Γ^p. In particular, associated to the relation R there is a distinguished 2-chain $c \in C_2(\mathbb{F})$ given by

$$c = \sum_{i=1}^{2g} (\partial R/\partial x_i, x_i). \tag{2.1}$$

(Here, $\partial/\partial x_i$ refers to the differential in the Fox free differential calculus: see [3], sections 3.1-3.3.) Goldman ([3], above Proposition 3.9) shows

$$\partial c = 1 - R. \tag{2.2}$$

3. The de Rham-bar bicomplex.

Weinstein [17] introduces a bicomplex $(C^{*,*}(G), \delta, d)$ whose p, q term is $C^q(G^p)$. The second coboundary is the exterior differential d, while the first is the differential δ appearing in group cohomology:

$$\begin{aligned} (\delta\beta)(g_0, \ldots, g_p) = \;& (-1)^{p+1}\beta(g_0, \ldots, g_{p-1}) \\ & + \sum_{i=1}^{p} (-1)^i \beta(g_0, \ldots, g_{i-1}g_i, \ldots, g_p) \\ & + \beta(g_1, \ldots, g_p). \end{aligned} \tag{3.1}$$

Let Y denote $\mathrm{Hom}(\mathbb{F}, G) = G^{2g}$. Then there is a second bicomplex $(\tilde{C}^{*,*}(G), \delta, d)$ whose p, q term is $\Omega^q(\mathbb{F}^p \times Y) = \Omega^0(\mathbb{F}^p) \otimes \Omega^q(Y)$. (The differential δ in the bicomplex $\tilde{C}^{*,*}(G)$ is the adjoint of the differential ∂ in the Eilenberg-Mac Lane group chain complex.) As in [17], the evaluation maps

$$E_p : \mathbb{F}^p \times Y \to G^p$$

give rise to maps

$$E_p^* : \Omega^q(G^p) \to \Omega^0(\mathbb{F}^p) \otimes \Omega^q(Y)$$

which combine to form a map of bicomplexes $E^* : C^{*,*}(G) \to \tilde{C}^{*,*}(G)$.

We recall the following elements of $C^{*,*}(G)$ defined in [17]. Let $\alpha \in \Omega^1(G) \otimes \mathfrak{g}$ denote the (left-invariant) Maurer-Cartan form on G, and $\bar{\alpha}$ the corresponding right-invariant form. Define projection maps $\pi_i : G^p \to G$ $(i = 1, \dots, p)$ to the i'th copy of G, and let $\alpha_i = \pi_i^* \alpha$ and $\bar{\alpha}_i = \pi_i^* \bar{\alpha}$. In terms of this notation, the following are introduced in [17]:

$$\lambda = \frac{1}{6} \alpha \cdot [\alpha, \alpha] \quad \in \Omega^3(G), \tag{3.2}$$

$$\Omega = \alpha_1 \cdot \bar{\alpha}_2 \in \Omega^2(G^2). \tag{3.3}$$

Given $\eta \in \mathfrak{g}$, Weinstein also introduces

$$\theta_\eta = \eta \cdot (\alpha + \bar{\alpha}) \in \Omega^1(G). \tag{3.4}$$

(Here, $\cdot$ denotes an invariant inner product on $\mathfrak{g}$.) These forms satisfy the following properties ([17], Lemmas 3.1, 3.3 , 4.1 and 4.4):

Proposition 3.1. *We have*

$$d\lambda = 0; \tag{3.5}$$

$$d\Omega = \delta\lambda; \tag{3.6}$$

$$\iota_{\tilde{\eta}} \lambda = d\theta_\eta; \tag{3.7}$$

$$\iota_{\tilde{\eta}} \Omega = -\delta\theta_\eta, \tag{3.8}$$

where $\tilde{\eta}$ is the vector field generated by η.

4. Two-forms on moduli spaces.

Let us now introduce

$$\omega = \langle c, E^*\Omega\rangle \in \Omega^2(Y). \tag{4.1}$$

We then have

Proposition 4.1. *In terms of the identification of $Y = \mathrm{Hom}(\mathbb{F}, G)$ with G^{2g}, we have*

$$d\omega = -{\epsilon_R}^* \lambda \tag{4.2}$$

where $\epsilon_R : \mathrm{Hom}(\mathbb{F}, G) \to G$ is the map given by evaluation on the element $R \in \mathbb{F}$:

$$\epsilon_R(g_1, \ldots, g_{2g}) = \prod_{i=1}^{g} [g_{2i-1}, g_{2i}]. \tag{4.3}$$

Proof: We have

$$\begin{aligned} d\langle c, E^*\Omega\rangle &= \langle c, dE^*\Omega\rangle \\ &= \langle c, E^* d\Omega\rangle = \langle c, E^*\delta\lambda\rangle \\ &= \langle c, \delta E^*\lambda\rangle = \langle \partial c, E^*\lambda\rangle \\ &= \langle 1 - R, E^*\lambda\rangle, \end{aligned}$$

where the last step follows from (2.2).

If t is any element of G, define $Y_t = {\epsilon_R}^{-1}(t) \subset Y$. The following is an immediate consequence of Proposition 4.1:

Proposition 4.2. *The 2-form ω restricts on Y_t to a closed form.*

Definition 4.3. *If t is an element of G, let $Z_t \subset G$ be the centralizer of t in G.*

Notice that Y_t carries an action of Z_t by conjugation.

Definition 4.4. *If $t \in G$, the* relative character variety *associated to t is the space $\mathcal{M}_t = Y_t / Z_t$, where Z_t acts on Y_t by conjugation.*

Some properties of the symplectic geometry of relative character varieties were given in [9] and [10]. Relative character varieties also arise in algebraic geometry where (under appropriate circumstances) they are identified with moduli spaces of semistable parabolic vector bundles on Riemann surfaces: see [13] (where the identification between relative character varieties and moduli spaces of parabolic bundles is given) or [14]. A gauge theory construction of a symplectic form on relative character varieties is given in [5]: it will follow from the Remark at the end of

Section 5 that this is essentially the same as the symplectic form we construct here (i.e., the two are the same under a natural map identifying the relevant Zariski tangent spaces).

By extending the calculations in [17], we obtain the following two propositions.

Proposition 4.5. *The form ω is invariant under the action of G on Y by conjugation.*

Proof: Let $\eta \in \mathfrak{g}$; we will show that the Lie derivative of ω with respect to the vector field $\tilde{\eta}$ generated by η is zero, in other words $\mathcal{L}_{\tilde{\eta}}\omega = (d\iota_{\tilde{\eta}} + \iota_{\tilde{\eta}}d)\omega = 0$. Now

$$\iota_{\tilde{\eta}}d\omega = -\iota_{\tilde{\eta}}{\epsilon_R}^*\lambda = -{\epsilon_R}^*\iota_{\tilde{\eta}}\lambda. \tag{4.4}$$

Also $d\iota_{\tilde{\eta}}\omega = d\langle c, E^*\iota_{\tilde{\eta}}\Omega\rangle$. But $\iota_{\tilde{\eta}}\Omega = -\delta\theta_\eta$ by (3.8), so we get

$$\begin{aligned} d\iota_{\tilde{\eta}}\omega &= -d\langle c, E^*\delta\theta_\eta\rangle \\ &= -d\langle c, \delta E^*\theta_\eta\rangle \\ &= -d\langle \partial c, E^*\theta_\eta\rangle = -d\langle 1 - R, E^*\theta_\eta\rangle \\ &= d{\epsilon_R}^*\theta_\eta = {\epsilon_R}^*d\theta_\eta. \end{aligned}$$

But by (3.7) we have $d\theta_\eta = \iota_{\tilde{\eta}}\lambda$, so $d\iota_{\tilde{\eta}}\omega = {\epsilon_R}^*\iota_{\tilde{\eta}}\lambda$, and $\mathcal{L}_{\tilde{\eta}}\omega = 0$.

Proposition 4.6. *We have the following identification of 1-forms on G^{2g}:*

$$\iota_{\tilde{\eta}}\omega = {\epsilon_R}^*\theta_\eta. \tag{4.5}$$

Thus, the restriction of ω to Y_t is horizontal with respect to the action of Z_t. (In other words, if $\eta \in \mathrm{Lie}(Z_t)$ generates the vector field $\tilde{\eta}$ on Y, then $\iota_{\tilde{\eta}}\omega|_{Y_t} = 0$, where $\iota_{\tilde{\eta}}$ denotes the interior product with respect to $\tilde{\eta}$.)

Proof: We have

$$\begin{aligned} \iota_{\tilde{\eta}}\omega = \langle c, \iota_{\tilde{\eta}}E^*\Omega\rangle = \langle c, E^*\iota_{\tilde{\eta}}\Omega\rangle &= -\langle c, E^*\delta\theta_\eta\rangle, \\ = -\langle \partial c, E^*\theta_\eta\rangle = -\langle 1 - R, E^*\theta_\eta\rangle &= {\epsilon_R}^*\theta_\eta. \end{aligned}$$

This form necessarily restricts to zero on the level sets of ϵ_R.

Propositions 4.5 and 4.6 imply that the form ω descends to a 2-form $\bar{\omega}$ on the space $\mathcal{M}_t$, which is closed by Proposition 4.2. Nondegeneracy will be established in Corollary 5.5.

In particular if t is a central element $\beta \in Z(G)$ then Z_β is the full group G. If $G = U(n)$ and $\beta = e^{2\pi i d/n}\mathrm{diag}(1, \ldots, 1)$ then the space $\mathcal{M}_\beta$ appears in algebraic geometry (see [1]) as the moduli space of semistable holomorphic vector bundles of rank n and degree d. When β is as above but $G = SU(n)$, the algebraic geometry interpretation of the space $\mathcal{M}_\beta$

is as the moduli space of semistable holomorphic vector bundles of rank n and degree d with fixed determinant.[1]

The constructions in this section exhibit a group cohomology construction of a symplectic form on the 'twisted' moduli spaces $\mathcal{M}_\beta$ associated to central elements β of G. It is shown in [6] that the form $\bar{\omega}$ is in the cohomology class of (a constant multiple of) the standard generator f_2 of $H^*(\mathcal{M}_\beta; \mathbb{R})$ (in the notation of Sections 2 and 9 of [1]). In fact in [6] we extend the construction which gives rise to the symplectic form $\bar{\omega}$, to give representatives in de Rham cohomology for all the generators of the ring $H^*(\mathcal{M}_\beta; \mathbb{R})$ given in [1]. Our applications in [8] will rely at least as heavily on the fact that $\bar{\omega}$ is a de Rham representative of the cohomology class f_2 as on its being nondegenerate: in any case, many results of the type we shall invoke for Hamiltonian G-manifolds generalize (cf. [12]) to manifolds where the symplectic form degenerates on a locus of measure 0.

5. The symplectic form on the extended moduli space.

Let β be an element of the center $Z(G)$. The associated *extended moduli space* X_β constructed in [5] may be described as a fibre product

$$X_\beta = (\epsilon_R \times e_\beta)^{-1}(\triangle) \subset G^{2g} \times \mathfrak{g}. \tag{5.1}$$

Here, $\triangle$ is the diagonal in $G \times G$ and $\epsilon_R : G^{2g} \to G$ was defined above, while $e : \mathfrak{g} \to G$ is the exponential map and $e_\beta = \beta \cdot e$. The space X_β is equipped with two canonical projection maps $\mathrm{pr}_1 : X_\beta \to G^{2g}$ and $\mathrm{pr}_2 : X_\beta \to \mathfrak{g}$, for which there is the following commutative diagram:

$$\begin{array}{ccc} X_\beta & \xrightarrow{\mathrm{pr}_2} & \mathfrak{g} \\ \mathrm{pr}_1 \downarrow & & \downarrow e_\beta \\ G^{2g} & \xrightarrow{\epsilon_R} & G \end{array} \tag{5.2}$$

A straightforward argument using the regular value theorem endows X_β with a smooth structure on an open dense set ${X_\beta}^s$ including $\mathrm{pr}_2^{-1}(0)$: see [5], Proposition 5.4, where an explicit characterization of the singular locus of X_β is given.

The space X_β carries an action of the group G. A gauge theoretic construction of a G-invariant closed 2-form on X_β was given in [5], and it was shown that this form is nondegenerate on an open dense set in X_β and that the action of G is Hamiltonian where the 2-form is

---

[1]When $G = U(n)$ or $SU(n)$ and d is coprime to n, the spaces $\mathcal{M}_\beta$ are smooth manifolds. This is in contrast to the spaces $\mathcal{M}$, which are singular except in a few very special cases.

nondegenerate. There is thus an open dense set in X_β which is a smooth finite dimensional Hamiltonian G-space such that the space $\mathcal{M}_\beta$ is given by symplectic reduction of this G-space at 0.

By extending the techniques of [17], we construct here a G-invariant closed 2-form on X_β which is nondegenerate on an open dense set, and show that the moment map μ is given by a constant multiple of pr_2, as was shown in [5]. The symplectic form will in fact turn out to be the same as the one constructed using gauge theory (see the Remark at the end of this section): in other words, one of these symplectic forms is the pullback of the other under a natural map. It follows from our construction that the symplectic quotient (at 0) with respect to the action of G on X_β is the twisted moduli space $\mathcal{M}_\beta$ described above. For $t \in G$, the relative character variety $\mathcal{M}_t$ from Section 4 is the symplectic reduction of X at the orbit $\mathcal{O}_\Lambda \subset \mathfrak{g}$ (under the adjoint action) that corresponds to an element $\Lambda \in \mathfrak{g}$ for which $\exp(\Lambda) = t$.

To construct the symplectic form, we first construct a form $\sigma \in \Omega^2(\mathfrak{g})$ for which

$$e^*\lambda = d\sigma. \tag{5.3}$$

The existence of such a form follows from the following standard result (see e.g. [16], Lemma 4.18):

Proposition 5.1. [*Poincaré Lemma*]
(a) If $\gamma \in \Omega^{p+1}(V)$ where V is a vector space, and $d\gamma = 0$, then there is a form $\sigma \in \Omega^p(V)$ with $\gamma = d\sigma$.
(b) Denote by I the map $\Omega^{p+1}(V) \to \Omega^p(V)$ sending γ to σ. Then $dI + Id = \mathrm{id}$.

For $\beta \in \Omega^*(V)$, $I\beta$ is given at $v \in V$ by

$$(I\beta)_v = \int_0^1 F_t^*(\iota_{\bar{v}}\beta)dt \tag{5.4}$$

where $\bar{v}$ is the vector field on V which takes the constant value v, and F_t is the map $V \to V$ given by multiplication by t. In our case the form

$$\sigma = I(e^*\lambda)$$

is G -invariant because λ is G-invariant and e is a G-equivariant map.

We now restrict to the fibre product $X_\beta \subset G^{2g} \times \mathfrak{g}$. For $(h, \Lambda) \in X_\beta$ we have $\epsilon_R(h) = e_\beta(\Lambda)$, so if $(H, \zeta) \in T_h G^{2g} \times \mathfrak{g}$ represents an element in the tangent space to X_β, we have

$$\epsilon_{R*}H = e_{\beta_*}\zeta. \tag{5.5}$$

We define a 2-form on X_β by

$$\tilde{\omega} = \mathrm{pr}_1^* \omega + \mathrm{pr}_2^* \sigma, \tag{5.6}$$

where ω was defined in (4.1). We find that

$$d(\mathrm{pr}_1^* \omega + \mathrm{pr}_2^* \sigma) = \mathrm{pr}_1^* d\omega + \mathrm{pr}_2^* d\sigma$$

so

$$\begin{aligned} d\tilde{\omega}(H,\zeta) &= -\epsilon_R{}^*\lambda(H) + d\sigma(\zeta) \qquad \text{(by (4.2))} \\ &= -\lambda(\epsilon_{R*}H) + \lambda(e_*\zeta) = -\lambda(\epsilon_{R*}H) + \lambda(e_{\beta_*}\zeta) = 0. \end{aligned} \tag{5.7}$$

(Here, we have used the fact that λ is invariant under multiplication by β, so $e^*\lambda = e_\beta{}^*\lambda$.) So we have

Proposition 5.2. *The 2-form $\tilde{\omega}$ on $G^{2g} \times \mathfrak{g}$ restricts on X_β to a closed form.*

The G-invariance of $\tilde{\omega}$ follows because σ and ω are G-invariant and the projection maps pr_1 and pr_2 are G-equivariant maps.

We may now identify the moment map for the action of G on X_β: in other words, we find a function $\mu : X \to \mathfrak{g}$ such that $\iota_{\tilde{\eta}}\tilde{\omega} = \eta \cdot d\mu$ where $\tilde{\eta}$ is the vector field on X_β generated by $\eta \in \mathfrak{g}$.[2] First we recall from (4.5) that $(\iota_{\tilde{\eta}}\omega) = \epsilon_R{}^*\theta_\eta$. Now since $\sigma = I(e^*\lambda) = I(e_\beta{}^*\lambda)$, we have

$$\begin{aligned} \iota_{\tilde{\eta}}\sigma &= \iota_{\tilde{\eta}}(Ie_\beta{}^*\lambda) = -I(\iota_{\tilde{\eta}}e_\beta{}^*\lambda) = -I(e_\beta{}^*\iota_{\tilde{\eta}}\lambda) \\ &= -I(de_\beta{}^*\theta_\eta) \ \ \text{(by (3.7))}\ . \end{aligned} \tag{5.8}$$

Combining (5.8) with Proposition 5.1 (b) we find

$$\iota_{\tilde{\eta}}\sigma = -e_\beta{}^*\theta_\eta + d(Ie_\beta{}^*\theta_\eta) \tag{5.9}$$

Adding (4.5) and (5.9) we have

$$\iota_{\tilde{\eta}}(\tilde{\omega}) = d(Ie_\beta{}^*\theta_\eta)$$

so that a moment map $\mu : X_\beta \to \mathfrak{g}$ for the action of G on X_β is given by

$$\eta \cdot \mu = Ie_\beta{}^*\theta_\eta. \tag{5.10}$$

Now

$$(Ie_\beta{}^*\theta_\eta)_\Lambda = \int_0^1 F_t^*\Big(e_\beta{}^*\theta_\eta(\Lambda)\Big) \ \text{ by (5.4)} \tag{5.11}$$

---

[2] Ordinarily the moment map is specified as a map into $\mathfrak{g}^*$. Here we have used the inner product to identify $\mathfrak{g}^*$ with $\mathfrak{g}$.

$$= \int_0^1 ({e_\beta}^*\theta_\eta)_{\Lambda t}(\Lambda) = \int_0^1 (\theta_\eta)_{e_\beta(\Lambda t)}({e_\beta}_*\Lambda)$$
$$= \int_0^1 \eta\cdot(\alpha+\bar{\alpha})_{e_\beta(\Lambda t)}({e_\beta}_*\Lambda)$$
$$= 2\eta\cdot\Lambda.$$

Thus we have explicitly specified a moment map for the action of G, which is equivariant with respect to the adjoint action of G:

Proposition 5.3. *A moment map for the action of G on X_β is given by the map $\mu = 2\,\mathrm{pr}_2 : \ (h,\Lambda)\mapsto 2\Lambda$.*

To complete the identification of the form $\tilde{\omega}$ as a symplectic form on an open dense set in X_β, one needs the following:

Proposition 5.4. *The form $\tilde{\omega}$ is a nondegenerate bilinear form on the Zariski tangent space $T_{(h,\Lambda)}X_\beta$, for any $(h,\Lambda)\in X_\beta$ for which $(d\epsilon_R)_h$ is surjective.*

Our proof of this Proposition parallels the gauge theory argument given in [5] (see Proposition 3.1 of [5] for the case $G = SU(2)$). This material is treated in [4] (Theorem 4.4 and Section 5) and [11] (Theorem 4): the proof we sketch is essentially the one given by Huebschmann [4], to whom the group cohomology proof of the nondegeneracy of the symplectic form on an open neighbourhood of the zero locus of the moment map in the extended moduli space is due.[3]

Proof of Proposition 5.4*:* To establish nondegeneracy of $\tilde{\omega}$ we proceed as follows. Proposition 5.3 establishes that if $\tilde{\eta}$ is the vector field associated to the action of η on X_β, and if $(H,\zeta)\in T_{(h,\Lambda)}X_\beta$ for $(h,\Lambda)\in X_\beta$ (in other words, $(d\epsilon_R)_h H = (de_\beta)_\Lambda\zeta$), then

$$\tilde{\omega}_{(h,\Lambda)}(\tilde{\eta},(H,\zeta)) = 2\eta\cdot\zeta. \tag{5.12}$$

Thus to establish nondegeneracy of $\tilde{\omega}$ at those (h,Λ) for which $(d\epsilon_R)_h$ is surjective (see (5.5)) it suffices to establish it on the orthocomplement

[3]The proof given in [4] applies to a suitable neighbourhood of the zero locus of the moment map in X_β which is contained in $\mathrm{pr}_2^{-1}(\mathcal{O}_{\mathrm{reg}})$, where $\mathcal{O}_{\mathrm{reg}}$ is the subset of $\mathfrak{g}$ where the exponential map is regular: the subset $\mathrm{pr}_2^{-1}(\mathcal{O}_{\mathrm{reg}})$ is a proper subset of the smooth locus of X_β. We have adapted the proof so it applies to the Zariski tangent space $T_{(h,\Lambda)}X_\beta$ for all points $(h,\Lambda)\in X_\beta$ for which $(d\epsilon_R)_h$ is surjective.

in $\operatorname{Ker}(d\,\mathrm{pr}_2) \subset T_{(h,\Lambda)}X_\beta$ of the image of the action of $\mathfrak{g}$. This means we must establish that ω is nondegenerate restricted to

$$\frac{T_{(h,\Lambda)}\Big(\mathrm{pr}_2^{-1}(\Lambda) \subset X_\beta\Big)}{\{\tilde{\eta} : \eta \in \mathrm{Stab}(\Lambda)\}}. \tag{5.13}$$

We have commutative diagrams

(5.14)

$$\begin{array}{ccccccccc}
 & & 0 & & 0 & & 0 & & \\
 & & \downarrow & & \downarrow & & \downarrow & & \\
0 & \longrightarrow & K_B & \longrightarrow & B^1(\mathbb{F};\mathfrak{g}_h) & \xrightarrow{\delta} & B^1(\mathbb{Z};\mathfrak{g}_{\epsilon_R(h)}) & \longrightarrow & \dots \\
 & & \downarrow & & \downarrow & & \downarrow & & \\
0 & \longrightarrow & K_C & \longrightarrow & C^1(\mathbb{F};\mathfrak{g}_h) & \xrightarrow{\delta} & C^1(\mathbb{Z};\mathfrak{g}_{\epsilon_R(h)}) & \longrightarrow & \dots \\
 & & \downarrow & & \downarrow & & \downarrow & & \\
0 & \longrightarrow & K_H & \longrightarrow & H^1(\mathbb{F};\mathfrak{g}_h) & \xrightarrow{\delta} & H^1(\mathbb{Z};\mathfrak{g}_{\epsilon_R(h)}) & \longrightarrow & \dots \\
 & & \downarrow & & \downarrow & & \downarrow & & \\
 & & 0 & & 0 & & 0 & &
\end{array}$$

in which the columns are short exact sequences. Here, if Γ is a discrete group (where $\Gamma = \mathbb{Z}$ or $\Gamma = \mathbb{F}$) equipped with a representation $\rho : \Gamma \to G$, the notation $C^*(\Gamma;\mathfrak{g}_\rho)$ refers to the Eilenberg-Mac Lane group cochain complex with coefficients in the Γ-module $\mathfrak{g}_\rho$ specified by ρ under the adjoint action of G on $\mathfrak{g}$. For $\Gamma = \mathbb{F}$ or $\Gamma = \mathbb{Z}$, we have $C^1(\Gamma;\mathfrak{g}_\rho) = Z^1(\Gamma;\mathfrak{g}_\rho)$. The maps δ are induced by $d\epsilon_R$, while K_B, K_C and K_H are the kernels of the maps δ on B^1, C^1 and H^1. Then the vector space $T_{(h,\Lambda)}(\mathrm{pr}_2^{-1}(\Lambda))$ is identified with $K_C \subset C^1(\mathbb{F};\mathfrak{g}_h)$, while $\{\tilde{\eta} : \eta \in \mathrm{Stab}(\Lambda)\}$ is identified with $K_B = K_C \cap B^1(\mathbb{F};\mathfrak{g}_h)$. Then we see from (5.14) that the quotient K_C/K_B is canonically identified with $K_H = \operatorname{Ker}\Big(\delta : H^1(\mathbb{F};\mathfrak{g}_h) \to H^1(\mathbb{Z};\mathfrak{g}_{\epsilon_R(h)})\Big)$.

We have the long exact sequence

(5.15)

$$0 \to H^0(\mathbb{F};\mathfrak{g}_h) \to H^0(\mathbb{Z};\mathfrak{g}_{\epsilon_R(h)}) \xrightarrow{\delta^*} H^1(\mathbb{F},\mathbb{Z};\mathfrak{g}_h) \to H^1(\mathbb{F};\mathfrak{g}_h) \to$$

$$\xrightarrow{\delta} H^1(\mathbb{Z};\mathfrak{g}_{\epsilon_R(h)}) \to H^2(\mathbb{F},\mathbb{Z};\mathfrak{g}_h) \to 0.$$

The vector spaces and maps in this sequence satisfy Poincaré duality. Furthermore, the pairing $\cdot : \mathfrak{g} \otimes \mathfrak{g} \to \mathbb{R}$ gives rise to a cup product pairing

$$H^1(\mathbb{F},\mathbb{Z};\mathfrak{g}_h) \otimes H^1(\mathbb{F};\mathfrak{g}_h) \to H^2(\mathbb{F},\mathbb{Z};\mathbb{R}) \cong \mathbb{R}, \tag{5.16}$$

and Poincaré duality in (5.15) implies that the restriction of this pairing to $\left(H^1(\mathbb{F},\mathbb{Z};\mathfrak{g}_h)/\mathrm{Im}(\delta^*)\right) \otimes (\mathrm{Ker}(\delta) \subset H^1(\mathbb{F};\mathfrak{g}_h))$ is nondegenerate.

Now one may show (see for instance [11] Theorem 4 or [4] Theorem 4.4) that this cup product is the restriction of the form ω to $\mathrm{Ker}(\delta) \subset H^1(\mathbb{F};\mathfrak{g}_h) \cong T_h\left({\epsilon_R}^{-1}(C_{\epsilon_R(h)})/G\right)$. (Here, $C_{\epsilon_R(h)}$ denotes the conjugacy class of $\epsilon_R(h)$ in G.) The nondegeneracy of the pairing arising from the cup product thus completes the proof of nondegeneracy.

Remark. Goldman ([3], proof of Proposition 3.7) has shown that $\mathrm{Im}(d\epsilon_R)_h = \mathrm{Lie}(\mathrm{Stab}(h))$.

Suppose $G = SU(n)$ and β is a generator of $Z(G)$. The subset of X_β where $(d\epsilon_R)_h$ is surjective then contains the zero level set of the moment map. Thus the following is an immediate consequence of Proposition 5.4:

Corollary 5.5. *Let $G = SU(n)$ and suppose β is a generator of $Z(G)$. Then 0 is a regular value of the moment map $\mu = 2\,\mathrm{pr}_2$. Further, $\mathcal{M}_\beta$ is a smooth manifold and the form $\bar{\omega}$ on $\mathcal{M}_\beta$ is nondegenerate.*[4]

Remark. At the end of the last Proof, we alluded to the identification of ω (on quotients of level sets of pr_2) with the bilinear form given by the cup product (5.16). It is easy to see that the vector spaces $H^1(\mathbb{F};\mathfrak{g}_h)$ and $H^1(\mathbb{F},\mathbb{Z};\mathfrak{g}_h)$ (arising from group cohomology with coefficients in the $\mathbb{F}$-module $\mathfrak{g}_h$ specified by $h \in \mathrm{Hom}(\mathbb{F},G)$) are the same as the vector spaces $H^1(\Sigma - D^2; d_A)$ and $H^1(\Sigma - D^2, \partial D^2; d_A)$ arising in the gauge theory description of X_β (cf. Section 2.2 of [5]). Here, A is a flat connection on the punctured surface $\Sigma - D^2$ whose holonomy gives rise to the representation h of the fundamental group $\mathbb{F}$. This identification comes from the identification between gauge equivalence classes of flat G connections and conjugacy classes of representations of the fundamental group into G, which arises from the map sending a flat connection to the representation given by its holonomy.[5] Furthermore, the pairing $H^1(\mathbb{F};\mathfrak{g}_h) \otimes H^1(\mathbb{F},\mathbb{Z};\mathfrak{g}_h) \to \mathbb{R}$ arising from the group cohomology cup product is the same as the pairing $(\alpha,\beta) \mapsto \int_{\Sigma - D^2} \alpha \cdot \wedge \beta$ that gives the

[4]When zero is a regular value of the moment map on X_β, standard arguments establish the smoothness of the reduced space $\mathcal{M}_\beta$.

[5]Using the holonomy representation, a homeomorphism from the extended moduli space (as defined gauge theoretically in Section 2.1 of [5]) to X_β (as defined above) is given in Section 2.3 of [5]. This map identifies the relevant Zariski tangent spaces and gives rise to the identification of the symplectic forms.

symplectic form in the gauge theory description. (Here, α and β are d_A-closed $\mathfrak{g}$-valued 1-forms on $\Sigma - D^2$, and $\alpha \cdot \wedge \beta$ is the element of $\Omega^2(\Sigma)$ that arises from the wedge product combined with the pairing $\mathfrak{g} \otimes \mathfrak{g} \to \mathbb{R}$ given by the invariant inner product $\cdot$.) Hence the symplectic form we have constructed on X_β is in fact the same as the one constructed in [6] using gauge theory.

References

1. Atiyah, M.F. and Bott, R., *The Yang-Mills equations over Riemann surfaces*, Phil. Trans. R. Soc. Lond. **A 308** (1982) 523-615.
2. Bott, R., *On the Chern-Weil homomorphism and the continuous cohomology of Lie groups*, Advances in Math. **11** (1973) 289-303.
3. Goldman, W., *The symplectic nature of fundamental groups of surfaces*, Advances in Math. **54** (1984), 200-225.
4. Huebschmann, J., *Symplectic forms and Poisson structures of certain moduli spaces*, Preprint hep-th/9312112 (1993); Duke Math. J., in press.
5. Jeffrey, L.C., *Extended moduli spaces of flat connections on Riemann surfaces*, Math. Annalen **298** (1994), 667-692.
6. Jeffrey, L.C., *Group cohomology construction of the cohomology of moduli spaces of flat connections on 2-manifolds*, Duke Math. J. **77** (1995) 407-429.
7. Jeffrey, L.C., Kirwan, F.C., *Localization for nonabelian group actions*, Preprint alg-geom/9307001; Topology **34** (1995) 291-327.
8. Jeffrey, L.C., Kirwan, F.C., *Intersection theory on moduli spaces of holomorphic bundles on a Riemann surface*, Elec. Res. Notices AMS **1** (1995) 57-71.
9. Jeffrey, L.C., Weitsman, J. *Toric structures on the moduli space of flat connections on a Riemann surface: volumes and the moment map*, Advances in Math. **109** (1994) 151-168.
10. Jeffrey, L.C., Weitsman, J. *Torus actions and the topology and symplectic geometry of flat connections on 2-manifolds*, Topology, Geometry and Field Theory: Proceedings of Taniguchi Foundation Symposium on Low Dimensional Topology and Topological Field Theory (Kyoto, January 1993), ed. K. Fukaya, M. Furuta, T. Kohno, D. Kotschick, World Scientific, 1994.
11. Karshon, Y., *An algebraic proof for the symplectic structure of moduli space*, Proc. Amer. Math. Soc. **116** (1992) 591-605.
12. Karshon, Y., Tolman, S., *The moment map and line bundles over presymplectic toric manifolds*, J. Diff. Geom. **38** (1993) 465-484.
13. Mehta, V., Seshadri, C.S., *Moduli of vector bundles on curves with parabolic structure*, Math. Annalen **248** (1980) 205-239.
14. C.S. Seshadri, *Fibrés Vectoriels sur les Courbes Algébriques,* Astérisque **96** (1982).
15. Shulman, H.B., *Characteristic classes and foliations*, Ph.D. Thesis, University of California, Berkeley (1972).

16. Warner, F.W., *Foundations of Differentiable Manifolds and Lie Groups*, Springer-Verlag, 1983.
17. Weinstein, A., *The symplectic structure on moduli space*, in The Andreas Floer memorial volume (ed. H. Hofer, C. Taubes, A. Weinstein, E. Zehnder), Birkhäuser (Progress in Math. **133** (1995) 627-635.
18. E. Witten, *Two dimensional gauge theories revisited*, Preprint hep-th/9204083; J. Geom. Phys. **9** (1992) 303-368.

MATHEMATICS DEPARTMENT PRINCETON UNIVERSITY PRINCETON, NJ 08544, USA

AMS/IP Studies in Advanced Mathematics
Volume **2**, 1997 (Part 1)

AUTOMORPHISMS OF THE E_8 FOUR-MANIFOLD

ALLAN L. EDMONDS

1. Introduction.

Any integral, unimodular, symmetric bilinear form is realized as the intersection form of a closed, simply connected topological 4-manifold, according to a fundamental result of M. Freedman [11]. Moreover, any automorphism of the intersection form is realizable by a homeomorphism of the manifold (see Freedman [11] and Freedman and F. Quinn [12]). (Both of these results become dramatically false in the differentiable category, according to results of S. Donaldson–see Donaldson and P. Kronheimer [5].) The group of automorphisms of a definite form is necessarily finite, so it is natural to ask whether the topological realization of such an automorphism can be chosen to have finite order.

We examine this question in some detail here for the special case of the E_8 form and the corresponding manifold $M^4(E_8)$, one of the fundamental building blocks of manifolds in dimension 4. We determine exactly which automorphisms of the E_8 form can be realized by locally linear, periodic homeomorphisms of the E_8 4-manifold. It is already known that no automorphism of even order can be realized by a locally linear periodic homeomorphism (A. Edmonds [8], S. Kwasik and P. Vogel [14]). The possible odd orders that occur in $Aut(E_8)$ are 3,5,7,9, and 15. It turns out that there are automorphisms of periods 3 and 5 that are not topologically realizable in this sense. No automorphism of period 9 or 15 is realizable. Meanwhile every automorphism of period 7 and some of the automorphisms of periods 3 and 5 are indeed realizable.

The approach here is to identify the automorphisms of E_8 in an effective way, and in particular to identify the corresponding rank 8 integral representations they induce. One then examines the g-Signature Formula to rule out realizability in many cases. For realizability in most other cases we refer to recent work of Edmonds and J. Ewing [10], except in the more troublesome cases of period 3, which require extra attention.

1991 *Mathematics Subject Classification.* 57.
Research supported in part by the National Science Foundation

Theorem 1.1. *The only automorphisms of E_8 that can be realized by locally linear periodic maps on the E_8-manifold are of orders 3, 5, and 7. Such an automorphism is realizable if and only if it induces a permutation representation on the underlying* $\mathbf{Z}$*-module of* E_8.

2. The E_8 form.

We need a convenient way to understand this basic symmetric inner product space. The form E_8 can be described as the inner product space obtained by restricting the standard Euclidean inner product on $\mathbf{R}^8$ to the lattice spanned by the vectors $e_i + e_j$ and $\frac{1}{2}\sum e_i$, where $\{e_1, \dots, e_8\}$ denotes the standard basis of $\mathbf{R}^8$ on which the standard form is given by $e_i \cdot e_j = \delta_{i,j}$. One checks that this form is unimodular, positive definite, and of even type. It is the unique such form of rank 8 and it does not split orthogonally into smaller forms.

With respect to the basis $\{f_1, \dots, f_8\}$ given by $f_i = e_i + e_{i+1}$ for $i = 1, \dots, 6$, $f_7 = e_7 - e_8$, and $f_8 = \frac{1}{2}(e_1 - e_2 + e_3 - e_4 + e_5 + e_6 - e_7 - e_8)$ the matrix of the E_8 form is given by a well known 8 by 8 symmetric, unimodular matrix.

One may conveniently describe the E_8 form with respect to the basis $\{f_1, \dots, f_8\}$ by a standard graph or "Dynkin diagram" in which the nodes correspond to the basis vectors, and two nodes are connected by an edge if and only if the corresponding basis vectors have inner product 1.

3. The automorphism group of E_8.

The group of automorphisms of E_8 is a finite group, and the orientation preserving subgroup, modulo its center (which has order 2), is a simple group (denoted by $O_8^+(2)$ in the *Atlas of Finite Groups* (Conway *et al.* [2])) of order $174,182,400 = 2^{12} \cdot 3^5 \cdot 5^2 \cdot 7$ (compare J. Milnor and D. Husemoller [15], p. 26), while the full automorphism group then has order $2^{14} \cdot 3^5 \cdot 5^2 \cdot 7$. Any evenly signed permutation of the standard basis vectors then produces an element of $Aut(E_8)$. This produces a subgroup of $Aut(E_8)$ of order $2^7 \cdot 8! = 2^{14} \cdot 3^2 \cdot 5 \cdot 7$. Thus we see the entire 2-Sylow and 7-Sylow subgroups of $Aut(E_8)$ appearing, but we are missing part of the 3-Sylow and 5-Sylow subgroups. One has to look a little deeper to see the rest of the automorphism group. (One reason to emphasize this part of $Aut(E_8)$, however, is that it turns out to capture *all* that is realizable by locally linear group actions on the E_8 4-manifold.)

The E_8 graph described above can also be used to describe $Aut(E_8)$, as laid out in Coxeter and Moser [3], Chapter 9. We let R_i denote the reflection in the vector f_i defined by

$$R_i(v) = v - (v \cdot f_i)f_i$$

Now it turns out that $Aut(E_8)$ is generated by the eight reflections R_i and one can read off the relations among the R_i from the graph: Reflections corresponding to adjacent nodes have product of order 3; Reflections corresponding to nonadjacent nodes commute with each other.

An inductive consideration of similar groups generated by reflections corresponding to smaller such graphs leads to these facts and the computation of the order of $Aut(E_8)$.

The elements we shall need are mostly so-called "Coxeter elements" (the products of reflections corresponding to the nodes of a connected subdiagram), or powers of Coxeter elements, corresponding to natural subroups of $Aut(E_8)$. We have R_1R_2 of order 3; $R_1R_2R_3R_4$ of order 5; $R_1R_2R_3R_4R_5R_6$ of order 7; $R_1R_2R_3R_4R_5R_6R_7R_8$ of order 30; $R_1R_2R_3R_4R_5$ of order 6; $R_2R_3R_4R_5R_6R_7R_8$ of order 18; etc.

In particular, we have two commuting elements of order 3: $X_3 = R_1R_2$ and $Y_3 = R_1R_2R_4R_5$. We have an element of order 9: $Z_9 = (R_2R_3R_4R_5R_6R_7R_8)^2$. We will see that the element $Z_3 = Z_9^3$ is *not* conjugate into the subgroup generated by X_3 and Y_3. There is also $W_3 = (R_1R_2R_3R_4R_5R_6R_7R_8)^{10}$, another element of order 3. These four elements represent the four distinct conjugacy classes of elements of order 3 in $Aut(E_8)$.

We have elements of order 5: $X_5 = R_1R_2R_3R_4$, $Y_5 = (R_1R_2R_3R_4R_5R_6R_7R_8)^6$. These represent the two distinct conjugacy classes of elements of order 5. (We have not in this way succeeded to find *explicit* non-conjugate elements that commute with each other.)

Then we have an element $X_7 = R_1R_2R_3R_4R_5R_6$, which represents the unique conjugacy class of elements of order 7. Finally we have $(R_1R_2R_3R_4R_5R_6R_7R_8)^2$, an element of order 15.

A convenient reference for enumeration of conjugacy classes of elements in Coxeter groups is Carter [1].

4. Integral representations.

We describe here the possible rank 8 integral representations induced by cyclic groups C_3, C_5, and C_7 in $Aut(E_8)$. By a result of Reiner (Curtis and Reiner [4]) each integral representation of these groups can be expressed as a sum of copies of the group ring $\mathbf{Z}[C_p]$ of $\mathbf{Z}$-rank p, the trivial representation $\mathbf{Z}$ of $\mathbf{Z}$-rank 1, and the augmentation ideal I, the kernel of the augmentation homomorphism $\mathbf{Z}[C_p] \to \mathbf{Z}$, of $\mathbf{Z}$-rank $p-1$, and isomorphic to $\mathbf{Z}[\zeta_p]$, where $\zeta_p = e^{2\pi i/p}$. (This uses the fact that the ideal class group of $\mathbf{Z}[\zeta_p]$ vanishes for small primes; in general one must allow arbitrary ideals in $\mathbf{Z}[C_p]$.)

Proposition 4.1. *The following are the only possibilities for integral representations induced by C_7, C_5, and C_3 in $Aut(E_8)$:*
C_7: $\mathbf{Z}[C_7] \oplus \mathbf{Z}$
C_5: $\mathbf{Z}[C_5] \oplus \mathbf{Z}^3$ *and* $\mathbf{Z}[\zeta_5]^2$
C_3: $\mathbf{Z}[C_3] \oplus \mathbf{Z}^5$, $\mathbf{Z}[C_3]^2 \oplus \mathbf{Z}^2$, $\mathbf{Z}[C_3] \oplus \mathbf{Z}[\zeta_3]^2 \oplus \mathbf{Z}$, *and* $\mathbf{Z}[\zeta_3]^4$.

Proof. For $p = 3, 5$, or 7 such a representation would be of the form $\mathbf{Z}[C_p]^r \oplus \mathbf{Z}[\zeta_p]^s \oplus \mathbf{Z}^t$, where $pr + (p-1)s + t = 8$. By an algebraic result of I. Hambleton and C. Riehm [1978] s must be even. (If such a representation arises from an action of C_p, then s gives the first betti number of the fixed point set.) If $r = 0$, then one of s or t must be 0 as well. For otherwise the two summands give an invariant *orthogonal* splitting of the nonsplittable form E_8. The result follows. □

The four permutation representations listed in the proposition are easily seen to arise from the permutations of basis vectors of $\mathbf{R}^8$ as described above. More generally, all seven representations are realized by X_7, X_5, Y_5, X_3, Y_3, Z_3, and W_3, respectively. This amounts to a calculation, but one needs a little guidance about how to proceed. The main difficulty is that over the rationals one cannot distinguish $\mathbf{Z}[C_p]$ from $\mathbf{Z}[\zeta_p] \oplus \mathbf{Z}$, for which an explicit "cohomology" calculation is required.

Suppose that X is an $n \times n$ integral matrix such that $X^p = XX \cdots X = I$. Form the sum $Y = I + X + X^2 + \cdots + X^{p-1}$. Then apply elementary *integral* row and column operations to diagonalize Y. Each $\mathbf{Z}$-summand in the corresponding representation of C_p yields a 1×1 block (p); each $\mathbf{Z}[\zeta_p]$ (or ideal in $\mathbf{Z}[\zeta_p]$) yields a $(p-1) \times (p-1)$ block of zeroes; and each $\mathbf{Z}[C_p]$ yields a $p \times p$ block $diag(1, 0, \ldots, 0)$. These kinds of calculations are conveniently carried out with the computer algebra system Maple in which the integral diagonalization process is accomplished by the command *ismith.*

In addition we note that it follows from an inspection of the tables in Carter [1] that each of the integral representations listed above determines a unique conjugacy class. In particular, two elements of $Aut(E_8)$ of given period equal to 3, 5 or 7 are conjugate in $Aut(E_8)$ if and only if they induce equivalent linear representations (over the integers, the rationals, the real numbers, or the complex numbers), if and only if they have the same characteristic polynomial.

5. Automorphisms realizable by pseudofree actions.

A group action is called *pseudofree* if the fixed point set of each nontrivial group element is discrete. The possible automorphisms of E_8 of periods $p = 5$ or 7 that contain no $\mathbf{Z}[\zeta_p]$ are all realizable by locally linear, pseudofree actions, using the work of Edmonds and Ewing [10]. (The point is that standard actions on connected sums of copies of $\mathbf{C}P^2$ yield appropriate fixed point data, so that the conditions for realizability given in Edmonds-Ewing [10] are clearly satisfied.) Here is one precise statement (recall that an odd prime p is *regular* if the second factor of the class number of p is odd, e.g. $p < 163$):

Theorem 5.1. *(Edmonds and Ewing [10]) Let p be a regular prime greater than 3. Let Φ be a C_p-invariant symmetric, positive definite unimodular bilinear form on the $\mathbf{Z}[C_p]$-module $V = \mathbf{Z}[C_p]^r \oplus \mathbf{Z}^t$. Then Φ is realizable as the equivariant intersection form of a closed, simply connected 4-manifold with locally linear, pseudofree C_p action.*

See Edmonds and Ewing [10], Application 8.2. When $p = 3$, the corresponding statement becomes the following:

Complement 5.2. *Let Φ be a C_3-invariant symmetric, positive definite unimodular bilinear form on the $\mathbf{Z}[C_3]$-module $V = \mathbf{Z}[C_3]^r \oplus \mathbf{Z}^t$. Then Φ is realizable as the equivariant intersection form of a closed, simply connected 4-manifold with locally linear, pseudofree C_3 action if and only if $t \leq 1$.*

Proof. Suppose we have such an action. Let x denote the number of fixed points with local representation of "type" $(1, 1)$ and y the number of type $(1, -1)$. Then the g-Signature Formula would state that $t = \frac{1}{3}y - \frac{1}{3}x$ where $x + y = t + 2$. It then follows from $y \geq 0$ that $t \leq 1$. For the cases $t = 0$ and $t = 1$ one can easily construct actions on connected sums of copies of $\mathbf{C}P^2$ yielding appropriate fixed point data, so that the approach above applies. □

In particular the previous results do not suffice to realize any of the period 3 automorphisms of E_8. Any such realizations will necessarily have two-dimensional fixed point set.

Proposition 5.3. *An automorphism of prime period p of the E_8 form is realizable by a locally linear, pseudofree periodic homeomorphism of period p on the closed, simply connected 4-manifold $M^4(E_8)$ with intersection form E_8 if and only if $p = 7$, or $p = 5$ and the corresponding integral representation is of the form $\mathbf{Z}[C_5] \oplus \mathbf{Z}^3$.*

6. Nonrealizable automorphisms of prime periods.

In this section we will show that no automorphism of E_8 can be realized by a locally linear action if its underlying integral representation contains a $\mathbf{Z}[\zeta_p]$-summand. There will be several cases to consider.

Proposition 6.1. *There is no locally linear C_5 action on $M^4(E_8)$ whose representation on second homology is equivalent to $\mathbf{Z}[\zeta_5]^2$.*

Proof. If there were such an action, then the g-signature of a generator would be -2, implying that the fixed point set must be nonempty. Therefore the fixed point set would be a torus with Euler class $e \geq 0$, by Edmonds [1989]. The g-Signature Formula applied to an appropriate generator then states

$$-2 = -e\frac{4\zeta_5}{(\zeta_5 - 1)^2},$$

which contradicts the independence of 1, ζ_5, and ζ_5^2 in $\mathbf{Z}[\zeta_5]$. □

Corollary 6.2. *There is no homologically faithful, locally linear action of $C_5 \times C_5$ on $M^4(E_8)$.*

Proposition 6.3. *There is no locally linear action of C_3 on $M^4(E_8)$ whose representation on second homology is equivalent to $\mathbf{Z}[\zeta_3]^4$.*

Proof. The g-signature of a generator would be -4; the fixed point set would be a surface of genus 2. Then the g-Signature Formula applied to an appropriate generator would yield

$$-4 = -e\frac{4\zeta_3}{(\zeta_3 - 1)^2}.$$

From this it follows that $e = -3$, contradicting both evenness and positive definiteness of E_8. □

Proposition 6.4. *There is no locally linear action of C_3 on $M^4(E_8)$ whose representation on second homology is equivalent to $\mathbf{Z}[C_3] \oplus \mathbf{Z}[\zeta_3]^2 \oplus \mathbf{Z}$.*

Proof. In this case the g-signature of a generator would be -1. Then the fixed point set would be nonempty, and by Edmonds [1989] it would consist of a torus (with Euler class ≥ 0) and one isolated fixed point with local representation of type $(a, b) = (1, 1)$ or $(1, -1)$. The g-Signature Formula then yields

$$-1 = -e\frac{4\zeta_3}{(\zeta_3 - 1)^2} + \frac{({\zeta_3}^a + 1)({\zeta_3}^b + 1)}{({\zeta_3}^a - 1)({\zeta_3}^b - 1)},$$

yielding

$$-1 = -\frac{4e}{3} \pm \frac{1}{3}.$$

Then $e = 1$, contradicting evenness of the form E_8. □

Corollary 6.5. *There are automorphisms of periods 3 and 5 of the E_8 form that cannot be realized by locally linear periodic maps of the E_8 4-manifold.*

It is interesting to ask whether these automorphisms can be realized by purely topological periodic homeomorphisms. The fixed point set of such a homeomorphism would be a wildly embedded surface.

7. Automorphisms of prime period not realizable with two-dimensional fixed point sets.

We now consider the question whether any of the automorphisms considered above can be realized by locally linear actions with two-dimensional fixed set. We make use of the fact that the two-dimensional components of the fixed set represent nonzero homology classes, except possibly in the case when the fixed point set is connected, by Edmonds [8]. In the case at hand, this means that the two-dimensional components of the fixed point set have even, strictly positive, normal Euler classes or self-intersection numbers.

Proposition 7.1. *There is no locally linear C_7 action on $M^4(E_8)$ whose representation on second homology is equivalent to $\mathbf{Z}[C_7] \oplus \mathbf{Z}$ and which has a two-dimensional fixed set.*

Proof. The fixed point set would have to consist of a 2-sphere and an isolated point by Edmonds [8]. Let the normal bundle of the 2-sphere have Euler class $e \geq 2$. The g-Signature Formula then becomes

$$1 = -e\frac{4\zeta_7}{(\zeta_7 - 1)^2} + \frac{({\zeta_7}^a + 1)({\zeta_7}^b + 1)}{({\zeta_7}^a - 1)({\zeta_7}^b - 1)}$$

It then follows easily that $e = 1$ and $a = b = 1$. (See Dovermann [6] and, more generally, Wilczyński [17], who proved this in the context of actions on $\mathbf{C}P^2$.) This contradicts $e \geq 2$. □

Proposition 7.2. *There is no locally linear C_5 action on $M^4(E_8)$ whose representation on second homology is equivalent to $\mathbf{Z}[C_5] \oplus \mathbf{Z}^3$ and which has a two-dimensional fixed set.*

Proof. There are two cases. (i) The fixed point set consists of a 2-sphere and 3 isolated points; or (ii) The fixed point set consists of two 2-spheres and an isolated point.

In the first case the g-Signature Formula takes the trigonometric form

$$3 = e \cdot \csc^2(\frac{\pi}{5}) - \sum_{i=1}^{3} \cot(\frac{\pi a_i}{5}) \cot(\frac{\pi b_i}{5})$$

where e is the Euler class of the 2-sphere, and we assume without loss of generality that the generator of the action rotates the normal bundle by $2\pi/5$, and the rotation numbers at the 3 isolated fixed points are given by (a_i, b_i), $i = 1, 2, 3$.

A quick computer/calculator search, testing all values of a_i, and b_i between 1 and 4 inclusive yields only one case (up to relabeling) for which e can be an integer, and in that case $e = 1$, contradicting $e \geq 2$.

Case (ii) is split into two subcases. The first is when a chosen generator rotates the normal bundles of the two 2-spheres by the same amount. In this case the g-Signature Formula takes the trigonometric form

$$3 = E \cdot \csc^2(\frac{\pi b}{5}) - \cot(\frac{\pi a}{5}) \cot(\frac{\pi b}{5})$$

where E denotes the sum of the two Euler classes. Again a trivial computer search of all possible (a, b) yields no integral value for E.

It remains to consider the second subcase when the fixed point set consists of two 2-spheres, rotated by differing amounts, plus an isolated point. Without loss of generality the g-Signature Formula takes the trigonometric form

$$3 = e_1 \cdot \csc^2(\frac{\pi}{5}) + e_2 \cdot \csc^2(\frac{2\pi}{5}) - \cot(\frac{\pi a}{5}) \cot(\frac{\pi b}{5})$$

A calculator shows that $\csc^2(\frac{\pi}{5}) \approx 2.89443$, $\csc^2(\frac{2\pi}{5}) \approx 1.10557$, $\cot^2(\frac{\pi}{5}) \approx 1.89443$, $\cot^2(\frac{2\pi}{5}) \approx 0.105573$, and $\cot(\frac{\pi}{5})\cot(\frac{2\pi}{5}) \approx 0.447214$. It follows that $3 \geq 2e_1 + e_2 - 2 \geq 4 + 2 - 2 = 4$, a contradiction. □

8. Automorphisms of period 3 and representation type $\mathbf{Z}[C_3]^2 \oplus \mathbf{Z}^2$.

In this section we shall show that the automorphisms of E_8 of period 3 that induce permutation representations are indeed realizable by locally linear periodic maps, but with two-dimensional fixed point sets. We begin by determining what the fixed point sets and local representations must be, ruling out many possibilities along the way.

First consider the case of $\mathbf{Z}[C_3]^2 \oplus \mathbf{Z}^2$. By Proposition 5.3 the fixed point set must be 2-dimensional, and it must then consist either of (i) 2 S^2s or of (ii) an S^2 and 2 points.

Proposition 8.1. *There is no locally linear C_3 action on $M^4(E_8)$ whose representation on second homology is equivalent to $\mathbf{Z}[C_3]^2 \oplus \mathbf{Z}^2$ and which has a two-dimensional fixed set consisting of two 2-spheres.*

Proof. The g-Signature Formula takes the form

$$2 = -e\frac{4\zeta_3}{(\zeta_3 - 1)^2}$$

where e denotes sum of the two normal Euler numbers. A little algebraic simplification yields $2e = 3$, contradicting the integrality of the Euler number. □

Complement 8.2. *There is data satisfying the g-Signature Formula corresponding to a possible locally linear C_3 action on $M^4(E_8)$ whose representation on second homology is equivalent to $\mathbf{Z}[C_3]^2 \oplus \mathbf{Z}^2$ and which has a two-dimensional fixed set consisting of a 2-sphere and 2 isolated points, in which the 2-sphere has euler class 2 and the two isolated fixed points both have local representations of type $(1, 1)$.*

Proof. The normal bundle of the 2-sphere may be assumed to be rotated by $2\pi/3$ and have unknown Euler class e. The rotation numbers at the 2 fixed points are either $(1, 1)$ or $(1, 2)$. If the rotation numbers at the two fixed points are different, then they "cancel" in the g-Signature Formula, and as in the proof of Proposition 8.1 we again conclude that $2e = 3$, a contradiction. If both points have type $(1, 2)$, the g-Signature Formula yields, after simplification $e = 1$, contradicting $e \geq 2$. Finally, if both points have type $(1, 1)$, then the g-Signature Formula yields, after simplification, $e = 2$, as claimed. □

We now show that the preceding data can be realized.

Proposition 8.3. *There is a locally linear C_3 action on $M^4(E_8)$ whose representation on second homology is equivalent to $\mathbf{Z}[C_3]^2 \oplus \mathbf{Z}^2$ and whose fixed point set consists of a 2-sphere and two isolated points, in which the 2-sphere has Euler class 2 and the two isolated fixed points both have local representations of type $(1, 1)$.*

Proof. The construction will be similar to one for certain C_2 actions presented in Edmonds [1988], and will only be sketched. We will construct an equivariant handlebody obtained from the 4-ball with a standard C_3 action that rotates around a fixed 2-disk, with two invariant 2-handles, one of which is attached along the boundary of the fixed 2-disk, and six freely permuted 2-handles, realizing the equivariant E_8 form. The boundary of the resulting handlebody will be a homology 3-sphere, by Poincaré Duality and the unimodularity of E_8, with a free action action (by the construction) of C_3. By Freedman [1982] this homology sphere bounds a contractible topological 4-manifold. The only obstruction to extending the action locally linearly over this contractible 4-manifold is the α-invariant (measuring the failure of the g-Signature Formula to hold for actions on manifolds with nonempty boundary), which must equal that of some lens space corresponding to the local representation that would occur in the action on the contractible piece. (In general there is an additional Reidemeister torsion condition, which follows from the g-signature condition for small prime periods). The g-signature analysis above guarantees that this smooth 4-manifold can be capped off locally linearly with an action on a contractible manifold.

First we have to describe the equivariant form explicitly. As before, let $\mathbf{R}^8$ have standard basis $\{e_1, \dots , e_8\}$ with the standard inner product, with E_8 realized as the span of all $e_i + e_j$, together with $\frac{1}{2}\sum e_i$. Let a given generator $T \in C_3$ act on $\mathbf{R}^8$ by the rules $e_1 \to e_2 \to e_3 \to e_1$, $e_4 \to e_5 \to e_6 \to e_4$, $e_7 \leftrightarrow e_7$, $e_8 \leftrightarrow e_8$. Then T obviously preserves the standard integer lattice $\mathbf{Z}^8$ on which the induced C_3 representation is clearly $\mathbf{Z}[C_3]^2 \oplus \mathbf{Z}^2$. In addition, T preserves the E_8 lattice. It is slightly less clear that the corresponding integral representation on E_8 is the same. To see this explicitly, one may check that $e_1 + e_8$, $e_2 + e_8$,$e_3 + e_8$, e_4+e_5, e_5+e_6, e_6+e_4, $\frac{1}{2}\sum e_i$, e_7-3e_8 forms a basis for E_8. It then follows that the form restricted to $\mathbf{Z}^2$ is given by the 2×2 matrix $\left(\begin{smallmatrix} 2 & -1 \\ -1 & 10 \end{smallmatrix}\right)$. (Of course, this fixed summand is not unique, and the integral restriction of the form is thus only unique *mod* 3.

Now consider a standard C_3 action on the 4-ball D^4 in $\mathbf{C}^2$, fixing the 2-disk $D^2_1 = \{((z, 0) : |z| \leq 1\}$ and such that a chosen generator of C_3 rotates the complementary 2-disk $D^2_2 = \{((0, w) : |w| \leq 1\}$ through an angle of $2\pi/3$. Let $A = \partial D^2_1$ denote the "axis" of the action on S^3 and give it framing 2. Let $C = \partial D^2_2$ and give it framing 10. Orient A and C so that they have linking number -1. We attach 2-handles along these two framed knots equivariantly, producing a compact smooth 4-manifold W^4 with a smooth action of C_3 such that $H_2(W^4; \mathbf{Z}) = \mathbf{Z}^2$ with matrix of intersection numbers given by $\left(\begin{smallmatrix} 2 & -1 \\ -1 & 10 \end{smallmatrix}\right)$ with respect to the basis corresponding to the 2-handles. The fixed point set consists of a 2-sphere with self-intersection number 2 and an isolated fixed point of type $(1, 1)$. (Compare Corollary 11.2 below.) The action is free on ∂W. One now easily adds two free orbits of permuted 2-handles to extend

W^4 to a compact smooth 4-manifold V^4 with $H_2(V^4; \mathbf{Z}) = \mathbf{Z}^8$ on which the intersection matrix gives the E_8 form with respect to a suitable C_3 equivariant basis corresponding to the handles. The boundary of V^4 is a homology 3-sphere with a free C_3 action. By Edmonds [1987], this action extends topologically with one new fixed point over a contractible 4-manifold. The obstruction to extending it locally linearly is the Atiyah-Singer α-invariant, which vanishes here because of the observations in the preceding Complement. □

9. Automorphisms of period 3 and representation type $\mathbf{Z}[C_3] \oplus \mathbf{Z}^5$.

Here we examine the remaining case when $H_2 = \mathbf{Z}[C_3] \oplus \mathbf{Z}^5$ and the fixed point set is two-dimensional. There are three subcases, corresponding to whether the fixed point set contains one, two, or three 2-spheres, and correspondingly 5, 3, or 1 isolated points. Without loss of generality we may assume the two-dimensional fixed point components are oriented so that a chosen generator of C_3 acts by rotation by $2\pi/3$ in the normal bundles of all components. Letting E denote the total Euler number, the g-Signature Formula takes the trigonometric form

$$5 = E \cdot \csc^2(\frac{\pi}{3}) - \sum_{i=1}^{r} \cot(\frac{\pi a_i}{3}) \cot(\frac{\pi b_i}{3})$$

where r is the number of isolated fixed points, each $a_i = 1$, and $b_i = 1$ or 2.

Now $\csc^2(\frac{\pi}{3}) = \frac{4}{3}$, $\cot^2(\frac{\pi}{3}) = \frac{1}{3}$, and $\cot(\frac{\pi}{3}) \cot(\frac{2\pi}{3}) = -\frac{1}{3}$. A case-by-case analysis of possibilities, using the fact that E must be an even integer, yields only the case $E = 4$. This means either we have one fixed 2-sphere with Euler number 4 and 5 fixed points including 3 of type $(1,1)$ and 2 of type $(1,2)$; or we have two fixed 2-spheres, each with Euler number 2 and 3 fixed points, two of type $(1,1)$ and one of type $(1,2)$.

We can summarize this discussion as follows:

Proposition 9.1. *There is no locally linear C_3 action on $M^4(E_8)$ whose representation on second homology is equivalent to $\mathbf{Z}[C_3] \oplus \mathbf{Z}^5$ and whose fixed point set consists of of three 2-spheres and one isolated point.*

Complement 9.2. *There is data satisfying the g-Signature Formula, corresponding to a possible locally linear C_3 action on $M^4(E_8)$ whose representation on second homology is equivalent to $\mathbf{Z}[C_3] \oplus \mathbf{Z}^5$ and which has a two-dimensional fixed set consisting either of (i) a 2-sphere and 5 isolated points, in which the 2-sphere has Euler class 4 and three isolated* *fixed points have type* $(1,1)$ *and two have type* $(1,2)$*; or of (ii) two 2-spheres, each with Euler class 2, and three isolated fixed points, two of type* $(1,1)$ *and one of type* $(1,2)$.

We have been unable to realize an action of the second sort above. The following result shows that case (i) can arise.

Proposition 9.3. *There is a locally linear* C_3 *action on* $M^4(E_8)$ *whose representation on second homology is equivalent to* $\mathbf{Z}[C_3] \oplus \mathbf{Z}^5$ *and whose fixed point set consists of a 2-sphere and five isolated points, in which the 2-sphere has Euler class 2, three isolated fixed points have local representations of type* $(1,1)$*, and two have type* $(1,2)$.

Proof. We proceed much as in the construction in the proof of Proposition 8.3. Again we describe the equivariant form explicitly, starting with $\mathbf{R}^8$ and its standard basis $\{e_1, \ldots, e_8\}$ with the standard inner product, with E_8 realized as the span of all $e_i + e_j$, together with $\frac{1}{2}\sum e_i$. Let a given generator $T \in C_3$ act on $\mathbf{R}^8$ by the rules $e_1 \to e_2 \to e_3 \to e_1$, $e_4 \leftrightarrow e_4$, $e_5 \leftrightarrow e_5$, $e_6 \leftrightarrow e_6$, $e_7 \leftrightarrow e_7$, $e_8 \leftrightarrow e_8$. Then T obviously preserves the standard integer lattice $\mathbf{Z}^8$ on which the induced C_3 representation is clearly $\mathbf{Z}[C_3] \oplus \mathbf{Z}^5$. In addition, T preserves the E_8 lattice. We need to see that the corresponding integral representation on E_8 is the same. To do this we check that $e_1 + e_7$, $e_2 + e_7$,$e_3 + e_7$, $\frac{1}{2}(e_1 + e_2 + e_3 + e_4 + e_5 + e_6 + 3e_7 + e_8)$, $e_1 + e_2 + e_3 - e_4$, $e_1 + e_2 + e_3 - e_5$, $e_1 + e_2 + e_3 - e_6$, $e_7 - e_8$ forms a basis for E_8. It then follows that the form restricted to $\mathbf{Z}^5$ is given by the 5×5 matrix

$$\begin{pmatrix} 4 & 1 & 1 & 1 & 1 \\ 1 & 4 & 3 & 3 & 0 \\ 1 & 3 & 4 & 3 & 0 \\ 1 & 3 & 3 & 4 & 0 \\ 1 & 0 & 0 & 0 & 2 \end{pmatrix}$$

(Of course, this fixed summand is not unique, and the integral restriction of the form is thus only unique *mod* 3.)

Once again consider a standard action of C_3 on D^4 fixing $D^2_1 = \{(z,0) : |z| \leq 1\}$. Give the axis $A = \partial D^2_1$ framing $+4$. Choose 4 invariant meridian curves K_1, K_2, K_3, K_4 to A, oriented to have linking number $+1$ with A. Give the meridian curves framings 4,4,4, and 2, respectively. This almost realizes the form restricted to the fixed summand, except that the curves $K_1, \ldots, K_3$ are supposed to have mutual linking numbers equal to 3, not 0. We can easily rectify this by equivariantly dragging the 3 translates of a short arc on K_1 across K_2, equivariantly dragging the 3 translates of another short arc on K_1 across

K_3, and, finally, equivariantly dragging the 3 translates of yet another short arc on K_2 across K_3. (K_1, K_2, K_3 then form the components of a suitable torus link.)

We attach 2-handles along these five framed knots equivariantly, producing a compact smooth 4-manifold W^4 with a smooth action of C_3 such that $H_2(W^4; \mathbf{Z}) = \mathbf{Z}^5$ with matrix of intersection numbers given by the matrix above, with respect to the basis corresponding to the 2-handles. The fixed point set consists of a 2-sphere with self-intersection number 4 and 3 isolated fixed point of type $(1,1)$ and 1 isolated fixed point of type $(1,2)$. (Compare Corollary 11.2 below.) The action is free on ∂W. One now easily adds one free orbit of permuted 2-handles to extend W^4 to a compact smooth 4-manifold V^4 with $H_2(V^4; \mathbf{Z}) = \mathbf{Z}[C_3] \oplus \mathbf{Z}^5$ on which the intersection matrix gives the E_8 form with respect to a suitable C_3 equivariant basis corresponding to the handles. The boundary of V^4 is a homology 3-sphere with a free C_3 action. Again, the obstruction to extending it locally linearly is the Atiyah-Singer α-invariant of the generator on the boundary homology sphere, which vanishes here because of the calculations in the preceding Complement. Thus we finally arrive at the required action. □

10. Nonrealizable automorphisms of composite periods.

In $Aut(E_8)$ the possible non-prime odd orders are precisely 9 and 15.

Proposition 10.1. *No cyclic subgroup of $Aut(E_8)$ of order 9 can be realized by a locally linear group action on $M^4(E_8)$.*

Proof. We argue in this case that the third power of a generator T would be an element of order 3 that we have already shown cannot be realized. Now $x^9 - 1 = (x-1)(x^2+x+1)(x^6+x^3+1)$. Therefore, over the rationals, a matrix representation of T would be expressible as a block sum of the companion matrix for x^6+x^3+1 plus either that for x^2+x+1 or a rank 2 identity matrix. By explicit calculation then T^3 would induce a representation of C_3 of the form $\mathbf{Q}[\zeta_3]^3 \oplus \mathbf{Q}^2$. Comparing this information with our previous analysis of integral representations of C_3 arising from E_8, we see that the corresponding integral representation must be $\mathbf{Z}[C_3] \oplus \mathbf{Z}[\zeta_3]^2 \oplus \mathbf{Z}$, a nonrealizable type. □

Proposition 10.2. *No cyclic subgroup of $Aut(E_8)$ of order 15 can be realized by a locally linear group action on $M^4(E_8)$.*

Proof. In this case, C_{15} arises as a permutation representation in such a way that the subgroups of orders 3 and 5 are realizable. Let $T \in C_{15} \subset Aut(E_8)$ be a generator of order 15. The Lefschetz Fixed Point

Formula $\Lambda(T) = \chi(Fix(T))$ shows that $\chi(Fix(T)) = 2$. Our previous analysis shows that the fixed point set of T^3 is exactly 5 isolated points, on which the element T^5 of order 3 acts by cyclically permuting 3 points and fixing 2. On the other hand the fixed point set for the element T^5 of order 3 is either two 2-spheres and 3 points or one 2-sphere and 5 points. Then T^3 acts on $Fix(T^5)$ fixing 2 points. Each of the 2-spheres must be invariant. This would lead to too many fixed points for T in the former case. We conclude that $Fix(T^5)$ consists of a 2-sphere and 5 points, such that T^3 cyclically permutes the 5 points. We further conclude that the local C_3 representations at all 5 points must be the same, contradicting our earlier analysis of the local representations via the g-Signature Formula. □

This completes the analysis of single automorphisms and their realizability by periodic maps of the same period. We finish up by showing that one gains nothing by allowing the possibility of periodic maps of higher periods.

11. Periodic maps of higher periods.

We showed in Section 6 that certain automorphisms of E_8 of periods 3 and 5 are not realizable by locally linear periodic maps of periods 3 and 5, respectively. In Section 10 we further showed that the automorphisms of periods 9 and 15 are not realizable by locally linear periodic maps of periods 9 and 15, respectively. In fact these automorphisms are not realizable by locally linear periodic maps of *any* order. The arguments involved are minor extensions of the proofs of Sections 6 and 10. We leave unresolved whether the realizable automorphisms of periods 3, 5, and 7 are also realizable by locally linear maps of higher finite periods. But do note that by Edmonds [1987] a manifold such as $M^4(E_8)$ does admit locally linear periodic maps inducing the identity on homology.

We use the fact that the fixed point set of a homologically trivial action on a simply connected 4-manifold must consist of points and 2-spheres, with euler characteristic equal to that of the ambient manifold.

Proposition 11.1. *There is no locally linear action of C_{5n} on $M^4(E_8)$ inducing the representation $\mathbf{Z}[\zeta_5]^2$ on second homology.*

Proof. In Proposition 6.1 we handled the case $n = 1$, so we may suppose that $n > 1$. Let T be a periodic map generating such an action such that $T^{5n} = identity$ and T^5 induces the identity on homology. The fixed set $Fix(T)$ has euler characteristic 0 (since it is equal to the Lefschetz number of T), and consists of points and surfaces. On the other hand, $Fix(T)$ is contained in $Fix(T^5)$, which consists of points and 2-spheres.

Therefore $Fix(T) = \emptyset$. But the g-signature is -2, which implies that the fixed point set must be nonempty. $\square$

Proposition 11.2. *There is no locally linear action of C_{3n} on $M^4(E_8)$ inducing the representation $\mathbf{Z}[\zeta_3]^4$ on second homology.*

Proof. In Proposition 6.3 we handled the case $n = 1$, so we may suppose that $n > 1$. Let T be a periodic map generating such an action such that $T^{3n} = identity$ and T^3 induces the identity on homology. The fixed set $Fix(T)$ has euler characteristic -2, and consists of points and surfaces. On the other hand, $Fix(T)$ is contained in $Fix(T^3)$, which consists of points and 2-spheres. Therefore $Fix(T) = \emptyset$. But the g-signature is -4, which implies that the fixed point set must be nonempty. $\square$

Proposition 11.3. *There is no locally linear action of C_{3n} on $M^4(E_8)$ inducing the representation $\mathbf{Z}[C_3] \oplus \mathbf{Z}[\zeta_3]^2 \oplus \mathbf{Z}$ on second homology.*

Proof. In Proposition 6.5 we handled the case $n = 1$, so we may suppose that $n > 1$ and odd. Let T be a periodic map generating such an action such that $T^{3n} = identity$ and T^3 induces the identity on homology. The fixed set $Fix(T)$ has euler characteristic $+1$, and consists of points and surfaces. On the other hand, $Fix(T)$ is contained in $Fix(T^5)$, which consists of points and 2-spheres. Therefore $Fix(T) =$ a single point. Meanwhile the g-signature is -1. This leads immediately to a contradiction to the g-signature formula. $\square$

Proposition 11.4. *There is no locally linear action of C_{9n} on $M^4(E_8)$ inducing a representation of period 9 on homology.*

Proof. As in Proposition 10.1, the representation of the third power would be of type $\mathbf{Z}[C_3] \oplus \mathbf{Z}[\zeta_3]^2 \oplus \mathbf{Z}$, which is already known to be unrealizable in the extended sense, by Proposition 11.3. $\square$

Proposition 11.5. *There is no locally linear action of C_{15n} on $M^4(E_8)$ inducing a representation of period 15 on homology.*

Proof. In Proposition 10.2 we dealt with the case $n = 1$, so we may suppose that $n > 1$. Let T be a periodic map generating such an action such that $T^{15n} = identity$ and T^{15} induces the identity on homology, but no smaller power of T induces the identity on homology. As before, the representation on homology is the direct sum of a permutation of order 3 and a permutation of order 5. The g-signature of T is clearly 0. We have $\chi(Fix(T)) = \Lambda(T) = 2$, and $Fix(T) \subset Fix(T^{15}$, which is a union of 2-spheres and points, with euler characteristic 10. It follows that $Fix(T)$ is either a 2-sphere or two points. If it were a 2-sphere, then it would be a component of $Fix(T^{15})$, and hence represents a nontrivial homology class, by Edmonds [1989], with nonzero self-intersection. But this would contradict the g-signature formula, since we already know the g-signature is 0. Therefore we conclude that $Fix(T)$ consists of exactly two points, with fixed point data (a, b), and (c, d), say. The g-Signature Formula says that

$$\cot(\frac{a\pi}{15n})\cot(\frac{b\pi}{15n}) + \cot(\frac{c\pi}{15n})\cot(\frac{d\pi}{15n}) = 0$$

Since the action is not semifree, this alone does not imply that the fixed point data "cancels". The plan now is to relate $Fix(T)$ to $Fix(T^{5n})$ to reach a contradiction. There are two cases, depending on whether $3|n$ or $3 \not| n$.

We first consider the easier case when 3 does not divide n. Then T^{5n} is a locally linear map of period 3 inducing the integral representation of type $\mathbf{Z}[C_3] \oplus \mathbf{Z}^5$. From Complement 9.2 we know that there are two possibilities for the fixed point data of T^{5n}:

1. $S^2(4) + 3(1,1) + 2(1,2)$ or
2. $2S^2(2) + 2(1,1) + 1(1,2)$

Here $aS^2(b)$ indicates a disjoint 2-spheres, each with normal Euler class b; $a(b,c)$ indicates a isolated fixed points with local representation of type (b, c). In each case T acts on $Fix(T^{5n})$ with fixed set exactly $Fix(T)$. Since n is known to be odd, this clearly leads to too many fixed points for T.

In the remaining case when $3|n$, and hence T^{5n} acts homologically trivially, we need to assemble the possibilities for $Fix(T^{5n})$. □

Lemma 11.6. *The following are the possibilities consistent with the g-Signature Formula for the fixed point data of a homologically trivial locally linear action of C_3 on $M^4(E_8)$:*

1. $S^2(4) + 8(1,2)$
2. $S^2(16) + 8(1,1)$
3. $3S^2(2) + 2(1,1) + 2(1,2)$
4. $S^2(2) + S^2(4) + 3(1,1) + 3(1,2)$
5. $S^2(6) + 4(1,1) + 4(1,2)$

Proof. The fixed point set consists of k 2-spheres, each nontrivial in homology, with positive even self-intersection, plus x isolated points of local representation type $(1,1)$ and y points of local representation type $(1,2)$, where $x + y + 2k = 10$. The g-Signature Formula then states, in this special case: $-\frac{1}{3}x + \frac{1}{3}y + \frac{4}{3}e = 8$ Here e denotes the sum of the self-intersections, hence is even and at least $2k$. It is now straightforward to analyze the possibilities. (It helps to note that $x \equiv y \ mod\ 8$.) □

Now we return to complete the proof of the second and last case of Proposition 11.5. The map T acts on $\mathrm{Fix}(T^{5n})$ with fixed point set $Fix(T)$. On this set it has period a divisor of $5n$, where $3|n$.

In Case 4, the two 2-spheres are necessarily T-invariant and give rise to at least 4 fixed points for T, contradicting our knowledge that $Fix(T)$ consists of just two points. In Case 3, the 3 2-spheres could conceivably be permuted, but the two points of type $(1,1)$ and the two points of type $(1,2)$ would necessarily be fixed, again leading to too many fixed points. In Case 5 the single invariant 2-sphere gives rise to 2 fixed points, and we also get at least one fixed point of each of the two groups of size 4–again a contradiction.

In Cases 1 and 2, however, the 2-spheres must be T-invariant, leading to the required two fixed points in each case. The map T would permute the 8 like fixed points in two cycles of lengths 3 and 5.

Because the two elements of $Fix(T)$ lie on an invariant 2-sphere, the fixed point data must be of the form (a,b) and $(-a,d)$, where a is divisible by 3. The g-signature formula

$$\cot(\frac{a\pi}{15n})\cot(\frac{b\pi}{15n})+\cot(\frac{-a\pi}{15n})\cot(\frac{d\pi}{15n})=0$$

from above implies that $\cot(b\pi/15n) = \cot(d\pi/15n)$, hence that $d \equiv b \bmod 15n$ (and the data cancels).

On the other hand, because the two fixed points lie on an invariant 2-sphere of Euler class 4 or 16, we have $d \equiv b - 4a \bmod 15n$ or $d \equiv b - 16a \bmod 15n$. In either case we conclude that $a \equiv 0 \bmod 15n$, since $15n$ is odd, contradicting the fact that T rotates the 2-sphere S nontrivially.

12. Concluding questions.

Can the data for C_3 actions with fixed point set containing two 2-spheres be realized? Does $C_3 \times C_3$ act locally linearly and homologically faithfully on $M^4(E_8)$? We have been able to show that no other non-cyclic group of automorphisms can be realized.

What can one say about purely topological actions? For example, is $\mathbf{Z}[\zeta_5]^2$ realizable topologically? In the case of purely topological actions one would want to consider elements of even order also, leading to many more possibilities.

Finally, we conjecture on the basis of the results here that realizable automorphisms of positive definite simply connected 4-manifolds induce (signed) permutation representations on second homology.

References

1. R.W. Carter, *Conjugacy classes in the Weyl group*, Comp. Math. **25** (1972) 1–59.
2. J.H. Conway, et al. *Atlas of Finite Groups*, Oxford Univ. Press, 1985.
3. H.S.M. Coxeter and W.J.R. Moser, *Generators and Relations for Discrete Groups*, 4th Ed., Springer, New York, 1980.
4. C. W. Curtis and I. Reiner, *Representation Theory of Finite Groups and Associative Algebras*, Wiley Interscience, New York, 1962.
5. S. K. Donaldson and P.B. Kronheimer, *The Geometry of Four-Manifolds*, Oxford Univ. Press, 1990.
6. K. H. Dovermann, *Rigid cyclic group actions on cohomology complex projective spaces*, Math. Proc. Camb. Phil. Soc. **101** (1987) 487-507.
7. A. L. Edmonds, *Construction of group actions on four-manifolds*, Trans. Amer. Math. Soc. **299** (1987) 155-170.
8. A. L. Edmonds, *Aspects of group actions on four-manifolds*, Topology Appl. **31** (1989) 109-124.
9. A. L. Edmonds, *Involutions on odd four-manifolds*, Topology Appl. **30** (1988) 43-49.
10. A. L. Edmonds and J. H. Ewing, *Realizing forms and fixed point data in dimension four*, Amer. J. Math. **114** (1992) 1103-1126.
11. M. H. Freedman, *The topology of four-dimensional manifolds*, J. Diff. Geom. **17** (1982) 357-453.
12. M. H. Freedman and F. Quinn, *Topology of 4-Manifolds*, Princeton Univ. Press, 1990.
13. I. Hambleton and C. Riehm, *Splitting of Hermitian forms over group rings*, Invent. Math. **45** (1978) 19-33.
14. S. Kwasik and P. Vogel, *Asymmetric four-dimensional manifolds*, Duke Math. J. **53** (1986) 759-764.
15. J. W. Milnor and D. Husemoller, *Symmetric Bilinear Forms*, Springer, New York, 1973.
16. J. P. Serre *Representations of finite groups*, Springer, New York, 1980.
17. D. W. Wilczyński *Group actions on the complex projective plane*, Trans. Amer. Math. Soc. **303** (1987) 707-731.

Department of Mathematics, Indiana University, Bloomington, IN 47405

E-mail address: edmonds@indiana.edu

AMS/IP Studies in Advanced Mathematics
Volume **2**, 1997 (Part 1)

ON THE STAR-CONSTRUCTION FOR TOPOLOGICAL 4-MANIFOLDS

PETER TEICHNER

ABSTRACT. We discuss existence and uniqueness of the $*$-construction which reverses the Kirby-Siebenmann invariant of a topological 4-manifold while fixing the homotopy type. In particular, we point out an error in Theorem 10.3 of [4] where a certain $*$-partner was mistakenly proven not to exist. The existence of this 4-manifold will follow from our Theorem 1 which proves that for fundamental group $\mathbb{Z}/2$ all possible isometries of the equivariant intersection form are realized by homeomorphisms.

1. INTRODUCTION.

One of the main surprises of Mike Freedman's classification of closed simply connected topological 4-manifolds in [3] was the existence of a manifold homotopy equivalent to $\mathbb{CP}^2$ but not homeomorphic to the same. In [4], Freedman and Quinn denote this manifold by $*\mathbb{CP}^2$ and use it to construct many more $*$-*partners* of 4-manifolds. More precisely, in [4, Thm.10.3] the authors prove that under the assumption that the universal covering of M is not spin (and the fundamental group is good) there exists a decomposition

$$(D) \qquad M\# * \mathbb{CP}^2 \cong *M\#\mathbb{CP}^2$$

which preserves the decomposition of π_2. By construction, M and $*M$ are homotopy equivalent and have opposite Kirby-Siebenmann invariants.

Freedman and Quinn proceed to define $*M := M$ if the universal covering of M is spin because they claim that in this case a decomposition as above does not exist, see the last sentence in [4, Thm.10.3(1)].

1991 *Mathematics Subject Classification.* Primary: 57N13 Secondary: 55N22.

Supported by a Feodor Lynen Fellowship from the Humboldt foundation

However, we will show that this sentence cannot be correct by proving the following result. (It is understood that a decomposition (D) always preserves the decomposition of π_2.)

Proposition 1. *Let E be the quotient of $S^2 \times S^2$ by the free involution sending (x,y) to $(-x,-y)$. Then a decomposition (D) as above exists for $M := E$.*

We would like to remark that in [12] Richard Stong found independently a mistake in the proof of [4, Thm.10.3] and gave a corrected statement which confirms our result on the existence of $*E$. Our methods differ considerably from those of Freedman, Quinn and Stong in that we make use of a modified surgery approach developed by Matthias Kreck in [11]. Proposition 1 will be implied by the following result which seems interesting for its own sake.

Theorem 1. *Let M_1 and M_2 be closed oriented 4-manifolds with fundamental group $\mathbb{Z}/2$ and with the same Kirby-Siebenmann invariants. Then any isometry $\pi_2 M_1 \to \pi_2 M_2$ of intersection forms which preserves the k-invariant is induced by an orientation preserving homeomorphism.*

Ian Hambleton pointed out to me that this result also follows from the classical surgery sequence, the homotopy theory developed in [5] and the classification result in [6]. Since this path of proof is very long it seems interesting to present the completely different approach via Kreck's modified surgery technique.

Let me mention the following subtlety in the definition of $*M$: In [7] we prove the existence of a manifold R homotopy equivalent to $\mathbb{RP}^4$ but with nontrivial Kirby-Siebenmann invariant. We also show that $R\#\mathbb{CP}^2$ is not homeomorphic to $\mathbb{RP}^4 \# *\mathbb{CP}^2$ which means that Theorem 1 is not true in the nonoriented case and that R is not a $*$-partner of $\mathbb{RP}^4$ in the sense of admitting a decomposition (D). However, in the classification of nonoriented 4-manifolds with fundamental group $\mathbb{Z}/2$ (see [7]), R plays a similar role as $*\mathbb{CP}^2$ does in the simply connected case. Similarly, in the known classification results for other fundamental groups (see [6] or [15]) it is not necessary to know that a decomposition (D) exists as long as one knows that one can reverse the Kirby-Siebenmann invariant while fixing the homotopy type of a given manifold. It seems safe to make the following

Definition. Set $*M := M$ if there exists no 4-manifold in the homotopy type of M which has the opposite Kirby-Siebenmann invariant.

It is wellknown that this definition applies if M is a spin manifold because in this case the homotopy type of M determines the Kirby-Siebenmann invariant by the formula (see [10])

$$ks(M) \equiv \sigma(M)/8 \mod 2$$

where $\sigma(M)$ denotes the signature of the intersection form on $H_2(M;\mathbb{Z})$.

It seemed not to be known whether our definition applies for any nonspin 4-manifold. The following examples answer this question.

Proposition 2. *If M is either the nontrivial S^3-bundle over S^1 or the nontrivial orientable S^2-bundle over a surface of genus > 0 then any manifold homotopy equivalent to M has trivial Kirby-Siebenmann invariant. In fact, it is stably homeomorphic to M, i.e. homeomorphic up to connected sum with copies of $S^2 \times S^2$.*

Let me now turn to the uniqueness question of the $*$-construction which is certainly one of the most desirable properties. If the fundamental group is good and the universal covering of M is not spin, Stong shows in [13] that the decomposition (D) determines $*M$ uniquely. (In the simply connected case this follows directly from Freedman's classification result [3].) Unfortunately, the following example shows that such a uniqueness statement cannot hold in general. First note that for $M := E\# * E$, E as in Proposition 1, the two manifolds $E\#E$ and $*E\# * E$ both satisfy a decomposition (D). Therefore, uniquess of the $*$-construction fails by the following result.

Proposition 3. *The manifolds $E\#E$ and $*E\#*E$ are not stably homeomorphic.*

I would like to thank Matthias Kreck for helpful discussions concerning Theorem 1.

2. Proofs of Propositions 2 and 3.

Propositions 2 and 3 are consequences of Kreck's stable classification theorem (see [11] or [15]). We will summarize the relevant definitions and results. Let M be a manifold with fundamental group π and universal covering spin. Choosing a 2-equivalence $u : M \to K(\pi,1)$, the homotopy fibration $\widetilde{M} \xrightarrow{p} M \xrightarrow{u} K(\pi,1)$ induces an exact sequence

$$0 \to H^2(\pi;\mathbb{Z}/2) \xrightarrow{u^*} H^2(M;\mathbb{Z}/2) \xrightarrow{p^*} H^2(\widetilde{M};\mathbb{Z}/2).$$

Since $0 = w_2\widetilde{M} = p^*(w_2M)$, there exists a unique element $w_2 \in H^2(\pi;\mathbb{Z}/2)$ with $w_2M = u^*(w_2)$. Clearly there also is a unique element $w_1 \in H^1(\pi;\mathbb{Z}/2)$ with $w_1M = u^*(w_1)$. The triple (π, w_1, w_2) is determined by M up to automorphisms of π. We call the isomorphism class of this triple the *w-type* of M and define the concept of (topological) (π, w_1, w_2)-*bordism groups* as follows. The classes w_1, w_2 give a fibration $(w_1, w_2) : K(\pi, 1) \to K(\mathbb{Z}/2, 1) \times K(\mathbb{Z}/2, 2)$ and we can form the pullback

$$\begin{array}{ccc} B(\pi, w_1, w_2) & \longrightarrow & K(\pi, 1) \\ {\scriptstyle \xi(\pi, w_1, w_2)} \downarrow & & \downarrow {\scriptstyle (w_1, w_2)} \\ BTOP & \xrightarrow{w_1(\gamma) \times w_2(\gamma)} & K(\mathbb{Z}/2, 1) \times K(\mathbb{Z}/2, 2) \end{array}$$

where γ denotes the stable universal micro-bundle over $BTOP$, the classifying space of stable micro-bundles. Now the bordism groups

$$\Omega_n(\pi, w_1, w_2) := \Omega_n(\xi(\pi, w_1, w_2))$$

can be defined as in [14]. They consist of bordism classes of triangles

$$\begin{array}{ccc} & {\scriptstyle \tilde{\nu}} & B(\pi, w_1, w_2) \\ & \nearrow & \downarrow {\scriptstyle \xi(\pi, w_1, w_2)} \\ M & \underset{\nu}{\to} & BTOP \end{array}$$

where M is a closed manifold and ν is the stable normal Gauß map of M given by some embedding into Euclidean space. The map $\tilde{\nu}$ is called a (π, w_1, w_2)-*structure* on M. If $\tilde{\nu}$ is a 2-equivalence, it is called a *normal 1-smoothing*. The following result is contained in [11].

Theorem. *Two closed 4-manifolds with the same w-type (π, w_1, w_2) are stably homeomorphic if and only if they admit (π, w_1, w_2)-bordant normal 1-smoothings. Moreover, any element in $\Omega_n(\pi, w_1, w_2)$ can be represented by a 4-manifold with w-type (π, w_1, w_2) together with a normal 1-smoothing.*

We remark that two such normal 1-smoothings always differ by a fibre homotopy equivalence of the fibration $\xi(\pi, w_1, w_2)$. Therefore, a 4-manifold with w-type (π, w_1, w_2) which is (π, w_1, w_2)-zero-bordant can never be stably homeomorphic to a 4-manifold which does not bound in $\Omega_4(\pi, w_1, w_2)$.
This applies in particular to the manifolds $E\#E$ and $*E\#*E$ of Proposition 3. Note that E is an S^2-bundle over $\mathbb{RP}^2$ and the corresponding D^3-bundle is a $(\mathbb{Z}/2, 0, x^2)$-bordism (here $x \in H^1(\mathbb{Z}/2; \mathbb{Z}/2)$ is a generator). The boundary connected sum of two copies of it provides a $(\mathbb{Z}/2 * \mathbb{Z}/2, 0, x_1^2 + x_2^2)$-bordism for $E\#E$. We will prove below that $*E\# * E$ is not zero in this bordism group which will finish the proof of Proposition 3.

But first we need to describe a method for computing the above bordism groups. The first step is to note that for any fibration $\xi : B \to BTOP$ the Pontrjagin-Thom construction gives an isomorphism

$$\Omega_n(\xi) \cong \pi_n^{st}(M\xi)$$

where $M\xi$ is the Thom spectrum for ξ with its bottom cell in dimension zero. In all the cases considered in this paper there exists a vector bundle E over $K(\pi,1)$ such that $w_i(E) = w_i$ for $i = 1,2$. Then the fibration $\xi(\pi, w_1, w_2)$ is fibre homotopy equivalent [15, Ch.5] to

$$\xi := \gamma \oplus E : B := BSpinTOP \times K(\pi,1) \longrightarrow BTOP$$

where γ is the universal spin micro-bundle over $BSpinTOP$. But this implies that we have a homotopy equivalence

$$M\xi \simeq M\gamma \wedge ME$$

and thus get an Atiyah-Hirzebruch spectral sequence (with respect to the homology theory associated to the spectrum $M\gamma$)

$$E^2_{p,q} = H_p(ME; \Omega_q^{SpinTOP}) \Longrightarrow \Omega_{p+q}^{SpinTOP}(ME) \cong \Omega_{p+q}(\xi).$$

Finally, using the Thom isomorphism (for ordinary homology), the E^2-term becomes $H_p(\pi; \Omega_q^{SpinTOP})$ where the coefficients are twisted via w_1. This also implies (compare [16]) that the d_2-differentials for $q \leq 2$ are dual to the homomorphisms sending $x \in H^p(\pi;\mathbb{Z}/2)$ to $Sq^2(x) + Sq^1(x) \cdot w_1 + x \cdot w_2$. This information implies the following

Lemma 2. *The signature and Kirby-Siebenmann invariant induce an isomorphism*

$$\Omega_4(\mathbb{Z}/2, 0, x^2) \cong 8 \cdot \mathbb{Z} \times \mathbb{Z}/2.$$

Moreover, together with an additional $\mathbb{Z}/2$-valued invariant they induce an isomorphism

$$\Omega_4(\mathbb{Z}/2 * \mathbb{Z}/2, 0, x_1^2 + x_2^2) \cong 8 \cdot \mathbb{Z} \times \mathbb{Z}/2 \times \mathbb{Z}/2.$$

Now fix $(\mathbb{Z}/2, 0, x^2)$-structures on E and $*E$ which are normal 1-smoothings and always use their connected sum as $(\mathbb{Z}/2 * \mathbb{Z}/2, 0, x_1^2 + x_2^2)$-structures on the connected sum manifolds. Consider the two inclusions $f_i : \mathbb{Z}/2 \to \mathbb{Z}/2 * \mathbb{Z}/2$. There are induced homomorphisms

$$f_{i*} : \Omega_4(\mathbb{Z}/2, 0, x^2) \longrightarrow \Omega_4(\mathbb{Z}/2 * \mathbb{Z}/2, 0, x_1^2 + x_2^2)$$

which send $*E$ to $E \# *E$ respectively $*E \# E$. A comparison of the corresponding spectral sequences implies that these two elements are generators of the two $\mathbb{Z}/2$-summands in this bordism group. Therefore, the sum of these two elements, which is bordant to $*E\# * E$, is nontrivial. $\square$

Now let $\pi(g)$ be the fundamental group of a surface of genus $g \geq 0$. Then $H^2(\pi(g);\mathbb{Z}/2) = \mathbb{Z}/2$ so there is a unique orientable but nonspin w-type for $\pi(g)$. Similarly, $H^1(\mathbb{Z};\mathbb{Z}/2) = \mathbb{Z}/2, H^2(\mathbb{Z};\mathbb{Z}/2) = 0$ implies that there is a unique nonspin w-type for fundamental group $\mathbb{Z}$. Proposition 2 immediately follows from the following

Theorem 3. *If M is one of the manifolds in Proposition 2, the bordism groups of its w-type are*

$$\Omega_4(\mathbb{Z}, \neq 0, 0) \cong \mathbb{Z}/2 \text{ and } \Omega_4(\pi(g), 0, \neq 0) \cong 8 \cdot \mathbb{Z} \times \mathbb{Z}/2$$

where the isomorphisms are given by the Kirby-Siebenmann invariant respectively the signature and the Kirby-Siebenmann invariant. Moreover, the intersection form on π_2 induces (additive) isomorphisms with Wall's L-groups

$$\Omega_4(\pi, w_1, w_2) \cong L_4(\pi, w_1).$$

In particular, the stable intersection form on π_2 determines the stable homeomorphism type of 4-manifolds in the w-type of M.

Proof. The computation of the bordism groups is a straightforward application of the methods described above. The second part splits naturally into two cases.

Case 1: M is the nontrivial S^3-bundle over S^1.

Then M and $M\#E_8$ constitute all stable homeomorphism classes in the w-type $(\mathbb{Z}, \neq 0, 0)$. Therefore, π_2 is always a stably free $\mathbb{Z}[\mathbb{Z}]$-module and the intersection form is nondegenerate. Since it uniquely determines a quadratic refinement, the class in $L_4(\mathbb{Z}, \neq 0)$ is welldefined. But one knows from [17] that this L-group is isomorphic to $\mathbb{Z}/2$, generated by $E_8 \otimes \mathbb{Z}[\mathbb{Z}] = \pi_2(M\#E_8)$.

Case 2: M is the nontrivial orientable S^2-bundle over a surface of genus g.

Then $\pi_2(M) = \mathbb{Z}$ is *not* a free $\pi(g)$-module but a trivial module. Also the intersection form vanishes for M. We claim that any 4-manifold with the given w-type has the trivial module $\mathbb{Z}$ as the radical of its intersection form on π_2 and the form on the quotient is nondegenerate on a stably free $\pi(g)$-module with a unique quadratic refinement. This defines our map to $L_4(\pi(g), 0)$. To prove our claim, let $X \to M$ be a degree 1 normal map with signature 0 and $ks(X) \neq 0$ (which always exists by [10, p.329] if M is orientable but nonspin). Then $M\#W$ and $X\#W$, W a simply connected closed spin 4-manifold, constitute all stable homeomorphism classes in the w-type $(\pi(g), 0, \neq 0)$. But the kernel on π_2 of a degree 1 normal map is always a nondegenerate form on a stably free module which proves our claim for X and thus for all manifolds in our w-type. Using the fact that the addition in the bordism group is realized by connected sum along the 1-skeleton [11] of manifolds with the correct w-type, it is not hard to check that our map to L_4 is a homomorphism. It is well known [1] that for surface groups the assembly map gives an isomorphism ($w_1 = 0$ in all cases and $g > 0$ as before)

$$L_4(\pi(g)) \cong L_4(1) \times L_2(1) \cong 8 \cdot \mathbb{Z} \times \mathbb{Z}/2$$

induced by the signature and the codimension 2 Arf invariant. But this Arf invariant equals $\sigma(N)/8 + ks(N)$ if N has the correct w-type. This finishes our proof. □

3. Proof of Theorem 1.

Our proof will not use the more difficult classification of oriented 4-manifolds with fundamental group $\mathbb{Z}/2$ in [6] in terms of the intersection form on $H_2(.; \mathbb{Z}/2)$. Instead, we will work along the lines of the method of Kreck in [11, Ch.13] who proved the analogue of Theorem 1 for certain *odd order* fundamental groups, e.g. odd order groups in which every Sylow subgroup has an abelian normal subgroup with cyclic quotient. However, our bordism calculations are much more involved in that we have to prove the existence of a nontrivial d_3-differential in a certain Atiyah-Hirzebruch spectral sequence.

Let me now show how Theorem 1 implies Proposition 1: Let M be a closed nonspin 4-manifold such that $\pi_1(M)$ is finite with a 2-Sylow subgroup of periodic cohomology. Then a result of [8] on the Wall obstruction for closed manifold surgery problems implies the existence of a manifold M' homotopy equivalent to M but with opposite Kirby-Siebenmann invariant. For details of this argument see [15, Ch.8]. In particular, we get a 4-manifold $E' \simeq E$ with $ks(E') \neq 0$. This gives a homotopy equivalence

$$E \# * \mathbb{CP}^2 \longrightarrow E' \# \mathbb{CP}^2$$

which preserves the decomposition on π_2. Since any homotopy equivalence preserves the k-invariant, Theorem 1 implies that the same decomposition of π_2 is induced by a homeomorphism. This proves Proposition 1 and shows the existence of $*E$. Note that by construction E' and $*E$ are homotopy equivalent and a second application of Theorem 1 proves that they are homeomorphic. □

Proof of Theorem 1. In order to control π_2 we have to climb up one more step in the Postnikov tower of the stable normal Gauß map of an oriented 4-manifold M. To do so, let X denote a CW-complex with trivial homotopy groups in dimensions 3 and higher and admitting a 3-equivalence $u : M \to X$. Then X is determined up to homotopy by M, in fact by the triple $(\pi_1 M, \pi_2 M, k(M))$ where $\pi_2 M$ is considered as a $\pi_1 M$-module and $k(M) \in H^3(\pi_1 M; \pi_2 M)$ is the first k-invariant of M classifying the fibration

$$(*) \qquad K(\pi_2 M, 2) \to X \to K(\pi_1 M, 1).$$

Using obstruction theory, it is easy to see that any automorphism of the above triple is induced by a self-homotopy equivalence of X. In particular, given an isomorphism of triples

$$(\pi_1 M_1, \pi_2 M_1, k(M_1)) \cong (\pi_1 M_2, \pi_2 M_2, k(M_2))$$

we may choose 3-equivalences $u_i : M_i \to X$ such that $u_{1*} \circ u_{2*}^{-1}$ induces the isomorphism. If we assume further that this isomorphism is an isometry of intersection forms on π_2 and that π_1 is cyclic then by [5] we also have

$$u_{1*}[M_1] = u_{2*}[M_2] \in H_4(X; \mathbb{Z}).$$

Let $\xi : BSpinTOP \times X \xrightarrow{\gamma \oplus L} BSTOP$ be a fibration coming from the classifying map of the Whitney sum of the universal spin micro-bundle γ and a complex line bundle L over X with $w_2(L) = u^*(w_2 M)$. By construction, the stable normal Gauß map $\nu : M \to BSTOP$ lifts over ξ and any such lift $\tilde{\nu}$ is called a *normal2-structure* of M. If $\tilde{\nu}$ is a 3-equivalence, it is called a *normal 2-smoothing*. Note that the difference bundle $\nu(M) - u^*(L)$ is spin and thus we obtain a normal 2-smoothing of the form $\tilde{\nu} = (s, u)$ with s some spin structure on this difference. We will use the following theorem from [11]. □

Theorem. *Let π_1 be a good group with $L_5^s(\pi_1) = 0$. If there exist ξ-bordant normal 2-smoothings $(M_i, \tilde{\nu}_i)$ then there exists an orientation preserving homeomorphism $h : M_1 \to M_2$ over X. In particular, h induces the given isometry on π_2.*

Noting that by [17] $L_5^s(\mathbb{Z}/2) = 0$, our Theorem 1 follows from the following

Proposition 4. *Let $\pi_1 = \mathbb{Z}/2$. Then signature, Kirby-Siebenmann invariant and image of the fundamental class induce a monomorphism*

$$(\sigma, ks, u_*) : \Omega_4(\xi) \rightarrowtail \mathbb{Z} \times \mathbb{Z}/2 \times H_4(X; \mathbb{Z}).$$

Proof. As in the previous section we make use of the Atiyah-Hirzebruch spectral sequence

$$E^2_{p,q} = H_p(ML; \Omega_q^{SpinTOP}) \Longrightarrow \Omega_{p+q}^{SpinTOP}(ML) \cong \Omega_{p+q}(\xi).$$

Since we are in the oriented case, we may use the Thom isomorphism to identify $H_p(ML; \Omega_q^{SpinTOP})$ with $H_p(X; \Omega_q^{SpinTOP})$ (ordinary homology, no twisting). Then the d_2-differentials for $q \leq 2$ are dual to the homomorphisms sending $x \in H^p(X; \mathbb{Z}/2)$ to $Sq^2(x) + x \cdot w_2(L)$. Moreover, the boundary homomorphism $\Omega_4(\xi) \to H_4(X; \mathbb{Z})$ is just the image of the fundamental class under u_* (see [16]) and controlling the signature means that we can work with the reduced theory $\widetilde{\Omega}_4(\xi)$. Therefore, our aim is to show that we have

$$E^\infty_{3,1} = 0 \text{ and } E^\infty_{2,2} = \mathbb{Z}/2.$$

This last $\mathbb{Z}/2$ will then necessarily have to be detected by the Kirby-Siebenmann invariant. The homology of X may be calculated using the Leray-Serre spectral sequence for the fibration (*) once one knows the structure of $\pi_2 := \pi_2 X \cong \pi_2 M$ as a π_1-module and the k-invariant. But for fundamental group $\mathbb{Z}/2$ this is easy since by [2] any $\mathbb{Z}$-torsionfree $\mathbb{Z}/2$-module has a unique decomposition as a direct sum of only three possible building blocks, namely $\mathbb{Z}^+$, $\mathbb{Z}^-$ and A. Here $A := \mathbb{Z}[\mathbb{Z}/2]$ is the free module of rank 1 and $\mathbb{Z}/2$ acts trivially on $\mathbb{Z}^+$ and as -1 on $\mathbb{Z}^-$. Therefore, the stable information on π_2 which one gets by the methods of the previous section imply the unstable isomorphism (see also [5])

$$\pi_2 \cong \mathbb{Z}^- \oplus \mathbb{Z}^- \oplus A^n \text{ and } 0 \neq k \in H^3(\mathbb{Z}/2; \pi_2) = \mathbb{Z}/2 \times \mathbb{Z}/2.$$

Since the nontrivial k-invariant implies a nontrivial differential d_3 in the Leray-Serre spectral sequence, it follows that for the generator $x \in H^1(X; \mathbb{Z}/2) = \mathbb{Z}/2$ we have the relation $x^3 = 0$. Moreover, a basis for $H^2(X; \mathbb{Z}/2)$ is given by

$$\{x^2, a, z_1, \ldots, z_n\}$$

with a restricting to the $\mathbb{Z}^- \oplus \mathbb{Z}^-$ part of π_2^* in the kernel of d_3 and z_i corresponding to the augmentation homomorphism of the i-th A-factor. We note also the relations $x^i \cdot z_j = 0 \quad \forall i > 0$ and the linear independence of all $z_i^k \cdot z_j^l, a^k \cdot z_j^l$ and $a^k \cdot x^m \quad \forall k, l \geq 0, m < 3$.

Corresponding to the three possible w_2-types of M, i.e. $\widetilde{M}$ nonspin, M spin and $\widetilde{M}$ spin but M nonspin, there are three possibilities for $w_2(L)$ (after a change of basis of A^n):

(I) $w_2(L) = x^2 + z_1$.
(II) $w_2(L) = 0$.

(III) $w_2(L) = x^2$.

In all these cases we have $E^3_{3,1} = 0$ because the differential

$$d_2 : H_5(X;\mathbb{Z}) \to H_3(X;\mathbb{Z}/2)$$

is onto: Since $H_5(X;\mathbb{Z})$ consists only of 2-torsion, it is enough to show that the composition

$$Sq^1 \circ (Sq^2 + w_2(L)) : H^3(X;\mathbb{Z}/2) \to H^6(X;\mathbb{Z}/2)$$

is injective. But $H^3(X;\mathbb{Z}/2)$ is generated by $a \cdot x$ which is mapped to $a^2 \cdot x^2 \neq 0$ under the above map. For $E^3_{2,2}$ we have to study the kernel of the map

$$Sq^2 + w_2(L) : H^2(X;\mathbb{Z}/2) \to H^4(X;\mathbb{Z}/2).$$

¿From the above information on $H^*(X;\mathbb{Z}/2)$ we can easily deduce that for w_2-type I this kernel is generated by x^2 and z_1 whereas for w_2-types II and III it is generated solely by x^2. In these last cases we are done since a single element must survive to $E^\infty_{2,2}$ because of the Kirby-Siebenmann invariant. For w_2-type I we have to show that the differential $d_3 : E^3_{5,0} \to E^3_{2,2}$ is nontrivial which will finish our proof. We concentrate on the important information by considering the commutative diagram of fibrations

$$\begin{array}{ccccc} K(\mathbb{Z}^-,2) & \longrightarrow & Y & \longrightarrow & K(\mathbb{Z}/2,1) \\ {\scriptstyle i}\downarrow & & {\scriptstyle f}\downarrow & & \| \\ K(\pi_2,2) & \longrightarrow & X & \longrightarrow & K(\mathbb{Z}/2,1) \end{array}$$

where Y has nontrivial k-invariant and the map i on the fibres is induced by the $\mathbb{Z}/2$-homomorphism sending the generator of $\mathbb{Z}^-$ to the sum $\kappa + (1-t) \cdot a_1$. Here $\kappa \in \mathbb{Z}^- \oplus \mathbb{Z}^-$ reduces to the k-invariant and a_1 is the free generator of the first component in A^n (which we decided to carry one part of $w_2(L)$). Also t denotes the generator of $\mathbb{Z}/2$. Since the class x is the pullback of the generator in $H^1(\mathbb{Z}/2;\mathbb{Z}/2)$ we can denote $f^*(x)$ again by x and it will generate $H^1(Y;\mathbb{Z}/2)$. We will now show the crucial property $f^*(z_1) = x^2$: This follows from the commutative diagram of Gysin-sequences where we consider the universal coverings ($\simeq$ fibres) of X and Y as S^0-bundles over X respectively Y (both having first Stiefel-Whitney class equal to x):

$$\begin{array}{ccccc} H^2(\widetilde{Y};\mathbb{Z}/2) & \xrightarrow{tr_Y} & H^2(Y;\mathbb{Z}/2) & \xrightarrow{\cdot x} & H^3(Y;\mathbb{Z}/2) \\ {\scriptstyle i^*}\uparrow & & {\scriptstyle f^*}\uparrow & & {\scriptstyle f^*}\uparrow \\ H^2(\widetilde{X};\mathbb{Z}/2) & \xrightarrow{tr_X} & H^2(X;\mathbb{Z}/2) & \xrightarrow{\cdot x} & H^3(X;\mathbb{Z}) \end{array}$$

Since both k-invariants are nontrivial, the above multiplication with x is identically zero. In $H^2(\widetilde{X};\mathbb{Z}/2)$ we know that there exists a class z corresponding to the nonequivariant homomorphism $A^n \to \mathbb{Z}$ which sends a_1 to 1, $t \cdot a_1$ to 0 and also all the other free generators and their t-translates to 0. Then $tr_X(z) = z_1$ and $i^*(z)$ generates $H^2(\widetilde{Y};\mathbb{Z}/2) = \mathbb{Z}/2$. Using the fact that tr_Y is an isomorphism and that $H^2(Y;\mathbb{Z}/2)$ is generated by x^2 we get

$$f^*(z_1) = f^*(tr_X(z)) = tr_Y(i^*(z)) = x^2.$$

This implies that the pullback under f of the line bundle L over X gives the trivial bundle and thus we get an induced map

$$\Omega_*^{SpinTOP}(Y) \cong \Omega_*^{SpinTOP}(Mf^*(L)) \longrightarrow \Omega_*^{SpinTOP}(ML) = \Omega_*(\xi).$$

Because the map $f_* : E^3_{2,2}(Mf^*(L)) \to E^3_{2,2}(ML)$ is nontrivial it suffices to show that the d_3-differential in question is nontrivial in the Atiyah-Hirzebruch spectral sequence for $\Omega_*^{SpinTOP}(Y)$. To do so, we have to ask the reader for some more patience because we have to make two more comparisons to reduce the problem to a d_2-differential where it finally will be solved. For the first comparison, let η be the real line bundle over Y with $w_1(\eta) = x$ and thus $S(\eta) = \widetilde{Y}$. Writing $T(\eta)$ for its Thom space we get a cofibration

$$\widetilde{Y} \xrightarrow{p} Y \xrightarrow{q} T(\eta).$$

This cofibration induces long exact sequences in any (generalized) homology theory. We conclude that q_* is an isomorphism in $H_2(;\mathbb{Z}/2)$ and $H_5(;gz)$ and thus it suffices to show the nontriviality of our d_3-differential in the Atiyah-Hirzebruch spectral sequence for $\Omega_*^{SpinTOP}(T(\eta))$. For the second comparison, we use the cofibration (compare [9])

$$T(p^*(\eta)) \to T(\eta) \to T(\eta \oplus \eta).$$

Since $p^*(\eta)$ is the trivial line bundle over $\widetilde{Y}$ we get the suspension isomorphism

$$\Omega_4(T(p^*(\eta))) \cong \Omega_3(\widetilde{Y}).$$

This group is easily checked to be trivial by the usual spectral sequence (only d_2-differentials are needed here!) We may conclude that $\Omega_4^{SpinTOP}(T(\eta))$ maps injectively into $\Omega_4^{SpinTOP}(T(\eta \oplus \eta))$. Since the filtration degree of an element in a spectral sequence can only go up, we are certainly done if $E^3_{2,2}(T(\eta \oplus \eta)) = 0$. But this follows from the injectivity of the map

$$Sq^2 : H^2(T(\eta \oplus \eta);\mathbb{Z}/2) \to H^4(T(\eta \oplus \eta);\mathbb{Z}/2)$$

which sends the Thom class u to $Sq^2(u) = w_2(\eta \oplus \eta) \cdot u = x^2 \cdot u \neq 0$. □

References

1. S. E. Cappell, *Unitary nilpotent groups and hermitian K-theory.* Bulletin of the AMS **80**, (1974) 1117-1122
2. C. W. Curtis, I. Reiner, *Representation Theory of Finite Groups and Associative Algebras,* Series of Texts and Monographs XI, Wiley and sons 1962.
3. M. Freedman, *The Topology of 4-dimensional Manifolds,* J. of Diff. Geom. **17** (1982) 357-453.
4. M. Freedman, F. Quinn, *Topology of* 4*-Manifolds,* Princeton Mathematical Series 39, Princeton University Press 1990.
5. I. Hambleton and M. Kreck, *On the Classification of Topological Four-Manifolds with Finite Fundamental Group,* Math. Annalen **280** (1988) 85-104.
6. I. Hambleton and M. Kreck, *Cancellation, Elliptic Surfaces and the Topology of certain Four-Manifolds,* Preprint of the Max-Planck Institut für Mathematik Bonn 1991, to appear in the Journal für reine und angewandte Mathematik.
7. I. Hambleton, M. Kreck, P. Teichner, *Nonorientable* 4*-Manifolds with Fundamental Group of Order* 2*,* Preprint 1992, to appear in the Transactions of the AMS.
8. I. Hambleton, R. J. Milgram, L. Taylor, B. Williams, *Surgery with Finite Fundamental Group,* Proc. London Math. Soc. (3) **56** (1988) 349-379.
9. I. M. James, *The Topology of Stiefel-Manifolds,* London. Math. Soc. Lecture Note Series 24, Cambridge University Press 1976.
10. R. C. Kirby, L. C. Siebenmann, *Foundational Essays on Topological Manifolds, Smoothings and Triangulations,* Ann. of Math. Studies 88, Princeton University Press 1977.
11. M. Kreck, *Surgery and Duality,* To appear as a book in the Vieweg-Verlag, Wiesbaden. As a preprint of the Johannes-Gutenberg-Universität Mainz 1985 available under the title: *An Extension of Results of Browder, Novikov and Wall about Surgery on Compact Manifolds.*
12. R. Stong, *Existence of* π_1*-Negligible Embeddings in* 4*-Manifolds. A Correction to Theorem 10. 5 of Freedman and Quinn,* UCLA Preprint 1992.
13. R. Stong, *Uniqueness of connected sum decompositions in dimension* 4, UCLA Preprint 1992.
14. R. M. Switzer, *Algebraic Topology-Homotopy and Homology,* Die Grundlehren der math. Wissenschaften 212, Springer-Verlag 1970.
15. P. Teichner, *Topological Four-Manifolds with Finite Fundamental Group,* PhD Thesis at the University of Mainz, Germany. Verlag Shaker 1992.
16. P. Teichner, *On the signature of four-manifolds with universal covering spin,* Math. Annalen **295** (1993) 745-759.

17. C. T. C. Wall, *Surgery on Compact Manifolds,* London Math. Soc. Monographs 1, Academic Press 1970.

UNIVERSITY OF MAINZ, DEP. OF MATHEMATICS, 55099 MAINZ,GERMANY
E-mail address: teichner@umamza.mathematih.uni-mainz.de

AMS/IP Studies in Advanced Mathematics
Volume **2**, 1997 (Part 1)

THE PROBLEM OF LAGRANGIAN KNOTS IN FOUR-MANIFOLDS

YAKOV ELIASHBERG AND LEONID POLTEROVICH

1. Introduction and Overview.

1.1. The Problem. Let (M^4, ω) be a symplectic 4-manifold, and L be a 2-dimensional surface without boundary. An embedding $f : L \to M$ is called *Lagrangian* if $f^*\omega \equiv 0$. The *problem of Lagrangian knots* can be formulated as follows. Assume that $f_1, f_2 : L \to M$ are two Lagrangian embeddings, which are *homotopic* as smooth maps. Is it true that

(1.1.A) (topological version): *f_1 and f_2 are smoothly isotopic?*

(1.1.B) (Lagrangian version): *f_1 and f_2 are isotopic through Lagrangian embeddings?*

(1.1.C) (Hamiltonian version): *there exists a Hamiltonian isotopy of M which takes f_1 to f_2?* (we refer the reader to [2] for the notion of Hamiltonian isotopy, as well as for basic definitions of symplectic geometry).

Notice that (1.1.C) implies (1.1.B) and (1.1.B) implies (1.1.A).

As far as we know, the Lagrangian knots problem was first formulated in the literature in [1]. The first answers appeared in [7], [13] (see also related papers [17] and [9]), [10], [11] and [6]. In the present short survey we give a brief description of the basic results and ideas of these works. Let us emphasize that all the approaches developed so far are based on Gromov's powerful theory of pseudo-holomorphic curves in symplectic manifolds [12] and the related technique of filling by holomorphic discs developed in [3], [4], [12], [5], and [8]. (See also the recent preprint [21].)

Authors supported by United States-Israel Binational Science Foundation grant 90-00150

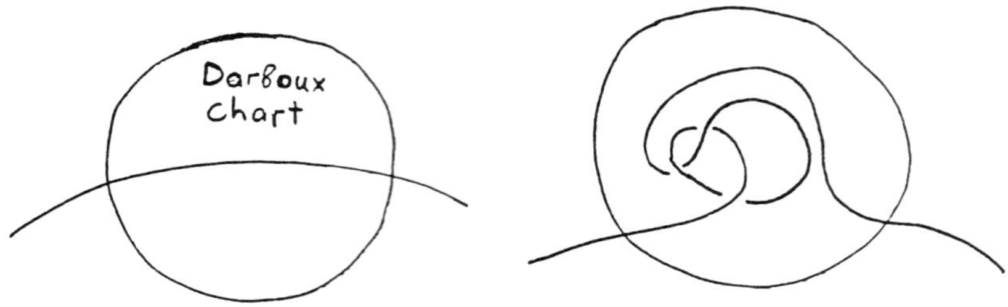

FIGURE 1

1.2. Local Knots. The simplest mechanism, which *a priori* might create Lagrangian knots is the following local knotting construction. Take a Lagrangian surface and modify it in a small neighborhood of a given point (see Fig. 1).

Our first result states that this is impossible!

Theorem 1.2.A ([11]) *Let* $(\mathbb{R}^4, \omega)$ *be the standard symplectic vector space, and let* $l \subset \mathbb{R}^4$ *be a Lagrangian plane. Consider the space* $\mathcal{L}$ *of all Lagrangian submanifolds of* $\mathbb{R}^4$ *which are diffeomorphic to* $\mathbb{R}^2$ *and which coincide with* l *outside a compact subset. Then* $\mathcal{L}$ *is contractible with respect to the* C^∞*-Whitney topology.*

Thus for the case of local knots this theorem gives the positive answer to all three questions 1.1.A–1.1.C.

1.3. Cotangent Bundles. In fact, the previous result deals with Lagrangian embeddings of $\mathbb{R}^2$ into its cotangent bundle endowed with the standard symplectic structure. It turns out that the *topological* unknottedness can be extended for some other surfaces, namely for S^2 and $\mathbb{T}^2$.

Theorem 1.3.A ([10]) *Let* L *be either sphere* S^2 *or torus* $\mathbb{T}^2$*, and let* $f : L \to T^*L$ *be a Lagrangian embedding homotopic to the zero section. Then* f *is smoothly isotopic to the zero section.*

The questions about Langrangian isotopies, as well as the situation with surfaces of higher genus are still far from being understood (see §5 below).

1.4. Lagrangian Tori in $\mathbb{R}^4$. In [13] K. Luttinger found restrictions on the knot types of embeddings $\mathbb{T}^2 \to \mathbb{R}^4$ which can be realized by Lagrangian ones. In order to formulate his result, recall that each embedded 2-torus $L \subset \mathbb{R}^4$ has a special framing, which we will call *topological.*

This framing is represented by a homotopically unique normal vector field v to L, which satisfies the following condition:

$$lk(L, \alpha + \varepsilon v) = 0 \text{ for all } \alpha \in H_1(L, \mathbb{Z}).$$

Here L is oriented, $lk(\cdot, \cdot)$ denotes the linking number, and $\alpha + \varepsilon v$ means a small shift of a cycle representing the class $\alpha \in H_1(L)$ along v. Denote by N a tubular neighbourhood of L, and set $\Sigma = \partial N$. The topological framing induces the splitting $\Sigma = S^1 \times \mathbb{T}^2$. Let θ and (x, y) be angular coordinates on S^1 and $\mathbb{T}^2$ respectively. Define the *Luttinger twist* $f_{m,n}$: $\Sigma \to \Sigma, (\theta, x, y) \mapsto (\theta, x + m\theta, y + n\theta)$, where $m, n, \in \mathbb{Z}$ and the *Luttinger surgery*

$$V_{m,n}(L) = (\mathbb{R}^4 \setminus N) \bigcup_{f_{m,n}:\Sigma \to \Sigma} N.$$

Theorem 1.4.A ([13]) *If L is a Lagrangian embedding, then $V_{m,n}$ is diffeomorphic to $\mathbb{R}^4$. In particular, $\pi_1(V_{m,n}(L)) = \{1\}$ for all $m, n, \in \mathbb{Z}$.*

It turns out that many topological knot types of 2-tori in $\mathbb{R}^4$ (for instance, spin knots) violate the conclusion of Luttinger's theorem, and hence *they cannot be represented by Lagrangian embeddings.*

1.5. Unknotedness of Lagrangian cylinders. The result we discuss in this section deals with a problem which lies between the problem of local knots and the problem of existence of knotted Lagrangan tori in $\mathbb{R}^4$.

Let us denote by U the domain $\{0 \le q_2 \le 1\} \subset \mathbb{R}^4$, and by S_0 and S_1 the circles

$$\{p_1^2 + q_1^2 = 1, p_2 = 0, q_2 = 0\}$$

and

$$\{p_1^2 + q_1^2 = 1, p_2 = 0, q_2 = 1\},$$

respectively. Let Λ be the Lagrangian cylinder

$$\{p_1^2 + q_1^2 = 1, p_2 = 0, 0 \le q_2 \le 1\}$$

and j be the inclusion $\Lambda \hookrightarrow U$. Denote by r the ray

$$\{p_1 = q_1 = p_2 = 0, q_2 > 1\}$$

and by D the disc

$$\{p_1^2 + q_1^2 \le 1, p_2 = 0, q_2 = 1\}$$

bounded by S_1.

Theorem 1.5.A ([7]) *Any Lagrangian embedding $\varphi : \Lambda \to U$ which coincides with j near $\partial\Lambda$ is isotopic to j relative $\partial\Lambda$. Moreover, φ and* *j are Lagrangian (and, therefore, Hamiltonian!) isotopic relative* $\partial\Lambda$ *as embeddings into* $\mathbb{R}^4_+ \setminus (r \cup D)$.

We do not know if φ and j are Lagrangian isotopic *inside* the domain U.

1.6. Hamiltonian vs. Lagrangian. In order to discuss the question (1.1.C) above we need to recall the following notions.

Consider, for simplicity, a Lagrangian embedding $f : \mathbb{T}^2 \to \mathbb{R}^4$, where $\mathbb{R}^4$ is endowed with the standard symplectic form $\omega = dp_1 \wedge dq_1 + dp_2 \wedge dq_2$. The cohomology class of the closed form $f^*\lambda$, where $\lambda = p_1 dq_1 + p_2 dq_2$ is invariant under an ambient Hamiltonian isotopy. We will call the class $[f^*\lambda] \in H^1(T^2; \mathbb{R})$ *the symplectic area class* of L.

Another cohomology class, this time in $H^1(T^2) = H^1(T^2; \mathbb{Z})$, which one can associate with the Lagrangian embedding $f : T^2 \to \mathbb{R}^4$ is its *Maslov class* $\mu(f)$. This is a characteristic class which measures the obstruction to the extension of the field of Lagrangian planes tangent to $L = f(T^2)$ along a cycle $\gamma \subset L$ to a surface spanned by γ in $\mathbb{R}^4$.

A Lagrangian embedding $f : T^2 \to \mathbb{R}^4$ is called k-monotone for $k \in \mathbb{R}_+$ if $[f^*\lambda] = k\mu(f)$. For instance, let $L_{a,b}$ be the *split Lagrangian torus*

$$L_{a,b} = \left\{ |z_1|^2 = \frac{2a}{\pi},\ |z_2|^2 = \frac{2b}{\pi} \right\}$$

in $\mathbb{R}^4$. Here we identify the standard symplectic $(\mathbb{R}^4, dp_1 \wedge dq_1 + dp_2 \wedge dq_2)$ with $\mathbb{C}^2$ endowed with the complex coordinates $z_1 = p_1 + iq_1$, $z_2 = p_2 + iq_2$. Then the torus $L_{a,a}$ is a-monotone. The tori $L_{a,b}$ with $a \neq b$ are not monotone.

Notice that the symplectic area class is invariant under a Hamiltonian isotopy and the Maslov class is invariant under a Lagrangian isotopy (and even a regular homotopy). So these invariants should be taken into account when formulating problems 1.1.B and 1.1.C.

In fact, especially in the Hamiltonian case, it is more convenient to talk about unparametrized Lagrangian tori. A reparametrization of the torus changes both classes via the induced action on $H^1(T^2)$. On the other hand the a-monotonicity is invariant under Hamiltonian isotopy. Thus there is no obvious obstruction to the existence of a Hamiltonian isotopy between two a-monotone Lagrangian tori in $\mathbb{R}^4$. However, we have

Theorem 1.6.A ([6]) *There exists a monotone Lagrangian embedded 2-torus in* $\mathbb{R}^4$ *which is Lagrangian isotopic but not Hamiltonian isotopic to the split monotone torus.*

In contrast with all other results described in this section, Chekanov's example is not specifically low-dimensional: it has analogues in $\mathbb{R}^{2n}$ for all $n \geq 2$.

1.7. Organization of the Paper. Results of 1.2, 1.3 and 1.5 above are based on the following fundamental idea: approximate a Lagrangian surface by a symplectic one, and then use pseudo-holomorphic curves machinery to study the obtained symplectic surfaces. We illustrate this idea in §2. In §3 we describe Luttinger's method (see 1.4 above) and mention some applications to Hamiltonian Mechanics. In §4 we give a new proof of 1.6.A. In constrast with Chekanov's original approach which was based on the theory of symplectic capacities, we use directly Gromov's theory of pseudo-holomorphic curves. In §5 we discuss some open questions.

1.8. Acknowledgements. This paper was written during our stay at the 1994 Symplectic Geometry program organized by Simon Donaldson, Dusa McDuff, Dietmar Salamon and Charles Thomas at the Isaac Newton Institute for Mathematical Sciences (Cambridge, UK). We warmly thank the Institute and the organisers for the hospitality and the remarkable research athmosphere. We thank also Yura Chekanov for critical remarks.

2. The Approximation Techniques.

In this section we illustrate the approximation techniques developed in [7], [10], and [11]. Namely, we derive the following result which is a topological version of the stronger Theorem 1.2.A.

Proposition 2.A *Local Lagrangian knots are topologically trivial.*

2.1. Notations. Consider the standard linear space $\mathbb{R}^4$ with coordinates (p_1, q_1, p_2, q_2) such that the symplectic form ω is written as $dp_1 \wedge dq_1 + dp_2 \wedge dq_2$. Let $l = \{p_1 = p_2 = 0\}$ be a Lagrangian plane, and let $L \subset \mathbb{R}^4$ be a Lagrangian submanifold which is diffeomorphic to $\mathbb{R}^2$ and which coincides with l outside a compact subset. We have to prove that L is isotopic to l via a compactly supported isotopy of $\mathbb{R}^4$.

2.2. Approximation (see [10], 2.1.A). Consider the family of symplectic forms $\omega_\varepsilon = \omega + \varepsilon dq_1 \wedge dq_2$, $\varepsilon > 0$.

Lemma 2.2.A *There exists a compactly supported isotopy of $\mathbb{R}^4$ which takes L to an ω_ε-symplectic surface, provided ε is small enough.*

Proof. One can easily find a closed 2-form, say τ, on a closed tubular neighborhood V of L with the following properties:

- $\tau = dq_1 \wedge dq_2$ outside a compact set;
- $\int_L (\tau - dq_1 \wedge dq_2) = 0$;
- L is τ-symplectic.

Then $\tau - dq_1 \wedge dq_2 = d\lambda$ for a compactly supported 1-form λ on V. Choose a bump function ρ on $\mathbb{R}^4$ which vanishes outside V and equals 1 near L. Set

$$\omega'_\varepsilon = \omega + \varepsilon(dq_1 \wedge dq_2 + d(\rho\lambda)).$$

Note that L is ω'_ε-symplectic, and ω'_ε coincides with ω_ε outside a compact subset of $\mathbb{R}^4$. Moreover, ω'_ε and ω_ε can be joined by a linear path of symplectic forms which coincide with ω_ε outside a compact set, provided ε is small enough. Then Moser's argument [16] implies that ω_ε and ω'_ε are isotopic. Clearly, this isotopy is compactly supported and takes L to an ω_ε-symplectic surface. □

2.3. Compactification. Denote by L' the ω_ε-symplectic submanifold of $\mathbb{R}^4$ which is constructed in 2.2. Choose a translationally invariant complex structure, say j, on $\mathbb{R}^4$ such that $\omega_\varepsilon(\cdot, j\cdot)$ is a Eucledian metric. Take a large ball B on $\mathbb{R}^4$ such that L' coincides with l outside B and compactify B to the projective plane $\mathbb{C}P^2$ by adding a line Γ_∞ at infinity. Denote by Ω and J_0 the compactifications of ω_ε and j, respectively. Thus, $(\mathbb{C}P^2, \Omega, J_0)$ is the usual Kähler structure on $\mathbb{C}P^2$. After such a compactification, the ω_ε-symplectic line l corresponds to a projective line Γ_0, and our knot L' is compactified to an Ω-symplectic embedded 2-sphere Σ. Notice that Σ is homologous to $\mathbb{C}P^1$ and Σ intersects Γ_∞ at the unique point $P = \Gamma_0 \cap \Gamma_\infty$.

2.4. Gromov's Pseudo-Holomorphic Curves. In order to prove Proposition 2.A stated at the beginning of this section, it is enough to show that Σ is isotopic to Γ_0 via an isotopy of $\mathbb{C}P^2$ which fixes Γ_∞. The construction of this isotopy is a standard application of Gromov's theory of pseudo-holomorphic curves. Recall that an almost complex structure J on $\mathbb{C}P^2$ is called *compatible* with Ω if $\Omega(\cdot, J\cdot)$ is a Riemannian metric. The set of all compatible J is contractible due to simple topological reasons. A basic observation of Gromov is that there exists an analogue of the classical theory of algebraic curves even in the case

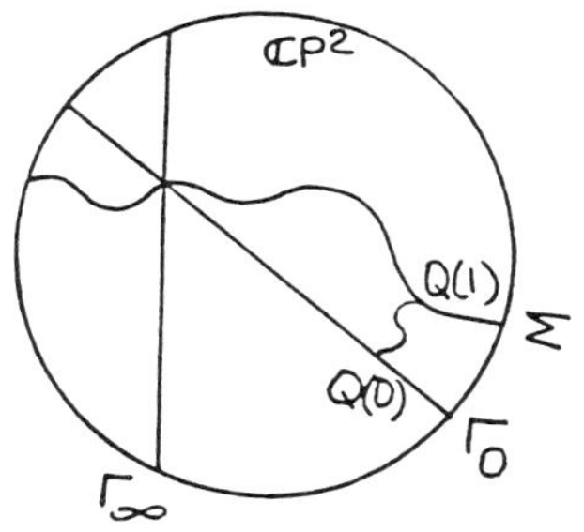

FIGURE 2

of non-integrable almost complex structures, provided these structures are compatible with a symplectic form.

More precisely, given a compatible J, a *J-holomorphic line* on $\mathbb{C}P^2$ is an embedded 2-sphere $C \subset \mathbb{C}P^2$ which is homologous to $\mathbb{C}P^1$ and whose tangent space T_xC is J-invariant for all points $x \in \mathbb{C}$.

Lemma 2.4.A [12] *For each compatible almost-complex structure J on $\mathbb{C}P^2$, and for each two distinct points $A, B \subset \mathbb{C}P^2$ there exists a unique J-holomorphic line which passes through A and B. Moreover, this line depends smoothly on J, A, B.*

Now apply this fact to our situation (see Fig. 2). Choose a smooth path $Q(t)$, $t \in [0;1]$ on $\mathbb{C}P^2 \setminus \Gamma_\infty$ such that $Q(0) \in \Gamma_0$ and $Q(1) \in \Sigma$. Let $J(t)$, $t \in [0,1]$ be a smooth family of compatible almost complex structures with the following properties:

- Γ_∞ is $J(t)$-holomorphic for all t;
- $J(0) = J_0$;
- Σ is $J(1)$-holomorphic.

Let C_t be the unique $J(t)$-holomorphic line passing through P and $Q(t)$. Since $C_0 = \Gamma_0$ and $C_1 = \Sigma$, the deformation $\{C_t\}$, $t \in [0,1]$ gives an isotopy between Γ_0 and Σ. The standard intersection index argument (which works perfectly for pseudo-holomorphic curves) shows that C_t intersects Γ_∞ transversally at the unique point P. Hence, the isotopy $\{C_t\}$ can be extended to an ambient isotopy of $\mathbb{C}P^2$ which preserves Γ_∞. $\square$

3. Luttinger Surgery and Quasi-periodic Motion.

In this section we explain 1.4.A and discuss its relations with classical mechanics.

3.1. Symplectic Framing. First observe that there exist an infinite number of different homotopy classes of non-singular vector fields on a 2-torus (see Fig. 3), and *only one of them* (see Fig. 3a) is $Diff$-invariant. We call this distinguished class *homotopically trivial.*

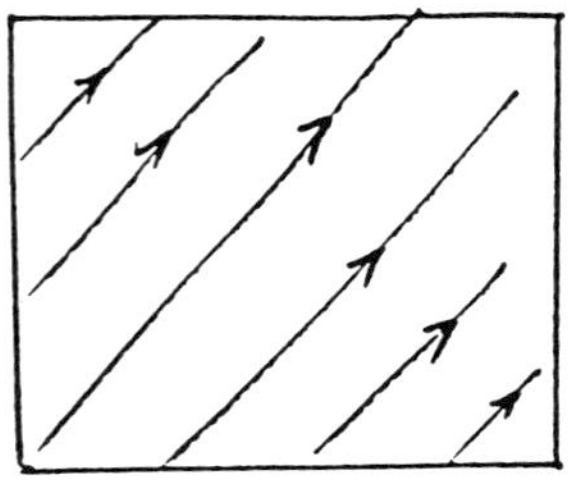

FIGURE 3.A.

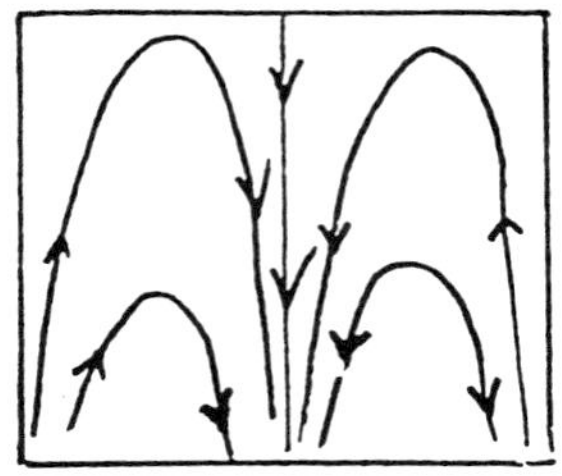

FIGURE 3.B.

Consider now an embedded Lagrangian torus $L \subset (\mathbb{R}^4, \omega)$. In section 1.4 we defined the topological framing of an embedded torus. Let us define another canonical framing of L which we call *symplectic.*

Choose a homotopically trivial vector field v on L and an ω-compatible almost complex structure J on $\mathbb{R}^4$. Then the required framing is defined by the non-vanishing vector field Jv.

Theorem 3.1.A ([9]) *For an embedded Lagrangian torus the topological and symplectic framings coincide.*

Notice that for a non-Lagrangian torus which is *totally real* with respect to the given almost complex structure J the symplectic framing can be also defined but it need not coincide with the topological one (see [17]).

3.2. Proof of 1.4.A. The main observation of Luttinger's paper [13] can be formulated as follows: *the twist* $f_{m,n}$ (see 1.4) *preserves the restriction of* ω *to the boundary of a small tubular neighbourhood of* L. Indeed, since the topological and symplectic framings coincide, we can write

$$\omega|_\Sigma = d(\cos\theta dx + \sin\theta dy),$$

and a trivial computation shows that $f^*(\omega|_\Sigma) = \omega|_\Sigma$. Thus the surgery described in 1.4 above leads to a *symplectic* manifold $V_{m,n}(L)$, which coincides with $(\mathbb{R}^4, \omega)$ outside a compact set. Hence Gromov's theory see ([12], [15]) predicts that $V_{m,n}(L)$ is symplectomorphic to $\mathbb{C}^2 \# k\overline{\mathbb{C}P^2}$ for some $k \geq 0$. In fact, Novikov's additivity of the signature theorem implies that $k = 0$ and, therefore, $V_{m,n}(L)$ is symplectomorphic to $\mathbb{C}^2$. (We owe this remark to R. Gompf.) In particular, we have $\pi_1(V_{m,n}(L)) = \{1\}$. □

Idea of the Proof of 3.1.A. It turns out that if the topological and symplectic framings were different, one could always perform a Luttinger surgery which would drastically change the topology of $V_{m,n}$ (see [9] for details). But this is impossible in view of the discussion above. □

3.3. Topological Universality of Quasi-periodic Motion. Theorem 3.1.A has the following consequence in Hamiltonian Mechanics. Let $W^3 \subset \mathbb{R}^4$ be a regular level of a Hamiltonian function $H : \mathbb{R}^4 \to \mathbb{R}$, and let $L \subset W$ be an invariant 2-torus of the Hamiltonian flow. Denote by v the Hamiltonian vector field of H. Obviously, v is tangent to L.

Corollary 3.3.A ([17], [9]) *Assume that L divides W. Then the restriction $v|_L$ is homotopically trivial.*

This follows from 3.1.A and the obvious fact that for a compatible J the field $J(v|_L)$ is transversal to W.

4. An Exotic Lagrangian Torus in $\mathbb{R}^4$.

In this section we prove Chekanov's theorem 1.6.A. Our proof is different from the one given by Chekanov in [6]. It suggests that one can use a relative version of Gromov-Ruan invariants (see [12], [19]) to distinguish between a-monotone Lagrangian tori which are non-isotopic through a Hamiltonian isotopy.

4.1. The Standard Torus. As in Section 1.6 above we denote by $L_{a,b}$ the split Lagrangian torus

$$L_{a,b} = \left\{ |z_1| = \frac{2a}{\pi},\ |z_2|^2 = \frac{2b}{\pi} \right\} \subset \mathbb{C}^2, z_1 = p_1 + iq_1,\ z_2 = p_2 + iq_2.$$

Set $\lambda = p_1 dq_1 + p_2 dq_2$. Notice that symplectic areas $\int_{e_1} \lambda$ and $\int_{e_2} \lambda$ of the cycles

$$e_1 = \{z_1 = \text{ const},\ z_2 = be^{i\tau}, \tau \in [0, 2\pi]\}$$

and

$$e_2 = \{z_2 = \text{ const },\ z_1 = ae^{i\tau}, \tau \in [0, 2\pi]\}$$

equal b and a, respectively.

Gromov proved in [12] the following remarkable property of the a-monotone torus $L_{a,a}$.

Proposition 4.1.A [12] *Let $f_t : L_{a,a} \to \mathbb{C}^2$ be a Hamiltonian isotopy. Then through each point of the torus $f_t(L_{a,a})$ pass at least two holomorphic discs whose boundaries are homologous to cycles $(f_t)_* e_1$ and $(f_t)_* e_2$, respectively.*

4.2. An oscillator. Let us call a monotone Lagrangian torus $L \subset \mathbb{C}^2$ *exotic* if it is not Hamiltonian isotopic to any of the tori $L_{a,a}$, $a \in \mathbb{R}$.

Consider a Hamiltonian function $H : \mathbb{C}^2 \to \mathbb{R}$,

$$H(z_1, z_2) = \frac{1}{2}|z_1|^2 - \frac{1}{2}|z_2|^2,$$

and set $M_a = \{H = a\} \setminus \{0\}$. The dynamics of this system is given by the flow

$$(z_1(t), z_2(t)) \mapsto (z_1(0)e^{it}, z_2(0)e^{-it}),$$

and hence it defines a free S^1-action on $\mathbb{C}^2 - \{0\}$. Consider the function $F : \mathbb{C}^2 \to \mathbb{C}$ defined by the formula $(z_1, z_2) \mapsto z_1 z_2$ and set $F_a = F|_{M_a}$.

Lemma 4.2.A *For all $a \in \mathbb{R}$ and $\gamma \in \mathbb{C}$ such that $a^2 + |\gamma|^2 \neq 0$ the set $F_a^{-1}(\gamma)$ consists of the unique orbit of the action.*

The proof is a trivial computation in symplectic polar coordinates and is left to the reader.

Let $D \subset \mathbb{C}$ be an embedded disc and $\Gamma = \partial D$ its boundary. Then the set $L_\Gamma^a = F_a^{-1}(\Gamma)$ is an embedded Lagrangian torus, provided $0 \notin \Gamma$ in the case $a = 0$.

Proposition 4.2.B *The torus L_Γ^0 is exotic if D does not contain zero.*

First we describe a basis (α, β) in $H_1(L_\Gamma^0, \mathbb{Z})$. The cycle α is represented by the map $\Gamma \to L_\Gamma^0$, defined by $w \mapsto (\sqrt{w}, \sqrt{w})$. Here $\sqrt{w}$ is a chosen branch of the square root ($0 \notin D$!). The cycle β is represented by an orbit of the S^1-action. One can easily verify that β has zero Maslov class as well as zero symplectic area, and hence the torus L_Γ^0 is monotone. The Maslov class takes the value 2 on the cycle α. Set $\int_\alpha \lambda = A_\Gamma$. Notice that the torus L_Γ^0 is foliated by circles homologous to

α which span holomorphic discs $\Delta_\theta, \theta \in [0, 2\pi)$. Since D does not contain 0, these discs are disjoint and can be parametrized by the family of maps

$$f^\theta : w \to \left(e^{i\theta}\sqrt{w},\, e^{-i\theta}\sqrt{w}\right), \quad w \in D,\, \theta \in [0, 2\pi).$$

It turns out, as the following Proposition 4.2.C shows, that all holomorphic discs with boundaries in L^0_Γ must coincide with one of these, maybe multiply covered, discs Δ_θ. In particular, 4.2.C together with 4.1.A implies 4.2.B.

Proposition 4.2.C *Suppose that D does not contain 0. Then any holomorphic disc in $\mathbb{C}^2$, whose area is equal to A_Γ and whose boundary is contained in L^0_Γ, coincides with one of the discs $\Delta_\theta, \theta \in [0, 2\pi)$.*

4.3. Proof of 4.2.C. Consider a holomorphic map $f : D \to L^0_\Gamma$ with $\int_D f^*\omega = A_\Gamma$ and $f(\Gamma) \subset L^0_\Gamma$.

Let us write $f(w) = (f_1(w), f_2(w)), w \in D$ and notice that $F \circ f : D \to \mathbb{C}$ is a holomorphic map which sends $\Gamma = \partial D$ to Γ. The condition $\int_D f^*\omega = A_\Gamma$ implies that the degree of $(F\circ f)|_\Gamma$ is equal to 1. Hence, after a conformal change of the variable w, we can assume that $(F\circ f)(w) = w$.

In other words,

$$f_1(w)f_2(w) = w. \tag{4.3.A}$$

On the other hand we have $L^0_\Gamma \subset M_0$, and hence,

$$|f_1(w)| = |f_2(w)| \text{ for } w \in \Gamma.$$

Notice that D does not contain 0, due to our assumption. In view of 4.3.A, $f_1(w) \neq 0$, $f_2(w) \neq 0$ for all w and, therefore, $\varphi(w) = f_1(w)/f_2(w)$ is a holomorphic function on D whose absolute value on the boundary equals 1. Thus φ maps D to the unit disc $\{|w| \leq 1\}$, and *does not* assume the value 0. We conclude that φ is a constant map, that is,

$$f_1(w) = e^{i\theta_0} f_2(w)$$

for all $w \in D$, where $\theta_0 \in [0, 2\pi)$.

In view of (4.3.A) this implies that $f = f^\theta$, where $\theta = \theta_0/2$ or $\theta = \theta_0/2 + \pi$. This completes the proof. □

4.4. Lagrangian Isotopy. We claim that L_Γ^0 is Lagrangian isotopic to a split Lagrangian torus. Indeed, L_Γ^0 is Lagrangian isotopic to L_Γ^a, $a > 0$. Let us deform Γ to a round circle $S \subset \mathbb{C}$ centered at the origin. Hence, L_Γ^0 is Lagrangian isotopic to L_S^a for some $a > 0$. The latter torus is a split one, as a simple computation shows. This proves our claim, and therefore completes the proof of 1.6.A. □

5. Problems and Discussions.

5.1. Lagrangian Knots in Cotangent Bundles of Surfaces. Let L be an orientable closed surface of genus g and $F \subset T^*L$ be a Lagrangian embedded surface homotopic to the 0-section.

5.1.A *Is L isotopic to the 0-section through smooth embeddings?*

Theorem 1.3.A above gives a positive answer for $g \leq 1$.

5.1.B *Does there exist a Lagrangian isotopy?*

This is unknown for all g. Notice that when $g \leq 1$ then any homologically non-trivial Lagrangian surface in T^*L has to be diffeomophic and homotopic to the 0-section (see [14]). This is not true when $g > 1$.

5.2. Lagrangian Knots in $\mathbb{R}^4$. The only orientable closed surface which admits a Lagrangian embedding in the standard symplectic $\mathbb{R}^4$ is the 2-torus T^2. Indeed, its normal bundle has to be trivial according to Whitney's theorem (see [20]) and has to be isomorphic to the tangent bundle.

5.2.A *Are all Lagrangian tori in $\mathbb{R}^4$ isotopic?*

This is unknown both in the smooth and Lagrangian categories. Theorem 1.4.A provides partial information. See 1.6 for the discussion of the problem of existence of a Hamiltonian isotopy.

Let us now consider a non-orientable closed surface F. If there exists a Lagrangian embedding of F into $\mathbb{R}^4$ then its normal bundle has to be isomorphic to the tangent one and hence Whitney's theorem ([20]) says that the Euler characteristic $\chi(F)$ is divisible by 4.

It is easy to construct Langrangian embeddings into $\mathbb{R}^4$ of all non-orientable surfaces with $\chi(F) = 4k$, $k > 0$ (see, for instance, [18]). On the other hand, the answer to the following question is still unknown.

5.2.B *Does there exist a Lagrangian embedding of the Klein bottle into* $\mathbb{R}^4$ *?*

It seems likely that for the case of non-orientable surfaces of higher genus, one should be able to construct non-isotopic Lagrangian embeddings.

5.3. Effect of Luttinger Surgery. As was explained in Section 1.4 above, one can perform Luttinger surgery along a Lagrangian torus $T \subset \mathbb{R}^4$ and obtain another symplectic manifold V^4 which coincides with the standard $\mathbb{R}^4$ at infinity and which is canonically symplectomorphic, via Gromov's theorem, to the standard symplectic $\mathbb{R}^4$. However, this symplectomorphism maps T onto a different Lagrangian torus $\tilde{T} \subset \mathbb{R}^4$.

5.3.A *Is* $\tilde{T}$ *isotopic to* T *through smooth, Lagrangian or Hamiltonian isotopies?*

5.4. Lagrangian embeddings in higher dimensions. All results and methods discussed in this paper, except those in §4, are strictly 4-dimensional. The problem of Lagrangian knots in higher dimensions is of different nature and requires, probably, different tools.

There are no local (in the sense of Section 1.2) topological knots in $\mathbb{R}^{2n}$ for $n > 2$. In other words, all embeddings of $\mathbb{R}^n \to \mathbb{R}^{2n}$, $n > 2$, which are flat at infinity, are isotopic in this class.

However, submanifolds of more complicated topology can be knotted and we expect that, for some n-dimensional manifolds several isotopy classes of their embeddings into $\mathbb{R}^{2n}$ can be realized by Lagrangian embeddings.

Similarly, we think that the problem of the existence of a Lagrangian isotopy has a negative solution even for local knots. Probably, the necessary invariant to distinguish Lagrangian isotopy classes could be provided by the theory of generating functions.

5.5. Lagrangian vs. Legendrian. An exact Lagrangian submanifold $L \subset T^*(M)$ admits a lift to a Legendrian submanifold $\widehat{L}$ of the space $J^1(M, \mathbb{R})$ of 1-jets of functions on M. Thus a Hamiltonian isotopy between two exact Lagrangian submanifolds lifts to a Legendrian isotopy between their lifts.

5.5.A *Do there exist exact Lagrangian embeddings into* $T^*(M)$ *whose Legendrian lifts are not Legendrian isotopic?*

Similarly, any a-monotone Lagrangian submanifold of $T^*(M)$ admits a Legendrian lift to the space $J^1(M, S^1)$ with an appropriate contact structure which depends on a.

5.5.B *Let $L_{a,a}$ and L^0_Γ be the split and the exotic a-monotone Lagrangian tori constructed in §4. Let $\Lambda = \widehat{L_{a,a}}$ and $\Lambda' = \widehat{L^0_\Gamma}$ be their Legendrian lifts to $J^1(\mathbb{R}, S^1) = \mathbb{R}^{2n} \times S^1$. Is there a Legendrian isotopy between Λ and Λ'?*

References

1. V. I. Arnold, *First steps in symplectic topology*, Russ. Math. Surv. **41 (6)** (1986) 1–21.
2. V .I. Arnold and A. B. Givental, *Symplectic geometry*, in Dynamical Systems –IV, Encyclopedia of Math. Sciences, Springer, 1990, 1–136.
3. E. Bishop, *Differentiable manifolds in complex Euclidean space*, Duke Math. J. **32** (1965) 1-22.
4. E. Bedford and B. Gaveau, *Envelopes of holomorphy of certain 2-spheres in $\mathbb{C}^2$*, Amer. Journal of Math. **105** (1983) 975–1009.
5. E. Bedford, W. Klingenberg, *On the envelope of homotopy of a 2-spheres in $\mathbb{C}^2$* , J. AMS, **4** (1993) 623–646.
6. Yu. Chekanov, *Lagrangian tori in a symplectic vector space and global symplectomorphisms*, Preprint, 1993.
7. Y. Eliashberg, *Symplectic geometry of 2-knots in $\mathbb{R}^4$*, Preprint, 1992, to appear in Floer Memorial Volume.
8. Y. Eliashberg, *Filling by holomorphic discs and its applications*, London Math. Soc. Lect. Notes Ser., **151** (1991) 45–67.
9. Y. Eliashberg and L. Polterovich, *New applications of Luttinger's surgery*, Comm.Math. Helvet. **69** (1994) 512–522.
10. Y. Eliashberg and L. Polterovich, *Unknottedness of Lagrangian surfaces in symplectic 4-manifolds*, Int. Math. Res. Notices., 1993, N 11, 295–301.
11. Y. Eliashberg and L. Polterovich, *Local Lagrangian 2-knots are trivial*, to appear in Ann. Math.
12. M. Gromov, *Pseudo-holomorphic curves in symplectic manifolds*, Invent. Math. **82** (1985) 307–347.
13. K. Luttinger, *Lagrangian tori in $\mathbb{R}^4$*, Preprint, 1992.
14. F. Lalonde and J.-C. Sikorav, *Sous-varietes lagrangiennes exactes de fibres cotangents*, Comment. Math. Helv. **66** (1991) 18-33.
15. D. McDuff, *The structure of rational and ruled symplectic manifolds*, Journal AMS, **3** (1990) 679–712.
16. J. Moser, *On the volume elements on a manifold*, Trans. AMS, **120** (1969) 286–294.
17. L. Polterovich, *New invariants of embedded totally real tori and one problem of Hamiltonian mechanics*, in "Methods of Qualitative Theory and Theory of Bifurcations", Gorkii, 84-90 (1988) in Russian.

18. L. Polterovich, *The surgery of Lagrange submanifolds*, Geom. and Funct. Analysis, **1** (1991) 198–210.
19. Y. Ruan, *Topological sigma model and Donaldson type invariants in Gromov theory*, Preprint, 1993.
20. H. Whitney, *On the topology of differentiable manifolds*, Univ. of Michigan Lect., (1941) 101–141.
21. R. Ye, *Filling by holomorphic discs in symplectic manifolds*, Preprint, 1994.

STANFORD UNIVERSITY
E-mail address: eliash@gauss.stanford.edu

TEL AVIV UNIVERSITY
E-mail address: polterov@math.tau.ac.il

AMS/IP Studies in Advanced Mathematics
Volume **2**, 1997 (Part 1)

ENERGY AND CAPACITIES IN SYMPLECTIC TOPOLOGY

FRANÇOIS LALONDE

1. Introduction.

Over the past years, the notion of energy has become a basic concept in symplectic topology. It is related to most of the previously defined symplectic invariants, and especially to the so-called capacities. This notion has led to an intrinsic geometry on the (infinite dimensional) group of Hamiltonian diffeomorphisms, introduced a few years ago by Hofer, and whose study has been recently undertaken. This is a Finslerian geometry which measures the length of a path in that group by a L^∞-norm on the tangent vectors to the path (these tangent vectors can be identified with the real valued functions that generate the Hamiltonian isotopy, and the norm is the total variation – or oscillation – of functions). That geometry has even more interest since recent works have shown that this norm leads to a metric which is essentially unique amongst bi-invariant ones.

It has also been proved that the cornerstone of that geometry – the energy-capacity inequality – is equivalent, by a some geometric transformation, to the Non-Squeezing Theorem. This is a crucial fact since this theorem can be considered as a fundamental result of symplectic topology, which moreover has been established for all symplectic manifolds, with no restriction.

The aim of this survey is to provide a non technical introduction to symplectic capacities and energy, with two goals: the first one is to explain as much as possible the basic ideas and relevant concepts with a minimum of technicalities. The second one is to provide an overview of the recent development of some aspects of the subject, which could serve as a guide to the literature.

The central theme of this article is the development of Hofer's geometry, and most topics have been chosen for this purpose. I present in Section 2 a very brief introduction to symplectic invariants. Section 3

Partially supported by NSERC grant OGP 0092913 and FCAR grant ER-1199.

is devoted to the pseudo-holomorphic approach: Gromov's width, the first rigidity results, the Non Squeezing theorem (area-width inequality) with a sharp local version of it, are described there. In the fourth one, I present the variational techniques, that have been successful in defining symplectic invariants of subsets of $\mathbf{R}^{2n}$ and of compactly supported Hamiltonian diffeomorphisms defined on euclidean spaces. In Section 5, Hofer's geometry on the group $\mathrm{Ham}(M)$ is introduced as well as the basic energy-capacity inequality. This central section contains the recent development of that geometry which reveals the unity, or at least the similarity, among different approaches. I mention there in particular the equivalence of the Non Squeezing Theorem and the energy-capacity inequality, the relation between the energy and the simplest C^0 data, and a deep result on the variational approach to stability due to Ustilovsky that enables one to decrease the energy needed to generate a given Hamiltonian diffeomorphism by making use in a clever way of a closed trajectory of the linearised flow. Actually, this section contains a quite complete account of the most basic properties of geodesics in Hofer's geometry. Finally, the last section on Viterbo's norm shows how generating functions provide a convenient way to retain much of the information of the action functional within a finite dimensional setting. The C^1-local equality of Hofer's and Viterbo's norms is described there, once again using either variational or holomorphic techniques.

I would like to thank the organizers of the Georgia Topology Conference for encouraging me to write this survey, and Dusa McDuff for reading carefully the first version of this article and making many useful comments. Some books have been recently published or will soon appear, which give a complete treatment of some aspects of symplectic topology, with much rigour (see the references in the text). I hope that this article will fill the need for a more informal presentation of the subject.

2. Symplectic invariants.

A symplectic manifold (M, ω) is a manifold endowed with a non-degenerate closed 2-form. Symplectic topology is the study of invariants of this structure, aiming ultimately at the classification of these spaces, at least of those in some reasonnably well limited families. It appeared both as a generalisation of Kahler geometry (where the integrability condition on the complex structure is abandonned) and as the framework of classical physics (with Hamiltonian systems and related problems of quantization). Very briefly speaking, it has emerged in connection with a deep theorem of Eliashberg [7] that showed that the behaviour of symplectic diffeomorphisms goes beyond the h-principle.

As with sectional curvature in Riemannian geometry, the basic invariants of symplectic geometry are 2-dimensional. But, unlike Riemannian geometry, these invariants are not local, they are only of global nature. Indeed, Darboux's theorem states that any sufficiently small neighbourhood of a point in a symplectic manifold (M, ω) is diffeomorphic to a neighbourhood of 0 in $\mathbf{R}^{2n}$ by a diffeomorphism which sends ω to the standard symplectic form $\omega_0 = \sum dx_i \wedge dy_i$ on $\mathbf{R}^{2n} = \{(x_1, y_1, \ldots, x_n, y_n)\}$. Thus symplectic geometry is the reign of topology, inasmuch as topology is "geometry in the large".

Before stating Eliashberg's theorem, let's briefly discuss symplectic invariants. Note that any symplectic diffeomorphism ϕ of a manifold can be C^0-approximated by symplectic diffeomorphisms which remain C^1-far from ϕ: simply compose with the time-one map of a Hamiltonian defined as the sum of high bump functions with small disjoint supports. Thus symplectic maps enjoy some flexibility. This and Darboux's theorem suggest that all symplectic invariants (of say closed domains of symplectic manifolds) should be global invariants defined by C^0-data. Now, any symplectic diffeomorphism is also a diffeomorphism preserving the volume form ω^n. The latter can be distinguished from smooth diffeomorphisms by a single invariant, the volume, which is the simplest example of a C^0-symplectic invariant. Do there exist other invariants, besides the volume, for symplectic maps ? Suppose that any volume-preserving diffeomorphism could be C^0-approximated by symplectic diffeomorphisms. Then the only symplectic invariant would be the volume, and the topology of symplectic spaces would reduce to volume topology, that is to ordinary differential topology. The following result due to Eliashberg shows that this is not the case:

Theorem 2.1. (C^0-rigidity) *In any symplectic manifold M, the group* $\mathrm{Diff}_\omega(M)$ *of symplectic diffeomorphisms is closed in* $\mathrm{Diff}_{\mathrm{vol}}(M)$ *(and therefore in* $\mathrm{Diff}(M)$*) for the C^0-topology.*

We will prove this theorem below (see Theorem 3.11) as a consequence of the Non-Squezing Theorem which establishes the existence of a C^0-invariant quite different from the volume.

3. Pseudo-holomorphic approach.

Global invariants might be quite difficult to define in situations where they cannot be obtained as integrals of local invariants. A fruitful approach to extract these global invariants has been introduced by Gromov [15] and consists in the study of (cobordism invariants of) the moduli spaces of pseudo-holomorphic curves in given homotopy classes. This approach is based on the following observation. Let Compl($\mathbf{R}^{2n}$) be the space of all complex structures on $\mathbf{R}^{2n}$, that is of all automorphisms $J \in Gl(2n, \mathbf{R})$ such that $J^2 = -id$. Since $Gl(2n, \mathbf{R})$ acts transitively on that space with stabilisator $Gl(n, \mathbf{C})$, it can be identified with the homogeneous space $Gl(2n, \mathbf{R})/Gl(n, \mathbf{C})$. The subspace $\mathcal{J}_c(\mathbf{R}^{2n})$ of all ω_0-*compatible* J is defined by the condition that $\omega_0(\cdot, J\cdot)$ defines an inner product on $\mathbf{R}^{2n}$. This can be identified with $Sp(2n, \mathbf{R})/(Gl(n, \mathbf{C}) \cap Sp(2n, \mathbf{R})) = Sp(2n, \mathbf{R})/U(n)$. Because $Sp(2n, \mathbf{R})$ retracts on $U(n)$, $\mathcal{J}_c(\mathbf{R}^{2n})$ is contractible. Thus we can endow $\mathbf{R}^{2n}$ with a unique, up to homotopy, complex structure compatible with ω_0. It is easy to check that the open subspace $\mathcal{J}(\mathbf{R}^{2n}) \subset$ Compl($\mathbf{R}^{2n}$) of all J such that $\omega_0(v, Jv) > 0$ for all $v \neq 0$, called ω_0-*tame* complex structures, retracts on $\mathcal{J}_c(\mathbf{R}^{2n})$ so that there is also, up to homotopy, a unique ω_0-tame complex structure on $\mathbf{R}^{2n}$.

The same conclusions also apply for a complex structure defined on each tangent space T_xM of a symplectic manifold and varying smoothly with x. Let (M, ω) be a $2n$-symplectic manifold. Denote by $\mathcal{J}$ the infinite dimensional space of C^∞ almost complex structures J on M, such that $J(x)$ is $\omega(x)$-tame for all $x \in M$. This space is also contractible: if $E \to M$ is the bundle whose fiber at $x \in M$ is the set of ω_x-tame complex structures on T_xM, and if σ is a section of E, there exists a smooth fiberwise retraction of E on σ. This implies that $\mathcal{J}$ is contractible and that any such structures given on any subset of M can be extended over all M (in a unique way, up to homotopy). Now endow M with anyone of its (homotopically unique) generic ω-tame almost complex structure. Gromov showed in [15] that a large part of the study of complex curves in Kahler geometry can be extended to almost complex manifolds tamed by a symplectic form. This gives rise, for each choice of a $J \in \mathcal{J}$, to finite dimensional moduli spaces of J-curves of given genus in given homology 2-classes of M. Since J is tame by ω, these represent symplectic surfaces of M, so that one actually obtains for each J a distinguished finite dimensional subset of all symplectic surfaces, which obey the (generalised) elliptic Cauchy-Riemann equations. The whole theory of elliptic PDE's can therefore be applied to show existence or to deform symplectic surfaces. Further, the topological invariants of the resulting J-moduli spaces furnish symplectic invariants of M (when one varies the J-structure in its contractible domain $\mathcal{J}$).

Let us explain this more precisely. Endow $\mathcal{J}$ with the C^∞-topology. This gives a Fréchet space. As explained in McDuff-Salamon [34], all relevant constructions can be carried out in this C^∞ framework (although one uses in the proofs the larger Banach manifolds endowed with some Sobolev norm), which is quite useful because some properties are only known to hold in the C^∞-case (this is the case of the positivity of intersections of a J-curve with a J-invariant $2n-2$ submanifold for instance; see [1] for the proof).

Now, for any given homology class $\alpha \in H_2(M, \mathbf{Z})$ and any non negative integer g, consider the space $\mathcal{M}_g(\alpha)$ of all non-multiply covered C^∞-maps $f : (\Sigma_g, J_0) \to (M, J)$ which are (J_0, J)-holomorphic and which realize the class α, where Σ_g is a real closed orientable surface of genus g, J_0 is a (necessarily integrable) complex structure on Σ_g and J belongs to $\mathcal{J}$ (or $\mathcal{J}'$). Here g and α are fixed but f, J_0 and J are allowed to take any value. Thus, up to the conformal group of reparametrisations, this is the space of J-holomorphic curves of genus g in class α, for any ω-tame J. A basic fact is that the projection

$$\mathcal{P} : \mathcal{M}_g(\alpha) \to \mathcal{J}$$

sending f to J is a Fredholm map between Fréchet manifolds, whose real index is $\dim_{\mathbf{R}} \mathcal{T}_g + 2(c(\alpha) - n(g-1))$ where $\mathcal{T}_g$ is the Teichmuller space of genus g and c is the first Chern class of T_*M (note that since $\mathcal{J}$ is contractible, the Chern classes of (TM, J) do not depend on the choice of $J \in \mathcal{J}$). Thus, if J is a regular value of $\mathcal{P}$ which is in the image of $\mathcal{P}$, the space $\mathcal{M}_g(\alpha, J) = \mathcal{P}^{-1}(J)$ is a smooth manifold of real dimension equal to Index$(\mathcal{P})$. When the group of conformal reparametrizations is independent of the structure J_0, the quotient is a manifold (of unparametrized curves) of dimension Index$(\mathcal{P}) - \dim_{\mathbf{R}}(\mathrm{Conf}(g))$ generically. Of course, the conformal group depends on the structure J_0 and therefore may vary from points to points in $\mathcal{M}_g(\alpha)$, but in many interesting cases the structure J_0 which appears with the maps f is fixed by some constraints. This is of course the case for rational curves (that is those of genus 0), but also for those which are sections of ruled manifolds.

For instance, in a 4-dimensional manifold M, the dimension of the unparametrized moduli space $\tilde{\mathcal{M}}_g(\alpha)$, for a class α that can be represented by an embedded real surface, is equal to $2(3g - 3 + \alpha \cdot \alpha + e_g - n(g-1) - \dim_{\mathbf{C}}\mathrm{Conf}(g))$ where e_g is the Euler characteristic of Σ_g. This is because the Chern number of α is the sum of the Euler characteristics of the tangent and normal bundles, and in the embedded case, the latter is equal to the self-intersection of the surface.

As in algebraic geometry, there is also here a compactness theorem which forms, with the positivity of intersection, the cornerstone of the holomorphic approach:

Theorem 3.1. (Gromov) *Let (M, ω) be a geometrically bounded symplectic manifold and $f_i : (\Sigma_g, (J_0)_i) \to (M, J_i)$, $i \geq 1$, a sequence of holomorphic maps of fixed genus g such that:*

1) the sequence $\{J_i\}$ belongs to a compact set K, and

2) there exists an upper bound A on the ω-areas of the curves:

$$\int_{\Sigma_g} f_i^* \omega \leq A \quad \text{for all } i \geq 1.$$

Then there exists a subsequence weakly converging to a J-cusp-curve, for some $J \in K$.

See [1] for a precise definition of geometric boundedness (which depends on the data (M, ω, J)). Any compact manifold or any cotangent bundle is geometrically bounded, and this property is stable under products. See [34, 1] for the definition of cusp-curves and weak convergence. In the ambient space M, this theorem essentially means that the curves C_i (the image of f_i, that we may consider as singular 1-dimensional complex submanifolds) converge to a connected union of J-curves, one of which has genus g (when considered as parametrized by a surface of genus g with perhaps some non homotopically trivial curves collapsed to points) and all other having genus 0 (these rational curves are produced by bubbling off). A particular case of this theorem is that any moduli space $\tilde{\mathcal{M}}_g(\alpha, J)$ has a natural compactification.

Now let's call a spherical class $\alpha \in H_2(M, \mathbf{Z})$ *decomposable* if there exist $k \geq 2$ and $J \in \mathcal{J}$ such that α decomposes in the form $\alpha = \sum_{1 \leq i \leq k} \alpha_i$ where each α_i can be realised by some J-curve. Then, for any indecomposable α, the moduli space $\tilde{\mathcal{M}}_0(\alpha, J)$ for J a regular value of $\mathcal{P}$, is either empty or a smooth compact manifold by the compactness theorem. By Smale's generalisation of Sard's theorem to Banach manifolds ([1, 34]), the cobordism class of the manifold $\mathcal{M}_g(\alpha, J)$ corresponding to a regular value J of $\mathcal{P}$ is well-defined, it does not depends on the particular choice of J. This is because in the connected space $\mathcal{J}$, any two regular values can be joined by a path Γ transverse to $\mathcal{P}$: then $\mathcal{P}^{-1}(\Gamma)$ defines a compact cobordism between the moduli spaces at the endpoints of the path. This gives a symplectic invariant of M attached to each indecomposable class α spherical. (Here again one has to use Sobolev norms in the proof as an intermediate step, but the result can then be shown to hold in the Fréchet case as well).

This is the Fredholm setting. A similar but more delicate analysis can be developed for holomorphic curves with boundary in a given Lagrangian submanifold $L \subset M$. This has led to the proofs of some well known conjectures as well as to new results in the study of the global ambient geometry of a pair $L \subset M$. See Chapter X by Audin-Lalonde-Polterovich in [1] for an exposition of this subject, which will not be further explained here.

One of the most useful results of this approach, proved in some particular cases by Gromov in [15], is the Non-Squeezing Theorem that has been established in [23] by Lalonde-McDuff for all symplectic manifolds. Write $B^{2m}(b)$ for the standard closed round ball in $\mathbf{R}^{2m}$ whose equatorial disc has area b (that is: $b = \pi r^2$ where r is the radius of the ball).

Theorem 3.2. (Non-Squeezing) *Let (M, ω) be any symplectic manifold. Then any symplectic embedding of the standard closed ball $B(c) \subset \mathbf{R}^{2n+2}$ in $(M \times B^2(a), \omega \oplus dx \wedge dy)$ satisfies*

$$c \leq a.$$

This is shown by first embedding $M \times B^2(a)$ in its S^2-compactification $M \times S^2(a+\varepsilon)$ and studying the moduli space of J-curves of genus 0 in the class $\{pt\} \times S^2$. The proof is based on the following two lemmas:

Lemma 3.3. *Let J_0 be the standard complex structure on $\mathbf{C}^n$, $B(c)$ the standard ball and C any J_0-holomorphic curve without boundary (the image of any open subset of a Riemann surface) which passes through the center 0 of the ball and which properly intersects B, that is $C \cap B$ is a closed subset of B. Then*

$$\text{area}(C \cap B) \geq c.$$

Here the area is the usual area, which coincides with the ω_0-area on all J_0-curves because the area and the ω_0-area coincides on any J_0-invariant 2-plane. See [1] for a proof of this lemma, which also holds for more general almost complex structures.

Definition 3.4. A manifold (M, ω) is *semi-monotone* if there exists a non-negative number κ such that $c = \kappa[\omega]$ on $\text{Im}(\pi_2(M) \to H_2(M, \mathbf{Z}))$, where c is the first Chern class of $T_* M$.

The second lemma needed, proved by holomorphic techniques, is the following ([23])

Lemma 3.5. *Let M be a semi-monotone manifold. Then, given any $J \in \mathcal{J}$ on $(M \times S^2, \Omega = \omega \oplus \sigma)$ such that at least one M-fiber is J-complex, and given any point $p \in M \times S^2$, there exists a J-holomorphic 2-sphere in class $\{pt\} \times S^2$ passing through p.*

Proof of Theorem 3.2. With this lemma, one can easily prove the Non-Squeezing Theorem when M is semi-monotone. Indeed, if $f : B(c) \to M \times B^2(a) \subset M \times S^2(a+\varepsilon)$ is any symplectic embedding, there exists a Ω-tame almost complex structure on $M \times S^2$ which extends the structure $f_*(J_0)$, where J_0 is the standard structure on $\mathbf{C}^n$, and which is split near some fiber $M \times \{pt\}$ outside $M \times B^2(a)$. Then the previous lemma gives the existence of a holomorphic sphere S passing through $f(0)$. Let S_0 be the connected component of $S \cap f(B)$ passing through $f(0)$. Applying Lemma 3.3 to the J_0-curve $C = f^{-1}(S_0)$ gives:

$$a + \varepsilon = \int_S \Omega > \int_{S_0} \Omega = \int_{f^{-1}(S_0)} \omega_0 = \text{ area}(f^{-1}(S_0)) \geq c.$$

The first relation is due to the fact that Ω is a closed form, the second one holds because J is Ω-tame, and the third one because f is a Kahler diffeomorphism onto its image. Since this relation holds for all ε, we conclude that $a \geq c$.

The proof of the Non-Squeezing Theorem for general M is much more delicate, see [23]. □

We now give a sketch of the proof of Lemma 3.5 that shows how holomorphic techniques can be used. It can be skipped.

Sketch of the proof of Lemma 3.5. Write $p = (p_1, p_2) \in M \times S^2$. Let $f : U \to B^{2n}$ be any Darboux map from an open neighbourhood U of p_1 in M to a standard ball of $\mathbf{C}^n$. Extend the structure $f^{-1}(J_0)$ to a ω-tame almost complex structure J_M on M. Writing J_0 for the complex structure on S^2, consider the section $f : S^2 \to \{p_1\} \times S^2$ as a $(J_M \oplus J_0)$-curve C of $M' = M \times S^2$. It is not difficult to see that this map is a regular point of the Fredholm projection $\mathcal{P}$ since regularity is a local condition which involves only a neighbourhood of C in M' that can be taken to be $B^{2n} \times S^2$. Indeed, by [29], the differential of this Fredholm projection is equivalent to the $\bar{\partial}$ operator on maps from S^2 to $\mathbf{C}^n$ with the standard complex structures, which is onto.

Thus there is a nearby structure J' in $\mathcal{J}(\Omega)$ which is both regular and in the image of $\mathcal{P}$.

Let Γ be a path in $\mathcal{J}$ from $J_{spl} = (J_M \oplus J_0)$ to the J given in the Lemma, which is such that each $\Gamma(t)$ admits a $\Gamma(t)$-M-complex fiber. Such a path can admit only very particular kinds of cusp-curves. Indeed write $C = \sum C_i$ for a $\Gamma(t)$-cusp-curve in class $\alpha = [\{pt\} \times S^2]$. Since there is a $\Gamma(t)$-complex fiber M, the intersection of any C_i with M must have positive sign, but their sum is constrained by the homological intersection $(\{pt\} \times S^2) \cdot M = 1$. Thus only one curve intersects M (positively): this readily implies that

$$C = (A - B) + B_1 + \ldots + B_k$$

where $A \in \alpha$, $[B] = \sum[B_i]$ and all $[B_i] \in H_2(M)$. Now by the compactness theorem, the same holds for any path in $\mathcal{J}$ in a neighbourhood of $\Gamma(t)$. In particular, it holds for J' and all paths that we will consider below. Call this homological cusp-curve decomposition a *special* type.

If J' is close enough to J_{split}, the evaluation map

$$\begin{array}{rcl} ev : \mathcal{M}_0(\alpha = [\{pt\} \times S^2], J') \times_G CP^1 & \to & M' \\ (f, z) & \mapsto & f(z) \end{array}$$

where G is the conformal reparametrization group of CP^1, is regular at some point p' near p, and has only one point in the inverse image of p'. We can then choose a path J_t, $0 \leq t \leq 1$ in $\mathcal{J}$ from J' to the structure J given in the hypothesis of the lemma, and which is everywhere transversal to $\mathcal{P}$, except perhaps at J. Thus $W = \mathcal{P}^{-1}(\{J_t\})$ is a smooth manifold (except maybe at time 1). Consider the map:

$$ev : W \times_G CP^1 \to M' \times [0, 1].$$

Let γ be a path from p' to p, tranverse to ev, and denote by N the $1_{\mathbf{R}}$-dimensional manifold $ev^{-1}(\gamma)$. If it is compact, and because it contains only one point at time 0, it must also contains at least one point at time 1 and the Lemma is proved. If it is not compact, then by the compactness theorem, there must be a sequence of curves in W weakly converging to a J_{t_0}-cusp-curve C of special type. But each B_i has positive Ω-area and therefore positive ω-area, which implies by the semi-monotonicity property that their Chern numbers are non-negative. A detailed analysis of the dimension of the submanifold of M' span by cusp-curves which can generically appear in view of the index formula, shows that one can perturb slightly the path J_t and the path γ so that the corresponding manifold N avoids all cusp-curves of special type, and therefore all cusp-curves (see [30, 23]). This implies that the manifold N must reach the time 1, which means that there must exist a J-curve passing through p in class α. $\square$

Definition 3.6. The *Gromov width* or *Gromov capacity* of a manifold (M, ω) is defined as

$$c_G(M) = \sup\{c : \text{there is a symplectic embedding } f : B^{2n}(c) \to M\}.$$

Of course, the width of a ball $B(c)$ is simply c.

Remarks. 1) The Non-Squeezing Theorem holds for any ball B in products $M \times S^2$ as soon as B misses at least one fiber $M \times \{pt\}$. It has been shown that higher genus versions of the Non-Squeezing Theorem do not hold (that is when one considers products $M \times \Sigma_g$ where Σ_g is a closed surface of genus ≥ 1). Indeed the product $T^2(a) \times S^2(2a)$ has capacity $2a$ (see Lalonde [22]). And Polterovich showed that $\mathbf{R}^4$ can be symplectically embedded in the product of the 2-torus with $\mathbf{R}^2$ (!!) which therefore has infinite width.

2) One can define in a similar way a sequence of symplectic invariants: for each integer k, $c_G(M,k)$ is the supremum over all c such that the disjoint union of k balls of capacity c can be symplectically embedded in M. See McDuff-Polterovich [32] where these higher widths have been studied. Besides the usefulness of new invariants, a basic interest in these higher capacities, pointed out by Hofer, was to obtain a drastic improvement of the Poincaré return map theorem. If $\phi : M \to M$ is any symplectic diffeomorphism, the dynamics induced by the iterates of ϕ could be quite influenced by the higher widths. Indeed, define $c_G^{-1}(M,c)$ as the greatest integer k such that the disjoint union of k balls of capacity c can be symplectically embedded in M. Then, for a given point $p \in M$ and a ball $B(c)$ containing p, $c_G^{-1}(M,c) - 1$ is an upper bound for number of iterates of ϕ on B which do not intersect B. Unfortunately, it can be shown that when c becomes smaller, $c_G^{-1}(M,c)$ becomes closer and closer to the number obtained by the volume constraint. Thus these capacities are useful only for small k.

The Non-Squeezing Theorem, as well as other results of this kind, are *rigidity* constraints or invariants, in the sense that they are symplectic invariants which cannot be derived from the h-principle or from other topological constraints such as the algebraic number of double points of Lagrangian immersions for instance. In this context, a large part of Symplectic Topology could be characterized as the study of the rigidity behaviour of the underdetermined PDE systems associated to symplectic (or Lagrangian) embeddings.

The Non-Squeezing Theorem shows that the volume is not the only constraint that determines the width of a manifold. Classically, in potential theory for instance, a capacity is a relative invariant satisfying a certain number of properties. The width is an intrinsic capacity in the following sense.

Definition 3.7. An *intrinsic capacity* c is a map from the set of all symplectic manifolds (with or without boundary) of given dimension $2n$ to $[0,\infty) \cup \{\infty\}$ which satisfies the following properties:

1) $c(M,\sigma) \leq c(N,\tau)$ whenever there exists a symplectic embedding of (M,σ) in (N,τ) (monotonicity);
2) $c(M,\lambda\omega) = |\lambda| c(M,\omega)$ for all non-zero real number λ (conformality);
3) $c(B^{2n}(1),\omega_0) = c(B^2(1) \times \mathbf{R}^{2n-2},\omega_0) = 1$ (non triviality and normalisation).

Definition 3.8. A *relative capacity* is a map defined on the set of all subsets of a given symplectic manifold M (if $\partial M \neq \emptyset$, we restrict to subsets disjoint from the boundary) with values in $[0,\infty) \cup \{\infty\}$ satisfying:
1) $c(S) \leq c(S')$ if $S \subset S'$ (monotonicity);
2) $c(\phi(S)) = |\lambda| c(S)$ for any $\phi: (M,\lambda\omega) \to (M,\omega)$ (conformality);
3) given any symplectic embedding $f: B^{2n}(a) \to M$, $c(f(B^{2n}(a'))) = a'$ for all sufficiently small a' (non triviality and normalisation).

Of course, when M is compact, property (2) has a meaning only for $\lambda = \pm 1$. Note that this is a relative capacity since we do not require that property (2) holds when ϕ is defined only on some open neighbourhood of S.

We will see later many examples of these capacities: Gromov's width and the c_0 capacity of Hofer-Zehnder are examples of intrinsic capacities; the capacity of Ekeland-Hofer and the disjunction energy are examples of relative capacities (see below), the latter being defined for any symplectic manifold.

Let's return to holomorphic methods for a moment. They also enable one to define a localised version of the Non-Squeezing Theorem associated to *local* rigidity.

Let $W = (\mathbf{C}^n \times D^2, \Omega)$ where $D^2 = D^2(1)$ is the closed disk of area 1, and Ω is the split form. We take the action-angle coordinates $(c,t) \in [0,1] \times [0,1]$ on the disc, with $2\pi t = \theta$ and $c = \pi r^2$. Let S be a closed $(2n+2)$-dimensional subset of W which contains the flat disk $D_0 = \{0\} \times D^2$. Assume that ∂S contains no point of D_0 other than ∂D_0, and that it is smooth near ∂D_0. Then ∂S is tangent to ∂W along ∂D_0 and there is some neighbourhood N of D_0 in S of the form:

$$S(F) = \{(x,c,t) \in V \times D^2 : c \leq F_t(x)\}$$

where V is a neighbourhood of $0 \in C^n$ and $F_t: V \to [0,1]$, $0 \leq t \leq 1$, is smooth and such that $F_t(0) = \max_V F_t = 1$ for all t. Denote by $\phi_t, 0 \leq t \leq 1$, the isotopy generated by F_t.

Definition 3.9. A set $S \subset W$ is *unsqueezable by isotopy inside* W if, for any symplectic isotopy $f_t : S \to W, 0 \leq t \leq 1$, beginning at the identity, the image of $\pi_{D^2(1)} \circ f_t$ equals $D^2(1)$ at all times. It is *locally unsqueezable inside* W if there exists an open neighbourhood $\mathcal{U}$ of the inclusion in the space $\mathrm{Emb}_\omega(S, W)$ endowed with the C^∞-topology, such that $D^2(1) = \mathrm{Im}(\pi_{D^2(1)} \circ f)$ for all $f \in \mathcal{U}$.

Theorem 3.10. (Local Non-Squeezing Theorem) *If $\psi_t, 0 \leq t \leq 1$ has no non-constant closed trajectory in time less or equal to 1, the set $S(F)$ is both locally unsqueezable and unsqueezable by isotopy inside the cylinder W. Conversely, if $\psi_t, 0 \leq t \leq 1$ has a non-constant closed trajectory in time less than 1, then $S(F)$ is locally squeezable by isotopy inside W. Actually, there exists in this case a smooth symplectic isotopy $f_t : S(F) \to W$ beginning with the inclusion and such that the image of $\pi_{D^2(1)} \circ f_t$ is strictly smaller than $D^2(1)$ for all $t > 0$.*

Remarks. (1) The first part of the theorem is the main result of the paper [26] by Lalonde-McDuff. It had been previously proved by Ustilovsky ([39]) under some non-degeneracy condition. The main arguments of the proof are briefly described in § 5 below. It is morally based on a Local Monotonicity Lemma concerning the behaviour of properly embedded holomorphic discs in S passing through 0, which are C^∞-close to the flat disc D_0. The second part is a consequence of a deep fact on the shortening of the Hofer length of a Hamiltonian path (see § 5 below, and [39, 24]).

(2) One can use the results of Hofer's geometry to produce many examples of subspaces S in cylinders $W = M \times D^2$, where M is any symplectic manifold, which are unsqueezable through isotopies with image inside W, but which are squeezable if one considers isotopies whose images are allowed to go in larger cylinders during intermediate times !! This leads to a variety of examples of embeddings $f_i : S \to W, 1 \leq i \leq k$ which are smoothly isotopic but not symplectically isotopic.

We will see in § 5 many other applications of holomorphic techniques. Let's close this section with a last application, the C^0-rigidity mentioned earlier, which is presented here as a corollary of the Non-Squeezing Theorem. The proof consists in showing that if a map ϕ is not symplectic, it cannot satisfy the Non-Squeezing Theorem. But because the width is a C^0-invariant, any map C^0-close to ϕ must also contradicts the Non Squeezing Theorem, and therefore cannot be symplectic.

Theorem 3.11. (C^0-rigidity) *In any symplectic manifold M, the group* $\mathrm{Diff}_\omega(M)$ *of symplectic diffeomorphisms is closed in* $\mathrm{Diff}(M)$ *for the C^0-topology.*

Proof. We follow Eliashberg [7]. Since the symplectic property of a diffeomorphism is a local property, the proof reduces to the following statement by Darboux's Theorem:

"Let $\phi_k : B^{2n}(1) \to \mathbf{R}^{2n}$ be a sequence of symplectic embeddings converging C^0-uniformly to a continuous map ϕ. Assume that ϕ is differentiable at 0. Then the differential of ϕ at 0 is symplectic."

To prove this, observe first that it is enough to show that $d\phi(0)^*\omega_0 = \lambda\omega_0$ for some real number λ. Indeed, we then get $\lambda = 1$ by considering $\phi_k \times id$ instead of ϕ_k. So assume by contradiction that $d\phi(0)^*\omega_0$ is not a multiple of ω_0. Let $A \in \mathrm{End}(\mathbf{R}^{2n})$ be the symplectic transpose of $d\phi(0)$, and denote by ω the bilinear form $A^*\omega_0$. Then ω is not a multiple of ω_0, and the skew-symmetry implies that there exists $x \in \mathbf{R}^{2n}$ such that $\omega(x,\cdot)$ is not a multiple of $\omega_0(x,\cdot)$. Thus there exists $y \in \mathbf{R}^{2n}$ such that

$$\omega_0(x,y) = 1 \quad \text{and} \quad \omega(x,y) = \frac{1}{4}.$$

Recalling that $\omega(x,y) = \omega_0(Ax, Ay)$, there exist then two symplectic bases $(e_1, f_1, \dots)$ and $(e_1', f_1', \dots)$ such that:

$$e_1 = x, \quad f_1 = y \;\; e_1' = 2Ax, \quad f_1' = 2Ay.$$

In these bases, we find $Ae_1 = \frac{1}{2}e_1'$ and $Af_1 = \frac{1}{2}f_1'$, which means that

$$d\phi(0)(x_1, y_1, \dots, x_n, y_n) = (\frac{1}{2}x_1', \frac{1}{2}y_1', \dots).$$

But we may suppose that $d\phi(0)$ has the above form for coordinates in the standard basis (both at the domain and the codomain). Hence, $d\phi(0)(B^{2n}(1)) \subset B^2(\frac{1}{4}) \times \mathbf{R}^{2n-2}$. Therefore, $\phi(B^{2n}(\varepsilon)) \subset B^2(\frac{1}{3}\varepsilon) \times \mathbf{R}^{2n-2}$ for all ε sufficiently small. Finally,

$$\phi_k(B^{2n}(\varepsilon)) \subset B^2(\frac{1}{2}\varepsilon) \times \mathbf{R}^{2n-2}$$

for all sufficiently large k, which contradicts the Non-Squeezing Theorem. □

Finally, we refer the reader to [34] by McDuff-Salamon and references there for other applications of holomorphic methods, such as invariants of Gromov-Witten (or Gromov-Donaldson) type. These are numbers defined with the use of the evaluation map: they count essentially the number (with sign) of rational J-curves in a given homology class of M which meet every element of a given finite collection of representants of homology classes of M of codimension d. The cardinality of the collection, and d, are chosen so that the result be a pure number, which is then shown to be independent of the choice of J. Closely related to these invariants is the *quantum cohomology.* This is a quantum deformation of the ring structure of the ordinary cohomology of M (mod torsion) and has been recently shown to be linked with function fields over Lagrangian manifolds (see Givental-Kim [13]).

Of course, there are also applications of J-curves to problems not directly related to symplectic invariants, thus beyond the scope of this paper. Amongst those, let's mention the classification of rational and ruled symplectic 4-manifolds, the classification of contact structures on S^3, and the existence of closed orbits of the flow generated by Reeb vector fields on contact 3-dimensional manifolds.

4. Variational approach.

Gromov's width is a geometrically defined capacity. The variational study of the action functional leads to other capacities, dynamically defined. The method is based on the possibility of choosing, in a clever way, a distinguished periodic orbit of any Hamiltonian system defined on $\mathbf{R}^{2n}$ (although there might be a way to define these capacities for more general manifolds, the theory works best in euclidean spaces). There are now a survey and a very recent monograph on that subject (see [21, 20]), we refer the reader to them, and summarize here what is relevant for our purposes.

Let's first explain roughly why there should be a relation between symplectic invariants of subsets of $\mathbf{R}^{2n}$ and periodic orbits of Hamiltonian systems. Let S be a smooth compact hypersurface of $\mathbf{R}^{2n}$, and let B be the closure of the compact component of $\mathbf{R}^{2n} - S$, that we equip with the restriction of the standard symplectic structure ω_0 of $\mathbf{R}^{2n}$. Thus $\partial B = S$. The restriction of the symplectic form ω_0 to S is a closed 2-form on an odd-dimensional manifold, and therefore the following kernel of $\omega_0 \mid_S$ cannot vanish:

$$L_S = \{(x, v) \in T_*S : \omega_0(v, w) = 0 \text{ for all } \ w \in T_xS\}.$$

It is easy to check that this is a smooth line bundle over S, which is determined by any autonomous Hamiltonian H admitting S as regular level-hypersurface:

$$X_H(x) \in L_S(x) \quad \text{for all } \ x \in S.$$

Note that because S is a regular level hypersurface of H, ∇H does not vanish on S and therefore the symplectic gradient $X_H = J\nabla H$ does not vanish either. Hence any Hamiltonian, regular on S, determines L_S. The latter determines a 1-dimensional foliation $\mathcal{F}_S$ of S. A *characteristic* of S is a leaf of $\mathcal{F}$. A *closed characteristic* is a leaf of $\mathcal{F}_S$

homeomorphic to S^1. Of course, one can endow L_S, $\mathcal{F}$, and all characteristics with an orientation, the one given by the symplectic gradient of a Hamiltonian, regular on S, whose gradient points outwards. Thus closed characteristics correspond to unparametrized closed orbits of H. Let us call "multiple closed characteristic" a covering by S^1 of a closed characteristic which respect the orientation of the characteristic (these are oriented characteristics with multiplicities in $\mathbf{N}^*$). Note that any multiple closed characteristic γ encloses a well-defined (symplectic) area

$$\operatorname{area}(\gamma) = \int_\gamma \lambda = \int_D \omega_0$$

where λ is any primitive of ω_0, and D any disc with boundary γ. Let us denote by $\sigma(S) \subset \mathbf{R}$ the set of these areas, called the *action spectrum* of S. These definitions can be generalized to any smooth boundary of a compact exact symplectic manifold $(M, d\lambda)$.

If B, B' are two compact domains of $\mathbf{R}^{2n}$ with smooth connected boundaries S, S', the existence of a symplectomorphism

$$\phi : (B, \omega_0 \mid_B) \to (B', \omega_0 \mid_{B'})$$

implies that the characteristic foliations of S and S' are conjugate (there exists a diffeomorphism, namely ϕ, which sends $\mathcal{F}_S$ to $\mathcal{F}_{S'}$). Further, if $H^1(B, \mathbf{R}) = 0$, the area of the corresponding closed characteristics are equal, as one can readily check. Thus conjugation classes of characteristic foliations (and in some cases action spectrum too) are symplectic invariants, which carry a great deal of information. Indeed, in $\mathbf{R}^{2n}$, it is not excluded that the following statement holds: if ∂B and $\partial B'$ are symplectomorphic by an exact symplectomorphism (that is $\phi^*\lambda - \lambda$ is an exact 1-form on ∂B) that can be extended to a diffeomorphism of B onto B', then B and B' are symplectomorphic. Note that the hypothesis of exactness is necessary here since two discs in $\mathbf{R}^2$ of different areas have symplectomorphic boundaries. A version of the above statement has been proved by Gromov and McDuff ([15, 31]) in the case of the standard 3-sphere but when one allows any filling: they proved that a compact symplectic 4-manifold with boundary equal to the standard 3-sphere, and which contains no blown up point (= exceptional curve, that is symplectic 2-sphere of self-intersection -1), must be symplectomorphic to the standard 4-ball.

However, it is known that the interior Int$B$ does not determine the boundary ∂B: Eliashberg (see [8] for instance) has constructed examples of open convex domains of $\mathbf{R}^4$ which are symplectomorphic (and therefore exactly symplectomorphic), but whose smooth boundaries do not have conjugate foliations. However, recently Floer, Hofer, and Wysocki

[12] have shown that symplectomorphic open convex domains (and more generally starshaped domains) have boundaries with the same action spectrum. This is actually true for any exact symplectomorphism between exact compact manifolds having contact type boundaries, under a mild hypothesis of genericity. Thus it seems that, at least in the contact case, the interior of a compact manifold determines some part of the behaviour of the boundary. See [30] for instance and references there for the definition and properties of contact type hypersurfaces: they are the symplectic analogue of convex hypersurfaces.

In any event, this shows that the area of a closed characteristic is a fine symplectic invariant, that one can consider as the area of a closed orbit of a Hamiltonian having S as regular energy level. Now, let $H : [0,1] \times \mathbf{R}^{2n} \to \mathbf{R}$ be any Hamiltonian of class C^∞, and $\Omega(\mathbf{R}^{2n})$ be the space of loops $\gamma : S^1 = \mathbf{R}/\mathbf{Z} \to \mathbf{R}^{2n}$ with some Sobolev norm H^k ($k = 1/2$ for instance). Then the action functional $\mathcal{A}_H : \Omega(\mathbf{R}^{2n}) \to \mathbf{R}$ is the continuous map defined by

$$\mathcal{A}_H(\gamma) = \int_\gamma \lambda - \int_0^1 H_t(\gamma(t))\,dt.$$

When all derivatives $H_t^{(j)}, j \geq 1$ are uniformly Lipschitz on $\mathbf{R}^{2n}$ with respect to the space variable, then $\mathcal{A}_H$ is of class C^∞ and all derivatives are obtained by differentiation under the sign integral. This is the case when H has compact support or more generally when it is equal to $a(t)|z|^2 + b(t)$ outside a compact set K not depending on t, where $(a,b) : [0,1] \to \mathbf{R}^2$ is (necessarily) smooth and therefore has compact image. The gradient of $\mathcal{A}$ with respect to the L^2-inner product induced by the standard Riemannian structure of $\mathbf{R}^{2n}$ is then the following vector field along γ:

$$\nabla \mathcal{A}_H(\gamma) = -J\gamma' - \nabla H_t(\gamma)$$

so that the critical points of $\mathcal{A}_H$ coincide with the closed trajectories of period 1 of the Hamiltonian flow. Thus critical values of $\mathcal{A}_H$ are the actions of closed trajectories of period 1, which approach the areas of closed trajectories when the values of H_t along γ are small.

Therefore one can hope to recover the action spectrum of some closed compact hypersurface S by looking at the critical values of $\mathcal{A}_H$ for autonomous Hamiltonians H which are everywhere non-negative, vanish on either the interior or the exterior of S, but which increase very quickly on the other side of S. Indeed, this is the case when the hypersurface S is *stable*, which means that there is an embedding $\Phi : S \times (-\varepsilon_0, \varepsilon_0) \to U$ onto a tubular neighbourhood U of S with $\Phi(\cdot, 0)$ the obvious inclusion which is such that the identification of $S_\varepsilon = \operatorname{Im}(\Phi_\varepsilon)$ with any other $S_{\varepsilon'}$ given by Φ preserves the characteristic foliations (as an example, any hypersurface of contact type is stable). Then consider an autonomous Hamiltonian $H_{c,\delta}$ whose restriction to U is equal to the function $g \circ f$ where $f : U \to \mathbf{R}$ is given by the projection on the ε-factor, and $g : (-\varepsilon_0, \varepsilon_0) \to \mathbf{R}$ is 0 on $(-\varepsilon_0, 0]$ and then increases to $c > 0$ on a small interval $(0, \delta)$ with $\delta < \varepsilon_0$. Then when $\delta \to 0$, the gradient of $H_{c,\delta}$ on the interval $(0, \delta)$ increases from 0 to an arbitrarily large value, which implies that any closed characteristic of S will eventually appear as the closed trajectory OF PERIOD 1 of H at some level S_ε for $\varepsilon \in (0, \delta)$. Since the value c of H can be made very small (while keeping the gradient large) on U, the actions of these closed trajectories approach their areas. Thus the whole spectrum of S will eventually appear, when δ is small enough, as critical values of the action functional $\mathcal{A}_H : \Omega(\mathbf{R}^{2n}) \to \mathbf{R}$ for $H = H_{c,\delta=c^2}$. This is a basic observation, which leads to a variational definition of symplectic invariants of domains of $\mathbf{R}^{2n}$ (and of more general manifolds). Of course, one has to be careful in the choice of the extension of H outside $(-\varepsilon_0, \delta)$. One also has the choice of approaching S from the inside or the outside, that is one can define the family $H_{c,\delta}$ as being zero either on the outside or the inside respectively. The full consequences of these observations have been formalised with the recent symplectic (local) homology ([11, 38]) which has succeeded in assigning to an open domain V of a compact symplectic manifold a homology defined as the direct limit as $c \to 0$ of the Floer homology associated to the functional $\mathcal{A}_{H_{c,c^2}} : \Omega(V) \to \mathbf{R}$. To encode the full information available, one also keeps track of the critical values in the definition of this homology (actually one considers here Hamiltonians which have support inside V instead of outside). For our purposes, it will be enough to consider a simpler invariant, which is linked to only one value of the action spectrum of S (when it is not empty), which turns out to be the smallest strictly positive value of this spectrum when the hypersurface is convex. This is the capacity c of Ekeland-Hofer ([5, 6]), which is defined for all subsets of $\mathbf{R}^{2n}$. It is the outcome of a longlasting endeavour in the study of critical points of indefinite functionals, like $\mathcal{A}_H$, which has infinite index and co-index, and for which the existence of critical points cannot be deduced from the usual methods of Morse theory. Following a method due to Benci and Rabinowitz ([4]), they defined c in the following way.

One first defines a kind of capacity for C^∞-smooth Hamiltonians $H : \mathbf{R}^{2n} \to \mathbf{R}$. We will refer to it as a *quasi-selector* for reasons that will become clear below. Let $\Omega^-, \Omega^+, \Omega^0 \subset \Omega = \Omega(\mathbf{R}^{2n})$ be the vector subspaces of the Hilbert space Ω generated by $\{xe^{2\pi ikt} : k < 0, x \in \mathbf{C}^n\}, \mathbf{C}^n$, and $\{xe^{2\pi ikt} : k > 0, x \in \mathbf{C}^n\}$ respectively.

Definition 4.1. A homeomorphism $\beta : \Omega \to \Omega$ is an *admissible deformation* if $\beta = \beta_1$ is the endpoint of a homotopy β_s beginning at $\beta_0 = id$ and such that:

1) $\beta_s(\Omega - (\Omega^- \oplus \Omega^0)) = \Omega - (\Omega^- \oplus \Omega^0)$, and

2) $\beta_s(x) = a(x,s)x^+ + b(x,s)x^0 + c(x,s)x^- + K(x,s)$, where $(a,b,c,K) : \Omega \times [0,1] \to (0,\infty)^3 \times \Omega$ is continuous and sends any bounded set to a relatively compact set.

Denote by $\mathcal{B}$ the set of all admissible deformations. Note that $\mathcal{B}$ is stable under composition. Then:

Definition 4.2. Let $H \in C^\infty(\mathbf{R}^{2n}, \mathbf{R})$. The *quasi-selector* $\delta(H)$ is defined by applying to $\mathcal{A}_H$ the following maximin:

$$\delta(H) = \sup_{\beta \in \mathcal{B}} \inf_{\beta(S^+)} \mathcal{A}_H \quad \in \mathbf{R} \cup \{-\infty, \infty\},$$

where S^+ is the unit sphere of Ω^+.

It follows directly from the definition that β is continuous for the C^0-topology and that it is monotone:

$$\delta(H) \geq \delta(K) \text{ if } H \leq K.$$

The important facts are summarized in the following theorem:

Theorem 4.3. *Denote by $\mathcal{H}_{ad}$ the subset of $C^\infty(\mathbf{R}^{2n}, [0,\infty))$ of* admissible *Hamiltonians, that is those which satisfy:*

1) Int$H^{-1}(0) \neq \emptyset$

2) there exist $a, b \in \mathbf{R}$ with $a > \pi$ and not belonging to $\mathbf{N}\pi$, such that $H(z) - (a|z|^2 + b)$ has compact support.

Then, for all $H \in \mathcal{H}_{ad}$, $\delta(H)$ is a strictly positive critical value of $\mathcal{A}_H$. Further, $\delta : \mathcal{H}_{ad} \to \mathbf{R}$ is invariant under composition with compactly supported symplectic diffeomorphisms of $\mathbf{R}^{2n}$.

See the nice exposition [37] by Sikorav for a proof of this theorem, which is based on the minimax theory of Lyusternik-Schnirelman and the fact that when H belongs to $\mathcal{H}_{ad}$, the functional $\mathcal{A}_H$ satisfies the Palais-Smale condition.

Finally, one defines the Ekeland-Hofer capacity of any subset B of $\mathbf{R}^{2n}$ by a procedure which consists, as we said above, in approaching the boundary of the set B by Hamiltonians which are zero on B and increase quickly outside B. More precisely, write $\mathcal{H}_{ad}(B)$ for the subset of $\mathcal{H}_{ad}$ of those H which vanish on some open subset containing the closure $\bar{B}$ of B. Then

Definition 4.4. The Ekeland-Hofer *capacity* of the set B is

$$c(B) = \inf_{H \in \mathcal{H}_{ad}(B)} \delta(H) \in [0, \infty)$$

when B is bounded and

$$c(B) = \sup_{B' \subset B \text{ is bounded}} c(B') \in [0, \infty]$$

otherwise.

Note that the monotonicity of $\delta(H)$ means that $c(B)$ is defined by looking at Hamiltonians that increase very quickly near the boundary of B. Then we have:

Proposition 4.5. *c is an relative capacity, which moreover satisfies $c(B^2(1) \times \mathbf{R}^{2n-2}) = 1$.*

All the properties are easy to prove, except $c(B^{2n}(1)) \leq 1$ (see [37], p.30).

One can use this capacity to prove the Non-Squeezing Theorem in the euclidean case. Indeed, if $\phi : B^{2n}(a) \to B^2(a') \times \mathbf{R}^{2n}$ is a symplectic embedding, then it can be extended to a Hamiltonian diffeomorphism. Hence, by monotonicity and ambient symplectic invariance,

$$c(B^2(a') \times \mathbf{R}^{2n}) \geq c(\phi(B^{2n}(a))) = c(B^{2n}(a)).$$

By normalisation and conformality, we conclude that $a' \geq a$. Hence, as in § 3, one can derive the C^0-rigidity too. One of the most elegant applications of this approach is the proof of the Weinstein conjecture in $\mathbf{R}^{2n}$, which states that any closed compact smooth hypersurface of $\mathbf{R}^{2n}$ of contact type has at least one closed characteristic. With Theorem 4.3, one can actually prove that any stable hypersurface has a closed characteristic. Indeed, let $\{S_\varepsilon\}$ define a foliated neighbourhood V of S by hypersurfaces having equivalent characteristic foliations. Define on $\mathbf{R}^{2n}$ a Hamiltonian $H \in \mathcal{H}_{ad}$ which is constant on the bounded component of $\mathbf{R}^{2n} - V$, which is equal to a monotone increasing function $f(\varepsilon)$ on V, which equals a constant from V to some large $(2n-1)$-sphere containing V, and equals a quadratic function $g(|z|^2)$ outside the sphere. It is easy to see that, if $f(\varepsilon_0)$ is taken large enough, one can choose g so that there is only one 1-periodic orbit of $g(|z|^2)$, and this orbit has negative action. By Theorem 4.3, there exists a periodic orbit of period 1 of H with positive action, which can only lie inside V.

Note that this argument, due to Hofer and Zehnder, shows that, for *any* hypersurface of $\mathbf{R}^{2n}$, the set of ε for which S_ε has a closed characteristic is dense (it is even of co-measure zero, by an improvement due to Struwe).

The above capacity is a very useful tool for the study of symplectic properties of subsets of $\mathbf{R}^{2n}$: its main advantage is that one can play with both aspects, the definition with the maximin procedure which implies many useful properies, and the property of being a critical value of some action functional which is easy to handle and to apply to many particular problems. But it has also some disavantages, for instance the fact that, since the whole geometry of $\mathbf{R}^{2n}$ intervenes in the definition, the capacity $c(B \subset \mathbf{R}^{2n})$ is not invariant under a symplectomorphism defined on an open neighbourhood of B. Also, it cannot be easily extended to subsets of more general manifolds. Recently, Hofer and Zehnder ([19]) have introduced a dynamically defined intrinsic capacity attached to each symplectic manifold M with or without boundary, that has some relations with the capacity c. It is defined by $c_0(M) = \sup_{H\in\mathcal{E}} \max_M H \in (0, \infty]$ where $\mathcal{E}$ is the set of all positive Hamiltonians $H : M \to [0, \infty)$ which vanish on some open subset U of M, which are equal to their maximum outside a compact subset K and which have no non-constant closed trajectory in time less or equal to 1. Here U and K depend on each Hamiltonian H, and are assumed to be disjoint from ∂M. With this capacity, the difficulty is to prove the finiteness of c_0, but once this is done, all applications above can be proved in a quite elegant way using the properties of a capacity for c_0. See [19, 20, 21] and references there for results in this direction.

We end this section with the definition of a very useful tool introduced by Hofer, the selector, which is a variant of the quasi-selector, but defined this time for compactly supported (not necessarily autonomous) Hamiltonians. We will need it in the next section.

Let $\mathcal{H}$ be the set of all compactly supported smooth functions $H : [0,1] \times \mathbf{R}^{2n} \to \mathbf{R}$ and denote by ϕ_H the time-1 flow.

Definition 4.6. A homeomorphism $\beta : \Omega \to \Omega$ is a *bounded admissible deformation* if:

1) β and β^{-1} map bounded sets into bounded sets, and

2) $\beta(x) = a(x)x^+ + x^0 + c(x)x^- + K(x)$, where $(a, c, K) : \Omega \to (0, \infty)^2 \times \Omega$ is continuous and sends any bounded set to a relatively compact set. Further, there exists a constant $R > 0$ depending on β such that $(a, c, K)(x) = (1, 1, 0)$ for all $x \in \Omega^+$ outside the ball of radius R. Let $\mathcal{B}_b$ denote the set of bounded admissible deformations.

Admissible deformations were appropriate when dealing with critical points of Hamiltonians quadratic at infinity; bounded admissible deformations are the ones needed for compactly supported Hamiltonians. Note that in both cases, the space of homeomorphisms of Ω is chosen so that it be sufficiently small to be manageable, but sufficiently large to ensure that the maximin of the action functional over the family of subsets corresponding to all admissible deformations gives a critical value of the action. It is chosen so that it satisfies the conditions under which the Lyusternik-Schnirelman minimax method for finding critical points works.

Definition 4.7. The *selector* $\gamma(H)$ is defined by the following maximin process:

$$\gamma(H) = \sup_{\beta \in \mathcal{B}_b} \inf_{\beta(\Omega^+)} \mathcal{A}_H \quad \in [0, \infty).$$

Proposition 4.8. (Hofer-Zehnder [21]) γ *is a continuous function* $\mathcal{H} \to [0, \infty)$ *which satisfies:*

1) $\gamma(H \equiv 0) = 0$, *and* $\gamma(H) > 0$ *if* $H \leq 0$ *and is not identically* 0*;*

2) $\gamma(H) \geq \gamma(K)$ *if* $H \leq K$*;*

3) $\gamma(H)$ *depends only on* ϕ_H, *it belongs to* $\sigma(\phi_H)$, *and is invariant under conjugation.*

Here $\sigma(\phi)$ is the *action spectrum* of the compactly supported Hamiltonian diffeomorphism ϕ defined in the following way. If $H \in \mathcal{H}$ is any Hamiltonian which generates ϕ, then

$$\sigma(\phi) = \{a \in \mathbf{R} : a \text{ is a critical value of } \mathcal{A}_H\}$$

which is of course not empty since $0 \in \sigma(\phi)$. It turns out that this set is independent of the choice of H such that $\phi = \phi_H$, and that $\sigma(\phi)$ is compact and nowhere dense as a subset of $\mathbf{R}$ (see [21], p.151). The property 3 above shows that γ induces a function on the group $\mathrm{Ham}(\mathbf{R}^{2n})$ of Hamiltonian diffeomorphisms with compact support, invariant under conjugation. Thus $\gamma : \mathrm{Ham}(\mathbf{R}^{2n}) \to [0, \infty)$ selects a distinguished element of the spectrum, that is to say a distinguished critical value of the action on the space of loops.

5. Hofer's geometry.

Hofer's geometry is the study of the group $\mathrm{Ham}(M)$ of compactly supported Hamiltonian diffeomorphisms with a norm defined as the total amount of energy needed to generate a given diffeomorphism. The interest in this geometry arose from a combination of factors: firstly, it turns out that this geometry is closely related to some important questions concerning the rigidity aspects of symplectic geometry, such as the classification at the π_0-level of symplectic embeddings for which there are no soft obstructions. Secondly, Eliashberg and Polterovich ([9]) showed that this norm is essentially the only bi-invariant (that is to say: intrinsic) metric on the group Ham(M) amongst those which are naturally defined as the infimum of length of paths between points. Since this group has an obvious importance in classical mechanics, as well as in thermodynamics (where it serves as the configuration space), this is another reason that prompted its development. Thirdly, it has the (pedagogical) advantage of expressing some of the most elusive aspects of rigidity in a geometric framework which leads naturally to the formulation of questions that would not have arisen so easily without it: the generalised Non-Squeezing Theorems of Lalonde-McDuff are examples of this (see below). Finally, Bialy and Polterovich recently showed that the Hofer norm is closely related to the classical problem of KAM-tori breaking mechanism ([2, 3]).

Let M be any symplectic manifold, and denote by Ham(M) the group of time-one maps of compactly supported C^∞-Hamiltonians $H : M \times [0,1] \to \mathbf{R}$. It is an infinite dimensional Lie group whose Lie algebra $\mathcal{H}am$ is the algebra of exact vector fields (that is: vector fields on M whose duals, defined by the non-degenerate form ω, are exact 1-forms), that we may identify with $C_0^\infty(M)/\mathbf{R}$, the Poisson algebra of smooth compactly supported functions modulo constants. The Hofer norm on $\mathcal{H}am(M)$ is the total variation of functions:

$$\|X\| = Totvar(H) = \max_M H - \min_M H$$

where X is the symplectic gradient of H. Because this norm is invariant under composition with Hamiltonian diffeomorphisms, it defines a bi-invariant norm on Ham(M). The Hofer length of a path $\gamma(t), 0 \le t \le 1$ between two points of Ham(M) is defined in the usual way:

$$\mathcal{L}(\gamma) = \int_0^1 \|\frac{d\gamma}{dt}(t)\|\, dt = \int_0^1 Totvar(H_t)\, dt$$

where the symplectic gradient of $H_t, a \le t \le b$ is $\frac{d\gamma}{dt}(t)$. The distance $d(\phi, \psi)$ between two points is the infimum of the length of all paths joining these two points. In particular, the Hofer norm is the distance from the identity:

$$\|\phi\| = \inf_H \int_0^1 Totvar(H_t)\, dt$$

where the infimum is taken over all compactly supported Hamiltonians $H : [0,1] \times M \to \mathbf{R}$ whose time-one maps are equal to ϕ. Of course, by bi-invariance,

$$d(\phi, \psi) = \|\phi\psi^{-1}\|.$$

All properties of a norm are easily proved ($\|\phi\| \geq 0$ and $= 0$ iff $\phi = id$, $\|\phi\| = \|\phi^{-1}\|$, $\|\phi \circ \psi\| \leq \|\phi\| + \|\psi\|$) , except the non-degeneracy ($\|\phi\| = 0 \Longrightarrow \phi = id$, see below). Note that these properties correspond to the properties of a metric for d.

So Hofer's metric is a Finslerian metric, defined as the length metric arising from the total variation of functions. It is L^1 on the time variable, and L^∞ on the space variable.

It was introduced by Hofer in [16], who used it to prove that orbits of a system of n independent harmonic oscillators persist under non-autonomous perturbations whenever the perturbation is C^0-bounded ! The development of this geometry appears a priori as a difficult task, since it presents at least two independent difficulties. The first one comes from the group itself: it is an infinite dimensional group whose exponential map is not onto, even locally. Thus, a fortiori, the tangent space at $\phi_t, 0 \leq t \leq 1$ of the space $\mathcal{P}$ of all paths between two fixed points $\phi_0, \phi_1 \in \mathrm{Ham}(M)$, does not exponentiate onto an open neighbourhood of the path. This means, for instance, that even if the second derivative of the Hofer length functional $\mathcal{L} : \mathcal{P} \to \mathbf{R}$ were say positive definite at ϕ_t, one could not conclude that that path is a local minimum of $\mathcal{L}$. The second, more serious difficulty, comes from the L^∞-nature of the Hofer metric which makes its analysis more awkward and necessitates the use of non smooth analysis.

Outline of the main results

The first steps in the development of that geometry were taken by Hofer, who showed in [16] that this pseudo-metric was indeed a non-degenerate metric on $\mathrm{Ham}(\mathbf{R}^{2n})$. Note that all properties of this pseudo-metric can be derived directly from the definitions, except the fact that $\phi \neq id$ implies that $\|\phi\| \neq 0$, which turned out to be uneasy to prove. For this, Hofer introduced another relative capacity, the *energy of disjunction* defined as follows on any subset $B \subset \mathbf{R}^{2n}$:

$$e(B) = \inf\{\|\phi\| : \phi \in \mathrm{Ham}(\mathbf{R}^{2n}) \text{ and } \phi(B) \cap B = \emptyset\}.$$

Of course, any Hamiltonian diffeomorphism, which is distinct from the identity, must disjoin some small ball from itself. Thus its energy is bounded below by the disjunction energy of that ball. So, in order to prove that $\| \quad \|$ is non-degenerate, it is enough to show that $e(B) > 0$ for all non-empty balls. This is a consequence of

Theorem 5.1. (Energy-capacity inequality) *For any subset $B \subset \mathbf{R}^{2n}$*

$$e(B) \geq c(B).$$

As a corollary, $e(B^{2n}(a)) = a$.

See Hofer [16] (or Sikorav [37]) for the proof of this inequality. We present below a proof of a more general energy-width inequality. Note that the proof of the corollary is immediate. Indeed, $e(B^{2n}(a)) \geq a$ by the theorem, and the inequality $\leq$ is obtained by constructing explicitely, for each $\varepsilon > 0$, a diffeomorphism of energy $\leq a + \varepsilon$ which disjoins the ball. For this, consider the ball inside a cylinder $B^2(a) \times \mathbf{R}^{2n-2}$, which is conjugate to the product $([0, \sqrt{a}] \times [0, \sqrt{a}]) \times \mathbf{R}^{2n-2}$. It is easy to see then that the square can be disjoined by a diffeomorphism of energy $a + \varepsilon$.

Note that this gives a lower bound of the norm in terms of C^0-data. The second step, in [37, 17] by Sikorav and Hofer, is an upper bound in terms of C^0-data:

Theorem 5.2. (Sikorav-Hofer) *For any* $\phi \in \mathrm{Ham}(\mathbf{R}^{2n})$,

$$\|\phi\| \leq 256 \cdot \mathrm{diam}(supp(\phi)) \cdot \|\phi\|_{C^0}$$

where the C^0*-norm is* $\sup|\phi(x) - x|$.

This implies in particular that this norm is continuous for the C^0-topology on $\mathrm{Ham}(\mathbf{R}^{2n})$, a non obvious result. Hofer's length is clearly continuous with respect to the C^0-norm on the generating Hamiltonians, but it is less evident how to show that C^0-small diffeomorphisms can always be generated by Hamiltonians with small mean total variations. Consider the following example: triangulate any compact domain K of a manifold M by small simplices, and define $H : M \to \mathbf{R}$ to be equal to the sum of high bump functions with support in each of the $2n$-dimensional simplices. Then the time-one map of H is C^0-close to the identity, but it is not evident how to decrease the total variation of H while keeping the time-one flow unchanged. Indeed, C^0-continuity is an open question for manifolds as elementary as S^2.

The following result exhibits an essential feature of that geometry, which relates it to important properties of Hamiltonian systems.

Theorem 5.3. (Hofer [17]) *On* $\mathrm{Ham}(\mathbf{R}^{2n})$, *any path* $\phi_t, t \leq 0$, *beginning at the identity and generated by an autonomous Hamiltonian* H *is absolutely length minimizing until time* t_0 *where the first non-constant closed orbit appears:*

$$\|\phi_t\| = t\|H\| = t\mathrm{Totvar}(H)$$

for all $t < t_0$.

This result on absolutely minimizing geodesics in Hofer's metric was proved by considering the selector of § 4: for the given Hamiltonian H, and any non-autonomous Hamiltonian K_t whose time t flow is also ϕ_t, Hofer defined an action functional by gluing the Hamiltonians H and K (H on the first half of the time interval, and K on the second one). He then used in a clever way the properties of this selector arising from both the minimax definition of the selector and from the fact that the value of the selector must be a critical value of the action functional, in order to show that

$$\int_0^t \|H\| \leq \int_0^t \|K_t\|.$$

See below a detailed sketch of the proof of this theorem. It has recently been generalised by Siburg [36] to *quasi-autonomous* Hamiltonians defined on $\mathbf{R}^{2n}$ (see below the definition of quasi-autonomous).

By a purely geometric construction, Lalonde-McDuff [23] showed that the energy-capacity inequality is equivalent to the Non-Squeezing Theorem in any symplectic manifold, and proved the Non-Squeezing Theorem in full generality, using holomorphic techniques (see § 3). Here the capacity is the Gromov's width. Note that it is obvious that if the energy-capacity inequality $e(B) \geq c_G(B)$ holds for any subset $B \subset M$ of a symplectic manifold, the Non-Squeezing Theorem must also holds because any cylinder $B^2(a) \times M \subset \mathbf{R}^2 \times M$ can be disjoined from itself by a diffeomorphism of energy as close as we wish to a. The converse is the consequence of a geometric construction where one uses the characteristic foliation on the graph of the Hamiltonian which generates the disjoining diffeomorphism in order to embed a large ball by a kind of stabilized version of the figure 8 trick. Thus:

Theorem 5.4. (Lalonde-McDuff) *For any subset B of any symplectic manifold:*

1) $e(B) \geq \frac{1}{2}c_G(B)$, and

2) $e_p(B) \geq c_G(B)$.

As a corollary, Hofer's pseudo-metric is non-degenerate on any manifold.

The proof in the case $M = \mathbf{R}^{2n}$ is given below. Here $e_p(B)$ is the proper disjunction energy, that can be easily defined on balls as the energy needed to disjoin a ball B of capacity c from a ball of capacity $2c$ containing B.

Very recently, a more detailed analysis of Hofer's geometry has been undertaken and apppeared in [2, 24, 25, 26, 27, 28, 36, 39]. We now present the results of these works, which give complete answers to many basic questions about the behaviour of geodesics in that geometry.

Definition 5.5. Let $\phi_t, 0 \leq t \leq 1$ be any path in $\mathrm{Ham}(M)$. It is a *geodesic* if it is a regular path (in the sense that its tangent vectors never vanish) and if it is a local minimum of the Hofer length at each moment.

Here, for a given path $\phi_t, 0 \leq t \leq 1$ from ϕ_0 to ϕ_1, denote by $\mathcal{P}_{\{\phi_t\}}$ the space of paths from ϕ_0 to ϕ_1 which are homotopic to $\{\phi_t\}$. Then $\phi_t, 0 \leq t \leq 1$, is a *local minimum* of the length if the Hofer length functional $\mathcal{L} : \mathcal{P}_{\{\phi_t\}} \to \mathbf{R}$ has a local minimum at $\{\phi_t\}$ with respect to the C^∞-topology. A path is a local minimum *at each moment* if each t belongs to a connected neighbourhood N such that the restriction of the path to that interval is a local minimum of the length.

Definition 5.6. (Bialy and Polterovich) A *fixed extremum* of a Hamiltonian $\{H_t\}_{t\in[0,1]}$ (or of the path that it generates) is a point $x \in M$ which is either a minimum of each H_t, or a maximum of each H_t. A path that has at least one fixed minimum and one fixed maximum is called *quasi-autonomous.*

Theorem 5.7. (Bialy-Polterovich) *A path is a geodesic if and only if it has at each moment at least one fixed minimum and one fixed maximum (that is to say: if and only if it is quasi-autonomous at each moment).*

This characterization was obtained by Bialy-Polterovich in $\mathbf{R}^{2n}$ using action spectrum analysis ([2]), and was generalised by Lalonde-McDuff to manifolds using holomorphic methods ([25]).

As above, "each moment" means that the property holds in the (time) neighbourhood of any $t \in [0, 1]$.

Let us now consider stability. A *stable* geodesic is a geodesic $\phi_t, 0 \leq t \leq 1$, which is a local minimum of the length on the corresponding space $\mathcal{P}_{\phi_t}$ of all paths, homotopic to ϕ_t, from ϕ_0 to ϕ_1. Of course, the notion of geodesic does not depend on the endpoints but the notion of stability does. The following fundamental theorem gives an essentially sharp characterisation of stability: it was proved by Ustilovsky under non-degeneracy hypotheses, and in general by Lalonde-McDuff. It is the main result of papers [39, 24, 26, 27]:

Theorem 5.8. *If a regular path $\phi_t, 0 \leq t \leq 1$ has a fixed minimum and a fixed maximum where the linearised isotopy has no non-constant closed trajectory in time less or equal to 1, then it is a stable geodesic. Conversely, a stable geodesic $\phi_t, 0 \leq t \leq 1$ must have at least one fixed minimum and one fixed maximum, and assuming that all fixed extrema* *are isolated, it must then have at least one fixed minimum p and one fixed maximum P where the linearised isotopies $d\phi_t, 0 \leq t \leq 1$, have no non-constant closed trajectory in time less than* 1.

It is not known whether there exist stable geodesics with non isolated fixed extrema such that the linearised isotopy has a non-constant closed trajectory in time < 1 at each fixed minimum or at each fixed maximum. But clearly this cannot happen in dimension 2 since any fixed extremum with a non-constant closed trajectory in time < 1 must be isolated. Also, this cannot happen, in any dimension, for a path whose endpoint is a generic Hamiltonian diffeomorphism. Note the following obvious corollary:

Corollary 5.9. *Any endpoint of a stable geodesic must have at least two fixed points.*

It is not known either whether a Hamiltonian diffeomorphism of a compact symplectic manifold must always have at least two fixed points. This corollary would provide a proof of that if one could show that any Hamiltonian diffeomorphism is the endpoint of a stable geodesic. But this is false ([24]):

Theorem 5.10. (Lalonde-McDuff) *There exist Hamiltonian diffeomorphisms of S^2 which cannot be joined to the identity by a stable geodesic. A fortiori, there is no shortest path joining them to the identity.*

Another basic question on Hofer's geometry is the diameter of $\mathrm{Ham}(M)$ in Hofer's norm for closed manifolds: in particular, is it always unbounded ? In [37], Sikorav showed, by an elementary but very tricky argument, that although the diameter of $\mathrm{Ham}(\mathrm{Int}B^{2n}(c))$ is infinite (a direct consequence of the Calabi invariant), it is contained in a bounded subset of $\mathrm{Ham}(\mathbf{R}^{2n})$ (here $\mathrm{Int}B^{2n}(c)$ is the interior of the standard ball of capacity c of $\mathbf{R}^{2n}$). In other words, there is some constant C such that any Hamiltonian diffeomorphism ϕ generated by a Hamiltonian with compact support in $\mathrm{Int}B^{2n}(c)$, satisfies

$$\|\phi\| \leq Cc$$

when ϕ is considered as an element of $\mathrm{Ham}(\mathbf{R}^{2n})$ (this is included in the statement of Theorem 5.2). Of course, this implies that the Hamiltonian of $\mathbf{R}^{2n}$ of length $Cc + \varepsilon$ whose time-one flow is ϕ must have very large support, much bigger than $B^{2n}(c)$. This suggests that the above question for closed manifolds might be difficult to answer in general. An easy argument based on the energy-width inequality on the universal cover of closed orientable surfaces Σ_g of genus ≥ 1 shows: $\mathrm{Ham}(\Sigma_g)$ is unbounded in Hofer's metric for all $g \geq 1$. A less obvious consequence, proved by holomorphic techniques [24], is:

Theorem 5.11. (Lalonde-McDuff) $\mathrm{Ham}(M \times \Sigma_g)$ *is unbounded in Hofer's metric for any* $g \geq 1$ *and any compact symplectic manifold* M.

Sketch of the proof of Theorem 5.4

We begin by proving Theorem 5.4, restricting ourselves to the case $M = \mathbf{R}^{2n}$. The proof for general manifolds is very similar (see [23]), at least if one can assume the general Non-Squeezing Theorem 3.2. We first need the following lemma:

Lemma 5.12. *Suppose that* $0 < \delta \ll c < C$, *and let* $Y = U_1 \cup S \cup U_2 \subset \mathbf{R}^2$ *be the union of two rectangles,* U_1 *of area* $C + \delta - c$ *and* U_2 *of area* c, *joined by a very thin strip* S *of area* δ. *Further, let*

$$Z_{C,c} = B^{2n}(C) \times U_1 \;\cup\; B^{2n}(c) \times (S \cup U_2).$$

Then, there is a symplectic embedding $B^{2n+2}(C) \hookrightarrow Z_{C,c}$.

Proof. Let π be the projection $B^{2n+2}(C) \to B^2(C)$ induced by projection onto the first two coordinates. This represents $B^{2n+2}(C)$ as a kind of fibration over the disc, with fibers equal to concentric balls of different capacities. Note that, for every c, the set

$$\{\, x \in B^2(C) : c(\pi^{-1}(x)) \geq c\}$$

is exactly $B^2(C - c)$. Clearly, because $\delta > 0$, there is an area preserving embedding $\phi : B^2(C) \hookrightarrow Z_{C,c}$ which takes $B^2(C - c)$ into U_1: actually, ϕ sends an open neighbourhood of

$$B^2(C - c) \cup (\,(-\infty, 0) \times \{0\} \cap B^2(C)\,) \subset B^2(C)$$

into U_1. The desired result follows immediately. □

Proposition 5.13. *For any compact subset* $A \subset M = \mathbf{R}^{2n}$,

$$e(A) \geq c_G(A).$$

Proof. Let $\phi_t, 0 \leq t \leq 1$ be any compactly supported path such that $\phi = \phi_1$ disjoins A from itself. We may assume that the Hamiltonian H_t which generates ϕ_t is 0 near $t = 0, 1$. Note that the hypersurface Q defined by

$$gr(H) = \{(x, t, s) \in M \times [0, 1] \times \mathbf{R} \mid s = H_t(x)\}$$

has a characteristic foliation whose monodromy defines a map $M \times (0, 0) \to M \times (1, 0)$ equal to ϕ. Define the energy e of Q as the area of its projection on $[0, 1] \times \mathbf{R}$, which is equal to $\mathcal{L}(\{H_t\})$. Let $B \subset A$

be the image of a standard ball of capacity c. We must then show that $e \geq c$. This will follow if, for any $\varepsilon > 0$, we can find an embedding Φ of the ball $B^{2n+2}(2c)$ of capacity $2c$ into the product $\mathbf{R}^{2n} \times B^2(e+c+\varepsilon)$, since the Non-Squeezing Theorem then tells us that $2c \leq e+c+\varepsilon$.

By the Lemma, it suffices to embed $Z_{2c,c}$ symplectically into $\mathbf{R}^{2n} \times X$, where X is an annulus of area $e + c + \varepsilon/2$. By hypothesis, there is a rectangle R in $[0,1] \times \mathbf{R}$ of area e such that $Q \subset \mathbf{R}^{2n} \times R$. Note that, by hypothesis, Q is flat near its ends, that is to say Q coincides with the hypersurface $\{s = 0\}$ near the boundary $t = 0, 1$. Let R' be another rectangle in $\mathbf{R}^2$ of area c with one edge along $t = 1$, so that $R \cup R'$ is a rectangle of area $e + c$ with one edge along $t = 0$ and another along $t = t_1 > 1$, say, and then form X by identifying these two edges and adding a little area to both R and R'.

Let $g : B^{2n}(c) \to B \subset A$ be a symplectic embedding, and extend g to an embedding, which we also call g, of $B^{2n}(2c) \to \mathbf{R}^{2n}$. (This is possible because the space of embedded symplectic balls of any given radius in $\mathbf{R}^{2n}$ is path connected so that any ball is isotopic to a standard one and thus can be extended as much as we wish.) Clearly, we may suppose that the ball $g(B^{2n}(2c))$ is disjoint from $\phi_Q(B)$, where ϕ_Q is the monodromy of Q. (To see this, note that because the space of symplectic embeddings of two balls in $\mathbf{R}^{2n}$ is path-connected, there is a symplectomorphism τ which is the identity on B and which moves $\phi_Q(B)$ far away, and so we may alter Q without changing its energy to a hypersurface Q' with monodromy $\tau \circ \phi_Q \circ \tau^{-1}$.) Then, we define the embedding $Z_{2c,c} \to \mathbf{R}^{2n} \times X$ as follows.

- $B^{2n}(2c) \times U_1$ goes to $\mathbf{R}^{2n} \times R'$ by $g \times \iota$, where $\iota : U_1 \to R'$.
- Think of the strip S as a slight thickening of an interval I, and map $B^{2n}(c) \times I$ to the hypersurface $Q \subset \mathbf{R}^{2n} \times R$ by taking each line $\{x\} \times I$ to the corresponding flow line of the characteristic flow on Q. Since X contains a neighborhood of R, this embedding extends to S provided that δ is sufficiently small by the symplectic neighborhood theorem.
- $B^{2n}(c) \times U_2$ goes onto $\phi_Q(B) \times R'$ by the map $(\phi \circ g) \times \iota$.

It is very easy to check that this gives the desired symplectomorphism. □

Sketch of the proof of Theorem 5.3

Proof. We give now a brief sketch of the main arguments needed to prove Theorem 5.3.

Suppose that two compactly supported Hamiltonians

$$H, K : \mathbf{R}^{2n} \times [0,1] \to \mathbf{R}$$

are given, and assume that the first one, H, is autonomous and has no non-constant periodic orbit of period less or equal to 1. Let ϕ_t, ψ_t denote the diffeotopies generated by H, K respectively, with $\phi_1 = \psi_1$. We wish to show that the Hofer length of ψ_t is greater or equal to the Hofer length of ϕ_t. Hofer actually proved more than that: he showed that

$$\begin{aligned} \|K\|^+ &= \int_{[0,1]} \max K_t dt \geq \|H\|^+ (= \max H) \quad \text{and} \\ \|K\|^- &= \int_{[0,1]} \min K_t dt \leq \|H\|^- (= \min H). \end{aligned}$$

So first let's concatenate the two Hamiltonians. Define the time dependent Hamiltonian $(-H) * K : \mathbf{R}^{2n} \times [0,1] \to \mathbf{R}$, by taking $-2H_{2t}$ during the first half interval $[0,1/2]$ and $2K_{2t-1}$ during the second half $[1/2,1]$. Then of course the time-one map $\psi_1 \circ \phi_1^{-1}$ of this Hamiltonian is the identity.

Now here is the main point: the selector $\gamma((-H) * K)$ can be interpreted both through the maximin procedure and through the choice of a particular value of the action of a closed trajectory of period 1 of the Hamiltonian $(-H) * K$. Let's write $F = (-H) * K$, and f_t for the corresponding Hamiltonian flow.

1) The second point of view leads immediately to $\gamma(F) = 0$ because the corresponding diffeotopy f_t is a loop in the Hamiltonian group. So, for each point $x \in \mathbf{R}^{2n}$, $f_t(x)$ is a closed trajectory of period 1. The value of the action on such a loop is

$$\int_{f_t(x)} \lambda - \int_{f_t(x)} F_t dt.$$

But the value of the first term is the area enclosed by the loop, which is equal to the area of the "cone" $\{f_t(\alpha) \mid t \in [0,1]\}$ where α is any path from a point p out of the support to the point x. The area of this cone is equal to the integral over time of $F_t(f_t(x)) - F_t(f_t(p)) = F_t(f_t(x))$. Thus the action is zero on all closed trajectories, so $\gamma(F) = \gamma((-H) * K) = 0$.

2) The first point of view gives: $\gamma((-H)*K) \geq \gamma((-H)*0) - \|0*K\|^+$. This inequality is immediate with the maximin definition: the infimum over any set of loops of $\int \lambda - \int_{[0,1/2]} (-2H) - \int_{[1/2,1]} 2K_{t-1}$ is clearly greater or equal to

$$\inf \left(\int \lambda - \int_{[0,1/2]} (-2H) \right) + \inf \left(- \int_{[1/2,1]} 2K_{t-1} \right)$$

which gives the inequality.

Thus (1) and (2) give:

$$\|K\|^+ \geq \gamma((-H) * 0).$$

Note that we have not made use yet of the fact that H is autonomous and of our assumptions about its closed orbits. The new point is:

If G is an autonomous Hamiltonian with no non-constant periodic orbit of period less or equal to 1, *then* $\gamma(G * 0) = \gamma(G) \geq -\min G = -\|G\|^-$.

The proof of this proceeds in two steps. The selector is the action of some periodic orbit of period 1. But this orbit must be a trivial one, so $\gamma(G) = -G(p)$ for some critical point p of G. To show that this point is the minimum of G is the second step: since the selector is continuous, this point p must be the same for all Hamiltonians sG where $s \in (0, 1]$. But for s small enough, the first and second derivatives of sG become so small that sG is uniformly bounded above by a quadratic function Q that has same minimum as sG. Apply now the monotonicity property of the selector:

$$\gamma(sG) \geq \gamma(Q).$$

Note that Q can be chosen quadratic with small second derivative on all the support of G and then smoothed out to 0 outside $supp(G)$. Then $\gamma(Q)$ is either $-\min Q$ or 0. But if it were 0, one could deform Q to a function with values in $(-\infty, 0]$ with no negative closed trajectory except at $\min Q$, and by continuity the selector would also be 0 at the end of the deformation, contradicting Proposition 4.8 (1). Hence:

$$\gamma(sG) \geq \gamma(Q) = -\min Q = -\min sG.$$

This concludes the proof of $\|K\|^+ \geq \|H\|^+$. The proof of the other (negative) statement is easily deduced from this one by considering the concatenation $H * \tilde{K}$ where $\tilde{K}$ is $-K_{1-t}$. □

Sketch of the proofs of the necessary conditions of Theorems 5.7 and 5.8. Let us explain the main arguments needed to prove Theorems 5.7 and 5.8. The necessary conditions for both statements are based on elementary (but tricky) arguments while the sufficient ones are based on local holomorphic techniques. We begin with the necessary conditions.

Proposition 5.14. ([24]) *A path $\phi_t, 0 \leq t \leq 1$ in* $\mathrm{Ham}(M)$ *from id to a given diffeomorphism ϕ which has either no fixed minimum or no fixed maximum is not a local minimum of the Hofer length.*

Sketch of the proof. Assume say that there is no fixed maximum. Then, for each point $p(t_0) \in M$ where H_{t_0} reaches its maximum, there is some small closed interval $I \subset [0,1]$ and some compact neighbourhood U of $p(t_0)$ in M, such that $H_t(x) < \max H_t$ for all $x \in U$ and all $t \in I$. Now let $f : M \to [0,1]$ be a bump function with support strictly inside U, and let $\alpha : [0,1] \to \mathbf{R}$ be a small function (defined on the time interval) which is negative on some small neighbourhood J of t_0, positive on I, zero elsewhere, and such that its integral on $[0,1]$ vanishes. Let $K_t = \alpha f$, and write $\psi_t, 0 \le t \le 1$, for the path generated by $\{K_t\}$. Then, because $\int_0^1 \alpha = 0$ and since any K_t commutes with any $K_{t'}$, we must have $\psi_1 = id$. Thus $\{\psi_t\}$ is a small contractible loop based at $id \in \mathrm{Ham}(M)$, and

$$\tilde{\phi}_t = \psi_t \circ \phi_t$$

is a path C^∞-close and homotopic to ϕ_t, joining the same endpoints (here the composition is the timewise composition of diffeomorphisms). But, if $\tilde{H}_t$ denotes the generating Hamiltonian of $\tilde{\phi}_t$, then:

$$\tilde{H}_t = K_t + H_t \circ \psi_t^{-1}.$$

By definition of K_t, this Hamiltonian has smaller values than H_t near $(p(t_0), t_0)$, but has same maximum values as H_t outside J, and same minimal values as H_t everywhere. If $p(t_0)$ were the only point where H_{t_0} reaches its maximum, this would show that $\mathcal{L}(\{\tilde{H}_t\}) < \mathcal{L}(\{H_t\})$. If there are other points where H_{t_0} reaches its maximum, one can use a similar argument with this time the help of a partition of unity. □

This establishes the necessary condition of Theorem 5.7, and shows that any stable geodesic must have at least one fixed maximum and one fixed minimum. The proof that, for a stable geodesic, the linearised flow has no closed trajectory in time < 1 at at least one fixed minimum and one fixed maximum is the consequence of a deeper analysis, which involves this time both the first and second variational formulas of $\mathcal{L} : \mathcal{P} \to \mathbf{R}$. This analysis was carried out in [39], and also in [24] with a slightly different approach.

Consider the length functional $\mathcal{L}(\gamma)$ on the space $\mathcal{P} = \mathcal{P}(\phi_0, \phi_1)$ of paths from ϕ_0 to ϕ_1.

Given $\gamma = \{\phi_t\}_{t\in[0,1]}$, generated by $\{H_t\}_{t\in[0,1]}$, let $\{G_t\}_{t\in[0,1]}$ be a tangent vector field along γ vanishing at both ends $t = 0, 1$. For any $\varepsilon \in \mathbf{R}$, set

$$\gamma_\varepsilon(t) = \phi_{\varepsilon G_t} \circ \phi_t$$

which is a 1-parameter family of paths with the given ends, where $\phi_{\varepsilon G_t}$ is the time 1 flow of εG_t. We wish to compute

$$\frac{d^k}{d\varepsilon^k}\Big|_{\varepsilon=0} \mathcal{L}(\gamma_\varepsilon) = \int_0^1 \frac{d^k}{d\varepsilon^k}\Big|_{\varepsilon=0} \|\frac{d}{dt}\gamma_\varepsilon\|_H dt$$

for $k = 1, 2$.

Proposition 5.15. ([24]) *The Taylor expansion of the vector field $\frac{d}{dt}\gamma_\varepsilon$ in powers of ε up to order 2 is*

$$\begin{aligned}\frac{d}{dt}\gamma_\varepsilon = \;& \textit{symplectic gradient of}\,[H_t + \varepsilon\,(G_t' + \{-G_t, H_t\}) \\ & + \frac{\varepsilon^2}{2}\,(\{-G_t, G_t'\} + \{-G_t, \{-G_t, H_t\}\}) + o(\varepsilon^2)].\end{aligned}$$

Here the notation $o(\varepsilon^k)$ denotes a term R satisfying:

$$\frac{\|R\|}{\varepsilon^k} \to 0 \quad \text{when} \quad \varepsilon \to 0$$

uniformly with respect to all other variables.

Let us denote by $K_{\varepsilon,t}$ the function in the right hand side of the above equation. To derive the first and second variation formulas, we first make the following assumption:

Hypothesis of continuity. The path $\phi_t, 0 \le t \le 1$, satisfies the hypothesis of continuity if, given any tangent vector field $\{G_t\}_{t\in[0,1]}$, it is possible to make a choice $p_t(\varepsilon) \in M$ of a point at which the minimum value of $K_{\varepsilon,t}$ is reached in such a way that $p_t(\varepsilon)$ is a smooth path for small values of ε and all $0 \le t \le 1$. In this case, $p_t(0) = p_t$ where p_t is a minimal point of H_t. We assume that the same holds for $P_t(\varepsilon)$ with maximum instead of minimum values.

A particular simple case is the following:

Lemma 5.16. *Suppose that there exist two points p, P such that for all t:*

1) minset $H_t = \{p\}$ *and* maxset $H_t = \{P\}$ *and*
2) *p, P are non-degenerate extrema of H_t.*

Then $H_{t\in[0,1]}$ satisfies the hypothesis of continuity. Here minset and maxset denote the set of points where the function reaches its minimum (maximum).

The proof is elementary. We now state the variation formulas.

Theorem 5.17. (First variation formula) *Suppose that the Hamiltonian $H_{t\in[0,1]}$ satisfies the hypothesis of continuity. Then the first variational formula is:*

$$\delta^1 \mathcal{L}(\{G_t\}) = \frac{d}{d\varepsilon}\Big|_{\varepsilon=0} \mathcal{L}(\gamma_\varepsilon) = \int_0^1 (\sup_{\text{maxset } H_t} G_t' - \inf_{\text{minset } H_t} G_t')\, dt$$

where the ′ denotes the derivative with respect to time. Given any Hamiltonian $H_{t\in[0,1]}$, the right hand side of this equation is non-negative for all tangent vector fields $G_{t\in[0,1]}$ if and only if there exist at least one fixed minimum and one fixed maximum of $\{H_t\}$.

The second variation at a fixed extremum q when evaluated at $\{G_t\}$ depends only on the loop $g(t)$ traced out by the gradient $\nabla G_t(q)$ of G_t at q. We will choose symplectic coordinates around q and then identify the tangent space T_qM with $\mathbf{R}^{2n}$ equipped with its standard symplectic form ω_0 and complex structure J. Here $Jx_{2i-1} = x_{2i}$, and $Jx_{2i} = -Jx_{2i-1}$, so that

$$\omega_0(u, v) = (Ju) \cdot v,$$

where $\cdot$ denotes the usual dot product. Then the symplectic area enclosed by a loop g in $\mathbf{R}^{2n}$ is

$$\operatorname{area} g = \int_{D_g} \omega_0 = \frac{1}{2}\int_0^1 (Jg)\cdot g'\, dt,$$

where D_g is a 2-disc with boundary along g. We will write

$$\langle u, v\rangle_t = ((\mathrm{d}^2 H_t)^{-1})u \cdot v$$

for the metric induced on T_qM by the inverse of the Hessian $\mathrm{d}^2 H_t$ of H_t at q. Note the sign convention that we will use: the symplectic gradient is defined by $X_H = +J\nabla H$.

Theorem 5.18. (Ustilovsky's Second variation formula [39], see also [24]) *Suppose that $\{H_t\}_{t\in[0,1]}$ satisfies the hypothesis of continuity and has at least one fixed minimum and one fixed maximum. Suppose further that each fixed extremum of $\{H_t\}$ is a non-degenerate critical point of all the functions $H_t, 0 \le t \le 1$. Let $\{G_t\}_{t\in[0,1]}$ be a tangent vector field along $\{\phi_t\}_{t\in[0,1]}$, and set $g(t) = \nabla G_t(p)$. Then the contribution of the fixed minimum p of $H_{t\in[0,1]}$ to the second variational formula, is*

$$\delta^2 \mathcal{L}(\{G_t\})(p) = \int_0^1 \langle g', g'\rangle_t\, dt \; - \; 2\operatorname{area}(g).$$

Similarly, the contribution of the fixed maximum P is

$$\delta^2 \mathcal{L}(\{G_t\})(P) = \int_0^1 \langle g', g' \rangle_t \, dt \; + \; 2 \, \mathrm{area}\,(g)$$

where this time $g(t) = \nabla G_t(P)$.

Let q be a fixed extremum of the path γ at which the Hessian $\mathrm{d}^2 H_t$ of H_t is non-degenerate for all t. We denote by $\mathcal{Q}_q$ the quadratic functional

$$\mathcal{Q}_q(g) = \int_0^1 \left(\langle g', g' \rangle_t \pm (Jg) \cdot g' \right) dt$$

on the space of smooth loops g based at the origin in $T_q M = \mathbf{R}^{2n}$ which appears in the second variation formula. Assume that the isotopy is generated by a Hamiltonian $H_t, 0 \leq t \leq 1$, such that there exist points p, P which are non-degenerate extrema of H_t for each t with minset $H_t = \{p\}$ and maxset $H_t = \{P\}$ for each t.

Theorem 5.19. (Ustilovsky [39]) *Let γ be a path with fixed non-degenerate extrema $q = p, P$ as above. Then the quadratic functional $\mathcal{Q}_q$ is positive definite if the linearized isotopy $d\phi_t$ at q, generated by the 2-jet $\tilde{H}_t, 0 \leq t \leq 1$ of H_t at q, has no non-constant closed trajectory α in time less or equal to 1. If such α does exist in time less than 1, then $\mathcal{Q}_q$ has non-vanishing index. If this is the case at either p or P, then the path γ is not a local minimum of $\mathcal{L}$.*

This can be proved by looking at the 1-parameter family of functionals

$$\mathcal{Q}_{t'}(g) = \int_0^{t'} \left(\langle g', g' \rangle_t \pm (Jg) \cdot g' \right) dt, \quad t' \in (0, 1]$$

defined on the space of closed loops $g : [0, t'] \to T_q M = \mathbf{R}^{2n}$ based at the origin. Note that these functionals are quadratic (and therefore have generically only the zero loop as critical point) and are invariant by translation and multiplication by -1. Ustilovsky showed that the values t' where the null space of $\mathcal{Q}_t$ is non-trivial are conjugate values. In other words, for $t < \min t'$, $\mathcal{Q}_t$ is positive definite, and the index of $\mathcal{Q}_t$ increases at the passage of a conjugate value t' by a quantity equal to the (finite) nullity of $\mathcal{Q}_{t'}$. The first statement in the theorem now follows from:

Lemma 5.20. *The loop g belongs to the null space of $\mathcal{Q}_{t'}$ if and only if Jg is the translate of a closed trajectory α of $d\phi_t$, $0 \leq t \leq t'$.*

Proof. Let us suppose that q is a minimum so that

$$\mathcal{Q}_{t'} = \int_0^{t'} (\langle g', g' \rangle_t - (Jg) \cdot g')\, dt.$$

Since we may normalise H_t so that its minimum value $H_t(p)$ is 0, its 2-jet $\tilde{H}_t$ may be written in local symplectic coordinates about $p = 0$ as

$$\tilde{H}_t(x) = \frac{1}{2} \sum B_{ij}(t) x_i x_j = \frac{1}{2} x \cdot B_t x,$$

for some symmetric matrix $B_t = B_{ij}(t)$. Then the inner product $\langle u, v \rangle_t = v \cdot (B_t)^{-1} u$, and the linearized flow $L_t = d\phi_t$ is generated by the vector field JBx.

Recall that the null space of a quadratic form $\mathcal{Q}$ on a vector space V is defined to be

$$\operatorname{null} \mathcal{Q} = \{g : \mathcal{Q}(g, h) = 0 \text{ for all } h \in V\}.$$

Thus $g \in \operatorname{null} \mathcal{Q}$ if and only if g is a critical point of $\mathcal{Q}$. Now,

$$\begin{aligned} \frac{\partial}{\partial s}\Big|_{s=0} \mathcal{Q}_{t'}(g + s\xi) &= \int_0^{t'} (2\xi' \cdot B_t^{-1} g' - J\xi \cdot g' - Jg \cdot \xi')\, dt \\ &= \int_0^{t'} 2\xi' \cdot (B_t^{-1} g' - Jg) dt. \end{aligned}$$

Hence, g is in the null space of $\mathcal{Q}_{t'}$ if and only if

$$B_t^{-1} g' - Jg = const,$$

or equivalently if

$$g'(t) = B_t(Jg(t) + c), \quad 0 \le t \le t'.$$

It follows that $Jg(t) + c$ is a closed trajectory of the linearized flow $L_t, 0 \le t \le t'$. □

The last part of Theorem 5.19 is proved by constructing an explicit deformation, given by a tangent vector field $G_t, 0 \le t \le 1$, with

$$\nabla G_t(p) = g(t), \quad \text{and} \quad \mathcal{Q}(g) < 0.$$

Such a loop is defined in the following way. Let $\bar{\alpha} : [0, t'] \to T_pM = \mathbf{R}^{2n}$ be a closed trajectory of L_t, $0 \le t \le t'$ and compose it with some slowing down function $f : [0, 1] \to [0, t']$ which is the identity on $[0, t' - \varepsilon]$ for $\varepsilon > 0$ sufficiently small and sends $[t' - \varepsilon, 1]$ onto $[t' - \varepsilon, t']$. If

$$\alpha(t) = \bar{\alpha}(f(t))$$

denotes this composition, define the loop g by requiring that

$$g(t) = -J(\alpha(t) - \alpha(0)).$$

Thus, Jg, which is the symplectic gradient of G, follows a path which is, up to translation, the same as the path of a closed trajectory of the linearised isotopy during the time interval $[0, t']$. Observe that the choice of $\bar{\alpha}$ is not unique: it may be replaced by $\rho\bar{\alpha}$ for any non-zero scaling factor ρ, positive or negative.

The corresponding deformation $\phi_{\varepsilon G_t} \circ \phi_t$ is the composition of ϕ_t with the time-1 map of εG_t. Thus, up to order 1 in ε,

$$\phi_{\varepsilon G_t} \circ \phi_t(p) = \varepsilon Jg(t) = \varepsilon(\alpha(t) - \alpha(0)).$$

The striking fact here is that the deformation which optimally reduces the length is given by composing the isotopy ϕ_t with a motion that moves p in the *same* direction as does the flow of $\phi_t, 0 \leq t \leq 1$, round p. Thus, if the linearised motion at p has a closed orbit, the path $\{\phi_t\}$ uses extra energy to keep the point p fixed rather than letting it move around p in the direction of this orbit.

This proves the necessary condition of Theorem 5.8 when minset H_t and maxset H_t are independent of t and consist of a single non-degenerate minimum and a single non-degenerate maximum. The general case is proved by a careful study of the dynamics of the linearised flow when the 2-jet of H_t is degenerate, which permits the extension of Ustilovsky's theorem to the degenerate case. One then uses this repeatedly together with the technique of Proposition 5.14 in order to decrease the maximal values of all H_t near each fixed maximum. Finally, one applies the technique of Proposition 5.14 again to decrease the maximal values of some H_t near points which belong to maxset H_t but are not fixed extrema (see [24]).

Sketch of the proofs of the sufficient conditions of Theorems 5.7 and 5.8. We consider now the sufficient conditions of Theorems 5.7 and 5.8. Of course, it is enough to prove the sufficiency of the condition of Theorem 5.8. A special case (the non-degenerate case) was proved by Ustilovsky [39], who used variational techniques.

The proof that we present below, which applies to all cases, is based on a geometric construction (the gluing of monodromies), on holomorphic methods (the local non-squeezing theorem), and on a careful study of "negative flows" in the symplectic linear group.

Let $\phi_t, 0 \leq t \leq 1$ be a path in $\mathrm{Ham}(M)$ joining the identity to ϕ and generated by $H_t, 0 \leq t \leq 1$, that we rescale so that $\min H_t = 0$ for all

t. We may assume that H_t vanishes near $t = 0, 1$. Denote by R_-H and R_+H the parts under and over the graph of H respectively:

$$\begin{aligned} R_-H &= \{(x,s,t) : 0 \leq s \leq H_t(x)\} \\ R_+H &= \{(x,s,t) : H_t(x) \leq s \leq m(t)\} \end{aligned}$$

where $m(t) = \max H_t$. Note that both R_-H and R_+H are regions comprised between the graph of H and an hypersurface of $M \times \mathbf{R}^2$ with trivial characteristic foliation, namely the hypersurfaces $\{s = 0\}$ and $\{s = m(t)\}$. Note also that the gluing of both halves R_-H and R_+H by the identification of their common side (which matches the characteristic foliations) gives back simply the set $M \times \{0 \leq s \leq m(t)\}$ which is symplectomorphic to a split cylinder of area $\mathcal{L}(\{H_t\})$.

Now let $K_t, 0 \leq t \leq 1$ be any other Hamiltonian generating a path ψ_t with same endpoints, homotopic and C^∞-close to ϕ_t. Assume also that $\min K_t = 0$ and that it vanishes near $t = 0, 1$. Denote by $R_{H,K}$ the space obtained by attaching R_-H to R_+K by the map which identifies the characteristic foliations of graph(H) and graph(K) in the following way: each characteristic curve of graph(H) beginning at some point $(x, 0, 0)$ and ending at $(\phi(x), 0, 1)$ is sent to the corresponding characteristic curve of graph(K) beginning at $(x, 0, 0)$ and ending at $(\phi(x), 0, 1)$ by the map which preserves time. Then the fact that the paths ϕ_t and ψ_t are homotopic rel endpoints implies:

1: the characteristic foliation around $\partial R_{H,K}$ closes up;

2: $M \times D^2(a)$ is diffeomorphic [1] to $R_{H,K}$ by a diffeomorphism which pulls back the form Ω on $R_{H,K}$ to a form Ω on $M \times D^2(a)$ which equals the split form $\omega \oplus ds \wedge dt$ near the boundary, and equals ω on each fiber $M \times \{pt\}$.

Call *quasi-cylinder* any space satisfying these conditions, and define its *area* as the quotient of its volume to the volume of M (restrict to a compact set of M containing both supports of H and K if M is not compact). Define similarly the quasi-cylinder $R_{K,H}$. Then obviously:

Proposition 5.21. *If area*$(R_{H,K})$ *and area*$(R_{K,H})$ *are both* $\geq \mathcal{L}(\{H_t\})$, *then* $\mathcal{L}(\{K_t\}) \geq \mathcal{L}(\{H_t\})$.

Assume that there is at least one fixed minimum p and one fixed maximum P where the linearised isotopy $d\phi_t$ have no non-constant closed

---

[1] Since the spaces $R_H^-, R_H^+, R_{H,K}$ have two cusps, one at $t = 0$ and one at $t = 1$, we need to thicken slightly these spaces in order to make the statement (2) rigourous.

trajectory in time ≤ 1. We wish to prove that, for all homotopic sufficiently C^∞-close isotopy K_t, area $(R_{H,K})$, area $(R_{K,H}) \geq \mathcal{L}(\{H_t\})$. But splitness is an open condition, so there exists some C^∞-neighbourhood of the path H_t such that both $R_{H,K}$ and $R_{K,H}$ are split, since this is true for $K = H$. Assume that what we wish to prove is false: there is then a sequence K_t^i of paths converging to H_t such that say $a_i = \text{area}\,(R_{H,K^i}) < \mathcal{L}(\{H_t\})$. On the other hand, for each i:

$$R_H^- \hookrightarrow R_{H,K^i} = (M \times D^2(a_i), \omega + \sigma)$$

is a symplectic embedding f_i which smoothly converges to the inclusion $\iota : R_H^- \hookrightarrow R_{H,H} = (M \times D^2(a), \omega + \sigma)$ where $a = \mathcal{L}(\{H_t\}) > a_i$. This means that the symplectic space R_H^- is locally squeezed inside smaller split cylinders. Thus, after some tight jiggling that we skip (see [26]), it is enough to prove the first part of the Local Non-Squeezing Theorem 3.10.

Sketch of the proof of Theorem 3.10. The second part is a consequence of the necessary condition of Theorem 5.8. We give a brief sketch of the proof of the first part (see [26, 27] for details). We use the same notations as in Theorem 3.10. Observe first that if the 2-jet of F_t at 0, $\tilde{F}_t = F_t(0) + \frac{1}{2}d^2F_t$, has no non-constant closed trajectory in time ≤ 1, then the same is true of $K_t = F_t(0) + G_t$, where G_t is a negative definite quadratic form defined on some neighbourhood B of $0 \in U$ satisfying $G_t < \frac{1}{2}d^2F_t$. Then, for B small enough, the set

$$S(K) = \{(x, s, t) : x \in B \text{ and } 0 \leq s \leq K_t(x)\}$$

is a subset of $S(F)$. Thus it is enough to prove the Local Non-Squeezing Theorem for K_t. Note that the path $\psi_t \in \mathrm{Sp}(2n, \mathbf{R})$ generated by the quadratic Hamiltonian G_t is a *negative* path beginning at the identity, negative in the sense that its tangent vector is everywhere negative definite. More precisely, the Lie algebra of $\mathrm{Sp}(2n, \mathbf{R})$ can be identified with the Poisson algebra of all quadratic forms (homogeneous polynomials of degree 2) on $\mathbf{R}^{2n}$. The symplectic gradient of such a form is a linear vector field, and the flow of a time dependent quadratic Hamiltonian G_t is then given by solutions of $X'(t) = A(t)X(t)$ for matrices $A(t) = X_{G(t)}$. Therefore solutions are linear symplectic maps, that is to say elements of $\mathrm{Sp}(2n, \mathbf{R})$. A path in $\mathrm{Sp}(2n, \mathbf{R})$ starting at the identity is negative if it is generated by a (time-dependent) Hamiltonian which is quadratic negative definite at each time. It is *short* if A_t, $t > 0$, has never 1 as eigenvalue. Thus $K_{t\in[0,1]}$ is a short negative path.

Lemma 5.22. ([26]) *The local non-squeezing theorem holds when K_t is a short negative autonomous path.*

Indeed, in this case, the corresponding set $S(K) = S(K_{aut})$ is an ellipsoid, up to a global linear change of coordinates. The theorem is then proved by a direct application of symplectic homology. Actually, one can even show more: define $W_\varepsilon = B \times D^2(1+\varepsilon)$, then for all $\varepsilon > 0$, there are compact sets in $S(K_{aut}) \cap \mathrm{Int}W$ which are unsqueezable by isotopy inside W to a set lying in $W_{-\varepsilon}$. This gives some extra room that will be used below.

Lemma 5.23. ([26]) *If all eigenvalues of ψ_1 are on the unit circle, then the above local non-squeezing theorem holds for K_t.*

Sketch of the proof of the lemma. In this case, there is a path ψ'_t from id to ψ_1 which is unitary in the sense that it is conjugate to an autonomous path generated by a (2×2)-block-diagonal negative definite Hamiltonian K_{aut}. Because both paths lie in the subset S_0^+ of Sp defined by $det(A-I) > 0$ and since by homotopy index theory any loop there is contractible in Sp (see [35]), the set $S(K)$ is symplectically equivalent to the corresponding set $S(K_{aut})$. Moreover, it is shown in [27] that there is a symplectic isotopy Ψ_s *inside* W between the inclusion $S(K_{aut}) \to W$ and the above symplectomorphism $\Psi : S(K_{aut}) \to S(K)$ (which essentially means that the paths K_t and $K_{aut,t}$ are homotopic rel endpoints in S_0^+ through negative paths). Note that, by the extension of Hamiltonian isotopies, one could define a symplectomorphism sending the pair $(S(K_{aut}), W_\varepsilon)$ to the pair $(S(K), W_\varepsilon)$, where $W_\varepsilon = B \times D^2(1+\varepsilon)$. Unfortunately, this does not hold for $\varepsilon = 0$, so we have to proceed differently. Instead, one shows that if $S(K)$ were locally squeezable in W, then for some small enough ε, every compact subset of $S(K) \cap \mathrm{Int}W$ would be locally squeezable by isotopy inside W to a subset lying in $W_{-\varepsilon}$. This would then imply that the same holds for the ellipsoid $S(K_{aut})$, a contradiction.

Thus the proof of Theorem 5.8 boils down to the following: □

Proposition 5.24. (Lalonde-McDuff) *Any short negative path in S_0^+ beginning at the identity can be extended to a short negative path lying in S_0^+ and ending at an element of Sp whose eigenvalues all lie on the unit circle.*

The proof of this proposition is more delicate (see [27]): it relies on a detailed analysis of the admissible bifurcations of the configurations of eigenvalues along a negative path, which extends the works of Bott on the Sturm-Liouville problem and of Krein ([14], Chapter VI) on Lyapunov stability. Note that because a path is a local minimum iff all subpaths are local minima too, this Proposition finishes the proof of Theorem 5.8.

6. Generating functions and Viterbo's norm.

In this last section, we present Viterbo's norm on Hamiltonian diffeomorphisms of $\mathbf{R}^{2n}$ and its relation to the Hofer norm. Its definition relies on the topology of generating functions. In this context, generating functions appear as a finite dimensional reduction of the action functional, as will be explained below.

Let V be any manifold, and consider its cotangent bundle T^*V endowed with the standard symplectic form $\omega = d\lambda$, where λ is the Liouville form. Given any Hamiltonian diffeomorphism ϕ of T^*V, let L be the Lagrange submanifold of T^*V defined by $L = \phi(L_0)$ where L_0 is the zero section.

Definition 6.1. A generating function for L is a smooth function $f : E \to \mathbf{R}$ defined on some real vector bundle $E \to V = L_0$ of finite rank, which has the following properties:

1) the set $S_v(f) = \{(x,v) \in E : \partial_v f(x,v) = 0\}$ is a smooth submanifold of E, where ∂_v is the differential of the restriction of f to the fibers of $E \to V$. Note that there corresponds then a well-defined cotangent vector $\alpha \in T^x V$ to each $(x,v) \in S_v(f)$, defined as the restriction of df to any horizontal linear subspace of $T_{(x,v)}E$.

2) The map $S_v(f) \to T^*V$ which assigns to each $(x,v) \in S_v(f)$ the element $\alpha \in T^*V$, embeds $S_v(f)$ onto L.

A generating function is *quadratic at infinity* (abbreviated by q.i. below) if it is equal to q_∞ outside some compact neighbourhood of the zero section, where the restriction of q_∞ to each fiber of $E \to V$ is a non-degenerate quadratic map. We will always restrict attention to q.i. generating functions such that the negative bundle E_- of q_∞ is trivial.

As Y.-G. Oh mentionned to me, one can consider the action functional associated to the Hamiltonian $H_t, 0 \leq t \leq 1$ that generates ϕ, as the canonical generating function for L. More precisely, let L_0 be the zero section of $M = T^*V$, and define

$$\Omega_{I,0} = \{\gamma : I \to M \mid \gamma(0) \in L_0\}$$

Heuristic principle: the action functional $\mathcal{A}_H : \Omega_{I,0} \to \mathbf{R}$ is the canonical generating function of $L = \phi(L_0)$. Let $\pi : \Omega_{I,0} \to V$ be the projection of $\gamma(1)$, which is a submersion, and indeed a fiber map when $V = \mathbf{R}^n$. Then one easily sees that the paths γ for which the vertical derivative of $\mathcal{A}_H$ vanishes are

$$\Sigma = \{\gamma \mid \gamma(0) \in L_0 \quad \text{and} \quad \gamma' = X_H(\gamma)\}$$

But the horizontal derivative of $\mathcal{A}_H$ at $\gamma \in \Sigma$ is simply $\gamma(1)$.

Since generating functions are essentially unique ([40], § 1), this shows that one can consider generating functions to be finite dimensional reductions of the action functional, so that they should also lead to invariants similar to the ones defined by the maximin procedure of § 4. This is indeed the case, as will be clear below.

If $\phi \in \mathrm{Ham}(\mathbf{R}^{2n})$ is any compactly supported Hamiltonian diffeomorphism, its graph $\Gamma(\phi)$ in $(W = \mathbf{R}^{2n} \times \mathbf{R}^{2n}, \omega \oplus (-\omega))$ is a Lagrange submanifold. Since the Weinstein tubular neighbourhood of the Lagrange submanifold $\Delta = \Gamma(id)$, the diagonal of W, can be extended to all W, $\Gamma(\phi)$ can be viewed as a Lagrange submanifold of $T^*\Delta$. (Note that we need here to work in $\mathbf{R}^{2n}$. This is what limits Viterbo's norm to euclidean spaces). Therefore one can consider q.i. generating functions for $\Gamma(\phi) \subset T^*\Delta$. Actually, it is better to compactify first Δ to the sphere S^{2n} by adding the point p_∞ at infinity, and then $T^*\Delta$ and $\Gamma(\phi)$ to T^*S^{2n} and $\Gamma'(\phi) \subset T^*S^{2n}$ respectively. This can be done because ϕ is compactly supported. Then Viterbo's norm $\|\phi\|_V$ is defined as follows:

Definition 6.2. (Viterbo) Let $S : E \to \mathbf{R}$ be any q.i. generating function for $L = \Gamma'(\phi) \subset T^*S^{2n}$ normalised so that $S(p_\infty, 0) = 0$. Denote by k the rank of E_-. Let $E^\lambda = S^{-1}(-\infty, \lambda]$ and note that, by the Thom isomorphism, $H^*(S^{2n}) \simeq H^{*+k}(E^\infty, E^{-\infty})$ where $H^{*+k}(E^\infty, E^{-\infty})$ is by definition the limit of $H^{*+k}(E^\infty, E^\lambda)$ as $\lambda \to -\infty$. Set

$$c_+(\phi) = \inf\{\lambda : H^{2n+k}(E^\infty, E^{-\infty}) \to H^{2n+k}(E^\lambda, E^{-\infty}) \text{ is not the zero map}\}$$

and

$$c_-(\phi) = \inf\{\lambda : H^k(E^\infty, E^{-\infty}) \to H^k(E^\lambda, E^{-\infty}) \text{ is not the zero map}\}.$$

Then $\|\phi\|_V = c_+(\phi) - c_-(\phi)$.

Viterbo showed ([40]) that this definition is invariant under the choice of the q.i. generating function for $\Gamma'(\phi)$ (and that actually both c_+ and c_- are invariants). The interest of this definition of norm is that it leads easily to important properties, which are much more difficult to establish for the Hofer norm (see in particular [40] Proposition 4.2, and [40] Corollary 4.5 which has not been proved yet for Hofer's norm). In particular, it is easy to show that $\| \quad \|_V$ is a non-degenerate norm on $\mathrm{Ham}(\mathbf{R}^{2n})$ (although the triangle inequality is more difficult to prove).

Thus, Viterbo's norm is the difference between two distinguished levels λ where special homotopic changes of $(E^\lambda, E^{-\infty})$ occur. (For $k = 0$ for instance, c_+ is the level λ at which E^λ does not contain the $2n$-sphere anymore, and c_- is the one at which E^λ becomes empty.) Of course, the normalization $S(p_\infty, 0) = 0$ implies that $c_+ \geq 0$ and $c_- \leq 0$. By Lyusternik-Schnirelman theory, both levels are critical (not only in the fiber direction). Thus they correspond to points of L on the zero section of T^*S^{2n}, that is to fixed points of ϕ. Denote by p_+ and p_- two such points (their choice is in general not unique). The following characterization is an (almost) immediate but crucial consequence of the definition:

Proposition 6.3. (Viterbo) *Let z be any point outside the support of an isotopy $\phi_t, 0 \leq t \leq 1$ ending at ϕ, and let γ_+, γ_- be paths from z to p_+, p_-. Then:*

$$c_\pm(\phi) = \int_{\gamma_\pm^{-1} \circ \phi(\gamma_\pm)} p\, dq$$

so that

$$\|\phi\|_V = \int_{\alpha^{-1} \circ \phi(\alpha)} p\, dq$$

where α is any path joining p_- to p_+. In other words, $c_\pm$ is the area swept out by the arc $\gamma_\pm$ during the isotopy (which equals the action of the fixed point $p_\pm$ as defined in § 4), and $\|\phi\|_V$ is the area swept out by the arc α.

In [40], it is also shown that $\|\ \ \|_V$ is a continuous norm for the C^0-topology: actually, in this approach, the only fact whose proof is less natural is the triangle inequality for $\|\ \ \|_V$.

Viterbo and Hofer norms coincide in a C^1-neighbourhood of *id*. Let us now show that the Viterbo norm coincides with the Hofer norm in some C^1-neighbourhood of the identity. As we will see, this easily implies that $\|\phi\|_H \geq \|\phi\|_V$ for all $\phi \in \mathrm{Ham}(\mathbf{R}^{2n})$. This inequality was first proved by Sikorav, and a simpler proof by Bialy-Polterovich appeared in [2] (see also [25]).

Consider the C^1-neighbourhood U of $id \in \mathrm{Ham}(\mathbf{R}^{2n})$ consisting of diffeomorphisms ϕ such that $\Gamma(\phi) \subset T^*\Delta$ project diffeomorphically onto the diagonal $\Delta \subset W = \mathbf{R}^{2n} \times \mathbf{R}^{2n}$. In this case, we can choose a very simple generating function for ϕ, namely the unique function $S : \Delta = \mathbf{R}^{2n} \to \mathbf{R}$ such that:

1) $S = 0$ at infinity, and

2) the graph of dS in $T^*\Delta$ coincides with $\Gamma(\phi)$.

Then obviously, $\|\phi\|_V = \|S\|$, the total variation of S. But this determines a distinguished path $\phi_t, 0 \leq t \leq 1$ joining id to ϕ, the one defined by $\Gamma(\phi_t) = \mathrm{graph}(d(tS))$. Let H_t be the generating Hamiltonian of this path.

Proposition 6.4. *The Hamilton-Jacobi equation states that:*

$$\frac{\partial(tS)}{\partial t}(x) = H_t(x + \frac{1}{2}J\nabla(tS)(x))$$

which boils down to:

$$S(x) = H_t(x + \frac{tJ}{2}\nabla S(x)).$$

Thus $\|H_t\| = \|S\|$ *for all* t, *which implies that*

$$\mathcal{L}(\{H_t\}) = \|S\| = \|\phi\|_V.$$

Note that, by the above formula, H_t is defined by composing S with a diffeomorphism depending on time and which is the identity at all critical points of S. In particular, the points p_-, p_+ where S reaches its minimum and maximum are fixed extrema of $\{H_t\}_{t\in[0,1]}$.

In order to show that $\| \quad \|_H$ and $\| \quad \|_V$ coincide C^1-locally, it remains to show:

Proposition 6.5. (Bialy-Polterovich) *The path* $H_t, 0 \leq t \leq 1$ *is a shortest path from id to* ϕ.

This is proved in [2] using a variational argument based on the selector of § 4. It is similar but more general than Hofer's argument showing that time-one flows of autonomous Hamiltonians with no non-constant closed trajectory in time ≤ 1 are absolute minima of the length. The above proposition is also proved in [25] by a holomorphic argument where the selector is replaced by the introduction of a Gromov width associated to diffeomorphisms and defined by an infimum of the width of the regions comprised under or over the graph of the generating Hamiltonians. (The argument in [25] extends this proposition to surfaces and other manifolds.)

Note that this implies that some C^1-neighbourhood U of the identity in $\mathrm{Ham}(\mathbf{R}^{2n})$ is isometric to some C^2-neighbourhood of 0 in $(C_0^\infty(\mathbf{R}^{2n}, \mathbf{R}), \| \quad \|)$ endowed with the total variation norm. Thus $(U, \| \quad \|_H)$ is flat.

An easy corollary is:

Theorem 6.6. (Bialy-Polterovich, Sikorav) $\|\phi\|_H \geq \|\phi\|_V$ *for all* $\phi \in \mathrm{Ham}(\mathbf{R}^{2n})$.

Proof. Take any path $\phi_t, 0 \leq t \leq 1$ from id to some ϕ, and let ϕ_{t_i}, $1 \leq i \leq N$, be sufficiently close so that $\phi_{t_i}^{-1} \circ \phi_{t_{i+1}}$ always belongs to the C^1-neighbourhood where both norms coincide. Then:

$$\begin{aligned}\mathcal{L}(\{\phi_t\}_{t\in[0,1]}) &\geq \sum_i d_H(\phi_{t_i}, \phi_{t_{i+1}}) \\ &\geq \sum_i d_V(\phi_{t_i}, \phi_{t_{i+1}}) \\ &\geq d_V(id, \phi)\end{aligned}$$

where the last inequality is due to the triangle inequality of d_V. Since this is true for all paths from id to ϕ, we conclude that $\|\phi\|_H \geq \|\phi\|_V$. □

Note that, besides the C^1-equality of both norms, we only needed in this proof the property that the Hofer norm is defined as the infimum of lengths of paths. If the same were known to hold for d_V, we would obtain a proof of the equality of both norms. Thus:

Proposition 6.7. $\| \ \|_V = \| \ \|_H$ *if and only if the Viterbo norm satisfies: for any* $\phi \in \mathrm{Ham}(\mathbf{R}^{2n})$ *and any* $\varepsilon, \delta > 0$, *there exist a sequence* $id = \phi_1, \dots, \phi_N = \phi$ *such that* $\|\phi_i^{-1} \circ \phi_{i+1}\|_{C^1} \leq \delta$ *and* $\sum_i d_V(\phi_i, \phi_{i+1}) \leq \|\phi\|_V + \varepsilon$.

Finally, this generating functions approach also leads to a definition of relative capacities. For any bounded subset $U \subset \mathbf{R}^{2n}$, define the capacity and the energy of disjunction as follows:

$$\begin{aligned}c_V(U) &= \sup\{c_+(\psi) : supp(\psi) \subset U\} \quad \text{and} \\ e_V(U) &= \inf\{\|\psi\|_V : \psi(U) \cap U = \emptyset\}.\end{aligned}$$

The advantage of this approach is again the simplicity of the proofs. Indeed the energy-capacity inequality:

Proposition 6.8. *For all bounded open subset* $U \subset \mathbf{R}^{2n}$, $e_V(U) \geq c_V(U)$.

is particularly easy to prove. It is an almost direct consequence of the properties of a norm and of Proposition 6.3 (see [40], Corollary 4.5).

Following an idea of Eliashberg, Lisa Traynor has just written down a presentation of symplectic homology via generating functions (see [38]): generating functions do to this homology exactly what they do for the capacities, that is to say the infinite dimensional analysis of the action functional is replaced by a finite dimensional topological study of generating functions. Thus it simplifies greatly the presentation although, in counterpart, its applications are limited to subsets of euclidean spaces.

REFERENCES

1. *Holomorphic curves in symplectic geometry*, M. Audin and J. Lafontaine eds, Progress in Mathematics, vol. 117, Birkauser, 1994.
2. M. Bialy and L. Polterovich, *Geodesics of Hofer's metric on the group of Hamiltonian diffeomorphisms*, Duke Math. J. **76** (1994) 273-292.
3. M. Bialy and L. Polterovich, *Hofer's metric and invariant tori*, Preprint, 1994.
4. V. Benci and P. Rabinowitz, *Critical point theory for indefinite functionals*, Invent. Math. **52** (1979) 336-352.
5. I. Ekeland and H. Hofer, *Symplectic topology and Hamiltonian dynamics I*, Math. Z. **200** (1990) 355-378.
6. I. Ekeland and H. Hofer, *Symplectic topology and Hamiltonian dynamics II*, Math. Z. **203** (1990) 553-567.
7. Y. Eliashberg, *A theorem on the structure of wave fronts and its applications to symplectic topology*, Funct. Anal. and Appl. **19** (1987) 227-231.
8. Y. Eliashberg and H. Hofer, *Unseen symplectic boundaries*, Preprint, Stanford, 1993.
9. Y. Eliashberg and L. Polterovich, *Biinvariant metrics on the group of Hamiltonian diffeomorphisms*, Inter. J. Math. **4** (1993) 727-738.
10. A. Floer, *The unregularized gradient flow for the symplectic action,* Commun. Pure Appl. Math. **41** (1988) 775-813.
11. A. Floer and H. Hofer, *Symplectic homology I: open sets in* $\mathbf{C}^n$, Preprint.
12. A. Floer, H. Hofer and K. Wysocki, *Applications of symplectic homology II*, in preparation.
13. A. Givental and B. Kim, *Quantum cohomology of flag manifolds and Toda lattices*, Preprint 1993.
14. I. C. Gohberg and M. G. Krein, *Theory and applications of Volterra operators in Hilbert space*, Translations of Mathematical Monographs vol. 24, American Mathematical Society, 1970
15. M. Gromov, *Pseudo-holomorphic curves in symplectic manifolds*, Invent. Math. **82** (1985) 307-347.
16. H. Hofer, *On the topological properties of symplectic maps*, Proc. Royal Soc. Edinburgh **115 A** (1990) 25-38.
17. H. Hofer, *Estimates for the energy of a symplectic map*, Comment. Math. Helv. **68** (1993) 48-72.
18. H. Hofer and E. Zehnder, *Periodic solutions on hypersurfaces and a result of C. Viterbo*, Invent. Math. **90** (1987) 1-7.
19. H. Hofer and E. Zehnder, *A new capacity for symplectic manifolds*, in: Analysis et cetera, P. Rabinowitz and E. Zehnder eds, Academic Press, 1990, pp. 405-428.
20. H. Hofer and E. Zehnder, *Lectures on symplectic invariants and Hamiltonian dynamics*, manuscript, 1993.
21. H. Hofer and E. Zehnder, *Symplectic invariants and Hamiltonian dynamics*, Birkhaüser Adv. Texts, 1994.

22. F. Lalonde, *Isotopy of symplectic balls, Gromov's radius, and the structure of ruled symplectic 4-manifolds*, Math. Ann. **300** (1994) 273-296.
23. F. Lalonde and D. McDuff, *The geometry of symplectic energy*, Ann. of Math. **141** (1995) 349-371.
24. F. Lalonde and D. McDuff, *Hofer's L^∞-geometry: energy and stability of Hamiltonian flows I*, Invent. Math. **122** (1995) 1-34.
25. F. Lalonde and D. McDuff, *Hofer's L^∞-geometry: energy and stability of Hamiltonian flows II*, Invent. Math. **122** (1995) 35-70.
26. F. Lalonde and D. McDuff, *Local non-squeezing theorems and stability*, J. Geom. Funct. Anal. **5** (1995) 364-386.
27. F. Lalonde and D. McDuff, *Positive paths in the linear symplectic group*, to appear in "The Gelfand Seminar", Birkhäuser.
28. Y. Long, *Geodesics in the compactly supported Hamiltonian diffeomorphisms group*, preprint, 1993.
29. D. McDuff, *Elliptic methods in symplectic geometry*, Bull. AMS **23** (1990) 311-358.
30. D. McDuff, *Symplectic manifolds with boundaries of contact-type*, Invent. Math. **103** (1991) 651-671.
31. D. McDuff, *Blow-ups and symplectic embeddings in dimension 4*, Topology **30** (1991) 409-421.
32. D. McDuff and L. Polterovich, *Symplectic packings and Algebraic Geometry*, Invent. Math. **115** (1994) 405-430.
33. D. McDuff and D. Salamon, *Introduction to Symplectic Topology*, monograph, Oxford press, 1994.
34. D. McDuff and D. Salamon, *J-holomorphic curves and Quantum Cohomology*, University Lecture Series, AMS, 1994.
35. D. Salamon and E. Zehnder, *Floer homology, the Maslov index and periodic orbits of Hamiltonian equations*, in: Analysis et cetera, P. Rabinowitz and E. Zehnder eds, Academic Press, 1990, pp. 573-600.
36. K. Siburg, *New minimal geodesics in the group of symplectic diffeomorphisms*, to appear in Calculus of Variations.
37. J.-C. Sikorav, *Systèmes hamiltoniens et topologie symplectique*, ETS Editrice, Pisa, 1990.
38. L. Traynor, *Symplectic homology via generating functions*, Preprint 1994.
39. I. Ustilovsky, *Conjugate points on geodesics of Hofer's metric*, Preprint, 1994.
40. C. Viterbo, *Symplectic topology as the geometry of generating functions*, Math. Ann. **292** (1992) 685-710.

Université du Québec à Montréal, Montréal H3C 3P8, Québec, Canada

E-mail address: flalonde@math.uqam.ca

AMS/IP Studies in Advanced Mathematics
Volume **2**, 1997 (Part 1)

PIECEWISE LINEAR MANIFOLDS AND CERF THEORY

W. B. R. LICKORISH

ABSTRACT. An exploration of the Pachner moves, that suffice to relate two different triangulations of a piecewise linear n-manifold, shows that the handle structures associated to different triangulations are related by a sequence of births and deaths of handle pairs. A *version* of this type of Cerf theory, for piecewise linear manifolds, is then proved for any two piecewise linear handle structures on the manifold.

1. INTRODUCTION.

Handle theory for smooth manifolds has been an important aspect of geometric topology, certainly ever since Smale [17] used it to solve the Poincaré conjecture for high dimensional differential manifolds. An extension of Morse theory shows that, for a smooth manifold M, handles correspond to the critical points of a smooth function $f : M \to \mathbb{R}$ having only non-degenerate critical points. Handle manoeuvres, births and deaths of handle pairs, correspond to the way that critical points come and go as f varies under a well-behaved homotopy. Cerf theory [2], [3] states that any two handle structures on M are related by such manoeuvres. Elegant though that may be in outline, the details are somewhat sophisticated. The theory of piecewise linear manifolds has the advantage that it is almost immediate that handle structures exist (take neighbourhoods of barycentres of simplexes), but there has been no direct piecewise linear analogue of Cerf theory, describing moves on handles that suffice to manoeuvre one handle structure to another on the same manifold. Although, for ease of intuition, much of low-dimensional geometric topology is conducted in the piecewise linear framework, when Cerf theory is needed recourse is had to a palliative of the form 'Transfer briefly to the smooth category and use transversality' [4], thereby concealing many technicalities. Familiar results that can be regarded as consequences of Cerf theory include the Reidemeister-Singer theorem [16], [14], [7] on the stability of Heegaard splittings of a 3-manifold, the Hatcher-Thurston technique for a presentation of the mapping class group of a surface [6] (that leads to Wajnryb's presentation [19]), and the (essentially 4-dimensional) theorem of Kirby [9] on moves between surgery presentations of a 3-manifold. Although for each of these results other methods of proof are available, it should be recalled that, in high dimensions, unsmoothable piecewise linear manifolds do exist [10] so that recourse to smooth Cerf theory is not always available.

The theory of piecewise linear topology rests on that of simplicial complexes. Manifolds are triangulated by complexes called combinatorial manifolds; a brief review of relevant traditional definitions will be given below. Very recently Pachner [12] [13] has proved a result to the effect that two triangulations of the same piecewise linear n-manifold are related by a finite sequence of 'moves'. There are at most $n+1$ types of move, each being of the form 'remove a certain standard pattern of simplexes and replace it with another'. It seems that this fundamental and elegant result was not known during the development of piecewise linear topology in the 1920's and 1930's nor during its renaissance in the 1960's. In fact, the introduction of Pachner's combinatorial work to most geometric topologists was consequent only upon the defining of a 'quantum' 3-manifold invariant by Turaev and Viro [18]. To define an invariant of piecewise linear n-manifolds it suffices to create some expression involving simplexes of a triangulation and to check it is unchanged by each of Pachner's $n+1$ moves.

The close connection between triangulations and handle structures, means that Pachner's result has consequences for handle theory. It is here shown that any two handle structures for a piecewise linear manifold that come from triangulations are related by births and deaths of pairs of handles, exactly as in Cerf theory. A similar result is obtained for arbitrary piecewise linear handle structures after first adjusting all the handles by the use of collars. The aim of this paper is not only to prove these rather technical, fundamental results, giving details of the definitions involved, but also to give some publicity to Pachner's result in a topological environment.

2. Piecewise linear terminology.

It is probably desirable to indulge in a very brief survey of the language of piecewise linear topology. A *(finite simplicial) complex* K is a finite collection of simplexes, contained (linearly) in some $\mathbb{R}^N$, that is closed under the operations of intersecting simplexes and taking faces. The notation '$A \leq B$' will mean that simplex A is a face of simplex B, the join of simplexes A and B will be denoted '$A \star B$' and '$Lk(A, K)$' will be used for the link of simplex A in complex K. Recall that

$$Lk(A, K) = \{B \in K : A \star B \in K\}.$$

A *simplicial isomorphism* between two complexes is a face-preserving bijection between them (as sets of simplexes). The *polyhedron* $|K|$ underlying the complex K is defined to be $\cup_{A \in K} A$ and K is called a *triangulation* of $|K|$.

A complex K' is a *subdivision* of the complex K if $|K'| = |K|$ and each A in K is the union of some of the simplexes in K'. Thus K' can be thought of as being obtained from K by subdividing (consistently) each simplex A of K into many 'smaller' simplexes (each linearly contained in A), but no specific tidy method is prescribed for this process. If a function $f : |K_1| \to |K_2|$ between polyhedra happens to induce a simplicial isomorphism $K_1' \to K_2'$ for some subdivisions K_1' and K_2' and f is affine (or 'linear') on each $A \in K_1'$, then f is called a *piecewise linear homeomorphism.* Piecewise linearly homeomorphic polyhedra will be called *equivalent.* If $|K_1| = |K_2|$ a standard result, [8], [15], asserts that the identity map is a piecewise linear homeomorphism. Let Δ^n denote the complex consisting of all faces of an n-simplex (including the n-simplex itself), and let $\partial\Delta^n$ be the subcomplex of all the *proper* faces of Δ^n. A piecewise linear n-ball B^n is any polyhedron eqivalent to $|\Delta^n|$, a piecewise linear $(n-1)$-sphere S^{n-1} is any polyhedron eqivalent to $|\partial\Delta^n|$. Thus, polyhedra should be thought of as complexes modulo the operation of taking arbitrarily complicated subdivisions and subpolyhedra as underlying sub-complexes of such subdivisions. It can be verified that products of polyhedra are polyhedra.

A complex K is a *combinatorial n-manifold* if the polyhedron underlying the link of each vertex of K is an S^{n-1} or B^{n-1}, a *piecewise linear n-manifold* M is the underlying polyhedron of such a K. Plausible reasons for this definition are in the literature [8], [15]. The required property of links is preserved under subdivision of K; those vertices with link a B^{n-1} are those in ∂M, the *boundary* of M.

Definition. A handle structure for a piecewise linear n-manifold M is a decomposition of the form $M = \bigcup_i h_i$, expressing M as a union of 'handles'. Each handle h_i is, for some r depending on i, a subpolyhedron equivalent to $B^r \times B^{n-r}$,

$$h_i \cap \bigcup_{j<i} h_j = h_i \cap \partial \bigcup_{j<i} h_j$$

with this intersection coresponding to $\partial B^r \times B^{n-r}$.

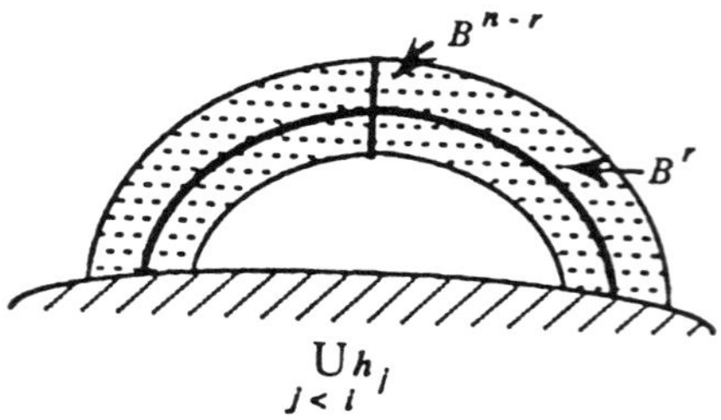

FIGURE 1

Note that the handles are ordered for this to make sense. Each r-handle $B^r \times B^{n-r}$ is thought of as being glued to the boundary of the union of all the preceeding handles by its *attaching tube* $\partial B^r \times B^{n-r}$ (see Figure 1). The handles may be reordered in any way provided that, for each handle, the union of the preceeding handles does always contain the set to which the attaching tube is to be glued.

The standard example of a handle structure on M is one obtained from a triangulation as follows. A derived subdivison $K^{(1)}$ of a complex K is defined by

$$K^{(1)} = \{\widehat{A}_0 \star \widehat{A}_1 \star \dots \widehat{A}_t : A_0 \leq A_1 \leq \cdots \leq A_t \in K\}$$

where each $\widehat{A}_i$ is any chosen point in the interior of A_i. The second derived, $K^{(2)}$, is $(K^{(1)})^{(1)}$. Suppose that $|K| = M$. Let $A_1, A_2, \dots$ be the simplexes of K listed in some order of increasing dimension and let $h_i = |\widehat{A}_i \star Lk(\widehat{A}_i, K^{(2)})|$. Then, for these h_i's, $M = \bigcup_i h_i$ is a handle structure for M, having an r-handle $|\widehat{A} \star Lk(\widehat{A}, K^{(2)})|$ corresponding to each r-simplex A of K. Of course, the precise subsets that constitute the handles depend on the choice of the A_i, but any two such handle structures so derived from K are clearly equivalent under a piecewise linear automorphism of M.

In favourable circumstances handles can be cancelled in pairs. Suppose that in a handle structure for a closed n-manifold M, h_j and h_{j+1} are an r-handle and an (r+1)-handle. Suppose that $h_j \cap h_{j+1}$ corresponds in h_j to $B^r \times \beta^{n-r-1} \subset B^r \times \partial B^{n-r}$ and in h_{j+1} to $\beta^r \times B^{n-r-1} \subset \partial B^{r+1} \times B^{n-r-1}$ where the β's are piecewise linear balls of the dimensions indicated. Then $h_j \cup h_{j+1}$ is an n-ball contained in M, its boundary is the union of two $(n-1)$-balls one contained in $\partial \bigcup_{i<j} h_i$ the other in $\partial \bigcup_{i>j+1} h_i$. The process of cancelling h_j and h_{j+1} consists of removing the interior of $h_j \cup h_{j+1}$, identifying together the above mentioned $(n-1)$-balls using *any* piecewise linear homeomorphism that is the identity on their common boundary, and identifying the result with M *via*

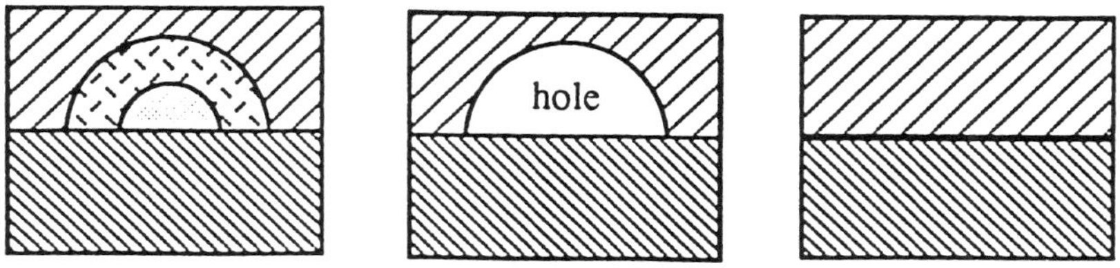

FIGURE 2

any piecewise linear homeomorphism fixed outside a ball neighbourhood of $h_j \cup h_{j+1}$. M thus acquires a new handle structure with two fewer handles. Informally this can be thought of as removeing the n-ball $h_j \cup h_{j+1}$, stretching the two hemispheres of the boundary thus created across the hole and gluing them together. This is a description of the *cancellation* or *death* of a handle pair, the *creation* or *birth* of a handle pair is the reverse process. The process is illustrated in Figure 2.

Note A. If A is an r-dimensional face of an $(r+1)$-dimensional simplex B in some triangulation of M then, in the associated handle structure the corresponding r- and $(r+1)$-handles are correctly situated for cancellation.

Note B. In any handle structure, the homeomorphism that glues together the union of the first N handles with the remainder can, using the above procedures, be changed by composition with any piecewise linear homeomorphism supported on an $(n-1)$-ball: just cut along this $(n-1)$-ball, create there a pair of handles, then immediately cancel them using the homeomorphism of interest when gluing the relevant $(n-1)$-balls together.

(Recall [8] that any piecewise linear automorphism of a manifold that is isotopic to the identity is the composition of automorphisms supported on balls.)

Definition. Handle structures on two piecewise linear n-manifolds are *handle-move equivalent* if they are related by a sequence of births and deaths of handle pairs and handle-preserving piecewise linear homeomorphisms.

Recall one more definition. A complex K *collapses* to a subcomplex L, written $K \searrow L$, if the simplexes of $K - L$ can be ordered $B_1, A_1, B_2, A_2, \ldots$ so that each $B_i < A_i$ and $B_i < C_i \in K$ implies that $C_i \in \{B_1, A_1, B_2, A_2, \ldots B_i, A_i\}$. Thus, if the piecewise linear manifold

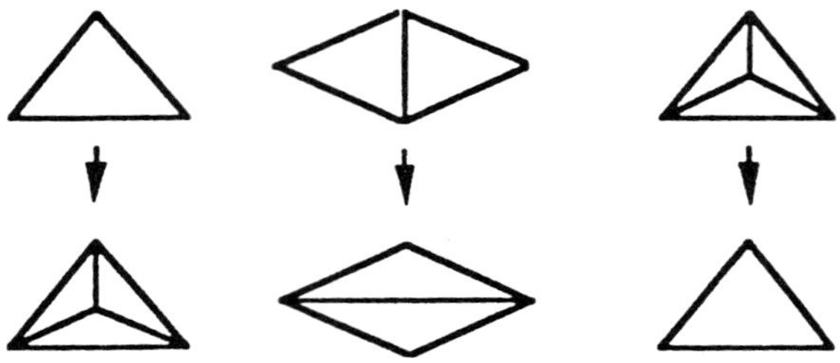

FIGURE 3

M is $|J|$ where $J \supset K \supset L$ and $K \searrow L$, then, in the associated handle structure, the handles corresponding to the simplexes of $K - L$ may be cancelled in pairs.

3. PACHNER'S MOVES.

Piecewise linear topology is always dominated by the idea of a join. Suppose an $(n+1)$-simplex C is regarded as the join of two of its disjoint faces A and B. Then $dimA + dimB = n$ and

$$\partial C = \partial(A \star B) = (\partial A \star B) \cup (A \star \partial B).$$

Assuming the pair (A, B) is ordered, this gives, up to isomorphism, $n+1$ ways of expressing the standard triangulation of the n-sphere (as the boundary of an $(n+1)$-simplex) as the union of two triangulations of n-balls glued along their boundaries. Suppose that A is an r-simplex in a combinatorial n-manifold K and that $Lk(A, K) = \partial B$ for some $(n-r)$-simplex $B \notin K$. Note that B is a *new* simplex not seen in K. Ignore the fact that this condition may seem unlikely. The *bistellar move* $\chi(A, B)$ consists of changing K to $\widetilde{K}$ by removing $A \star \partial B$ and inserting $\partial A \star B$. In dimension two, the three types of bistellar move are shown in Figure 3.

Note that the move $\chi(A, B)$ has $\chi(B, A)$ as an inverse. The move $\chi(A, B)$ might be thought of as moving K across a prototype 'cobordism' formed by gluing $A \star B$ to K along $A \star \partial B$. Pachner's result is that for closed manifolds bistellar moves suffice.

Theorem 1. (Pachner [12], [13]) *Closed combinatorial n-manifolds are piecewise linearly homeomorphic if and only if they are related by a finite sequence of bistellar moves and simplicial isomorphisms.*

Although the situation for bounded manifolds will not be explored here, it is worth recording Pachner's result for them too. The basic move to be considered is an *elementary shelling*. This consists of removing, from a combinatorial n-manifold K, an n-simplex $A \star B$ together with $\partial A \star B$, in the circumstance where the interior of A is in the interior of K and $\partial A \star B$ is in the boundary of K. Pachner [13] proves that two connected combinatorial n-manifolds with non-empty boundary are piecewise linearly homeomorphic if and only if they are related by a finite sequence of elementary shellings, inverses of elementary shellings and simplicial isomorphisms.

Bistellar moves, being very simple changes in simplicial structure, can be transferred into handle theory in the following straightforward way.

Lemma 2. *Suppose that two combinatorial n-manifolds K_1 and K_2 are related by a bistellar move, then the corresponding handle structures of $|K_1|$ and $|K_2|$ are handle-move equivalent.*

Proof. Suppose that K_2 is created from K_1 by the process of removing $A \star \partial B$ and inserting $B \star \partial A$ where $A \in K_1$, $B \notin K_1$, $dim A = r$ and $dim B = n - r$. If C is an $(n-r)$-simplex in ∂B, then $\partial B - C$ is a cone with vertex v (the vertex of ∂B not in C). Hence $\partial B - C \searrow v$ and this means that

$$(A \star \partial B) - (A \star C) \searrow (\partial A \star \partial B) \cup (A \star v) \searrow (\partial A \star \partial B).$$

The handles of $|K_1|$ corresponding to the K_1-triangulation may be ordered so that handles corresponding to $K_1 - int(A \star \partial B)$ come first, then the handles corresponding to the remaining simplexes in any order of increasing dimension. The above collapsing procedure shows that those last handles all cancel in pairs except for the n-handle corresponding to $A \star C$. Thus the handles corresponding to K_1 become those corresponding to $K_1 - int(A \star \partial B)$ together with one n-handle added on to the resulting $(n-1)$-sphere boundary. The handles corresponding to K_2 can be treated in exactly the same way, with the rôles of A and B interchanged, to obtain exactly the same result. □

Theorem 3. *Let M_1 and M_2 be closed piecewise linear n-manifolds that are piecewise linearly homeomorphic. Then any handle structure on M_1 corresponding to a triangulation is handle-move equivalent to any handle structure on M_2 corresponding to a triangulation.*

Proof. This follows at once from the theorem of Pachner quoted above, and from Lemma 2. □

4. Arbitrary piecewise linear handle structures.

Are any two piecewise linear handle structures on a closed piecewise linear manifold M handle-move equivalent? It will be shown that they are so equivalent after a collaring adjustment. In the light of the last result it is required only to prove that any handle structure on M is handle-move equivalent to one corresponding to a triangulation. Firstly consider a preparatory lemma.

Lemma 4. *Suppose a piecewise linear ball B^n is triangulated as $|v \star K|$ where v is a vertex and $|K|$ is a piecewise linear $(n-1)$-sphere. Then the handle structure for B^n obtained from $v \star K$ is handle-move equivalent to a handle structure consisting of all the handles corresponding to simplexes in K (which are not changed in the equivalence) together with one n-handle.*

Proof. Order the handles of B^n by first taking those corresponding to K, then the 0-handle corresponding to v, then the handles corresponding to the simplexes, in order of increasing dimension, of the form $v \star A$ where $A \in K$. Each of these latter handles can be regarded as a copy of $h_i \times I$ (I being the interval $[0,1]$) where h_i is the handle of $|K|$ corresponding to $A \in K$. If h_i is an r-handle $h_i \times I$ is an $(r+1)$-handle. Now suppose that h_i and h_{i+1} are such handles of $|K|$ correctly situated for the process of cancellation. Then $h_i \times I$ and $h_{i+1} \times I$ are likewise correctly situated for cancellation. Cancel them in the following way, in a manner modelled on any given method of cancelling h_i and h_{i+1}. Suppose that in $|K|$ the cancellation of h_i and h_{i+1} takes place as follows. Handles h_i and h_{i+1} are removed, and (see Figure 4 for notation) and the resultant hole is closed up using a homeomorphism $\phi : N \to N \cup h_i \cup h_{i+1}$ that identifies the $(n-2)$-balls β_+ and β_- together, ϕ being the identity on $\partial N - \beta_+$. Now, from the handle structure of B^n, remove $h_i \times I$ and $h_{i+1} \times I$ and glue $\beta_+ \times I$ to $((h_i \cup h_{i+1}) \times \{0,1\}) \cup (\beta_- \times I)$ in any way. Then, using Note B, cut along $((h_i \cup h_{i+1} \cup N) \times \{0,1\}) \cup (\beta_- \times I)$ and readjust the gluing so that the resulting hole made in B^n is closed up by the homeomorphism

$$\phi \times 1 : N \times I \to (N \cup h_i \cup h_{i+1}) \times I.$$

This procedure changes the handle structure on B^n to one that has the original handles corresponding to K and v, together with handles that are the products of I with handles of $|K|$, corresponding to the simplexes of K, with one pair cancelled. Now the handles of $|K|$ are, by Theorem 3, handle-move equivalent to handles corresponding to simplexes

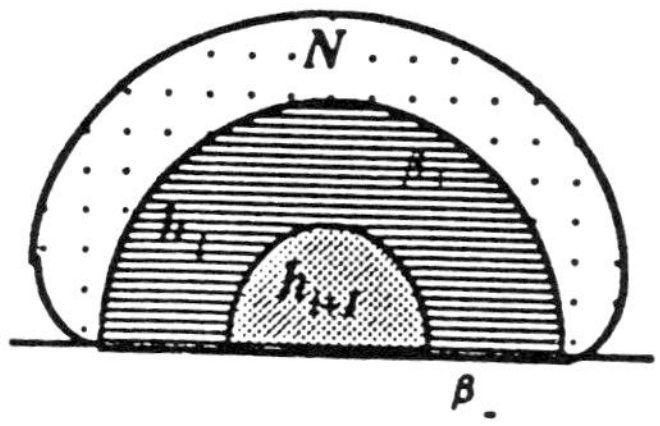

FIGURE 4

of a triangulation by $\partial\Delta^n$, the boundary of a single simplex. Because $\partial\Delta^n$ less an n-simplex collapses to a vertex, handles corresponding to $\partial\Delta^n$ cancel in pairs to become a 0-handle and an $(n-1)$-handle. Hence the above handle structure on B^n, associated to $v \star K$, is equivalent to one having the original handles corresponding to the simplexes of K, a 1-handle connecting these to the 0-handle corresponding to v, together with an n-handle. Cancelling the 0-handle and the 1-handle gives the required result. □

Suppose now that M is a closed piecewise linear n-manifold with an arbitrary piecewise linear handle structure. Suppose that $M = |K|$ and that every handle in the structure is the underlying polyhedron of some subcomplex of K (this can always be achieved, [8], after subdivison of any triangulation of M). Construct a new handle structure by adding to each handle its second derived neighbourhood (in $K^{(2)}$), subtracting the second derived neighbourhood of its attaching tube, and taking the closure. This has the effect of replacing a handle $B^r \times B^{n-r}$ by a handle $B^r_- \times B^{n-r}_+$ where the suffices indicate augmentation or depletion by a collar. This will be called the *K-adjustment* of the given handle structure.

Conjecture. *A handle structure and any K-adjustment of it are handle-move equivalent.*

Theorem 5. *The K-adjustment of any piecewise linear handle structure on a closed piecewise linear n-manifold M (where each handle underlies a subcomplex of K) is handle-move equivalent to the handle structure corresponding to the simplexes of K.*

Corollary. *Any adjustments of any two handle structures on M are handle-move equivalent.*

Proof. The result of the corollary follows at once from Theorems 3 and 5. Let a complex J be chosen so that $M = |J|$ and each handle underlies a subcomplex. A theorem of J.H.C.Whitehead [20, Theorem 6] asserts that, if a polyhedron X contains a finite collection $\{(Y_i, Z_i)\}$ of pairs of subpolyhedra such that, for each i, Y_i collapses to Z_i with respect to *some* triangulation in the structure of X, then there is a triangulation of X such that every Y_i and Z_i underly subcomplexes, the first of which collapses to the second for *every* i. Such a triangulation can be taken to be a subdivision of any given triangulation. Thus, let J be chosen so that for each handle of the form $B^r \times B^{n-r}$, the subpolyhedra $\partial B^r \times B^{n-r}$ and $\partial B^r \times 0$ (for 0 a given point in the interior of B^{n-r}) underly subcomplexes, the first of which collapses to the second. Now, starting with J, replace the triangulation of each handle with the *cone* on the J-triangulation of the boundary of the handle. As the interiors of the handles are mutually disjoint n-cells, this produces a new complex K that triangulates M within the same piecewise linear structure. For each handle $B^r \times B^{n-r}$, using (abusively) the same notation for a polyheron and the subcomplex of K that triangulates it,

$$B^r \times B^{n-r} \searrow v \star (\partial B^r \times B^{n-r}) \searrow (v \star (\partial B^r \times 0)) \cup (\partial B^r \times B^{n-r}).$$

Here, v is the cone point in $B^r \times B^{n-r}$, the first of the collapses is conewise, and the second uses coning on the given collapse $\partial B^r \times B^{n-r} \searrow \partial B^r \times 0$.

Consider now the handle structure of M corresponding to K; handles of this will be called K-handles. These K-handles are to be ordered in the following way. The subset of the K-handles that lie in a K-adjusted $B^r \times B^{n-r}$ (one of the original handles) are ordered amongst themselves by first taking those corresponding to the simplexes, in some order of increasing dimension, in the interior of $v \star (\partial B^r \times 0)$, then taking the remaining K-handles in K-adjusted $B^r \times B^{n-r}$ (again in an order corresponding to increasing dimension). K-handles in K-adjustments of distinct original handles are ordered according to the ordering of those original handles.

Now consider the handle structure of the core of one of the original r-handles corresponding to the triangulation by $v \star (\partial B^r \times 0)$. It is, by Lemma 4, handle-move equivalent to the handles corresponding to the triangulation of $(\partial B^r \times 0)$ with one extra (r-dimensional) r-handle added. Thus the K-handles of M that corresponding to simplexes in $(v \star (\partial B^r \times 0)) - (\partial B^r \times 0)$ (which are just the above mentioned handles in the interior of $B^r \times 0$ producted with an $(n-r)$-ball) are handle-move equivalent to a single (n-dimensional) r-handle h which occupies the same set as the union of the K-handles that it replaces.

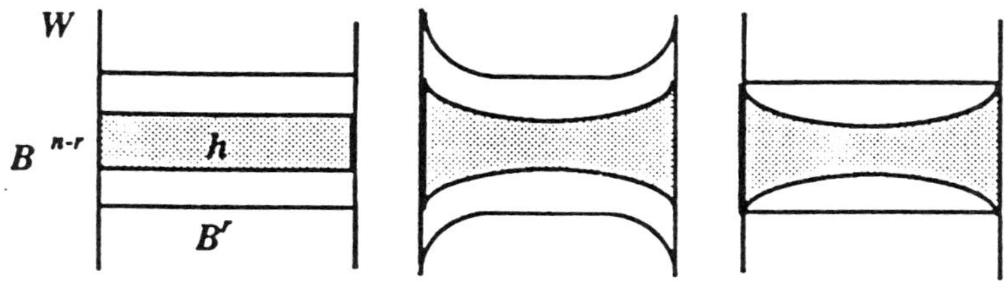

FIGURE 5

The remaining K-handles of the adjusted $B^r \times B^{n-r}$ would cancel in pairs following the simplicial collapse

$$B^r \times B^{n-r} \searrow (v \star (\partial B^r \times 0)) \cup (\partial B^r \times B^{n-r})$$

but, do *not* cancel them yet. Instead, use Note B to readjust the gluing of h to the union, W say, of all the K-handles that preceed h, by expanding $h \cap W$ to be the set previously occupied by $\partial B^r \times B^{n-r}$, thus sliding $(\partial B^r \times B^{n-r}) - (h \cap W)$ into a collar of $\partial B^r \times \partial B^{n-r}$ in ∂W . The existence of such a sliding follows from the standard result [8], [15], on uniqueness of regular neighbourhoods up to ambient isotopy applied to neighbourhoods of $\partial B^r \times 0$ in ∂W. Now use Note B again to slide that collar back into its original position, this time sliding handles that are attached to the union of W and the new h. Figure 5 is intended to illustrate this procedure.

This last sliding can be chosen so that, at the end of it, all K-handles that follow those of the adjusted $B^r \times B^{n-r}$ are back in their original positions. The K-handles of the adjusted $B^r \times B^{n-r}$ less those in h have been moved, but only by a homeomorphism. Thus they still in a position to cancel in pairs. Cancel them now, removing them and stretching h to occupy precisely the set occupied by the adjusted $B^r \times B^{n-r}$. This procedure has manoeuvred all the K-handles in the adjusted $B^r \times B^{n-r}$ to become one r-handle occupying the same set, the attaching tubes exactly corresponding. Repeating this procedure on all the original handles shows that the adjusted handles of the original structure are handle-move equivalent to the handles of the structure corresponding to the simplexes of K. The Theorem now follows as any two different adjustments of the given handle structure are, by regular neighbourhood theory, equivalent under a handle-preserving automorphism. □

5. Epilogue.

Of course, the last result of this paper would be improved if the use of *adjusted* handles could be avoided. Such avoidance would need some piecewise linear technicalities in a 'classical' vein. Probably the results here could be sharpened a little by insisting that many of the piecewise linear automorphisms be isotopic to the identity; simplicity has been the aim rather than generality. Similarly, Pachner's result on bounded manifolds could be interpreted in terms of handle structures.

Whereas it seemed desirable to have on record a version of Cerf theory in the piecewise linear category, it should be emphasised that the proofs given here depend on significant pieces of piecewise linear theory. Theorems available in books have been regarded as 'standard', but that does not make them trivial. In addition, the theorem of J.H.C.Whitehead, used in the proof of Theorem 5, is quite sophisticated. Pachner's work that is central to this paper is not understood without effort. It depends heavily on the theorem of Alexander that asserts that any subdivision of a complex K is related to K by 'stellar moves'. The standard reference books [15], [8], for piecewise linear topology omit that result, preferring to develop the subject by way of another route. Proofs of Alexander's result can be found in [1], [5] and [11]. In principle perhaps the basic theory of piecewise linear topology should be rewritten to accommodate Pachner's result, but market forces would inhibit such a project.

References

1. J. W. Alexander, *The combinatorial theory of complexes*, Ann. of Math. **31** (1930) 292-320.
2. J. Cerf, *Isotopie et pseudo isotopie*, Proc. Int. Congress of Math., Moscow (1966).
3. J. Cerf, *La stratification naturelle des espaces de fonctions différentiables réeles et la théorème de la pseudoisotopie* , Publ. Math. I.H.E.S. **39**, (1970).
4. R. A. Fenn and C. P. Rourke, *On Kirby's calculus of links*, Topology **18** (1979) 1-15.
5. L. C. Glaser, *Geometric combinatorial topology*, Van Nostrand Reinhold, New York (1970).
6. A. Hatcher and W. Thurston, *A presentation for the mapping class group of a closed orientable surface*, Topology **19** (1980) 221-237.
7. J. Hempel, *3-manifolds* , Ann. of Math. Studies **86**, Princeton University Press, Princeton, N.J. (1976).
8. J. F. P. Hudson, *Lecture notes on piecewise linear topology*, Benjamin, New York (1969).
9. R. C. Kirby, *A calculus for framed links in* S^3, Invent. Math. **45** (1978) 35-56.

10. R. C. Kirby and L. C. Siebenmann, *Foundational essays on topological manifolds, smoothings and triangulations* , Ann. of Math. Studies **88**, Princeton University Press, Princeton, N.J. (1977).
11. W. B. R. Lickorish, *Combinatorial topology. Notes on seminars by E.C.Zeeman*, Cambridge University Notes (1961).
12. U. Pachner, *Konstruktionsmethoden und das kombinatorische Homöomorphieproblem für Triangulationen kompakter semilinearer Mannigfaltigkeiten*, Abh. Math. Sem. Hamb. **57** (1987) 69-86.
13. U. Pachner, *P.L. homeomorphic manifolds are equivalent by elementary shellings*, Europ. J. Combinatorics **12** (1991) 129-145.
14. K. Reidemeister, *Zur dreidimensionalen Topologie*, Abh. Math. Sem. Univ. Hamberg **9** (1933) 189-194.
15. C. P. Rourke and B. J. Sanderson, *Introduction to piecewise-linear topology*, Ergebnisse der mathematik **69**, Springer-Verlag, Berlin Heidelberg New York (1972).
16. J. Singer, *Three-dimensional manifolds and their Heegaard diagrams*, Trans. Amer. Math. Soc. **35** (1933) 88-111.
17. S. Smale, *Generalised Poincaré's conjecture in dimensions greater than 4*, Ann. of Math. **74** (1961) 391-406.
18. V. G. Turaev and O. Y. Viro, *State sum invariants of 3-manifolds and quantum 6j-symbols*, Topology **31** (1992) 865-902.
19. B. Wajnryb, *A simple presentation for the mapping class group of an orientable surface*, Israel J. Math. **45** (1983) 157-174.
20. J. H. C. Whitehead, *Simplicial spaces, nuclei and m-groups*, Proc. London Math. Soc. **45** (1939) 243-327.

University of Cambridge, Department of Pure Mathematics and Mathematical Statistics, 16, Mill Lane, Cambridge, CB2 1SB, England

E-mail address: wbrl@pmms.cam.ac.uk

AMS/IP Studies in Advanced Mathematics
Volume **2**, 1997 (Part 1)

LOCAL HOMOTOPY PROPERTIES OF TOPOLOGICAL EMBEDDINGS IN CODIMENSION TWO

GERARD A. VENEMA

ABSTRACT. Suppose N is an $(n-2)$-manifold topologically embedded in an n-manifold M. Chapman and Quinn have shown that N is locally flat in M provided that the local homotopy groups of $M-N$ at points of N are infinite cyclic and that all the higher local homotopy groups of $M-N$ at points of N vanish. The main theorem in this paper asserts that it is only necessary to assume the homotopy conditions in dimensions below the middle dimension; i.e., if the local fundamental groups are good and if $M-N$ is locally k-connected at points of N for $2 \leq k < n/2$, then N is locally flat.

1. INTRODUCTION

Let N^{n-2} be a topological $(n-2)$-submanifold of the topological n-manifold M^n. (All manifolds are without boundary.) In this paper we wish to study the problem of characterizing local flatness for N in terms of local homotopy properties of the complement $M-N$. We begin with the relevant definitions.

Definitions. Let $x \in N \subset M$. We say that N is *locally flat* at x if there exists a neighborhood U of x in M such that the pair $(U, U \cap N)$ is homeomorphic with the standard pair $(\mathbb{R}^n, \mathbb{R}^{n-2} \times \{0\})$. We say that N is *locally 1-alg* at x if for every neighborhood U of x in M there exists a neighborhood V of x in U such that the image of $\pi_1(V-N)$ in $\pi_1(U-N)$ is abelian. We say that N is *locally k-co-connected* (abbreviated k-LCC) at x if for every neighborhood U of x in M there exists a neighborhood V of x in U such that the image of $\pi_k(V-N)$ in $\pi_k(U-N)$ is trivial.

1991 *Mathematics Subject Classification.* 57M30, 55Q05, 55M05 .
Key words and phrases. Local homotopy groups, local π_1, locally flat, topological embedding, codimension two, duality .

We say that N is *locally homotopically unknotted* at x if N is locally 1-alg at x and N is k-LCC at x for every $k \geq 2$.

Remark. The "alg" in 1-alg stands for "abelian local groups." Other authors (e.g. [4]) use the terminology "good local π_1" to mean the same thing. It is not difficult to see that if N is locally 1-alg at x, then the image of $\pi_1(V - N)$ in $\pi_1(U - N)$ is infinite cyclic for sufficiently small U and V. (See Lemma 2.1, below.)

The basic theorem relating these concepts is due to Chapman and Quinn. Chapman first proved that locally homotopically unknotted codimension two embeddings are locally flat as long as $n \geq 5$ [2] but Quinn later gave a different proof of this result [10] and also proved the theorem in case $n = 4$ [11]. The theorem is true in dimension 3 as well [1], but that case is not of interest to us here.

Theorem [Chapman-Quinn]. *Let N^{n-2} be a topological $(n-2)$-submanifold of the n-manifold M^n. If N is locally homotopically unknotted at x for every $x \in N$, then N is locally flat in M.*

Note that the definition of locally homotopically unknotted involves assumptions about the local homotopy groups in all dimensions. In this paper we wish to investigate the question of whether or not it might be possible to weaken the hypotheses in the Chapman-Quinn Theorem and still conclude that the embedding is locally flat. In order to understand the question it is helpful to recall the analogous global problem for codimension two knots: If $\Sigma^{n-2} \subset S^n$ is a smooth knot and $\pi_i(S^n - \Sigma^{n-2}) \cong \pi_i(S^1)$ for $i < n/2$, then Σ^{n-2} is unknotted [8]. On the other hand, for each $k < n/2$ there exists a smooth slice knot $\Sigma^{n-2} \subset S^n$ such that $\pi_i(S^n - \Sigma^{n-2}) \cong \pi_i(S^1)$ for $i < k$ but $\pi_k(S^n - \Sigma^{n-2}) \ncong \pi_k(S^1)$ [12].

The analogy above leads us to conjecture that it should suffice to assume that N is k-LCC for k in the range $2 \leq k < n/2$. This is the best we can hope for because taking the suspension of one of the knots mentioned at the end of the previous paragraph gives the following example.

Example 1.1. For each n and k with $1 \leq k < (n-1)/2$ there exists a piecewise linear $(n-2)$-sphere $\Sigma \subset S^n$ such that Σ is locally 1-alg and j-LCC at x for every $x \in \Sigma$ and for every $j < k$ but there exists a point $x_0 \in \Sigma$ such that Σ is not k-LCC at x_0. The embedding is locally flat except at two points.

But we must be careful with this analogy because we are dealing with *topological* embeddings, not piecewise linear or smooth ones. In the topological category we can have the following more surprising example.

Example 1.2. [13] For each n and k with $1 \leq k \leq n-2$ there exists a topological $(n-2)$-sphere $\Sigma \subset S^n$ and a point $x \in \Sigma$ such that Σ is locally 1-alg at x and Σ is j-LCC at x for every $j < k$ but Σ is not k-LCC at x. The embedding is locally flat except at a sequence of points; the sequence of nonlocally flat points has x as its only limit point.

These examples show that we must be very careful about how we ask our question. In particular, Example 1.2 shows that if we focus our attention on a single point $x \in N$, all the k-LCC conditions are independent of each other in the range $2 \leq k \leq n-2$. So it is not the case that locally 1-alg at x together with k-LCC at x for $2 \leq k < n/2$ imply that N is locally homotopically unknotted at x. However, the examples constructed in [13] do not have just one point at which they fail to be locally flat. If $k \geq n/2$, then the examples are forced to have a sequence of points converging to x at which the codimension two sphere is not j-LCC for some $j < n/2$. Thus there is still hope for a positive answer to the question. In fact, if we assume the condition at all points in a neighborhood of x, then the answer to the question follows the expected pattern. That fact is the main result of this paper. Since the conclusion we seek is local we may as well assume that the open set on which the hypothesis holds is all of N.

Theorem 1.3. *Let N^{n-2} be a topological $(n-2)$-submanifold of the n-manifold M^n. If N is locally 1-alg at x for every $x \in N$ and N is k-LCC at x for every $x \in N$ and for $2 \leq k < n/2$, then N is locally flat in M.*

Notice that the inequality $k < n/2$ in the theorem is a strict inequality. This is the real novelty of the theorem. The fact that it suffices to assume k-LCC for k in the range $2 \leq k \leq n/2$ was apparently known to Hollingsworth and Rushing [5]. Although they did not explicitly state a theorem about this, they incorporated that fact into their definition of locally homotopically unknotted [5, page 393].

A particularly interesting special case of the theorem is the case $n = 4$. That is the case which inspired the work on the present paper. In the 4-dimensional case it is only necessary to assume a π_1 condition.

Corollary. *Let Σ be a surface (without boundary) topologically embedded in the 4-manifold M^4. If Σ is locally 1-alg at x for every $x \in \Sigma$, then Σ is locally flat in M.*

This result has already appeared as part of Theorem 9.3A in [4] but the proof given in [4] contains a gap. In particular, [4, Lemma 9.3B] purports to prove that, for a fixed $x \in \Sigma$, if Σ is locally 1-alg at x, then Σ is 2-LCC at x. Example 1.2 and the example in [7] show that this is not the case.

Acknowledgment. The author wishes to express his thanks to Vo Thanh Liem for many helpful conversations related to the results in this paper.

2. Homotopy properties of codimension two embeddings

Let $N^{n-2} \subset M^n$ be a pair of topological manifolds-without-boundary as in the Introduction. In this section we develop the basic homotopy properties enjoyed by such pairs if they have good local homotopy properties at every point. We begin by quoting a lemma of Chapman which allows us to specify exactly which neighborhoods of points of N are good for π_1 of $M - N$. The proof is an application of Alexander Duality and can be found in [2, Lemma 3.1]. All homomorphisms of homotopy and homology groups are assumed to be inclusion-induced unless otherwise indicated.

Lemma 2.1. [2] *Suppose N is locally 1-alg at $x \in N$. Let $V_1 \supset V_2 \supset V_3$ be three neighborhoods of x such that*

(1) *$V_i \cap N \cong \mathbb{R}^{n-2}$ for each i,*
(2) *V_i is homotopic to a point in V_{i-1} for $i = 2, 3$, and*
(3) *V_i satisfies the 1-alg hypothesis with respect to V_{i-1} for $i = 2, 3$.*

Then both $\operatorname{im}[\pi_1(V_2 - N) \to \pi_1(V_1 - N)]$ and $\operatorname{im}[\pi_1(V_3 - N) \to \pi_1(V_2 - N)]$ are infinite cyclic and the inclusion induces an isomorphism from $\operatorname{im}[\pi_1(V_3 - N) \to \pi_1(V_2 - N)]$ to $\operatorname{im}[\pi_1(V_2 - N) \to \pi_1(V_1 - N)]$.

We next prove three technical lemmas which show the key way in which our hypothesis of k-LCC at *every* point is used. The techniques of proof are very similar to those in [2] and [5]. The first of these lemmas (Lemma 2.2) says that homotopies of j-dimensional polyhedra can be pushed off N. The second (Lemma 2.3) says roughly that if N is locally j-co-connected, then it is globally j-co-connected. The final lemma (Lemma 2.4) says that j-dimensional polyhedra in the complement of N can be pushed close to N with homotopies supported in the complement of N.

Lemma 2.2. *Fix $j \geq 1$ and an open set V_0 such that $V_0 \cap N \cong \mathbb{R}^{n-2}$. Suppose N is locally 1-alg at x for every $x \in N$ and N is k-LCC at x for every $x \in N$ and for all k in the range $2 \leq k \leq j$. For every positive function ϵ defined on V_0 there exists a neighborhood V of $V_0 \cap N$ such that* *if K^j is a compact j-dimensional polyhedron and $f : K \times [0,1] \to V$ is a map with $f(K \times \{0\}) \subset V - N$, then there is a map $g : K \times [0,1] \to V_0 - N$ such that $g(x,0) = f(x,0)$ for every $x \in K$ and dist($f(x,t), g(x,t)$) $< \epsilon(f(x,t))$ for every $(x,t) \in K \times [0,1]$.*

Proof. First consider the case $j = 1$. For each $x \in N \cap V_0$ there exists a neighborhood V_x of x such that any loop in $V_x - N$ which is null-homologous in $V_x - N$ bounds a disk of diameter less than ϵ in $V_0 - N$. Let $V = \cup_{x \in V_0 \cap N} V_x$. Suppose $f : K^1 \times [0,1] \to V$ with $f(K \times \{0\}) \subset V - N$. Take a triangulation of K with small mesh and a partition $0 = t_0 < t_1 < \cdots < t_m = 1$ such that for each 1-simplex σ in K and each i, the diameter of $f(\sigma \times [t_i, t_{i+1}])$ is small relative to a Lebesgue number for the cover $\{V_x\}$ of $f(K \times [0,1])$. Let $K^{(0)}$ denote the 0-skeleton of K. Use the fact that N does not separate any open set to shift $f(K^{(0)} \times [0,1])$ off N with a very small move. This defines $g|K^{(0)} \times [0,1]$. Let σ^1 be a 1-simplex in K. We next push each of the arcs $f(\sigma^1 \times \{t_i\})$ off N in such a way that each $g(\partial(\sigma^1 \times [t_i, t_{i+1}]))$ is null-homologous in some $V_x - N$. This is done by pushing them off one at a time, starting with $f(\sigma^1 \times \{t_1\})$. Push $f(\sigma^1 \times \{t_1\})$ off with a small move to define $g|\sigma^1 \times \{t_1\}$. If $g(\partial(\sigma^1 \times [t_0, t_1]))$ links N, add a small loop to $g(\sigma^1 \times \{t_1\})$ to make $g(\partial(\sigma^1 \times [t_0, t_1]))$ null-homologous missing N. Then proceed to $f(\sigma^1 \times \{t_2\})$. Push it off and add a loop to it, if necessary, to make $g(\partial(\sigma^1 \times [t_1, t_2]))$ null-homologous. This process is continued inductively. Once $g(\partial(\sigma^1 \times [t_i, t_{i+1}])) \cap N = \emptyset$ for every i, we can use the choice of V_x to extend g to each of the 2-cells $f(\sigma^1 \times [t_i, t_{i+1}])$.

Next consider the case $j = 2$. For each $x \in N \cap V_0$ there exists a neighborhood V_x of x such that any singular 2-sphere in $V_x - N$ bounds a singular 3-cell of diameter less than $\epsilon/2$ in $V_0 - N$. Let δ be a Lesbegue function for the cover $\{V_x\}$. For each $x \in V_0 \cap N$ we can use Lemma 2.1 to choose a neighborhood V'_x such that any loop in $V'_x - N$ that is null-homologous in $V_0 - N$ bounds a singular disk of diameter less than $\delta(x)/2$ in $V_0 - N$. Let δ' be a Lesbegue function for the cover $\{V'_x\}$. Finally use the case $j = 1$ of the Lemma to choose a neighborhood V such that any homotopy of a 1-dimensional polyhedron in V is $(\delta'/2)$-homotopic to a homotopy in the complement of N. Now consider a map $f : K^2 \times [0,1] \to V$ such that $f(K^2 \times \{0\}) \subset V - N$. Take a triangulation of K with small mesh and a partition $0 = t_0 < t_1 < \cdots < t_m = 1$ such that for each simplex σ in K and each i, the diameter of $f(\sigma \times [t_i, t_{i+1}])$ is small relative to $\delta'/2$. Use the choice of V to define $g|K^{(1)} \times [0,1]$. Consider a 2-simplex σ^2 in K. For each i, $g((\partial\sigma^2) \times \{t_i\})$ is null-homotopic in the complement of N (it bounds $g(\sigma^2 \times \{0\} \cup (\partial\sigma^2) \times [0, t_i])$), so the choice of V'_x allows us to extend g to $\sigma^2 \times \{t_i\}$. Then the choice of V_x allows us to extend g to each of the 3-cells $\sigma^2 \times [t_i, t_{i+1}]$.

The cases $j \geq 3$ are handled inductively. They are a little easier than either of the cases $j = 1$ or $j = 2$ because there is no problem of linking in these higher dimensions. $\square$

Lemma 2.3. *Fix $j \geq 1$. Suppose N is locally 1-alg at x for every $x \in N$ and N is k-LCC at x for every $x \in N$ and for every k in the range $2 \leq k \leq j$. If V_1 is an open set such that $V_1 \cap N \cong \mathbb{R}^{n-2}$, then there exists an open set $V_2 \subset V_1$ with $V_1 \cap N = V_2 \cap N$ such that any map $f : S^j \to V_2 - N$ is null-homotopic in $V_1 - N$. (If $j = 1$, we must also assume that f is null homologous in $V_1 - N$.)*

Proof. Consider the case $j = 1$. For each $x \in V_1 \cap N$ we can use Lemma 2.1 to find a positive number $\epsilon(x)$ such that any loop in the ϵ neighborhood of x that misses N and is null-homologous in $V_1 - N$ is null-homotopic in $V_1 - N$. Choose V' using this ϵ and Lemma 2.2. Then choose $V_2 \subset V'$ so that V_2 strong deformation retracts to N in V'. Suppose $f : S^1 \times \{0\} \to V_2 - N$ and f is null-homologous in $V_1 - N$. The choice of V_2 allows us to extend f to $f : S^1 \times [0,1] \to V'$ so that $f(S^1 \times \{1\})$ is a single point. Using Lemma 2.2 we can find an ϵ approximation $g : S^1 \times [0,1] \to V_1 - N$. Now $g(S^1 \times \{1\})$ has diameter less than ϵ and $g(S^1 \times \{1\})$ is null-homologous in $V_1 - N$ (since $g(S^1 \times \{0\})$ was), so the choice of ϵ allows us to find a singular 2-cell which $g(S^1 \times \{1\})$ bounds in $V_1 - N$. Thus $f(S^1 \times \{0\})$ bounds a singular 2-cell in $V_1 - N$.

The proof of the higher dimensional cases is similar. $\square$

Lemma 2.4. *Fix $j \geq 1$ and an open set V_0 such that $V_0 \cap N \cong \mathbb{R}^{n-2}$. Suppose N is locally 1-alg at x for every $x \in N$ and N is k-LCC at x for every $x \in N$ and for all k in the range $2 \leq k \leq j$. For every neigborhood U_1 of $N \cap V_0$ in V_0 there exists a neighborhood V_1 of N in U_1 having the following property: for every neighborhood U_2 of N there exists a neighborhood V_2 of $N \cap V_0$ in U_2 such that if $f : K^j \to V_1 - N$ is a map of a j-dimensional polyhedron into $V_1 - N$, then f is homotopic in $U_1 - N$ to a map of K into $U_2 - N$. Furthermore, if L is a subpolyhedron of K and $f(L) \subset V_2 - N$, then the homotopy can be chosen to keep L fixed.*

Proof. Again we first consider the case $j = 1$. Given U_1, we choose V_1 in such a way that any loop in $V_1 - N$ that is null-homologous in $U_1 - N$ is null-homotopic in $U_1 - N$ (Lemma 2.3). Given U_2, we choose V_2 to be any connected neighborhood of N contained in U_2. Consider a map $f : K^1 \to V_1 - N$. It is easy to pull the 0-skeleton in, so we may assume that $f(K^{(0)}) \subset V_2$. For each 1-simplex σ^1 in K, we choose an arc α

in $V_2 - N$ joining the two points $f(\partial\sigma^1)$ and then add a loop to it, if necessary, so that $\alpha \cup f(\sigma^1)$ is null-homologous in $U_1 - N$. By the choice of V_1, this loop is null-homotopic in $U_1 - N$, so there is a homotopy from $f(\sigma^1)$ to α that keeps the endpoints fixed.

Now consider the case $j = 2$. Let U_1 be given. Use Lemma 2.3 to choose $U' \subset U_1$ such that any singular 2-sphere in $U' - N$ bounds a singular 3-cell in $U_1 - N$. Then use the previous case of this lemma to choose $V_1 \subset U'$. Let U_2 be given. Use Lemma 2.3 to choose $V' \subset U_2$ so that any loop in $V' - N$ that is null-homologous in $U_2 - N$ is null-homotopic in $U_2 - N$. Finally choose $V_1 \subset V'$ using the previous case of this lemma. Suppose $f : K^2 \to V_1 - N$. By the previous case of this lemma, we can pull the image of the 1-skeleton into $V' - N$ so we assume that $f(K^{(1)}) \subset V' - N$. For any 2-simplex σ^2 in K, the choice of V' allows us to find a singular 2-cell $C \subset U_2 - N$ whose boundary is $f(\partial\sigma^2)$. But the choice of U' says that the singular 2-sphere $f(\sigma^2) \cup C$ bounds in $U_1 - N$, so there is a homotopy of $f(\sigma^2)$ to C that keeps $\partial\sigma^2$ fixed.

The higher dimensional cases are similar. □

3. Homology properties of codimension two embeddings

We begin this section by fixing some notation which will be assumed for the remainder of this paper. For the remainder of this paper it will be assumed that $N \subset M$ satisfies the hypotheses of Theorem 1.3, even though this will not be explicitly stated in the lemmas which follow. Specifically, we assume the following.

Hypothesis. N^{n-2} is a topological $(n-2)$-submanifold of the n-manifold M^n such that N is locally 1-alg at x for every $x \in N$ and N is k-LCC at x for every $x \in N$ and for $2 \leq k < n/2$.

Consider one point of $x \in N$ and fix a neighborhood U_0 of x. Let $U_1 \subset U_0$ be a second neighborhood of x such that $U_1 \cap N \cong \mathbb{R}^{n-2}$ and U_1 satisfies the 1-alg condition relative to U_0. We will denote the infinite cyclic group $\operatorname{im}[\pi_1(U_1 - N) \to \pi_1(U_0 - N)]$ by $\mathbb{J}$. We use t to denote a fixed generator of $\mathbb{J}$ and write elements of $\mathbb{J}$ multiplicatively; thus $\mathbb{J} = \{t^i \mid i = 0, \pm 1, \pm 2, \dots\}$. Let $W = U_1 - N$ and let $p : \widetilde{W} \to W$ denote the infinite cyclic cover determined by the natural homomorphism of $\pi_1(W)$ to $\mathbb{J}$. We will use the same symbol t to denote the generator of the group of deck transformations of $\widetilde{W}$ which corresponds to t. If P is any subset of W, we use P^* to denote $p^{-1}(P) \subset \widetilde{W}$; we use the notations $H_i(P^*; \mathbb{Z})$ and $H_i(P; \mathbb{Z}[\mathbb{J}])$ interchangeably. From now on we will work with W as our ambient manifold; the reader should be certain to note that W is a deleted neighborhood of N and does not include any points of N. The reason for doing this is so that we can take the $\mathbb{J}$-cover of W.

We are assuming that N is k-LCC at each point for $2 \leq k < n/2$. We wish to conclude that N is k-LCC at x for $k \geq n/2$. Fix $x \in N$ and let $B_2 \subset \operatorname{int} B_1$ be two concentric n-balls centered at x. If we can show that it is possible to choose B_1 and B_2 in such a way that $\pi_k(B_2 - N) \to \pi_k(B_1 - N)$ is the trivial homomorphism for $k \geq n/2$, then the proof will be complete. Since each map of S^k into $B_2 - N$ has compact image, it suffices to show that there exist arbitrarily close neighborhoods $W_1 \subset W_2$ of N such that if $P_i = B_i - W_i$, then $\pi_k(P_2) \to \pi_k(P_1)$ is trivial. By the Eventual Hurewicz Theorem ([3, Proposition 3.1] or [10, Theorem 5.2]), it suffices to prove this in homology with local coefficients. In other words, we need to prove the following Proposition.

Proposition 3.1. *Fix $x \in N$. For each sufficiently small n-ball B_1 centered at x and contained in V_1 there exists a concentric n-ball B_2 such that there are arbitrarily close neighborhoods $W_1 \subset W_2$ of N for which the inclusion-induced homomorphism $H_k(B_2 - W_2; \mathbb{Z}[\mathbb{J}]) \to H_k(B_1 - W_1; \mathbb{Z}[\mathbb{J}])$ is trivial for $k \geq n/2$.*

The proof of Proposition 3.1 is broken into three parts. First, in the remainder of this section we prove some lemmas about related homology inclusions that will be used later. Second, in §4 we will prove that there are arbitrarily small neighborhoods of x having cell decompositions with no cells above the middle dimension. This leaves only the middle dimensional case to be completed. In §5 we prove the middle dimensional case using coefficients in a field. Finally, we observe in the last section that the results of the preceding two sections suffice to complete the proof of Proposition 1.3.

Fix an open set V_0 such that $V_0 \cap N = U_1 \cap N$ and there is a strong deformation retraction of V_0 to $V_0 \cap N$ in U_1. The let V_1 be a neighborhood of $N \cap U_1$ such that polyhedra in $V_1 - N$ of dimension less than $n/2$ can be pulled near N via homotopies whose tracks lie in $V_0 - N$ (Lemma 2.4).

Lemma 3.2. *Let B_1 be an n-ball centered at x and contained in V_1, let C be an $(n-2)$-ball in the interior of $N \cap B_1$, and let B_2 be a second n-ball centered at x such that $B_2 \cap N$ is contained in the interior of C. There exist arbitrarily close neighborhoods $W_1 \subset W_2$ of N such that if*

$P_i = B_i - W_i$, then

$$\ker[H_k(W - P_1; \mathbb{Z}[\mathbb{J}]) \to H_k(W; \mathbb{Z}[\mathbb{J}])] \subset \ker[H_k(W - P_1; \mathbb{Z}[\mathbb{J}]) \to H_k(W - P_2; \mathbb{Z}[\mathbb{J}])]$$

for each $k < n/2$.

In other words, if a k-dimensional class in $W - P_1$ is null-homologous in W, then it is null-homologous missing P_2.

Proof. Let W_2 be any connected open neighborhood of N in V_1. By Lemma 2.3 there exists a neighborhood W_1' of N in W_2 such that $\pi_k(p^{-1}(W_1' - N)) \to \pi_k(p^{-1}(W_2 - N))$ is trivial for every $k < n/2$. Choose ϵ to be a positive number that is smaller than either $\frac{1}{3}\text{dist}(C, W - B_1)$ or $\frac{1}{3}\text{dist}(N - C, B_2)$. Then use ϵ as input to Lemma 2.2 to choose W_1'' so that homotopies in W_1'' can be ϵ-pushed off N. Finally W_1 should be chosen so that there is an ϵ-homotopy deformation of W_1 to N that stays inside W_1''.

Now consider $\gamma \in \ker[H_k(W - P_1; \mathbb{Z}[\mathbb{J}]) \to H_k(W; \mathbb{Z}[\mathbb{J}])]$. We can represent γ with a map $g : (K, L) \to (\widetilde{W}, \widetilde{W} - P_1^*)$, where K is a compact $(k+1)$-dimensional polyhedron and L is a k-dimensional subpolyhedron. We trim away the part of $g(K)$ that is outside of $p^{-1}(V_1 - N)$ so that $p(g(K)) \subset V_1 - N$.

We first wish to push $p(g(L))$ out of B_2. The choice of W_1 allows us to find an ϵ-homotopy of $p(g(L))$ that pushes the part of $p(g(L))$ in the ϵ-neighborhood of C into N and is the identity outside the 2ϵ-neighborhood of C. The image of L under that homotopy can be further pushed along N out of the $(n-2)$-cell C because $\dim L < \dim C$. We then use Lemma 2.2 to approximate by a homotopy in the complement of N. This last homotopy may be lifted to $\widetilde{W}$, so it allows us to assume that $p \circ g : (K, L) \to (V_1 - N, V_1 - (N \cup B_2))$.

We next cut $p(g(K))$ off on ∂B_2. In other words, put $p \circ g$ in general position relative to ∂B_2. Then $(p \circ g)^{-1}(B_2)$ is a subpolyhedron of K. So $p \circ g | (p \circ g)^{-1}(\partial B_2)$ can be extended (using the Tietze Extension Theorem) to a map of $(p \circ g)^{-1}(B_2)$ into ∂B_2. Thus there is a map $g' : K \to (V_1 - (N \cup B_2)) \cup \partial B_2$ that agrees with $p \circ g$ outside $(p \circ g)^{-1}(B_2)$.

By Lemma 2.4 and the choice of V_1, there exists a neighborhood W_2' of N such that k-dimensional polyhedra can be pulled into W_1' keeping W_2' fixed. Let Q be a compact subpolyhedron of K such that $g'(Q) \subset W_2'$, $g'(K - Q) \subset V_1 - (N \cup B_2)$, and $Q \cap L = \emptyset$. We may assume that Q does not separate K. (If it does, adjust g' slightly so that arcs in Q which connect up components of $K - Q$ are moved off N and then remove neighborhoods of the arcs from Q.) Then there exists a k-dimensional subpolyhedron R of K such that $K \searrow Q \cup R$. By the choice of W_2' there is a homotopy that pulls $g'(R)$ into W_1', keeping $g'(R \cap Q)$ fixed. This defines a new map $g'' : Q \cup R \to W_1'$ such that $g''|Q = g'|Q$ and $g''(R) \cap N = \emptyset$. Let S denote a close neighborhood of $Q \cup R$ in K and extend g'' to S. Then $p(g(L))$ is homologous to $g''(\partial S)$ since the difference bounds $g'(\overline{K - Q}) \cup g''(\overline{S - Q})$. All those homologies are supported in W. Furthermore, the fact that $g' \simeq g''$ in W allows us to lift the homologies to $\widetilde{W}$.

Let γ'' denote the homology class represented by a lift of $g''(\partial S)$. We have that $\gamma'' \sim \gamma$ in $\widetilde{W} - P_2^*$ and γ'' is represented by a map into $p^{-1}(W_1' - N)$. The choice of W_1' now gives $\gamma'' \sim 0$ in $p^{-1}(W_2 - N)$, so the proof is complete. □

Note: The entire construction described in the proof above takes place in V_0, so we may assume that the homology classes we have constructed are supported in $p^{-1}(V_0 - N)$. This observation will be important in the proof of Lemma 3.4, below.

Lemma 3.3. *If P_1 and P_2 are as in Lemma 3.2, then*

$$H_k(W, W - P_1; \mathbb{Z}[\mathbb{J}]) \to H_k(W, W - P_2; \mathbb{Z}[\mathbb{J}])$$

is the trivial homomorphism for $k < n/2$.

Proof. Suppose $(c, \partial c)$ is a class in $H_k(W, W - P_1; \mathbb{Z}[\mathbb{J}])$. We use the notation $|c|$ to denote the support of a representative of c. Lemma 3.2 says $\partial c = \partial d$, where $|d| \subset \widetilde{W} - P_2^*$. But Lemma 2.4 says that $c - d$ is homologous to e where $|e| \subset W_2$. Hence c is homologous to $d + e$ and the proof is complete because $|d| \cup |e| \subset \widetilde{W} - P_2^*$. □

Downstairs we can improve the dimension by 1. The reason for this is the fact that the local homology of the complement downstairs is always trivial.

Lemma 3.4. *If P_1 and P_2 are as in Lemma 3.2 and P_1 is sufficiently small, then*

$$H_k(W, W - P_1; \mathbb{Z}) \to H_k(W, W - P_2; \mathbb{Z})$$

is the trivial homomorphism for $k \leq n/2$.

Proof. (The reader should probably review the definitions of V_0 and V_1 at this point; those definitions can be found just above the statement of Lemma 3.2) By Alexander Duality, $H_{k+1}(U_1, U_1 - N) \cong \bar{H}_c^{n-k-1}(U_1 \cap N)$ and $\bar{H}_c^{n-k-1}(U_1 \cap N) \cong \bar{H}_c^{n-k-1}(\mathbb{R}^{n-2}) = 0$ since $k \geq 2$. The following diagram with exact rows shows that $H_k(V_0 - N) \to H_k(U_1 - N)$ is trivial.

$$\begin{array}{ccccc} H_{k+1}(V_0, V_0 - N) & \longrightarrow & H_k(V_0 - N) & \longrightarrow & H_k(V_0) \\ \downarrow & & \downarrow & & \downarrow 0 \\ 0 = H_{k+1}(U_1, U_1 - N) & \longrightarrow & H_k(U_1 - N) & \longrightarrow & H_k(U_1) \end{array}$$

Now suppose $(c, \partial c)$ is a class in $H_k(W, W - P_1; \mathbb{Z})$. We can trim off part of $|c|$ so that $|c| \subset V_1$. Then the proof of Lemma 3.2 shows that $\partial c = \partial d$, where $|d| \subset V_1 - P_2$. Since $\partial(c - d) = 0$, $c - d$ represents a class in $H_k(V_0 - N)$ and $c - d$ is null-homologous in W by the paragraph above. Thus $c = d$ in $H_k(W, W - P_1; \mathbb{Z})$ and the proof is complete. □

4. The structure of neighborhoods

In this section we use the homology results of the previous section to construct neighborhoods of x having improved homotopy properties. Before we can do that we need a result about π_1.

Lemma 4.1. *Fix $x \in N$ and suppose $n \geq 5$. For each sufficiently small n-ball B_1 centered at x there exists a concentric n-ball B_2 in B_1 such that for sufficiently small n-balls $B_3 \subset B_2$ there are arbitrarily small neighborhoods $W_1 \subset W_2 \subset W_3$ of N such that if $P_i = B_i - W_i$, then the inclusion-induced homomorphism $\pi_1(\partial P_2^*) \to \pi_1(P_1^* - P_3^*)$ is trivial.*

Proof. The outline of the proof is as follows. Step 1 consists of choosing B_2 and B_3. Step 2 is the construction of W_2. Step 3 is the proof that $\pi_1(\partial P_2^*) \to \pi_1((B_1 - N)^* - P_3^*)$ is trivial. The final step is the observation that W_1 can be chosen so that the conclusion of the Lemma holds.

Let B_1 be given. We first explain how to choose B_2 and B_3. Begin by choosing an $(n-2)$-cell C_1 in N such that $x \in \operatorname{int} C_1$ and $C_1 \subset \operatorname{int} B_1$. Choose B_2 to be an n-cell centered at x such that $B_2 \cap N \subset \operatorname{int} C_1$ and the inclusion $B_2 - N \hookrightarrow B_1 - N$ satisfies the 1-alg hypothesis. Let C_2 be an $(n-2)$-cell in N such that $x \in \operatorname{int} C_2$ and $C_2 \subset \operatorname{int} B_2$. Let U be a neighborhood of $\overline{C_1 - C_2}$ in B_1 whose closure misses x. For each $y \in \overline{C_1 - C_2}$, choose a neighborhood V_y such that any loop in $V_y - N$ which is null-homologous in W is null-homotopic in $U - N$. Let ϵ be a Lesbegue number for this cover. Also choose ϵ so that $\epsilon < \frac{1}{3}\min\{\operatorname{dist}(B_2, N - C_1), \operatorname{dist}(\partial B_2, C_2)\}$. Use this ϵ and Lemma 2.2 to determine a neighborhood V of N. Choose another neighborhood V' such that V' ϵ-deforms to N in V. Let U' be the preimage of $\overline{C_1 - C_2}$ under this deformation. Define $Q = \overline{\partial B_2 - V'}$ and $\partial Q = Q \cap \overline{V}'$. We may assume that both Q and ∂Q are finite polyhedra. Let $\ell_1, \ell_2, \ldots, \ell_q$ be a set of generators for $\pi_1(Q, \partial Q)$. For each arc ℓ_i there exists an arc

m_i in $U' - N$ such that $\ell_i \cup m_i$ is null-homologous in W. Hence $\ell_i \cup m_i$ bounds a singular 2-cell $\Delta_i \subset B_1 - N$ by the choice of B_1. Choose B_3 to be an n-ball centered at x which misses $(\cup_{i=1}^q \Delta_i) \cup \overline{U}$.

We next explain how to choose W_2. Let W_3 be any connected neighborhood of N which is contained in V'. Use Lemma 2.2 to choose a neighborhood W_2' of N such that homotopies in W_2' can be pushed off N staying in W_3. Then choose W_2 so that W_2 deforms to N in W_2'.

Define $P_2 = B_2 - W_2$ and $P_3 = B_3 - W_3$. Our next goal is to show that $\pi_1(\partial P_2^*) \to \pi_1((B_1 - N)^* - P_3^*)$ is trivial. Let γ be a loop in ∂P_2^* and define $\gamma' = p(\gamma)$. We show that γ' is null-homotopic in $(B_1 - N) - P_3$ and then lift this homotopy to $\widetilde{W}$ to obtain the desired result. We first use the choice of W_2 to push γ' to the boundary of C_2 and out of B_3 in exactly the same way we pushed out of B_2 in the proof of Lemma 3.2. Next put γ' in general position with respect to ∂Q. Then $\gamma' \cap Q$ consists of a finite number $\gamma_1, \gamma_2, \ldots, \gamma_m$ of subarcs. The choice of B_3 allows us to find an arc μ_i in U' joining the endpoints of γ_i such that $\gamma_i + \mu_i$ is null-homotopic missing $B_3 \cup N$. This means that γ' may be replaced by a loop γ'' in $U' - N$. There exists an ϵ-deformation of γ'' to $\overline{C_1 - C_2}$. Since $n \geq 5$, $\overline{C_1 - C_2}$ is simply connected. So γ'' is homotopic to a point in the ϵ neighborhood of $\overline{C_1 - C_2}$. We can use the choice of ϵ and V along with Lemma 2.2 to ϵ-push this homotopy into the complement of N. This means that γ'' is homotopic to γ''' with $\mathrm{diam}(\gamma''') < \epsilon$. The choice of ϵ then shows us that γ'' is null-homotopic in U. So γ' is null-homotopic in $(B_1 - N) - B_3$. Let $g_1, g_2, \ldots, g_n, t$ be a set of generators for $\pi_1(\partial P_2)$. We may assume that each g_i lifts to ∂P_2^*. (If not, replace g_i with $g_i t^\ell$ for some ℓ.) Then each element of $\pi_1(\partial P_2^*)$ can be written in terms of $\{t^j \tilde{g}_i\}_{i,j}$. We fix a homotopy shrinking each g_i in $(B_1 - N) - P_3$ and choose W_1 so close to N that it misses the tracks of all these homotopies. Then we will have $\pi_1(\partial P_2^*) \to \pi_1(P_1^* - P_3^*)$ trivial. □

Lemma 4.2. *Fix $x \in N$ and suppose $n \geq 5$. For each $q > 0$ there exists a sequence $B_1 \supset B_2 \supset \cdots \supset B_q$ of n-balls centered at x such that there are arbitrarily close neighborhoods $W_1 \subset W_2 \subset \cdots \subset W_q$ so that if $P_i = B_i - W_i$ then the inclusion induced homomorphisms $\pi_1(P_i^*) \to \pi_1(P_{i-1}^*)$ and $\pi_1(P_i^* - P_{i+1}^*) \to \pi_1(P_{i-1}^* - P_{i+2}^*)$ are trivial.*

Proof. Construct the sequence $B_1 \supset B_2 \supset \cdots \supset B_q$ inductively, using Lemma 4.1 to construct B_{i+1} from B_i. Using the 1-alg hypothesis we can also arrange that $\pi_1(p^{-1}(B_i - N)) \to \pi_1(p^{-1}(B_{i-1} - N))$ is trivial. Choose $W_1 \subset W_2 \subset \cdots \subset W_q$ to satisfy the conclusion of Lemma 4.1 as well.

Consider a loop $\ell \subset P_i^*$. There exists a map $f : \Delta^2 \to p^{-1}(B_{i-1} - N)$ such that $f|\partial\Delta^2 = \ell$. Put f in general position with respect to ∂P_i^*. Then $f^{-1}(\partial\Delta^2)$ consists of a finite collection of circles in int Δ^2. Each outermost circle shrinks in $P_{i-1}^* - P_{i+1}^*$ by Lemma 4.1. Thus we can redefine f to map the interior of this circle into $P_{i-1}^* - P_{i+1}^*$. We do this for each outermost circle. The result is a map $f' : \Delta^2 \to P_{i-1}^*$ such that $f'|\partial\Delta^2 = \ell$. So $\pi_1(P_i^*) \to \pi_1(P_{i-1}^*)$ is the trivial homomorphism.

The proof of the second part of the Lemma is similar. We start with a loop in $P_i^* - P_{i+1}^*$. It bounds a singular disk in $p^{-1}(B_{i-1} - N)$. We use Lemma 4.1 to cut this disk off on $\partial P_i^* \cup \partial P_{i+1}^*$. The result is a singular disk in $P_{i-1}^* - P_{i+2}^*$. □

Proposition 4.3. *Fix $x \in N$. For each sufficiently small n-ball B_1 centered at x there exists a concentric n-ball B_2 and arbitrarily small neighborhoods $W_1 \subset W_2$ of N such that there exists a compact PL manifold P with $B_2 - W_2 \subset P \subset B_1 - W_1$ and P has a handle decomposition with no handles of index greater than $n/2$.*

Proof. First suppose $n \geq 5$. Fix $q \gg 2^n$. Let $P_1 \supset P_2 \supset \cdots \supset P_q$ be a sequence as in Lemma 4.2. We may also assume that each $P_{i+1} \subset P_i$ satisfies the hypotheses of Lemma 3.2. We may then apply the Relative Eventual Hurewicz Theorem [10, Theorem 5.2] to the inclusions $(P_i, P_i - P_{i+j}) \hookrightarrow (P_{i-1}, P_{i-1} - P_{i+j+1})$ since Lemma 4.2 gives us exactly the π_1-conditions we need and Lemma 3.3 gives us the higher homology hypotheses we need. Thus there is an integer s such that for all $k < n/2$, $\pi_k(P_i, P_i - P_{i+1}) \to \pi_k(P_{i-s}, P_{i-s} - P_{i+s+1})$ is trivial. (The number s is independent of the particular P_i used; it depends only on k. In fact $s = 2^{k+2}$ will work; see [10, Theorem 5.2].) This means that relative k-dimensional polyhedra in $(P_i, P_i - P_{i+1})$ can be pushed off P_{i+s+1} with a homotopy and hence (using engulfing) can be pushed off some P_{i+t} with an ambient isotopy whose support lies in the interior of P_i.

The PL manifold P can now be constructed as follows. Start with P_1 and form a handle decomposition of P_1. Each j-dimensional handle has an $(n-j)$-dimensional cocore. If $j > n/2$, then $n - j < n/2$, so the paragraph above implies that the cocore of each j-handle can be isotoped off P_{i+t+1} keeping $P_1 - P_i$ fixed. We construct P by starting with P_1, pushing the cocore of each j-handle off P_q for $j > n/2$, and then removing a small regular neighborhood of the cocore of each of the j-handles, $j > n/2$. The remaining manifold has only handles of index $n/2$ or less.

In case $n = 4$ the proof is much easier. Each P_i has a handle decomposition with handles of dimensions ≤ 3, so we need to eliminate the 3-handles. It is enough to show that $\pi_1(P_i, P_i - P_{i-1}) \to \pi_1(P_{i+1}, P_{i+1})$

is trivial. The proof proceeds as follows: Given a path α in P_i with endpoints in $P_i - P_{i-1}$, use the fact that P_{i+1} does not separate to join the endpoints with a path β in $P_i - P_{i-1}$. Add a loop, if necessary, so that $\alpha \cup \beta$ is null-homotopic in P_{i+1}. □

Corollary 4.4. *Proposition 3.1 holds for $k > n/2$.*

Proof. This is clear since the inclusion factors through a space with trivial homology in dimensions $k > n/2$. □

5. A DUALITY PRINCIPLE FOR HOMOLOGY WITH COEFFICIENTS IN A FIELD

In this section we prove a weak version of Proposition 3.1. It is weaker than the version we need in that it only works for coefficients in a field. It uses a duality principle for infinite cyclic covers that is similar to the one used by Milnor in [9]. Let F be a field and let Λ denote the group algebra $F[\mathbb{J}]$. Notice that for any $U \subset W$ the homology group $H_k(\widetilde{W}, U^*; F)$ can be thought of either as a vector space over F or as a module over Λ. It is important to distinguish between "finitely generated over F" and "finitely generated over Λ." We use $H_k(W, U; \Lambda)$ as a shorthand for $H_k(\widetilde{W}, U^*; F)$.

Fix a sequence of three PL manifolds $P_3 \subset P_2 \subset P_1$ such that each of the inclusions $P_3 \subset P_2$ and $P_2 \subset P_1$ satisfies the hypotheses of Lemmas 3.2 and 3.4. Fix $k \leq n/2$ and let $\alpha_i : H_k(W, W - P_i; \Lambda) \to H_k(W, W - P_{i+1}; \Lambda)$ denote the inclusion-induced homomorphism.

Lemma 5.1. $\operatorname{im}(\alpha_2) = \operatorname{im}(\alpha_2 \circ \alpha_1)$.

Proof. Consider the following commutative diagram in which each row is exact.

$$\begin{array}{ccccc}
H_k(W;\Lambda) & \xrightarrow{\beta_1} & H_k(W, W-P_1;\Lambda) & \longrightarrow & H_{k-1}(W-P_1;\Lambda) \\
\| & & \downarrow{\scriptstyle \alpha_1} & & \downarrow \\
H_k(W;\Lambda) & \xrightarrow{\beta_2} & H_k(W, W-P_2;\Lambda) & \xrightarrow{\gamma} & H_{k-1}(W-P_2;\Lambda) \\
\| & & \downarrow{\scriptstyle \alpha_2} & & \downarrow{\scriptstyle \delta} \\
H_k(W;\Lambda) & \xrightarrow{\beta_3} & H_k(W, W-P_3;\Lambda) & \longrightarrow & H_{k-1}(W-P_3;\Lambda)
\end{array}$$

By Lemma 3.2, we have $\delta \circ \gamma = 0$, so $\operatorname{im}(\alpha_2) \subset \operatorname{im}(\beta_3)$. On the other hand, $\operatorname{im}(\alpha_2) \supset \operatorname{im}(\beta_3)$ is obvious so we have $\operatorname{im}(\alpha_2) = \operatorname{im}(\beta_3)$. Now the inclusion $P_1 \subset P_3$ has all the same properties that the inclusion $P_2 \subset P_3$ has, so we have $\operatorname{im}(\alpha_2 \circ \alpha_1) = \operatorname{im}(\beta_3)$ as well. □

Lemma 5.2. $\operatorname{im}(\alpha_2)$ *is finite dimensional over* F.

Proof. For each i the exact sequence

$$0 \to C_*(W, W - P_i; \Lambda) \xrightarrow{t-1} C_*(W, W - P_i; \Lambda) \xrightarrow{p_*} C_*(W, W - P_i; F) \to 0$$

of chain complexes gives rise to a long exact sequence

$$\cdots \to H_k(W, W - P_i; \Lambda) \xrightarrow{t-1} H_k(W, W - P_i; \Lambda) \xrightarrow{p_*} H_k(W, W - P_i; F) \to \cdots$$

of homology groups. Therefore we have the following commutative diagram with exact rows.

$$\begin{array}{ccccc}
H_k(W, W - P_1; \Lambda) & \xrightarrow{t-1} & H_k(W, W - P_1; \Lambda) & \xrightarrow{p_*} & H_k(W, W - P_1; F) \\
\downarrow{\scriptstyle \alpha_1} & & \downarrow{\scriptstyle \alpha_1} & & \downarrow{\scriptstyle 0} \\
H_k(W, W - P_2; \Lambda) & \xrightarrow{t-1} & H_k(W, W - P_2; \Lambda) & \xrightarrow{p_*} & H_k(W, W - P_2; F) \\
\downarrow{\scriptstyle \alpha_2} & & \downarrow{\scriptstyle \alpha_2} & & \downarrow{\scriptstyle 0} \\
H_k(W, W - P_3; \Lambda) & \xrightarrow{t-1} & H_k(W, W - P_3; \Lambda) & \xrightarrow{p_*} & H_k(W, W - P_3; F)
\end{array}$$

The vertical arrows on the right are 0 by Lemma 3.4. The upper half of the diagram shows that $\operatorname{im}(\alpha_1) \subset (t-1)(H_k(W, W - P_2; \Lambda)) = \operatorname{im}(t-1)$. Thus $\operatorname{im}(\alpha_2) = \operatorname{im}(\alpha_2 \circ \alpha_1) \subset \operatorname{im}(\alpha_2 \circ (t-1)) = (t-1)\operatorname{im}(\alpha_2)$. It follows that $\operatorname{im}(\alpha_2) = (t-1)\operatorname{im}(\alpha_2)$. Now $\operatorname{im}(\alpha_2)$ is a subspace of $H_k(W, W - P_3; \Lambda)$ which is finitely generated over Λ. Since $\operatorname{im}(\alpha_2)$ is finitely generated over Λ and $\operatorname{im}(\alpha_2) = (t-1)\operatorname{im}(\alpha_2)$, $\operatorname{im}(\alpha_2)$ is a finitely generated Λ-module whose order ideal in non-trivial. It follows that $\operatorname{im}(\alpha_2)$ is finite dimensional over F. (See the proof of Assertion 5, page 118, of [9]. This is the point in the proof at which it is important to be working over a field.) □

Let K be a connected subpolyhedron of $\widetilde{W}$ such that $p(K) = W$ and for each compact subpolyhedron L of W, $p^{-1}(L) \cap K$ is compact. Notice that, for each compact $C \subset \widetilde{W}$, $C \cap t^i(K) = \emptyset$ for all but finitely many i. Define

$$N_q = \bigcup_{i \geq q} t^i(K), \quad \text{and}$$

$$N'_q = \bigcup_{i \leq -q} t^i(K).$$

Lemma 5.3. $\operatorname{im}[H_{k-1}(N_0, N_0 - P_i^*; F) \to H_{k-1}(N_0, N_0 - P_{i+1}^*; F)]$ *is finite dimensional over* F *for* $i = 1, 2$.

Proof. Define $N = N_0 \cap N_0'$. Consider the following commutative diagram in which each row is part of an exact Mayer-Vietoris sequence. (Coefficients in F are understood.)

$$\begin{array}{ccccc}
H_{k-1}(N, N - P_i^*) & \longrightarrow & H_{k-1}(N_0, N_0 - P_i^*) \oplus H_{k-1}(N_0', N_0' - P_i^*) & \longrightarrow & H_{k-1}(\widetilde{W}, \widetilde{W} - P_i^*) \\
\downarrow & & \downarrow \gamma & & \downarrow \beta \\
H_{k-1}(N, N - P_{i+1}^*) & \longrightarrow & H_{k-1}(N_0, N_0 - P_{i+1}^*) \oplus H_{k-1}(N_0', N_0' - P_{i+1}^*) & \longrightarrow & H_{k-1}(\widetilde{W}, \widetilde{W} - P_{i+1}^*)
\end{array}$$

Since $H_{k-1}(N, N - P_i^*; F)$ is finite dimensional (by excision and the choice of K) and the image of β is finite dimensional, it follows that the image of γ is finite dimensional. □

Lemma 5.4. $\operatorname{im}[H_k(\widetilde{W}, N_0 \cup (\widetilde{W} - P_1^*); F) \to H_k(\widetilde{W}, N_0 \cup (\widetilde{W} - P_3^*); F)]$ *is finite dimensional over* F.

Proof. Consider the following diagram in which each row is part of the long exact sequence of a triple.

$$\begin{array}{ccccc}
H_k(\widetilde{W}, \widetilde{W} - P_1^*; F) & \longrightarrow & H_k(\widetilde{W}, N_0 \cup (\widetilde{W} - P_1^*); F) & \longrightarrow & H_{k-1}(N_0 \cup (\widetilde{W} - P_1^*), \widetilde{W} - P_1^*; F) \\
\downarrow & & \downarrow \delta_1 & & \downarrow \beta_1 \\
H_k(\widetilde{W}, \widetilde{W} - P_2^*; F) & \longrightarrow & H_k(\widetilde{W}, N_0 \cup (\widetilde{W} - P_2^*); F) & \longrightarrow & H_{k-1}(N_0 \cup (\widetilde{W} - P_2^*), \widetilde{W} - P_2^*; F) \\
\alpha_2 \downarrow & & \downarrow \delta_2 & & \downarrow \\
H_k(\widetilde{W}, \widetilde{W} - P_3^*; F) & \longrightarrow & H_k(\widetilde{W}, N_0 \cup (\widetilde{W} - P_3^*); F) & \longrightarrow & H_{k-1}(N_0 \cup (\widetilde{W} - P_3^*), \widetilde{W} - P_3^*; F)
\end{array}$$

Since both β_1 and α_2 have finite dimensional images (by Lemmas 5.2 and 5.3), an elementary linear algebra argument shows that $\delta_2 \circ \delta_1$ has finite dimensional image as well. □

Lemma 5.5. *There exists an integer* $s \geq 0$ *such that*

$$H_k(\widetilde{W}, N_s \cup (\widetilde{W} - P_1^*); F) \to H_k(\widetilde{W}, N_0 \cup (\widetilde{W} - P_3^*); F)$$

is the trivial homomorphism.

Proof. By Lemma 5.4, the image of $H_k(\widetilde{W}, N_0 \cup (\widetilde{W} - P_1^*); F)$ in $H_k(\widetilde{W}, N_0 \cup (\widetilde{W} - P_3^*); F)$ is finitely generated. We choose s large enough so that N_{-s} contains a representative of each of these generators. Then the homomorphism $H_k(\widetilde{W}, N_0 \cup (\widetilde{W} - P_1^*); F) \to H_k(\widetilde{W}, N_{-s} \cup (\widetilde{W} - P_3^*); F)$ is trivial. We shift everything by t^s in order to reach the desired conclusion. □

We end this section with a proof of a weak version of Proposition 3.1. It assumes coefficients in a field, but is valid in the crucial middle dimension.

Proposition 5.6. *If the sequence of PL manifolds $P_5 \subset P_4 \subset P_3 \subset P_2 \subset P_1$ is such that each of the inclusions $P_{i+1} \subset P_i$ satisfies the hypotheses of Lemmas 3.2 and 3.4, then the inclusion-induced homomorphism $H_k(P_1; \Lambda) \to H_k(P_5; \Lambda)$ is trivial for $k \leq n/2$*

Proof. The cases $k < n/2$ are covered by Corollary 4.4, so we focus on the case $k = n/2$. By Poincaré Duality, $H_k(P_i; \Lambda) \cong H^k(P_i, \partial P_i; \Lambda)$ and, by excision, $H^k(P_i, \partial P_i; \Lambda) \cong H^k(W, W - P_i; \Lambda)$. Furthermore, the inclusion induced homomorphism $H_k(P_1; \Lambda) \to H_k(P_3; \Lambda)$ is dual to $H^k(W, W - P_1; \Lambda) \to H^k(W, W - P_3; \Lambda)$, so we show that this last homomorphism is trivial.

Note that $H^k(W, W - P_i; \Lambda) = H^k_c(\widetilde{W}, \widetilde{W} - P_i^*; F)$ and that

$$H^k_c(\widetilde{W}, \widetilde{W} - P_i^*; F) = \lim_{q \to \infty} H^k(\widetilde{W}, (\widetilde{W} - P_i) \cup N_q \cup N'_q; F).$$

Let s be the integer given by Lemma 5.5 and consider the following diagram. (Each row is part of an exact Mayer-Vietoris sequence; coefficients in Λ are understood; $E_q = N_q \cup N'_q$.)

$$\begin{array}{ccccc}
H^{k-1}(\widetilde{W}, \widetilde{W} - P_1^*) & \longrightarrow & H^k(\widetilde{W}, (\widetilde{W} - P_1^*) \cup E_q) & \longrightarrow & H^k(\widetilde{W}, (\widetilde{W} - P_1^*) \cup N_q) \oplus H^k(\widetilde{W}, (\widetilde{W} - P_1^*) \cup N'_q) \\
{\scriptstyle 0}\downarrow & & \downarrow & & \downarrow{\scriptstyle 0} \\
H^{k-1}(\widetilde{W}, \widetilde{W} - P_3^*) & \longrightarrow & H^k(\widetilde{W}, (\widetilde{W} - P_3^*) \cup E_{q+s}) & \longrightarrow & H^k(\widetilde{W}, (\widetilde{W} - P_3^*) \cup N_{q+s}) \oplus H^k(\widetilde{W}, (\widetilde{W} - P_3^*) \cup N'_{q+s}) \\
{\scriptstyle 0}\downarrow & & \downarrow & & \downarrow{\scriptstyle 0} \\
H^{k-1}(\widetilde{W}, \widetilde{W} - P_5^*) & \longrightarrow & H^k(\widetilde{W}, (\widetilde{W} - P_5^*) \cup E_{q+2s}) & \longrightarrow & H^k(\widetilde{W}, (\widetilde{W} - P_5^*) \cup N_{q+2s}) \oplus H^k(\widetilde{W}, (\widetilde{W} - P_5^*) \cup N'_{q+2s})
\end{array}$$

The vertical arrows on the right are 0 by Lemma 5.5 (translated by t^q and t^{q+s}). The vertical arrows on the left are 0 by Lemma 3.3. It follows that the composite vertical arrow in the center is 0 as well. Taking a limit as $q \to \infty$ gives the desired conclusion. □

6. Proof of Theorem 1.3

As observed in §3, it suffices to prove Proposition 3.1. Most cases of Proposition 3.1 were proved in §4. The only remaining case is that in which n is even and $k = n/2$.

Use Propositions 4.3 and 5.6 to choose PL manifolds $P_2 \subset P_1$ satisfying the hypotheses of Lemmas 3.2 and 3.4 such that

(1) P_1 has a cell decomposition with no cells of dimension greater than k, and
(2) $H_k(P_2^*; \mathbb{Q}) \to H_k(P_1^*; \mathbb{Q})$ is trivial.

Since $H^{k+1}(P_1^*;\mathbb{Z})=0$, the Universal Coefficient Theorem for Cohomology shows that $\operatorname{Ext}(H_k(P_1^*;\mathbb{Z}))=0$. Thus $H_k(P_1^*;\mathbb{Z})$ is torsion-free [6, Theorem III.6.1]. Consider the commutative diagram

$$\begin{array}{ccc} H_k(P_2^*;\mathbb{Z}) & \xrightarrow{\alpha} & H_k(P_1^*;\mathbb{Z}) \\ \downarrow{\scriptstyle \otimes\mathbb{Q}} & & \downarrow{\scriptstyle \otimes\mathbb{Q}} \\ H_k(P_2^*;\mathbb{Q}) & \xrightarrow{0} & H_k(P_1^*;\mathbb{Q}). \end{array}$$

Let $x\in H_k(P_2^*;\mathbb{Z})$. If $\alpha(x)\in H_k(P_1^*;\mathbb{Z})$ is not 0, then its image in $H_k(P_1^*;\mathbb{Q})$ is not 0 either by [6, Exercise III.7.5]. But the bottom arrow is the trivial homomorphism by the choice of P_1 and P_2. Hence $\alpha=0$ and the proof is complete.

References

1. J. W. Cannon, *ULC properties in neighborhoods of embedded surfaces and curves in E^3*, Canadian J. Math. **25** (1973) 31–73.
2. T. Chapman, *Locally homotopically unknotted embeddings of manifolds in codimension two are locally flat*, Topology **18** (1979) 339–348.
3. S. C. Ferry, *Homotoping ϵ-maps to homeomorphisms*, Amer. J. Math. **101** (1979) 567–582.
4. M. H. Freedman and F. Quinn, *Topology of 4-manifolds*, Princeton University Press, Princeton, 1990.
5. J. G. Hollingsworth and T. B. Rushing , *Homotopy characterization of weakly flat codimension 2 spheres*, Amer. J. Math. **98** (1976) 385–394.
6. P. J. Hilton and U. Stammbach, *A course in homological algebra*, 1971, Springer-Verlag, New York, Graduate Texts in Mathematics, vol. 4.
7. V. T. Liem and G. A. Venema, *On the asphericity of knot complements*, Canadian J. Math. **45** (1993) 340–356.
8. J. Levine, *Unknotting spheres in codimension two*, Topology **4** (1965) 9–16.
9. J. Milnor *Infinite cyclic coverings*, in Conference on the Topology of Manifolds, Prindle, Weber and Schmidt, Boston, 1968 115–133.
10. F. Quinn, *Ends of maps, I*, Annals of Math. **110** (1979) 275–331.
11. _____, *Ends of maps, III: dimensions 4 and 5*, J. Differential Geometry **17**(1982) 503–521.
12. D. W. Sumners, *Homotopy torsion in codimension two knots*, Proc. Amer. Math. Soc. **24** (1970) 229–240.
13. G. A. Venema, *Duality on noncompact manifolds and complements of topological knots*, Proc. Amer. Math. Soc. to appear.

CALVIN COLLEGE, GRAND RAPIDS, MICHIGAN 49546 USA
Current address: University of Michigan, Ann Arbor, Michigan 48109-1003
E-mail address: venema@calvin.edu

AMS/IP Studies in Advanced Mathematics
Volume **2**, 1997 (Part 1)

$SL(n,\mathbb{Z})$ CANNOT ACT ON SMALL TORI

SHMUEL WEINBERGER

We prove the following result (answering a question of Zimmer):

Theorem. *The discrete group $SL(n,\mathbb{Z})$, $n \geq 3$, can act smoothly on the torus T^m, $m < n$, only trivially.*

Of course, one expects that no subgroup of finite index so acts either, but our proof does not give this. The smoothness, however, can be removed at the cost of making the argument (even) less elementary.

Lemma 1. *If G is a compact group acting smoothly on a compact manifold M , then there is an affine G action on a torus T, and an equivariant map $\phi : M \longrightarrow T$ which induces an isomorphism on $H^1(-;\mathbb{Z})$.*

Proof. (from [3]). The torus is $T = \mathrm{Hom}(H^1(M;\mathbb{Z}) : \mathbb{R}/\mathbb{Z})$. We define a G action on the torus by combining the action of G on $H^1(M;\mathbb{Z})$ with the set of translations defined as follows: Pick a base point b. To a homomorphism α and a cohomology class c represented by the harmonic form ω, one adds $\int_b^{gb} \omega$ to the action of G on H^*. Since this is a crossed homomorphism, it defines an affine G action on T. The map $M \longrightarrow T$ is defined by $\phi(p)(c) = \int_b^p \omega$ (analogous to the Jacobi map from a curve to its Jacobian). It is readily seen to have the advertised properties. □

Lemma 2. *If G acts homologically trivially on a torus, then the action is equivariantly homotopy equivalent to an action that factors through the group of translations of the torus. In particular, if G is nonabelian, the action is not effective.*

Proof. We apply the previous lemma to get an equivariant map from our G-torus to an affine G-torus. It is unequivariantly a homotopy equivalence. Observe that for the target the action is necessarily by translations, and that furthermore, in the target each element either acts freely or trivially. Certainly, an element which acts freely in the target acts freely in the domain. Conversely, consider an element g

Partially supported by an NSF grant.

which acts trivially on range. If g is a p-primary element, it must act trivially on the domain as well, by Smith theory (see e.g. [2]). (Smith theory asserts that an equivariant map between spaces on which a p-group acts, that induces an isomorphism on cohomology, induces an isomorphism on mod p cohomology of fixed sets. Since the fixed set of g has top dimensional cohomology in the range, it must in the domain, which means it must be the whole torus, i.e. that g acts trivially in the domain.) Now, it is straightforward to induct on the number of primes dividing the exponent of g. □

Remark. As the referree observed, lemma 2 implies that many groups, such as those whose p-ranks are larger than that of $(\mathbb{Z}/p\mathbb{Z})^k \rtimes SL(k,\mathbb{Z})$ cannot act effectively on T^k.

Lemma 3. *If $2 < n > m$, then every homomorphism $SL(n,\mathbb{Z}) \longrightarrow SL(m,\mathbb{Z})$ is trivial.*

Proof. Superrigidity (see [4]) implies that the homomorphism necessarily has finite image. The congruence subgroup property [1] implies that the action factors through $SL(n,\mathbb{Z}/k\mathbb{Z})$ for some k. However, it is well known and quite simple to see that for $k > 1$ and every quotient of $SL(n,\mathbb{Z}/k\mathbb{Z})$ there is a prime p such that it does not have a nontrivial map to $PSL(m,\mathbb{Z}/p\mathbb{Z})$. □

According to lemma 3, the purported action would be homologically trivial. Lemma 2 then implies that we can find some elements which are acting trivially on the torus (use a finite nonabelian subgroup). The subgroup normally generated by such an element in $SL(n,\mathbb{Z})$ is necessarily finite index (see the "essential simplicity" theorem of Kazhdan and Margulis [4]) the action is now known to factor through $SL(n,\mathbb{Z}/k\mathbb{Z})$. Since this group has no abelian quotients, it cannot act by translations, and the theorem is established.

An alternative way to obtain the conclusion of Lemma 2 sufficient for our application (and which does not require smoothness) is to look at all the lifts of the group elements to the universal cover, and extend continuously to the boundary. If one can show that there is an element of SL_n whose lift is of order p acting trivially on homology (which the calculation of $K_2(\mathbb{Z})$ permits) then one sees that its lift is trivial on the boundary (by "coarse" tricks) and then Smith theory [2] informs us that this element acts trivially.

Remark. If one assumes appropriate conditions on the representation of $SL(n,\mathbb{Z})$ on homology, one can prove theorems regarding actions on higher dimensional tori, as well. The hard case regards actions whose induced representations factor through nontrivial finite quotients.

Remark. One can show that the action of $SL(n,\mathbb{Z})$ on T^n is *not* C^0 unique.

Remark. If this note has a theme, it is the importance of examining the compact subgroups when one has a large group acting. (In the measure theoretic approaches, these do not seem to have any value.) I do not know any torsion free group which can can be proven not to act on any manifold (without assuming that some geometric structure is preserved).

I would like to thank Alex Lubotzky and Bob Zimmer for very useful conversations.

References

1. H. Bass, J. Milnor, and J.P. Serre, *The congruence subgroup property for* SL_n $(n > 3)$ *and* Sp_{2n} $(n > 2)$, Publ. Math. d'IHES. **33** (1967) 59-137.
2. G. Bredon, *Introduction to compact transformation groups*, Academic Press, 1972.
3. J. Rosenberg and S. Weinberger, with an appendix by J.P. May, *An equivariant Novikov conjecture*, K-theory, **4** (1990) 29-53.
4. R. Zimmer, *Ergodic theory and semisimple groups*, Birkhauser, 1988.

Department of Mathematics, University of Chicago
E-mail address: shmuel@math.upenn.edu

AMS/IP Studies in Advanced Mathematics
Volume **2**, 1997 (Part 1)

MORSE HOMOTOPY AND ITS QUANTIZATION

KENJI FUKAYA

0. INTRODUCTION

Morse theory, which is one of the roots of differential topology, is a method to study topology of manifolds using functions on it.

One of its main applications is Morse inequality, which gives a relation between Betti number and the number of critical points of Morse function. The original proof (due to Morse) involves something more. That is it gives a way to find a homology group of a manifold using a Morse function. Variants of this idea appeared repeatedly in many important works by Thom, Smale, Milnor, etc. This point again appeared in Witten's famous paper [30] and calls attentions of many mathematicians. Floer [5] (generalizing [4]) uses it to built a new infinite dimensional homology theory. There the moduli space of gradient lines played an essential role.

One main point in [30] is to regard Morse theory as a topological field theory. In fact, one may regard it as one dimensional topological field theory or quantum mechanics, while topological σ-model, Gauge field theory etc. are regarded as (topological) quantum field theories.

The purpose of this paper is two fold. One is to push forward this point of view and shows that we can regard various operators of algebraic topology as n-point functions. We use several functions for this purpose while in finding homology one uses a single Morse function.

The other is to show a relation of (finite dimensional) Morse theory to topological σ-model and two dimensional gravity. Namely we regard topological σ-model as a quantization of Morse theory. (Furthermore they are related also to Gauge theory as we discussed in [7],[8],[9].) This construction generalize A-model [32], which is introduced to study Mirror symmetry of Calabi-Yau manifold. There it was already realized that the construction of A-model is a quantization of the ring structure of cohomology. Here we quantize many other operators in algebraic topology. They are genus zero case, in the sense that it is related to the moduli space of pseudo holomorphic map from Riemann sphere or disk. We will also discuss some ideas how to handle higher genus case. Then, we find that the triangulation of Teichmüller space due to Mumford

[18]-Strebel [28]-Harer [11]-Penner[22]-Kontsevitch [13], arises naturally in Morse theory. It is remarkable that it appears even in the situation before we quantize the story. In other words, from this point of view we can find easily what is the natural way to quantize "higher genus Morse theory", though there is a serious analytic trouble to make those arguments rigorous.

Here is a bit more detailed outline and contents of each sections.

In section 1 we consider cup product, Massey product and its higher analogue. There we study the moduli space of trees with metric such that the number of vertices with one edge is given. It is a classical result by Stasheff that such a moduli space is a cell. Then we consider a finitely many functions on manifolds. By using these functions (whose number is equal to the number of vertices with one edge), we construct a moduli space of Feynman diagrams. (Here the edges are regarded as gradient lines of difference of two functions.) One can prove (using Stasheff's result) that this moduli space is a manifold. And one can find a dimension formula for it. Then using the case when the dimension is zero one finds several maps, which give cup product, Massey product, etc.

Since there are many kinds of trees one can use, it is not clear what this construction describes. Then we introduce the notion of A^∞-category and find that we constructed an A^∞-category whose object is a function on M.

This construction is a "classical limit" of one discussed in [7],[8]. There we considered a symplectic manifold X such that the Chern class is proportional to the symplectic form and found an A^∞-category whose object is a Lagrangian or Bohr-Sommerfeld orbit of it. In fact they are related to each other as follows. We take a symplectic manifold M satisfying the above assumption. And put $X = M \times M$. Given a function f on M we find an exact perturbation of the diagonal $M \subset X$ by the Hamiltonian flow generated by f. Hence a smooth function corresponds to a Lagrangian. In this way we can find a relation between two constructions. However the rigorous proof that they really coincide to each other is not yet known. (Except for homology group itself (without considering product etc.) which is due to Floer [5].) There is also a quantum correction that is related to the bubbling phenomenon in the moduli space of pseudo holomorphic curves. We recall these facts briefly in section 3 and refer [8] for further detail.

The construction we mentioned above corresponds open string of zero loop, since we consider pseudo holomorphic maps from 2 disk whose boundary condition is given by Lagrangians. In section 2 we discuss it in a different way by using pseudo holomorphic spheres in place of pseudo holomorphic disks. There we work on a symplectic manifold M (and do not embed it to $M \times M$.) For this version it is obvious that its classical limit is the same as one discussed in section 1. So we get quantum Massey product etc. What is not obvious and leads us a delicate singular bifurcation problem is the conjecture that this version is related to the special case of open string version we mentioned above. This conjecture is discussed at the end of section 3 but the author do not have a proof of it yet.

Also it is easy to show that the quantum cup product we obtain in that way coincides with Gromov-Ruan invariant [23]. Using symmetries of graphs we also define a quantum Steenrod square in section 2. This discussion depends much on the work by M.Betz-R.Cohen [2].

Thus the discussion of sections 1,2,3 concern the case of tree (zero loop) or equivalently the case of Riemann surface of genus zero. Sections 4,5 are devoted to the case of higher genus. Here we need to borrow another important idea by M.Betz and R.Cohen. ([2]) That is to introduce a graph which is not a tree. (However we do it in a way different from theirs.) See the remark at the end of this section the relation between their work and this paper.

In section 4, we again consider the "classical limit". Namely Morse theory. Here we use the moduli of metric ribbon graph, that is a direct generalization of the construction of section 1 and also closely related to the 2 dimensional (topological) gravity. But at this point we do not use any complex or symplectic structure on the manifold. Also the moduli theory of Riemann surface itself does not appear at this point. Our construction is based on quite elementary moduli spaces of ribbon graphs embedded into our manifold such that each edge is a gradient line of Morse function. We also introduce a variant of mapping class group as the symmetry of moduli space of metric ribbon graphs. So using cohomology classes of this group coupled with the cohomology classes of the manifold, we obtain various numbers (or maps.) They should coincide to the classical limit of the n point functions of the (open) string theory as we will discuss in section 5. Since we use only elementary moduli spaces, the construction of section 4 is rigorous and do not require nonlinear PDE etc. The author however do not know an appropriate algebraic machinery to describe what we obtain in that way. (We do have such an algebraic machinery in the case of trees, that is A^{∞}- category.) So the discussion of section 4 is rather a preliminary sketch of the theory which we hope to develop in future.

Section 5 is devoted to a quantization of the higher genus Morse homotopy. Again there is two versions. But we discuss mainly open string version (one similar to section 3.) Our discussion in section 4 is organized so that it is quite immediate to see what we need to do for open string quantization. However we meet a serious trouble to make this argument mathematically rigorous. The major problem is that we do not know well about the compactification of the moduli space of pseudo holomorphic maps from Riemann surfaces of higher genus. (The author would like to thank Prof. Y. Ruan who explained him some essential points of this problem.) But it seems to the author that the problem is somewhat easier in this open string situation.

We also meet the same kind of singular bifurcation problems as we meet in genus zero case.

Because of these difficulties, we can not quantize higher genus Morse homotopy in mathematical level of rigor. So this section should be regarded as a collection of remarks than results.

Finally we remark here that related ideas are employed by M.Betz and R.Cohen [2]. So the author mentions here which one he learned from their paper. First the idea that cup product is obtained by considering Feynman diagrams as in section 1 is found independently by the author [7], [8] and them. (Although the proof of this fact is rather trivial the fact itself I believe is new and interesting.) (However Floer [5] may have known this fact already.) Next to consider the symmetry of graph to obtain Steenrod square is discovered by [2]. Also they first considered the graph which is not necessary a tree. These two facts the author learned from their paper. The idea to use moduli parameter of the graph and to patch the moduli spaces of graphs with different combinatorial types in order to obtain secondary invariants is due to the author [8]. Also we use a ribbon graph in place of a graph. This looks more natural from the point of quantization. Moreover the symmetry we obtain in that way (the mapping class group) looks more interesting than the symmetry of a graph. The observation that one can relate the Mumford-Strebel-Harer-Penner-Kontsevitch's triangulation of Teichmüller space to Morse theory in that way is discussed in this paper for the first time.

1. Morse homotopy.

We first recall briefly the construction of Witten complex. Let M be a closed oriented manifold of finite dimension and f be a Morse function on it. Let $Cr(f)$ denote the set of all critical points of f, namely

$$Cr(f) = \{p \in M | df(p) = 0\}.$$

For $p, q \in Cr(f)$ we put

$$\mathcal{M}(p,q) = \left\{ \ell : (-\infty,\infty) \to M \;\middle|\; \begin{matrix} \frac{d\ell}{dt} = -grad\ f, \\ \ell(-\infty) = p, \quad \ell(+\infty) = q. \end{matrix} \right\} /\!\sim.$$

Here $\ell \sim \ell'$ if and only if $\ell(t) = \ell'(t+c)$ for some constant c.

For $p \in Cr(f)$, let $\mu(f)$ be the Morse index of f. (Namely the number of negative eigenvalues of $Hess_p(f)$.) The following lemmas are proved by an easy transversality argument.

Lemma 1.1. *For a generic function f, the space $\mathcal{M}(p,q)$ is a smooth manifold of dimension $\mu(p) - \mu(q) - 1$.*

Lemma 1.2. *For a generic function f, the space $\mathcal{M}(p,q)$ is compactified to $\mathcal{CM}(p,q)$ such that*

$$\partial \mathcal{CM}(p,q) = \bigcup_r \mathcal{CM}(p,r) \times \mathcal{CM}(r,q).$$

Lemma 1.3. *For a generic function f, the space $\mathcal{M}(p,q)$ has an orientation such that they are compatible with respect to the compactification in Lemma 1.2*

Here orientation of a 0-dimensional manifold means that we assign $+1$ or -1 to each point. Then for a 0-dimensional oriented manifold X we can define $\sharp X$ as the number of point counted with sign. Now we define the Witten complex $C(M,f)$ for our Morse function f as follows.

$$C_k(M,f) = \bigoplus_{\mu(p)=k, p \in Cr(f)} \mathbf{Z}[p].$$

$$\partial[p] = \sum \sharp \mathcal{M}(p,q)[q].$$

Theorem 1.4. (Morse,Thom,Smale,Milnor,Witten,Floer,etc.) *$(C(M,f),\partial)$ is a chain complex. Its homology is canonically isomorphic to the homology of M.*

The proof of $\partial\partial = 0$ is based on Lemma 1.2. For the rest of this paper we shift to cohomology notation from homology. We can identify them by Poincaré duality. So we put

$$C^k(M,f) = C_{n-k}(M,f).$$

and $d = \partial$. Here n is the dimension of M. We hereafter write $\mu(p) = n -$ Morse index of p.

Our purpose here is to deduce more detailed structure on M from Morse theory. For this purpose we use several functions. First we define a moduli space of metric trees $\widetilde{\mathcal{T}}_{0,k}$. We consider the trees (one dimensional compact and simply connected simplicial complex). We assume that it has k vertices with one edges and no vertex with two edges. We let $\mathcal{V}_e$ the set of vertices with one edge. We say that an edge is *interior edge* if its boundary is disjoint to $\mathcal{V}_e$. Otherwise it is called an *exterior edge.* We fix an order of $\mathcal{V}_e = \{v_1, \cdots, v_k\}$. (Then an isotopy type of the embedding of our tree to $\mathbf{R}^2$ is automatically fixed.) We assign a positive real number to each interior edges. This number is regarded as the length of the edge. (We do *not* assign numbers to exterior edges. The length of them are regarded as infinity.)

We say such an object a *metric tree with k-exterior vertices.* Let $\widetilde{\mathcal{T}}_{0,k}$ be the set of such metric trees with k-exterior vertices. We can define a topology on this space in an obvious way. Now we recall the following :

Theorem 1.5. (Stasheff [26]) *$\widetilde{\mathcal{T}}_{0,k}$ is homeomorphic to $k-3$-dimensional (open) disk.*

See Figure 1 for the case when $k = 5$. Compactification of $\widetilde{\mathcal{T}}_{0,k}$, (which is also essential for our discussion) is studyed in [26]. The compactification is homeomorphic to a closed disk.

For an element Γ of $\widetilde{\mathcal{T}}_{0,k}$ we embed it to $D^2 = \{x \in \mathbf{R}^2 | |x| \le 1\}$, such that $\partial D^2 \cap \Gamma = \mathcal{V}_e$. Then there are k connected components of $D^2 - \Gamma$. We number it so that the closure of i-th component contains v_i and v_{i+1}. (See Figure 2.) Its boundary is a subgraph of Γ, and is called the *i-th (exterior) circle.*

Now let us take k smooth functions $f_1, \cdots, f_k$ on a manifold M. Let $x_i \in Cr(f_i - f_{i-1})$. We define a moduli space $\mathcal{M}(M, 0, k; f_1, \cdots, f_k; x_1, \cdots, x_k)$ as follows. (Sometimes we write $\mathcal{M}(M; f_1, \cdots, f_k; x_1, \cdots, x_k)$ or $\mathcal{M}(x_1, \cdots, x_k)$ to save notations.) $\mathcal{M}(x_1, \cdots, x_k)$ is the set of all maps $I : \Gamma \to M$, where $\Gamma \in \widetilde{\mathcal{T}}_{0,k}$ such that the following holds.

(1.6.1) $I(v_i) = x_i$.

(1.6.2) Let e is an interior edge of length ℓ such that it is contained in i-th and j-th circles. We assume that its orientation is as in Figure 3. We identify $e \simeq [0, \ell]$. Then we assume that the restriction of I to e is a gradient line of $f_i - f_j$.

(1.6.3) Let e is the k-th exterior circle. We identify $e - \{v_i\} = (-\infty, 0)$. We then assume that $I_{e-\{v_i\}}$ is a gradient line of $f_i - f_{i-1}$

There is a map $\pi : \mathcal{M}(x_1, \cdots, x_k) \to \widetilde{\mathcal{T}}_{0,k}$. Now the following is a consequence of a standard transversality argument. (And hence its proof is omitted.)

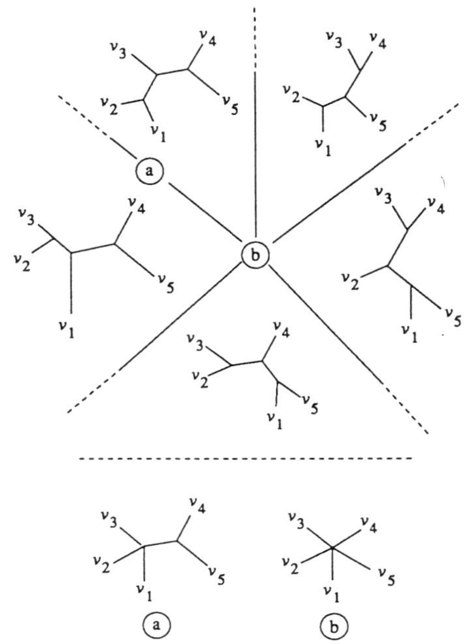

FIGURE 1

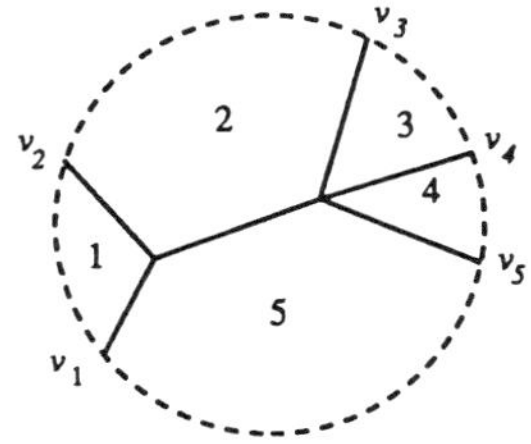

FIGURE 2

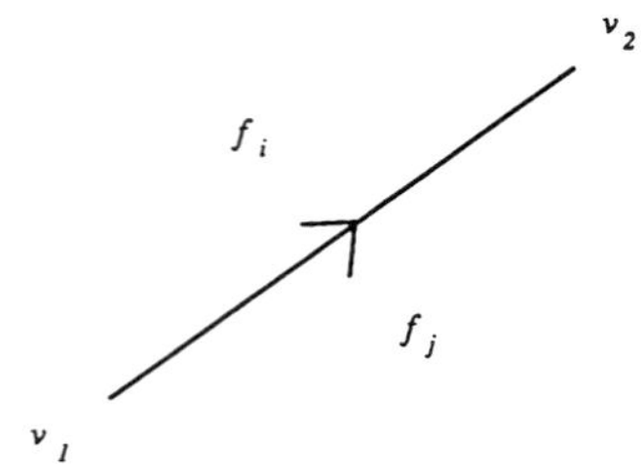

FIGURE 3

Theorem 1.6. *For generic $f_1, \cdots, f_k$, the space $\mathcal{M}(M; f_1, \cdots, f_k; x_1, \cdots, x_k)$ is a manifold of dimension $n - \sum \mu(x_i) + k - 3$*

Here $\mu(x_i)$ is the Morse index of x_i as a critical point of $f_i - f_{i-1}$. (Note that we consider x_1 as a critical point of $f_1 - f_k$ and not one of $f_k - f_1$. The Morse index depends on the sign of the Morse function.) We need also the compactification of our moduli space.

Theorem 1.7. *One can find a compactification $\mathcal{CM}(M; f_1, \cdots, f_k; x_1, \cdots, x_k)$ of $\mathcal{M}(M; f_1, \cdots, f_k; x_1, \cdots, x_k)$ which is a manifold with boundary and corner. The codimension one stratum (the boundary) of this compactification is the union of the following spaces.*

(1.8.1) $\mathcal{M}(M; f_1, \cdots, f_k; x_1, \cdots, x_i', \cdots, x_k) \times \mathcal{M}(x_i, x_i')$. *Here* $x_i' \in Cr(f_i - f_{i-1})$.

(1.8.2) $\mathcal{M}(M; f_1, \cdots, f_i, f_j, \cdots, f_k; x_1, \cdots, x_i, x, x_{j+1}, \cdots, x_k) \times \mathcal{M}(M; f_i, \cdots, f_j; x, x_{i+1} \cdots, x_j)$

This theorem is an immediate consequence of the proof of Lemma 1.2 and Stasheff's compactification of $\widetilde{\mathcal{T}}_{0,k}$. Roughly speaking (1.8.1) corresponds to the compactification of the fibre of π and (1.8.2) corresponds to the compactification of $\widetilde{\mathcal{T}}_{0,k}$.

Now we define a map

$$\begin{aligned}\eta_{k-1} : C^{a_1}(M; f_2 - f_1) \otimes & \cdots \otimes C^{a_{k-1}}(M; f_k - f_{k-1}) \\ & \to C^{\sum a_i + k - 3}(M; f_k - f_1)\end{aligned}$$

by

$$\eta_{k-1}([x_1] \otimes \cdots \otimes [x_{k-1}]) = \sum_{x_k} \sharp \mathcal{M}(M; f_1, \cdots, f_k; x_1, \cdots, x_k)[x_k].$$

(We write sometimes $\eta_{k-1}([x_1], \cdots, [x_{k-1}])$ in place of $\eta_{k-1}([x_1] \otimes \cdots \otimes [x_{k-1}])$.) A consequence of Theorem 1.8 is

Corollary 1.8.

$$\begin{aligned}(\partial\eta_k) \quad & ([x_1], \cdots, [x_k]) \\ & = \textstyle\sum_{i<j} \epsilon \cdot \eta_{k-j+i+1}([x_1], \cdots, \eta_{j-i}([x_{i+1}], \cdots, [x_j]), \cdots, [x_k]).\end{aligned}$$

Here $\epsilon = (-1)^{(j-i)(degx_1 + \cdots + degx_i)}$. Upto sign Corollary 1.9 is immediate from Theorem 1.8. The verificatin of the sign is omitted.

Let us consider the case when $k = 2$. Corollary 1.9 implies that η_2 is a chain map. Hence by Theorem 1.4 it induces a map

$$H^i(M; \mathbf{Z}) \otimes H^j(M; \mathbf{Z}) \to H^{i+j}(M; \mathbf{Z}) \tag{1.10}$$

Proposition 1.9. *The map (1.10) is the cup product.*

Sketch of the proof. Let $\sum a_m[x_m] \in C^i(M, f_2 - f_1)$, $\sum b_m[y_m] \in C^j(M, f_3 - f_2)$ be cocycles. The Poincaré dual of the corresponding cohomology class is represented by $\sum a_m X_m \in C^i(M, f_2 - f_1)$, and $\sum b_m Y_m \in C^j(M, f_3 - f_2)$ where X_m is the closure of the unstable manifold of the critical point x_m and Y_m is the closure of the unstable manifold of the critical point y_m. Hence the Poincaré dual of the cup product of $[\sum a_m[x_m]]$ and $[\sum b_m[y_m]]$ is represented by $\sum a_m b_{m'}[X_m \cap Y_{m'}]$.

On the other hand by definition we have, for $z \in Cr(f_3 - f_1)$,

$$\mathcal{M}(x_m, y_{m'}, z) = X_m \cap Y_{M'} \cap Z.$$

Where Z is the stable manifold of the critical point z of $grad(f_3 - f_1)$. The proposition now follows from the definition.

We next consider the case when $k = 3$. We have

$$\begin{aligned} 0 = {} & d(\eta_3(x_1, x_2, x_3)) - \eta_3(dx_1, x_2, x_3) \\ & -(-1)^{degx_1}\eta_3(x_1, dx_2, x_3) - (-1)^{degx_2+degx_3}\eta_3(x_1, x_2, dx_3) \\ & -\eta_2(\eta_2(x_1, x_2), x_3) - \eta_2(x_1, \eta_2(x_2, x_3)). \end{aligned} \tag{1.12}$$

There is two consequences of this formula. One is that the cup product is associative. In fact our map η_3 gives a chain homotopy for the cup product to be associative.

The other is that we can define Massey product using it. Let $x_i \in C(M, f_i - f_{i-1})$ be cocycles such that $[x_1] \cup [x_2] = [x_2] \cup [x_3] = 0$. We put $\eta_2(x_1, x_2) = dy_1$, $\eta_2(x_2, x_3) = dy_2$. Then we find

$$d(\eta_2(y_1, x_3) + (-1)^{degy_1}\eta_2(x_1, y_2) + \eta_3(x_1, x_2, x_3)) = 0.$$

Proposition 1.10. *The cocycle* $\eta_2(y_1, x_3) + (-1)^{deg x_1}\eta_2(x_1, y_2) + \eta_3(x_1, x_2, x_3)$ *represents the Massey product of three elements* $[x_1]$,$[x_2]$, $[x_3]$.

The proof is similar to one of Proposition 1.9 and is omitted. In this way one can find the higher Massey product using our maps η_k. One can summarize these propositions by using the notion of (topological) A^∞-category.

Definition 1.11. A topological A^∞-*category* is a collection of the topological space, the set of object $\mathfrak{Ob}$, a cochain complex $C(a,b)$ for Baire subset of $(a,b) \in \mathfrak{Ob}^2$, and maps $\eta_k : C(a_0,a_1) \otimes \cdots \otimes C(a_{k-1},a_k) \to C(a_0,a_k)$ of degree $k-3$ for Baire subset of $(a_0,\cdots,a_k) \in \mathfrak{Ob}^{k+1}$, such that Formula (1.9) holds in a Baire subset.

Now it is easy to see that our discussion so far proves the following :

Theorem 1.12. *For each oriented manifold* M*, there exists a (topological)* A^∞*-Category whose object is a smooth function on it, whose morphism is an element of Witten complex, and whose (higher) composition is* η_k.

Problem 1.13. Can one construct Sullivan's minimal model [29], using our A^∞-category ?

Remark 1.14. We remark that our operator η_k is defined everywhere and gives (higher) Massey product after appropriate modification. The idea that one can construct everywhere defined operator from secondary operator (which is defined only in an appropriate subset and is well defined modulo some indeterminacy) using something like A^∞-category is found independently by V.Smirnov [25].

2. Quantization

In this section we join the discussion of section 1 with topological σ-model or Gromov-Ruan invariant [23]. It seems to the author that one can define quantum rational homotopy type of Calabi-Yau manifold in this way.

In physical literature, there is a construction know as quantum ring or A-model. ([32].) This is a quantization of cup product. In this section we also quantize (higher) Massey product and Steenrod square.

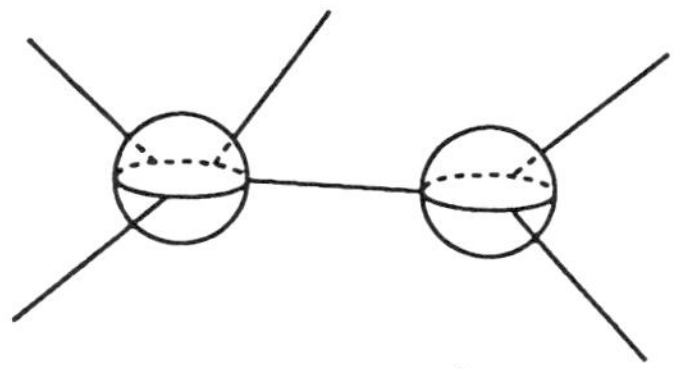

FIGURE 4. An element of $\hat{\mathcal{T}}_{0,5}$

We first consider the moduli space $\widehat{\mathcal{T}}_{0,k}$ which modify a bit the moduli space $\widetilde{\mathcal{T}}_{0,k}$ we used in section 1. For a positive integer m we put

$$\mathcal{T}_{0,\partial,m} = \left\{ (D^2; x_1, \cdots, x_k) \;\middle|\; \begin{array}{l} x_i \in \partial D^2,\ x_1, \cdots, x_k \text{ are dis-} \\ \text{joint to each other, cyclic or-} \\ \text{dering of them is respected} \end{array} \right\} / \sim .$$

Here $\sim$ denotes the biholomorphic equivalence. We can compactify the moduli space $\mathcal{T}_{0,\partial,m}$ by adding the configuration where $x_i = x_{i+1}$ etc., and denote it by $\mathcal{CT}_{0,\partial,m}$. We may and will regard the element of $\mathcal{T}_{0,\partial,m}$ as Riemann sphere with m point on it which is contained in one circle, (the set conformal to equator.) We put

$$\widehat{\mathcal{T}}_{0,k} = \left\{ (\Gamma; D_1, \cdots, D_a) \;\middle|\; \begin{array}{l} \Gamma \in \widetilde{\mathcal{T}}_{0,k} \text{ with } a \text{ inner vertices} \\ D_b \in \mathcal{CT}_{0,\partial,m} \text{ where } b\text{-th vertex of } \Gamma \\ \qquad\qquad \text{has } m\text{-edges.} \end{array} \right\}$$

(See Figure 4.) We define a topology as in Figure 5. Now the following can be proved in a similar way as Theorem 1.5.

Theorem 2.1. *$\widehat{\mathcal{T}}_{0,k}$ is diffeomorphic to the $k-3$-dimensional open disk.*

See Figure 6 for the case when $k = 5$. Compactification of this moduli space is obtained in a similar way.

We regard an element of $\widehat{\mathcal{T}}_{0,k}$ as 2 dimensional space as in Figure 4. We now define a moduli space $\mathcal{M}^m(M, 0, k; f_1, \cdots, f_k; x_1, \cdots, x_k)$ as follows. Let M be a symplectic manifold. We choose a compatible almost complex structure on M. We assume that first Chern class of this almost complex structure vanishes. (In fact one can discuss in a similar way the case when the first Chern form is a multiplication of symplectic form by a positive number . See [8] Chapter 4.) Let f_i be functions on it. x_i is a critical point of $f_i - f_{i-1}$. Let $P \in \widehat{\mathcal{T}}_{0,k}$. Then we consider a continuous map $I : P \to M$ such that they satisfy (1.6.1),(1.6.2),(1.6.3) on graph part of P, and

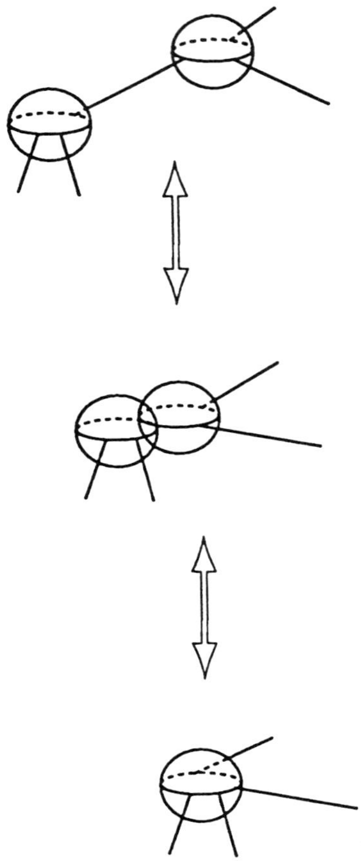

FIGURE 5

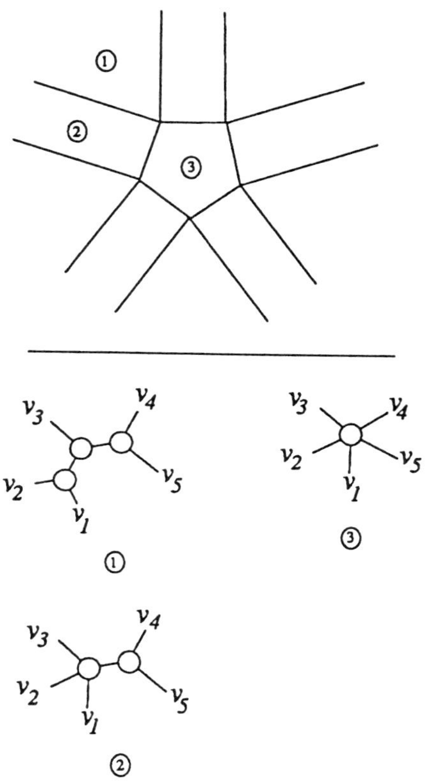

FIGURE 6

(2.2.1) I is pseudo-holomorphic on each sphere.

(2.2.2)

$$\int_P I^*\omega = m.$$

Here ω denotes the symplectic form. We let $\mathcal{M}^m(M, 0, k; f_1, \cdots, f_k; x_1, \cdots, x_k)$ be the space of all such objects. One can define a topology on it in a similar way. There is again a map $\pi : \mathcal{M}^m(M, 0, k; f_1, \cdots, f_k; x_1, \cdots, x_k) \to \widehat{\mathcal{T}}_{0,k}$.

To be more precise, we have to modify (2.2.1) and consider the equation

$$\overline{\partial} I = g \tag{2.3}$$

in place of Caucy-Rieman equation. (See [23].) But we do not discuss this point since the argument one needs for it is the same as Ruan's one. Now we have :

Theorem 2.2. *For generic f_i, almost complex structure, and perturbation of Caucy Riemann equation, the space $\mathcal{M}^m(M, 0, k; f_1, \cdots, f_k; x_1, \cdots, x_k)$ is a manifold of dimension $n - \sum \mu(x_i) + k - 3$.*

Here n is the real dimension of our manifold. The proof is a minor modification of one by Ruan [23] and is omitted.

(We remark that our assumption that the first Chern class vanishes implies that the dimension is independent of m. In case when $c^1 = N\omega$, the dimension will be $n - \sum \mu(x_i) + k - 3 + nmN$.)

We can also prove the following analogy of Theorem 1.8.

Theorem 2.3. *One can find a compactification $\mathcal{CM}^m(M; f_1, \cdots, f_k; x_1, \cdots, x_k)$ of $\mathcal{M}^m(M; f_1, \cdots, f_k; x_1, \cdots, x_k)$ which is a manifold with boundary and corner. The codimensions one stratum (the boundary) of this compactification is one of the following spaces.*

(2.4.1) $\mathcal{M}^m(M; f_1, \cdots, f_k; x_1, \cdots, x_i', \cdots, x_k) \times \mathcal{M}(x_i, x_i')$. *Here* $x_i' \in Cr(f_i - f_{i-1})$.

(2.4.2) $\mathcal{M}^{m'}(M; f_1, \cdots, f_i, f_j, \cdots, f_k; x_1, \cdots, x_i, x, x_{j+1}, \cdots, x_k) \times \mathcal{M}^{m-m'}(M; f_i, \cdots, f_j; x, x_{i+1}, \cdots, x_j)$

We now define

$$\eta_{k-1}^m : C^{a_1}(M; f_2 - f_1) \otimes \cdots \otimes C^{a_{k-1}}(M; f_k - f_{k-1}) \to C^{\sum a_i + k - 3}(M; f_k - f_1)$$

by the same formula as η_{k-1}. Namely

$$\eta^m_{k-1}([x_1]\otimes\cdots\otimes[x_{k-1}]) = \sum_{x_k} \sharp\mathcal{M}^m(M;f_1,\cdots,f_k;x_1,\cdots,x_k)[x_k].$$

We put $\widehat{C}^a(M;f) = C^a(M;f)\otimes \mathbf{Z}[[T]]$. (The ring $\mathbf{Z}[[T]]$ arises here by the same reason as Novikov ring is used in symplectic Floer theory ([24] [12] [16] [21] [8]). Then we define

$$\widehat{\eta}_{k-1} : \widehat{C}^{a_1}(M;f_2-f_1)\otimes\cdots\otimes\widehat{C}^{a_{k-1}}(M;f_k-f_{k-1}) \to \widehat{C}^{\sum a_i+k-3}(M;f_k-f_1)$$

by

$$\widehat{\eta}_{k-1} = \sum T^m\eta^m_{k-1}.$$

Hence using Theorem 2.4 we obtain :

Corollary 2.4.

$$\begin{aligned}(\partial\widehat{\eta}_k)\quad &([x_1],\cdots,[x_k])\\ &=\textstyle\sum_{i<j}\epsilon\cdot\widehat{\eta}_{k-j+i}([x_1],\cdots,\widehat{\eta}_{j-i}([x_{i+1}],\cdots,[x_j]),\cdots,[x_k]).\end{aligned}$$

Here ϵ is as in Corollary 1.9. It is a direct consequence of the definition that η^m_2 coincides to Gromov-Ruan invariant [23]. Thus $\widehat{\eta}_2$ is the same as quantum ring in physical literature. This coincidence is pointed out by M.Kontsevitch to the author in the case of open string quantization. Kontsevitch proposed the following :

Conjecture 2.5. $\sum_m \sharp\mathcal{M}^m(M;f_1,\cdots,f_k;x_1,\cdots,x_k)\,t^m$ *converges for small* t.

If Conjecture 2.6 is true, then one can consider $\widehat{\eta}_k$ over real coefficient in place of formal power series ring.

We can define the quantum (higher) Massey product by exactly the same formula as in section 1. Namely

$$< x_1,x_2,x_3 >= \widehat{\eta}_2(y_1,x_3) + (-1)^{degx_1}\widehat{\eta}_2(x_1,y_2) + \widehat{\eta}_3(x_1,x_2,x_3).$$

Next we discuss the quantum Steenrod square based on Betz-Cohen's result. For this purpose we first recall their idea to relate Steenrod square to Morse theory. Our treatment is basically the same as Betz-Cohen [2]. However we modify and organize the construction so that it is easy to see the way of quantization. Let M be an oriented manifold. (We do not need to assume that it is symplectic at this point.) First we need to modify a bit our moduli space. $\mathcal{M}(M;f_1,\cdots,f_k;x_1,\cdots,x_k)$ and define $\mathcal{M}(M;f_1,\cdots,f_k)$. We recall that in the definition of $\mathcal{M}(M;f_1,\cdots,f_k;x_1,\cdots,x_k)$ we regard the length of exterior edges to be infinity. Here we take an arbitrary but a fixed finite number (we choose 1) and suppose the length of exterior edges to be that number. More precisely we define $\mathcal{M}(M; f_1, \cdots, f_k)$ to be the set of all maps $I : \Gamma \to M$ such that it satisfies (1.6.2) and

(2.7) Let e be the i-th exterior edge. We identify $e = [0,1]$. Then $I|_e$ is a gradient line of $f_i - f_{i-1}$.

(Here we do not need to take the critical points x_i.) There is a map :

$$\pi : \mathcal{M}(M; f_1, \cdots, f_k) \to \widetilde{\mathcal{T}}_{0,k}$$

and

$$\pi_2 : \mathcal{M}(M; f_1, \cdots, f_k) \to M^k.$$

Here π_2 is defined by $\pi_2(\Gamma, I) = (I(v_1), \cdots, I(v_k))$. We consider the union of $\mathcal{M}(M; f_1, \cdots, f_k)$ for all f_i and denote it by $\mathcal{M}_{0,k}$. Then we have

$$\pi_1 : \mathcal{M}_{0,k} \to \widetilde{\mathcal{T}}_{0,k} \times (C^\infty(M))^k. \tag{2.8}$$

Here we remark that the cyclic group $\mathbf{Z}_k$ of order k acts as a symmetry group of this map. We choose an appropriate subspace $\overline{(C^\infty(M))^k}$ of $(C^\infty(M))^k$ such that $\mathbf{Z}_k$ act freely there and that $(C^\infty(M))^k - \overline{(C^\infty(M))^k}$ is of infinite codimension. We obtain

$$\pi_1 : \overline{\mathcal{M}}_{0,k} \to \frac{\widetilde{\mathcal{T}}_{0,k} \times \overline{(C^\infty(M))^k}}{\mathbf{Z}_k}.$$

By dividing the map (2.8). We also obtain a map

$$\pi_2 : \overline{\mathcal{M}}_{0,k} \to M^k / \mathbf{Z}_k.$$

Now we can prove by a standard transversality argument that

Theorem 2.6.

$$\pi_1 : \overline{\mathcal{M}}_{0,k} \to \frac{\widetilde{\mathcal{T}}_{0,k} \times \overline{(C^\infty M)^k}}{Z_k}.$$

is a Fredholm map of index n.

We need one more map. We remark that there is a universal family

$$Univ \to \frac{\widetilde{\mathcal{T}}_{0,k} \times \overline{(C^\infty M)^k}}{\mathbf{Z}_k},$$

whose fiber is a tree. Since a tree is contractible, we can choose a section s to this map. Then we get a map

$$\pi_3 : \overline{\mathcal{M}}_{0,k} \to M$$

by

$$\pi_3([\Gamma, I]) = I(s([\Gamma, I])).$$

Now let p be an arbitrary prime. We remark that $\frac{\widetilde{\mathcal{T}_{0,p}} \times \overline{(C^\infty M)^p}}{\mathbf{Z}_p}$ is a $K(\mathbf{Z}_p, 1)$ space. Hence its homology over $\mathbf{Z}_p$ has a canonical generator $[X_i]$ in each degree i. We choose a cycle X_i which represents this class. Next let U be a cycle of degree $n - j$ which represents a cohomology class $H^j(M; \mathbf{Z}_p)$. (To be precise one needs to use the notion of geometric cycle of degree i. See [6].) Then in case we choose everything transversal we get a cycle

$$\pi_1^{-1} X_i \cap \pi_2^{-1}(U^p/\mathbf{Z}_p).$$

This is a cycle of dimension $n - pj + i$. Hence by using π_3 it gives an element of

$$H_{n-pj+i}(M; \mathbf{Z}_p) = H^{pj-i}(M; \mathbf{Z}_p).$$

Theorem 2.7. $\pi_3(\pi_1^{-1} X_i \cap \pi_2^{-1}(U^p/\mathbf{Z}_p))$ *represents the element* $D_i([U])$. *Where* D_i *is as in* [27]

The proof is rather immediate from the constructions and hence omitted. We recall, for example, $Sq^k = D_{j-k}$ for an element of degree j.

Now it is quite obvious how to quantize the construction. We can construct the moduli space $\mathcal{M}^m(M; f_1, \cdots, f_k)$ etc. Hence we obtain a map

$$D^m_{i,(p)} : C^j(M; \mathbf{Z}_p) \to C^{pj-i}(M; \mathbf{Z}_p),$$

in case M is a symplectic manifold with trivial first Chern class. We define

$$\widehat{D}_{i,(p)} : C^j(M; \mathbf{Z}_p[[T]]) \to C^{pj-i}(M; \mathbf{Z}_p[[T]])$$

by

$$\widehat{D}_{i,(p)} = \sum_m T^m D^m_{i,(p)}.$$

Thus we get quantum Steenrod square and higher reduced power.

We close this section by proposing some open questions.

Problem 2.8. Does Adem and Cartan relations hold for quantum Steenrod square and higher reduced power ? If not what is a quantized version of it ?

Problem 2.9. Can one define secondary operations by combining ideas of this and last sections ?

We remark that the idea of section 1 is to use Teichmüller parameter (moduli parameter of graphs) to define a secondary operator (Massey product). We remark that in the definition of quantum Steenrod square we do not in fact need a moduli parameter since there is an element of $\widetilde{T}_{0,p}$ which is fixed by $\mathbf{Z}_p$. (But this does not the case for higher genus.)

Problem 2.10. Can one quantize Adams spectral sequence ?

Solving these problems should be the way to find "quantum homotopy type" of, say, Calabi-Yau manifold.

3. Remarks

First we describe another way to quantize Morse homotopy. Let X be a symplectic manifold. For simplicity we again assume that the first Chern class of X vanishes. We assume also that the cohomology class of the symplectic form ω is integral. We choose and fix a complex line bundle L with connection ∇ on it such that its Chern form is equal to symplectic form (not only as cohomology class but also as forms.) We say a Lagrangian $\Lambda \subset X$ to be a Bohr-Sommerfeld orbit (BS-orbit) if the restriction of (L, ∇) to it is trivial. For two oriented BS-orbits Λ_1,Λ_2 we define its (Lagrangian intersection) Floer homology (roughly) as follows. (See [8] for detail.) Let $p, q \in \Lambda_1 \cap \Lambda_2$. We consider the moduli space

$$\mathcal{M}(p,q) = \left\{ h : D^2 \to X \;\middle|\; \begin{array}{ll} h(-1) = p, h(1) = q, & \\ h(e^{\sqrt{-1}\theta}) \in \Lambda_1, & \text{if } \theta \in [0,\pi], \\ h(e^{\sqrt{-1}\theta}) \in \Lambda_2, & \text{if } \theta \in [\pi, 2\pi]. \end{array} \right\}$$

We split the space using symplectic area. Namely we put

$$\mathcal{M}_m(p,q) = \left\{ h \in \mathcal{M}(p,q) \;\middle|\; \int_{D^2} h^*\omega = m + c(p,q) \right\}.$$

Here $c(p,q)$ is a number depending only on p and q. (See [8] Chapter 4.) Because of the presence of the fundamental group of Λ_i and of Maslov index, the dimension of this space depends on the components. We can split $\mathcal{M}_m(p,q)$, as

$$\mathcal{M}_m(p,q) = \bigcup_{\ell > \ell_o} \mathcal{M}_{m,\ell}(p,q)$$

and that $M_{m,\ell}(p,q)$ is a manifold of dimension $\mu(p) - \mu(q) + 2\ell$. Here $\mu(p)$ is an integer depending only on p. We divide the moduli space by

$\mathbf{R} = Aut(D^2; -1, +1)$ and let $\overline{\mathcal{M}}_{m,\ell}(p,q)$ be the quotient space. Now we put

$$\widehat{C}(\Lambda_1, \Lambda_2) = \bigoplus_{x \in \Lambda_1 \cap \Lambda_2} \mathbf{Z}[[T]][T^{-1}][x].$$

Here $[x]$ is regarded to have degree $\mu(x)$ and T is regarded to be of degree 2. The boundary operator is defined by

$$\partial[x] = \sum_y \sharp \overline{\mathcal{M}}_{m,\ell}(x,y) T^m [y].$$

Here we choose ℓ such that $dim \overline{\mathcal{M}}_{m,\ell}(x,y) = 0$. Thus we obtain a chain complex.

From now on we again use cohomology notation. Hence we put

$$\widehat{C}^k(\Lambda_1, \Lambda_2) = \widehat{C}_{n-k}(\Lambda_1, \Lambda_2).$$

Here $2n = dim_{\mathbf{R}} X$.

Let $\Lambda_1, \cdots, \Lambda_k$ be BS-orbits and $p_i \in \Lambda_i \cap \Lambda_{i-1}$. ($\Lambda_0 = \Lambda_k$.) We put

$$\begin{aligned} &\mathcal{M}(\Lambda_1, \cdots, \Lambda_k; p_1, \cdots, p_k) = \mathcal{M}(p_1, \cdots, p_k) \\ &= \left\{ (h; x_1, \cdots, x_k) \;\middle|\; \begin{array}{l} [D^2; x_1, \cdots, x_k] \in \mathcal{T}_{0,\partial,k} \\ h : D^2 \to X \ \text{ is pseudo holomorphic,} \\ h(x_i) = p_i, \quad h(\overline{x_i x_{i+1}}) \subset \Lambda_i \end{array} \right\} / \sim \end{aligned}$$

Here $(h; x_1, \cdots, x_k) \sim (h\varphi; \varphi^{-1}(x_1), \cdots, \varphi^{-1}(x_k))$ for $\varphi \in Aut(D^2)$ and $\overline{x_i x_{i-1}}$ denotes the subset of ∂D^2 which bounds $\{x_i, x_{i-1}\}$. We let $\mathcal{M}_k(p_1, \cdots, p_k)$ be the subset of $\mathcal{M}(p_1, \cdots, p_k)$ such that $\int h^* \omega = m + c(p_1, \cdots, p_k)$. (See [8] Chapter 4.)

We again split $\mathcal{M}_k(p_1, \cdots, p_k)$ into $\mathcal{M}_{k,\ell}(p_1, \cdots, p_k)$ such that

$$dim \mathcal{M}_{k,\ell}(p_1, \cdots, p_k) = n - \sum \mu(p_i) + (k-3) + 2\ell.$$

We count its number and obtain

$$\eta_k^\ell([p_1], \cdots, [p_{k-1}]) = \sharp \mathcal{M}_{k,\ell}(p_1, \cdots, p_k) \cdot [p_k].$$

We put

$$\widehat{\eta}_k = \sum T^\ell \eta_k^\ell.$$

Theorem 3.1. ([7],[8])

$$\begin{aligned} (\partial \widehat{\eta}_k) \quad &([x_1], \cdots, [x_k]) \\ &= \textstyle\sum_{i<j} \epsilon \cdot \widehat{\eta}_{k-j+i}([x_1], \cdots, \widehat{\eta}_{j-i}([x_{i+1}], \cdots, [x_j]), \cdots, [x_k]). \end{aligned}$$

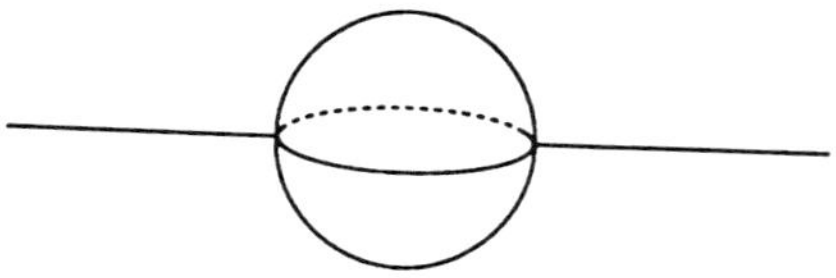

FIGURE 7

Here ϵ is as in Corollary 1.9. This is a consequence of a result similar to Theorem 2.3.

Let us mention here a relation of this construction to one in section 2.

First we mention one difference between them. In the construction of section 2, we do *not* consider the quantum correction, (which is defined by using the moduli space of pseudo holomorphic maps with nonzero symplectic area), in the definition of boundary operator. (But it appeared in cup and Massey products.)

The reason for it is as follows. (This observation is due to Floer [5].) The quantum correction of the boundary operator, if exists, should be defined by counting the order of the moduli space of the map from the space of Figure 7.
The point is that the 2-sphere with two points is *not* a stable curve. That is it has an automorphism group S^1. So if such a moduli space is nonempty then its (virtual) dimension is at lease one. Since we count the moduli space of dimension zero, the nontrivial contribution comes only from the degenerate case, which corresponds to the usual boundary operator.

On the other hand, in our situation of Lagrangian intersection, we are considering a pseudo holomorphic map from D^2 with two points on its boudary. Its automorphism group is $\mathbf{R}$, which corresponds, in Morse theory, the reparametrization of the gradient line. Hence there is no reason that the quantum correction is 0 in this situation, even for boundary operator. However the author do not have an explicit example that it is really nonzero.

We next discuss how the version of this section is related to one in section 2. Let M be a symplectic manifold of dimension n. Suppose that its first Chern class vanishes. We embed M to $X = M \times M$ as diagonal. Then M is a BS-orbit. One can prove that a neighborhood of M in X is symplectic ally diffeomorphic to a neighborhood of zero section in T^*M, the cotangent bundle. Let $f \in C^\infty(M)$. We consider a graph of

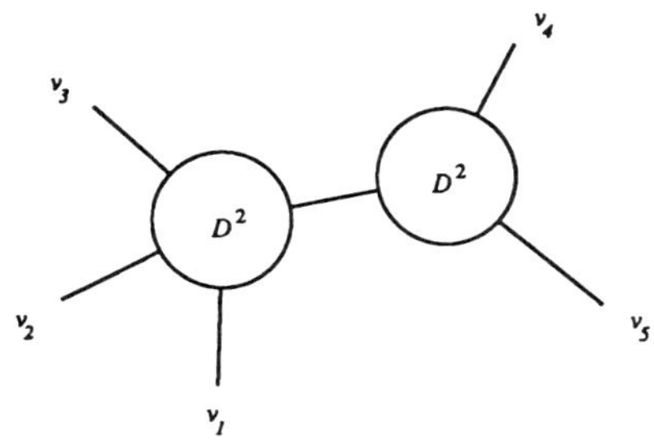

FIGURE 8

ϵdf in T^*M. Then for ϵ small, we may regard it as a submanifold in X. We denote it by Λ_f. We remark that $\Lambda_f \cap \Lambda_g = Cr(f-g)$. Let $p_i \in \Lambda_{f_i} \cap \Lambda_{f_{i-1}} = Cr(f_i - f_{i-1})$. We may choose $c(p_1, \cdots, p_k) \sim \epsilon$. Namely for

$$h \in \mathcal{M}_{m,\ell}(\Lambda_{f_1}, \cdots, \Lambda_{f_k}; p_1, \cdots, p_k)$$

we have

$$\int h^*\omega = m + O(\epsilon).$$

Then $\mathcal{M}_{m,\ell}(\Lambda_{f_1}, \cdots, \Lambda_{f_k}; p_1, \cdots, p_k)$ is nonempty only if $m \geq 0$. Hence we can take $\mathbf{Z}[[T]]$ as coefficient ring rather than $\mathbf{Z}[[T]][T^{-1}]$. Moreover we can verify that the Maslov index $\pi_1(M) \to \mathbf{Z}$ defined in [8] Chapter 4 is trivial in this case where M is diagonal. As a consequence we do not need to split using ℓ.

Conjecture 3.2. *For small ϵ the space $\mathcal{M}_m(\Lambda_{f_1}, \cdots, \Lambda_{f_k}; p_1, \cdots, p_k)$ is diffeomorphic to $\mathcal{M}^m(M; f_1, \cdots, f_k; p_1, \cdots, p_k)$.*

Here the first moduli space is introduced in this section and the second one is introduced in section 2.

Idea of the proof. We write $\mathcal{M}^\epsilon_m(\Lambda_{f_1}, \cdots, \Lambda_{f_k}; p_1, \cdots, p_k)$ in place of $\mathcal{M}_m(\Lambda_{f_1}, \cdots, \Lambda_{f_k}; p_1, \cdots, p_k)$. We consider elements $h_\epsilon \in \mathcal{M}_m(\Lambda_{f_1}, \cdots, \Lambda_{f_k}; p_1, \cdots, p_k)$ and study what happens when ϵ goes to zero. We want to describe the moduli space which we expect to be the set of limit of the images $h_\epsilon(D^2)$, (say, in Haudsorff topology.)

Let $(\Gamma; D_1, \cdots, D_a) \in \widehat{T}_{0,\partial,k}$. In place of regarding it as a two dimensional space like Figure 4. we regard it as the space of Figure 8. Then we consider the maps $h : P \to X$ such that on the graph part it satisfies (1.6.1), (1.6.2),(1.6.3), on disks it satisfies (3.3.1),(3.3.2) and boundary condition (3.3.3).

(3.3.1) P is pseudo holomorphic on each disk,
(3.3.2) $\int_P h^*\omega = m$.
(3.3.3) $h(\partial P) \subset M =$ diagonal.

We write moduli space of such maps $\mathcal{M}^0_m(\Lambda_{f_1}, \cdots, \Lambda_{f_k}; p_1, \cdots, p_k)$.

Lemma 3.3. *$\mathcal{M}^0_m(\Lambda_{f_1}, \cdots, \Lambda_{f_k}; p_1, \cdots, p_k)$ is diffeomorphic to $\mathcal{M}^m(M; f_1, \cdots, f_k; p_1, \cdots, p_k)$.*

This is an immediate consequence of reflection principle. Thus to show Conjecture 3.2 we need only to prove

$$\begin{aligned}\lim_{\epsilon\to 0} \mathcal{M}^\epsilon_m(\Lambda_{f_1}, \cdots, \Lambda_{f_k}; p_1, \cdots, p_k) \\ = \mathcal{M}^0_m(\Lambda_{f_1}, \cdots, \Lambda_{f_k}; p_1, \cdots, p_k)\end{aligned} \tag{3.5}$$

The author do not know the proof of it. (See also Hofer-Salamon [12].)

In fact we can prove that the left hand side is contained in right hand side. But the opposite is more delicate. (It is an analogy of Taubes' construction in Gauge theory.) To use Conjecture 3.2 to find relations between two versions we need probably Ono's idea in [21].

Finally we give a remark about our assumption that $x_1, \cdots, x_k$ is on one circle, in the definition of $\mathcal{T}_{0,\partial,m}$. From the point of view of moduli space of Riemann surface with points, this assumption looks strange and it looks natural to remove it. (Of course one needs it for Conjecture 3.2 to be true.) So it seems not appropriate to regard the construction of section 2 as closed string theory.

However there is another reason why we put that assumption. Under this assumption the moduli space $\mathcal{T}_{0,\partial,m}$ has codimension one boundary which we use to construct our moduli space $\widehat{\mathcal{T}}_{0,k}$ by patching moduli spaces of configurations of different combinatorial types. If we remove that assumption in the definition of $\mathcal{T}_{0,\partial,m}$, it will have only codimension 2 boundary. This is inconvenient to patch it to other spaces. The author do not understand well have to handle various components in the case of closed string.

4. Higher genus Morse homotopy

In this section we generalize the construction of section 1 by using graphs (more precisely Ribbon graphs) which is not necessary a tree.

Let $g \geq 0, k \geq 2$ be integers. We define a space $\widetilde{\mathcal{T}}_{g,k}$ as follows.

A ribbon graph Γ is a one dimensional simplicial complex together with cyclic order of the set of the edges containing a vertex. Any ribbon graph is embedded in a unique way to a Riemann surface Σ such that each connected component of $\Sigma - \Gamma$ is a cell and that the cyclic order is respected. The genus of Γ is by definition the genus of Σ.

A k-marked ribbon graph is a ribbon graph Γ and a vertex v_0 on it with exactly k-edges containing v_0 together with an order of the set of the edges containing v_0 compatible with the cyclic order. (Here we only consider the case with one marked point. One can generalize it to the case when there are several marked points. But we do not try to do it here.) In case when genus is zero and the graph minus v_0 is a tree, we go back to the situation in section 1 by removing the vertex v_0 and taking its completion.

For a k-marked ribbon graph, we call an edge to be an interior edge if it is disjoint to v_0. Otherwise the edge is called an exterior edge. A length function ℓ is a function from the set of interior edges to the set of positive numbers. We assume that the number of edges containing each vertex is not smaller than 3. A circle of a k-marked ribbon graph Γ is by definition a subcomplex $\subset \Gamma$ which is a boundary of a connected component of $\Sigma - i(\Gamma)$. A circle is called an interior circle if it is disjoint from v_0. Otherwise it is called an exterior circle. By our assumption there are exactly k-exterior circles. In other words each exterior circle is embedded to Γ in a neighborhood of v_0.

We put

$$\widetilde{\mathcal{T}}_{g,k} = \left\{ ((\Gamma, v_0), \ell, [i]) \;\middle|\; \begin{array}{l} (\Gamma, v_0) \text{ is a } k\text{-marked ribbon graph,} \\ \ell \text{ is a length function,} \\ [i] \text{ is an isotopy class of an embedding of} \\ \;\Gamma \text{ to a Riemann surface together with} \\ \text{an order of the set of interior circles.} \end{array} \right\}$$

Let $\widetilde{\mathcal{T}}_{g,m,k}$ be the set of all elements of $\widetilde{\mathcal{T}}_{g,k}$ with m interior circles and k exterior circles.

We remark that we can define a topology of these spaces $\widetilde{\mathcal{T}}_{g,k}$, in an obvious way.

The space $\widetilde{\mathcal{T}}_{g,k,m}$ is almost the same as the triangulation of Teichmüller space based on Strebel's quadratic differentials. This fact is important for quantization of the construction of this section. However in this section we use this space in a different way. We remark that one can prove that $\widetilde{\mathcal{T}}_{g,k}$ is a smooth manifold in exactly the same way as the case of the triangulation of Teichmüller space. (The argument by Stasheff quoted in section 1 can be applied to this case.) But contrary to the case of section 1 this space is not contractible. For example in case of $g = 0$ $m \neq 0$ the fundamental group of it is the braid group.

Now let us take an oriented manifold M (of finite dimension). We choose an ordered set $(f_1, \cdots, f_n)$ of smooth functions on M. We assume that $f_i - f_j$ is a Morse function for each $i \neq j$.

We take the point $x_1, \cdots, x_k$ of M such that x_i is a critical point of $f_i - f_{i-1}$.

We now generalize the construction of section 1 and define a moduli space

$$\mathcal{M}(M, g, m, k; f_1, \cdots, f_k, f_1', \cdots, f_m', x_1, \cdots, x_k),$$

together with a map from there to $\widetilde{\mathcal{T}}_{g,m,k}$ as follows. (We remark that

$$\mathcal{M}(M, 0, 0, k; f_1, \cdots, f_k; x_1, \cdots, x_k) = \mathcal{M}(M; f_1, \cdots, f_k; x_1, \cdots, x_k)$$

where the right hand side is the space defined in section 1.)

Let $(\Gamma, (v_1, \cdots, v_n), \ell, [i]) \in \widetilde{\mathcal{T}}_{g,m,k}$. Let $\mathcal{V}$ be the set of all vertices of Γ. We consider the set of all maps $I : \Gamma - \{v_0\} \to M$ such that the following is satisfied for each edge e. We put $\partial e = \{v_1, v_2\}$ and the circle containing e is i-th and j-th one such that the orientation is as in Figure 3.

(4.1.1) $\quad \partial e \cap \{v_0\} = \emptyset$.

We assume $I|_e$ (together with its parameterization) is a gradient line of $f_j - f_i$. (We identify the edge e with $[0, \ell(e)]$.) (It can happen that $i = j$. But it does not matter.)

(4.1.2) $\quad v_1 = v_0$.

Then e is an i-th exterior edge for $i \in \{1, \cdots, k\}$. We identify $e - v_1$ by $(-\infty, 0]$. We assume $I|_{e-v_1}$ is a gradient line of $f_i - f_{i-1}$. Such that $\lim_{t \to -\infty} I(t) = x_i$. (We remark that x_i is a critical point of $f_i - f_{i-1}$.)

We define

$$\begin{aligned} \pi^{-1}((\Gamma, \ & (v_1, \cdots, v_n), \ell, [i])) \\ & \subset \mathcal{M}(M, g, m, k; f_1, \cdots, f_k, f_1', \cdots, f_m', x_1, \cdots, x_k), \end{aligned}$$

to be the set of all maps $I : \Gamma \to M$ satisfying the above condition for each edges.

Taking the union of all $\pi^{-1}((\Gamma, (v_1, \cdots, v_n), \ell, [i]))$ we obtain the moduli space

$$\mathcal{M}(M, g, m, k; f_1, \cdots, f_k, f_1', \cdots, f_m', x_1, \cdots, x_k)$$

and the map

$$\mathcal{M}(M, g, m, k; f_1, \cdots, f_k, f_1', \cdots, f_m', x_1, \cdots, x_k) \to \widetilde{\mathcal{T}}_{g,m,k}.$$

Theorem 4.1. *For generic choice of* $f_1, \cdots, f'_m$ *the moduli space*

$$\mathcal{M}(M, g, m, k; f_1, \cdots, f_k, f'_1, \cdots, f'_m, x_1, \cdots, x_k)$$

is a manifold of dimension

$$n - \sum \mu(x_i) + k - 3 + (m + 2g)(3 - n).$$

Proof. The fact that our moduli space is a smooth manifold is a consequence of standard transversality argument. To count its dimension, we first remark that we can forget ribbon structure to consider the dimension of the fibre. (The ribbon structure come to play when we patch the moduli spaces which belong to the graphs of different combinatorial types.) To count the dimension we need only the case when each interior vertex has three edges. In that case to add one edge decreases the dimension of the fibre by n.

We next count the dimension of the Teichmüller parameter. We consider the dimension of the moduli space of ribbon graphs of each combinatorial type. Among them the stratum of the largest dimension is one when each interior vertex has three edges. Since we fix combinatorial type we can forget again the ribbon structure. Then we can find easily from Euler's formula that the number of internal edges is $k - 3 + 3(m + 2g)$. The theorem follows. □

We can also consider the case when we take the length of exterior edges to be one. (Namely we assume (2.7) in place of (4.1.2).) (We do not need to take the critical point in this case by the same reason as in section 2.) By moving functions $f_1, \cdots, f'_m$, we obtain $\mathcal{M}(M, g, m, k)$ and a map $\pi : \mathcal{M}(M, g, m, k) \to (C^\infty(M))^{k+m}$. Also there is a map π_2 : $\mathcal{M}(M, g, m, k) \to \widetilde{\mathcal{T}}_{g,m,k}$ and a map $\pi : \mathcal{M}(M, g, m, k) \to (C^\infty(M))^{k+m} \times \widetilde{\mathcal{T}}_{g,m,k}$.

Theorem 4.2. $\pi : \mathcal{M}(M, g, m, k) \to (C^\infty(M))^{k+m}$ *is a Fredholm map of index* $n + k - 3 + (m + 2g)(3 - n)$.

We omit the proof. One may be able to find an "invariant" of manifold by counting the number of the moduli space when it is 0-dimensional. However one finds from the formula that the dimension decreases quite rapidly as m or g grows, unless $n = 3$. So one might worry that there is only a few invariant obtained in that way. Of course by taking k large there is always something. But the author do not know how much nontrivial it is.

In case dimension is 3 there is infinitely many invariant obtained in this way.

In higher dimensional case, there are two ways to get something of positive dimension. (We can do them at the same time also.) One is to use symmetry (extended mapping class group), as we will discuss soon. The other is to try to construct an invariant of M fibre bundle. Namely we consider the following situation. Let $p : E \to B$ be a smooth oriented fibre bundle with M as a fibre. We consider the mapping $F : E \to \mathbf{R}$ and regard it as a family of smooth function on M parameterized by B . By applying the construction fibrewise we obtain $\pi : \mathcal{M}(E,B,g,m,k) \to (C^\infty(E))^{k+m}$, $\pi_2 : \mathcal{M}(E,B,g,m,k) \to E^k$, $\pi_1 : \mathcal{M}(E,B,g,m,k) \to (C^\infty(E))^{k+m} \times \widetilde{\mathcal{T}}_{g,m,k}$. The map $\pi : \mathcal{M}(E,B,g,m,k) \to (C^\infty(E))^{k+m}$ is Fredholm map of index $n+k-3+(m+2g)(3-n)+\dim N$. In this way we can get extra dimension and may find a nontrivial invariant.

This construction and the critical dimension 3 is similar to Kontsevitch's construction in [14]. Namely in his case, one obtains an invariant in 3 dimensional case, and an invariant of fibre bundle in higher dimension. (However Kontsevitch used it in that case graph in place of ribbon graph.) Kontsevitch's construction is a generalization of one obtained for Chern-Simons perturbation theory by Axelrod-Singer [1], Kontsevitch, and is based on De-Rham theory. There might be a relation of it to this construction (together with mapping class group we will soon discuss). Witten [33] might be related to it.

We now consider the action of mapping class group. We first take a neighborhood U of p_0 which are disjoint to each other and take a differomorphism $\varphi : U \to D^2$. We put

$$K = \varphi^{-1}\left\{re^{2\pi\ell\sqrt{-1}/k} | \ell \in \mathbf{Z}, r \in \mathbf{R}_+\right\}.$$

For an element (Γ, i) of $\widetilde{\mathcal{T}}_{g,m,k}$ we may assume that $i(\Gamma)\cap U = K$. (Since we fixed the order of the edges around marked point the isotopy class satisfying, this condition is same as one without assuming it.)

$$\mathfrak{M}_{g,k} = \{\Psi : \Sigma_g \to \Sigma_g | \Psi(p_0) = p_0, \Psi(K_i) = K_i, \Psi \text{ is a diffeomorphism.}\}$$

This group acts on our space $\widetilde{\mathcal{T}}_{g,m,k}$. There is a group homomorphism

$$\sigma : \mathfrak{M}_{g,k} \to \mathbf{Z}_k$$

(Namely an element of $\mathfrak{M}_{g,k}$ acts as a cyclic permutation of exterior edges around the marked point.) The kernel of this homomorphism is isomorphic to the usual mapping class group $\mathfrak{M}_g$. (The group of isotopy classes of homeomorphisms of surface of genus g which fix the based point.)

One the other hand, the symmetric group $\mathfrak{S}_m$ of order $m!$ acts on $\widetilde{\mathcal{T}}_{g,m,k}$ by changing the order of interior circles.

We put

$$\mathcal{T}_{g,k,m} = \widetilde{\mathcal{T}}_{g,m,k}/(\mathfrak{M}_g \times \mathfrak{S}_m),$$
$$\overline{\mathcal{T}}_{g,k,m} = \widetilde{\mathcal{T}}_{g,m,k}/(\mathfrak{M}_{g,k} \times \mathfrak{S}_m),$$

It seems to the author that the fundamental group of $T_{g,k,m}$ is $\pi_0(Diff(\Sigma_g; p_1, \cdots, p_{n+1}))$, and this space is a $K(\pi, 1)$ space. Now we can divide the moduli space and maps in Theorem 4.2 and Theorem 4.3, by these groups. In the case of Theorem 4.3, we can simply divide everything by the group $\mathfrak{M}_{g,k} \times \mathfrak{S}_m$ and obtain the spaces $\overline{\mathcal{M}}(M, g, m, k) = \mathcal{M}(M, g, m, k)/(\mathfrak{M}_{g,k} \times \mathfrak{S}_m)$, and maps

$$\pi_1 : \overline{\mathcal{M}}(M, g, m, k) \to ((C^\infty(M))^{k+m} \times M^k)/(\mathbf{Z}_k \times \mathfrak{S}_m)$$
$$\pi_2 : \overline{\mathcal{M}}(M, g, m, k) \to \overline{\mathcal{T}}_{g,k,m}.$$

In the case of Theorem 4.2, situation is a bit more complicated. First we fix $f_1, \cdots, f_k$, functions corresponding to exterior circles, and $x_1, \cdots, x_k$, the critical points. Then we take the union of moduli spaces

$$\mathcal{M}(M, g, m, k; f_1, \cdots, f_k, f_1', \cdots, f_m', x_1, \cdots, x_k)$$

moving $f_1', \cdots, f_m'$. Let

$$\mathcal{M}(M, g, m, k; f_1, \cdots, f_k, x_1, \cdots, x_k)$$

be the space we obtained. We can divide this space by $\mathfrak{M}_g \times \mathfrak{S}_m$, to obtain

$$\overline{\mathcal{M}}(M, g, m, k; f_1, \cdots, f_k, x_1, \cdots, x_k).$$

There are maps

$$\pi_1 : \overline{\mathcal{M}}(M, g, m, k; f_1, \cdots, f_k, x_1, \cdots, x_k) \to (C^\infty(M))^m/\mathfrak{S}_m,$$
$$\pi_2 : \overline{\mathcal{M}}(M, g, m, k) \to \mathcal{T}_{g,k,m}.$$

One can use these spaces and maps (and the homology classes of $\overline{\mathcal{T}}_{g,k,m}$ and $\mathcal{T}_{g,k,m}$) to find an "invariant" of M, as we did in sections 1,2. We remark that $K(\mathbf{Z}_k \times \mathfrak{S}_m, 1)$ has nonzero torsion homology class in arbitrary high degree but has finite $\mathbf{Q}$-homology dimension. As we mentioned in introduction, it is obscure for the author what we get in that way, since we do not have algebraic machinery to describe it. We do not have an explicit calculation of these invariants yet so there is still a gloomy possibility that everything is trivial, though it is quite unlikely.

5. Preliminary ideas for quantization of higher genus Morse homotopy

Now we try to "quantize" the construction of section 4. Let $g \geq 0$,$k \geq 2$,$m > 0$ be integers. We consider a compact Riemann surface Σ of genus g and $m+1$ boundaries. Let $(p_1, \cdots, p_k)$ be points in first component of the boundary. We assume that the cyclic order $(p_1, \cdots, p_k)$ is respected in $S^1 \subset \partial\Sigma$. Let $\widetilde{T}'_{g,m,k}$ be the set of all such objects modulo biholomorphic isomorphisms isotopic to identity. In a way similar to [18], [28], [11], [22], we can prove that $\widetilde{T}'_{g,m,k}$ is diffeomorphic to $\widetilde{T}_{g,m,k}$. Where the later is defined in section 4.

Let X be a symplectic manifold with trivial first Chern class and $\Lambda_1, \cdots, \Lambda_k$,$\Lambda'_1, \cdots, \Lambda'_m$ be BS-orbits in it. Let $x_i \in \Lambda_i \cap \Lambda_{i-1}$. We define the space

$$\mathcal{M}_{g,m,k}(X; \Lambda_1, \cdots, \Lambda_k, \Lambda'_1, \cdots, \Lambda'_m; x_1, \cdots, x_k).$$

Let $(\Sigma, p_1, \cdots, p_m) \in \widetilde{T}'_{g,m,k}$. We consider the map $h : \Sigma \to X$ such that

(5.1.1) h is pseudo holomorphic.
(5.1.2) $h(p_i) = x_i$.
(5.1.3) $h(\partial_i\Sigma) \subset \Lambda'_i$, for $i \geq 2$.
(5.1.4) $h(\overline{x_i x_{i-1}}) \subset \Lambda_i$.

Here $\partial_i\Sigma$ denote the i-th component of $\partial\Sigma$. Let $\mathcal{M}_{g,m,k}(X; \Lambda_1, \cdots, \Lambda_k, \Lambda'_1, \cdots, \Lambda'_m; x_1, \cdots, x_k)$ be the set of all such objects. We can split it as in section 2 using symplectic area a and Maslov index ℓ . Then a similar dimension formula can be proved. Namely

$$\begin{aligned} &dim\mathcal{M}_{g,m,k;a,\ell}(X; \Lambda_1, \cdots, \Lambda_k, \Lambda'_1, \cdots, \Lambda'_m; x_1, \cdots, x_k) \\ &\qquad = n - \sum \mu(x_i) + (k-3) + (m+2g)(3-n) + 2\ell. \end{aligned}$$

(Where $2n = dim_{\mathbf{R}}X$.)

Conjecture 5.1.

$$\begin{aligned} &\lim_{\epsilon \to 0} \mathcal{M}^{\epsilon}_{g,m,k;a=0}(X;\Lambda_{f_1}, \cdots, \Lambda_{f_k}, \Lambda_{f'_k}, \cdots, \Lambda_{f'_m}; x_1, \cdots, x_k) \\ &\qquad = \mathcal{M}(M, g, k, m; f_1, \cdots, f_k, f'_1, \cdots, f'_m; x_1, \cdots, x_k). \end{aligned}$$

Here Λ_{f_i} is the graph of ϵdf_i. And in the left hand side we can omit ℓ by the same reason as section 3. The right hand side is the moduli space we introduced in section 4.

We remark that the compactification of $\mathcal{M}_{g,m,k}(X; \Lambda_{f_1}, \cdots, \Lambda_{f_k}, \Lambda_{f'_k}, \cdots, \Lambda_{f'_m}; x_1, \cdots, x_k)$ has various troubles.

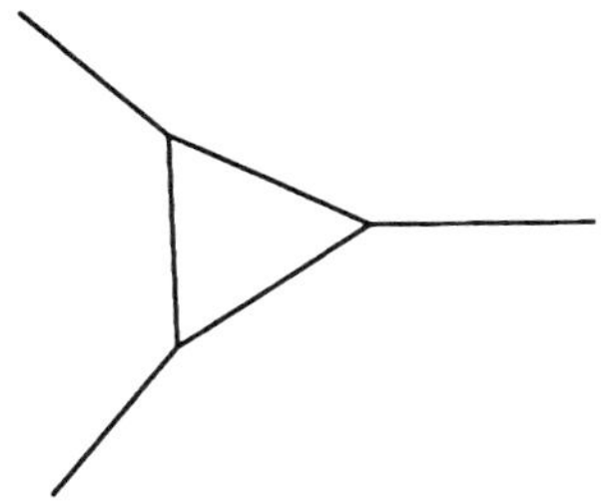

FIGURE 9

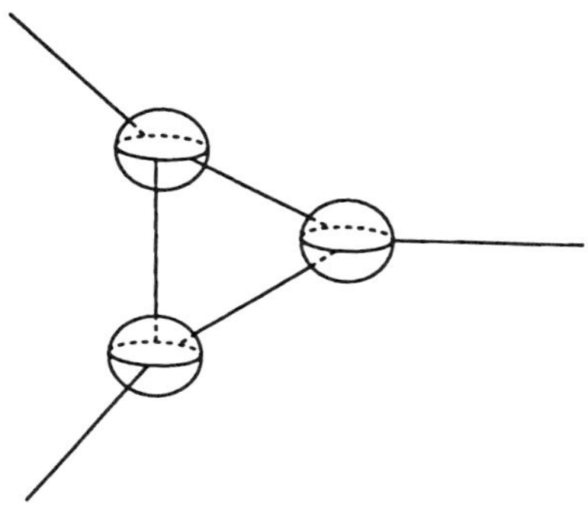

FIGURE 10

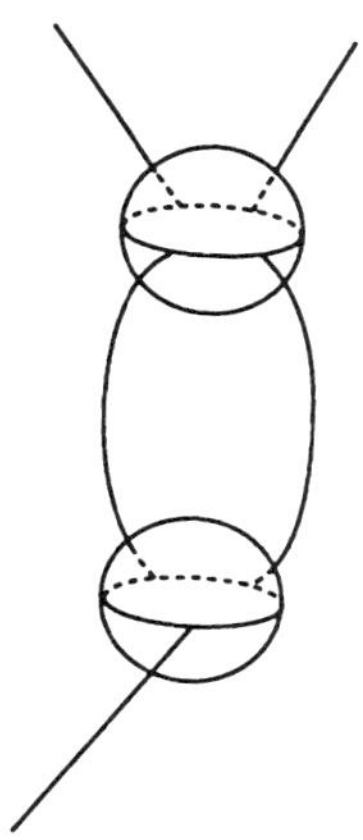

FIGURE 11

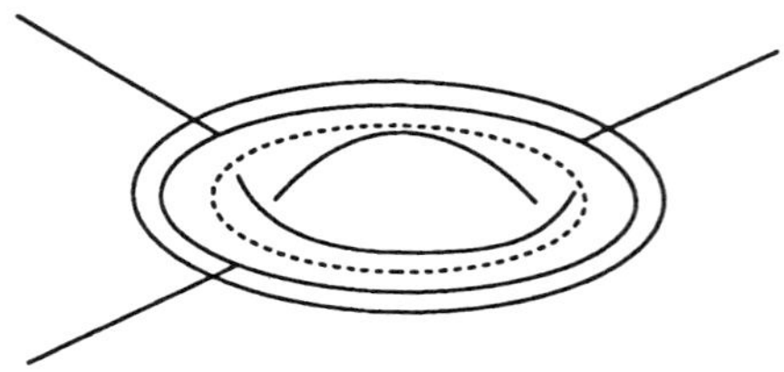

FIGURE 12

Finally we remark a bit the quantization similar to section 2. Let us consider the ribbon graph as in Figure 9. Naturally to quantize it in a way similar to section 2, one needs to consider the configuration in Figure 10. Deforming it continuously we get the space in Figure 11, and finally we get one in Figure 12. Thus we need to study pseudo holomorphic map from Riemann surface of higher genus.

In this say we can guess what is the limit

$$\lim_{\epsilon\to 0}\mathcal{M}^{\epsilon}_{g,m,k}(X;\Lambda_{f_1},\cdots,\Lambda_{f_k};\Lambda_{f'_1},\cdots,\Lambda_{f'_m};x_1,\cdots,x_k).$$

There are a lot of difficulties to prove it.

As we remarked at the end of section 3, Riemann surface we can use in this way is restricted. Namely it should have an anti holomorphic involution.

REFERENCES

1. S. Axelrod and I.Singer, *Chern Simons perturbation theory*, I. Proc. XXth DGM Conf. (S Catto, A. Rocha ed.) World Scientific, Singapore, II. J. Differential Geometry **39** (1994) 173–213.
2. M. Betz and R. Cohen, *Graph Moduli space and Cohomology operations*, Preprint.
3. P. Candelas, P. Green, L. Parke and dela Ossa, *A pair of Calabi Yau manifolds as an exactly soluble supper conformal field theory*, Nucl. Phys., **B 359** (1991) 21.
4. C. Conley and E. Zender, *The Birkoff-Lews fixed point theorem and a conjecture of V.I. Arnold*, Invent. Math. **73** (1983) 33–49.
5. A. Floer, *Symplectic fixed point and holomorphic spheres*, Commun. Math. Phys., **120** (1989) 575–611.
6. K. Fukaya, *Floer homology for connected sum of homology 3-spheres*, to appear in Topology.
7. _____, *Floer homology for 3 manifolds with boundary (abstract)*, in "Topology Geometry and field theory" ed by K. Fukaya, M. Furata, T. Kohno, D. Kotchik, World Scientific, Singapore.

8. _____, *Morse homotopy, A^∞-category, and Floer homologies*, Proc. Garc. Workshop on Geometry and Topology'93, ed. H. J. Kim, Seoul National University.
9. _____, *Floer homology for 3 manifolds with boundary*, in preparation.
10. M. Gromov, *Pseudo holomorhpic curves in symplectic manifolds*, Invent. Math., **82** (1985) 307–347.
11. H. Harer, *The cohomology of the moduli space of curves*, Lecture note in Math., Springer, **1337** 138–221.
12. H. Hofer and D. Salamon, *Floer homology and Novikov ring*, Preprint.
13. M. Kontsevitch, *Intersection theory on the Moduli space of curves and the Matrix Airy Functions* , Commun. Math. Phys., **147** (1992) 1–23.
14. _____, *Feynman diagram and low dimensional topology*, First European Congress of Mathematics, Vol II, Birkhauser, 97–121.
15. _____, *A^∞ algebras in Mirror symmetry*, Preprint.
16. V. Lê and K. Ono, *Symplectic fixed points, Calabi invariant, and Novikov ring*, Preprint.
17. D. MacDuff, *Example of symplectic structures*, Invent. Math., **89** (1987) 13–36.
18. D. Mumford, *Toward an enumerative geometry of moduli spaces of curves*, in book Arithmetric and Geometry, Birkhauser, info ed. S.T. M.Artin and J.Tate, 271 – 326, 1983.
19. S. Novikov, *Multivalued functions and functionals - an analogue of Morse theory*, Soviet Math. Dokl., **24** (1981) 222–225.
20. Y. Oh, *Floer cohomology of Lagrangian intersections and Pseudo-holomorphic disks*, Comm. Pure Appl. Math. **46** (1993) 949–994.
21. K. Ono, *The Arnold conjecture for weakly monotone symplectic manifolds*, Invent. Math. **119** (1995) 519–537.
22. R. Penner, *The decoretad Teichmüller space of punctured surface*, Comm. Math. Phys., **113** (1987) 299–339.
23. Y. Ruan, *Topological Sigma model and Donaldson type invariant in Gromov Theory*, Preprint.
24. T. Sikorov, *Homologie de Novikov associe e a une classe de cohomologie reelle de degree un*, Thesis Univ. Orsay.
25. V. Smirnov, *Lecture at Geogia international conference.*
26. J. Stasheff, *Homotopy comutative H-spaces I*, Trans. Amer. Math. Soc., **108** (1963) 295–292.
27. N. Steenrod and D. Epstein, *Cohomology operations*, Princeton Univ. Press.
28. K. Strebel, *Quadratic differentials*, Springer, 1985, Berlin.
29. D. Sullivan, *Infinitesimal calculations in Topology*, Publ. IHES, **74** (1978) 269 – 361.
30. E. Witten, *Super symmetry and Morse theory*, J. Differential Geom., **17** (1982) 661–692.
31. _____, *Topological sigma model*, Commun. Math. Phys., **118** (1988) 441.
32. _____, *Mirror manifolds and topological field theory*, in "Essays on Mirror

manifolds", International Press, Hong-Kong, info ed. S.T. Yau, 120 – 156.
33. _____, *Chern Simons Gauge theory as a string theory*, Preprint.

DEPARTMENT OF MATHEMATICAL SCIENCES, UNIVERISTY OF TOKYO, HONGO, BUNKYO-KU TOKYO, JAPAN
E-mail address: fukaya@kusm.kyoto-u.ac.jp

AMS/IP Studies in Advanced Mathematics
Volume **2**, 1997 (Part 1)

CAT(0) REFLECTION MANIFOLDS

F. D. ANCEL, M. W. DAVIS AND C. R. GUILBAULT

Let M^n be a compact, contractible manifold. Its boundary Σ^{n-1} is a homology sphere. We assume that Σ^{n-1} admits a PL triangulation. (This is automatic for $n \neq 5$.)

In [7], Davis showed how to construct an action of a reflection group on an open contractible manifold with fundamental chamber M. In [10], Gromov showed that a modified version of this construction could be given a piecewise Euclidean, CAT(0) metric. (Roughly, a "CAT(0) metric" is the generalization to singular metric spaces of the notion of a complete Riemannian metric of nonpositive sectional curvature on a simply connected manifold. The precise definition is given in [10, §2.4.C].) In Gromov's version, M is replaced by $C\Sigma$, the cone on Σ, and the CAT(0) space on which the group acts is generally, no longer a manifold, but only a polyhedral homology manifold. In [1], Ancel and Guilbault showed that any contractible M^n, with $n \geq 5$, can be written as the union of two cones along the cone on a homology $(n-2)$-sphere. This allows us to use Gromov's idea to put CAT(0) metrics on many of the original examples of [7]. In particular, we get the following result.

Theorem. *Let M^n be a compact, contractible manifold, with $n \geq 5$ and with boundary a PL homology sphere. Then there is an open contractible n-manifold $\mathcal{X}$ with a piecewise Euclidean CAT(0) metric and an isometric action of a reflection group W on $\mathcal{X}$ with fundamental chamber M^n.*

Basically, the proof consists of recalling the constructions of [7], [10] and [1]

The construction of [7] Triangulate Σ and denote the resulting simplicial complex again by Σ. Let S be the set of vertices in Σ and let

Partially supported by NSF Grant DMS 9208071

W be the right-angled Coxeter group generated by S with relations:

$$\begin{aligned} s^2 &= 1\,, &&\text{for all } s \in S \\ (ss')^2 &= 1\,, &&\text{if } s \text{ and } s' \text{ span an edge of } \Sigma\,. \end{aligned}$$

We identify S with its image in W. "Right-angled" refers to the fact that if s and s' are distinct elements of S, then the order of ss' in W is either 2 or ∞. For each s in S, let Σ_s denote the closed star of s in the barycentric subdivision of Σ. For each $x \in M$, let W_x denote the subgroup of W generated by $\{s \in S \mid x \in \Sigma_s\}$. (If this set is empty, then $W_{\mathrm{x}} = \{1\}$.) Set

$$\mathcal{X}(W, M) = (W \times M)/\sim$$

where the equivalence relation $\sim$ is defined by $(w, x) \sim (w', x')$ if and only if $x = x'$ *and* $w^{-1}w' \in W_x$. Then W acts naturally on $\mathcal{X}(W, M)$. It is proved in [7] that $\mathcal{X}(W, M)$ is a manifold (since M is a manifold with boundary).

More generally, if X is any space and $\{X_s\}_{s\in S}$ is a family of closed subspaces, then we can define a W-space $\mathcal{X}(W, X)$ in exactly the same way. By a *reflection group* we mean an action of a Coxeter group on a space which is equivariantly homeomorphic to some $\mathcal{X}(W, X)$. The space X is called the *fundamental chamber*.

In particular, the construction $\mathcal{X}(W, M)$ could be modified by replacing M by $C\Sigma$. The resulting space $\mathcal{X}(W, C\Sigma)$ is a polyhedral homology manifold, which is W-equivariantly homotopy equivalent to $\mathcal{X}(W, M)$. However, $\mathcal{X}(W, C\Sigma)$ is generally not a topological manifold since there will be singularities at the cone point and its W-translates whenever Σ is not simply connected (and of dimension greater than 1).

A simplicial complex is a *flag complex* if any finite set of vertices, which are pairwise joined by edges, span a simplex.

It is proved in [7] that $\mathcal{X}(W, M)$ is contractible if and only if Σ is a flag complex. (This condition is easy to achieve, for example, the barycentric subdivision of any cell structure on Σ is a flag complex.)

The construction of [10] The cone on the barycentric subdivision of a k-simplex can be identified with a standard simplicial subdivision of a $(k+1)$-cube in a natural way. This gives an identification of the cone on a k-simplex with a $(k+1)$-cube, well-defined up to symmetries. The picture for $k = 2$ is given below.

Gromov puts a piecewise Euclidean structure on $C\Sigma$ by identifying the cone on each simplex of Σ with a regular Euclidean cube of edge length 1. Each translate of $C\Sigma$ in $\mathcal{X}(W, C\Sigma)$ is given a cubical structure

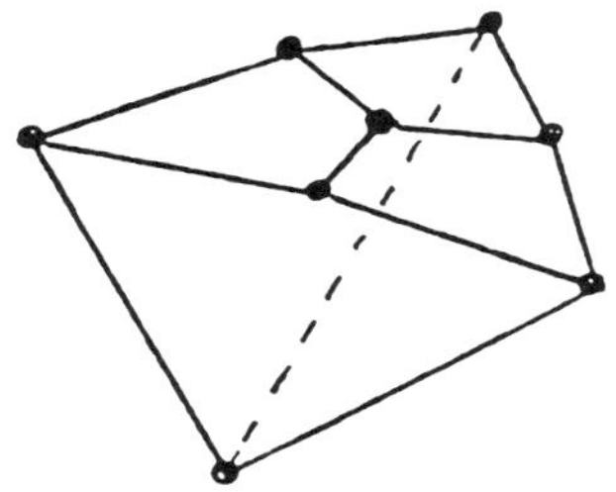

FIGURE 1

in a similar fashion. Further details on this cubification can be found in [5, §6].

The distance between two points in $\mathcal{X}(W, C\Sigma)$ is then defined to be the length of the shortest path between them. This metric gives $\mathcal{X}(W, C\Sigma)$ the structure of a "geodesic space" (also called a "length space"). Since $\mathcal{X}(W, C\Sigma)$ is simply connected, to prove that this metric is CAT(0), it sufficies to show that it satisfies CAT(0) locally (i.e., that it is nonpositively curved), cf., [10, p. 119] and [3, p. 195]. To prove this it is necessary and sufficient to show that the link of each cubical face of $C\Sigma$ in $\mathcal{X}(W, C\Sigma)$ is a flag complex, cf., [10, pp. 120-122] and [4, Lemma 1.3]. But this is true if and only if Σ is a flag complex. (Proof: The link of the cone point is Σ. The link of any other cubical face in the interior of $C\Sigma$ can be identified with the link of a simplex in Σ and this is a flag complex if Σ is. Finally, the link of a cubical face on the boundary of $C\Sigma$ can be identified with an iterated suspension of a link of a simplex in Σ and the suspension of a flag complex is a flag complex.) In the case when the Coxeter group W is not right-angled there is a more refined version of this construction, due to Moussong, [11].

The construction of [1]. Let $\Sigma_0^{n-2} \subset \Sigma^{n-1}$ be a PL-embedded homology sphere of codimension one. Then Σ_0 divides Σ into two homology $(n-1)$-cells, call them N_1 and N_2. It is proved in [1, Lemma 1] and [6, Proposition 2] that, for $n \geq 5$, one can always find a Σ_0 so that the induced homomorphism $\pi_1(\Sigma_0) \to \pi_1(\Sigma)$ is surjective. It follows from van Kampen's Theorem that, for $i = 1, 2$, $\pi_1(N_i)$ is normally generated by the image of $\pi_1(\Sigma_0)$. This result is used in [1] to prove that M can be written as a mapping cylinder of some map from Σ to an arc and in [2] to prove that the interior of M can be given a complete

CAT(ϵ) metric, for any $\epsilon \leq 0$.

Proof of the theorem. As above, let Σ_0 be a PL homology sphere of codimension one in Σ so that, for $i = 1, 2$, $\pi_1(N_i)$ is normally generated by the image of $\pi_1(\Sigma_0)$. Choose a PL triangulation of Σ as a flag complex so that Σ_0 is a full subcomplex. For $i = 1, 2$, set

$$\Sigma_i = N_i \underset{\Sigma_0}{\cup} C\Sigma_0$$

where, as before, CY denotes the cone on Y. By van Kampen's Theorem, Σ_i is simply connected. Let F denote the closed star of the cone point in the barycentric subdivision of $C\Sigma_0$. Glue $C\Sigma_1$ to $C\Sigma_2$ along F and call the result X. Then X is clearly a contractible, polyhedral homology n-manifold with boundary. Since the complement of the interior of F in $C\Sigma_0$ is a collared neighborhood of Σ_0, we have $\partial X = N_1 \cup (\Sigma_0 \times I) \cup N_2$, which is PL-homeomorphic to Σ. For $i = 0, 1, 2$, let v_i denote the cone point of $C\Sigma_i$ and let e denote the union of the edge from v_1 to v_0 and the edge from v_0 to v_2. The triangulation of X (as a union of two cones) has PL singularities only along e. According to a well-known theorem of Edwards [9], a polyhedral homology manifold of dimension $n \geq 5$ (with boundary a manifold) is a topological manifold if and only if the link of each vertex in its interior is simply connected. This holds in our case. (The link of v_0 is the suspension of Σ_0; for $i = 1, 2$, the link of v_i is Σ_i.) Hence, X is a contractible n-manifold. By the h-cobordism Theorem, it is homeomorphic rel boundary to M.

Now apply Gromov's cubification to $C\Sigma_1$ and $C\Sigma_2$, separately. This defines a cubical structure on their union, X. For $i = 0, 1, 2$, let S_i be the set of vertices in Σ_i so that $S_0 = S_1 \cap S_2$ and $S = S_1 \cup S_2$. For $s \in S_i - S_0$, set

$$X_s = (\Sigma_i)_s,$$

the closed star of s in the barycentric subdivision of Σ_i. For $s \in S_0$, set

$$X_s = (\Sigma_1)_s \cup (\Sigma_2)_s,$$

In both cases, X_s is an $(n-1)$-cell in ∂X and a union of cubical faces. Let W be the right-angled Coxeter group generated by S defined as before. Put

$$\mathcal{X} = \mathcal{X}(W, X)$$

It has a cubical structure induced from that of X. One checks, just as in Gromov's construction, that the link of each cubical face is a flag complex. Hence, we see that $\mathcal{X}$ is CAT(0) and the proof is complete. □

Remarks. 1) If Σ^{n-1} is not simply connected, $n > 2$, then the CAT(0) manifold $\mathcal{X}$ is not simply connected at infinity and hence, not homeomorphic to $\mathbb{R}^n$.

2) A very similar construction using reflection groups was mentioned in Remark 5b.2 of [8, p. 384]. The idea there was to use only one of the cones, say $C\Sigma_1$, as the fundamental chamber. Let W' be the right-angled Coxeter group generated by the vertices of Σ_1. It follows, as above, that $\mathcal{X}(W', C\Sigma_1)$ is a CAT(0) manifold. Moreover, if the double of N_1 (along Σ_0) is not simply connected, then $\mathcal{X}(W', C\Sigma_1)$ is not simply connected at infinity. The point of our theorem is that *any* compact, contractible manifold (with PL boundary) can occur as a fundamental chamber.

References

1. F. D. Ancel and C. R. Guilbault, *Compact contractible n-manifolds have arc spines* ($n \geq 5$), Pacific J. Math., to appear.
2. ———, *Interiors of compact contractible n-manifolds are hyperbolic* ($n \geq 5$), to appear.
3. W. Ballman, *Singular spaces of non-positive curvature*, in "Sur les Groupes Hyperboliques d'apres Mikhael Gromov," edited by E. Ghys and P. de la Harpe, Progress in Math. 83, Birkhaüser, Boston, Basel, Berlin, 1990, 189-202.
4. R. Charney and M. Davis, *Singular metrics of nonpositive curvature on branched covers of Riemannian manifolds*, Amer. J. Math.
5. ———, *The Euler characteristic of a nonpositively curved, piecewise Euclidean manifold*, Ohio State Univ. preprint, 1993.
6. R. J. Daverman and F. C. Tinsley, *Laminations, finitely generated perfect groups, and acyclic maps*, Michigan Math. J. **33** (1986), 343-351.
7. M. Davis, *Groups generated by reflections and aspherical manifolds not covered by Euclidean space*, Ann. of Math. (2)**117** (1983) 293-325.
8. M. Davis and T. Januszkiewicz, *Hyperbolization of polyhedra*, J. Differential Geometry **34** (1991) 347-388.
9. R. D. Edwards, *The topology of manifolds and cell-like maps*, Proc. ICM Helsinki, 1978, 111-127.
10. M. Gromov, *Hyperbolic groups* in "Essays in Group Theory" (S. M. Gersten, ed.), M.S.R.I. Publ. 8, Springer-Verlag, New York and Berlin, 1987, 75-264.
11. G. Moussong, *Hyperbolic Coxeter groups*, Ph.D. thesis, Ohio State Univ., 1988.

AMS/IP Studies in Advanced Mathematics
Volume **2**, 1997 (Part 1)

ON ORIENTED BUNDLE THEORIES AND APPLICATIONS

DUANE RANDALL

ABSTRACT. The structural groups for the classical oriented CAT and block bundle theories with fiber $\mathbb{R}^n$ have distinct homotopy types for all $n \geq 5$, except possibly for SO(5) and SPL(I^5). Those groups which can be finite H-spaces are determined from low-dimensional homotopy and H-space theorems of Lin. Normal bundles to codimension 4 embeddings of homotopy spheres are determined. Relevance to the smooth 4-dimensional Smale Conjecture is explained.

1. STRUCTURAL GROUPS.

We recall briefly the definitions from [11] or [21] of the structural groups for the classical oriented bundle theories which we consider in this paper. Let CAT = PL or TOP. SCAT(n) denotes the semi-simplicial group whose i-simplices consist of all fiber-preserving and orientation-preserving CAT homeomorphisms $h : \Delta^i \times R^n \to \Delta^i \times R^n$ such that the restriction of h to $\Delta^i \times \{0\}$ is the identity map. Replacing R^n by I^n yields SCAT(I^n). The i-simplices of the block bundle structural groups $\widetilde{\mathrm{SCAT}}(n)$ omit the fiber-preserving property for orientation-preserving CAT homeomorphisms h of $\Delta^i \times R^n$ which are the identity map on $\Delta^i \times \{0\}$. Let SG(n) denote the H-space of degree one homotopy equivalences of S^{n-1} while SF($n-1$) denotes the subset preserving a fixed basepoint. Now SCAT(I^{n+1}) can be regarded, up to homotopy, as the group of orientation-preserving CAT homeomorphisms of the standard CAT n-sphere Σ^n via the Alexander construction. The CAT concordance group $C^{\mathrm{CAT}}(S^m)$ is the semi-simplicial space of CAT homeomorphisms of $S^m \times I$ restricting to the identity map on $S^m \times \{0\}$. All groups, including quotient groups, should be understood in the semi-simplicial category. The fundamental diagram (1.1) displays these structural groups.

The semi-simplicial group complexes in (1.1) have the same homotopy type for dimension $n = 1$ and also for $n = 2$ by [24, p.121]. Hatcher

[7] proved that the inclusions $\mathrm{SO}(3) \to \mathrm{SCAT}(I^3) \to \mathrm{SCAT}(3)$ and also $\mathrm{SO}(4) \to \mathrm{SCAT}(I^4)$ are homotopy equivalences for CAT $=$ PL or TOP. However, $\widetilde{\mathrm{SPL}}(3)$, $\widetilde{\mathrm{STOP}}(3)$ and $\mathrm{SG}(3)$ have distinct homotopy types. The open problems concerning *distinctness* of homotopy types and the existence of *finite* H-space structures on the structural groups in (1.1) occur only in dimensions 4 through 6 by the results of this paper. Before discussing the existence of finite H-space structures, we first give elementary arguments for some known properties of the groups in (1.1). The first property states that PL bundle theory differs from vector bundle theory in every fiber dimension ≥ 5.

$$
\begin{array}{ccccc}
 & & \mathrm{SO}(n) & & \\
 & \swarrow & & \searrow & \\
\mathrm{SPL}(I^n) & & \longrightarrow & & \mathrm{STOP}(I^n) \\
\downarrow & & & & \downarrow \\
\mathrm{SPL}(n) & & \longrightarrow & & \mathrm{STOP}(n) \\
\downarrow & & & & \downarrow \\
\widetilde{\mathrm{SPL}}(n) & & \longrightarrow & & \widetilde{\mathrm{STOP}}(n) \\
 & \searrow & & \swarrow & \\
 & & \mathrm{SG}(n) & &
\end{array}
\tag{1.1}
$$

Proposition 1.2. *The inclusion* $i : \mathrm{SO}(m) \to \mathrm{SPL}(m)$ *is not a homotopy equivalence for each* $m \geq 5$.

Proof. The following commutative diagram consists of homotopy exact sequences for the pairs $(\widetilde{\mathrm{SPL}}(m), \mathrm{SO}(m))$ and $(\widetilde{\mathrm{SPL}}(m), \mathrm{SPL}(m))$ where i, j and ℓ denote the inclusion morphisms.

The natural map induces an *isomorphism* from $\pi_7(\widetilde{\mathrm{SPL}}(m)/\mathrm{SO}(m))$ to $\pi_7(\mathrm{SPL}/\mathrm{SO}) \approx Z/28$ for $m \geq 6$ and an *epimorphism* for $m = 5$ by [4] while $\pi_6(\mathrm{SO}(m)) \approx 0$ for all $m \geq 5$. Consequently, j_* cannot be surjective. However, $\pi_7(\widetilde{\mathrm{SPL}}(m)/\mathrm{SPL}(m))$ is trivial for all $m \geq 5$ by Proposition 5.6 of [4]. Thus ℓ_* is surjective so i_* cannot be surjective for each $m \geq 5$. □

Whether the inclusion $i : \mathrm{SO}(4) \to \mathrm{SPL}(4)$ is a homotopy equivalence is an important open problem. Let $\mathrm{SDiff}(S^4)$ denote the group under composition of all orientation-preserving diffeomorphisms of the standard 4-sphere endowed with the C^∞ topology. By [1] and [11] we know that $\mathrm{SDiff}(S^4)$ has the homotopy type of $\mathrm{SO}(5) \times \Omega^5(\mathrm{SPL}(4)/\mathrm{SO}(4))$. Since $\mathrm{SPL}(4)/\mathrm{SO}(4)$ is at least 4-connected by [4], $\Omega^5(\mathrm{SPL}(4)/\mathrm{SO}(4))$ is contractible if and only if $\mathrm{SPL}(4)/\mathrm{SO}(4)$ is contractible. So i :

$\mathrm{SO}(4) \to \mathrm{SPL}(4)$ is a homotopy equivalence if and only if the smooth 4-dimensional Smale Conjecture is true, that is, the inclusion $\mathrm{SO}(5) \subset \mathrm{SDiff}(S^4)$ is a homotopy equivalence. Various equivalences are given in [17, Theorem 3.5]. The profound result that $i : \mathrm{SO}(3) \to \mathrm{SPL}(3)$ is a homotopy equivalence is again due to Hatcher [7]. A direct proof that $i : \mathrm{SO}(2) \to \mathrm{STOP}(2)$ is a homotopy equivalence is given in [11, p. 254].

$$
\begin{array}{ccc}
0 & & \\
\downarrow & & \\
\pi_7(\mathrm{SO}(m)) & \xrightarrow{i_*} & \pi_7(\mathrm{SPL}(m)) \\
\downarrow j_* & & \downarrow \ell_* \\
\pi_7(\widetilde{\mathrm{SPL}}(m)) & = & \pi_7(\widetilde{\mathrm{SPL}}(m)) \\
\downarrow & & \downarrow \\
\pi_7(\widetilde{\mathrm{SPL}}(m)/\,\mathrm{SO}(m)) & \longrightarrow & \pi_7(\widetilde{\mathrm{SPL}}(m)/\,\mathrm{SPL}(m)) \\
\downarrow & & \downarrow \\
0 & &
\end{array}
\qquad (1.3)
$$

Proposition 1.4. *The oriented structural groups in diagram (1.1) have distinct homotopy types for all dimensions $n \geq 5$, except possibly for* $\mathrm{SO}(5)$ *and* $\mathrm{SPL}(I^5)$.

Sketch of proof. Now $\pi_*(\mathrm{SPL}(I^5), \mathrm{SO}(5)) \approx \pi_*(\mathrm{SPL}(4), \mathrm{SO}(4))$ by [21] so triviality of the relative homotopy is equivalent to the truth of the smooth 4-dimensional Smale Conjecture. We claim that distinctness of the remaining homotopy types for all dimensions $n \geq 5$ can be obtained from the following fibrations

$$
(1.5)\qquad
\begin{array}{lr}
\mathrm{SO}(n) \to \mathrm{SO}(n+1) \to S^n & \\
\mathrm{SCAT}(n) \to \mathrm{SCAT}(I^{n+1}) \to \Sigma^n & [11] \text{ or } [21] \\
\mathrm{SPL}(n) \to \mathrm{STOP}(n) \to K(Z/2,3) & [11] \\
\widetilde{\mathrm{SPL}}(n) \to \widetilde{\mathrm{STOP}}(n) \to K(Z/2,3) & [19] \\
\mathrm{SF}(n-1) \to \mathrm{SG}(n) \to S^{n-1} & [22] \\
\widetilde{\mathrm{SCAT}}(n) \to \mathrm{SG}(n) \to \mathrm{SG}/\mathrm{SCAT} & [18] \text{ or } [24] \\
C^{\mathrm{CAT}}(S^{n-1}) \to \mathrm{SCAT}(I^n) \to \mathrm{SCAT}(n) & [12] \text{ or } [5] \\
C^{\mathrm{CAT}}(S^n) \to \mathrm{SCAT}(n)/\mathrm{SO}(n) & \\
\qquad \to \mathrm{SCAT}(n+1)/\mathrm{SO}(n+1) & [4],[5] \text{ or } [17].
\end{array}
$$

The fiber inclusion i in the fibration $\mathrm{SG}/\mathrm{SCAT} \xrightarrow{i} \mathrm{BSCAT} \to \mathrm{BSG}$ is clearly a rational homotopy equivalence since the stable homotopy of spheres in positive stems is torsion. Moreover, $\mathrm{SG}/\mathrm{SCAT}$ is the homotopy fiber of the fibration $\widetilde{\mathrm{BSCAT}}(n) \to \mathrm{BSG}(n)$ for all $n \geq 3$ by [18] or [24]. Consequently, all the rational Pontrjagin classes live in $H^*(\widetilde{\mathrm{BSCAT}}(n); Q)$ for every $n \geq 3$, a result of Maunder. So the rational homotopy of the pair $(\widetilde{\mathrm{SCAT}}(n), \mathrm{SCAT}(n))$ is nontrivial for every $n \geq 3$. We remark that $C^{\mathrm{CAT}}(S^n)$ has nontrivial homotopy for every $n \geq 4$. Indeed, $\pi_{n+2}(C^{\mathrm{CAT}}(S^n))$ contains $Z/2$ for every $n \geq 7$ by [8] while $\pi_7(C^{\mathrm{CAT}}(S^5))$ contains Z by [17, Theorem 5.1]. Nontriviality of the homotopy for the fibrations in (1.5) together with Proposition 1.2 and the above observations yield distinctness of homotopy types for the structural groups in (1.1) with the possible exception of SO(5) and $\mathrm{SPL}(I^5)$. □

2. Finite H-Spaces.

A finite H-space is an H-space having the homotopy type of a finite CW complex. Which structural groups in (1.1) are finite H-spaces? This question is partially motivated by the work of Antonelli, Burghelea and Kahn. Theorem A of [1] states that the homotopy commutative H-spaces $\Omega^{n+1}(\mathrm{SPL}(n)/\mathrm{SO}(n))$ are neither finite H-spaces nor even dominated by a finite CW complex for every $n \geq 7$, since the fundamental groups contain non-zero torsion elements. We first show that the structural groups in (1.1) are *finitely dominated* if and only if they are *finite* H-spaces. Then we apply H-space theory, especially theorems of Lin in [13] and [14], to deduce that only $\mathrm{SPL}(n)$ and $\mathrm{SPL}(I^{n+1})$ for $n = 4, 5$ could be *finite* H-spaces in addition to the already known finite H-spaces. We recall that the smooth 4-dimensional Smale Conjecture implies that SPL(4) and SPL(I^5) are homotopically equivalent to SO(4) and SO(5), respectively.

Theorem 2.1. *Let Y denote any structural group in diagram (1.1) with dimension $n \geq 3$. Then Y is a finite H-space if and only if $H_*(Y; Z)$ is finitely generated as a graded abelian group.*

Proof. This result follows from an observation of [17]. The nilpotent CW complex Y is *finitely dominated* by [16] if and only if $H_*(Y; Z)$ is finitely generated as a graded group. Now $\pi_1(Y) \approx Z/2$ for all dimensions $n \geq 3$. Since the reduced projective class group for the group ring $Z(Z/2)$ is trivial by [23], the Wall obstruction for Y to have the homotopy type of a *finite* complex must be zero whenever defined. □

Theorem 2.2. *The only structural groups in diagram (1.1) in all dimensions $n \geq 3$ which can possibly admit the compatible structure of a finite H-space are the following:* SO(n) *for* $n \geq 3$, SPL(I^n) *for* $3 \leq n \leq 6$, SPL(n) *for* $3 \leq n \leq 5$, STOP(3) *and* STOP(I^4).

Proof. The proof of Proposition 1.4 showed that $H^t(\widetilde{\mathrm{SCAT}}(n); Q)$ is nonzero for every $t \equiv 3 \bmod 4$ and for all $n \geq 3$. Thus the integral homology of $\widetilde{\mathrm{SCAT}}(n)$ is not finitely generated for every $n \geq 3$. By (2.1) $\widetilde{\mathrm{SCAT}}(n)$ cannot be a finite H-space for every $n \geq 3$. From the homotopy exact sequence for the fibration $\mathrm{SF}(n-1) \to \mathrm{SG}(n) \to S^{n-1}$, we deduce that $\pi_2(\mathrm{SG}(n))$ contains $\pi_2(\mathrm{SF}(n-1)) \approx \pi_{n+1}(S^{n-1}) \approx Z/2$ for all $n \geq 3$. Since $\pi_2(X)$ is trivial for any path-connected H-space X such that $H_*(X; Z)$ is a finitely generated group by [2], SG(n) cannot be a finite H-space for every $n \geq 3$. By Theorem 6.1 of [17], neither STOP(n) nor STOP(I^{n+1}) for every $n \geq 4$ can be a finite H-space for the following reason. The third homotopy group of any connected finite H-space must be free abelian by [10]; however, π_3 of both STOP(n) and STOP(I^{n+1}) contain a $Z/2$-summand, the Kirby-Siebenmann class in homotopy, for every $n \geq 4$.

Let Spin PL(n) denote the universal cover of SPL(n). The quotient complex
Spin PL(n)/ Spin(n) = SPL(n)/ SO(n) is 6-connected for all $n \geq 5$ by [4]. We consider the commutative diagram (2.3). By stability [11] the vertical morphisms are *isomorphisms* for every $n \geq 9$. By naturality the Hurewicz morphism $\pi_7(\text{Spin PL}(n)) \to H_7(\text{Spin PL}(n); Z)$ must be a *monomorphism* for every $n \geq 9$. But the torsion in the homotopy of every simply connected finite H-space X with an associative ring multiplication on $H_*(X; Z/2)$ constitutes the kernel of the Hurewicz

morphism by [13]. Consequently, neither Spin PL(n) nor SPL(n) can be a finite H-space for every $n \geq 9$.

(2.3)

$$\begin{array}{cccccccc} & \pi_7(\mathrm{Spin}(n)) & \longrightarrow & \pi_7(\mathrm{Spin\ PL}(n)) & \longrightarrow & \pi_7(\mathrm{SPL}(n)/\mathrm{SO}(n)) & \longrightarrow & \\ & \downarrow & & \downarrow & & \downarrow & & \\ \xrightarrow{0} & \pi_7(\mathrm{SO}) & \longrightarrow & \pi_7(\mathrm{SPL}) & \longrightarrow & \pi_7(\mathrm{SPL}/\mathrm{SO}) & \xrightarrow{0} & \\ & \| & & \| & & \| & & \\ & Z & \longrightarrow & Z \oplus Z/4 & \longrightarrow & Z/7 \oplus Z/4 & & \end{array}$$

The natural morphisms $\mathrm{SPL}(6)/\mathrm{SO}(6) \to \mathrm{SPL}(7)/\mathrm{SO}(7) \to \mathrm{SPL}/\mathrm{SO}$ induce *isomorphisms* on π_7 by [4]. However, the inclusions $\mathrm{SPL}(6) \subset \mathrm{SPL}(7) \subset \mathrm{SPL}$ induce *monomorphisms*, not isomorphisms, on π_7 which is isomorphic to $Z \oplus Z/4$, since multiplication by 2 occurs repeatedly for $\pi_7(\mathrm{SO}(6)) \to \pi_7(\mathrm{SO}(7)) \to \pi_7(\mathrm{SO})$. By naturality, the Hurewicz morphism is again a *monomorphism* on $\pi_7(\mathrm{Spin\ PL}(6)) \approx \pi_7(\mathrm{SPL}(6))$ and $\pi_7(\mathrm{Spin\ PL}(7)) \approx \pi_7(\mathrm{SPL}(7))$, each isomorphic to $Z \oplus Z/4$. Lin's theorem [13] guarantees that neither Spin PL(6) nor Spin PL(7), and consequently neither SPL(6) nor SPL(7), can be a *finite* H-space, since the torsion subgroup of π_7 does not belong to the kernel of the Hurewicz morphism. Now the pair $(\mathrm{Spin\ PL}(n)), \mathrm{Spin\ PL}(I^n))$ is $(n+1)$-connected for $n \geq 6$ by [12] and [4]. Replacing Spin PL(n) by Spin PL(I^n) in the previous argument shows also that Spin PL(I^n), and thus SPL(I^n) also, cannot be a *finite* H-space for every $n \geq 9$ and also for $n = 7$. However, this argument does not apply to SPL(n) and SPL(I^n) for $n = 5$ and 8 and to SPL(I^6), since π_7 of these spaces is torsion-free.

We now present a different argument that neither SPL(n) nor SPL(I^n) can be a finite H-space for every $n \geq 8$, and so establish the desired result for dimension 8. We consider the fibration $\mathrm{SPL} \to \mathrm{SG} \to \mathrm{SG}/\mathrm{SPL}$. By [3] the homotopy boundary of the generator of $\pi_{10}(\mathrm{SG}/\mathrm{SPL}) \approx Z/2$ yields a class in $\pi_9(\mathrm{SPL})$ with non-zero image under the mod 2 Hurewicz morphism. Since the inclusion $\mathrm{SG}(m)/\widetilde{\mathrm{SPL}}(m) \to \mathrm{SG}/\mathrm{SPL}$ is a homotopy equivalence for all $m \geq 3$ by [18], naturality gives a class of order 2 in $\pi_9(\widetilde{\mathrm{SPL}}(m))$ with non-zero image under the mod 2 Hurewicz morphism for all $m \geq 3$. The inclusions

$$\mathrm{Spin\ PL}(I^m) \to \mathrm{Spin\ PL}(m) \to \widetilde{\mathrm{SPL}}(m)$$

induce epimorphisms on π_9 for all $m \geq 8$ by [4]. So the mod 2 Hurewicz morphism

$$\pi_9(X) \to H_9(X; Z) \to H_9(X; Z/2)$$

is nontrivial for $X = \text{Spin PL}(m)$ or $\text{Spin PL}(I^m)$ for every dimension $m \geq 8$. On the other hand, Theorem 9 of [14] asserts that the mod 2 Hurewicz morphism $\pi_j(Y) \to H_j(Y; Z/2)$ is always trivial in dimension j with $j+1$ not a power of 2 for all simply connected *finite* H-spaces Y having associative mod 2 homology rings. Consequently, neither $\text{SPL}(m)$ nor $\text{SPL}(I^m)$ can be a *finite* H-space for every $m \geq 8$. □

3. Codimension 4 spherical embeddings.

Numerous classification problems for immersions and embeddings were reduced to the calculation of homotopy groups of the structural groups in (1.1) by Haefliger [6], and Rourke and Sanderson [18]. Massey [15] asked whether it is possible to embed a homotopy 7-sphere smoothly in $\mathbb{R}^{11}$ with a nontrivial normal bundle. We show that there are 4 distinct normal bundles for smooth embeddings of 7-spheres in $\mathbb{R}^{11}$. The bundles are classified by the 2-primary component of $\pi_6(\text{SO}(3)) \approx Z/4 \oplus Z/3$. Every smooth embedding of a 7-sphere in $\mathbb{R}^m$ for $m \neq 11$ must have a trivial normal bundle by [9]. Following [6], let Γ_n^q denote the group of concordance classes of smoothed (lisses) embeddings of S^n in S^{n+q} which are piecewise differentiably unknotted. (This condition holds for $q > 2$.) Let N_n^q denote the group of normal bundles associated to Γ_n^q.

Theorem 3.1. *Γ_7^q is isomorphic to Z for $3 \leq q \leq 5$ and to $Z/28$ for $q \geq 6$ while N_7^q is trivial for $3 \leq q \neq 4$ and $N_7^4 \approx Z/4$.*

Proof. By [6], $\Gamma_n^q \approx \pi_n(\widetilde{\text{SPL}}(q), \text{SO}(q))$ while $N_n^q \approx \text{image}\, \partial \subseteq \pi_{n-1}(\text{SO}(q))$ where ∂ is the boundary morphism for the homotopy exact sequence of $(\widetilde{\text{SPL}}(q), \text{SO}(q))$. Since $\Gamma_n^q \approx \pi_n(\text{SPL}/\text{SO})$ for $2q > n+3$ by [18], $\Gamma_7^q \approx Z/28$ for $q \geq 6$. Now N_7^q is trivial for $q \geq 5$ since $\pi_6(\text{SO}(q)) \approx 0$ for $q \geq 5$. Further, N_7^3 is trivial by Theorem 1.10 of [9].

Theorem 5.1 of [17] shows that $\pi_7(\widetilde{\text{SPL}}(5)) \approx Z \oplus Z$. A *careful* analysis of the homotopy sequences for $(\widetilde{\text{SPL}}(5), \text{SO}(5))$ and $(\widetilde{\text{SPL}}(6), \text{SO}(6))$ and morphisms between them shows that generators can be chosen so that $i_* : \pi_7(\text{SO}(5)) \to \pi_7(\widetilde{\text{SPL}}(5))$ is given by $i_*(1) = (14, 1) \in Z \oplus Z$. Thus $\pi_7(\widetilde{\text{SPL}}(5), \text{SO}(5)) \approx Z \approx \Gamma_7^5$. Direct calculation shows that the exact sequence from [18, p. 476]

$$\begin{aligned} \to \pi_8(\text{SF}(4), \text{SG}(4)) \to \pi_7(\widetilde{\text{SPL}}(4), \text{SO}(4)) &\to \pi_7(\widetilde{\text{SPL}}(5), \text{SO}(5)) \\ &\to \pi_7(\text{SF}(4), \text{SG}(4)) \to \end{aligned}$$

is given by $\to Z/2 \xrightarrow{0} Z \to Z \to 0 \to$. Also, calculations using [22] show that the exact sequence

$$\begin{aligned}\to \pi_8(\mathrm{SF}(3), \mathrm{SG}(3)) \to \pi_7(\widetilde{\mathrm{SPL}}(3), \mathrm{SO}(3)) &\to \pi_7(\widetilde{\mathrm{SPL}}(4), \mathrm{SO}(4)) \\ &\to \pi_7(\mathrm{SF}(3), \mathrm{SG}(3)) \to\end{aligned}$$

is given by

$$Z/2 \oplus Z/2 \xrightarrow{0} Z \xrightarrow{\times 20} Z \to Z/20 \oplus Z/2 \to \ .$$

Now $\mathrm{SG}(4) = \mathrm{SF}(3) \times S^3$ so $\pi_6(\mathrm{SG}(4)) \approx Z/3 \oplus Z/12$ by [22]. The morphism
$i_* : \pi_6(\widetilde{\mathrm{SPL}}(4)) \to \pi_6(\mathrm{SG}(4))$ is an *isomorphism* from the homotopy exact sequence for $\widetilde{\mathrm{SPL}}(4) \to \mathrm{SG}(4) \to \mathrm{SG}/\mathrm{SPL}$, since $\pi_6(\mathrm{SG}(4)) \to \pi_6(\mathrm{SG}(5))$ is the trivial morphism while $\pi_6(\mathrm{SG}(5)) \to \pi_6(\mathrm{SG}/\mathrm{SPL}) \approx Z/2$ is an isomorphism. We can deduce that $\pi_6(\widetilde{\mathrm{SPL}}(4), \mathrm{SO}(4))$ is trivial from exact sequences in [6] or by the following observation. The groups $\pi_7(\mathrm{SF}(4), \mathrm{SG}(4))$ and $\pi_6(\widetilde{\mathrm{SPL}}(5), \mathrm{SO}(5))$ are both trivial by direct calculation using [22] and by [4], respectively. Consequently, the middle group is trivial in the exact sequence of [18, p. 476]

$$\to \pi_7(\mathrm{SF}(4), \mathrm{SG}(4)) \to \Gamma_6^4 \to \Gamma_6^5 \to$$

We can now conclude that the homotopy exact sequence

$$\to \pi_7(\widetilde{\mathrm{SPL}}(4), \mathrm{SO}(4)) \xrightarrow{\partial} \pi_6(\mathrm{SO}(4)) \xrightarrow{j_*} \pi_6(\widetilde{\mathrm{SPL}}(4)) \to \pi_6(\widetilde{\mathrm{SPL}}(4), \mathrm{SO}(4))$$

is given by $\to Z \to Z/12 \oplus Z/12 \to Z/3 \oplus Z/12 \to 0$. Thus the kernel of j_* is the $Z/4$ summand in $\pi_6(\mathrm{SO}(3))$. Those are the 4-plane bundles over S^7 which are both stably trivial and fiber homotopy trivial. □

An important problem concerns codimension 4 spherical embeddings in the PL category. We conjecture that only the trivial SPL(4) bundle can occur as the SPL(4) normal bundle associated to the concordance class of any PL open tube of Σ^6 embedded in Σ^{10}. We express our gratitude to Francisco Gonzalez-Acuna for information about the codimension 4 embedding problem and also to Allen Hatcher and Peter Kahn for informative conversations at the GITC.

References

1. P. L. Antonelli, D. Burghelea, and P. J. Kahn, *The non-finite homotopy type of some diffeomorphism groups*, Topology **11** (1972) 1–49.
2. W. Browder, *Torsion in H-spaces*, Ann. of Math. (2) **74** (1961) 24–51.
3. G. Brumfiel, I. Madsen and R. J. Milgram, PL *characteristic classes and cobordism*, Ann. of Math. (1) **97** (1973) 82–159.

4. D. Burghelea and R. Lashof, *The homotopy type of the space of diffeomorphisms II*, Trans. Amer. Math. Soc. **196** (1974) 37–50.
5. D. Burghelea and R. Lashof, *Stability of concordances and the suspension homomorphism*, Ann. of Math. (2) **105** (1977) 449–472.
6. A. Haefliger, *Lissage des immersions – I*, Topology **6** (1967) 221–239.
7. A. Hatcher, *A proof of the Smale Conjecture* $\mathrm{Diff}(S^3) \simeq \mathrm{O}(4)$, Ann. of Math. **117** (1983) 553–607.
8. A. Hatcher, *Concordance spaces, higher simple-homotopy theory, and applications*, Proc. Sympos. Pure Math., Vol. 32 Part I, 3–21, Amer. Math. Soc., Providence, 1978.
9. W. C. Hsiang, J. Levine and R. H. Szczarba, *On the normal bundle of a homotopy sphere embedded in euclidean space*, Topology **3** (1965) 173–181.
10. J. Hubbuck and R. Kane, *On* π_3 *of a finite* H*-space*, Trans. Amer. Math. Soc. **213** (1975) 99–105.
11. R. Kirby and L. Siebenmann, *Foundational essays on topological manifolds smoothing and triangulations*, Annals of Math. Studies 88, Princeton Univ. Press, Princeton, 1977.
12. N. H. Kuiper and R. K. Lashof, *Microbundles and bundles II. Semisimplicial theory*, Invent. Math. **1** (1966) 243–259.
13. J. Lin, *Two torsion and the loop space conjecture*, Ann. of Math. **115** (1982) 35–91.
14. J. Lin, *Steenrod connections and connectivity in* H*-spaces*, Memoirs of Amer. Math. Soc. **68** (1987) no. 329.
15. W. S. Massey, *On the normal bundle of a sphere imbedded in Euclidean space*, Proc. Amer. Math. Soc. **10** (1959) 959–964.
16. G. Mislin, *Finitely generated nilpotent spaces* , Ann. of Math. **103** (1976) 547–556.
17. D. Randall and P. A. Schweitzer, *On foliations, concordance spaces, and the Smale conjectures*, Differential Topology, Foliations and Group Actions, Contemp. Math., Vol. 161 Amer. Math. Soc. Providence, 235–258, 1994.
18. C. P. Rourke and B. J. Sanderson, *Block bundles III, homotopy theory*, Ann. of Math. (2) **87** (1968) 431–483.
19. C. P. Rourke and B. J. Sanderson, *On topological neighborhoods*, Compositio Math. **22** (1970) 387–424.
20. B. L. Sharma, *Topologically invariant integral characteristic classes*, Topology Appl. **21** (1985) 135–146.
21. R. J. Stern, *On topological and piecewise linear vector fields*, Topology **14** (1975) 257–269.
22. H. Toda, *Composition methods in homotopy groups of spheres*, Annals of Math. Studies 49, Princeton Univ. Press, Princeton, 1962.
23. C. T. C. Wall, *Finiteness conditions for CW complexes*, Ann. of Math. **8** (1965) 56–69.

24. _____, *Surgery on compact manifolds*, Academic Press, London, 1970.

DEPARTMENT OF MATHEMATICS AND COMPUTER SCIENCE, LOYOLA UNIVERSITY, NEW ORLEANS, LA 70118

E-mail address: randall@beta.loyno.edu

AMS/IP Studies in Advanced Mathematics
Volume **2**, 1997 (Part 1)

ON MINIMAL COVERINGS IN THE SENSE OF LUSTERNIK-SCHNIRELMANN FOR SPACES AND MANIFOLDS

JEAN-CLAUDE HAUSMANN

1. Introduction

Let $f : Y \longrightarrow X$ be a continuous map between topological spaces. We assume that X is endowed with a base point $* \in X$ (Y has no base point). A subset $A \subset Y$ is called *categorical for* f (or f*-categorical*) if $f|_A$ is homotopic to the map sending A onto the base point. An open covering $\mathcal{R} = \{U_i\}_{i\in\mathcal{I}}$ of X is called f*-categorical* if each U_i is so. The *category* $\operatorname{cat} f$ of f (in the sense of Lusternik-Schnirelmann and Berstein-Ganea; see [5], (§7)) is the minimal cardinality of a f-categorical open covering of Y. For instance, $\operatorname{cat} f = 1$ if and only if f is homotopic to a constant map. It is obvious that $\operatorname{cat} f$ depends only on the homotopy class of f. One writes $\operatorname{cat} X$ for $\operatorname{cat}(\mathrm{id}_X)$.

A f-categorical open covering $\mathcal{R}$ is called *minimal* if the cardinality of $\mathcal{R}$ is equal to $\operatorname{cat} f$. The aim of this paper is to study these minimal coverings and to try to classify them. Denote by $\widetilde{\mathcal{M}}_f$ the set of minimal coverings of X which are categorical for f. Our first result is

(1.1) Proposition A. *Let $f : Y \longrightarrow X$ be a continuous map between topological spaces, with Y paracompact. Suppose that* $\operatorname{cat} f = k+1$. *Let $\mathcal{R} \in \widetilde{\mathcal{M}}_f$. Then the nerve of $\mathcal{R}$ is isomorphic to the simplex Δ^k.*

Suppose $\operatorname{cat} f = n < \infty$. Two (ordered) coverings $\mathcal{R} = \{U_1, \ldots, U_n\}$ and $\mathcal{R}' = \{U_1', \ldots, U_n'\}$ in $\widetilde{\mathcal{M}}_f$ are called *concordant* if there is an open covering $\{V_1, \ldots, V_n\}$ of $Y \times [0,1]$ such that the following properties hold.

a) $V_i \cap Y \times \{0\} = U_i$ and $V_i \cap Y \times \{1\} = U_i'$.
b) V_i is categorical for the map $(y,t) \mapsto f(y)$.

"Concordance" might not be a transitive relation. Taking its transitive closure, it becomes an equivalence relation. Denote by $\mathcal{M}_f$ the set of equivalence classes of coverings in $\widetilde{\mathcal{M}}_f$. The following problem is open:

(1.2) Problem. *Find a map f such that $\mathcal{M}_f$ has more than one element*

To study $\mathcal{M}_f$, we introduce the set $\widetilde{\mathcal{H}}_f$ whose elements are pairs $(\mathcal{R}, h)$, where $\mathcal{R} = \{U_1, \dots, U_n\} \in \widetilde{\mathcal{M}}_f$ and $h = (h_1, \dots, h_n)$ is a n-tuple of homotopies

$$h_i : U_i \times [0,1] \longrightarrow X$$

such that $h_i(y,0) = f(y)$ and $h(y,1) = *$, the base point of X. The relation of concordance between $(\mathcal{R}, h)$ and $(\mathcal{R}', h')$ is defined as above, with in addition homotopies $H_i : V_i \times [0,1] \longrightarrow X$ such that

$$H_i(y,t,0) = f(x), \quad H_i(y,t,1) = *, \quad H_i(y,0,s) = h_i(y,s)$$

$$\text{and} \quad H_i(y,1,s) = h_i'(y,s)$$

This relation of concordance is now an equivalence relation. The set of concordance classes of elements of $\widetilde{\mathcal{H}}_f$ is denoted by $\mathcal{H}_f$. One has obvious surjections $\widetilde{\mathcal{H}}_f \twoheadrightarrow \widetilde{\mathcal{M}}_f$ and $\mathcal{H}_f \twoheadrightarrow \mathcal{M}_f$

Let G be a topological group and BG its Milnor's classifying space. Recall that $BG = \lim_k B_kG$, where B_kG is the quotient of the joint $E_kG := G * \cdots * G$ (k times) by the right action of G. Denote by $i_k : \widetilde{B_kG} \longrightarrow BG$ the inclusion $B_kG \subset BG$ seen as a fibration by Serre's construction. Now, if X is homotopy equivalent to a connected CW-complex, one can chose a topological group G_X together with a homotopy equivalence $\tau_X : X \xrightarrow{\simeq} BG_X$ ([6], (§3)). One denotes by L_f^k the space of maps $\tilde{f} : Y \longrightarrow \widetilde{B_kG_X}$ such that $i_k \circ \tilde{f} = \tau_X \circ f$. Set $\mathcal{L}_f^k := \pi_0(L_f^k)$. Recall that $\operatorname{cat} f \leq k$ if and only if $L_f^k \neq \emptyset$ (see [2, Lemma 1.1] or [5, Proposition 8.4]).

(1.3) Theorem B. *Suppose that Y is a paracompact space and that X is homotopy equivalent to a connected CW-complex. Let $n := \operatorname{cat} f$. Then there is a bijection*

$$\Phi : \mathcal{H}_f \xrightarrow{\simeq} \mathcal{L}_f^n$$

Theorem B gives a unicity result for minimal categorical coverings when the set $[Y, B_kG(X)]_f$ is reduced to a single element (Corollary B1 below). Recall that if X is $(r-1)$-connected, one has the inequality $\operatorname{cat} f \leq \frac{\operatorname{hdim} Y}{r} + 1$, *where* $\operatorname{hdim} Y$ *is the homotopy dimension of* Y. [5], *(Proposition 7.4).*

(1.4) Corollary B1. *Assume the hypothesis of Theorem A. Suppose that X is $(r-1)$-connected, with $r > 1$ and that* $\operatorname{cat} f = \frac{\operatorname{hdim} Y}{r} + 1$. *Then $\mathcal{H}_f$ and $\mathcal{M}_f$ both contain one element.*

One will also draw the following result from Theorem B :

(1.5) Corollary B2. *Assume the hypothesis of Theorem A. Suppose that X is 1-connexted and that the reduced homology groups $\tilde{H}_*(X;\mathbb{Z})$ are all finite for $* \leq \operatorname{hdim} Y - \operatorname{cat} f + 2$. Then $\mathcal{H}_f$ and $\mathcal{M}_f$ are both finite.*

The paper is organized as follows : in §2, we give the proof of Proposition A while Theorem B is proven in §3. Section 4 is devoted to examples. In §5 and 6, we give a geometric version of the Lusternik-Schnirelmann theory in the category of manifold, where the open cover is replaced by the notion of a "patchwork". A patchwork on a manifold M^n is a special type of stratification of M. The codimension 0 strata are called the "colors" of the patchwork and are manifolds with certain types of corners (a patchwork can also be described by a structure on M for some pseudo-group of diffeomorphisms of $\mathbb{R}^n$). These patchworks might be useful in trying to decompose a manifold M into pieces endowed with some structure which does not exist globally on M. In §7, for instance, we discuss spin-patchworks on non-spin-manifolds of dimension 4.

The author is grateful to the referee for correcting mistakes and suggesting improvement in the first version of the paper. I also would like to thank Shmuel Weinberger for a useful discussion.

2. Proof of Proposition A

Let $\mathcal{R} = \{U_0, \dots, U_k\}$ be an open covering of a topological space Y and let $\mu_i : Y \longrightarrow [0,1]$ $(i = 0, \dots, k)$ be a partition of unity subordinate to $\mathcal{R}$. As $\sum \mu_i = 1$, one may consider $\mu := (\mu_0, \dots, \mu_k)$ as a continuous map from Y into the standard simplex Δ^k. We shall prove :

(2.1) Proposition. *Let $f : Y \longrightarrow X$ be a continuous map between topological spaces. Let $\mathcal{R} = \{U_0, \dots, U_k\}$ be a f-categorical open covering of Y and let $\mu : Y \longrightarrow \Delta^k$ be a partition of unity subordinate to $\mathcal{R}$. Suppose that* $\operatorname{cat} f = k+1$. *Then μ is surjective.*

Let us show that Proposition (2.1) implies Proposition A. Let $\mathcal{R} := \{U_0, \dots, U_k\} \in \widetilde{\mathcal{M}}_f$. As Y is paracompact, there exists a partition of unity $\mu : Y \longrightarrow \Delta^k$ subordinnate to $\mathcal{R}$. By (2.1), μ is onto. We then have

$$\emptyset \neq \mu^{-1}(\operatorname{int} \Delta^k) \subset U_0 \cap \cdots \cap U_k$$

The nerve $\mathcal{N}$ of $\mathcal{R}$ is then of dimension k. As $\mathcal{N}$ has $k+1$ vertices, it is isomorphic to Δ^k.

Proof of Proposition (2.1). Let us first introduce a few notions of simplicial geometry [3], (pp. 5–9). Let K be a simplicial complex. For instance, $K = \Delta^k = \Delta$, the standard simplex of dimension k. The underlying topological space of K is denoted $|K|$ (elswhere in the paper, Δ^k denotes in fact $|\Delta^k|$). If σ is a simplex of K, we denote its barycenter by $b(\sigma)$. The *open-star* $\mathrm{ost}(\sigma; K)$ is the subspace of $|K|$ defined by

$$\mathrm{ost}(\sigma; K) := |\overline{\mathrm{star}}(\sigma; K)| - \mathrm{link}(\sigma; K)|$$

where $\overline{star}$ and link are the usual notions [3], (p. 7).

Let α be a face of the standard simplex $\Delta = \Delta^k$ (abreviation : $\alpha \leq \Delta$). One denotes by $\Delta[\alpha]$ the minimal subdivision of Δ which contains the first barycentric subdivision of α. The barycenters of the faces of α are then 0-simplices of $\Delta[\alpha]$. For $0 \leq j \leq \dim \alpha$, one defines

$$M_j(\alpha) := \bigcup_{\substack{\sigma < \alpha \\ \dim \sigma = j}} \mathrm{ost}(b(\sigma), \Delta[\alpha]) \tag{2.2}$$

which is an open set of $|\Delta|$. One checks that $|\Delta| - \{b(\alpha)\}$ admits the open covering

$$|\Delta| - \{b(\alpha)\} = M_0(\alpha) \cup \cdots \cup M_{\dim \alpha - 1}(\alpha) \cup M(\alpha) \tag{2.3}$$

where $M(\alpha) := \mathrm{ost}(\mathrm{link}(\alpha; \Delta), \Delta)$.

Let us suppose that $\mu : Y \longrightarrow |\Delta|$ is not surjective. Let $u \in |\Delta|$ be a point which is not in the image of μ. By composing μ with a PL-isomorphism of Δ, we may assume that u corresponds to the barycenter $b(\alpha)$ of a face $\alpha \leq \Delta$. If $(e_0, \ldots, e_k)$ are the vertices of Δ, we may, by reordering the open sets U_i if necessary, assume that the vertices of α are $(e_0, \ldots, e_m)$. Using (2.3), we have :

$$Y = \mu^{-1}(M_0(\alpha)) \cup \cdots \cup \mu^{-1}(M_{\dim \alpha - 1}(\alpha)) \cup \mu^{-1}(M(\alpha))$$

Observe that $\mu^{-1}(M(\alpha)) \subset U_{m+1} \cup \cdots \cup U_k$. Moreover, $\mu^{-1}(M_j(\alpha))$ is f-categorical. Indeed, this follows from the following three observations:

a) if σ is a face of α of dimension j containing the vertex e_i, $\mu^{-1}(\mathrm{ost}(b(\sigma), \Delta[\alpha]) \subset U_i$;
b) the union, in Formula (2.3) is a *disjoint* union;
c) as Y admits a f-categorical covering, $f(Y)$ is contained in the path-connected component of the base point of X.

This then produces a f-categorical open covering of Y containing k elements, which contradicts the hypothesis $\mathrm{cat}\, f = k + 1$. □

Remark. The use of the open sets $\mu^{-1}(M_j(\alpha))$ is classical in the case $\alpha = \Delta$. Indeed, $\mu^{-1}(M_j(\Delta))$ is nothing else than the space W_{j+1} of [4], (proof of 12.1) or of [5], (proof of 2.1).

3. Proof of Theorem B

Let $G = G_X$. We shall identify X with BG_X via the homotopy equivalence $\tau_X : X \longrightarrow BG_X$. Replacing the map f by $\tau_X \circ f$ enables us to consider f as a map from Y to BG. Let $\xi : EG \xrightarrow{\pi} BG$ be the universal G-principal bundle. Let $\eta : P \xrightarrow{p} Y$ be the G-principal bundle induced on Y by the map f. We start with a couple of lemmas.

(3.1) Lemma. *Let* $(\mathcal{R}, h) = \{(U_1, h_1), \ldots, (U_n, h_n)\} \in \widetilde{\mathcal{H}}_f$. *Let* $(\mathcal{R}', h') = \{(U_1', h_1'), \ldots, (U_n', h_n')\}$ *be finer than* $(\mathcal{R}, h)$, *namely* $U_i' \subset U_i$ *and* $h_i' = h_i | U_i'$. *Then* $(\mathcal{R}, h)$ *is concordant to* $(\mathcal{R}', h')$.

Proof. For $i = 1, \ldots, n$, define

$$V_i := U_i \times [0, 1[\cup U_i' \times \{1\}, \qquad H_i(x, t, s) := h_i(x, t)$$

Then $\{(V_1, H_1), \ldots, (V_n, H_n)\}$ makes $(\mathcal{R}, h)$ concordant to $(\mathcal{R}', h')$. □

Denote by $\mathcal{B}_f^k$ the set of homotopy classes of G-bundle maps from η to $\xi_k : E_k G \xrightarrow{\pi_k} B_k G$. The elements of $\mathcal{B}_f^k$ are thus homotopy classes of couples $(\tilde{g}, g)$ where

$$\begin{array}{ccc} P & \xrightarrow{\tilde{g}} & E_k G \\ {\scriptstyle p}\downarrow & & \downarrow{\scriptstyle \pi_k} \\ Y & \xrightarrow{g} & B_k G \end{array}$$

is a strictly commutative diagram and $\tilde{g}$ is a G-equivariant map.

(3.2) Lemma. *For any integer* k, *there is a bijection*

$$\mathcal{L}_f^k \xrightarrow{\simeq} \mathcal{B}_f^k$$

Proof. By the definition of Serre's construction, an element of $\mathcal{L}_f^k$ is represented by a pair (g, H) where $g : Y \longrightarrow B_k G$ and $H : Y \times [0, 1] \longrightarrow BG$ is a homotopy from f to g. Lifting the homotopy H gives rise to a G-map $\tilde{g}$ over g. It is easy to check that this correspondance defines a map $\mathcal{L}_f^k \xrightarrow{\alpha} \mathcal{B}_f^k$.

On the other hand, let $(\tilde{g}, g)$ represent an element of $\mathcal{B}_f^k$. Composing with the inclusions $(E_k G \subset EG, B_k G \subset BG)$ gives a bundle map from η into ξ. The space of such bundles maps is contractible (use [4], (12.3 and 12.4)). Therefore, there is a homotopy H from f to g and two such H are homotopic. Thus, the correspondance $(\tilde{g}, g) \mapsto (g, H)$ defines a map $\mathcal{B}_f^k \xrightarrow{\beta} \mathcal{L}_f^k$ which is an inverse of α. □

(3.3) Lemma. *Let $f : A \longrightarrow BG$ be a map with* $\operatorname{cat} f = 1$. *Then there is a bijection between $\mathcal{H}_f$ and the set of homotopy classes of G-equivariant maps from P onto G.*

Proof. By Lemma (3.2), it suffices to construct a bijection $\mathcal{H}_f \xrightarrow{\simeq} \mathcal{B}^1_f$. But B_1G is contractible, and thus there is a G-equivariant map $E_1G \xrightarrow{\simeq} G$, unique to homotopy. □

Proof of Theorem B. By Lemma (3.2), it suffices to construct a bijection $\mathcal{H}_f \xrightarrow{\Phi'} \mathcal{B}^n_f$ (observe that Lemma (3.3) gives the case $n = 1$). Let $(\mathcal{R}, h) = \{(U_1, h_1), \ldots, (U_n, h_n)\} \in \widetilde{\mathcal{H}}_f$. The homotopy h_i represents an element in $\mathcal{H}_{f|U_i}$. By Lemma (3.3), this produces a G-equivariant map $\alpha_i : p^{-1}(U_i) \longrightarrow G$, unique up to homotopy.

As Y is paracompact, there is a partition of unity $\mu_i : Y \longrightarrow [0,1]$ with $\mu_I^{-1}(]0,1]) \subset U_i$. The formula

$$\tilde{g}(z) := (\alpha_1(z), \mu_1(p(z)), \ldots, \alpha_n(z), \mu_n(p(z))$$

defines a G-equivariant map $\tilde{g} : P \longrightarrow E_nG$. Let $g : Y \longrightarrow B_nG$ be the induced map on the quotient spaces. The class of $(\tilde{g}, g)$ in $\mathcal{B}^n_f$ will, by definition, represent $\Phi'(([\mathcal{R}, h)])$. It is easy to check that this definition makes sense using Lemma (3.3) and the fact that the partitions of unity subordinate to $\mathcal{R}$ form a convex space.

We shall define a map $\Psi : \mathcal{B}^n_f \longrightarrow \mathcal{H}_f$ which will be an inverse to Φ'. Let $(\tilde{g}, g)$ represent an element $\gamma \in \mathcal{B}^n_f$. Recall that B_nG is equipped with a partitions of unity $(t_1, \ldots, t_n)$ such that ξ_n has a canonical section over $t_i^{-1}(]0,1])$ (mapping $u \in B_nG$ to the only element $(T_jg_j) \in E_nG$ above u such that $g_i = 1$). The map $\tilde{g}$ giving an isomorphism between η and $g * \xi_n$, we get, for $i = 1, \ldots, n$, sections s_i of $\eta|V_i$, where $V_i := g^{-1}(t_i^{-1}]0,1])$. By Lemma (3.3), each s_i produces a well defined homotopy class of contractions h_i of $f|V_i$. It is easy to check that the class of $\{V_i, h_i)\}$ in $\mathcal{H}_f$ depends only of the class of $(\tilde{g}, g)$ in $\mathcal{B}^n_f$. By definition, $\{V_i, h_i)\}$ represents $\Psi(\gamma)$.

Checking that $\Psi \circ \Phi' = \mathrm{id}$ is easy. The equality $\Phi' \circ \Psi = \mathrm{id}$ uses Lemmas (3.1) ($\{\mu_i^{-1}([0,1]), h_i\}$ being finer than $(\mathcal{R}, h)$) and Lemma (3.3). □

Proof of Corollaries B1 and B2. Let $m := \dim Y$ and $k := \operatorname{cat} f$. Let F_k be the fibre of $\widetilde{B_kG} \xrightarrow{i_k} BG$. The space F_k is homotopy equivalent to the k-fold join $G * \cdots * G$. The group $G = G_X$ being $(r-2)$-connected and $r - 2 \geq 0$, the space F_k is ν-connected with

$$\nu = k(r-2) + 2(k-1)$$

According to our hypotheses, one has $k = \frac{m}{r} + 1$ and $r \geq 2$. It is easy to check that one then has $\nu \geq m$. By obstruction theory, this implies that $\mathcal{L}_f^k$ has only one element, which proves Corollary B1.

For Corollary B2, we observe that our hypotheses imply, by the Serre theory for the class of finite abelian groups, that $\pi_*(X)$ is finite for $* \leq m-k+2$. Therefore, $\pi_*(G)$ is finite for $* \leq m-k+1$. The classical extension of the above Serre theory to simple spaces implies that $H_*(G)$ is finite for $* \leq m-k+1$. The formula for the homology of a join (see [7], (Lemma 2.1)) then tells us that $H_*(F_k)$ is finite for $* \leq m$. The same holds then for $\pi_*(F_k)$. By obstruction theory, one deduces that $\mathcal{L}_f^k$ is finite. □

4. Examples

Except for Corollary A1, the author could compute $\mathcal{M}_f$ only in special situations. In all cases, it turned out that $\mathcal{M}_f$ has only one element. Here is an exemple :

(4.1) Proposition. *Let $f : Y \longrightarrow X$ be a continuous map. Suppose that* $\operatorname{cat} Y \leq 2$. *Then $\mathcal{M}_f$ has only one element.*

Proof. Note that $\operatorname{cat} f \leq \operatorname{cat} Y \leq 2$. As the case $\operatorname{cat} f = 1$ is trivial, one may assume that $\operatorname{cat} f = \operatorname{cat} Y = 2$.

If $\operatorname{cat} Y = 2$, the homotopy equivalence $Y \longrightarrow BG_Y$ factors through $B_1 G_Y \simeq \Sigma G_Y$. Therefore, there are maps $Y \xrightarrow{j} \Sigma G_Y \xrightarrow{p} Y$ such that $p \circ j$ is homotopic to id_Y. The map p then induces a bijection $p_* : \mathcal{M}_f \xrightarrow{\simeq} \mathcal{M}_{p\circ f}$. Therefore, one may assume that Y is a suspension. Denote by a, b be the two cone points of Y. Let $\{U_1, U_2\} \in \widetilde{\mathcal{M}}_f$.

Case 1 : $a \notin U_1$. Then $U_1 \subset Y - \{a\}$ and $a \in U_2$. As in Lemma (3.1), one proves that if $\mathcal{R} \in \widetilde{\mathcal{M}}_f$, and if $\mathcal{R}'$ is finer then $\mathcal{R} \in \widetilde{\mathcal{M}}_f$ and $\mathcal{R} \sim \mathcal{R}'$, where "$\sim$" means "being concordant". One then has

$$\begin{aligned} \{U_1, U_2)\} &\sim \{Y - \{a\}, U_2\} \sim \{Y - \{a\}, U_2 \cap (Y - \{b\})\} \\ &\sim \{Y - \{a\}, (Y - \{b\})\} \end{aligned}$$

Case 2 : general case. There exists $u \in Y$ with $u \notin U_1$ (otherwise, $\operatorname{cat} f = 1$). The inclusion of $\{a\} \subset Y$ being a cofibration, there is a map $\alpha : Y \longrightarrow Y$, homotopic to id_Y, such that $\alpha(a) = u$. As α is in the homotopy class of id_Y, one has $\{U_1, U_2)\} \sim \{\alpha^{-1}(U_1), \alpha^{-1}(U_2))\}$. On the other hand, $a \notin \alpha^{-1}(U_1)$ and then Case 1 applies. □

(4.2) Example. Let $f : S^1 \longrightarrow \mathbb{R}P^\infty = B\mathbb{Z}_2$ represent the generator of $\pi_1(\mathbb{R}P^\infty) \simeq \mathbb{Z}_2$. Then, $\operatorname{cat} f = 2$. The space $B_2\mathbb{Z}_2$ is equal to

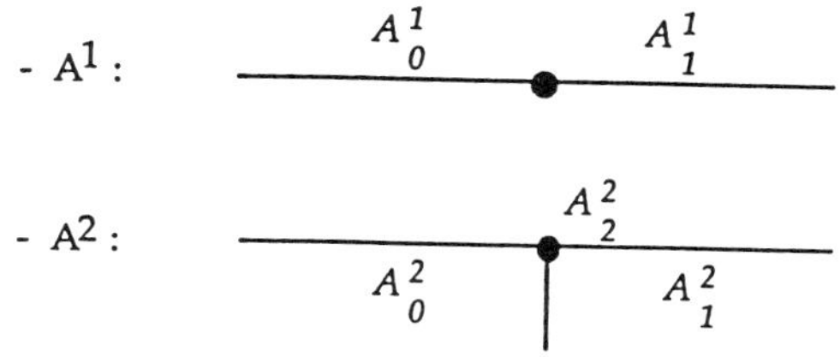

FIGURE 1

$\mathbb{R}P^1 \simeq S^1$ and the map which associates to a lifting $\tilde{f} : S^1 \longrightarrow \widetilde{B_2\mathbb{Z}_2}$ of f its degree gives a bijection between $\mathcal{L}_f^2$ and the set of odd integers. Thus $\mathcal{H}_f$ is infinite. But, as S^1 is a suspension, the set $\mathcal{M}_f$ contains only one element by (4.1).

5. PATCHWORKS ON MANIFOLDS

Local models and definitions. Denote by A^q the space $\mathbb{R}^q$ decomposed into $q+1$ *colors*:

$$A^q := A_0^q \cup \cdots \cup A_q^q$$

where the colors A_i^q are, by induction on q, defined by the following three properties:

a) $A^0 = A_0^0 := \{0\}$
b) if $i < q$, then $A_i^q = A_i^{q-1} \times \mathbb{R}_{\leq 0}$
c) $A_q^q := \{(x_1, \ldots, x_q) \in \mathbb{R}^q \mid x_q \geq 0\}$

For instance, Figure 1 for A^1 and A^2.

For $q \leq n$, let us consider the pseudo-group $\mathcal{P}(\mathbb{R}^{n-q} \times A^q)$ composed of all diffeomorphisms $f : U \longrightarrow U'$, where U and U' are open sets of $\mathbb{R}^{n-q} \times A^q$, such that f satisfies $f((\mathbb{R}^{n-q} \times A_i^q) \cap U) \subset \mathbb{R}^{n-q} \times A_i^q$ for all $i \in \{0, \ldots, q\}$.

Let M be a (smooth) manifold of dimension n. A *patchwork* with $q+1$ *colors* is an atlas $\mathcal{Q}$ such that

a) the charts of $\mathcal{Q}$ are diffeomorphisms.
b) all changing charts of $\mathcal{Q}$ are in $\mathcal{P}(\mathbb{R}^{n-q} \times A^q)$.
c) $\mathcal{Q}$ is maximal for the properties a) and b).

A patchwork $\mathcal{Q}$ gives a decomposition of M in $q+1$ *colors* :

$$M = P_0 \cup \cdots \cup P_q$$

where the color $P_i = P_i(\mathcal{Q})$ is defined by

$$P_i := \bigcup_{\varphi \in \mathcal{Q}} \varphi^{-1}(\mathbb{R}^{n-q} \times A_i^q)$$

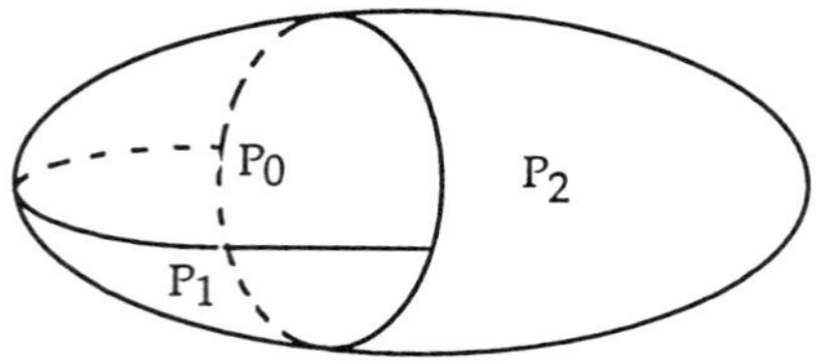

patchwork with 3 colors

FIGURE 2

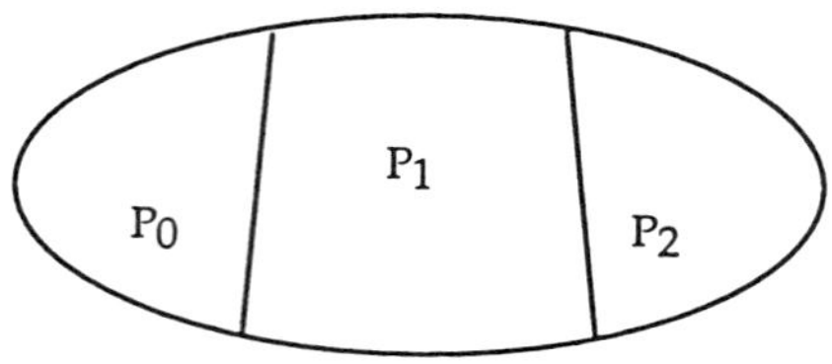

A non-complete patchwork with 3 colors

FIGURE 3

Observe that P_i is a manifold "with corners" and a "piece with corners" of V in the sense of [1].

By replacing $\mathcal{P}(\mathbb{R}^{n-q} \times A^q)$ by the pseudo-group $\mathcal{P}(\mathbb{R}_{\leq 0} \times \mathbb{R}^{n-1-q} \times A^q)$ (one needs $q \leq n-1$), we get the notion of a patchwork with $(q+1)$ colors on a manifold of dimension n M with boundary. Such a structure naturally induces on the boundary of M a patchwork with $(q+1)$ colors in the previous sense. We may then speak of *concordance* or *cobordism* of patchworks.

More precisely, a patchwork $\mathcal{Q}$ determines a stratification $\mathcal{S}_{\mathcal{Q}}$ of its underlying manifold M. The *strata* of $\mathcal{S}_{\mathcal{Q}}$ are the equivalence classes defined by the following relation on M : $x \sim x'$ if there exists $\varphi, \varphi' \in \mathcal{Q}$ such that $\varphi(x) = \varphi'(x')$. The colors are the closure of the top-dimensional strata. As an immediate consequence of our definitions, two patchworks $\mathcal{Q}$ and $\mathcal{Q}'$ on the same manifold are equal if and only if $\mathcal{S}_{\mathcal{Q}} = \mathcal{S}_{\mathcal{Q}'}$ or if and only if they have the same colors.

A patchwork $\mathcal{Q}$ on a connected manifold M is *complete* if there exists $x \in M$ and a chart $\varphi \in \mathcal{Q}$ defined in a neighborhood of x such that $\varphi(x) = 0$. Such a point x then belongs to all the colors of $\mathcal{Q}$ and a patchwork is complete if and only if the intersection of all its colors is non-empty. On a non-connected manifold, one says that a patchwork is complete if its restriction to each connected component is complete.

(5.2) Examples–Remarks.

a) A^q determines the standard patchwork on $\mathbb{R}^q$, with $q+1$ colors.
b) A patchwork with 2 colors on M corresponds to a decomposition $M = P_0 \cup P_1$ of M into two smooth sub-manifolds with boundary, with $P_0 \cap P_1 = \partial P_0 = \partial P_1$. If M is connected, this patchwork is complete when P_0 and P_1 are both not empty.
c) Let $f : E \longrightarrow M$ be a submersion. A patchwork $\mathcal{Q}$ with $q+1$ colors on M may be pulled back into a patchwork with $q+1$ colors $f^*\mathcal{Q}$ on E. The colors of $f^*\mathcal{Q}$ are the pre-images of the colors of $\mathcal{Q}$. The patchwork $f^*\mathcal{Q}$ is complete if and only if $\mathcal{Q}$ is complete.
d) Denote by $\mathbb{K}$ one of the fields $\mathbb{R}$, $\mathbb{C}$ or $\mathbb{H}$. The projective space $\mathbb{K}P^n$ admits the structure of a complete patchwork with $n+1$ colors $\mathcal{Q}(\mathbb{K}P^n)$ which is inductively defined as follows :
 –: $\mathbb{K}P^0 = \{*\} = P_0$.
 –: Recall that $\mathbb{K}P^n = E \cup D$, where $p : E \longrightarrow \mathbb{K}P^{n-1}$ is the disk bundle associated to the Hopf bundle on $\mathbb{K}P^{n-1}$ and D is a disk of codimension 0. The colors $P_0, \ldots, P_{n-1}$ of $\mathcal{Q}(\mathbb{K}P^n)$ are defined as the pre-images by p of the corresponding colors in $\mathcal{Q}(\mathbb{K}P^{n-1})$ and we set $P_n := D$.

For instance, if one describes points of $\mathbb{K}P^1$ by their homogenous coordinates $[z_1, z_2]$ ($(z_1, z_2) \in \mathbb{K}^2 \setminus \{0\}$), the two colors of $\mathbb{K}P^1$ are just the disks :

$$P_0 := \{[z_1, z_2] \mid z_2 \neq 0 \text{ et } \|z_1 z_2^{-1}\| \leq 1\},$$
$$P_1 := \{[z_1, z_2] \mid z_1 \neq 0 \text{ et } \|z_2 z_1^{-1}\| \leq 1\}$$

6. Minimal categorical patchworks

Let $f : M \longrightarrow X$ be a continuous map from a connected manifold M into a topological space X. A patchwork on M is called *f-categorical* if each color is f-categorical. Denote by $p(f)$ the minimal integer r such that M admits a f-categorical patchwork with r colors $P_0, \ldots, P_{r-1}$ (*a priori*, $p(f) \in \mathbb{N} \cup \{\infty\}$). Such a patchwork is called *minimal f-categorical.*

(6.1) Theorem C.

a) $p(f) = \operatorname{cat} f$.
b) *Any minimal f-categorical patchwork on M is complete.*

The proof of theorem C requires some techniques of manifolds with corners that we shall develop now. Our notations are inspired by those of [1], but are slightly different.

(6.2) *The sector of index k in $\mathbb{R}^n$ is $S_k^n := \mathbb{R}^{n-k} \times \mathbb{R}^k_{\leq 0}$.* An *atlas of manifold with corners* on a (metrisable) space M is an open covering $\{U_i \mid i \in I\}$ of M and a collection of homeomorphisms (charts) $\varphi_i : U_i \longrightarrow V_i$, where V_i is an open set of $S^n_{k(i)}$, such all the changing charts are smooth (C^∞) maps. The data of such a maximal atlas endows M with a structure of a *manifold with corners*.

Let x be a point in a manifold with corners M. There exists a chart φ defined in a neighbourhood of x such that $\varphi(x) = 0$. If this chart takes value in an open set of S_k^n, one says that k is the *index* of x. The set of points of index k is denoted by $\partial_k M$: it is the space of *corners of index* k in M. The *boundary* ∂M of M is the union of $\partial_i M$ for $i \geq 1$; it is also the closure of $\partial_1 M$ (in [1], ∂M is denoted $M^{(1)}$). The manifold M is "without boundary" if $\partial M = \emptyset$ and has a "smooth boundary" if $\partial_2 M = \emptyset$. The space $\partial_k M$ is a smooth manifold of dimension $n-k$.

Let $f : M \longrightarrow Z$ be a continuous map from a manifold with corners M into a manifold (without boundary) Z. A point $z \in Z$ is a *super-regular value* of f if z is a regular value of $f|\partial_k M$ for all k. As there are only finitely many non-empty $\partial_k M$, Sard's theorem implies that the set of super-regular values of f is dense in Z. The proof of the following lemma, using only elementary techniques, is left to the reader :

(6.3) Lemma. *Let $M = K_0 \cup \cdots \cup K_s$ be a patchwork with $s+1$ colors on a manifold M. Let $g : M \longrightarrow \mathbb{R}$ be a smooth map and let $t_0 \in \mathbb{R}$ super-regular value for $g|K_0$. Then, if $s < \dim M$, the manifold M admits a patchwork structure $M = L_0 \cup \cdots \cup L_{s+1}$ with $s+2$ colors, where the colors are :*

$$L_0 := K_0 \cap g^{-1}(]-\infty, t_0]),\ L_1 := K_0 \cap g^{-1}([t_0, \infty[),\ L_i := K_{i-1},\ (i \geq 2)$$

We can now start the proof of Theorem C.

(6.4) *Proof that* $p(f) \geq \operatorname{cat} f$ Let $M = P_0 \cup \cdots \cup P_{r-1}$ be a patchwork with r colors on M whose colors are f-categorical. Each color P_i being a piece with corners in M, there exists, according Proposition (4.1) and (4.2) of [1], an open set U_i of M containing P_i and such that the inclusion $P_i \subset U_i$ extends to a homeomorphism of $P_i \cup_{\partial P_i} (\partial P_i \times \mathbb{R}_{\geq 0})$. The collection $\{U_i\}$ is then an open covering of M by r f-categorical open sets.

(6.5) *Proof that* $p(f) \leq \operatorname{cat} f$: Set $r := \operatorname{cat} f$. The proof is actually an induction argument on r, the case $r = 1$ being trivial. However, the induction step being technically heavy, we present the proof as a sequence of steps.

Step 1 : Let $Q_{r-1} := M$. One sees Q_{r-1} as a manifold with corners ($\partial_2 Q_{r-1} = \emptyset$ and $\partial_1 Q_{r-1} = \emptyset$ if M is without boundary). Let $\{U_0, \ldots, U_{r-1}\}$ be an open f-categorical covering of M. Let $\mu = (\mu_0, \ldots, \mu_{r-1}) : M \longrightarrow \Delta^{r-1}$ be a smooth partition of unity subordinated to $\mathcal{R}$. As $r = \operatorname{cat} f$, the map μ is surjective by Proposition (2.1). Thus there exists $z_{r-1} \in]0,1[$ which is a super-regular value of μ_{r-1}. By composing μ with a PL-automorphism of Δ^{r-1} if necessary, one may assume that $z_{r-1} = \frac{1}{2}$. We set :

$$P_{r-1} := \mu_{r-1}^{-1}([\tfrac{1}{2}, 1]), \quad Q_{r-2} := \mu_{r-1}^{-1}([0, \tfrac{1}{2}])$$

The decomposition $M = Q_{r-2} \cup P_{r-1}$ determines a patchwork with 2 colors on M. As $\mu_{r-1}|P_{r-1} > 0$, the color P_{r-1} is f-categorical.

Step 2 : Observe that : $\operatorname{cat}(f|Q_{r-2}) = r-1$. The open sets $U_i \cap Q_{r-2}$ $(i = 0, \ldots, r-2)$ cover Q_{r-2} and are f-categorical; this shows that $\operatorname{cat}(f|Q_{r-2}) \leq r-1$. On the other hand, suppose that $\{V_0, \ldots, V_k\}$ is a covering of Q_{r-2} by f-categorical open sets, with $k < r-2$. One then would have an open covering of Q_{r-1} with the sets $V_i \cap \operatorname{int} Q_{r-2}$ together with U_{r-1}. Such covering is f-categorical, which contradicts $r = \operatorname{cat} f$.

Let us choose a new partition of unity (still called μ) :

$$\mu : U_0 \cup \cdots \cup U_{r-2} \longrightarrow \Delta^{r-2}$$

which is subordinated to the covering $\{U_0, \ldots, U_{r-2}\}$. As $\operatorname{cat}(f|Q_{r-2}) = r-1$, Proposition (2.1) says that μ is surjective. One can then, as in Step 1, suppose that $1/2$ is a super-regular value for $\mu_{r-2}|Q_{r-2}$. One applies Lemma (6.3) with $K_0 = Q_{r-2}$, $K_1 = P_{r-1}$ and $g = \mu_{r-2}$. This cuts the color Q_{r-2} into two parts and one obtains a patchwork with 3 colors

$$M = \underbrace{Q_{r-3} \cup P_{r-2}}_{Q_{r-2}} \cup P_{r-1}.$$

The color P_{r-2}, extracted from Q_{r-2} is f-categorical as was P_{r-1}. Moreover, $\operatorname{cat}(f|Q_{r-3}) = r-2$.

Step 3 and further : One extracts from Q_{r-3} a color P_{r-3} which is f-categorical, etc. Recall that

$$r \leq LS(\mathrm{id}_M) \leq \operatorname{hdim}(M) + 1$$

where hdim is the homotopy dimension (see [5], (§2)). One then has $r \leq n+1$ when M is closed and $r \leq n$ when ∂M is non-empty. Our process may then be pursued till Step r, providing us an f-categorical patchwork with r colors.

(6.6) *Proof of b).* Let $M = P_0 \cup \cdots \cup P_{r-1}$ be an f-categorical patchwork $\mathcal{Q}$ with r colors on M. Suppose that $r = p(f)$ and that $\mathcal{Q}$ is not complete. We then have $(P_0 \cap \cdots \cap P_{r-2}) \cap P_{r-1} = \emptyset$. The colors being closed sets of M, there are two open sets U and V of M such that $P_0 \cap \cdots \cap P_{r-2} \subset U$, $P_{r-1} \subset V$ and $U \cap V = \emptyset$. As in (6.4), one constructs for each i an open set U_i of M which is the union of P_i with a collar on ∂P_i (then U_i is f-categorical). The open sets U_i may be chosen so that $U_i \subset U$ for $i < r - 1$ and $U_{r-1} \subset V$. One then has a minimal f-categorical covering of M whose nerve is of dimension $\leq r - 2$. As $r = \operatorname{cat} f$ by the already proven Point a), this contradicts Proposition A. □

As in the definition of $\mathcal{M}_f$, note that concordance of minimal f-categorical patchworks might not be a transitive relation. However, taking its transitive closure, we define $\mathcal{P}_f$ to be the set equivalence classes of minimal f-categorical patchworks on M. An improvement of the above technique, not hard but technical, will give :

(6.7) Theorem D. *There is a bijection* $\psi : \mathcal{P}_f \xrightarrow{\simeq} \mathcal{M}_f$

Idea of proof. Starting with a minimal f-categorical patchwork, we can construct a covering in $\widetilde{\mathcal{M}}_f$ by adding collars to the colors, as was done in (6.4). It is easy to see that this provides a well defined map $\psi : \mathcal{P}_f \longrightarrow \mathcal{M}_f$. The argument of (6.5) proves that ψ is surjective. It remains to prove the injectivity of ψ which is technically harder; we just give the idea of the proof.

The first step is to establish that, given a covering in $\widetilde{\mathcal{M}}_f$ together with a partition of unity subordinate to it, the concordance class of the patchwork constructed in (6.5) does not depend on the choice of the sequence of super-regular values. We give the argument in the special case $\operatorname{cat} f = 2$. Let us consider a covering $\{U_0, U_1\} \in \widetilde{\mathcal{M}}_f$ together with a partition of unity $\mu = (\mu_0, \mu_1)$ subordinate to it. Let $z < z'$ be two super-regular values for μ_1. Define

$$P_0 := \mu_1^{-1}([0,z]) \ , P_1 := \mu_1^{-1}([z,1]) \ , P_0' := \mu_1^{-1}([0,z']) \ , P_1' := \mu_1^{-1}([z',1])$$

The patchworks constructed as in Step 1 of (6.5) would be $\mathcal{Q} := \{P_0, P_1\}$ and $\mathcal{Q}' := \{P_0', P_1'\}$. Let $W := \mu_1^{-1}([z,z']) = \operatorname{adh}(P_0' - P_0)$. The manifold $M \times [0,1]$, decomposed as shown in Figure 4.
provides an f-categorical concordance of patchworks between $\mathcal{Q}$ and $\mathcal{Q}'$. In the general case, this has to be done successively in each color.

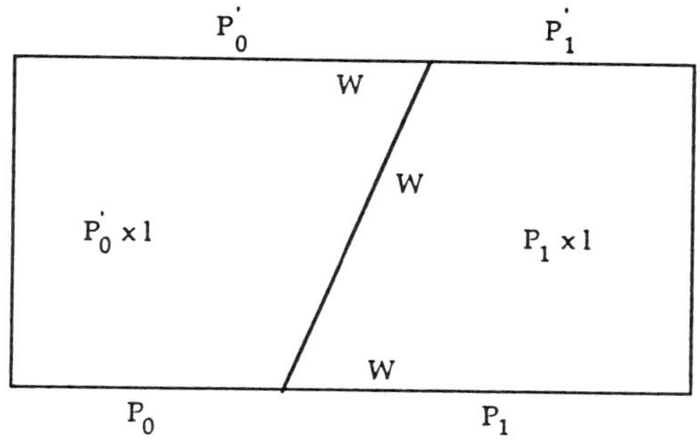

FIGURE 4

Now, let $\mathcal{R}$ and $\mathcal{R}'$ be two concordant minimal f-categorical coverings of M. Let μ and μ' be partitions of unity subordinate to $\mathcal{R}$ and $\mathcal{R}'$ respectively. Let $\mathcal{Q}$ and $\mathcal{Q}'$ be minimal f-categorical patchworks obtained as in (6.5), using super-regular values of the μ_i's and μ_i''s. Let $\overline{\mathcal{R}}$ be a minimal f-categorical open covering of $M \times [0,1]$ making $\mathcal{R}$ and $\mathcal{R}'$ concordant. As partitions of unity subordinate to a given covering form a convex set, there is a partition of unity subordinate to $\overline{\mathcal{R}}$ whose restriction to $M \times \{0,1\}$ is $\mu \cup \mu'$. Chose a minimal f-categorical patchwork on $M \times [0,1]$ as in (6.5), using $\overline{\mu}$. Its restriction to $M \times \{0,1\}$ gives minimal f-categorical patchworks $\mathcal{Q}_0$ and $\mathcal{Q}_1$ on these manifolds, which are obtained as in (6.5), using μ and μ'. By what has been done above, $\mathcal{Q}_0$ is concordant to $\mathcal{Q}$ and $\mathcal{Q}_1$ is concordant to $\mathcal{Q}'$. $\square$

Finally, we shall consider minimal f-categorical patchworks on M together with given homotopies

$$h_i : P_i \times [0,1] \longrightarrow X \ , \quad h_i(x,0) = f(x) \ , \quad h_i(x,1) = *$$

for each color P_i. The set of concordance classes of such objects is denoted by $\mathcal{HP}_f$ (the additional data of h_i makes concordance an equivalence relation). As for Theorem D, using Theorem B, one can prove

(6.8) Theorem E. *There are a bijections* $\mathcal{HP}_f \xrightarrow{\simeq} \mathcal{H}_f \xrightarrow{\simeq} \mathcal{L}_f^n$

Example. Let us consider the patchwork with $n+1$ colors on the projective space $\mathbb{K}P^n$ (Example d) of (6.2)). This is a representative of $\mathcal{P}(\mathbb{K}P^n) := \mathcal{P}_f$ for f the inclusion of $\mathbb{K}P^n$ into $\mathbb{K}P^\infty$. The sets $\mathcal{P}(\mathbb{C}P^n)$ and $\mathcal{P}(\mathbb{H}P^n)$ both contain one element by Theorem C and Corollary B1. For $\mathbb{K} = \mathbb{R}$, this conclusion can only be deduced for $n = 1$ by Proposition (4.1).

As for the set $\mathcal{HP}(\mathbb{K}P^n) := \mathcal{HP}_f$, it consists of concordance classes of minimal patchworks on $\mathbb{K}P^n$ together with a trivialization over each color of the canonical line bundle η (the map f being characteristic for

η). Again, there is only one element in $\mathcal{HP}(\mathbb{K}P^n)$ in the cases $\mathbb{K} = \mathbb{C}$ or $\mathbb{H}$, by Theorem D and Corollary B1. On the other hand, the set $\mathcal{HP}(\mathbb{R}P^1)$ is infinite by Example (4.2).

7. Spin and p_1-patchworks

Let W^n be an oriented manifold and let $\tau : W \longrightarrow BSO_n$ be a characteristic map for the tangent bundle of W. Call $\tilde{w}_2(W)$ the composed map $W \xrightarrow{\tau} BSO_n \xrightarrow{w_2} K(\mathbb{Z}_2, 2)$. The map $\tilde{w}_2(W)$ represents the second Stiefel-Whitney class $w_2(W) \in H^2(W; \mathbb{Z}_2)$. Recall that a *spin-structure* on W is a homotopy class of liftings of τ through $BSpin_n$. As $BSpin_n$ is the homotopy fiber of w_2, a spin-structure exists on W if and only if $\operatorname{cat} \tilde{w}_2(W) = 1$ and the set of spin-structures is then in bijection with $\mathcal{H}_{\tilde{w}_2(W)}$ as well as with $H^3(W; \mathbb{Z}_2)$.

A *spin-patchwork* on a manifold M^n is a patchwork such that each color is endowed with a spin-structure. We are interested in the set of concordance classes of spin-patchworks on a manifold. Here is a result in dimension 4.

(7.1) Proposition. *Let M^4 be a closed oriented smooth manifold with $w_2(M) \neq 0$. Then*

a) *If $w_2(M)^2 \neq 0$, the manifold M admits a spin-patchwork with 3 colors. Such a patchwork is unique up to concordance.*
b) *If $w_2(M)^2 = 0$, the manifold M admits a spin-patchwork with 2 colors. There are at least two concordance classes of such patchworks.*

Proof. Let $f = \tilde{w}_2(M)$. By what was said above, the set of concordance classes of spin-patchworks is in bijection with $\mathcal{HP}_f$. The later is in 1-1 correspondance with $\mathcal{H}_f$ by Theorem D and therefore contains 1 element by Corollary B1. This proves Part a).

To establish Part b), note that $w_2(M)^2$ is the unique obstruction for $w_2 : M \longrightarrow X = K(\mathbb{Z}_2, 2)$ to be lifted through $B_2G_X = \Sigma K(\mathbb{Z}_2, 1)$ [5], (5.3). Therefore, if $w_2(M)^2 = 0$, the manifold M admits a spin-patchwork with 2 colors. The concordance classes of such patchworks are in bijection with $\mathcal{L}_f^2$. The obstruction groups to decide whether two liftings are homotopic are $H^3(M; \pi_3(F))$ and $H^4(M; \pi_4(F)) \simeq \pi_4(F)$, where F is the fiber

$$F \longrightarrow \Sigma K(\mathbb{Z}_2, 1) \longrightarrow K(\mathbb{Z}_2, 2) \tag{7.2}$$

By obstruction theory, $\mathcal{L}_f^2$ surjects onto $H^3(M; \pi_3(F))$. If the latter is zero, then $H^4(M; \pi_4(F)) \simeq \pi_4(F)$ is the first (and last) obstruction group and then $\mathcal{L}_f^2$ is in bijection with $\pi_4(F)$. Therefore, it is enough to prove that $\pi_4(F)) \neq 0$ (incidentally, observe that $\pi_4(F)) = \pi_4(\Sigma K(\mathbb{Z}_2, 1)))$.

The Serre spectral sequence of the fibration in (7.2) gives rise to the exact sequence for the homology with integral coefficients :

$$0 = H_5(\Sigma K(\mathbb{Z}_2, 1)) \longrightarrow H_5(K(\mathbb{Z}_2, 2)) \longrightarrow H_4(F)$$

As $H_5(K(\mathbb{Z}_2, 2)) \approx \mathbb{Z}_2$, this proves that $H_4(F) \neq 0$. The space F being 2-connected, the Hurewicz homomorphism $\pi_4(F) \longrightarrow H_4(F)$ is onto and therefore $\pi_4(F) \neq 0$. $\square$

Let W^n as above. Let us now consider the map

$$\tilde{p}_1(W) : W \xrightarrow{\tau} BSO_n \xrightarrow{p_1} K(\mathbb{Q}, 4)$$

representing the first rational Pontrjagin class $p_1(W) \in H`4(W; \mathbb{Q})$ of W. We call a (rational) *p_1-structure* on W a homotopy class of liftings of $\tilde{p}_1(W)$ through the homotopy fiber of the map $BSO_n \longrightarrow K(\mathbb{Q}, 4)$.

A *p_1-patchwork* on a manifold M^n is a patchwork such that each color is endowed with a p_1-structure. As for Spin-patchworks, the set of concordance classes of p_1-patchworks on M can sometimes be estimated, for example when $n = 8$:

(7.3) Proposition. *Let M^8 be a closed oriented smooth manifold with*
$p_1(M) \neq 0$. Then

a) *If $p_1(M)^2 \neq 0$, the manifold M admits a p_1-patchwork with 3 colors. Such a patchwork is unique up to concordance.*
b) *If $p_1(M)^2 = 0$, the manifold M admits a p_1-patchwork with 2 colors. The set of concordance classes of such patchworks is in bijection with $H_1(M; \mathbb{Q})$.*

Proof. The proof is the same as for Prosition (7.1), replacing $f = \tilde{w}_2(M)$ by $f = \tilde{p}_1(M)$. Part a) is identical. For Part b), note that $K(\mathbb{Q}, 3)$ is a rational homology 3-sphere and one then has the principal fibration :

$$K(\mathbb{Q}, 7) \longrightarrow \Sigma K(\mathbb{Q}, 3) \longrightarrow K(\mathbb{Q}, 4)$$

Therefore, $\mathcal{L}_f^2$ is in bijection with $[M, K(\mathbb{Q}, 7)] \simeq H^7(M; \mathbb{Q}) \approx H_1(M; \mathbb{Q})$. $\square$

References

1. Douady A. & Hérault L., *Arrondissement des variétés à coins (appendice à un article de Borel-Serre)*, Comm. Math. Helv. **48** (1973) 484–489.
2. Hausmann J-C.& Vogel P., *Manifold categories of Poincaré duality spaces*, Differential Topology–Siegen, Springer. Lect. Notes **1350** (1987) 241–258.
3. Hudson J. F. P., *Piecewise linear topology*, Benjamin,1969.
4. Husemoller D., *Fibre bundles*, 2nd ed., Springer-Verlag, 1966.
5. James. I. M., *On category, in the sense of Lusternik-Schnirelmann*, Topology **17** (1978) 331–348.
6. Milnor J., *Construction of universal bundle* I, Annals of Math. **63** (1956) 272–284.
7. _____, *Construction of universal bundle* II, Annals of Math. **63** (1956) 430–436.

MATHÉMATIQUES-UNIVERSITÉ, B.P. 240, CH-1211 GENEVA 24, SWITZERLAND

E-mail address: hausmann@ibm.unige.ch

AMS/IP Studies in Advanced Mathematics
Volume **2**, 1997 (Part 1)

THE CENTERS OF FIBERED TWO-KNOT GROUPS

KATSUYUKI YOSHIKAWA

1. Introduction

An *n-knot* K is a smooth n-sphere in the $(n+2)$-sphere S^{n+2}. The fundamental group $\pi_1(S^{n+2}-K)$ of the complement of K is called the *group* of K.

We consider the centers of n-knot groups. For $n=1$, L. Neuwirth [9] showed that the center of a 1-knot group is trivial or infinite cyclic, and G. Burde and H. Zieschang [2] proved that the torus 1-knot groups are the only 1-knot groups with nontrivial center. For $n \geq 3$, J. C. Hausmann and M. Kervaire [3] showed that any finitely generated abelian group is the center of an n-knot group. In case $n=2$, some results on the centers of 2-knot groups have been obtained. For example, it is known that there exist 2-knot groups with center $1, Z_2, Z, Z \oplus Z_2$ or $Z \oplus Z$ ([10], [11], [12]), and J. A. Hillman [6] proved that if the center of a 2-knot group has rank greater than 1, then the center is torsion free of rank 2. However, it still remains open to characterize the centers of 2-knot groups. In this paper, we completely determine the centers of the groups of 2-knots with minimal Seifert manifold and those of the groups of fibered 2-knots.

2. Preliminaries

Let D and H be groups with monomorpihsms ϕ_1 and ϕ_2 of D to H, and let $H=<S:R>$. Then the *HNN extension (Higman-Neumann-Neumann extension)* of H with respect to ϕ_1 and ϕ_2 is the group defined by

$$G=<S,t:R,t^{-1}\phi_1(d)t=\phi_2(d),\forall d\in D> .$$

The group H is called the *base* of G. We denote the group G by $<H,t: t^{-1}\phi_1(D)t=\phi_2(D)>$. For a group G, denote the center of G by $C(G)$.

Let K be an n-knot. Then there exists a compact orientable submanifold V^{n+1} in S^{n+2} such that $\partial V=K$. This is called a *Seifert manifold* for K. Let Y be the $(n+2)$-manifold obtained from S^{n+2} by cutting along V. Then Y is a compact manifold with boundary $\partial Y=V_1\cup V_2$, where V_1 and V_2 are two copies of V with $V_1\cap V_2=\partial V_1\cap\partial V_2=K$. Let

$\nu_i : V \to V_i \hookrightarrow Y, i = 1, 2$, be the compositions. If both of the induced homomorphisms $\nu_{i*} : \pi_1(V) \to \pi_1(Y), i = 1, 2$, are monomorphisms, we say that V is a *minimal* Seifert manifold for K. If an n-knot K has a minimal Seifert manifold V, then the group G of K is expressed as an HNN extension

$$< H, t : t^{-1}\nu_{1*}(D)t = \nu_{2*}(D) >,$$

where $H = \pi_1(Y)$ and $D = \pi_1(V)$. For any n-knot group G, it is known that $G/[G, G] \cong Z$ and $H_2(G) = 0$, where $[G, G]$ is the commutator subgroup of G and $H_2(G)$ is the second homology group of G with integral coefficients.

We will prove two lemmas.

Lemma 1. *Let G be a group. If $G/[G, G] \cong Z$, then we have*

$$C(G) \cong C(G) \cap [G, G] \text{ or } (C(G) \cap [G, G]) \oplus Z.$$

Proof. Let $A = C(G) \cap [G, G]$. Since $G/[G, G] \cong Z$, it is easy to see that $C(G)/A \cong 0$ or Z. Therefore we obtain the split exact sequence of abelian groups

$$0 \to A \to C(G) \to C(G)/A \to 0.$$

It follows that $C(G) \cong A \oplus (C(G)/A)$. The proof is completed. $\square$

Lemma 2. *Let G be the HNN extension $< H, t : t^{-1}\phi_1(D)t = \phi_2(D) >$ of a group H. If $G/[G, G] \cong Z$ and $H_2(G) = 0$, then we have*

$$C(G) \cap [G, G] \subset C(\phi_1(D)) \cap [\phi_1(D), \phi_1(D)].$$

Proof. To prove that $C(G)\cap[G, G] \subset C(\phi_1(D))$, it suffices to show that $C(G) \cap [G, G] \subset \phi_1(D)$. Since $G/[G, G] \cong Z$, we have $[G, G] =<< H >>^G$, where $<< H >>^G$ is the normal closure of H. Therefore the commutator subgroup $[G, G]$ is the amalgamated free product of infinitely many factors

$$\cdots *_{D_{-2,2}=D_{-1,1}} H_{-1} *_{D_{-1,2}=D_{0,1}} H_0 *_{D_{0,2}=D_{1,1}} \cdots,$$

where $H_i = t^{-i}Ht^i$ and $D_{i,j} = t^{-i}\phi_j(D)t^i$, $i = 0, \pm1, \ldots$, $j = 1, 2$. If $\phi_1(D) = H$, then we have the descending sequence

$$\cdots \supset H_{-1} \supset H_0 \supset H_1 \supset \cdots.$$

Therefore we see that $[G, G] = \cup_{i=-\infty}^{+\infty} H_i$. Hence, for each element x of $C(G) \cap [G, G]$, there exists an integer k such that $x \in H_k = t^{-k}Ht^k$. Since $x \in C(G)$, it follows that $x = t^kxt^{-k} \in H$, i.e., $x \in \phi_1(D)$. Therefore we obtain $C(G) \cap [G, G] \subset \phi_1(D)$. Similarly, if $\phi_2(D) = H$, then we have $C(G)\cap[G, G] \subset \phi_2(D)$. Therefore we see that $C(G)\cap[G, G] = t(C(G)\cap[G, G])t^{-1} \subset t\phi_2(D)t^{-1} = \phi_1(D)$. Hence, suppose that $\phi_1(D) \neq$

H and $\phi_2(D) \neq H$. Let $A = \cdots *_{D_{-3,2}=D_{-2,1}} H_{-2} *_{D_{-2,2}=D_{-1,1}} H_{-1}$ and $B = H_0 *_{D_{0,2}=D_{1,1}} H_1 *_{D_{1,2}=D_{2,1}} \cdots$. Then $[G,G]$ is also expressed as the amalgamated free product $A *_{D_{-1,2}=D_{0,1}} B$ of A and B with amalgamated subgroup $D_{0,1}(= \phi_1(D))$. Hence, from [7, Corollary 4.5, p.211], we obtain

$$C(G) \cap [G,G] \subset C([G,G]) \subset \phi_1(D).$$

Therefore we conclude that $C(G) \cap [G,G] \subset C(\phi_1(D))$.

Next we prove that $C(G) \cap [G,G] \subset [\phi_1(D), \phi_1(D)]$. FFrom [1], we obtain the exact sequence

$$\cdots \to H_e2(G) \to H_1(D) \stackrel{\phi_{1*}-\phi_{2*}}{\longrightarrow} H_1(H) \to H_1(G) \to \cdots.$$

Therefore, since $H_1(D) = D/[D,D], H_1(H) = H/[H,H]$ and $H_2(G) = 0$, we have the exact sequence

$$0 \to D/[D,D] \stackrel{\phi_{1*}-\phi_{2*}}{\longrightarrow} H/[H,H].$$

Hence $\phi_{1*} - \phi_{2*} : D/[D,D] \to H/[H,H]$ is a monomorphism. Let c be an element of $C(G) \cap [G,G]$, and let $\tilde{c} = \phi_1^{-1}(c)$ $(\in D)$. Then, we have $\phi_1(\tilde{c}) = t^{-1}\phi_1(\tilde{c})t = \phi_2(\tilde{c})$. Therefore we see that

$$\begin{aligned}(\phi_{1*} - \phi_{2*})(\alpha(\tilde{c})) &= \phi_{1*}\alpha(\tilde{c}) - \phi_{2*}\alpha(\tilde{c}) \\ &= \phi_{1*}\alpha(\tilde{c}) - \phi_{1*}\alpha(\tilde{c}) = 0,\end{aligned}$$

where $\alpha : D \to D/[D,D]$ is the abelianization of D. Since $\phi_{1*} - \phi_{2*} : D/[D,D] \to H/[H,H]$ is one-one, it follows that $\alpha(\tilde{c}) = 0$. Therefore we have $\tilde{c} \in D/[D,D]$, that is, $c \in [\phi_1(D), \phi_1(D)]$. The proof is completed. □

3. The centers of the groups of 2-knots with minimal Seifert manifold.

Concerning the centers of the groups of 2-knots with minimal Seifert manifold , we obtain the following:

Theorem 1. *A finitely generated abelian group A is isomorphic to the center of some 2-knot group with minimal Seifert manifold if and only if $A \cong 1, Z_2, Z, Z \oplus Z_2$ or $Z \oplus Z$.*

Proof. The necessity follows from [10] and [11]. Conversely, suppose that K is a 2-knot with minimal Seifert manifold V. Let $D = \pi_1(V)$ and $H = \pi_1(Y)$. Then the group G of K is expressed as an HNN extension

$$< H, t : t^{-1}\nu_{1*}(D)t = \nu_{2*}(D) > .$$

If $C(D) = 1$, then, by Lemma 2, $C(G) \cap [G, G] = 1$. Therefore, from Lemma 1, we have $C(G) \cong 1$ or Z. Assume that $C(D)$ is finitely generated and nontrivial. Then, by [4, p.131], the group D of an orientable compact 3-manifold which has finitely generated nontrivial center is one of the following cases:

Case (1) : If $C(D)$ is finite, then so is D.
Case (2) : $C(D) \cong Z$.
Case (3) : $D \cong Z \oplus Z \cong C(D)$.
Case (4) : $D \cong Z \oplus Z \oplus Z \cong C(D)$.

In case (2), by Lemma 2, we have $C(G) \cap [G, G] \cong 1$ or Z. Therefore, Lemma 1 shows that $C(G) \cong 1, Z$ or $Z \oplus Z$. In cases (3) and (4), we have

$$C(D) \cap [D, D] \cong 1.$$

Thus , by Lemmas 1 and 2, we have $C(G) \cong 1$ or Z. In case (1), from Milnor's list [8] of finite 3-manifold groups, it is not difficult to verify that

$$C(D) \cap [D, D] \cong 1 \text{ or } Z_2.$$

Therefore, by Lemmas 1 and 2, we obtain $C(G) \cong 1, Z_2, Z$ or $Z \oplus Z_2$. The proof is completed. □

4. The centers of the groups of fibered 2-knots.

It is known that there exist fibered 2-knots whose groups have the centers $1, Z, Z \oplus Z_2$ and $Z \oplus Z$ respectively ([10], [12]). In this section, we prove the following:

Theorem 2. *A finitely generated abelian group A is isomorphic to the center of the group of some fibered 2-knot if and only if $A \cong 1, Z, Z \oplus Z_2$ or $Z \oplus Z$.*

Proof. The centers of the groups of the spun figure-eight 2-knot and the 1-, 3- and 6-twist spun trefoil 2-knots, which are fibered, are $1, Z, Z \oplus Z_2$ and $Z \oplus Z$, respectively [10]. Conversely suppose that K is a fibered 2-knot whose group has finitely generated center. Then K has a minimal Seifert manifold V. Therefore, by Theorem 1, it suffices to show that if the finitely generated center $C(G)$ of a fibered 2-knot group G has an element of order two, then $C(G)$ is isomorphic to $Z \oplus Z_2$. If $C(G)$ has an element of order two, then it follows from the proof of Theorem 1 that $D(= \pi_1(V))$ is finite. Since K is fibered, we have $\nu_{1*}(D) = \nu_{2*}(D) = H = [G, G]$. Therefore the commutator subgroup $[G, G]$ of G is finite. Hence we see that $C(G)$ is isomorphic to $Z \oplus Z_2$ ([5], [11]). The proof is completed. □

Remark. If the center of a 3-manifold group is not finitely generated, then it is isomorphic to a subgroup A of the additive group Q of the rationals [4]. Therefore, in the same way as in the above, it can be shown that if the center of the group of a 2-knot with minimal Seifert manifold (or a fibered 2-knot) is not finitely generated, then it is isomorphic to A or $A \oplus Z$, where A is a subgroup of the additive group Q of the rationals. It is not known whether such a 2-knot group exists.

References

1. R. Bieri, *Mayer-Vietoris sequences for HNN-groups and homological duality*, Math. Z. **143** (1975) 123–130.
2. G. Burde and H. Zieschang , *Eine Kennzeichnung der Torusknoten*, Math. Ann. **167** (1966) 169–176.
3. J. C. Hausmann and M. Kervaire , *Sur le centre des groupes de noeuds multidimensionnels*, C. R. Acad. Sc. Paris, t. **287** (1978) 699–702.
4. J. Hempel, *3-manifolds*, Ann. of Math. Studies 86, Princeton Univ. Press, 1976.
5. J. A. Hillman, *High dimensional knot groups which are not two-knot groups*, Bull. Austral. Math. Soc. **16** (1977) 449–462.
6. J. A. Hillman, *2-knot and their groups*, Cambrige Univ. Press, 1989.
7. W. Magnus, A. Karrass and D. Solitar, *Combinatorial Group Theory*, New York, Wiley, 1966.
8. J. Milnor, *Groups which act on S^n without fixed points*, Amer. J. Math. **79** (1957) 623–630.
9. L. Neuwirth, *The algebraic determination of the genus of knots*, Amer. J. Math. **82** (1960) 791–798.
10. K. Yoshikawa, *On 2-knot groups with the finite commutator subgroups*, Math. Sem. Notes Kobe Univ. **8** (1980) 321–330.
11. K. Yoshikawa, *On a 2-knot group with nontrivial center*, Bull. Austral. Math. Soc. **25** (1982) 321–326.
12. E. C. Zeeman, *Twisting spun knots*, Trans. Amer. Math. Soc. **115** (1965) 471–495.

Faculty of Science, Kwansei Gakuin University, Nishinomiya, Hyogo 662, Japan
E-mail address: yoshikawa@kgupyr.kwansei.ac.jp

AMS/IP Studies in Advanced Mathematics
Volume **2**, 1997 (Part 1)

EQUIVARIANT PERIODICITY IN SURGERY FOR ACTIONS OF SOME NONABELIAN GROUPS

MIN YAN

ABSTRACT. The isovariant homotopy classification was proved to have equivariant periodicity $S_G(M, \text{rel } \partial) \cong S_G(M \times DV, \text{rel } \partial)$ for the following cases: V is the four fold permutation representation of an odd order group [14], or V is twice of any complex representation of a compact abelian group [13]. In this paper, the equivariant periodicity is proved for twice of the natural complex representation of $SU(2)$, and twice of any complex representation of $O(2)$, providing further evidence that the equivariant periodicity should be true for twice of any complex representation of any compact Lie group.

Let $S(M, \text{rel } \partial M)$ be[1] the homeomorphism classes of topological manifolds that are homotopy equivalent to M and restrict to homeomorphisms on the boundary. Siebenmann proved that the homotopy classification of topological manifolds has the four fold periodicity (Appendix C of [5]):

$$S(M, \text{rel } \partial M) \cong S(M \times D^4, \text{rel } \partial(M \times D^4)).$$

As pointed out by Nicas, this is in fact not quite correct in the context of manifolds [7]. The deviation may be a copy of $\mathbf{Z}$. On the other hand, this is indeed true if the consideration is enlarged to ANR-homology manifolds [2]. Our purpose is to extend the periodicity to the isovariant (i.e., preserving the isotropy groups) homotopy classification of equivariant topological manifolds

$$S_G(M, \text{rel } \partial M) \cong S_G(M \times DV, \text{rel } \partial(M \times DV)),$$

where DV is the unit ball of some unitary G-representation V. Because of the fact that G may act nontrivially on DV, such kind of

[1]Research is supported by RGC grant HKUST 632/94P

generalization is theoretically and practically quite useful (see [13] for a more detailed discussion). Again the exact isomorphism is true only in the context of equivariant ANR-homology manifolds. However, the deviation in the context of equivariant topological manifolds is several copies of $\mathbf{Z}$ and is well understood. Therefore we will not be concerned with the deviation and will pretend that the exact statement is true for topological manifolds.

Under some mild combinatorial and small gap conditions, the equivariant periodicity has been proved for the following cases:

(1) G is a finite group. $V = \mathbf{R}S \otimes \mathbf{R}^4$ is a permutation representation, where S is a finite G-set such that all its orbits are of odd order [14].

(2) G is a compact abelian group, including the torus group. $V = W \oplus W$ is twice of some complex representation W [13].

Moreover, it was suggested in [13] that the equivariant periodicity should be true for twice of a complex representation of any (abelian or non-abelian) compact Lie group. This paper provides some evidence in this direction. The main result is the following theorem.

Theorem 0.1. *Suppose that M is a homotopically stratified G-manifold with codimension ≥ 3 gap. Suppose that V is a G-representation such that M and $M \times V$ have the same isotropy everywhere. Then there is a natural isomorphism*

$$S_G(M, \mathrm{rel}\ \partial M) \cong S_G(M \times DV, \mathrm{rel}\ \partial(M \times DV))$$

for the following cases:

(1) $V = \mathbf{C}^2 \oplus \mathbf{C}^2 = \mathbf{R}^8$ is twice the natural representation of $G = SU(2)$ on $\mathbf{C}^2$;

(2) $V = \mathbf{H} \otimes_{\mathbf{R}} \mathbf{R}^2 = \mathbf{R}^8$ is the quaternionization of the natural representation of $G = O(2)$ on $\mathbf{R}^2$.

The result remains true for the representations of G' induced from a group homomorphism $G' \to G$.

The codimension ≥ 3 gap condition is the following: If $H \subset K$ are subgroups of G, and $M_\alpha^H \supset M_\beta^K$ are two connected components of fixed point subsets. Then either $M_\alpha^H = M_\beta^K$ or $\dim M_\alpha^H \geq \dim M_\beta^K + 3$. The condition will make sure that the isovariant and equivariant connected fixed point components are in one-to-one correspondence, and the corresponding components have isomorphic fundamental groups. The assumption is useful because the surgery and K-obstructions are basically determined by such combinatorial data. Moreover, such gap condition is often satisfied in applications.

The condition that M and $M \times V$ have the same isotropy everywhere was introduced in [14] and means the following: For any $x \in M$, there is an equivariant neighborhood U of x, such that $\text{iso}(U) = \text{iso}(U \times V)$, where "iso" means the collection of isotropy groups. The condition makes sure that the isovariant fixed point components of M and $M \times V$ are in one-to-one correspondence, a naturally necessary combinatorial condition for the structures on M and $M \times DV$ to be equivalent. Moreover, if M has codimension ≥ 3 gap, then the same isotropy everywhere condition implies that $M \times V$ also has codimension ≥ 3 gap.

The statement in the theorem about representations induced from group homomorphisms may be applied to subgroups. For example, we get periodicity for the dihedral groups as subgroups of $O(2)$.

From the representation theory, any complex representation of $O(2)$ is induced either from a representation of $\mathbf{Z}_2$ via the group homomorphism $\det : O(2) \to \{\pm 1\}$, or from the complexification of the natural $O(2)$-representation on $\mathbf{R}^2$ via a homomorphism $O(2) \to O(2)$. In fact, this is also true for the dihedral subgroups of $O(2)$. Therefore by repeatedly making use of the periodicity for the abelian group $\{\pm 1\}$ and the above theorem, we may conclude the following:

Corollary 0.2. *The equivariant periodicity holds for twice of any complex representation of $O(2)$ or D_{2n}.*

The proof of the theorem is similar to the one in [13]. The key is to construct a periodicity space that contains the representation, so that away from the representation, we have a stratified $\pi - \pi$ structure. The Wall's $\pi - \pi$ theorem [11] can be generalized to show that the surgery obstructions always vanish for the product of the such $\pi - \pi$ structure with any equivariant manifold. Moreover, the Tate cohomology of the K-theoretical obstructions also vanish for such products. By making use of Weinberger's stratified surgery theory [12], we are then able to prove the periodicity.

This paper is organized as follows: In the first part, the key properties that enable us to prove the periodicity in [13],[14] are analysed. This motivates the definition 1.1 of stratified isovariant $\pi - \pi$ structures and the definition 1.8 of equivariant periodicity spaces. Some useful properties of these objects are also proved. In the second part, the periodicity spaces for actions by $SU(2)$ and $O(2)$ are constructed. The $SU(2)$-periodicity space is a simple straightforward generalization of the S^1-periodicity space in [13]. The construction of $O(2)$-periodicity space requires much more effort. In the third part, the general periodicity theorem 3.1 is proved. Therefore if one can find periodicity spaces (according to the definition 1.8) for other group actions, then one would obtain corresponding periodicity statement.

The author would like to point out that the definition 1.8 is very likely to be superseded by better ones. I hope that with more advanced machinery, one may come up with a definition of periodicity spaces that enables us to prove the following conjecture posed in [13]: Twice of any complex representation of any compact Lie group is a periodicity representation.

1. Definition of Periodicity Spaces.

The periodicity in the classical (nonequivariant) surgery theory was induced from the following isomorphisms

$$L(X) \overset{\times \mathbf{C}P^2}{\cong} L(X \times \mathbf{C}P^2) \overset{\text{incl}}{\cong} L(X \times D^4, \text{rel } S^3). \tag{1}$$

The reason for $\times \mathbf{C}P^2$ to be isomorphic is that $\mathbf{C}P^2$ is a closed manifold of signature 1. The reason for the inclusion to be isomorphic is that $\mathbf{C}P^2$ is connected and simply connected. Therefore if we replace $\mathbf{C}P^2$ by any closed, connected, simply connected manifold P^{4p} of signature 1, and replace (D^4, S^3) by (D^{4p}, S^{4p-1}), then (1) is still isomorphic. Such a manifold P is a *periodicity manifold* for the classical surgery theory.

The *equivariant periodicity manifolds* for the equivariant surgery theory were introduced in [4] and [14] as a straightforward generalization of the classical periodicity manifolds. A G-manifold P is a G-periodicity manifold if for any subgroup H of G, P^H is closed, connected, simply connected, and has equivariant $\pi_0 WH$-signature 1. Here the signature condition means that the $\pi_0 WH$-invariant intersection form of P^H at the middle dimension is isomorphic to the form

$$\text{multiplication} : \mathbf{Z} \otimes \mathbf{Z} \to \mathbf{Z}, \quad \text{trivial group action}$$

up to adding equivariant hyperbolic symmetric forms.

Unfortunately, such equivariant periodicity manifolds do not arise easily. The candidate considered in [14] is the following

$$P = \times^S \mathbf{C}P^2 = \mathbf{C}P^2 \times \mathbf{C}P^2 \times \cdots \times \mathbf{C}P^2 \quad (S \text{ copies}),$$

where S is a finite G-set, and G acts on P through permutation. It was then proved (theorem 3.16 of [14]) that if the orbits in S are all of odd orders, then P is a periodicity manifold. Dovermann and Schultz considered the special case $S = G$ for a finite group G acting by left multiplication. In addition to the fact that P is a periodicity manifold for odd order G, they also showed that P is not periodic for even order

G (theorem 3.7(ii) of [4], even for the general case that $\mathbf{C}P^2$ is replaced by any classical periodicity manifold).

It should also be noted that Browder suggested the construction of $P = \times^G \mathbf{C}P^2$ with the permutation action by a finite group G long time ago (1976 AMS Summer Symposium on Algebraic and Geometric Topology at Stanford). The purpose was to increase the dimensional gap between fixed point components of various subgroups to a big one by crossing with (maybe several copies of) P. The periodicity property of P implies that the process preserves much interesting geometric information, while the big gap enables people to apply many geometric machineries. The purpose is well served for odd order groups and will evidently not be achieved for even order ones. In fact, Dovermann and Schultz proved that one may never obtain big gaps by crossing with $\mathbf{Z}_2$-periodicity manifolds (theorem 3.7(i) of [4]).

It appears that there is no useful periodicity for actions of even order groups by using only periodicity *manifolds*. However, the stratified approach to group actions motivates the construction in [13] of a periodicity *space* for actions of the circle group S^1 and all cyclic groups. The resulting periodicity may be further extended to actions of all compact abelian groups. Moreover, such periodicity often increases the gap to a big one, for actions by abelian groups of odd as well as even orders.

The S^1-periodicity space considered in [13] is

$$P = \mathbf{C}P^2 \cup D^3, \tag{2}$$

where D^3 is attached to $\mathbf{C}P^2$ via the identification $S^2 = \mathbf{C}P^1$. The action of S^1 is trivial on D^3, and by the formula

$$\lambda[z_1, z_2, z_3] = [\lambda z_1, \lambda z_2, z_3] \tag{3}$$

on $\mathbf{C}P^2$.

By $\mathbf{C}P^2 = D^4 \cup_{S^3 \nearrow} S^2$ (the map $S^3 \to S^2$ is the Hopf bundle map), P is an S^1-stratified space with three strata

$$P = D^4 \cup_{S^3 \nearrow} S^2_{\nwarrow S^2} \cup D^3. \tag{4}$$

S^1 acts on $D^4 \subset \mathbf{C}^2$ (the first two coordinates of (3)) by complex multiplication, and acts trivially on D^3 and S^2.

The following two properties enable us to prove the S^1-periodicity:

(1) The signature of $\mathbf{C}P^2$ is 1, and the euler characteristic is 3, an odd number;
(2) D^3 is a manifold with boundary S^2. Both are fixed by S^1 , and have isomorphic fundamental groups (the trivial group).

Strictly speaking, we should consider the $\pi_0 S^1$-equivariant signature and euler characteristic. However, these are the same as the nonequivariant one because $\pi_0 S^1$ is trivial.

The S^1-periodicity is similarly obtained by first considering (compare with (1)):

$$(5) \qquad L_{S^1}(X) \overset{\times P}{\to} L_{S^1}(X \times P) \overset{\text{incl}}{\leftarrow} L_{S^1}(X \times D^4, \text{rel } S^3),$$

where the S^1-stratification on $X \times P$ is induced from that of P. The inclusion map fits into a long exact sequence in which the third term is $L_{S^1}(X \times (D^3, S^2))$. Because of the second property of P, each isovariant fixed point component of $X \times (D^3, S^2)$ is a $\pi - \pi$ pair. By the equivariant version of Wall's $\pi - \pi$ theorem [3], $L_{S^1}(X \times (D^3, S^2))$ vanishes. Consequently, the inclusion is an isomorphism. It is a little more complicated to show that $\times P$ is isomorphic. We simply point out that this essentially is a consequence of the first property of P. The fact about the euler characteristic is used in studying the map $\times P$ in the "destablization" stage of the homotopically stratified surgery theory.

It is quite conceivable that periodicity spaces may be constructed for actions of other compact Lie groups. In order to clarify the structure of such spaces, we first elaborate on the second property of P.

Let G be a compact Lie group. A G-stratified space X is generally indexed by a partially ordered set A. The closed strata of X will be denoted by X_α, while the (pure) open ones by X^α. X^α is a G-space.

X/G is a stratified space doubly indexed by A and the collection $\text{iso}(X)$ of isotropy subgroups of X. We say X is *homotopically G-stratified* if X/G is a homotopically stratified space.

Let B be a partially ordered set. Let ∂B be a copy of B, and $\beta \in B$ correspond to $\partial\beta \in \partial B$. Then the double $2B = B \coprod \partial B$ is partially ordered by keeping the original partial orders in B and ∂B, and adding the extra orders $\partial\beta < \beta$.

Definition 1.1. A homotopically G-stratified space X is an *isovariant $\pi - \pi$ structure* if X is indexed by the double $2B$ of some partially ordered set B, such that for any $\beta \in B$, $X^\beta \cup X^{\partial\beta}$ is a manifold with boundary $X^{\partial\beta}$, and the pair $(X^\beta \cup X^{\partial\beta}, X^{\partial\beta})$ satisfies the *isovariant $\pi - \pi$ condition*: the connected components of X^β and $X^{\partial\beta}$ are in one-to-one correspondence; and the corresponding ones have isomorphic fundamental groups.

The classical $\pi - \pi$ theorem has the following straightforward elaboration.

Proposition 1.2. *Suppose a homotopically G-stratified space X is an isovariant $\pi - \pi$ structure. Then $L_G(X)$ vanishes.*

Proof. Let β be a minimal element of B. Then $L_G(X)$ fits into a long exact sequence in which the other two terms are $L_G(X^\beta \cup X^{\partial\beta}, X^{\partial\beta})$ and $L_G(X - (X^\beta \cup X^{\partial\beta})) = L_G(X, \text{rel } X^\beta \cup X^{\partial\beta})$. Since $(X^\beta \cup X^{\partial\beta}, X^{\partial\beta})$ satisfies the isovariant $\pi - \pi$ condition, $L_G(X^\beta \cup X^{\partial\beta}, X^{\partial\beta})$ vanishes by the equivariant $\pi - \pi$ theorem. Since $X - (X^\beta \cup X^{\partial\beta})$ is also an isovariant $\pi - \pi$ structure with smaller index set $2(B - \{\beta\})$, its surgery obstructions may be assumed to vanish by induction. Then we may conclude that $L_G(X)$ vanishes. □

Remark 1.3. We will need the similar proposition for the other obstruction theories. By inspecting the proof of the proposition, we see that the following are the sufficient conditions for the proposition 1.2 to be valid for a stratified obstruction theory Λ:

(1) (classical $\pi - \pi$ theorem) For an unstratified manifold pair $(X, \partial X)$ with X, ∂X connected, and $\pi_1 X = \pi_1 \partial X$, the obstruction $\Lambda(X, \partial X)$ is trivial;
(2) (decomposition along singularities) For a stratified space X and a closed union Y of strata of X, $\Lambda(X)$ fits into a long exact sequence in which the other two terms are $\Lambda(Y)$ and $\Lambda(X - Y) = \Lambda(X, \text{rel } Y)$.

The (nonstratified) equivariant $\pi - \pi$ theorem used in the proof of the proposition 1.2 is a consequence of the two conditions.

Because we will take product with periodicity spaces, we need to know the behavior of isovariant $\pi - \pi$ structures with respect to products.

Proposition 1.4. *Suppose X and Y are homotopically G-stratified spaces. If X is an isovariant $\pi - \pi$ structure, then $X \times Y$ is also an isovariant $\pi - \pi$ structure.*

Proof. Suppose the $\pi - \pi$ structure of X is indexed by $2B$, and Y is indexed by C. Then $X \times Y$ has a stratification indexed by $2(B \times C)$ with strata pairs $(X \times Y)^{(\beta,\gamma)} = X^\beta \times Y^\gamma$ and $(X \times Y)^{\partial(\beta,\gamma)} = X^{\partial\beta} \times Y^\gamma$. However, it is not immediately clear that the pair satisfies the isovariant $\pi - \pi$ condition.

The problem is that the isovariant components behave badly with respect to products ($X_H = \{x \in X : G_x = H\}$):

$$(6) \qquad (X \times Y)_H = \cup_{H_1 \cap H_2 = H} X_{H_1} \times Y_{H_2}.$$

However, we may view the above expression as a stratification of $(X \times Y)_H$ with pure strata $X_{H_1} \times Y_{H_2}$. Therefore to complete the proof of the proposition, it suffices to show the following result and then apply it to the inclusion $X^{\partial\beta} \times Y^{\gamma} \subset (X^{\beta} \cup X^{\partial\beta}) \times Y^{\gamma}$. □

Proposition 1.5. *Suppose that $f : X \to Y$ is a homotopically transverse map between homotopically stratified spaces. If f is a one-to-one correspondence between the components of pure strata, and induces surjections (or isomorphisms) between fundamental groups of the corresponding components, then $f_* : \pi_1 X \to \pi_1 Y$ is surjective (or isomorphic).*

Proof. We may assume that $X = \bar{X} \cup E_X \times I \cup X_0$, where $\bar{X}$ is a top stratum of X, X_0 is a closed union of all strata except $\bar{X}$, and $E_X \to X_0$ is a stratified system of fibrations.

We may further assume that Y and f have similar decompositions, such that $f_E : E_X \to E_Y$ is a map of stratified systems of fibrations over the homotopically stratified map f_0.

By Vam-Kampen theorem, we have pushouts of fundamental groups and maps between them

$$(7) \qquad \begin{array}{ccc} \pi_1 E_X & \to & \pi_1 X_0 \\ \downarrow & & \downarrow \\ \pi_1 \bar{X} & \to & \pi_1 X \end{array} \Longrightarrow \begin{array}{ccc} \pi_1 E_Y & \to & \pi_1 Y_0 \\ \downarrow & & \downarrow \\ \pi_1 \bar{Y} & \to & \pi_1 Y \end{array}$$

Suppose that f induces surjections on the fundamental groups of pure strata. Then $\bar{f}_* : \pi_1 \bar{X} \to \pi_1 \bar{Y}$ is surjective. Moreover, we may assume by induction that $f_{0*} : \pi_1 X_0 \to \pi_1 Y_0$ is surjective. Therefore, we conclude that $f_* : \pi_1 X \to \pi_1 Y$ is surjective.

To prove the isomorphism part of the statement, we observe that the homotopy stratifications of X_0 and Y_0 induce homotopy stratifications of E_X and E_Y, such that the restrictions of f_E on pure strata are the fibration maps over the corresponding restrictions of f_0. The assumption that f_0 induces isomorphisms between the fundamental groups of the pure strata of X_0 and Y_0 implies that f_{E*} induces surjections (not necessarily isomorphisms!) between the fundamental groups of the pure strata of E_X and E_Y. By the surjectivity part of the proposition that we just proved, $f_{E*} : \pi_1 E_X \to \pi_1 E_Y$ is surjective.

Now in the map (7) between pushouts, f_{E*} is surjective, $\bar{f}_*$ is assumed to be isomorphic, and f_{0*} may be assumed to be isomorphic by induction. It then follows that $f_* : \pi_1 X \to \pi_1 Y$ is isomorphic. □

Remark 1.6. Strictly speaking, if some spaces in (6) are not connected, then we need to use the term groupoid instead of groups in the proof. In the language of fundamental groups, there may be some extra generators and relations that produce HNN-extensions. The one-to-one correspondence between the components makes these HNN-extensions equivalent for X and Y.

The following useful result says that isovariant $\pi - \pi$ structures may be restricted and induced.

Proposition 1.7. *Suppose that a homotopically G-stratified space X is an isovariant $\pi - \pi$ structure. Then for any subgroup H of G, X is an isovariant $\pi - \pi$ structure as an H-space, and X^H is an isovariant $\pi - \pi$ structure as a WH-space.*

Proof. The statement about X^H follows from the definition. In proving the statement about X as an H-space, one faces the complication of writing down the isovariant components of the H-action in terms of the components of the G-action:

$$X_{K,\text{for } H\text{-action}} = \cup_{K' \cap H = G} X_{K',\text{for } G\text{-action}}, \qquad K \subset H. \tag{8}$$

As in the proof of the proposition 1.4, we think of the expression (8) as a stratification of the isovariant components of the H-space X, with certain $X_{K',\text{for } G\text{-action}}$ as pure strata. Then by the proposition 1.5, the inclusion

$$X^{\partial\beta}_{K,\text{for } H\text{-action}} \subset (X^{\partial\beta} \cup X^{\beta})_{K,\text{for } H\text{-action}}$$

induces an isomorphism between fundamental groups. □

Now we are ready to present our definition of the equivariant periodicity spaces.

Definition 1.8. A *G-periodicity space* with *periodicity representation* V is a homotopically G-stratified space P satisfying the following properties:

(1) The representation disk DV is a closed stratum of P. Moreover, a closed union of DV with some other strata of P is a closed G-manifold Q after forgetting the stratification. Q has the property that for any subgroup $H \subset G$, the connected component of Q^H that contains DV^H is a WH-manifold with $\pi_0 WH$-signature 1 and $\pi_0 WH$-euler chracteristic an odd number;
(2) $\overline{P - DV}$ is a G-stratified isovariant $\pi - \pi$ structure.

Remark 1.9. The definition is very likely to be superseded by better ones as more technical machineries are available. Here we simply specify the basic properties our machinery requires.

By making use of the proposition 1.4, it is easy to prove the following result.

Lemma 1.10. *The products of periodicity spaces are periodicity spaces.*

It also immediately follows from our definition and the proposition 1.7 that periodicity spaces can be induced and restricted.

Lemma 1.11. *If P is a G-periodicity space with G-periodicity representation V, then*

(1) *For any homomorphism $G' \to G$, the induced action makes P a G'-periodicity space with the induced G'-periodicity representation V;*

(2) *For any subgroup $H \subset G$, P^H is a WH-periodicity space with WH-periodicity representation V^H.*

2. Periodicity Spaces for $SU(2)$ and $O(2)$.

We will make the free use of the following facts in constructing the periodicity spaces for $SU(2)$ and $O(2)$.

S^1 is the circle group of complex numbers of norm 1. $SO(2) \cong S^1$ and the natural representation of $SO(2)$ on $\mathbf{R}^2 = \mathbf{C}$ is the same as the scalar multiplication by complex numbers in S^1.

Similarly, let S^3 be the Lie group of quaternionic numbers of norm 1. Then $SU(2) \cong S^3$ and the natural representation of $SU(2)$ on $\mathbf{C}^2 = \mathbf{H}$ is the same as the scalar multiplication *on the right* by quaternions in S^3.

Motivated by (2), we consider the $SU(2)$-stratified space

$$P = \mathbf{H}P^2 \cup D^5, \tag{9}$$

where D^5 is attached to $\mathbf{H}P^2$ via the identification $S^4 = \mathbf{H}P^1$. The action of $SU(2) = S^3$ is trivial on D^5, and by the formula

$$\lambda[z_1, z_2, z_3] = [z_1\lambda, z_2\lambda, z_3] \tag{10}$$

on $\mathbf{H}P^2$.

By $\mathbf{H}P^2 = D^8 \cup_{S^7 \nearrow} S^4$ (the map $S^7 \to S^4$ is the quaternionic Hopf bundle map). P is an $SU(2)$-stratified space with three strata

$$P = D^8 \cup_{S^7 \nearrow} S^4_{\nwarrow S^4} \cup D^5, \tag{11}$$

where $SU(2)$ acts on $D^8 \subset \mathbf{C}^4$ (the first two coordinates of (10)) via twice of the natural representation of $SU(2)$ on $\mathbf{C}^2$, and acts trivially on D^5 and S^4. By taking $Q = \mathbf{H}P^2$, it is easy to see that P is $SU(2)$-periodic.

Lemma 2.1. (11) *is an $SU(2)$-periodicity space with twice of the natural $SU(2)$-representation as the periodicity representation.*

Our $O(2)$-periodicity space is much more complicated. Its stratification contains four isovariant $\pi - \pi$ pairs.

Let

$$V = \mathbf{H} \otimes_{\mathbf{R}} \mathbf{R}^2 = \mathbf{H}^2$$

be the quaternionization of the natural representation of $O(2)$ on $\mathbf{R}^2$. Motivated by the way the periodicity spaces are constructed for $SO(1)$ and $SU(2)$, we start with the projective space

$$\mathbf{C}P(V \oplus \mathbf{C}) = DV \cup_{SV \nearrow \text{Hopf}} \mathbf{C}P(V) \tag{12}$$

with the induced $O(2)$-action. Next we list all the subgroups of $O(2)$ and study their fixed points in $\mathbf{C}P(V \oplus \mathbf{C})$.

The group $O(2)$ consists of two types of elements: The rotations

$$\rho_\theta = \begin{pmatrix} \cos\theta & -\sin\theta \\ \sin\theta & \cos\theta \end{pmatrix} \qquad \theta \in \mathbf{R}\text{mod}2\pi$$

by angle θ and the flippings

$$\tau_\theta = \begin{pmatrix} \cos 2\theta & \sin 2\theta \\ \sin 2\theta & -\cos 2\theta \end{pmatrix} \qquad \theta \in \mathbf{R}\text{mod}\pi$$

with respect to the subspace $\rho_\theta(\mathbf{R} \oplus 0)$.

The only nontrivial infinite closed subgroup of $O(2)$ is the circle of rotations:

$$S^1 = SO(2) = \{\rho_\theta\}.$$

The finite subgroups of $O(2)$ are the cyclic subgroups of rotations:

$$C_n = < \rho_{\frac{2\pi}{n}} >, \quad n > 1,$$

the 2 element subgroups of flippings:

$$B_\theta = < \tau_\theta > = \{1, \tau_\theta\},$$

and the dihedral subgroups of rotations and flippings:

$$A_{\theta,n} = < \rho_{\frac{2\pi}{n}}, \tau_\theta > = B_\theta \propto C_n, \quad n > 1.$$

The normalizers and the Weyl groups of the subgroups are

$$\begin{array}{ll} N(S^1) = O(2), & W(S^1) = \{1, \bar{\tau}_0\} \cong \mathbf{Z}_2; \\ N(C_n) = O(2), & W(C_n) = O(2)/C_n \cong O(2); \\ N(B_\theta) = A_{\theta,2}, & W(B_\theta) = \{1, -1\} \cong \mathbf{Z}_2; \\ N(A_{\theta,n}) = A_{\theta,2n}, & W(A_{\theta,n}) = \{1, \bar{\rho}_{\frac{\pi}{n}}\} \cong \mathbf{Z}_2. \end{array}$$

The fixed points of the subgroups are described by the following proposition.

Proposition 2.2. *Suppose that W is a unitary complex H-representation. Then*

$$\mathbf{C}P(W)^H = \coprod_{\kappa} \mathbf{C}P(W^{\kappa}),$$

where

$$W^{\kappa} = \{w \in W : hw = \kappa(h)w \text{ for all } h \in H\}$$

is the eigenspace of a homomorphism (complex character) $\kappa : H \to S^1$.

We also observe that

$$\mathbf{C}P(W \oplus \mathbf{C})^H = \mathbf{C}P(W^H \oplus \mathbf{C}) \coprod [\coprod_{\kappa \text{ nontrivial}} \mathbf{C}P(W^{\kappa})]. \tag{13}$$

$\mathbf{C}P(W^H \oplus \mathbf{C})$ is the component of $\mathbf{C}P(W \oplus \mathbf{C})^H$ containing DW^H. If the group action preserves orientation and $\dim_{\mathbf{C}} W^H$ is always even (e.g., W is twice of some other complex representation), then the equivariant signature of $\mathbf{C}P(W^H \oplus \mathbf{C})$ is 1, and the equivariant euler characteristic is $1 + \dim_{\mathbf{C}} W^H$, an odd number. What prevents $\mathbf{C}P(W \oplus \mathbf{C})$ to become a periodicity manifold is the potential existence of nontrivial "eigencharacter" κ, making the fixed points nonconnected.

ρ_θ has two complex eigenvalues $e^{i\theta}$, $e^{-i\theta}$ for the representation V. If $\theta \neq 0 \bmod \pi$, then the eigenspaces in V are respectively

$$(1, i)\mathbf{H} = \{(w, iw) : w \in \mathbf{H}\}, \quad (1, -i)\mathbf{H} = \{(w, -iw) : w \in \mathbf{H}\}.$$

Note that these are independent of the choice of the angle. Therefore they are actually the eigenspaces of the rotation subgroup S^1, corresponding to the two characters:

$$id : SO(2) \to S^1, \quad \rho_\theta \mapsto e^{i\theta}; \qquad conj : SO(2) \to S^1, \quad \rho_\theta \mapsto e^{-i\theta}.$$

By (13),

$$\begin{aligned} \mathbf{C}P(V \oplus \mathbf{C})^{S^1} &= \mathbf{C}P(V \oplus \mathbf{C})^{C_n} \\ &= 0 \coprod \mathbf{C}P((1, i)\mathbf{H}) \coprod \mathbf{C}P((1, -i)\mathbf{H}), \quad n > 2. \end{aligned} \tag{14}$$

The action of the Weyl group $W(S^1) = \mathbf{Z}_2$ fixes 0 and exchanges the other two components. The action of the Weyl group $W(C_n) = O(2)$ factors through $O(2) \to \mathbf{Z}_2$.

τ_θ has two eigenvalues 1 and -1, with corresponding eigenspaces in V:

$$\mathbf{H}_\theta = \rho_\theta(\mathbf{H} \oplus 0), \quad \mathbf{H}_{\theta + \frac{\pi}{2}} = \rho_{\theta + \frac{\pi}{2}}(\mathbf{H} \oplus 0).$$

Thus by (13) again,

$$\mathbf{C}P(V \oplus \mathbf{C})^{B_\theta} = \mathbf{C}P(\mathbf{H}_\theta \oplus \mathbf{C}) \coprod \mathbf{C}P(\mathbf{H}_{\theta + \frac{\pi}{2}}). \tag{15}$$

The Weyl group $\mathbf{Z}_2$ acts by $-1 \oplus 1$ on the first component and trivially on the second component.

The fixed points of $A_{\theta,n}$, $n > 2$, are the intersection of the fixed points of C_n, $n > 2$, and those of B_θ. However, the intersections

$$\mathbf{C}P((1,i)\mathbf{H}) \cap \mathbf{C}P(\mathbf{H}_\theta \oplus \mathbf{C}), \qquad \mathbf{C}P((1,-i)\mathbf{H}) \cap \mathbf{C}P(\mathbf{H}_\theta \oplus \mathbf{C})$$

are easily seen to be empty. Therefore

$$\mathbf{C}P(V \oplus \mathbf{C})^{A_{\theta,n}} = 0, \quad n > 2. \tag{16}$$

The action of the Weyl group $\mathbf{Z}_2$ has to be trivial.

It remains to study the cyclic group C_2 and the 4 element groups $A_{\theta,2}$. We have

$$\mathbf{C}P(V \oplus \mathbf{C})^{C_2} = 0 \coprod \mathbf{C}P(V), \tag{17}$$

The action of the Weyl group $O(2)/C_2$ is trivial at 0 and is the one coming from the induced action of $O(2)$ on $\mathbf{C}P(V)$ (such action is trivial when restricted to C_2). Moreover, by taking the intersection of the fixed points of C_2 and B_θ, we obtain

$$\mathbf{C}P(V \oplus \mathbf{C})^{A_{\theta,2}} = 0 \coprod \mathbf{C}P(\mathbf{H}_\theta) \coprod \mathbf{C}P(\mathbf{H}_{\theta+\frac{\pi}{2}}). \tag{18}$$

The Weyl group $\mathbf{Z}_2$ acts trivially at 0 and exchanges the other two components.

Although the fixed point subspaces of $\mathbf{C}P(V \oplus \mathbf{C})$ are often not connected, the components not containing 0 are all contained in the submanifold $\mathbf{C}P(V)$. Therefore we may think of $\mathbf{C}P(V \oplus \mathbf{C})$ as a two strata space (12) with the equivariant submanifold $\mathbf{C}P(V)$ as the smaller stratum (later on we need to single out another submanifold $S^2 \times \{\pm 1\} \subset \mathbf{C}P(V)$, making $\mathbf{C}P(V \oplus \mathbf{C})$ a three strata space). If there is an $SU(2)$-manifold E with $\mathbf{C}P(V)$ as boundary, such that the pair $(E, \mathbf{C}P(V))$ satisfies the isovariant $\pi - \pi$ condition (see the definition 1.1), then $DV \cup_{SV \to \mathbf{C}P(V)} E$ is an $SU(2)$-periodicity space.

Our attempt of constructing E makes use of the quaternionic structure of the representation V. Consider the Hopf bundle map

$$S^2 \to \mathbf{C}P(V) \to \mathbf{H}P(V).$$

This is an equivariant $SU(2)$-bundle. Since the fibre is a sphere, the mapping cylinder

$$E = \mathbf{C}P(V) \times I \cup \mathbf{H}P(V)$$

is an $SU(2)$-manifold with boundary $\mathbf{C}P(V)$.

To check whether $(E, \mathbf{C}P(V))$ satisfies the isovariant $\pi - \pi$ condition, we need to know the structure of the fixed points in $\mathbf{H}P(V)$ and its relation to the structure of the fixed points of $\mathbf{C}P(V)$. First we need the quaternionic version of the proposition 2.2.

Proposition 2.3. *Suppose that W is a unitary quaternionic H-representation. Then*

$$\mathbf{H}P(W)^H = \coprod_{conj.\ class\ of\ \kappa} \mathbf{H}P(W^\kappa),$$

where

$$W^\kappa = \{w \in V : hw = \kappa(h)w \text{ for all } h \in H\}$$

is the set of eigenvectors of a homomorphism (quarternionic character) $\kappa : H \to S^3$.

Remark 2.4. If $hw = \kappa w$, then $h(\lambda w) = \lambda\kappa\lambda^{-1}(\lambda w)$. Therefore by the noncommutativity of the quaternionic multiplication, W^κ may not be a quaternionic subspace of W. However, we do have

$$\mathbf{H}W^\kappa = \cup_\lambda W^{\lambda\kappa\lambda^{-1}}.$$

The quaternionic projective space of W^κ then makes sense for the conjugacy class of κ.

The general relation between $\mathbf{C}P(W)^H$ and $\mathbf{H}P(W)^H$ is not very clear. However, we observe that if H is generated by a single element h, then the charater κ is given by a complex or quaternionic eigenvalue a of h, and we will use the notation

$$W^{h=a} = W^\kappa = \{w \in W : hw = aw\}.$$

As further pointed out in the remark above, only the conjugacy class of the eigenvalue matters in the quaternionic case. The following proposition summarizes the facts in this respect.

Proposition 2.5. *The conjugacy action of $S^3 \subset \mathbf{H}$ on itself has the following properties:*

(1) *The action has two fixed points: the real numbers* 1 *and* -1*;*

(2) *Any nonreal quaternion is conjugate to exactly two complex numbers a and $\bar{a}$;*

(3) *The isotropy subgroup of any nonreal complex number is S^1, the complex numbers in S^3.*

We are now ready to study the relation between $\mathbf{C}P(W)^h$ and $\mathbf{H}P(W)^h$.

Proposition 2.6. *The projection $\mathbf{C}P(W)^h \to \mathbf{H}P(W)^h$ is surjective. A component of $\mathbf{H}P(W)^h$ is of the form $\mathbf{H}P(W^{h=a})$ for some complex number a. The components of $\mathbf{C}P(W)^h$ over $\mathbf{H}P(W^{h=a})$ are the following:*

(1) *If a is real, then $W^{h=a}$ is a quaternionic subspace of W, and there is only one corresponding component $\mathbf{C}P(W^{h=a})$. Moreover, $\mathbf{C}P(W^{h=a}) \to \mathbf{H}P(W^{h=a})$ is a Hopf bundle with fibre S^2;*

(2) *If a is not real, then there are exactly two corresponding components $\mathbf{C}P(W^{h=a})$ and $\mathbf{C}P(W^{h=\bar{a}})$. Moreover, the projections $\mathbf{C}P(W^{h=a}) \to \mathbf{H}P(W^{h=a}) \leftarrow \mathbf{C}P(W^{h=\bar{a}})$ are homeomorphism.*

Proof. By the proposition 2.3, a component of $\mathbf{H}P(W)^h$ is of the form $\mathbf{H}P(W^{h=b})$ for some quaternion b. By the proposition 2.5, b is conjugate to a complex number a. Thus by the remark above, $\mathbf{H}P(W^{h=b}) = \mathbf{H}P(W^{h=a})$. This proves the statement about the components of $\mathbf{H}P(W)^h$.

The components of $\mathbf{C}P(W)^h$ over $\mathbf{H}P(W^{h=a})$ have to be of the form $\mathbf{C}P(W^{h=c})$ with c being a complex number conjugating to a via quaternions. By the first and the second properties of the proposition 2.5, we have the two situations listed in this proposition.

If a is real (i.e., $a = 1$ or -1), then the fact that $W^{h=a}$ is a quaternionic space implies that the projection $\mathbf{C}P(W^{h=a}) \to \mathbf{H}P(W^{h=a})$ is a Hopf bundle.

If a is not real, then we need to show that the projection $\mathbf{C}P(W^{h=a}) \to \mathbf{H}P(W^{h=a})$ is homeomorphism. It is obviously surjective. For the injectivity, we consider vectors $u, v \in W^{h=a}$ of length 1, such that $[u]_{\mathbf{H}} = [v]_{\mathbf{H}} \in \mathbf{H}P(W^{h=a})$. Thus $v = \lambda u$ for some quaternion $\lambda \in S^3$. Therefore $\lambda a u = \lambda h u = h \lambda u = hv = av = a\lambda u$. Since $u \neq 0$, we see that $\lambda^{-1} a \lambda = a$. By the third property of the proposition 2.5, λ is a complex number. Consequently, $[u]_{\mathbf{C}} = [\lambda u]_{\mathbf{C}} = [v]_{\mathbf{C}} \in \mathbf{C}P(W^{h=a})$. This proves the injectivity.

Finally, the projection $\mathbf{C}P(W)^h \to \mathbf{H}P(W)^h$ is surjective because in both cases, the projections on components are surjective. □

Now we are ready to study the fixed points in E.

First we observe that $C_2 = \{1, -1\}$ fixes E. Therefore the action of $O(2)$ factors through $O(2)/C_2 \cong O(2)$.

If $\theta \neq 0 \bmod \pi$, then the two complex eigenvalues $e^{i\theta}$, $e^{-i\theta}$ of ρ_θ are not real. Therefore the fixed points of $C_n = < \rho_{\frac{2\pi}{n}} >$, $n > 2$, and S^1 are described by the second case of the proposition 2.6. The following proposition describes the S^1-fixed points of E. The same description applies to the fixed points of C_n, $n > 2$, with the action of the Weyl group factoring through $O(2) \to \mathbf{Z}_2$.

Proposition 2.7. $E^{S^1} \cong S^2 \times [-1,1]$, *with the action of the Weyl group* $\mathbf{Z}_2$ *given by the antipodal on* S^2 *and the flipping* $t \mapsto -t$ *on the interval.*

Proof. E^{S^1} is the mapping cylinder of

$$\begin{aligned} &\mathbf{C}P(V)^{S^1} = \mathbf{C}P((1,i)\mathbf{H}) \coprod \mathbf{C}P((1,-i)\mathbf{H}) \\ \to\ &\mathbf{H}P(V)^{S^1} = \mathbf{H}P((1,i)\mathbf{H}) = \mathbf{H}P((1,-i)\mathbf{H}), \end{aligned}$$

From the natural identifications

$$\mathbf{H}P((1,i)\mathbf{H}) \cong \mathbf{C}P((1,i)\mathbf{H}) \cong \mathbf{C}P(\mathbf{H}) = S^2, \quad [w, iw] \to [w], \tag{19}$$

$$\mathbf{H}P((1,-i)\mathbf{H}) \cong \mathbf{C}P((1,-i)\mathbf{H}) \cong \mathbf{C}P(\mathbf{H}) = S^2, \quad [w, -iw] \to [w], \tag{20}$$

we see that

$$E^{S^1} = S^2 \times [-1,1], \tag{21}$$

where we put $\mathbf{C}P(V)^{S^1}$ at the two ends $S^2 \times \{\pm 1\}$ and $\mathbf{H}P(V)^{S^1}$ at the middle $S^2 \times 0$.

As for the action of the Weyl group $W(S^1) = \mathbf{Z}_2$, we observe that the action is the flipping $t \to -t$ on the interval $[-1,1]$ because the Weyl group exchanges the two components $\mathbf{C}P((1,i)\mathbf{H})$ and $\mathbf{C}P((1,-i)\mathbf{H})$. In the S^2 direction, we consider the action at the level $t = 0$:

$$\tau_0[w, iw] = [w, -iw] = [jw, i(jw)] \in \mathbf{H}P((1,i)\mathbf{H}).$$

If we use the identification (19), then

$$\tau_0[w] = [jw] \in \mathbf{C}P(\mathbf{H}) = S^2.$$

This is the antipodal map. $\square$

The eigenvalues of τ_θ over V are 1 and -1. Therefore the fixed points of $B_\theta = <\tau_\theta>$ are described by the first case of the proposition 2.6. In particular, $\mathbf{H}_\theta$ is always a quaternionic subspace of V, and E^{B_θ} is the mapping cylinder of

$$\begin{aligned} &\mathbf{C}P(V)^{B_\theta} = \mathbf{C}P(\mathbf{H}_\theta) \coprod \mathbf{C}P(\mathbf{H}_{\theta+\frac{\pi}{2}}) \\ \to\ &\mathbf{H}P(V)^{B_\theta} = \mathbf{H}P(\mathbf{H}_\theta) \coprod \mathbf{H}P(\mathbf{H}_{\theta+\frac{\pi}{2}}). \end{aligned}$$

In fact, the projection is simply $S^2 \coprod S^2 \to 2$ points. Therefore

$$E^{B_\theta} = D^3 \coprod D^3. \tag{22}$$

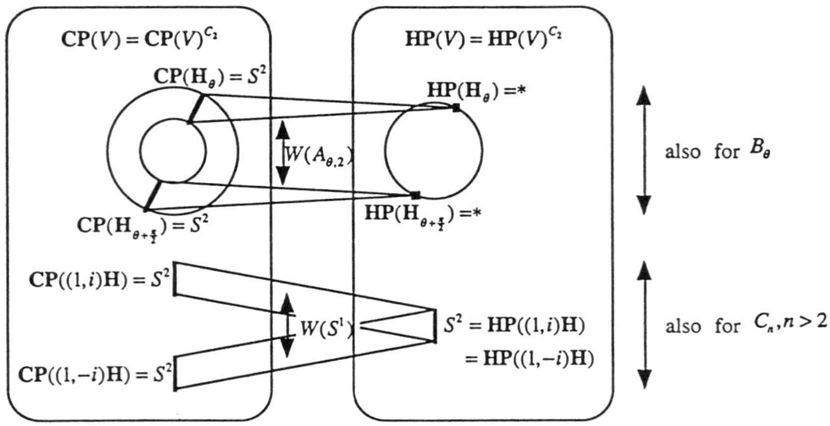

FIGURE 1

The action of the Weyl group $\mathbf{Z}_2$ is trivial.

The fixed points of $A_{\theta,n}$ is the intersection of the fixed points of C_n and B_θ. Because $\mathbf{H}P((1,i)H) \cap \mathbf{H}P(\mathbf{H}_\theta)$ and $\mathbf{H}P((1,-i)H) \cap \mathbf{H}P(\mathbf{H}_\theta)$ are easily seen to be empty, we have

$$E^{A_{\theta,n}} = \emptyset, \quad n > 2. \tag{23}$$

Moreover, since C_2 acts trivially on $\mathbf{C}P(V)$,

$$E^{A_{\theta,2}} = E^{B_\theta} = D^3 \coprod D^3. \tag{24}$$

The Weyl group $\mathbf{Z}_2$ of $A_{\theta,2}$ exchanges the two disks (in contrast to the behavior of the Weyl group of B_θ).

Figure 1 is the picture of all fixed points in E.

In order that the $O(2)$-stratified space $P_1 = \mathbf{C}P(V \oplus \mathbf{C}) \cup_{\mathbf{C}P(V)} E$ to be periodic, $(E, \partial E) = (E, \mathbf{C}P(V))$ has to satisfy the isovariant $\pi - \pi$ condition. We see from the above picture that $(E, \partial E)$ has codimension ≥ 3 gap. Therefore the inclusion from the isovariant components to the equivariant components is one-to-one, and the corresponding components have isomorphic fundamental groups. Consequently, the isovariant $\pi - \pi$ condition is the same as the analogous equivariant $\pi - \pi$ condition. In what follows, we check such equivariant condition.

First of all, the whole space E is fixed by the subgroup $C_2 = \{1, -1\}$. Both E and ∂E have trivial fundamental groups. Hence the equivariant $\pi - \pi$ condition is satisfied by C_2.

Second, the fixed points $E^{B_\theta} = E^{A_{\theta,2}}$ consists of two components, each of which is homeomorphic to D^3, with boundary homeomorphic to

S^2. Both have trivial fundamental groups. Hence the equivariant $\pi - \pi$ condition is satisfied by B_θ and $A_{\theta,2}$.

Third, the equivariant $\pi - \pi$ condition is trivially satisfied by $A_{\theta,n}$, $n > 2$, since no points in E is fixed by the subgroup.

The problem arises with the fixed points of S^1 and C_n, $n > 2$. The proposition 2.7 shows that $E^{S^1} = E^{C_n} = S^2 \times [-1,1]$ is connected, while $\partial E^{S^1} = \partial E^{C_n} = S^2 \times \{\pm 1\}$ has two components.

The failure of the $\pi - \pi$ condition for $(E, \partial E)^{S^1} = (E, \partial E)^{C_n}$ may be fixed by a $\mathbf{Z}_2$-equivariant nullcobordism (F, F^∂) of the pair $(E, \partial E)^{S^1}$ such that both (F, E^{S^1}) and $(F^\partial, \partial E^{S^1})$ satisfy the isovariant $\pi - \pi$ condition. With such (F, F^∂),

$$P_2 = P_1 \cup F = \mathbf{C}P(V \oplus \mathbf{C}) \cup_{\mathbf{C}P(V)} E \cup_{E^{S^1}} F.$$

is a periodicity space (Strictly speaking, we should take the union of P_1 with $O(2) \times_{N(S^1)} F$. However, $O(2) = N(S^1)$.).

We choose $F^\partial = D^3 \times \{\pm 1\}$ as the $\mathbf{Z}_2$-equivariant nullcobordism of $\partial E^{S^1} = S^2 \times \{\pm 1\}$. Then $E^{S^1} \cup_{\partial E^{S^1}} F^\partial = S^2 \times [-1,1] \cup D^3 \times \{\pm 1\}$ is a sphere S^3, and the $\mathbf{Z}_2$-action is antipodal. Therefore we may consider the Hopf bundle

$$S^1 \to S^3 \to S^2.$$

This is a principal S^1-bundle, with the antipodal action on S^3 induced by the fibrewise S^1-action. The associated disc bundle

$$F = S^3 \times I \cup S^2$$

is then a $\mathbf{Z}_2$-equivariant nullcobordism of the antipodal S^3.

Figure 2 is the picture of F.

Clearly, $(F^\partial, \partial E^{S^1}) = (D^3, S^2) \times \{\pm 1\}$ is a $\mathbf{Z}_2$-isovariant $\pi - \pi$ structure.

As for the pair (F, E^{S^1}), we have

$$F^{\mathbf{Z}_2} = S^2 \neq \emptyset = (E^{S^1})^{\mathbf{Z}_2}.$$

On the $\mathbf{Z}_2$-free part, however,

$$F - F^{\mathbf{Z}_2} = S^3 \times [0,1) \supset S^2 \times [-1,1] = E^{S^1} - (E^{S^1})^{\mathbf{Z}_2}$$

are all connected and simply connected. Thus the free part of (F, E^{S^1}) satisfies the isovariant $\pi - \pi$ condition, while the nonfree part fails to satisfy.

The failure on the nonfree part may again be fixed by another null-cobordism D^3 of $F^{\mathbf{Z}_2} = S^2$. The pair (D^3, S^2) satisfies the $\pi - \pi$ condition.

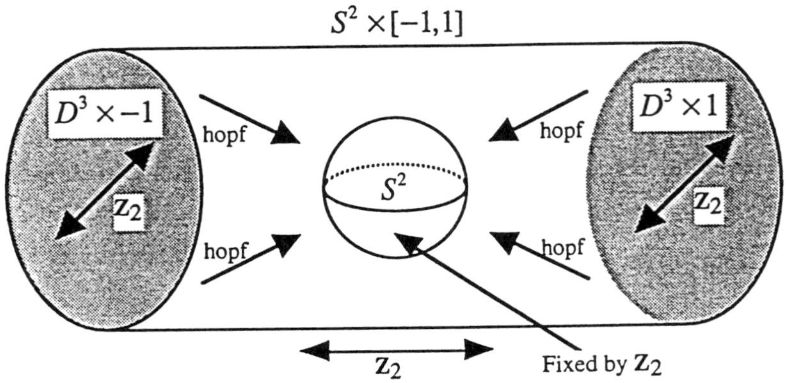

FIGURE 2

In conclusion, we build our periodicity space by successively equivariantly nullcobording fixed point components, eliminating all the nullcobordisms that do not satisfy the $\pi - \pi$ condition. The space we end up with is (recall $V = \mathbf{H}^2$)

$$P = \mathbf{C}P(V \oplus \mathbf{C}) \quad \begin{array}{l} \cup_{\mathbf{C}P(V)}(\mathbf{C}P(V) \times I \cup \mathbf{H}P(V)) \\ \cup_{S^2\times[-1,1]}(S^3 \times I \cup S^2) \\ \cup_{S^2} D^3, \end{array} \tag{25}$$

The action of $O(2)$ on P is the induced one on the projective spaces. It is the antipodal action on S^3 via the homomorphism $O(2) \to \mathbf{Z}_2$. The action is trivial on D^3.

The stratification of P is illustrated by the following diagram.

$$\begin{array}{ccccccc}
\mathbf{C}P(V \oplus \mathbf{C})_{(5)} & & & & & & \\
\cup & & & & & & \\
\mathbf{C}P(V)_{(\partial 4)} & \subset & E_{(4)} & & S^2_{(\partial 2)} & \subset & D^3_{(2)} \\
& & \cup & & \cap & & \\
& & E^{S^1}_{(\partial 3)} & \subset & F_{(3)} & & \\
& & \cup & & \cap & & \\
& & \partial E^{S^1}_{(\partial 1)} & \subset & F^{\partial}_{(1)} & &
\end{array}$$

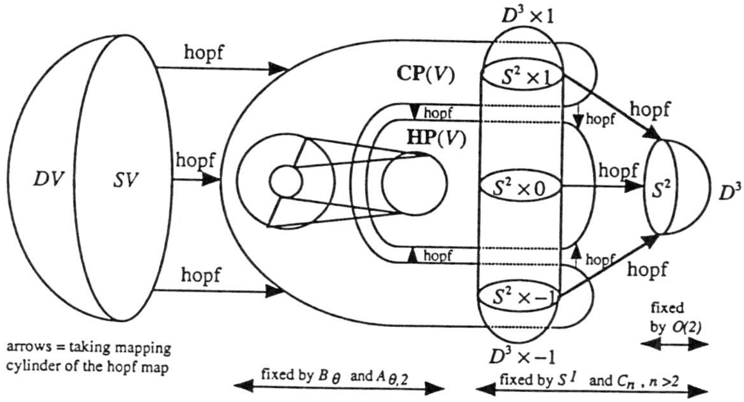

FIGURE 3

where

$$\begin{aligned}
&\mathbf{C}P(V \oplus \mathbf{C}) = DV \cup_{\text{Hopf}} \mathbf{C}P(V), \\
&E = \mathbf{C}P(V) \times I \cup_{\text{Hopf}} \mathbf{H}P(V), \\
&E^{S^1} = S^2 \times [-1,1], \\
&\partial E^{S^1} = S^2 \times \{\pm 1\}, \\
&F = S^3 \times I \cup_{\text{Hopf}} S^2, \\
&F^{\partial} = D^3 \times \{\pm 1\}.
\end{aligned}$$

The terms in the diagram are closed strata. The numerical footnotes denote the indices. The index set is obtained from the double of $1 < 2 < 3 < 4 < 5$ by deleting $\partial 5$.

Figure 3 is the picture of P.

Lemma 2.8. (25) *is an $O(2)$-periodicity space with the four fold natural $O(2)$-representation as the periodicity representation.*

3. PROOF OF THE PERIODICITY.

The theorem 0.1 is special cases of the following theorem.

Theorem 3.1. *Suppose that M is a homotopically stratified G-manifold with codimension ≥ 3 gap. Suppose that P is a G-periodicity space with periodicity representation V, such that M and $M \times V$ have the same isotropy everywhere. Then there is a natural equivalence*

$$S_G(M, \text{rel } \partial M) \cong S_G(M \times DV, \text{rel } \partial(M \times DV)).$$

Remark 3.2. The periodicity map in the theorem may depend on more than the representation V. If the $\pi - \pi$ structures in the periodicity spaces are cobordant through $\pi - \pi$ structures (of one dimension higher), then the corresponding periodicity maps are homotopy equivalent.

Remark 3.3. The periodicity equivalence is natural with respect to transverse isovariant G-maps, and the restriction to fixed points of subgroups. Such naturalities follow from the formal nature of the proof of the theorem.

Remark 3.4. For the special periodicity space for the abelian group actions, it was further proved in [13] that the periodicity is natural with respect to the induction $S_G(M) \to S_{G'}(M)$ (provided that $[G : im(G')]$ is finite) and the restriction $S_G(M) \to S_{WH}(M^H)$. Such naturality remains true in general.

The proof of the theorem 3.1 is similar to the one in [13].

We will use spacified version of the surgery theory. In the classical case, this was done in [8]. The formalism may be adopted to the isovariant and more generally, stratified case (see [12] for more details). Therefore $L(X)$ will be a space whose homotopy groups are the surgery obstruction groups of X at various dimensions (In particular, the proposition 1.2 may be interpreted as the contractibility of the space $L_G(X)$ for a G-isovariant $\pi - \pi$ structure X). Similarly, $S(X)$ will be a space whose homotopy groups are the structure sets of $X \times \Delta^i$ relative to the boundary for various i. We will also use $K^{\leq 1}$ to denote an involutive spectrum with

$$\pi_i K^{\leq 1} = \begin{cases} 0 & i > 1 \\ \text{whitehead torsion} & i = 1 \\ \text{finiteness obstruction} & i = 0 \\ \text{negative } K\text{-obstruction} & i < 0 \end{cases}$$

Finally we note that long exact sequences of obstruction groups are usually the long exact sequences of homotopy groups of fibrations of obstruction spaces.

According to Weinberger [12], the computation of the isovariant structure $S_G(M, \text{rel } \partial)$ involves several fibrations. First we compute the stable isovariant structure through the surgery fibration:

$$S_G^{-\infty}(M, \text{rel } \partial) \to H(M/G; L_G^{-\infty}(locM)) \to L_G^{-\infty}(M). \tag{26}$$

$L_G^{-\infty}$ is the stablization of Browder-Quinn's isovariant surgery obstruction [3]. The difference between L_G and $L_G^{-\infty}$ is given by the Rothenberg fibration

$$L_G \to L_G^{-\infty} \to \hat{H}(\mathbf{Z}_2; K_G^{\leq 1}). \tag{27}$$

$K_G^{\leq 1}$ is Browder-Quinn's isovariant K-obstruction [3]. $\hat{H}(\mathbf{Z}_2; ?)$ is the (spacified) Tate cohomology, which may be applied to involutive spectra and converts an involutive fibration into a fibration.

The unstable structure $S_G(M, \text{rel } \partial)$ can then be computed from another Rothenberg fibration

$$S_G \to S_G^{-\infty} \to \hat{H}(\mathbf{Z}_2; Wh_G^{top, \leq 0}). \tag{28}$$

$Wh_G^{top,\leq 0}$ is the topological isovariant K-obstruction [9][10]. It is related to $K_G^{\leq 1}$ in a way similar to the stable surgery fibration (26)

$$Wh_G^{top}(M) \to H(M/G; K_G^{\leq 1}(locM)) \to K_G^{\leq 1}(M). \tag{29}$$

There is however one catch: $Wh_G^{top,\leq 0}$ is the truncation of Wh_G^{top} at dimension 0. This troublesome catch is responsible for our requirement on the equivariant euler charateristic in the definition of periodicity spaces.

The theorem 3.1 is proved by showing succesively that

$$M \overset{\times P}{\Longrightarrow} M \times P \overset{\text{incl}}{\Longleftarrow} M \times (DV, \text{rel } SV), \tag{30}$$

induces periodicity equivalences on L_G, $L_G^{-\infty}$, $S_G^{-\infty}$, and S_G.

Proof of the Periodicity on L_G. Consider the maps

$$L_G(M) \xrightarrow{\times P} L_G(M \times P) \xleftarrow{\text{incl}} L_G(M \times DV, \text{rel } SV)$$

induced by (30). The inclusion fits into a fibration

$$L_G(M \times DV, \text{rel } SV) \xrightarrow{\text{incl}} L_G(M \times P) \to L_G(M \times \overline{P - DV}), \tag{31}$$

Since $\overline{P - DV}$ is an isovariant $\pi - \pi$ structure, we conclude from the propositions 1.2 and 1.4 that $L_G(M \times \overline{P - DV})$ is contractible. Therefore the inclusion is an equivalence.

The proof that $\times P$ also induces an equivalence is more complicated. We first consider the case that G acts on M freely. Construct the diagram

$$\begin{array}{ccc} L_G(M \times Q) & \xleftarrow{\text{incl}} & L_G(M \times DV, \text{rel } SV) \\ \times_Q \uparrow & \nwarrow \phi & \downarrow \text{incl} \\ L_G(M) & \xrightarrow{\times P} & L_G(M \times P) \end{array} \tag{32}$$

where the map ϕ first restricts to $Q \subset P$ and then forgets the stratification structure inside Q. The two triangles are commutative by the geometric meaning of the maps. Because M is a free G-manifold, $\times Q$ may be identified with the map

$$L(M/G) \to L((M \times Q)/G)$$

of classical surgery obstructions obtained as the transfer of the bundle

$$Q \to (M \times Q)/G \to M/G. \tag{33}$$

Lück and Ranicki [6] proved that such map depends only on the $\pi_0 G$-equivariant signature of Q. Since the signature is 1, $\times Q$ is an equivalence. Moreover, since G acts freely on M, the horizontal inclusion is the inclusion $L((M \times DV)/G) \subset L((M \times Q)/G)$ of classical surgery obstructions. By the simple connectivity of Q, we have $\pi_1((M \times DV)/G) = \pi_1(M/G) = \pi_1((M \times Q)/G)$. Consequently, the horizontal inclusion is an equivalence. Finally, we have already proved that the vertical inclusion is also an equivalence. Therefore we may conclude that $\times P$ at the bottom of the diagram is an equivalence.

For the general case, we induct on the isotropy subgroups of M. Let H be a maximal isotropy subgroup of M. Then we consider the diagram

(34)

$$\begin{array}{ccccc}
L_G(M - GM^H) & \to & L_G(M \times P - G(M^H \times P^H)) & \overset{\text{incl}}{\leftarrow} & L_G(M \times DV - G(M^H \times DV^H), \text{rel } SV) \\
\downarrow & & \downarrow & & \downarrow \\
L_G(M) & \overset{\times P}{\to} & L_G(M \times P) & \overset{\text{incl}}{\leftarrow} & L_G(M \times DV, \text{rel } SV) \\
\downarrow & & \downarrow & & \downarrow \\
L_{WH}(M^H) & \overset{\times P^H}{\to} & L_{WH}(M^H \times P^H) & \overset{\text{incl}}{\leftarrow} & L_{WH}(M^H \times DV^H, \text{rel } SV)
\end{array}$$

where the columns are fibrations. Observe that WH acts freely on M^H, and P^H is a WH-periodicity space by the lemma 1.11. We have just proved that the bottom maps are equivalences. Therefore in order to prove the middle maps are equivalences, it suffices to prove that the top maps are equivalences. We apply (30) to $M - GM^H$ and compare it with the top of (34):

(35)

$$\begin{array}{ccccc}
L_G(M - GM^H) & \overset{\times P}{\to} & L_G((M - GM^H) \times P) & \overset{\text{incl}}{\leftarrow} & L_G((M - GM^H) \times DV, \text{rel } SV) \\
\| & & \downarrow \text{incl} & & \downarrow \text{incl} \\
L_G(M - GM^H) & \to & L_G(M \times P - G(M^H \times P^H)) & \overset{\text{incl}}{\leftarrow} & L_G(M \times DV - G(M^H \times DV^H), \text{rel } SV)
\end{array}$$

By induction, we may assume that the upper row consists of equivalences. Therefore it remains to show that the vertical inclusions are equivalences.

The condition that M and $M \times DV$ have the same isotropy everywhere and the codimension ≥ 3 condition imply that the vertical inclusion on the right does not introduce any new strata or isovariant components and induce isomorphisms on the fundamental groups of corresponding isovariant components in pure strata. Therefore the right inclusion is an equivalence.

We have proved that the upper inclusion is an equivalence because $\overline{P - DV}$ is an isovariant $\pi - \pi$ structure. Similarly, the lower inclusion is an equivalence because both $\overline{P - DV}$ and $\overline{P^H - DV^H}$ are isovariant $\pi - \pi$ structures. As a consequence, the vertical inclusion at the middle is an equivalence. □

Proof of the Periodicity on $L_G^{-\infty}$. By Rothenberg fibration (27), the equivalences on the stable surgery obstruction $L_G^{-\infty}$ will follow from the compatible equivalences on L_G and $\hat{H}(\mathbf{Z}_2; K_G^{\leq 1})$. Thus we need to repeat the argument for the equivalence on L_G again, this time on $\hat{H}(\mathbf{Z}_2; K_G^{\leq 1})$.

$K^{\leq 1}$ is a functor over stratified (and in particular, equivariant) spaces. It has the following decomposition property: If Y is a closed union of strata of X, then there is a involutive fibration

$$K^{\leq 1}(X - Y) \to K^{\leq 1}(X) \to K^{\leq 1}(Y) \tag{36}$$

with a natural (and often noninvolutive) splitting $K^{\leq 1}(Y) \to K^{\leq 1}(X)$. Since the Tate cohomology functor $\hat{H}(\mathbf{Z}_2; ?)$ converts involutive fibrations to fibrations, the decomposition property implies that the functor $\hat{H}(\mathbf{Z}_2; K^{\leq 1})$ satisfies the second condition (decomposition along singularities) in the remark after the proposition 1.2. If we can show that $\hat{H}(\mathbf{Z}_2; K^{\leq 1})$ also satisfies the first condition in the remark (classical $\pi - \pi$ theorem, which $K^{\leq 1}$ alone does not satisfy), then the proposition 1.2 is also valid for $\hat{H}(\mathbf{Z}_2; K_G^{\leq 1})$.

For a manifold pair $(X, \partial X)$, we have

$$K^{\leq 1}(X, \partial X) = K^{\leq 1}(X, \mathrm{rel}\ \partial X) \times K^{\leq 1}(X), \tag{37}$$

with the involution

$$(\alpha, \beta)^* = (\alpha^* - (i_*\beta)^*, \beta^*), \tag{38}$$

where $i_* : K^{\leq 1}(\partial X) \to K^{\leq 1}(X, \mathrm{rel}\ \partial X)$ is the natural inclusion map and satisfies

$$(i_*\beta)^* = -i_*(\beta^*). \tag{39}$$

Suppose X, ∂X are connected, and $\pi_1 X = \pi_1 \partial X$. Since $K^{\leq 1}$ depends only on fundamental groups, we see that i_* is an equivalence. It was easy to prove (see [13], for example) that the involution (38) in which i_* is an isomorphism satisfying (39) has to have the vanishing Tate cohomology. Therefore the classical $\pi - \pi$ theorem is valid for $\hat{H}(\mathbf{Z}_2; K^{\leq 1})$.

With the proposition 1.2 and the subsequent remark at hand, we are able to show that the inclusion in (30) induces an equivalence on $\hat{H}(\mathbf{Z}_2; K_G^{\leq 1})$, similar to the equivalence on L_G.

In further repeating the argument for the periodicity of L_G, we encounter the following difference between L_G and $K_G^{\leq 1}$: By Anderson [1], the transfer of the K-obstructions is given by multiplying the equivariant euler characteristic, instead of Lück and Ranicki's equivariant signature. In the definition of periodicity spaces, we have assumed that the equivariant euler characteristic is an odd number (perhaps different for different isotropy groups). Therefore the product with the number induces equivalences after localizing at 2. Since the Tate cohomology is 2-torsion, localization at 2 does not change the Tate cohomology. Therefore by working with $\hat{H}(\mathbf{Z}_2; K_G^{\leq 1} \otimes \mathbf{Z}_{(2)}) = \hat{H}(\mathbf{Z}_2; K_G^{\leq 1})$, we may still carry out the argument.

Despite the difference between L_G and $K_G^{\leq 1}$, we have all the properties we need to repeat the proof of the periodicity on L_G for the functor $\hat{H}(\mathbf{Z}_2; K_G^{\leq 1})$. Consequently, (30) induces equivalences on the stable isovariant surgery obstructions $L_G^{-\infty}$. □

Proof of the Periodicity on $S_G^{-\infty}$. To prove the periodicity for the stable structure $S_G^{-\infty}$, we view the periodicity on $L_G^{-\infty}$ as a natural equivalence between functors:

$$L_G^{-\infty}(?) \xrightarrow{\times P} L_G^{-\infty}(? \times P) \xleftarrow{\mathrm{incl}} L_G^{-\infty}(? \times DV, \mathrm{rel}\ SV).$$

We may take the assemblies of their homologies over M/G to obtain a commutative diagram

$$\begin{array}{ccc} H(M/G; L_G^{-\infty}(locM)) & \to & L_G^{-\infty}(M) \\ \simeq\downarrow \times P & & \simeq\downarrow \times P \\ H(M/G; L_G^{-\infty}((locM) \times P)) & \to & L_G^{-\infty}(M \times P) \\ \simeq\uparrow \mathrm{incl} & & \simeq\uparrow \mathrm{incl} \\ H(M/G; L_G^{-\infty}((locM) \times DV, \mathrm{rel}\ SV)) & \to & L_G^{-\infty}(M \times DV, \mathrm{rel}\ SV) \end{array} \tag{40}$$

The homotopy fibre of the top map is $S_G^{-\infty}(M, \text{rel } \partial M)$. To work out the homotopy fibre of the bottom map, we rewrite the homology:

$$\begin{array}{lll} & & H(M \times DV/G; L_G(loc(M \times DV))) \\ (41) & \overset{\text{Fubini}}{\simeq} & H(M/G; H(DV/G_x; L_G(loc(M \times DV)))) \\ & \overset{\alpha}{\simeq} & H(M/G; L_G((locM) \times DV, \text{rel } SV)). \end{array}$$

The Fubini equivalence is a basic property of the stratified homology. α is obtained by applying the homology to the natural transformation (of functors of $locM$)

$$\alpha_0 : H(DV/G_x; L_G^{-\infty}(loc(M \times DV))) \to L_G^{-\infty}((locM) \times DV).$$

α_0 is an assembly over a cone $DV/G_x = \text{cone}(SV/G_x)$, which is always an equivalence (lemma 3.21 of [14]). Consequently α is an equivalence. By the naturality of the Fubini equivalence with respect to the assembly, the bottom of (40) may be identified with the assembly of $L_G^{-\infty}(?)$ over $(M \times DV)/G$. The homotopy fibre of this assembly is $S_G^{-\infty}(M \times DV, \text{rel } \partial(M \times DV))$. Hence we obtain the equivalence between $S_G^{-\infty}(M, \text{rel } \partial M)$ and $S_G^{-\infty}(M \times DV, \text{rel } \partial(M \times DV))$. This proves the periodicity of the stable structure.

Proof of the Periodicity on S_G. The last step is destablization. By the Rothenberg fibration (28), the periodicity on S_G will follow from compatible periodicities on $S_G^{-\infty}$ and on $\hat{H}(\mathbf{Z}_2; Wh_G^{top,\leq 0})$. This is the most intricate step.

The first intricacy is the meaning of the periodicity on $\hat{H}(\mathbf{Z}_2; Wh_G^{top,\leq 0})$. We are not studying the maps

$$Wh_G^{top,\leq 0}(M) \overset{\times P}{\to} Wh_G^{top,\leq 0}(M \times P) \overset{\text{incl}}{\leftarrow} Wh_G^{top,\leq 0}(M \times DV, \text{rel } SV)$$

induced from (30). In fact, these maps generally do not induce equivalences on the Tate cohomologies.

To clarify our problem, we start by considering a diagram similar to (40)

(42)

$$\begin{array}{ccc} H(M/G; K_G^{\leq 1}(locM)) & \to & K_G^{\leq 1}(M) \\ \downarrow \times P & & \downarrow \times P \\ H(M/G; K_G^{\leq 1}((locM) \times P)) & \to & K_G^{\leq 1}(M \times P) \\ \uparrow \text{incl} & & \uparrow \text{incl} \\ H(M/G; K_G^{\leq 1}((locM) \times DV, \text{rel } SV)) & \to & K_G^{\leq 1}(M \times DV, \text{rel } SV) \end{array}$$

We also introduce the notation $Wh_G^{top}(M; A, \text{rel } B)$ as the homotopy fibre of the assembly

$$H(M/G; K_G^{\leq 1}((locM) \times (A, \text{rel } B))) \to K_G^{\leq 1}(M \times (A, \text{rel } B)).$$

Then we have the induced maps between the homotopy fibres of the horizontal maps in (42):

$$Wh_G^{top}(M) \xrightarrow{\times P} Wh_G^{top}(M; P) \xleftarrow{\text{incl}} Wh_G^{top}(M; DV, \text{rel } SV). \tag{43}$$

As in the proof of the periodicity on $S_G^{-\infty}$, we may use Fubini equivalence to show that

$$Wh_G^{top}(M; DV, \text{rel } SV) \simeq Wh_G^{top}(M \times DV, \text{rel } SV). \tag{44}$$

Because the proof makes use of the fact that $DV = \text{cone}(SV)$, we cannot conclude $Wh_G^{top}(M; P) = Wh_G^{top}(M \times P)$ (which is generally false). The situation is similar to what happens for the periodicity on $S_G^{-\infty}$. We may introduce a notation $S_G^{-\infty}(M; P)$ for the homotopy fibre of the assembly at the middle row of (40). Then we obtain equivalences

$$S_G^{-\infty}(M) \overset{\times P}{\simeq} S_G^{-\infty}(M; P) \overset{\text{incl}}{\simeq} S_G^{-\infty}(M \times DV, \text{rel } SV).$$

If we replace $S_G^{-\infty}(M; P)$ by $S_G^{-\infty}(M \times P)$, then we generally will not obtain equivalences.

Combining (43) and (44) together, we obtain the maps

$$Wh_G^{top}(M) \xrightarrow{\times P} Wh_G^{top}(M; P) \xleftarrow{\text{incl}} Wh_G^{top}(M \times DV, \text{rel } SV) \tag{45}$$

induced by (30) on the topological K-theoretical obstructions, compatible with the stable periodicity. Therefore the problem boils down to proving that the truncation

(46)
$$Wh_G^{top, \leq 0}(M) \xrightarrow{\times P} Wh_G^{top, \leq 0}(M; P) \xleftarrow{\text{incl}} Wh_G^{top, \leq 0}(M \times DV, \text{rel } SV)$$

of (45) induces equivalences on Tate cohomologies.

Here arises the second intricacy. The maps in (43) are not equivalences themselves. It is not hard to show that the maps in (43) indeed induce equivalences on Tate cohomologies. However, this does not imply that, after the truncation, (46) still induces equivalences on Tate cohomologies.

We get around the difficulty by repeating the proof of the periodicity on L_G for the second time.

We will often make use of the following trick: Suppose that $K_1 \to K_2 \to K_3$ is a natural fibration of involutive functors, such that there is a (not necessarily involutive) splitting $K_3 \to K_2$. Then we may take the fibres of the assemblies of homologies of the functors to obtain a new fibration of involutive functors $W_1 \to W_2 \to W_3$ with a (still not necessarily involutive) splitting $W_3 \to W_2$. The splitting enables us to truncate and still obtain a fibration of involutive functors: $W_1^{\leq 0} \to W_2^{\leq 0} \to W_3^{\leq 0}$. Moreover, since the Tate cohomology converts involutive fibrations to fibrations, we see that $\hat{H}(\mathbf{Z}_2; W_1^{\leq 0}) \to \hat{H}(\mathbf{Z}_2; W_2^{\leq 0}) \to \hat{H}(\mathbf{Z}_2; W_3^{\leq 0})$ is still a fibration. Of course, because the splittings was not involutive, there may no longer be splitting $\hat{H}(\mathbf{Z}_2; W_3^{\leq 0}) \to \hat{H}(\mathbf{Z}_2; W_2^{\leq 0})$ to this fibration.

In particular, if we apply our trick to (36), then we see that the functor $\hat{H}(\mathbf{Z}_2; Wh_G^{top,\leq 0}(M; ?))$ satisfies the second condition (decomposition along singularities) in the remark after the proposition 1.2. As in the proof of the periodicity of $L_G^{-\infty}$, we will check whether the functor $\hat{H}(\mathbf{Z}_2; Wh_G^{top,\leq 0}(M; ?))$ also satisfies the first condition ($\pi - \pi$ theorem) in the remark.

Thus we consider a G-manifold pair $(X, \partial X)$ satisfying isovariant $\pi - \pi$ condition. We have decomposition of functors

$$(47)\quad K_G^{\leq 1}(? \times (X, \partial X)) = K_G^{\leq 1}(? \times (X, \mathrm{rel}\ \partial X)) \times K_G^{\leq 1}(? \times \partial X)$$

with the involution described by (38), in which i_* satisfies (39). By the proposition 1.4, $M \times (X, \partial X)$ still satisfies the isovariant $\pi - \pi$ condition for any G-manifold M. Therefore i_* is an equivalence.

We take the fibre of the assembly of (47) over M/G and then truncate to obtain a decomposition

(48)
$$Wh_G^{top,\leq 0}(M; X, Y) = Wh_G^{top,\leq 0}(M; X, \mathrm{rel}\ Y) \times Wh_G^{top,\leq 0}(M; Y).$$

The description about the involution still holds, and i_* is still an equivalence. Such description implies that the Tate cohomology of (48) is trivial. As in the proof of the periodicity on $L_G^{-\infty}$, we see that the equivariant $\pi - \pi$ theorem is valid for $\hat{H}(\mathbf{Z}_2; Wh_G^{top,\leq 0}(M; ?))$. Consequently, we are able to show that the inclusion in (46) induces equivalence on the Tate cohomologies.

The next step is to prove that $\times P$ in (46) induces equivalence on the Tate cohomologies. In proving the periodicity of $L_G^{-\infty}$, we first proved the free action case. The corresponding case here is $\hat{H}(\mathbf{Z}_2; Wh_G^{top,\leq 0}(M, \mathrm{rel}\ M_s; ?))$, where

$$M_s = \cup_{\{1\} \neq H \subset G} M^H$$

is the nonfree part of M, so that G acts freely on $M - M_s$.

Consider the commutative diagram of functors similar to (32)

$$
(49)\qquad
\begin{array}{ccc}
K_{\bar{G}}^{\leq 1}((?-?_s)\times Q) & \xleftarrow{\text{incl}} & K_{\bar{G}}^{\leq 1}((?-?_s)\times DV, \text{rel } SV) \\
\times_Q \uparrow & \nwarrow \phi & \downarrow \text{incl} \\
K_{\bar{G}}^{\leq 1}(?-?_s) & \xrightarrow{\times P} & K_{\bar{G}}^{\leq 1}((?-?_s)\times P)
\end{array}
$$

In proving the periodicity on $L_G^{-\infty}$, we have shown that $\times Q$ is an equivalence after localizing at 2, and the horizontal inclusion is an equivalence.

After taking the fibres of the assemblies over M/G of the functors in (49) and then truncating, we obtain

$$
(50)\qquad
\begin{array}{ccc}
 & & Wh_G^{top,\leq 0}((M, \text{rel } M_s)\times(DV, \text{rel } SV)) \\
 & & \simeq\downarrow (44) \\
Wh_G^{top,\leq 0}(M, \text{rel } M_s; Q) & \xleftarrow{\text{incl}} & Wh_G^{top,\leq 0}(M, \text{rel } M_s; DV, \text{rel } SV) \\
\times_Q \uparrow & \nwarrow \phi & \downarrow \text{incl} \\
Wh_G^{top,\leq 0}(M, \text{rel } M_s) & \xrightarrow{\times P} & Wh_G^{top,\leq 0}(M, \text{rel } M_s; P)
\end{array}
$$

The map $\times Q$ is still an equivalence after localizing at 2, and the horizontal inclusion is also still an equivalence. Moreover, we have proved that the vertical inclusion induces an equivalence on the Tate cohomologies. These imply that $\times P$ induces an equivalence on the Tate cohomologies.

To prove that $\times P$ induces an equivalence on $\hat{H}(\mathbf{Z}_2; Wh_G^{top,\leq 0}(M; ?))$ (not relative to the nonfree part), we consider the more general problem of the equivalence on the functor $\hat{H}(\mathbf{Z}_2; Wh_G^{top,\leq 0}(M, \text{rel } M^{\mathcal{H}}; ?))$, for a collection $\mathcal{H}$ of subgroups satisfying

$$g^{-1}Kg \supset H \in \mathcal{H} \Longrightarrow K \in \mathcal{H}.$$

In case $\mathcal{H}$ is all but the trivial subgroups, we have $M^{\mathcal{H}} = M_s$, and the equivalence has been proved. The case $\mathcal{H} = \emptyset$ is what we want at the end.

Let $K \subset G$ be a maximal subgroup not in $\mathcal{H}$. Let $\mathcal{K} = \mathcal{H} \cup \{g^{-1}Kg\}$. Then we compare the theory $K^{\leq 1}$ for $?-?^{\mathcal{H}}$ and $?-?^{\mathcal{K}}$ through a diagram similar to (34)

$$
(51)\qquad
\begin{array}{ccccc}
K_{\bar{G}}^{\leq 1}(?-?^{\mathcal{K}}) & \rightarrow & K_{\bar{G}}^{\leq 1}(?\times P-(?\times P)^{\mathcal{K}}) & \xleftarrow{\text{incl}} & K_{\bar{G}}^{\leq 1}(?\times DV-(?\times DV)^{\mathcal{K}}, \text{rel } SV) \\
\downarrow & & \downarrow & & \downarrow \\
K_{\bar{G}}^{\leq 1}(?-?^{\mathcal{H}}) & \xrightarrow{\times P} & K_{\bar{G}}^{\leq 1}((?-?^{\mathcal{H}})\times P) & \xleftarrow{\text{incl}} & K_{\bar{G}}^{\leq 1}((?-?^{\mathcal{H}})\times DV, \text{rel } SV) \\
\downarrow & & \downarrow & & \downarrow \\
K_{\bar{WK}}^{\leq 1}(?^K-(?^K)_s) & \xrightarrow{\times P^K} & K_{\bar{WK}}^{\leq 1}((?^K-(?^K)_s)\times P^K) & \xleftarrow{\text{incl}} & K_{\bar{WK}}^{\leq 1}((?^K-(?^K)_s)\times DV^K, \text{rel } SV)
\end{array}
$$

where the columns are involutive fibrations with (noninvolutive splittings). We also have the comparison similar to (35)

(52)

$$\begin{array}{ccccc}
K_G^{\leq 1}(?-?^{\mathcal{K}}) & \overset{\times P}{\rightarrow} & K_G^{\leq 1}((?-?^{\mathcal{K}})\times P) & \overset{\text{incl}}{\leftarrow} & K_G^{\leq 1}((?-?^{\mathcal{K}})\times DV, \text{rel } SV) \\
\| & & \downarrow \text{incl} & & \downarrow \text{incl} \\
K_G^{\leq 1}(?-?^{\mathcal{K}}) & \rightarrow & K_G^{\leq 1}(?\times P-(?\times P)^{\mathcal{K}}) & \overset{\text{incl}}{\leftarrow} & K_G^{\leq 1}(?\times DV-(?\times DV)^{\mathcal{K}}, \text{rel } SV)
\end{array}$$

If the functors in (52) are applied to the open pieces in M, then the same isotropy everywhere condition, the codimension ≥ 3 condition, and the isovariant $\pi-\pi$ structure in P imply that the all inclusions in (52) are equivalences.

Now we replace the top row of (51) with the top row of (52), take the fibres of the assemblies over M/G of the functors in the resulting diagram, and then truncate. The trick described before produces a commutative diagram

(53)

$$\begin{array}{ccccc}
\alpha(M, \text{ rel } M^{\mathcal{K}}) & \rightarrow & \alpha(M, \text{ rel } M^{\mathcal{K}}; P) & \leftarrow & \alpha(M, \text{ rel } M^{\mathcal{K}}; DV, \text{ rel } SV) \\
\downarrow & & \downarrow & & \downarrow \\
\alpha(M, \text{ rel } M^{\mathcal{H}}) & \rightarrow & \alpha(M, \text{ rel } M^{\mathcal{H}}; P) & \leftarrow & \alpha(M, \text{ rel } M^{\mathcal{H}}; DV, \text{ rel } SV) \\
\downarrow & & \downarrow & & \downarrow \\
\beta(M^K, (M^K)_s) & \rightarrow & \beta(M^K, (M^K)_s; P^K) & \leftarrow & \beta(M^K, (M^K)_s; DV^K, \text{ rel } SV^K)
\end{array}$$

where α is $Wh_G^{top,\leq 0}$, and β is $Wh_{WK}^{top,\leq 0}$, in which the columns are involutive fibrations. By the lemma 1.11, P^K is a WK-periodicity space with WK-periodicity representation V^K. Therefore by what we just proved, the bottom row of (53) induces equivalences on the Tate cohomologies. In the top row of (53), there are fewer isotropy groups in $M-M^{\mathcal{K}}$ than in $M-M^{\mathcal{H}}$. Therefore we may assume by induction that it also induces equivalences on the Tate cohomologies. Consequently, the middle row induces equivalences on the Tate cohomologies. □

This completes the proof of the periodicity on the unstable structure S_G. □

References

1. Anderson, D., *The Whitehead torsion of the total space of a fibre bundle*, Topology **11** (1972) 179-194.
2. Bryant, J., Ferry, S., Mio, W., Weinberger, S.,*Topology of homology manifolds*, Bull. A.M.S. **28** (1993) 324-328.
3. Browder, W. and Quinn, F., *A surgery theory for G-manifolds and stratified sets*, In Proceedings of International Conference on Manifolds, Tokyo, 1973, Hattori, A.(ed), 27-36. Tokyo, University of Tokyo Press 1975.

4. Dovermann, K.H. and Schultz, R., *Equivariant surgery theories and their periodicity properties*, Lecture Notes in Mathematics, vol.**1443**. Berlin-Heidelberg-New York, Springer-Verlag 1990.
5. Kirby, R. and Siebenmann, L., *Foundational essays on topological manifolds, smoothings, and triangulations*, Princeton, Princeton University Press 1977.
6. Lück, W. and Ranicki, A., *Surgery obstructions of fibre bundles*, J. Pure Appl. Algebra **81** (1992) 139-189.
7. Nicas, A., *Induction theorems for groups of homotopy manifold structure sets*, Mem.A.M.S. **267** (1982).
8. Quinn, F., *A geometric formulation of surgery*, Thesis, Princeton University, 1969.
9. Quinn, F., *Homotopically stratified sets*, Jour. A.M.S. **1** (1988) 441-499.
10. Steinberger, M., *The equivariant s-cobordism theorem*, Invent. Math. **91** (1988) 61-104.
11. Wall, C.T.C., *Surgery on compact manifolds*, New York, Academic Press 1971.
12. Weinberger, S., *The topological classification of stratified spaces*, Chicago, University of Chicago Press 1994.
13. Weinberger, S. and Yan, M., *Exotic products and equivariant periodicity for abelian group actions*, (Preprint).
14. Yan, M., *The periodicity in stable equivariant surgery*, Comm. in Pure and Applied Math. **46** (1993) 1013-1040.

DEPARTMENT OF MATHEMATICS HONG KONG UNIVERSITY OF SCIENCE AND TECHNOLOGY CLEAR WATER BAY, HONG KONG

AMS/IP Studies in Advanced Mathematics
Volume **2**, 1997 (Part 1)

MICROSURGERY ON STRATIFIED SPACES

SHMUEL WEINBERGER

The purpose of this paper is to give an overview of a general scheme for the classification of stratified spaces, at least assuming that all of the strata are high-dimensional manifolds. The main result is a kind of "local-global principle" that extends ordinary surgery theory [2, 34, 25] for the case of a manifold (= stratified space with one stratum). I will start by explaining what the classification theorem means in a number of concrete cases (such as manifolds with boundary, orbifolds, algebraic varieties) where the terms arising in the theory already have a conventional meaning. After this, we will discuss the theorem in full generality. A fuller account of this material has recently appeared as a volume in the Chicago Lectures in Mathematics series.

1. Background and Setup

We assume that all manifolds have dimension at least five. The prototypical classification theorem is the following, which is a rapid summary of the conclusions of simply connected surgery for topological manifolds:

Theorem. (See [2, 34, 25]**)** *If M is a simply connected manifold, then associated to any homotopy equivalence $M' \longrightarrow M$ there is an element in $H_n(M; \mathbb{L}(e))$, which vanishes by the time one pushes forward into $H_n(*; \mathbb{L}(e)) = L_n(e)$ ($\mathbb{Z}$, 0, $\mathbb{Z}_2$, or 0 depending on n mod 4) is zero. Every element of this kernel occurs for some M'*[1]*, and the map is injective. Furthermore, this element can be identified, away from 2, with the difference of the KO homology classes associated to the signature operators of M' and M.*

Partially supported by an NSF grant.

[1]This is not quite true as stated, because $\mathbb{L}(e)$ is not identical to F/Top (see immediately below). One has to remove an extra $\mathbb{Z}$ from this alleged calculation, or allow M' to be an ANR homology manifold (see [4] and the first remark of section three for more discussion) for the calculation to be correct as stated.

The spectrum $\mathbb{L}(e)$ was first defined by Quinn [22]. It is totally computed as $\mathbb{Z} \times F/Top$ and F/Top was computed by Sullivan (see [17])[2]. Rationally, all spectra are products of Eilenberg-MacLane spaces; here the basic invariants are the rational Pontrjagin classes (or better, the L-classes).

I should also point out that it takes quite a bit of work to develope the signature operator in the topological case, [33, 31]. However, the associated class in $KO \otimes \mathbb{Z}[1/2]$ homology was first constructed purely topologically by Sullivan.

As is well known, this theory extends to nonsimply connected manifolds (as well as to manifolds with boundary, to noncompact manifolds, as well as other variations). The set of homotopy equivalent manifolds is then part of a long exact sequence, rather than just the kernel of a homomorphism. The characteristic class part is identical, but the "constant term" is now a functor of the fundamental group (together with its orientation character) and the dimension mod 4.

For dealing with stratified spaces one is immediately forced to deal with the problem of lack of homogeneity of the spaces, so that conventional characteristic classes which measure the twisting of the tangent bundle, that is of a constant linear approximation to the space, are not available.

It turns out that one can nontheless define a characteristic class that lives in a "cosheaf homology group", where the local structure of the cosheaf reflects the singular nature of the space[3]. However, before getting to this, which can be, at first, somewhat hard to digest, I would like to point out how in special cases where characteristic classes have already been defined by other methods, the general machinery provides meaningful classification theories.

Definition. A *manifold stratified space* is a filtered space (X, P, Xp), that is a space, a finite partially ordered set P, and a covariant functor from P into closed subsets of X, subject to certain additional conditions. The first is that the *pure strata*, which are the differences of consecutive strata, are all manifolds. The second demands a certain tameness condition for the embedding of the strata in one another. This condition can be phrased purely in terms of fundamental groups and homologies of the system of open subsets of one pure strata that approach points of a lower stratum. We refer the reader to [21],[14] for precise definitions.

---

[2]In more traditional treatments, F/Top is viewed as the more basic object (which is natural, as this version makes perfect sense for the smooth and PL categories).

[3]Even in the manifold case, orientability is understood by the cosheaf.

It is important to note that in this definition we do not assume that points in pure strata have links (in the sense of PL topology). Indeed, Milnor's counterexamples to the hauptvermutung [19] (or Stallings': the cones on h-cobordant manifolds are always homeomorphic) are based on the topological ill-definedness of links, when they exist. Nonexistence of links for isolated singularities arise in one-point compactifications of manifolds with nonvanishing Siebenmann obstruction [29].

Definition. A stratified map is map between stratified spaces (with the same index set) that preserves the pure strata. A stratified homotopy equivalence is a stratified map between stratified spaces with an "inverse map" whose compositions are stratified homotopic to the identity maps.

Statement of Problem. *Classify all of the manifold stratified spaces stratified homotopy equivalent to a given one.*

Remark. The rather boring definitions given above, in situations of geometric interest, involve real decisions, and can be somewhat unnatural. Usually, it is more interesting to study the stratified spaces, with some additional condition or other, filtered homotopy equivalent to a given one. However, it seems that the general theory works out more naturally with the choices that we have made, and the filtered theory can often (or, at least, sometimes), by homotopical methods, be related to the stratified one.

Another problem not accessible, directly, to the machinery we'll present is whether stratified spaces are homeomorphic in a nonstratified way (i.e. the double suspension of a homology sphere is here viewed as distinct from a sphere). However, the stratification dependence can sometimes be turned into an advantage for some problems (see example 3 in section 2).

2. Examples

The most trivial example comes from manifolds with boundary, viewed as a space with two strata[4]. There are here two theories, where one works rel ∂, and where one allows the boundary to change. In the former case, the L-theory is the absolute one (since surgeries only occur in the interior, the absolute situation) while in the second, one obtains the relative L-groups.

Correspondingly, there are two characteristic class theories. The rel ∂ case leads one to the ordinary $KO_*(M)$, but the not rel ∂ case gives one characteristic classes living in $KO(M, \partial M) \cong KO(\text{Int}M)$. (We

[4]We leave it as an exercise to the reader to identify the stratified and filtered categories in this most unusual case.

make the convention that all unlabelled homology groups for noncompact spaces are the (Borel-Moore) locally finite homology groups.) The natural invariant lives in the latter group (consider the symbol class in the interior), but it seems a priori remarkable that one deals with the boundary simply by deleting it! The rel ∂ case is actually the more subtle one from the point of view of characteristic classes because one only gets into the absolute group via the identification of the signature operators on the boundaries via the homeomorphism.

In section 3, we will see that these parallels between K-theories and L-groups and calculations are entirely natural, and ultimately a consequence of the formula $\mathbb{L}(e, e) \cong *$.

It is also interesting to observe that it is only by using homology rather cohomology characteristic classes that the type of group matches the type of surgery. Cohomologically, absolute cohomology corresponds to not rel ∂ problems and hence to the relative L-group and vice versa.

Example 1. (Due to [18] for odd order groups acting locally linearly.) For any finite group G acting orientation preservingly we can still define analogous characteristic classes that now lie in $KO^G[1/2]$. (These are defined by analytic mathods in [28, 26] and by topological methods in [5].) *We assume that all fixed sets of group actions that arise throughout are never codimension two in one another.* This is sometimes called the small gap hypothesis in the literature. Without this assumption[5] one can perform calculations similar to the ones described below, but the results are more complicated to state.

Definition. If $f : X \longrightarrow Y$ is an equivariant map between G-spaces, then X is said to be isovariant if $G_x = G_{f(x)}$ for all x, i.e. if f preserves isotropy. There is an obvious notion of isovariant homotopy between isovariant maps, and isovariant homtopy equivalence, etc.

Our methods, involving stratified spaces, most naturally lead to isovariant results[6].

The main result is similar to that of the usual surgery, and of the previous example:

Theorem. *Suppose that M is a simply connected G-space where G is a finite group which acts orientation preservingly (all fixed sets are also*

[5]At least in some simple cases.

[6]However, it is a result of Browder (see [9]) that in the presence of a large gap hypothesis, equivariant homotopy equivalences are homotopic to isovariant homotopy equivalences. Without the gap hypothesis, one can sometimes reduce the equivariant classification theory to the isovariant one discussed here and some additional unstable homotopy theory.

assumed to be simply connected), and such that all fixed sets are locally flat submanifolds in one another. Then associated to any isovariant homotopy equivalence $X' \longrightarrow X$ we associate the difference of the $KO^{NH/H}[1/2]$ classes of each pair of corresponding closed strata. These elements detect, away from 2, the equivariant homeomorphism class of the map. All these classes can be prescribed arbitrarily, and independantly of each other, except for the restriction that each class pushes forward to the G-signature of the stratum[7].

Note that one can make sense of all of the terms in this theorem, even more simply, for smooth G-manifolds, but in this case the classification theorem is quite false; unlike the unequivariant situaion, even in high dimensions there can be infinitely many smooth structures on a given homeomorphism type of group action.

Also, while the local geometry is a bit simpler for locally linear actions, the theorem becomes a little more awkward to state.

The situation at 2 can be analysed, but is more complicated. Even in the case of manifolds, there is an interesting refinement of the class of the signature operator at 2 that lies in ordinary homology! (See [27].) However, here as elsewhere in this note, we have concentrated on what goes on at odd primes. Furthermore, the case of sonsimply connected strata is also a parallel to the manifold case where nonsimply connected equivariant surgery obstruction groups enter.

Example 1 (cont.). In addition to calculational results in the spirit of surgery theory, the machinery also leads to purely conceptual structural conclusions. Here we mention a couple of them. This work has been pioneered by my student Min Yan.

Theorem. [40][8] *For any odd order group G acting on any manifold M and $V = 4\times$Regular representation, there is an isomorphism $S(M) \cong S(M \times D(V)$ rel boundary).*

Here $S(M)$ denotes the set of G-manifolds isovariantly homotopy equivalent to M up to G-homeomorphism, and $D(V)$ is the unit disk bundle of V.

In some cases, one can prove more precise results:

---

[7]See the footnote regarding simply connected surgery on manifolds.

[8]I have skipped some technical assumptions necessary for the theorem to be entirely correct, involving dimensions and isotropy groups. Also there is a slight deviation in the isomorphism by $\mathbb{Z}$'s as usual, see also [20].

Theorem. [38] *If G is an arbitrary compact abelian Lie group, then the same holds for $V \cong 2W$, where W is an arbitrary complex representation.*

It is interesting to note that these periodicity results depend on parallel results for L-groups, of the sort considered in [10]. That book deals with only odd order groups, and, as they point out, for good equivariant bordism theoretic reasons. However, by enlarging one's category to include stratified spaces, one can handles some even order and even positive dimensional groups.

These results imply the following corollary:

Corollary. *For appropriate G, the classification of isovariant homotopy equivalent manifolds is an equivariant homotopy invariant.*

Using this corollary, I proved:

Theorem. *For every finite G there are G actions on some torus that are equivariantly homotopy equivalent to an affine G action on the torus, but not equivariantly homeomorphic (or even isovariantly homotopy equivalent) to the affine torus.*

These provide counterexamples to the "equivariant topological rigidity conjecture". See [36] for more history and details.

Example 2. Let us move to the classification of connected compact Lie groups, and the circle in particular. Here we will see some new phenomena, i.e. that the most conventional extensions of Pontrjagin classes or signature operators do not yield the "usual" classification theorem.

In the locally free case, i.e. even if G is an arbitrary compact Lie group acting with finite isotropy on M, the "usual" result holds: The equivariant signature operator is the primary invariant.

However, if there is fixed set for a positive dimensional isotropy subgroup then strange things can happen. We will only describe the semifree case; the general case is a combination of the phenomena visible in the locally free and semifree cases and is related to some splittings of equivariant K-theory that were first produced for other purposes in [1].

Theorem. *If $G = T$ is the circle group acts on M semifreely with connected and simply connected locally flat fixed sets, then $S(M, \text{ rel } M^T)$ is isomorphic to $KO_{*-1}(M/T)[1/2]$ (modulo a Hirzebruch formula type condition).*

For the not rel fixed set calculation, one has either $S(M) \cong S(M \text{ rel } F) \times S(F)$ *or* $S(M) \cong S(M - F)$[9] *for a component of fixed set* F*, depending on whether the codimension is 0 or 2 mod 4.*

The general case requires the construction of "rigid K-theory" which has ordinary K theory at a fixed point, as opposed to $KO^T(*) \cong RO(T)$. More precisely, rigid K-theory is the fiber (remember, we make spaces out of all groups!) of the composite

$$KO^T_*(M - F) \longrightarrow KO^T_{*-1}(F) \cong KO_{*-1}(F) \otimes RO(T) \longrightarrow KO_{*-1}(F)$$

The signature operator naturally lives here, because the G-signature of a manifold is a constant multiple of the trivial representation for $G = T$, and to get a general theory this additional universal bit of data must be encoded into the characteristic classes.

For more general Lie group actions, it is very challenging to analyze how the geometry of homogenous spaces enters into the characteristic class theory.

Example 3. (Embedding): If one manifold is embedded in another $M \subset W$, then we can consider W as a stratified space with two strata. The classification theorem asserts that the pair of forgetful maps:

$$S(W, M) \longrightarrow S(W) \times S(M)$$

gives an isomorphism if M is of codimension at least three. (This is a classical theorem of Browder, Casson, Haefliger, Sullivan, and Wall, see [34].) This geometrically means that if one has the underlying homotopy data of an embedding of M in W, a *Poincare embedding*, then there is a unique isotopy class of locally flat embedding of M in W realising it. (There will also be an embedding of any manifold homotopy equivalent to M in any manifold homotopy equivalent to W.)[10]

The way this is proven using the stratified theory is by proving local and global decompositions for (cosheaves of) L-spectra. This method works identically if M and W are not manifolds themselves, but only stratified spaces, with the proviso that each stratum of M maps into only one stratum of W, and is codimension at least three there. Thus there is an embedding theorem much more generally.

---

[9] Up to a s that arises, as before, and in Siebenmann periodicity [20].

[10] Here, too, one sees the difference between filtered and stratified homotopy. The former would enable one to study the embeddings homotopic to a given map, a much harder and, a priori, more interesting problem than the one of realizing Poincare embeddings.

Example 4. (See [6].) Let X be a stratified space with only even codimension strata. We also assume that the fundamental groups of all links of one stratum in another are trivial. These are examples of *supernormal Witt spaces.*

For spaces with even codimensional strata, Goresky and MacPherson have defined L-classes in rational homology [12], which were refined by Siegel to be $KO[1/2]$ classes. This class is also the class of the signature operator defined by Cheeger [8] using L^2 cohomological methods.

The main result asserts that using these characteristic classes one can extend the classification theory, almost verbatim.

Theorem. *If X is a simply connected space with even codimensional strata, all of which are also assumed to be simply connected, and all of whose links are simply connected as well, then associated to any stratified homotopy equivalence $X' \longrightarrow X$ we associate the difference of the $KO[1/2]$ classes of each pair of corresponding closed strata. These elements detect, away from 2, the homeomorphism class of the map. All these classes can be prescribed arbitrarily, and independantly of each other, except for the restriction that each class pushes forward to the Cheeger-Goresky-MacPherson signature of the stratum*[11].

3. Main Theorem

As promised, in this section we present a general theorem that "includes" all of the previous examples as special cases. The reason we write "includes" is because to apply the theory to examples one must do certain calculations, and in the above cases we are able to do these by making use of the special geometry present in each of the examples.

To understand the general theorem, we must again discuss ordinary surgery theory for manifolds. The reason for this is simple: if one considers the cones on manifolds, these are stratified homotopy equivalent iff the manifolds were homotopy equivalent, and are homeomorphic iff the manifolds are h-cobordant. The characteristic class theory in this case must therefore encode all of the information of surgery theory for the manifolds, not just the part encoded there within the characteristic classes.[12]

---

[11] As usual one must correct for the extra $\mathbb{Z}$'s.

[12] To be a little more precise, aside from torsion considerations, these spaces do not become homeomorphic even after crossing with Euclidean spaces, and it is an old phenomenon, first discovered by Mazur for manifolds, that characteristic classes are what remain beyond homotopy type after crossing with Euclidean space.

As I mentioned before, the statement that the set of manifolds homotopy equivalent to a given one is the kernel of a certain map, is replaced by a long exact sequence[13]. The new terms are of the form $L_n(\pi)$ where n is the dimension of π is the fundamental group(oid). It is much more convenient for the proofs in the stratified case to also replace this long exact sequence by a fibration of spaces (spectra) whose homotopy groups are the terms in the surgery exact sequence. (See [22, 20].)

The homotopy types of all of these spaces are understood to be functors on the homotopy groups, so, in some ways, this idea is largely a book-keeping device (See [32]). However, certain geometric problems naturally give rise to maps into these spaces, and the homotopy classification of the maps is not quite as directly dependant just on the homotopy groups (one must use homotopy with coefficients, as well).

The main result is the following calculation of $\mathbf{S}(X)$ the space whose homotopy groups measures the difference between spaces stratified simple homotopy equivalent to X and those homeomorphic to it:

Theorem. *One has two fibrations (of infinite loop spaces) for the classification of stratified spaces that are already homeomorphic on strata of dimension at most four:*

$$\boldsymbol{S}^{-\infty}(X) \longrightarrow H_0(X; \mathbb{L}^{BQ-\infty}(\textit{Neighborhood of } x)) \longrightarrow \mathbb{L}^{BQ-\infty}(X) \times \mathbb{Z}^s$$

$$\boldsymbol{S}(X) \longrightarrow \boldsymbol{S}^{-\infty}(X) \longrightarrow H(\mathbb{Z}_2; \boldsymbol{Wh}'(X)),$$

where $\boldsymbol{Wh}$*' denotes the* $\boldsymbol{Wh}(X)$ *spectrum with its homotopy groups above dimension one killed, s is the number of components of pure strata, and* $H(\mathbb{Z}_2; \boldsymbol{Wh}'(X))$ *is the Tate cohomology of this spectrum.*

The spectra $\mathbb{L}^{BQ-\infty}(X)$ are variants of the groups defined in [3]. These are built up by a sequence of fibrations from the terms coming from the pure strata. In the special case where X is a manifold the local cosheaf is $\mathbb{L}(\mathbb{R}^n)$, (where these twist according to the orientation character of X) and the stalk spaces are looped versions of $\mathbb{L}(e)$. The cosheaf homology then is just the normal invariant term in the usual surgery exact sequence (as it must be).

As another illustration, if X is a manifold with boundary, then the cosheaf is of course $\mathbb{L}(\mathbb{R}^n)$ at the interior points, but for the boundary points, one obtains instead $\mathbb{L}(\mathbb{R}^{n-1} \times [0,1)) \cong *$ (by the $\pi - \pi$ theorem = theorem 3.3 of [34]). This accounts for the fact that the characteristic class for not rel ∂ surgery problems lies in the theory where we just remove the boundary.

---

[13]It is really a long exact sequence of abelian groups only in the topological and PL categories, as far as we know.

The spectrum $\mathbf{Wh}(X)$ occurs in a topological h-cobordism theorem, and was studied in [21] and [30].

Remark. The extra $\mathbb{Z}$'s that arise in this theorem are an aesthetic mar. Using [4][14], they can be removed at the expense of classifying stratified spaces whose strata are ANR homology-manifolds. (The equivalence relation is then s-cobordism; unfortunately, we do not yet know an s-cobordism theorem for ANR homology manifolds.) The $\mathbb{Z}$'s measure on the various strata Quinn's resolution obstructions [23] and just check whether or not that stratum corresponds to a manifold.

The method of proof is to first work stably, that is cross with some unspecified high dimensional torus. This simplifies the local geometry to a large extent (as was explained and exploited in [28]), and also causes "controlled surgery" to become a homology theory, see [11]. Using either of these ideas, one can then do an inductive proof of the first sequence. In this inductive procedure one typically does an infinite number of surgeries, but all take place in carefully chosen open manifold subsets of the singular space. This can be viewed as being surgery performed on tiny handles in open subsets of normal tubes, which is what suggests the name microsurgery. The fact that the local cosheaf terms are of the same sort as the global spectrum that arises is perfect for the type of induction that arises: again, global geometry for k strata arises as the local geometry of $k+1$ strata spaces.

Remark. There is another approach that provides more precise information about unstable structures directly. This follows an idea of [16] and later extensions of this by Bruce Hughes. The main point is a generalized tubular neighborhood theorem for the structure of germs of neighborhoods of manifolds that are bottom strata of stratified spaces, and an analysis of this type of structure.

It turns out that for such a manifold bottom stratum B, there is a neighborhood $\mathfrak{U}$ with a map $\mathfrak{U} \longrightarrow B \times \mathbb{R}$ (morally the projection and radial coordinates for the neighborhood) with the property that the homeomorphism type of the inverse image of an open ball is independent of the ball. One should think of this as some kind of a "shape" analogue of being a fiber bundle. This beautiful structure is a generalization of the notion of a CE map and can be quite effectively studied. (In the manifold case, these were studied by Chapman, Quinn, Hughes,

[14] This was the topic of my talk at the 1993 Georgia International Topology conference. For an exposition of the classification of homology manifolds up to s-cobordism and a current conjectural view of their geometric topology, see [4, 37]

and Hughes-Taylor-Williams, see [15] and the references there.) In this approach to the induction, one ends up stabilizing just to get a nice closed form for the answer.

The method of proof of desuspension is a combination of controlled technology as in [21][15] (and [16] for realization of h-cobordisms) and the method of proof of a rather similar desuspension theorem in [39]. Roughly speaking, one is sewing together the difficulty in producing a boundary for the infinite cyclic cover of a stable structure with some fixed number of toral factors, and modding out by the indeterminacy involved, to produce an object with one fewer circle. It is a result of Siebenmann [29] in the case of manifolds that algebraic K-theory enters in the problem of putting boundaries on open manifolds.

Remark. One can use the middle term of this sequence as a useful general place to put the "signature class" of an arbitrary stratified space. There seems to be a connection from the geometric topology studied here to microlocal analysis involved in forming a suitable K-theory for receiving such signature operators.

References

1. M. Atiyah, *Elliptic operators and compact groups*, Springer LNM 401, 1974.
2. W. Browder, *Surgery on Simply Connected Manifolds*, Springer-Verlag, 1970.
3. W. Browder and F. Quinn, *A surgery theory for G-manifolds and stratified sets*, Manifolds, Tokyo, 1973, Univ. of Tokyo Press, 1975, 27-36.
4. J. Bryant, S. Ferry, W. Mio, and S. Weinberger, *The topology of homology manifolds*, (Preprint, see also BAMS 1993)
5. S. Cappell, J. Shaneson, and S. Weinberger, *Classes topologiques caracteristiques pour les actions de groupes sur les espaces singuliers*, Comptes Rendus, **313** (1991) 293-295.
6. S. Cappell and S. Weinberger, *Classification de certaines espaces stratifies*, Comptes Rendus, **313** (1991) 399-401.
7. S. Cappell and S. Weinberger, *Replacement theorems for fixed point sets and normal representations of group actions*, (to appear) .
8. J. Cheeger, *Spectral geometry of singular Riemannian spaces*, J. Diff. Geo., **18** (1983) 575-657.
9. K. H. Dovermann, *Almost isovariant normal maps*, Amer. J. of Math., **111** (1989) 851-904.
10. K.H.Dovermann and R.Schultz, *Periodicity in equivariant surgery*, LNM 1443, 1990.
11. S.Ferry and E.Pederson, *Epsilon surgery theory*, (to appear).

[15] See also [30] for the orbifold case.

12. M. Goresky and B. MacPherson, *Intersection homology I, II*, Topology **19** (1980), 135-162, Inven. Math., **72** (1983) 77-130.
13. B.Hughes, *Controlled homotopy topological structures*, Pacific J. of Math., **133** (1988) 69-97.
14. B.Hughes, *Geometric topology of stratified spaces*, (to appear).
15. B.Hughes, L.Taylor, and B.Williams, *Bundle theories for topological manifolds*, Trans. AMS, **319** (1990) 1-65.
16. B.Hughes, L.Taylor, S.Weinberger, and B.Williams, *The classification of neighborhoods in a stratified space*, (in preparation).
17. I.Madsen and R.J.Milgram, *The Classifying spaces for surgery and cobordism of manifolds*, Princeton University Press, 1979.
18. I. Madsen and M. Rothenberg, *On the classification of G-spheres I, II, III*, Acta Math. 160 (1988) 65-104. Math Scand. 64 (1989) and several related Aarhus preprints. See in particular Cont. Math. 19 (1983) 193-226 for a summary.
19. J.Milnor, *Two complexes which are homeomorphic but combinatorially distinct*, Ann of Math., **74** (1961) 575-590.
20. A. Nicas, *Induction theorems for groups of manifold homotopy structure sets*, Memoirs AMS, **267** 1982.
21. F. Quinn, *Homotopically stratified spaces*, Jour. AMS, **1** (1988) 441-499.
22. F. Quinn, *A geometric formulation of surgery*, 1969 Princeton PhD thesis.
23. F.Quinn, *An obstruction to the resolution of homology manifolds*, Mich. Math. J., **301** (1987) 285-292.
24. A. Ranicki, *The total surgery obstruction*, LNM, **763** (1979) 275-316.
25. A. Ranicki, *Algebraic L-theory and Topological Manifolds*, Cambridge University Press, 1992.
26. J. Rosenberg and S. Weinberger, *Higher G-indices on Lipschitz manifolds*, K-theory, 1993.
27. J. Rosenberg and S. Weinberger, *The signature operator at 2*, (in preparation).
28. M. Rothenberg and S. Weinberger, *Group Actions and Equivariant Lipschitz analysis*, BAMS, **17** (1987), 109-112. (long version in preparation).
29. L.Siebenmann, *The obstruction to finding a boundary for an open manifold of dimension greater than five*, 1965 Princeton, PhD thesis.
30. M. Steinberger, *The equivariant topological s-cobordism theorem*, Inven. Math., **91** (1988) 61-104.
31. D. Sullivan and N. Teleman, *An analytic proof of Novikov's theorem on rational Pontrjagin classes*, Publ. Math d'IHES, **58** (1983) 291-293.
32. L.Taylor and B.Williams, *Surgery spaces: structure and formulae*, LNM, **741** (1979) 170-195.
33. N.Teleman, *The index of the signature operator on Lipschitz manifolds*, Publ Math d'IHES, **58** (1983) 251-290.
34. C. T. C. Wall, *Surgery on Compact Manifolds*, Academic Press, 1971.
35. S. Weinberger, *The Topological Classification of Stratified Spaces*, University of Chicago Press, (to appear).

36. S. Weinberger, *Fake G-tori*, (to appear).
37. S. Weinberger, *Nonlocally linear manifolds and orbifolds*, Proc. 1994 ICM (Zurich).
38. S. Weinberger and M. Yan, *Exotic Products and Equivariant Periodicity*, (in preparation).
39. M. Weiss and B. Williams, *Automorphism of Manifolds and Algebraic K-theory I*, K-theory, **1** (1988) 575-626, II, J Pure and Applied Algebra, **62** (1989) 47-107.
40. M. Yan, *Periodicity for equivariant structure sets*, Comm. Pure Appl. Math., 1993.

DEPARTMENT OF MATHEMATICS, UNIVERSITY OF CHICAGO
E-mail address: shmuel@math.upenn.edu

AMS/IP Studies in Advanced Mathematics
Volume 2, 1997 (Part 1)

LARGE SCALE HOMOLOGY THEORIES AND GEOMETRY

JONATHAN BLOCK AND SHMUEL WEINBERGER

In recent years there has been an explosion of interest in the application of "large scale" or "coarse" methods in geometry, topology, and group theory. The modest aim of this paper is to describe some "homological" methods that can be (and have been) applied to questions of this ilk, and to explain some of the geometry involved in these theories.

There are two seperate strands that oft intertwine in this subject. The first type of question is itself coarse: we want to understand how someone with blurry vision would view a discrete metric space. Thus, bounded metric spaces are entirely indistinguishable from points.

The second type of question is almost complementary in its force: What kinds of restrictions on conventional geometric questions are forced by large scale considerations. In other words, one studies smooth questions on manifolds while insisting that the large scale structure of the space, or a map, remain intact.

Our choice of material has been dictated by a few considerations. Firstly, we tried to avoid topics that seem to be very well known (excluding therefore, bounded cohomology and L^2 cohomology) or recently surveyed (thus excluding the theories of hyperbolic and automatic groups, or, much more generally, the contents of [58]). Secondly, we have tried to cover topics that have simple to understand geometric meanings or that can be used to answer questions posed by friends. And, finally, we have tried not to stray too far from things we actually understand.

The contents are as follows:

1. Definitions and examples.
2. HX_* and R_*X.
3. Metric rigidity phenomena I (packing).
4. Metric rigidity phenomena II (topological rigidity).
5. Connection to large scale index theory.

The first author partially supported by an NSF-YI grant, the second author partially supported by an NSF grant.

We would like to thank Oliver Attie, Mladen Bestvina, Steve Ferry, Hillel Furstenberg, Steve Gersten, Misha Gromov, Steve Hurder, Greg Kuperberg, Erik Pederson, and John Roe for useful conversations and correspondence. We would also like to thank the anonymous referee for his careful job as a readers advocate and for suggesting references of which we were unaware. Inevitably, there will be some overlap between the theorems presented here and others in papers of theirs (and other people); we believe that the points of view, however, are somewhat different, and apologize if we are wrong.

1. Definitions and Examples

Let us start with some common definitions:

Definition. A *Lipschitz* map between metric spaces is a map $\phi : Y \longrightarrow Z$ such that $d(\phi x, \phi x') \leq Cd(x, x')$, for some constant C. A *coarse-Lipschitz* map satisfies $d(\phi x, \phi x') \leq Cd(x, x') + D$, for some C, D. Note that coarse Lipschitz maps need not be continuous. Note, too, that for *effectively* discrete metric spaces (e.g. metric spaces with a lower bound on the distance between distinct points) Lipschitz and coarse Lipschitz are the same. We use the term *bi-Lipschitz* to denote a coarse Lipschitz map with a coarse Lipschitz inverse.

Definition. A *quasilattice* is an effectively discrete subset of Z, which is C-*dense*[1], that is which is of some bounded distance of any point of Z. Thus $3n + (-)^{g(n)}$ is a quasilattice in $\mathbb{R}$ for any integer function g (e.g. the n-th base 2 digit of π).

Definition. Two metric spaces are *coarsely equivalent* (or *coarsely quasiisometric*) if they contain quasilattices which are homeomorphic by bi-Lipschitz homeomorphisms. There is a similar notion of *quasi-isometric embedding*.

The by now standard example of these notions is the relationship between a finitely generated group with its word metric and the as-

[1] Also called *syndectic* in the literature. Quasilattices are also called nets.

sociated cover of a compact manifold[2] (or polyhedron). This famous observation of Milnor and Svarc implies (as they noted) that, say, fundamental groups of negatively curved manifolds must have exponential growth.

The basic question to be asked under these circumstances is how can we distinguish coarse equivalence classes of spaces?

There are a number of subsidiaries to this question. Is there a "nicest" space associated to the quasiisometry type which captures the type, or at least which can be used effectively in constructing invariants? We have introduced the quasilattices, which of course, have the simplest structure in some sense in that they are discrete! Can one find other "best" spaces by focusing on other features?

Perhaps the most concrete way to think about this question is to ponder some individual metric spaces.

Example 1. *Finitely generated groups and coset spaces.*

Let π be a finitely generated group. Given a finite generating set, one can construct its *Cayley graph* which has a vertices the group elements, and as edges pairs of the form $(g, \Gamma g)$ where Γ or Γ^{-1} lies in the generating set. The associated distance between group elements is called the *word distance* or *length* or *metric.* If Δ is a subgroup of π then the coset space π/Δ also has a natural word metric. The study of groups (and to a lesser extent, coset spaces) via the coarse properties of the Cayley graph is the focus of the popular subject called "geometry of groups". The great theorems that started this subject and remain major sources of inspiration are probably Mostow's rigidity theorem (see [84, 85]) and Gromov's theorem on groups of polynomial growth [59], which we will state below.

Remark. It is natural to consider infinitely generated situtations like infinitely generated subgroups of finitely generated groups. One can build a Cayley graph from an infinite set of generators with weights on the generating set so that the resulting metric has a finitely many vertices in any ball. However, the quasiisometry class depends on the choice of generators and weights.

Definition. A metric space has *coarse bounded geometry* if it posesses a quasilattice with the property that the number of elements in arbitrary balls of a given radius is uniformly bounded. Such a quasilattice will be called a *bounded quasilattice.*

[2]This naturally generalizes to groups acting properly discontinuously and isometrically, provided there is a bound on the size of isotropy.

Groups and coset spaces are good examples of spaces of bounded geometry. A (simplicial) tree with vertices with growing numbers of neighbors is the prototypical example of a space with unbounded geometry.

Definition. The growth rate of a metric space is the "growth rate" of the number of points in a fixed quasilattice within distance R of a fixed base point. (Two nondecreasing functions f and $g : \mathbb{N} \longrightarrow \mathbb{R}$ have the same growth rate if there are constants such that $Af(Br) - C < g(r) < Df(Er) + F$.)

It is not hard to see that the growth rate of a metric space just depends on the coarse quasiisometry type, and that (for metric spaces of bounded geometry) it is at most exponential. In particular, the notion of polynomial growth makes sense.

Theorem. (Gromov) *A finitely generated group is virtually* [3] *nilpotent iff it has polynomial growth.*

Of course, one cannot do better than a virtual result because *commensurable* groups (i.e., virtually isomorphic groups) are quasiisometric. A central problem is to determine to what extent the converse holds, i.e., how far quasiisometry is from virtual isomorphism (and quasiisometries are from virtual isomorphisms). This is the core of one of the approaches to rigidity, but the "answer" is that there are many examples of noncommensurable quasiisomteric groups. For instance, Malcev's theorem on nilpotent groups together with Remark 2.15 in Raghunathan [91] gives examples of noncommensurable quasiisometric nilpotent groups. More explicitly, let Sol be the irreducible three dimensional solvable Lie group given as the semidirect product of $\mathbb{R}$ on $\mathbb{R}^2$ by the action $t \cdot (x, y) = (e^t x, e^{-t} y)$. Then any uniform lattice Γ intersects the normal subgroup in a free abelian subgroup of rank 2 and projects to a discrete subgroup of $\mathbb{R}$. Commensurable groups will have rationally dependent images and conversely. For any matrix A with $tr(A) > 2$ in $SL_2(\mathbb{Z})$ the semidirect product of $\mathbb{Z}$ acting by A on $\mathbb{Z}^2$ gives a lattice in Sol whose image in $\mathbb{R}$ is generated additively by

$$\log(\frac{tr(A) + \sqrt{tr(A)^2 - 4}}{2})$$

[3]For a property P, recall that a group is vitrually-P if it contains a finite index subgroup which is P. This is most convenient for properties P which are passed on from a group to its finite index subgroups.

It is an easy exercise that by varying A one obtains an infinite set of rationally independent images and therefore noncommensurable groups.[4]

Gersten [54], using a bounded cohomology trick, gave the first examples of uniform lattices in essentially different Lie which are quasi-isomteric. A "simple" question at the heart of rigidity is whether every group quasiisometric to a Lie group is a virtual uniform lattice. In the semi simple rank greater than one case, at least, this is a recent theorem of Kleiner and Lieb. For rank one nonuniform lattices Richard Schwartz has even shown that quasiisometry implies virtual isomorphism. (See also [48] for the Hilbert modular group.)

Other examples of noncommensurable groups can be obtained by using different lattices in the same Lie group and using invariants like volume.

Bass has given a nice formula for the growth of a virtually nilpotent group. (For a proof, see the appendix, by Tits, of [59].) If the dimensions of the successive free abelian quotients are denote by a_i, then growth $= \sum ia_i$. In a beautiful paper, Pansu [86] has shown that the individual a_i's are actually coarse invariants as well. For instance, we have the following consequence that is indeed much more elementary.[5]

Corollary. *"Virtually abelian" is a coarse notion.*

Corollary. *The length and cohomological dimension of nilpotent groups are coarse invariants.*

Problem. *When are virtually nilpotent groups quasiisometric?*

Problem. *Is virtual solvability coarse?*

Another important coarse algebraic notion is amenabilty which we will discuss in considerable detail in the Part II. We will also show that for a large class of groups virtual cohomological dimension[6] and the

---

[4]For subgroups of a given group, there are two meanings to commensurable. One means that literally that they have a common finite index subgroup, while the other notion (the one we are using) means that they are virtually isomorphic. This difference is significant in the case of *Sol* because lattices have deformations, detected by covolume change.

[5]By the end of section 2 the reader should also be able to prove this.

[6]The corollary on being virtually abelian follows from coarseness of cohomological dimmension as follows: Virtually abelian groups are those with polynomial growth of degree equal to their virtual cohomological dimension. By Bass's theorem, if the nilpotent group is nonabelian, its growth rate is larger than this.

property of being a virtual duality group are coarse. (This had been independently observed by Gersten [55] as well.)

Now we introduce a class of spaces whose proper homotopy type reflect features of its quasiisometry type.

Definition. A space A is uniformly contractible[7] if there is a function $\phi : \mathbb{R}^+ \longrightarrow \mathbb{R}^+$ such that any ball of radius r, $B_x(r)$ is contractible in the ball of radius $\phi(r)$ $B_x(\phi(r))$. Note, ϕ need not be continuous or have $\phi(0) = 0$ or anything like that. We refer to ϕ as a contractibility function for A.

Proposition. *Finite dimensional coarsely equivalent uniformly contractible polyhedra are proper homotopy equivalent.*

This follows easily from the following proposition, which shows that if one thinks of the quasiisomtery type as being like the fundamental group, uniformly contractible spaces will then play the roles of $K(\pi, 1)$'s.

Proposition. (Connect the Dots) *If $f : Z \longrightarrow A$ is a coarse Lipschitz map from a finite dimensional polyhedron to uniformly contractible space, then f is of bounded distance*[8] *from a continuous coarse Lipschitz map $f' : Z \longrightarrow A$.*

The proof is by induction on the skeleta of Z, extending maps over simplices using the contractibility function.

Proposition. *The universal cover on any compact aspherical space is uniformly contractible.*

For many groups one knows compact $K(\pi, 1)$'s and so the above proposition gives very convenient models for the group as a metric space. The most basic examples of this to keep in mind are uniform lattices π of real connected Lie groups G, which act (freely, when π is torsion free) on the symmetric space G/K, K a maximal compact subgroup of G; G/K is uniformly contractible so it is coarsely quasiisometric to π. In the S-arithmetic case, one can use products of Tits buildings to replace the symmetric space. (See [23].) We will often call the universal

[7] There is a similar notion of uniformly k-connected. It is also of some interest to examine the k-contracitiblity functions.

[8] The distance depends only on the dimension of Z, the coarse Lipschitz constants, the density of the quasilattice of Z that f is defined on, and the contractibility function of A.

cover of a compact aspherical space a *coarse-homogenous space* for the fundamental group[9].

The cohomological dimension of $\pi = \dim G/K$ which is of course a proper homotopy invariant (determined as the maximum dimension for which the cohomology with compact supports is nonzero.) In the next section we will make this reasoning more directly homological. In Part II we will also discuss how to build homology theories that take volume-type considerations and the like into account.

Remark. There are more delicate geometric properties of the "coarse homogenous" space implicated in the above discussion, if one looks at it more carefully. For instance, one can easily check that the growth types of contractibility functions are coarse. The 1-contractibility function has been called the *isodiametric function* by Gersten [56]; it is recursive iff the group has solvable word problem.

Example 2. *Cones on manifolds and polyhedra.*

A rather different class of examples comes about by taking cones on metric spaces. If (X, d) is a (typically compact) metric space, we can form its *(Euclidean) cone.* This is done (for finite dimensional X) by viewing X as isometrically embedied in the unit sphere of some Euclidean space (with say a box metric) and the looking at the open cone as a subspace of Euclidean space.

Other types of cones can be formed by replacing Euclidean space by other spaces. Alternatively, one can think of a process where one weights the levels of the cone by a function ϕ, and then forms the metric given by paths of shortest lengths (see [43] for some information about this construction, e.g. when it produces a metric!). This will be called the *ϕ-cone.* Using rapidly growing ϕ, one gets many examples of hyperbolic metric spaces in Gromov's sense [61].

Let us start with Euclidean coning. The basic question is to what extent one can recover (X, d) from the Euclidean cone?

Example. [10] *Consider $X = \{0\} \cup \{1/n^\alpha | n \in \mathbb{N}\}$. The growth rate of cX is $r^{(2\alpha+1)/(\alpha+1)}$ so that for varying α these metric spaces are not coarsely equivalent. As a consequence, the topological type of X does not determine the Lipschitz type of the cone.*

---

[9]The reader should realize immediately always that coarse-homogenous spaces are not at all spaces coarsely equivalent to homogenous spaces.

[10]This is a variant of an example shown to us by Greg Kuperberg which simplified our more complicated examples based on metrics of varying Hausdorff dimension on the circle.

In light of this example we will restrict attention to situations where X has some preferred class of metrics, e.g. X a polyhedron, or smooth compact manifold (all Riemannian metrics are biLipschitz) or even a topological manifold (by Sullivan's theorem [100][11].)

Proposition. *If M and N are smooth (or topological) closed manifolds (with Sullivan's metric, that is any metric compatible with Sullivan's Lipschitz structure, see* [101]*) of dimension at least four then cM and cN are coarsely equivalent iff the M and N are topologically h-cobordant.*

This is probably true in all dimensions, but we have not thought through the low dimensional case in enough detail to see how hard it would be to prove this. Using the ideas of [98] and [105] one can also extend this to "Lipschitz stratified spaces" with the suitable notion of stratified s-cobordism. For an application of this proposition, see [106]

Proof. We can restrict attention to quasiisometric quasilattices on which the map is genuinely bi-Lipschitz. Now rescale the metric spaces and maps by a factor of $1/n$. This gives a sequence of maps on denser and denser subsets which are *uniformly* bi-Lipschitz. The Arzela-Ascoli theorem (as extended to Gromov-Hausdorff space) gives a convergent subsequence of "bi-Lipeomorphisms" on the whole cone. Thus cM and cN are bi-Lipschitz homeomorphic in the non-coarse sense! This immediately implies that M and N are h-cobordant, since M embeds as a deformation retraction in $N \times \mathbb{R}$. □

Conversely, if they are topologically h-cobordant, then $M \times S^1 \cong N \times S^1$ by a bi-Lipschitz homeomorphism (according to Sullivan's uniqueness theorem). The infinite cyclic cover of this homeomorphism with the 0-points added gives a Bi-Lipschitz homeomorphism between these cones, as has been verified in [106].

Thus for Euclidean cones[12] one cannot hope for a well defined "space at infinity"; there is at most a well defined h-cobordism class. However for hyperbolic cones, there is the "Gromov boundary", made up

[11]Sullivan's theorem requires that the dimension of the manifold be other than 4. (See [101].) However, one can use the discussion below to show that homeomorphic 4-dimensional Lipschitz manifolds have biLipschitz cones. Even for 4-manifolds without Lipschitz structures, one can produce the Lipschitz structure on cX that would have come from its Lipschitz structure, had it had one.

[12]It seems that even with slower growing coning functions one still gets the h- cobordism class at infinity by a more complicated argument; if the coning function is bounded then the coarse type is of course always the same as that of a ray.

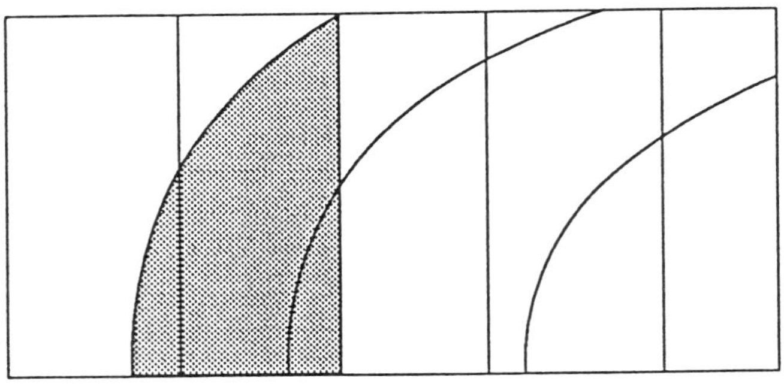

FIGURE 1. These represent two product structures on the same manifold, the shaded region an h-cobordism between them.

by equivalence classes of "coarse quasigeodesic rays" and the space at infinity not only includes the homeomorphism type of X, but also the conformal class of metric! (See [61, 57, 36, 97] for more on this.)

There is a weird type of coning construction which preserves exactly the homotopy type of a given polyhedron at infinity. Essentially one uses an iterated mapping cylinder of regular neighborhoods of embeddings of the polyhedron in higher and higher dimensional Euclidean spaces. This is like replacing the polyhedron with its product with the Hilbert cube and then takes the Euclidean cone. Then one can make use of regular neighborhood characterization of simple homotopy types and the previous arguments to prove the equivalence. (We note that what is produced on the ball of radius n in this coned metric space can be viewed as what would have been constructed from $P \times I^n$, so there is nothing "locally infinite dimensional" in this cone.)

Moral of the example: The conclusion is that there are various natural construcitons from compact metric spaces to coarse metric spaces which preserve more and more structure: from the mere homotopy type to h-cobordism type to homeomorphism and conformal type of the metric! It is not hard to see that there are manifolds (even Riemannian manifolds, see [50]) approximating these spaces. Thus "coarse considerations" can lead one to much more precise or fine problems in the geometry of complicated spaces than geometric topologists had been comfortable in studying in the past!

We note that it seems difficult to figure out what the overlap between the quasiisometry types described by cones and those described by groups (and coset spaces) are.

Example 3. *Leaves of foliations*

The leaves of a foliation of a compact manifold have natural quasiisometry classes of metrics. If Γ is a group of diffeomorphisms of a compact manifold M, then it is possible to get a foliation with many leaves quasiisometric to Γ, via the fiberred product construction $M \times_\Gamma \tilde{N}$, where N is a manifold with a surjection $\pi_1 N \longrightarrow \Gamma$, which produces the cover implicit in the notation $\tilde{N}$.

Problem. *Is every finitely generated group quasiisometric to a leaf?*[13]

There are similar questions that can be asked about coset spaces. (Their quasiisometry types would arise if Δ were the isotopy subgroup of some element for a Γ action on M.)

In any case, it is interesting to try to restrict the coarse quasiisometry type of leaves of foliations on manifolds. In Part II we will explain the method of Phillips-Sullivan-Janskiewicz that sometimes does this. There is also a very nice entropy method of Attie-Hurder that uses coarse/fine invariants of the sort to be discussed in the next section to produce leaves bounded-geometry homotopy equivalent to leaves that are not leaves (of at least smooth foliations).

Nonetheless, there are also many vary complicated leaves of foliations known that are not at all related to the previous examples. (As far as we know, the most detailed information on spaces at infinity of leaves is the realization of compact totally disconnected spaces as the space of ends [27].) It would be very interesting to understand how foliation invariants are related to the coarse geometry of the leaves. One beautiful result is Hurder's [76] that gives the vanishing of the Godbillon-Vey invariant of a foliation all of whose leaves have subexponential growth.

[13]One can produce certain, fairly unnatural, groups that do not act by diffeomorphisms on any manifold; however, as far as we know, no such torsion free example is known.

2. HX_* AND R_*X

In the previous section, we explored the utility of having a uniformly contractible space within a given coarse quasiisometry type, especially its use in defining coarse invariants. Now we shall introduce invariants without having a uniformly contractible model around.

Examples. *If an increasing sequence $\{a_n\}$ of real numbers is viewed as a metric space, it is coarsely (equivalent to a) uniformly contractible (metric space) iff the successive differences are uniformly bounded.*

One can similarly string together spheres of dimension i and radii given by the $\{a_n\}$, and one obtains a uniformly i-connected space iff the sequence is bounded. If one varies the dimensions of the spheres one can get exmples that are coarsely uniformly k-connected for each k, but not uniformly contractible.

Depending on the metric on X, the open cone will or will not be uniformly contractible. Finite dimensional ANR's have metrics for which the warped open cones are uniformly contractible [43]. We do not know whether the Euclidean cone can always be taken UC.

A finitely generated group is coarsely equivalent to a uniformly k-connected space iff $B\pi$ has a finite $k+1$ skeleton. (Gersten, Gromov, Block-Weinberger, and probably others). Abels [1] has given many examples of groups with varying finiteness conditions, which are relevant here.

The basic method here is a variant of an idea of Rips (and has also been discovered on many occasions by different authors). Rips observed (see [61, 36, 57, 97]) that for X a ∂-hyperbolic metric space, for all sufficiently large k the complex obtained by considering the Cech nerve of the covering of X by k-balls is contractible[14].

However, for many non-word hyperbolic groups or yet more general metric spaces, even those posessing uniformly contractible models, there is no fixed size at which this Cech nerve is contractible. The solution to this problem is to consider the whole sequence of nerves and work with this.

More precisely, let Z be a metric space, which we assume to be (effectivly) discrete. Then one has the locally finite cover of X (a quasilattice

[14]Rips' application was to prove beautiful general facts regarding the cohomology of hyperbolic groups. Bestvina and Mess have given further properties of the compactified Rips complex [18]. These are applied in [52, 30] to give a geometric approach to the Novikov type theorems for hyperbolic groups, see [28, 39] and below.

in Z) by balls of radius n. We call $R_n(Z)$ the nerve of this cover[15]. Since these covers become coarser as we increase n, there are natural (proper) maps $R_n(Z) \longrightarrow R_{n+1}(Z)$, which are, in fact, coarse equivalences. The pro-homotopy type of this system is invariant of the choice of X.

The point of this construction is that topological structure at scale n is removed in $R_n(Z)$. We denote by R_*Z this system[16]. It is "pro-functorial" for proper Lipschitz maps ($f_* : R_n(Z) \longrightarrow R_{Cn+D}(Z)$ for a (C, D)-Lipschitz map) and its "pro-homotopy-type" is a well defined quasi-isometry invariant of the metric space. Note that if V is a simplicial complex with simplices of bounded diameter, and f is Lipschitz, one obtains a continuous map $f : V \longrightarrow R_n(Z)$ for n sufficiently large.

We can therefore define invariants of coarse metric spaces by taking invariants of R_*X.

Definition. $HX_*(Z) \cong H_*(R_*(Z))$ (or by the direct limit, if we are not interested in the details of the pro-system.)

Proposition. (Roe [92]**)** *If Z is uniformly contractible then the natural map $H_*^{lf}(Z) \longrightarrow HX_*(Z)$ is an isomorphism.*

This is not too hard. Now, if E is any homology theory (or homotopy functor!) one can now make:

Definition. $EX_*(Z) \cong E_*(R_*(Z))$.

Furthermore, there is a dual cohomological notion. These were in fact what were introduced in [58, 92]. Roe's interest in this cohomology was that it could be used to pair with elliptic operators on noncompact manifolds to form "exotic indices". We will return to this in a later section.

For us convenient choices of E are K-theory and K-theory with coefficients. It turns out that for nonconnective spectra there is no need for the proposition to hold, and there is different information in the EX-theory than the E-theory of a uniformly contractible model! Remarkably, it seems that EX-theory is actually more suitable for coarse geometric applications than the homology it was invented to model!

The following is the result of tracing definitions:

Proposition. *If $B\Gamma$ is a finite complex*[17], *then* $HX_*(\Gamma) \cong H_*(\Gamma; \mathbb{Z}\Gamma)$.

---

[15] Metrized simplicially.

[16] The use of the system rather than taking the limit is suggested by shape theory. In this paper, though, we will only be studying limit type invariants.

[17] Under suitable finiteness condition on Γ, this proposition will be true in the appropriate range.

In particular the latter group is a coarse invariant, in this setting. Since, the cohomological dimension is the last degree in which the r.h.s. is nonzero, it is also coarse (as promised above). Finally, we have the following:

Proposition. *If $B\Gamma$ is a finite complex, then either*

$$(1) HX_*(\Gamma) \text{ is infinitely generated in some dimension}$$
$$\text{or } (2) HX_*(\Gamma) = 0 \text{ for } * \neq n, \text{ and } HX_*(\Gamma) = \mathbb{Z} \text{ for } * = n, \text{ for some } n.$$

In that case $B\Gamma$ is an n-dimensional Poincaré duality group.

Results similar to the proposition have been proven by Bieri-Eckmann, and Farrell (all though not with the coarse terminology); see Brown [23] for some discussion and references to their work.

Corollary. *No uniform lattice is coarsely equivalent to a nonuniform lattice.*

(Because $HX_*(\Gamma)$ is infinite dimensional in its top nonvanishing dimension for nonuniform lattices as one can see using the results of Borel and Serre [21].)

This proposition is probably known to those who work in cohomology of groups in light of the previous one [18]. We will give an alternate proof that has additional topological implications.

Proof. We shall make use of the following result: □

Theorem. (Quinn, see [31]) *Suppose $F \longrightarrow E \longrightarrow B$ is a fibration with all three space finite complexes, then E is a Poincaré complex iff F and B are.*

This follows from a consideration of boundaries of regular neighborhoods of embeddings in Euclidean space, and the following characterizations of Poincaré duality (which is essentially a special case[19] of the theorem, as well)

Theorem. (Mischenko and Quinn) *X is a Poincaré space iff when X is PL embedded in Euclidean space, the homotopy fiber of the map from the boundary of a regular neighborhood of X into X is a homotopy sphere.*

---

[18]The corollaries to the proposition and the coarseness of Poincaré (or Bieri-Eckmann) duality were observed by Gersten [55] without reference to HX.

[19]In light of Spivak's thesis [99].

Now we can prove the proposition. Let $N, \partial N$ be a regular neighborhood of $B\Gamma$. (We assume without loss of generality that the maps $\partial N \longrightarrow N \longrightarrow B\Gamma$ are isomorphisms on π_1.) The homotopy fiber of the map $\partial N \longrightarrow B\Gamma$ is the universal cover of ∂N. Thus, $B\Gamma$ is a Poincaré space iff this is a homotopy sphere. Since ∂N is a manifold, if this fiber is a finite complex, $B\Gamma$ is a Poincaré space, by the previous theorem. Thus, if the fiber (=cover) is finite, it is a homotopy sphere.

Since $B\Gamma$ is aspherical, N is contractible, and therefore $H_*^{lf}(N) \cong HX_*(B\Gamma)$. However $H_*^{lf}(\tilde{N}) \cong \tilde{H}^{n-*-1}(\partial\tilde{N})$ (where $n = \dim N$), so if HX is finite dimensional, it is the reduced cohomology (shifted) of a sphere, proving the proposition.

Corollary (of proof). *If the universal cover of a manifold with fundamental group of type virtual-FP has finitely generated homology, then the group is a virtual duality group.*

Remark. If $\mathbb{R}^n \times M$[20] has a proper discontinuous cocompact Γ action, M compact, then we have $HX^*(\Gamma) = 0$ for $* \neq n$, and $HX^*(\Gamma) = s$ for $* = n$, since Γ and a uniformly contractible copy of $\mathbb{R}^n$ have the same coarse type. Consequently, if Γ is torsion free, it is a duality group. For M a sphere, this is a theorem in [40].

This is nice because the coarse theory describes an analogue of the duality condition that is necessary even for non-residually torsion free groups (see [48] for such an example). It is quite easy to show that for torsion free groups a converse holds (with an extra condition[21]); for virtually torsion free groups the converse still holds if we are allowing arbitrary compact M—if we insist M is a sphere, there is at least an additional condition on periodicity of Farrell cohomology, and it is a difficult problem to characterize the groups that can act (see e.g. [68, 3]).

Problems. *Which countable groups (i.e. no sissy conditions) can act freely and cocompactly on $\mathbb{R}^n \times M$ for some compact M, on $\mathbb{R}^n \times S^k$ for some (specific?) sphere, or on some space of finite type? Which Γ act on a sphere in a way that is free and cocompact outside of its domain of discontinuity?*[22]

---

[20] Or even $X \times M$, for X a contractible manifold and M merely a compact space.

[21] Basically, one can define the fundamental group of the end of Γ in a coarsely invariant fashion; it must be trivial for the group to act cocompactly on a space that is coarsely simply connected at infinity. A contractible manifold that is simply connected at infinity is Euclidean space by a theorem of Stallings.

[22] For torsion free hyperbolic groups the answer is given entirely by the Poincaré duality condition.

3. Metric rigidity phenomena. I (Packing)

The coming sections contain a miscelleny of coarse rigidity theorems of a coarse /topological/homological nature. As we mentioned in the introduction, the literature is very large, (after all one of the main inspirations of coarse geometry was Mostow's work) and we would surely be remiss if we did not at least mention Gromov's bounded cohomological proof of Mostow rigidity for compact hyperbolic manifolds (see [66, 102]). However, our focus will be on phenomena that seem at least philosophically related to the types of considerations of the previous section[23].

Definition. A map $\phi : X \longrightarrow Y$ is *effectively proper* if there is a function real valued function f such that if $d(x, x') \geq f(r)$, then $d(\phi x, \phi x') \geq r$. A map which is both coarsely Lipschitz and effectively proper will be called EPL.

Remark. This condition has appeared in [1, 19] and a number of other investigations. However, this condition now seems too strong, as it encodes "asymptotic coarse injectivity". Probably one should modify the definition so that $\phi : \mathbb{C} \longrightarrow \mathbb{C}$ given by $\phi(z) = z^2$ is effectively proper[24].

Problem 1 (A packing problem). *Suppose that $\phi : X \longrightarrow Y$ is EPL, what can one say about ϕ? Sometimes, we will use the (hopefully) evocative terminology that ϕ packs X into Y, or that ϕ is a packing.*

Of course, in this generality, nothing. We first thought about this question[25] in the special case where $X = Y = \mathbb{Z}^n$.

A metric space will be called overeuclidean if there is a (uniformly and coarsely) Lipschitz homeomorphism from it to Euclidean space. For instance any simply connected manifold of non-positive curvature is over-Euclidean as well as any simply connected nilpotent Lie group.

Theorem. [26] *Any packing of $\mathbb{Z}^n$ into itself is a quasiisometry. In particular, the image is always C-dense. In fact, any packing of a quasi-lattice in an over-Euclidean space into a uniformly contractible Riemanninan manifold of the same dimension is a quasiisometry.*

---

[23]Indeed, these phenomena were our main motivations for the introduction of these homological techniques.

[24]The reader who insists on Lipschitz should modify ϕ to be $\phi(z) = z^2/|z|$.

[25]Hillel Furstenberg brought it to the attention of the second author in 1990.

[26]With some extra condition on ϕ, this was proved by Furstenberg as well, by ergodic methods, at least in low dimensions.

The proof we'll give implies that one can't assemble simplices in any coherent way using far apart vertices and get a PL homeomorphism:

The theorem can be restated that any of these packings necessarily has a linear effective propriety function.

Note however that it is critically important that the packings be quite tight: It's trivial to pack the line into the plane in many exotic ways. However, it seems possible that in the case of higher rank lattices there are relatively few packings of spaces even into rather larger ones.

We will prove the theorem geometrically. However, the argument (perhaps by being more perspicacious in applying the techniques of [49] coarsely—rather than just the theorem) could possibly be reorganized purely homologically. Intuitively, one wants to say that EPL maps are asymptotically injective, so that they have HX-degree one, which implies that they are coarsely onto.

We shall instead go back to uniformly contractible manifolds and use the following amazing theorem of Ferry [49] (See also [51] for the noncompact version we're using here):

Theorem. *For any manifold M^n, $n \geq 5$[27], there is a δ such that any continuous map $\phi : M \longrightarrow N^n$ with point inverses of diameter less than δ are within ϵ of a homeomorphism*[28].

Addendum. *If M is noncompact, δ might have to be taken to be a function which decays at infinity. However, if M has bounded geometry*[29]*, δ can be chosen to be a small constant.*

However, many M have *shrinking self-homeomorphisms* [30] (e.g., Euclidean space has dilation), in which case obviously for any δ, any "δ-map" can be homotoped to a homeomorphism.

Remark. It is remarkable that one can use shrinking self homeomorphisms, in that their inverses are not Lipschitz.

In any case the theorem is now fairly direct. If one connects the dots on a packing one abtains a continuous-Lipschitz map between Euclidean spaces. Ferry's theorem puts it a finite distance from a homeomorphism. The image of the lattice under this homeomorphism is C-dense. One can also now define a coarse inverse map, g, which has a "bornology property", $d(gy, gy') \leq F(d(y, y'))$ for some function F (by the effective

[27]4 suffices according to [16, 51].

[28]This is true equivariantly as well for the locally linear case.

[29]The injectivity radius is bounded below and the curvature is bounded (see [11] for some discussion).

[30]This is related to "combing conditions" as studied in [45, 6, 5, 56] and elsewhere.

propriety). (The bornological terminology is due to Roe, [92]). However, since $\mathbb{R}^n$ is a coarse path space (i.e., the distance between points is proportional to the length of shortest paths connecting them), obviously, bornological implies Lipschitz, so the original map was coarse bi-Lipschitz.

The result on spaces comes from reflecting on Ferry's theorem, local contractibility of homeomorphism spaces [44], and Sullivan's theorem [100].

Regarding other uniform lattices, it is a little tricky to deduce the C-density result directly from Ferry's theorem because of the non-Lipschitzness of the inverse of the squeezing map (although one could directly try to prove a variant of Ferry's theorem for arbitrary δ on a symmetric space). Here homology theory enters. Let $\Gamma \longrightarrow \Delta$ be a packing and let X and Y be symmetric spaces for these groups. We also denote by $s : X \longrightarrow X$ a Lipschitz squeeze homeomorphism. Consider the diagram

$$\begin{array}{ccc}
HX_*(\Gamma) & \longrightarrow & HX_*(\Delta) \\
\downarrow & & \downarrow \\
HX_*(X) & \longrightarrow & HX_*(Y) \\
\uparrow & & \uparrow \\
H_*^{lf}(X) & \longrightarrow & H_*^{lf}(Y) \\
{\scriptstyle s^{-1}}\uparrow & & \\
H_*^{lf}(X) & &
\end{array}$$

where the bottom up arrow is induced by s^{-1}. This, and all vertical arrows are known to be isomorphism. The composite of the connect the dots map with s^{-1} is near a homeomorphism, by Ferry's theorem, so it is degree one, and we deduce that all the arrows are isomorphisms. The result then follows as before from the following lemma:

Lemma. *If X and Y are uniformly contractible manifolds and $f : X \longrightarrow Y$ is a coarsely proper Lipschitz map, and is nonzero on HX, then the image $f(X)$ is C-dense in Y.*

This is proven by a modification of the standard proof that a map of nonzero degree is onto There one assumes by way of contractiction that an open subset $\mathfrak{O}$ of Y is missed by f and then argues using the isomorphism $H_n^{lf}(Y) \longrightarrow H_n^{lf}(\mathfrak{O})$.

The trouble is that we only have compact balls in Y that are assumed missed by f, and HX_* of any such set vanishes. To deal with this, one uses the pro-system that defines HX. The idea is this, for each C, the balls of much greater size have "scale C homology" with the fundamental class of Y mapping nontrivially; that is, in the notation introduced before, $H_n^{lf}(Y) \longrightarrow H_n^{lf}(R_C(Y)) \longrightarrow H_n(R_C(B, \partial B))$ is injective. (In other words, for a fixed scale, the larger compact sets do not disappear; it is on changing scale that they get lost.) Having done this, the usual argument goes through.

Remark. Subsequent to the circulation of an early draft of this paper, Farb and Schwartz [48] applied packing theorems to gain control at infinity for maps between spaces that contain flat directions where the usual negative curvature ideas do not seem to directly apply. Their paper also contains a more direct approach to packing.

Exercise: Assuming the balls of Y to be convex, estimate how large the C in C-density is in terms of the diameter of the simplices in Δ needed to represent the fundamental class and the lipschitz constant of the packing.

We suspect that, at least overeuclidean groups, but probably Poincaré duality groups, and most other "nondegenrate" groups (maybe: fundamental groups of aspherical pseudomanifolds) cannot have exotic self-packings and cannot be packed into finite dimensional uniformly contractible metric spaces whose HX vanishes above the dimension of the group. (Indeed, some of this might follow from the proof of Ferry's theorem combined with the above argument in place of application of the theorem itself.) It would be interesting to know about other packing problems where the shapes of the spaces involved aren't as directly comparable. We note though that there are some "surprising" packings which raise rank, like horospherical packing of Euclidean space within Hyperbolic space.

4. Metric rigidity phenomena II (Topological rigidity)

In this section we will briefly discuss a statement which, through no fault of his, is referred to the Borel conjecture, and its metric analogue.

(Borel) Conjecture. *A group is the fundamental group of a closed aspherical manifold, which is unique up to homeomorphism iff it satisfies Poincaré duality.*

This conjecture can be viewed as one of the organizing problems of topology[31]. The uniqueness statement includes the three-dimensional Poincaré conjecture. Its analysis for tori [73, 103] was a key step in triangulation theory for manifolds [81]. It is known to be true for non-positively curved manifolds and other symmetric spaces [47].

As for the existence, there are hardly any nontrivial cases of aspherical manifolds produced via the group theory and proving the conjecture. (One can show that certain extensions of manifolds are realized by manifolds by producing geometric block bundles via a proof of the conjecture; superrigidity suggests that such examples rarely occur where not predicated by Lie theory.)

Our goals in this section are to (1) explain metric analogues of this conjecture (as in [52], (2) show that these lead to counterexamples to the analogue of existence for groups with torsion (but also provide evidence for a more reasonable version of existence discussed in [26]), and (3) give counterexamples to the metric rigidity conjecture following [43].

(Coarse Borel) Conjecture. *If M is a uniformly contractible manifold, then any bounded homotopy equivalence $M' \longrightarrow M$ is a bounded distance from a homeomorphism.*

If we view coarse types as the analogues of fundamental groups, then the uniformly contractible spaces are the analogues of the aspherical ones (they are both the terminal objects in the category of spaces with the given coarse type or fundamental group). In addition, the universal cover of a manifold to which the Borel conjecture applies is a metric space to which the coarse version applies. The Borel conjecture then asserts that in the absence of additional algebraic topology, there can be only unique geometric topology.

We note that Ferry's theorem (discussed in the previous section) implies that the Coarse Borel Conjecture is true for Euclidean space. In a suitable sense, it is true for weighted cones. It is also true for Hadamard manifolds, for instance. If it were true for groups then one could deduce the Novikov conjecture for all groups for which $B\Gamma$ is a finite complex (the principle of descent, see [52, 30])

To just indicate this direction of ideas, let us begin with an interesting theorem regarding a classical theme in smooth transformation groups.

Theorem. *Suppose a group G acts affinely on a torus T with a fixed point p, and that G acts continuously on T' as well. Let $\phi : T \longrightarrow T'$ be an equivariant map which is a homotopy equivalence. Then, there*

[31] There are, of course, many generalizations and analogues. We will just work with the closed manifold and metric versions here.

is an equivariant topological embedding of the tangent space TTp in a neighborhood of $\phi(p)$.

In short, if the action of G on T' were assumed smooth, one would be asserting the topological conjugacy of the tangential representations at the fixed points. One can modify the examples of [29] to show that even in the smooth category, linear conjugacy is not possible.

The point of view that this suggests to me is that if one thinks of the Novikov conjecture as a statement about tangentiality, one had best work in the topological category, and then one can enlarge the context some to obtain interesting information regarding even the restrictions of the tangent bundle to points! (In the classical (i.e. nonequivariant, manifold) case there is no information here except the dimension ... but we will soon see that here, too, there is more than meets the eye.)

Also, since this theorem is really part of the "Novikov package", one can go much beyond the torus case to at least, nonpositively curved spaces and groups (in some interpretation or other). *We will be using this extension shortly.* The following exercise describes one of the key ideas in this extension.

Exercise. *After reading the proof below, modify it to hold if the affine action on T is replaced by a G-aspherical manifold whose universal cover admits an equivariant homeomorphism that is Lipschitz to a G-representation (G-overeuclidean).*

Having increased the suspense by not giving the proof instantly, we give it now, for simplicity in the case that the action of G in a neighborhood of p has just p as a fixed point[32]. Consider $\phi : \tilde{T} \longrightarrow \tilde{T}'$. Clearly, the map is equivariant and has bounded size point inverses. This passes to the quotient. Apply Ferry's theorem near ∞, to produce isomorphisms outside of compact sets[33]. Now engulfing allows one to extend the homeomorphism to the complements of p and $\phi(p)$. These we extend by continuity! Now, by restricting to a small enough neighborhood of p, one still has an embedding of TTp on projection $\tilde{T}' \longrightarrow T'$. This completes the proof.

Now let's consider the existence part of the Borel conjecture for groups with torsion. We will see that not all reasonable candidates can be realized by aspherical orbifolds.

---

[32]The general case only requires more equivariant or stratified machinery, but is no different in conception.

[33]This essentially gives a proof of the bounded Borel conjecture for open cones of manifolds, with the understanding that an isolated singularity is allowed, if one works relatice to a neighborhood of the singularity.

Proposition. *There are virtually PD^n groups that are not realized by aspherical orbifolds. Moreover, these groups can be constructed so that all torsion is of order p and all torsion elements generate their own normalizers.*

We emphasize the simple structure of the finite subgroup lattice to eliminate any homotopical obstructions. The groups considered are true analogs of PD^n groups with torsion if one studies the orbifold question.

To construct these groups, we start with a PL (topologically) nonlinear semifree action of $\mathbb{Z}_p$ on a sphere with, say, two fixed points. These can be obtained by suspending suitable free actions on the sphere. (The early ones produced by Milnor and Hirsch will not do; ones detected by the (ρ-invariant [103] are perfect—there are such examples constructible through the Brieskorn method of isolated hypersurface singularities that are invariant under a group action. For instance, for involutions, where every Brieskorn singularity is preserved by refelction through the origin, Hirzebruch had first proved the nonlinearity of the action on the link sphere. For a thorough discussion of several points of view on these involutions and the interrelationships among the invariants involved in their analysis, see [82].)

Since the action is PL, we can apply a hyperbolization technique [61, 41, 78]. This will produce an aspherical manifold with a PL $\mathbb{Z}_p$ action. The group of lifts of this action (the orbifold fundamental group!!) is not realized by an orbifold!

Firstly, how can this be? Orbifolds are spaces that are modelled on linear representations. Here the process of hyperbolozation will produce an action that is not locally linear at the fixed points; the action there will be modelled on the free PL action on the sphere we started with.

The proof is not so hard. Any other action will be equivariant homotopy equivalent to the action constructed. The hyperbolic analog of the previous proposition shows that the local structure must be (topologically) the same as the one we started with, yielding a contradiction.

Remark. In [107] this same strategy of using true rigidity theorems as a method for disproving false ones was applied to producing counterexamples to the equivariant Borel conjecture for affine tori which violate the gap hypothesis.

On reflection it's easy to understand the source of these examples. The definition of orbifold is too narrow. One really wanted to say there was an appropriate group action on a contractible manifold. The local linearity is just some extra feature of a group action that we should not have assumed.

The same is probably true for PD^n groups. In [26] it was shown that surgery theory holds better for the existence (and classification) of homology manifolds than it does for manifolds! (That is, s-cobordism classes of homology manifolds simple homotopy equivalent to a given simple Poincaré complex are in a 1-1 correspondence with the fiber of the 4-periodic surgery obstruction assembly map, and exist iff the corresponding "total surgery obstruction" vanishes.) For instance, all of the evidence for uniqueness Borel is equally evidence for existence Borel if we modify it to include homology manifolds (i.e. if we remove the assumption of local linearity from our homology manifolds), but not for the existence BC as we've phrased it.

More precisely, there is a $\mathbb{Z}$-obstruction to resolving homology manifolds, see [90]. Assuming Novikov, for $B\pi$ the $\mathbb{Z}$-obstruction for any realizing homology manifold is the same as that of any other one. (The Borel conjecture would say than any two realizations were s-cobordant.) It is unthematic, and in violation of the analogy between group actions and the theory of homology manifolds [108], to assume that the only realized homology manifolds will be resolvable.

Indeed, there is another generalization of the Borel conjecture where one can also see the failure of the existence half if one does not include homology manifolds. Following [105], say that X is haspherical if X has the same integral homology as $B\pi_1(X)$. The Borel conjecture for the fundamental group of X implies that if X is a topological manifold, it is rigid. However, it would only imply for Poincaré spaces realizability by homology manifolds.

Conjecture. *If Γ is a torsion free hyperbolic (or "nonpositively curved") group then there is an aspherical homology manifold with fundamental group Γ iff it is a PD^n group (iff $HX_i(\Gamma) = 0$ for $i \neq n$ and $HX_n(\Gamma) = \mathbb{Z}$). The realization is unique up to s-cobordism.*

There is some direct evidence for this in [52, 30, 47, 74] and probably elsewhere. More indirect evidence can also be adduced. However, most importantly, the all values of the realization obstruction should occur (at least for $n \geq 5$).

Now, let us return to the large scale Borel conjecture itself

Theorem. [43] *There is a uniformly contractible manifold which is not boundedly rigid, i.e., there are coarse q.i. between uniformly contractible manifolds M' and M, that are not a finite distance from a homeomorphism.*

The construction begins with a remarkable result of Dranishnikov [42] as refined in subsequent work of [43] (and depends importantly on other constructions of Edwards and Walsh [104]):

Theorem. *There is a cell-like map $S^n \longrightarrow X$ which is neither an injection nor surjection on mod p K-theory.*

This X cannot be a finite dimensional space. (CE maps induce isomorphisms on all connective homology theories, and on all theories if the target is finite dimensional.) The basic idea is to build a metric on $\mathbb{R}^{n+1}$ that asymptotically resembles a sufficiently quickly growing cone cX. Doing this one can achieve uniform contractibility of the metric on $\mathbb{R}^{n+1}$ but the natural map $K^{lf}(\mathbb{R}^{n+1}) \longrightarrow KX(\mathbb{R}^{n+1})$ is neither injective nor surjective[34]. This means that there are a different class of "large scale bundles" on this manifold than predicted by ordinary K-theory. These are used to construct invariants that surgery theory realizes. We will discuss this somewhat in the next section.

5. Connection to large scale index theory

As far as we can tell, the first major paper in "large scale index theory" and its application to the geometry of compact manifolds was the beautiful paper [GL] that contains within it many of the ideas that have been developed from any number of different points of view over the past decade. Let me start with one of their main theorems whose proof contains elements which should already resonate with reader.

Theorem. *No compact manifold of nonpositive sectional curvature possesses any Riemannian metric with positive scalar curvature.*

This is deduced from a "large scale" result:

Theorem. *If a Riemannian manifold M^n of bounded geometry*[35] *with positive scalar curvature, then M^n does not have a Lipschitz degree one map to $\mathbb{R}^n$.*

Remark. Gromov has asked [62] whether every uniformly contractible M^n has a coarse Lipschitz degree one map to $\mathbb{R}^n$. In [43] this is shown

[34]Nothing like this could happen for homology as we had seen earlier; the point is that for nonconnective theories the construction of large-scale homology theories produces a quite different theory than that of the (assuming it exists) uniformly contractible representative of the coarse type.

[35]We will discuss the significance of this condition in chapter 2.

not to be the case[36]. However, it is still conceivable (likely?) that no uniformly contractible manifold is coarsely equivalent to a manifold with positive scalar curvature.

The proof is a coarse variant of a theorem of Lichnerowicz (see [10]) which asserts that any spin manifold with positive scalar curvature has $\hat{A}$-genus. This theorem is proven using the index theorem and a Bochner-Weitzenbock formula

$$D^*D = \Delta + \kappa/4$$

where D is the Dirac operator, Δ is the Laplacian (which is positive semidefinite) and κ is the scalar curvature. If κ is positive, then neither D nor D^* can have kernel, so the index is 0, but the index is $\hat{A}$-genus, and Lichnerowicz's result follows.

Gromov and Lawson next prove an index theorem for open manifolds. The setting is that one has two open manifolds with elliptic operators on them, and a positivity condition on these operators at ∞. They also presume that the manifolds and operators are identified outside of some compact set. This situation leads them to a relative Fredholm index, which they compute as being the difference of the integral of the Atiyah-Singer integrand on the corresponding compact pieces of the two manifolds.

In the applications, there is only one manifold, but one will have two operators on it, which are identified near ∞. These operators will just be Dirac twisted by appropriate bundles (with connection). It is in the construction of these bundles that the coarse geometry enters.

Note now that if a bundle is *almost flat*, say, has sufficiently small curvature, then when we tensor the Dirac operator with it, the Weitzenbock formula still gives the same vanishing of the index. In this noncompact setting, we can define a bundle E with a family of connections ∇_t to be almost flat if (1) the connections are each flat outside of compact sets (depending on t) and (2) the curvatures of ∇_t are going to 0 as $t \longrightarrow \infty$.

If we apply the relative index theorem to (E, ∇_t) (relative to the trivial bundle trivially trivialized), we get the conclusion that $\int_M(\hat{A} \wedge ch(E))dVol = 0$.

Now we invoke the hypothesis on Lipschitz maps to construct the almost flat bundles. Suppose without loss of generality that the dimension is even. Then we can consider $\mathbb{R}^{2k} \cong \mathbb{C}^k$ which has the Bott bundle on it (pick some connection on it). By rescaling one gets a sequence of

[36]This follows simply from the construction outlined in the previous section. If there were such a Lipschitz map then one could use it to see that K^{lf} would at least split off KX, which it does not, by construction.

connections with decaying curvatures on $\mathbb{C}^k$, which when pulled back give the desired bundle on M. It is easy to see that for this bundle $\int_M(\hat{A} \wedge ch(E))dVol \neq 0$, and the theorem is proven.

Remark. In the nonpositive curvature case the inverse of the exponential map (at some base point) is a convenient choice of Lipschitz map to Euclidean space. The theory of almost flat bundles is further elaborated in [39, 70].

It is also worth pointing out here that in the last section of this same paper Gromov and Lawson show how to use the families version of the same index theorem (coupled with a family of Bott bundles) to show that if M is a compact spin manifold with the fundamental group that of a nonpositively curved manifold, then the $\hat{A}$-genus must be 0. One considers the family $\tilde{M} \times_\Gamma E\Gamma \longrightarrow B\Gamma$ and considers its family relative index in $K(B\Gamma)$ and obtains its vanishing.

This method of obtaining restrictions on invariants of compact manifolds by considering families of coarse problems on the universal cover is used, in one way or another, in [80, 39, 52, 30, 75]. We like to call it the method (or principle) of descent. It is quite remarkable that a good enough understanding of what is often a single large scale obstruction in $HX^n(M)$, is enough to get information on obstructions lying in a whole range of dimensions and of essentially unkown groups $H^i(B\Gamma)$.

In a very nice memoir, Roe [92], building on work of Higson conceptually clarified the situation greatly[37]. His idea is this: One considers on complete Riemannian manifolds elliptic operators of bounded propagation speed (see [35, 92] for this concept and some of its implications; essentially one wants approximations to the defining kernel of e^{itD} the wave group associated to the given elliptic operator to be zero outside a neighborhood of uniformly bounded girth around the diagonal in $X \times X$; it holds for "geometric" operators like laplacians and Dirac operators on manifolds).

Now, according to [11, 25, 80], an elliptic operator gives rise to an element of $K^{lf}(X)$. The business of constructing an index is finding a C^*-algebra over which the projections to kernel and cokernel live, so that the index can then be defined to be the difference of these projective modules in K_0 of this algebra.

This is only intuitively correct. There are several types of complications that can arise in this program. The simplest and perhaps most familiar example is the index theorem for families. Here the appropriate C^*-algebra is $C(B)$, the agebra of continuous complex valued functions

---

[37]It still is somewhat unclear what the precise connection between almost flat bundles and bundles on the corona is.

on the parameter space B. The kernel and cokernel of this family of operators do not themselves form vector bundles, i.e., projective modules over this algebra, because the ranks of these vector spaces can change from point to point. Nonetheless, a well known argument produces perturbations such that the kernel and cokernel become constant rank, and the difference between these projective modules (i.e. vector bundles) is a well-defined element in K-theory. In the general situation that we need, there is a simple abstract algebraic K-theoretic construction that gives us our index, see [92, 93].

Here what is relevant is the completion of the bounded propagation speed kernels, $C^*(X)$. Indices with values in C^*-algebras have all of the properties that you would expect from an index: one can get vanishing from a Bochner formula, or homotopy invariance from a signature operator[38].

The index theorem really has something to do with the map $K^{lf}(X) \longrightarrow K(C^*(X))$. $K(C^*(X))$ is essentially a coarse object. For X compact, it boils down to $K(\mathbb{C})$ which is the familiar alternating sequence of $\mathbb{Z}$'s and 0's. In general, one doesn't know that much about it[39], but one can get some information using an idea of Higson.

Definition. The corona of X is the set of ideal points in the completion defined as follows: Define the variation of a function f on X by $V_f(x,r) = sup|f(x) - f(y)|$ where $y \in B_x(r)$. We then consider the algebra of bounded functions with decaying variations, i.e., for all r, $V_f(x,r) \longrightarrow 0$ as $x \longrightarrow \infty$. By the Gelfand-Naimark theorem, this corresponds to a compact space, and on reflection one sees that this contains X as a dense open subset.

In practice one always uses quotients of this completion, but even before doing that, one does obtain a nice "space at infinity" by Higson's construction. If Z is a quotient of the corona and is a nice space, then there is a map $K_*(C^*(X)) \longrightarrow K_{*-1}(Z)$. The large scale index theorem states that the image of the elliptic operator is given by the topological pushforward of the "symbol cycle" under the ∂ map $K_*^{lf}(X) \longrightarrow K_{*-1}(Z)$.

Dually, thinking in terms of pairing, one gets to pair intelligently an operator with a bundle that's coming from $K^*(cZ)$. These are precisely Bott type bundles in the sense we just used them: they live on a compact

[38] We are being a little glib here. See [87] for a discussion of what's involved and cases where this can currently be proven.

[39] See [71] for the current state of knowledge (which contains all Hadamard manifolds).

part of a cone, and rescaling them makes them "almost flat". The Gromov-Lawson argument is associated to compactifications using the sphere, but one can now systematically use other spaces (like, e.g., Tits boundaries).

Of course, the space at infinity, and cones on it, are ways at getting at a more basic invariant: $KX_*(X)$. One has an assembly map $\alpha : K_*^{lf}(X) \longrightarrow K_*(C^*(X))$, which we will will see factors through the large scale theory $K_*^{lf}(X) \longrightarrow KX_*(X) \longrightarrow K_*(C^*(X))$. α is constructed by a variant of the usual construction of the assembly map, see [17, 92, 80]. One chooses a quasilattice $Z \longrightarrow X$ and a Borel map $\phi : Z \longrightarrow X$ with $\phi^{-1}(z)$ of measure 1. Then $C^*(X)$ can be identified (not quite canonically) with compact operator valued kernels on Z of bounded propagation speed $C^*(Z)$. Then one constructs maps $K_*(R_n(Z)) \longrightarrow K_*(C^*(Z))$ by realizing elements of $K_*(R_n(Z))$ by "geometric operators" and taking the index of these operators. The maps commute with the system $R_n(Z) \longrightarrow R_{n+1}(Z)$ producing α and its factorization.

Those who are optimistic could conjecture that $KX_*(X) \longrightarrow K_*(C^*(X))$ is an isomorphism. This would be compatible with all known calculations and with the examples in [43].

Note that in the [43] example the map $K_*^{lf}(X) \longrightarrow KX_*(X)$ is not an isomorphism. It is a secondary invariant associated to the cokernel $K_{*+1}^{lf}(X) \longrightarrow KX_{*+1}(X)$ that distinguishes these manifolds from one another; it is morally like an η invariant associated to an almost flat bundle that is being used[40]. Unfortunately, one does not have enough spectral theory available for such a manifold and such bundles to make real mathematical sense out of this.

The geometry implicit in these examples, and in the theory of stratified space [105] suggests the extension of these ideas to a more sheaf theoretic context.

The idea is quite simple: Note that $R_*(X)$ is built by the same construction that produces Cech homology. We may reverse the usual arrows and construct the homology associated to a "cosheaf" (even of spectra, if we are discussing generalized cohomology theories). (See [105] for a quick treatment.) Since we are coarsening all the time, one doesn't need the cosheaf defined on all "open" sets (or balls), only sufficiently large ones. Thus, one can discuss "coarse cosheaves" if one liked.

The idea of these is that we can use them to record some part of

[40]This is parallel to the use of η-invariants and flat bundles in [9] to distinguish homotopy equivalent lens spaces from one another.

the information in the small, or in a semilocal fashion, that we do not want to lose as we go large scale. One useful example would be to take $B\pi \wedge K(\mathbb{Z}, 0)$ (where π is the fundamental groupoid). This lets one take higher signatures that keep track of both fundamental groups on some scale, and the coarse geometry. Indeed, using this, one can see that the punctured torus T^n does not have any complete metric of positive scalar curvature: Note though, that the metric could be quasiisometric to just the ray $[0, \infty)$, for which $KX \cong 0$, and the previous methods would not detect. Here we are keeping track of the fact that even by being coarse, the part at ∞ does not carry enough fundamental group for some of the compact obstruction to escape through. However, by taking a Lipschitz map ϕ to a ray, and giving it the coarse cosheaf described, there would be an obstruction in $KX_*([0, \infty); \phi * B\pi) \cong K_*^{lf}(T^{n'})$, (here $'$ denotes a puncture) which contains the nontrivial obstruction.

By choosing coarse cosheaves carefully one can do some more coarse analysis, in the same way that sheaves help out in complex manifold theory. The coarse cosheaves could, for instance, be things that keep track of growth of functions (or cycles) on chunks of the space. Some of these will be studied in the next chapter. But there seem to be many more possibilities, such as using Rips complexes of one growing scale in forming space fragments whose invariants are then assembled into a larger scale coarse theory.

6. Uniformly Finite Homology

Beginning with this section, we will put growth conditions of various sorts on the (co)cycles of our coarse (co)homology theories and show how these can be used to study more delicate coarse properties of metric spaces and manifolds. The theory we study in this section, H_*^{uf} was introduced in [19], and is useful for studying spaces with bounded geometry.

Let Z be a space with coarse bounded geometry. This means that Z contains a bounded quasilattice. Riemannian manifolds of bounded geometry (i.e., injectivity radius bounded from below and curvatures bounded in absolute value), or uniformly locally finite simplcial complexes (with simplicial metric) and finitely generated groups with the word metric all provide examples.

If Γ is a quasilattice in Z (Γ does not necessarily connote a group), then the Rips complexes $R_n(\Gamma)$ are uniformly[41] locally finite simplicial complex. Now consider simplicial chains on $R_n(\Gamma)$ which, while

[41] The uniformity is not in n.

perhaps having infinite support, have uniformly bounded coefficients[42], $c = \sum a_\sigma \sigma$, with $|a_\sigma| \le K$. These form a chain complex $C^{uf}(R_n(\Gamma))$, and we can consider (as we did earlier for HX) the pro-system of the homology groups of these chain complexes. We will denote the direct limit of these by $H_i^{uf}(X;R)$; it is independent of the bounded quasilattice chosen[43]. We will concentrate on the cases $R = \mathbb{Z}$ or $\mathbb{Q}$.

It is also possible to define uniformly finite homology with coefficients in other spectra, see [12]. These are important in studying the geometry and the surgery theory of manifolds of bounded geometry.

To illustrate, let $X = \mathbb{N}$, and $R = \mathbb{Z}$. A typical chain is $\Sigma[n]$. In HX_0 this cycle is 0 since $\Sigma[n] = \partial(\Sigma n[n-1,n])$. Since the coefficients (n) are unbounded, this 1-chain is not uniformly finite. In fact one can easly check that

$$H_0^{uf}(\mathbb{N};\mathbb{Z}) \cong \{\phi : \mathbb{N} \longrightarrow \mathbb{Z} | \Delta\phi \text{ is bounded}\}/\{\text{bounded } \phi\},$$

where the isomorphism is given by summing up the coefficients of the chain from 0 to n. Note that this is a $\mathbb{R}$ vector space! This group tends to be very large if it is nonzero.

For another example consider the free group on 2 generators F_2. The 0-cycle $\Sigma[\gamma]$ vanishes in $H_0^{uf}(F_2;\mathbb{Z})$ as is demonstrated in the following figure.

The intuition involved in this group is that we are trying to produce geometric one chains whose boundaries are a given collection of points (with multiplicities), in such a way that there is only a bounded number of arcs passing through any ball. From this point of view $\Sigma[\gamma]$ is the hardest cycle to kill, since there is no room for cancellation. The "spaghetti" that this bounds can be used to produce escape routes for any 0-cycle.

Similarly, by partitioning the cosets of F_2 in the Cayley graph of a discrete group containing F_2, one has enough spaghetti to kill any 0-cycle. This, and the unfortunately disproved Von Neumann conjecture suggest parts of the following theorem:

Theorem. [19] *For a metric space X of coarse bounded geometry with bounded quasilattice Γ, the following are equivalent:*

---

[42]We are envisioning a ring with norm. Other possibilities for the uniform finiteness are to allow only finitely many elements of the ring to appear as coefficients for any given chain. Note that for $\mathbb{Q}$, these chain complexes obtained in this way are quite different.

[43]These were first defined in [19]; similar cohomology groups were introduced in somewhat less generality in [54]. Obviously, this is an L^∞ notion and one can form coarse L^p groups as well.

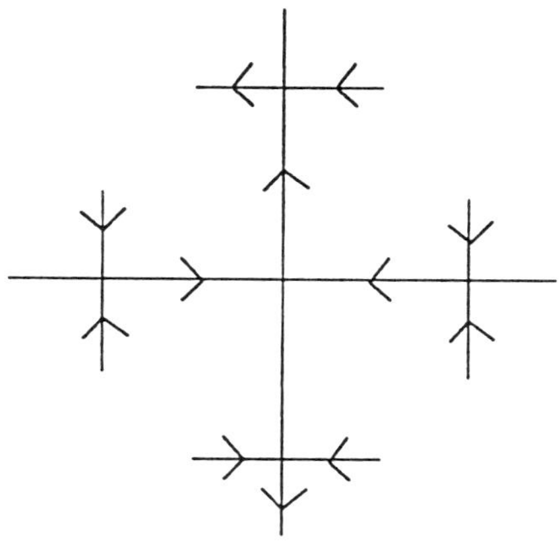

FIGURE 2. The pattern visible at the identity element is everywhere repeated. Two edges enter, one leaves, and one just has coefficient -Q. An easy induction show this is possible.

- $\Sigma[\gamma] = 0$ *in* $H_0^{uf}(X;\mathbb{Z})$
- $H_0^{uf}(X;\mathbb{Z}) \cong 0$
- $H_0^{uf}(X;\mathbb{R}) \cong 0$
- $\bar{H}_0^{uf}(X;\mathbb{R}) \cong 0$
- X *is nonamenable.*

The group $\bar{H}_0^{uf}(X;\mathbb{R})$ is defined similarly to the groups introduced before, except one kills the closure of ∂, not just the image of ∂. The condition of nonamenability requires a little preparation to state:

Definition. For a set S in Γ, we denote by $\partial_r S$ the points that are within r of both S and $\Gamma - S$.

Definition. If V is a finite set in Γ, then vol(V) = the cardinality of V.

Now, we say that X is nonamenable if for some r, and positive c, there is an inequality $\text{vol}(S) < c\text{vol}(\partial_r S)$ for all finite sets S.

The alternative is amenability: there is a sequence of subsets S_i for which $\text{vol}(\partial_r S_i)/\text{vol}(S_i) \longrightarrow 0$ for each r. Such a sequence will be called a regular sequence. It is more common, and, in fact, equivalent to suppose that $S_i \subset S_{i+1}$, and that $X = \cup S_i$. When these conditions hold, the sequence will be called a regular exhaustion. These arose in Folner's original characterisation of amenability of groups (see [88]) and in function theory (see [4]). Thus a discrete group is amenable iff in its word metric, it is an amenable metric space. In any case, as we will see, regular sequences are as easy to work with as the exhaustions, and have greater flexibility for studying certain geometric problems.

The original idea of using regular exhaustions was to be able to measure the average of a function on the manifold. One integrates a function[44] and rescales according to the volume of the region[45]. For this to be well defined in cohomology (i.e., that the boundary term in integration by parts to be negligable) one uses (boundedness of the function and) the regularity condition. This can be done equally well using regular sequences, except that one should then intuitively think of this as a process that samples the function in, perhaps carefully chosen, regions.

If the space is amenable then $\Sigma[\gamma] \neq 0$ in $H_0^{uf}(X;\mathbb{Z})$ as the average value of the coefficients is 1 on any region, and so for any regular exhaustion (or sequence).

It is also important to realize that although the vanishing of $H_0^{uf}(X;\mathbb{Z})$ is equivalent to that of $H_0^{uf}(X;\mathbb{R})$, the obvious map is far from injective. There seems to be a lot of subtle information located in this kernel, and we hope to unravel some of this. As an example, though, we point out that $\Sigma[n^2]$ vanishes in $H_0^{uf}(\mathbb{N};\mathbb{R})$ although it is nontrivial in $H_0^{uf}(\mathbb{N};\mathbb{R})$ (as is any infinite set of points with the same sign).

The following result is a more precise version of the relation of amenability to the vanishing of $H_0^{uf}(X;\mathbb{R})$.

Theorem. [20] *For a uniformly discrete metric space X and any cycle $c \in C_0^{uf}(X)$, $[c] = 0 \in \tilde{H}_0^{uf}(X;\mathbb{R})$ iff for any regular sequence S_i,*

$$\int_{S_i} c = o(vol(Si)),$$

where the integral of a cycle over a set is the sum of the coefficents the vertices in the cycle that lie in S.

The elements of the kernel of $H_0^{uf}(X;\mathbb{R}) \longrightarrow \tilde{H}_0^{uf}(X;\mathbb{R})$ are represented by very thinly sprinkled cycles. The relation of $\tilde{H}_0^{uf}(X;\mathbb{R})$ to integration over regular sequences means that the reduced theory is much more susceptible to direct analysis. The following theorem gives a complete hold on the meaning of 0-dimensional homology.

---

[44]We view the coefficients of a 0-cycle as a function in this intuitive discussion.

[45]This should be done in a suitable function space, and a weak $*$ limit chosen for the process.

Theorem. *Given a uniformly discrete metric space X and a cycle $c \in C_0^{uf}(X)$, $[c] = 0 \in H_0^{uf}(X;\mathbb{R})$ iff for any regular sequence S_i,*

$$|\int_{S_i} c\,| = O(vol(\partial S_i)).$$

We have applied this to proving the bounded geometry homotopy invariance of the top Hirzebruch L-form.

7. Applications of H_0^{uf}

The results explained in the previous section have a number of interesting applications, especially to the geommetry of manifolds of bounded geometry. The details of some of these depend on the connection to index theory that we'll only get to in Section 9. Nonetheless, aspects of these can be apprehended immediately.

The most important application that provides the pattern for most of the others was pioneered early this century by the Italian American immigrant Ponzi in the area of financial services. The method was essentially to consider a set of clients each of which was trying to earn 1 unit and find a 1-cycle which can be viewed as giving instructions for how these clients should redistribute wealth. The idea is that if the boundary of the 1-cycle is the 0-cycle, then the redistribution will result in all points having gained, with no loss. Now Ponzi knew that this is impossible with a finite set, but he was dealing with *investors*, so he used time to allow the set to become infinite. However, the trouble was the one that occurred with the 1-chain on $\mathbb{N}$ whose boundary is $\Sigma[n]$: the coefficients were growing, and ultimately one couldn't find investors to provide for the large transfers required. Ponzi was jailed for this failure.

Now on the 4-valent tree, if each node would follow the instructions given in figure in the previous section, each would give a dollar to one neighbor and get two from two of its neighbors: universal profits for all![46] Moreover, this plan can be executed on any nonamenable space.

For the more mathematical applications, one is more often trying to remove resources, or actually, obstructions from all parts of a space, in a controlled fashion (which prevents the blow-up of the 1-cycle used for this purpose).

[46]As a moral lesson, we should point out that it is necessary for the agents to perform acts of kindness to neighbors who will not reciprocate it to them for the universal happiness to ensue.

Application 1. *We will follow* [93] *and* [19]. *Roe considers manifolds with bounded geometry*[47] *to to be complete Riemannian manifolds with positive injectivity radius and with uniformly bounded curvature tensor together with its its covariant derivatives. (Cheeger and Gromov* [32, 33], *see also Abresch* [2], *have shown that there is no need to worry about the covariant derivative condition. We will not discuss this and cognate points as they arise.) A strict quasi-isometry is a diffeomeorphism with upper and lower bounds on the lengths of images of unit vectors and for which the difference of the Levi-Civita connection of one and the pullback of the other is a bounded form.*

The PL version of these is that one is looking at PL manifolds of bounded combinatorial complexity, and the maps are allowed to only use uniformly finite subdivisions before being defined. See [11] for more on this and the connection to bounded distortion.

Covers of compact manifolds and leaves of foliations have natural bounded geometry structures and strict quasi-isometry classes of metrics; their study is a primary motivation for our work in this direction.

Roe used regular exhaustions and a modification of Atiyah's L^2 index theorem for regular covers to prove the following:

Theorem. [93] *If M is a spin manifold with $\int \hat{A}(M)dVol \neq 0$, then no amenable cover of M has a nonnegative scalar metric in its natural strict quasi-isometry class.*

We will discuss this result and refinements of it in Section 9. Using Ponzi schemes one can prove a sort of converse:

Theorem. [19] *For any nonamenable group Γ, there is a spin manifold M with $\int \hat{A}(M)dVol \neq 0$ and a regular Γ cover of M strictly quasi-isometric to manifold with positive scalar curvature.*

The method of proof is to use the fact that positive scalar curvature obstructions are known to be ultimately obstruction of spin bordism class (see [65]). One uses a Ponzi scheme to move all of these obstructions for the universal cover of a manifold $N^{\#}W$, where N is a positive scalar curvature manifold with fundamental group Γ and W a simply connected spin manifold with $\hat{A}$-genus, off to infinity. (One can use the "spaghetti", i.e. the arcs that go off to Y that have the bounded "intersection and near-miss property", discussed last section, to build a bounded geometry cobordism from the cover of $N^{\#}W$ to N.)

---

[47] This is intimately related to the uniform category considered by Januskiewicz [77].

Application 2. *A similar example occurs in* [14], *where the question of to what extent characteristic classes of a compact manifold obstruct the construction of strict quasi-isometries of for the universal cover.*

Theorem. [14] *For a group* Γ, *manifolds* M *and* M', *and a map* $\phi\colon M' \longrightarrow M$ *for which* $\phi^*(p(M')) \neq p(M) \in H^*(M;\mathbb{Q})$, *yet which is a finite distance from a strict quasi-isometry of* Γ *covers, exist iff* Γ *is nonamenable.*

In other words, for nonamenable groups characteristic classes downstairs do not necessarily prevent strict quasi-isometries upstairs, while for amenable groups they do. The obvious PL version of this theorem holds as well.

The construction of examples is an application of surgery together with a Ponzi scheme. That no examples exist in the amenable case is a reflection of properties of lower bounds on the higher H_i^{uf} for amenable groups. This will be discussed in the next section.

Application 3. *Another theorem that comes out of the Ponzi method is the following (which we call "Gerrymandering"):*

Theorem. *Let* M *be a manifold with a volume form, then there is a measurable* ϕ *a bounded distance from the identity, such that* $\phi^*(Vol) = 2 \cdot Vol$ *iff* M *is nonamenable.*

Application 4. *There is a pretty connection between tilings and* H_0^{uf}. *We consider a triangulated manifold* M *of bounded combinatorial complexity and tilings by codimension 0 submanifolds with oriented triangulated boundaries. A semibalanced tiling is one where weights are assigned to each oriented face, so that each tile has nonnegative total weight. The impurity set of a tiling, is a set of points in* M *taken one from each tile with positive total weight. An unbalanced tiling is one where all tiles have positive weight.*

Theorem. [19] *Any nonamenable manifold has an unbalanced tiling. A discrete set* S *of vertices of* M *is the impurity set of a semibalanced tiling iff* $\Sigma[s] = 0 \in H_0^{uf}(M)$ *(the sum taken over all* $s \in S$*).*

Thinking about tilings purely combinatorially leads to some insights about the nature of cycles in H_0. The following is a standard result, see Tilings and Patterns [67]:

Proposition. *Let* M *be an almost homogenous manifold (i.e., one which possesses a cocompact group of isometries). A finite set of prototiles which tiles arbitrarily large balls of* M, *tiles* M.

So, if one can tile, say, the upper half plane with a given set of tiles, one can tile the plane! (Note, tiles are oriented, so one can't reflect them!) Philosophically, this indicates that when H^{uf} is used to obstruct a structure, the violations of this property will have to be C-dense. In Section 9 such a refinement of Roe's theorem on the existence of negative scalar curvature will be given.

The proof of the proposition is an application of Konig's lemma, which finds an infinite path from a point through a finite valence directed graph provided that paths of arbitrarily large length exist (Exercise: verify this.). We build a tree with nodes collections of prototiles that fill up the ball of radius r, and such that no subset does, and connect two of these if one contains the other. This is a graph with finite valence, and the hypothesis gives paths of arbitrary length, and the conclusion of Konig's lemma is a tiling of M.

Application 5. *The previous example implicates H_0^{uf} in a geometric problem. The following results implicate H_0^{uf}.*

Theorem. *Let $\phi : M' \longrightarrow M$ be a bounded geometry homotopy equivalence, then $\phi^*(L(M') \cap [M']) = L(M) \cap [M] \in H_0^{uf}(M; \mathbb{Q})$.*

Theorem. *Suppose that M is a spin manifold with bounded geometry and positive scalar curvature, $\hat{A}(M) \cap [M] = 0 \in \tilde{H}_0^{uf}(M; \mathbb{Q})$.*

We conjecture these are true without the reduction, and for higher homology groups as well, but we have only been able to verify this in special cases related to where the Novikov conjecture can be verified by the principle of descent. The above theorems are proven using index theory, estimates on η-invariants by [32] and the criterion for vanishing of elements of $\tilde{H}_0^{uf}(M; \mathbb{Q})$ mentioned in the previous section.

Application 6. *The following is a reformulation of the mean ergodic theorem (again in light of the vanishing theorem):*

Theorem. *Suppose that Γ acts unitarily on a Hilbert space H, and that $P : H \longrightarrow H^\Gamma$ is orthogonal projection to the fixed point set. Then for $\xi, \nu \in H$, we have:*

$$\Sigma < \gamma\xi, \nu > [\gamma] =< P\xi, \nu > \Sigma[\gamma] \in \tilde{H}_0^{uf}(\Gamma; \mathbb{R}).$$

This does not hold if the unreduced version is used. (One can see this using the shift acting on l^2: for almost any pair of vectors the equality is false in $H_0^{uf}(\mathbb{Z}; \mathbb{R})$.) We hope that there are higher homological ergodic theorems that are useful when Γ is nonamenable.

More conventional forms of the ergodic theorem can be obtained by assuming that the action on H comes from an ergodic action on a probability space. Then it is intelligent to take for ξ and ν characteristic functions, and the theorem directly encodes recurrence properties of the group action. One can quickly deduce (using adapted regular sequences as we will explain in another context in Section 9) that for an action of an amenable group, the "return times" of a set of nonzero measure is always a C-dense subset of the group (Khinchine's theorem).

8. Higher Homology

The higher homology groups are much more mysterious. However, a few cases have been somewhat analyzed.

Theorem. ([14, 54, 60]) *If Γ is amenable, then there is an injection $H_*(B\Gamma;\mathbb{R}) \longrightarrow \tilde{H}_*^{uf}(\Gamma;\mathbb{R})$.*

There is a useful variant of H_*^{uf}. Define $H_*^{\infty}(Z)$ for a simplicial complex of bounded geometry Z by using uniformly finite chains on Z (i.e. just like uf, but not coarsified using Rips complexes)[48].

Theorem. ([14]) *If Γ is amenable acting proper discontinuously, isometrically, and simplicially[49] on Z, a polyhedron of bounded geometry, then there is an injection $H_*(Z/\Gamma) \longrightarrow H_*^{\infty}(Z)$.*

This can be proven most directly[50] by defining a reverse map $H_*^{\infty}(Z) \longrightarrow H_*(Z/\Gamma)$: Given a chain c, assign to the simplex k in Z/Γ the average value (using the invariant mean on $l^{\infty}(\Gamma)$) of the coefficients of c of the simplices that lie above k. Clearly the composite $H_*(Z/\Gamma) \longrightarrow H_*^{\infty}(Z) \longrightarrow H_*(Z/\Gamma)$ is the identity.

Problem. *How is the image of $H_*(Z/\Gamma)$ related to $H_*^{\infty}(Z)^{\Gamma}$?*

Now using (Poincaré duals of) characteristic classes for manifolds of bounded geometry (or using the related cohomology theory defined using bounded differential forms, see [77, 60, 93] or below), and observing that they are invariant under strict quasiisometry, one obtains the other half of Example 2.

For symmetric spaces, one can get a certain amount of information on $H_*^{uf}(\Gamma;\mathbb{R})$. This leads to the statement that rank is a quasiisometry

---

[48]For uniformly contractible Z, these theories coincide.

[49]This includes the proper discontinuity, but that hypothesis is included for clarity.

[50]This is an oblique criticism of [14], although this argument is implicit inthe more complicated one given there.

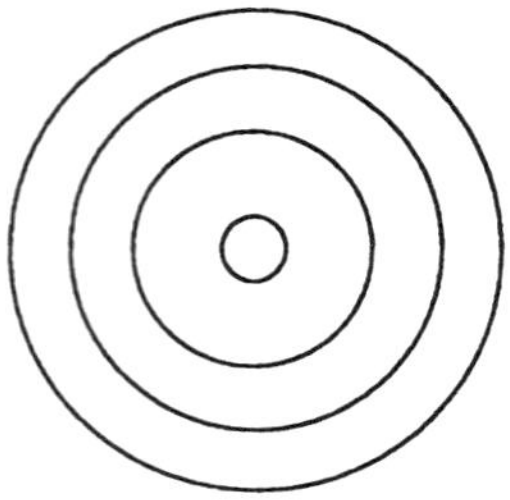

FIGURE 3. These concentric circles extending to the boundary give a nontrivial 1-cycle on the hyperbolic plane.

invariant, but to be honest, the proof depends on a key estimate of Mostow's that captures a critical aspect of the rank.

Theorem. ([20, 60]) *For X^n a symmetric space of real rank k, $H_*^{uf}(\Gamma;\mathbb{R}) = 0$ for $* \leq n-k-1$ and $* > n$; it is uncountable for $* = n-k$ and $\cong \mathbb{Z}$ in dimension n. (The same hold for $\tilde{H}_*^{uf}(\Gamma;\mathbb{R})$ as well.)*

So for $\mathbb{H}^2$, the hyperbolic plane, H_1^{uf} is very large. For later reference, we give some (undoubtedly incomplete) invariants. (Since the X's we are considering are uniformly contractible, one can use conventional cycles to represent the homology.)

A first, simplest, invariant is to take a geodesic and intersect the one cycle with it. A little thought shows that transversality enables us to take this intersection to be a uniformly finite cycle in the geodesic, i.e., an element of $H_0^{uf}(\mathbb{R})$. Using regular sequences on these one gets copies of the enormous $H_0^{uf}(\mathbb{R})$ in $H_1^{uf}(\mathbb{H}^2)$. It isn't hard to get different copies of this using different geodesics. As in Figure 3. Note that for the obvious geodesic, and regular exhaustions symmetric around the origin the concentric circle cycle is nontrivial.

One can do more creative things than just use a single geodesic. Using a collection of geodesic fragments with a regular sequence in this union one has greater flexibility. One way to do this is to find a bunch of balls of larger and larger size and use diametric (or other) geodesic chords in these balls. Then one can hopefully analyze cycles in terms of positions in hyperbolic space where one already has information.

The nonvanishing result for symmetric space comes via intersection with flats (or unions of fragments of these).

The vanishing result, as we've already mentioned, is a consequence of a basic estimate of Mostow [85] regarding projection from a symmetric

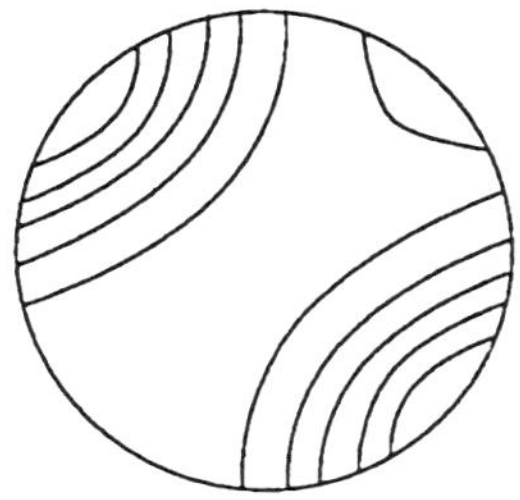

FIGURE 4. This collection of arcs gives a rather different looking nontrivial 1-cycle.

space down to a maximal flat. One can see that given a low dimensional cycle, the "shadow" of the cycle based on projection from a generic flat is a uniformly finite cycle with boundary the given cycle. Alternatively, one can argue cohomologically and use the proof of homotopy invariance of DeRham cohomology together with the exponential deacy of the deformation retraction to the flat (along the projection line) to see that for high degree bounded differential form ω the form ν with $d\nu = \omega$ is bounded.

The cognate cohomology theory to $H_*^\infty(Z)$ just applied is used in [77, 60, 93, 54, 14]. This is defined using cochains that are uniformly bounded on the simplices of Z or differential k-forms that are uniformly defined on k-tuples of unit vectors. We shall denote this by $H^*_\infty(Z)$ and by $H^*_\beta(Z)$ in the coarse version (using *inverse* limits of theories for Rips complexes). One can redo all of the arguments made for the homology theory for these groups as well. For n-manifolds one has the following useful relation:

Theorem. [13] *If Z is an n-manifold of bounded geometry, then $H^*_\infty(Z) \cong H^\infty_{n-*}(Z)$.*

In any case our last two applications in this section are most easily framed in terms of $H^*_\infty(Z)$.

The main advantage formally of $H^*_\infty(Z)$ is that it has a cup product structure. One also can put characteristic classes directly in it.

Proposition. [77] *If L is the leaf of a foliation of M and c is a characteristic class of L, then $c \in imr^* : H^*(M) \longrightarrow H^*_\infty(L)$.*

Now for $c =$ Euler class, Januskiewicz recovers the result of Phillips-Sullivan [PhS] that the average Euler characteristic of a(n amenable) leaf must vanish if it is a codimension one leaf in a simply connected manifold. This excludes Jacob's ladder or the infinite jungle gym. Using the previous results, one can exclude some exponentially growing quasiisometry classes from arising as leaves by taking universal covers of suitable manifolds with fundamental group a lattice. (We remark that there is a pretty paper of Attie and Hurder [15] that excludes certain quasi-isometry types as being leaves of C^1 foliations in any compact manifold based on entropic ideas.)

The final application of these methods that we'd like to mention is due to Gersten and Gromov [54, 58]. The Dehn function for a finitely presented group is the number of conjugates of relators one must multiply together to obtain any word of length n that represents the trivial element. Basically (but not exactly, as one can see quickly for the free group) its growth is an invariant of the group. It is recursive iff the group has a solvable word problem and is a first measure of the logical complexity of a group. In the riemannian setting, the parallel idea is to try to estimate the area of a bounding disk for nullhomotopic curves in terms of the length of the curve.

Gromov showed that a group is hyperbolic iff the Dehn function is linear. In [45], it is shown that automatic groups have quadratic Dehn functions. Interestingly, one can sometimes give lower bounds on the Dehn function (and the higher "filling functions" of Gromov) using $H^2_\beta(\Gamma)$ (and $H^i_\beta(\Gamma)$).

Elementary Illustration: The idea is quite simple. We try to find closed curves of small length which have explicit bounding disks. Integrating a representative from $H^2_\infty(\Gamma)$ over this coboundary gives a lower bound to its area.

The simplest case is $\mathbb{Z}^2$. One can use the 2-form $dx \wedge dy$ and the boundary of the square with size n to see that the Dehn function is quadratic.

More interesting is the Heisenberg group $H = 3 \times 3$ integral matrices with 0's below the diagonal and 1's on the diagonal. It has a presentation $\langle x, y | [x, [x, y]] = [y, [x, y]] = e \rangle$. ($x$, y, and $[x, y] = z$ are all elementary matrices.) It can also be viewed as $\mathbb{Z}^2 \times \mathbb{Z}$, where $\mathbb{Z}$ acts on $\mathbb{Z}^2$ by the shear map $(z, x) \longrightarrow (z, z+x)$. This is volume preserving, so one obtains a 2-form on Γ, which lifts to the (the universal cover of the nilmanifold, the) contractible Nilpotent group. Note that the word of linear length $[x^n, [x^n, y^n]]$ is trivial in the group because $[x^n, y^m] = [x, y]^{nm}$, and thus lies in the center. This reasoning produces a bounding 2-disk, for which one easily computes the integral to be cubic. It is easy to then check Gersten's assertion that for the Heisenberg group the Dehn function has precisely cubic growth.

Remark. The cohomological method gives a good first approximation to Dehn functions, but us far from the last word. For instance [22] give examples where one can compute the Dehn function despite vanishing $H^2(\Gamma)$; I do not know whether $H^2_\beta(\Gamma)$ can be better applied in those cases, but in any case, there do not seem to be obvious cocycles to use. Gromov [58] has a lengthy discussion of both deeper upper and lower bounds, especially in the case of nilpotent groups.

9. Index Theory Again

Very similar to the manner in which HX and KX entered in the study of large scale index theory, the uniformly finite homology arises in the analysis of elliptic operators on manifolds of bounded geometry. This relation is necessary for some of the applications already mentioned (e.g., Example 1 in Section 7) and is the topic of this section.

This topic was initiated in Roe's thesis [93], which itself was motivated by the wish to remove equivariance from Atiyah's L^2 index theorem for coverings. It also shares some in spirit with Connes' foliation index theorem (see [38, 85]); our interest was spurred by the philosophical issue of whether Conne's theorem was one about foliations or really one about leaves. Our current view is that Connes' theorem really measures average phenomena about leaves, but requiring more of a "recurrence structure" than is merely encoded by the bounded geometry. In any case, the theorems here well fit into the Connes program wherein geometry is encoded by an operator algebra and indices take value in its K-theory, with geometric characteristic class theory living in the cyclic cohomology of dense subalgebras of the algebra. (See [37].)

To put the matter very bluntly, the connection roughly goes like this: On a compact connected manifold one assigns an integer to each elliptic operator, namely its index. For a disconnected compact manifold, one still has the index, but there is certainly more information in the pair of integers, namely the index of the restriction of the operator to each component. This is an element of H_0. For manifolds with bounded geometry, we have an index in H_0^{uf} and interesting information contained in the higher homology as well (the "fine theory" containing "symbol" information and the coarse theory containng "higher indices" a la the Novikov philosophy and that discussed in Section 5).

The relevant algebra for this problem is the algebra $\mathfrak{U}_{-\infty}(X)$ of smoothing operators of bounded propagation speed that are also unifirmly bounded. (N.B. A smoothing kernel need not be uniformly bounded to define a bounded operator.) The usual geometric elliptic operators will have inverses modulo the C^*-completion of this algebra, and thus define indices in the K-theory of that C^*-algebra.

Rather than pair cyclic cocycles of $\mathfrak{U}_{-\infty}(X)$ with K-theory, we take a naturally defined Chern character into HC_*, and this has a natural map into $H_*^{uf}(X;\mathbb{C})$. The computation of this class is the index theorem. Formally, the index theorem looks identical to, say, the original one in [10]. For the Dirac operator, it says:

$$A(\hat{X}) \cap [X] = \text{ind} D.$$

In degree zero this is essentially a restatement of the index theorem in [93]. However, even here there are advantages to the restatement in terms of uniformly finite homology.

Corollary. *Suppose M is a spin manifold with amenable fundamental group and $\int A(M)dVol \neq 0$, then any metric g on $\tilde{M}$ strict quasi-isometric to the pullback metric has C-dense negative scalar curvature set, i.e., $N = x|s_g(x) < 0$ is C-dense.*

Proof. Suppose not. Then we can find balls of arbitrarily large size missing N. We can translate using the covering group copies of Folner (=regular) sequence into these balls. Now pair the index class with the average associated with this modified Folner sequence. The Bochner formula shows this must be 0, but, after a moments reflection one sees that this quantity is precisely $\int A(M)dVol$ and the conclusion follows. □

Remark. Even if the original Folner sequence was an exhaustion, the modified one introduced in the course of the proof was not. It is true that the index associated to any regular sequence is the same as the index associated to some other regular exhaustion; however, that exhaustion will contain parts of N, which would complicate the argument.

The higher classes enter as well in the problem of positive scalar curvature on universal covers. (Indeed it would not be hard to describe a "Gromov-Lawson-Rosenberg" conjecture for this problem.) The following theorem is interesting in this context.

Theorem. *Suppose M^n is a spin manifold with fundamental group that of a closed hyperbolic k-manifold $k > 2$. Then*

- *$\tilde{M}$ admits a complete metric of positive scalar curvature.*
- *$\tilde{M}$ has the coarse quasiisometric metric of positive scalar curvature iff ind $D = 0 \in KOX_n(M) = KO_{n-k}(*)$.*
- *$\tilde{M}$ has a strict quasiisometric metric of positive scalar curvature iff, in addition $\int \hat{A}(M) \cup \alpha \, dVol = 0$ for all closed 1-forms α. (What??)*

Addendum. *If* $\int \hat{A}(M) \cup \alpha \, dVol \neq 0$, *then the negative scalar curvature set of any strictly quasiisometric manifold is* C*-dense.*

Let us ignore the constructive issue. The exotic condition follows from the type of machinery discussed in Section 5. (See also [64], and combine the basic idea with [72].) The necessity is a consequence of pairing the index class with the cycle arising from the lift of appropriate closed geodesics from the hyperbolic manifold.

The C-density follows the same pattern as the case of amenable fundamental group. The geodesic used for detection of the index class has translates that have large intersections with the balls that would putatively be missing N.

10. Final Remarks

This paper has dealt with some of the geometry and index theory related to two particular large scale homology theories. There are many variants of these constructions, which have yet to be explored ....

The most obvious variant of uniformly finite homology would allow growth conditions. That is one could examine i-cycles with the property that the sum of the coefficients of the cycles that touch $B_x(r)$ is $\leq f(d(x,*))\phi(r)$ for some ϕ and function f of the distance from x to an arbitrary base point. There are different versions according to whether or not ϕ's growth type is specified.

Furthermore, nothing requires that the same function be used for different dimensional chains. One restricts attention to i-chains whose boundaries are $i-1$ allowable in defining homology.

Yet another variant just keeps track of balls around some base point, and insists that the amount of chain that one sees in $B(r)$ is bounded by $D(r)$.

We believe that all of these growth homologies, say even for polynomials, are an important invariant. It would be interesting to do the calculations even in the case of nilpotent groups. Presumably, by examining the growths for which the homologies of nilpotent groups vanish, one could gain some insight into Pansu's theorem that the dimensions of the associated graded pieces are coarse invariants.

What is nice about the growth homology (and cohomology) theories is that they seem likely to have nice connections to geometry, and to index and surgery theories. They can be used to give information to the questions raised by H^{uf}: If it is not possible to achieve a quasiisometric metric with some property, how must distortion must there in fact be in producing metrics with the given property.

In addition to these, one can, of course, develop Lp theories, and again with various weights, but here the geometric intuition is more remote. (Analytially $p = 2$ has many advantages: for some beautiful applications of this, see, e.g. [34].)

More possiblities for yet more theories arise for any quantifiable adjective. For instance the entropic method of [15] suggests building homology theories where one insists that the entropy of the cycles are not too large. Many of these possible theories will be impossible to compute; it seems that one usually needs some kind of functional analytic device for these to be useful.

Another theme that has arisen, most clearly in Sections 1,4,5 is the interconnection between the microscopic scale and the largest scale. The tangential representation of a group action is reflected in the large scale motion of a very thin quasilattice in the universal cover. For hyperbolic groups with homology spheres at ∞, one discover the local structure of any aspherical homology manifold with that fundamental group.

In the future one would hope for a large scale variant of Newtonian calculus, so that Euclidean lattices produce the conventional subject, but that other structures would yield different nonlinear best approximating functions.

Less speculatively, there are the techniques of variable conings. These can be thought of as being the microscopes that are (simultaneously) magnifying a space to many different scales, and encoding (partially) how these scales fit together to be indeed the different scales of one object. It seems certain that the small scale structure of fractals can be studied by using the growth homologies of their cones. As we saw, depending on the growth of the cone, more or less of the metric geometry of the space will be reflected in the coarse type.

REFERENCES

1. H. Abels *Finiteness properties of certain arithmetic groups in the function field case,* 1988, preprint.
2. U. Abresch *Uber das glatten Riemann'sher metriken*, Habilitationsschrift, Rheinischen Friedrisch-Wilhelms-Universität Bonn, 1988.
3. D. Anderson and F. Connolly, *Finiteness obstructions and cocompact actions on $S^m \times \mathbb{R}^n$*, Comm. Math. Helv. **68** (1993) 85-110 .
4. L. Ahlfors and L. Sario, *Riemann Surfaces*, Princeton University Press.
5. J. Alonso, *Combings of groups*, MSRI Pub. **23** (1992). .
6. J. Alonso and M. Bridson, *Semihyperbolic groups*, Preprint .
7. M. Atiyah, *Elliptic operators and K-homology*, Global theory of elliptic operators, Proc. Int. Conf. on Functional Analysis, University of Tokyo Press, 1969, 21-30.

8. M. Atiyah, *Elliptic operators, discrete groups, and Von Neumann algebras*, Asterisque 32 (1976) 43-72.
9. M. Atiyah, V. Patodi and I. Singer, *Spectral asymmetry and Riemannian geometry, I*, Math. Proc. Cambridge Phil. Soc. **77** (1975) 43-69.
10. M. Atiyah and I. Singer, *The Index of Elliptic Operators, I, III*, Ann. of Math **87** (1968) 484-530, 546-604.
11. O. Attie, *The classification of some manifolds of bounded geometry*, 1992, Thesis, Courant Institute.
12. O. Attie, *Bounded geometric surgery*, 1993, preprint.
13. O. Attie and J. Block, paper in preparation.
14. O. Attie, J. Block, and S. Weinberger, *Characteristic classes, distortion of diffeomorphisms, and amenability*, J. AMS **5** (1992) 919-921.
15. O. Attie and S. Hurder, *Manifolds which are not leaves*, 1993, preprint.
16. K. K. Au, Thesis, University if California, San Diego, 1990.
17. P. Baum, CBMS lecture series (notes in preparation).
18. V. Bestvina and G. Mess, *The boundary of negatively curved groups*, JAMS **4** (1991) No. 3, 469-481.
19. J. Block, and S. Weinberger, *Aperiodic tilings, positive scalar curvature and amenability*, JAMS **5** (1992) 907-918.
20. J. Block, and S. Weinberger, (in preparation).
21. A. Borel and J. P. Serre, *Corners and arithmetic groups*, J. Comm. Math. Helv. **48** (1974) 244-297.
22. M. Bridson and S. Gersten, *The optimal isoperimetric inequality for* $\mathbb{Z}^n \times \mathbb{Z}$, 1993, preprint.
23. K. Brown, *Cohomology of Groups*, Springer, 1982.
24. K. Brown and R. Geoghegan, *An infinite dimensional torsion free FPY group*, Inven. Math. **77** (1984) 367-381.
25. L. Brown, R. Douglas and Fillmore, *Extensions of* C^**-algebras and* K*-homology*, Ann. Math. **105** (1977) 265-324.
26. J. Bryant, S. Ferry, W. Mio and S. Weinberger, *The topology of homology manifolds*, 1993, preprint. (See also BAMS 1993).
27. Cantwell and Conlon, *Leaves with isolated ends in foliated three manifolds*, Topology **16** (1977) 311-322.
28. A. Connos and H. Moscovici, *Cyclic cohomology, the Novikov conjecture and hyperbolic groups*, Topology **29** (1990) No. 3, 345-388.
29. S. Cappell and J. Shaneson, *Representations at fixed points*, Inven. Math. **68** (1982) 1-19.
30. G. Carlsson and E. Pederson, *Controlled algebra and the Novikov conjecture for* K*- and* L*- theory*, 1993, preprint.
31. A. Casson and D. Gottlieb, *Fibrations with compact fibers*, Amer. J. of Math. **99** (1979) 159-189.
32. J. Cheeger and M. Gromov, *On the characteristic numbers of complete manifolds of bounded curvarue and finite volume*, Rauch Memorial volume (Chavel and Farkas, eds).
33. J. Cheeger and M. Gromov, *Chopping riemannian manifolds*, Differential

Geometry, Pitman lecture series **52** (1990) 85-94.

34. J. Cheeger and M. Gromov, *L^2 cohomology and group cohomology*, Topology **25** (1986) 189-215.
35. J. Cheeger, M. Gromov and M. Taylor, *Finite propagation speed, kernel estimates for functions of the Laplace operator, and the geometry of complete Riemannian manifolds*, J. Differential Geometry **17** (1982) 15-54.
36. M. Coornaert, T. Delzant and A. Papadapolous, *Notes sur les groupes hyperboliques de Gromov,* LNM **1441** (1990).
37. A. Connes, *Noncommutative Differential Geometry*, Publ. Math. d'IHES **62** (1986) 41-03.
38. A. Connes, *A survey of foliations and operator algebras,* Proc. Symp. Pure Math. **38** part I (1982) 521-628.
39. A. Connes, M. Gromov and H. Moscovicci, *Conjecture de Novikov et Fibres Presque Plats*, Comptes Rendus (1990) 273-282.
40. F. Connolly and S. Prassidis, *Groups which act freely on* $\mathbb{R}^{m-1} \times S^n$, Topology **28** (1989) 133-148.
41. M. Davis and T. Januskiewicz, *Hyperbolizations of polyhedra*, J. Differential Geometry **31** (1991) 347-388.
42. A. Dranishnikov, *Homological dimension theory*, Russian Math Surveys **43** (1988) 11-63.
43. A. Dranishnikov, S. Ferry and S. Weinberger, *Large Riemannian manifolds which are flexible,* 1993, preprint.
44. R. Edwards and R. Kirby, *Deformations of spaces of embeddings*, Ann of Math **93** (1971) 63-88.
45. D. Epstein, J. Cannon, D. Holt, S.Levy, M.Paterson and W.Thurston, *Word Processing in Groups*, Jones and Bartlett, 1992.
46. B. Farb and R. Schwartz, *The large scale geometry of Hilbert modular groups*, Preprint, 1994. .
47. T. Farrell and L. Jones, *Classical Aspherical Manifolds*, CBMS lecture notes **75** 1990.
48. T. Farrell and C. Stark, *Cocompact spherical-Euclidean spaceform groups of infinite vcd*, Bull. LMS **25** (1983) 189-192.
49. S. Ferry, *Homotoping ϵ-maps to homeomorphisms*, Am. J. of Math. **101** (1979) 567-582.
50. S. Ferry and B. Okun, *Approximating topological metrics by Riemannian metrics*, Proc AMS, (to appear).
51. S. Ferry and S. Weinberger, *Curvature, tangentiality, and controlled topology*, Inven. Math. **105** (1991) 401-414.
52. S. Ferry and S. Weinberger, *A Coarse approach to the Novikov conjecture,* 1992, preprint .
53. S. Ferry, J. Rosenberg and S. Weinberger, *Equivariant topological rigidity phenomena*, Comptes Rendus **306** (1988) 777-782.
54. S. Gersten, *Bounded cohomology and combings of groups,* 1992, preprint .
55. S. Gersten, *Quasiisometry invariance of ends and other homological invariants*, Comptes Rendus, 1992.

56. S. Gersten, *Isoperimetric and Isodiametric functions of finite presentations,* Geometric Group Theory, (Nibling and Roller, eds.), LMS lecture note serie **181** (1993).
57. E. Ghys and P. de la Harpe, *Sur les groupes hyperboliques de Mikael Gromov*, Birkhauser.
58. M. Gromov, *Asymptotic invariants of infinite groups,* Geometric Group Theory, (Nibling and Roller, eds.), LMS lecture note serie **182** (1993).
59. M. Gromov, *Groups of polynomial growth and expanding maps*, Publ Math d'IHES **53** (1981) 53-73.
60. M. Gromov, *Kahler hyperbolicity and L^2 Hodge theory*, J. Differential Geometry **33** (1991) 263-292.
61. M. Gromov, *Hyperbolic groups,* Essays on Groups, (Gersten ed.) MSRI series number 8, Springer Verlag.
62. M. Gromov, *Large riemannian manifolds*, LNM 1201 (1985) 108-122.
63. M. Gromov, *Volume and Bounded Cohomology*, Publ. Math. de I.H.E.S., No. 56, (1982) 5-100.
64. M. Gromov and H. B. Lawson, *Positive scalar curvature and the dirac operator on complete Riemannian manifolds*, Publ Math d'IHES **58** (1983) 295-408.
65. M. Gromov and B. Lawson, *The classification of simply connected manifolds of positive scalar curvature*, Ann. of Math.**111** (1980) 423-424.
66. M. Gromov and P. Pansu, *Rigidity of lattices: an introduction*, Geometric Topology: Recent Developments, LNM 1504, 39-137 (1991).
67. Grunbaum and Branko, *Tilings and Patterns*, Freeman, New York 1987.
68. I. Hambleton and E. Pederson, *Bounded surgery and dihedral group actions on spheres*, JAMS **4** (1991) 105-126.
69. N. Higson, *On the relative homology of Baum and Douglas,* 1988, preprint.
70. M. Hilsum and G. Skandalis, *Invariance par homotopie de la signature a coefficients dans un fibre presque plat*, Crelle J. **423** (1992) 73-99.
71. N. Higson and J. Roe, Preprint.
72. N. Hitchin, *Harmonic Spinors*, Advances in Math. **14** (1974) 1-55.
73. W. C. Hsiang and J. Shaneson, *Fake tori*, Topology of Manifolds, (Cantrell and Edwards, eds.), Markham Press, 1969, 18-51.
74. B. Hu, *Whitehead groups of finite polyhedra of nonpositive curvature*, J. Differential Geometry **38** (1993) 501-518, (N.B. There is a sequel to this which will do the Borel conjecture in the same cases.).
75. S. Hurder, *Coarse geometry of foliations,* 1993, preprint.
76. S. Hurder, *The Godbillon measure of amenable foliations*,J. Differential Geometry **23** (1986) 347-355.
77. T. Januskiewicz, *Characteristic invariants of noncompact Riemannian manifolds*, Topology **23** (1984) 289-301.
78. T. Januskiewicz, *Hyperbolizations,* Group Theory from a geometric viewpoint, (Short ed.), World Scientific.
79. G. Kasparov, *Topological invariants of elliptic operators I: K-homology*, Math USSR, Izvestia **39** (1975) 751-792.

80. G. Kasparov, *Equivariant KK theory and the Novikov conjecture*, Inven. Math. **91** (1988) 147-201.
81. R. Kirby and L. Siebenmann, *Foundational Essays on Topological Manifolds, Smoothings, and Triangulations*, Princeton University Press, 1977.
82. S. Lopez de Medrano, *Involutions on Manifolds*, Springer-Verlag, 1972.
83. C. Moore and C. Schochet, *Global Analysis of Foliated Spaces*, MSRI series **9** 1988.
84. G. D. Mostow, *Quasiconformal mappings in n-space and the rigidity of hyperbolic spaceforms*, Publ Math d'IHES 34 (1968) 53-104.
85. G. D. Mostow, *Strong rigidity for symmetric spaces*, Ann of Math Studies.
86. P. Pansu, *Metriques de Carnot-Caratheadory et quasi-isometries des espaces symetriques de rang un*, Ann. of Math. **129** (1989) 1-61.
87. E. Pedersen, J. Roe and S. Weinberger, Preprint.
88. A. Paterson, *Amenability*, AMS Publications.
89. A. Phillips and D. Sullivan, *Geometry of leaves*, Topology **20** (1981) 209-218.
90. F. Quinn, *Resolution of homology manifolds*, Inven. Math. **72** (1983) 267-284, Corrigendrum **85** (1986) 653.
91. M. S. Raghunathan, *Discrete Subgroups of Lie Groups*, Springer, 1972.
92. J. Roe, *Coarse cohomology and index theory on complete Riemannian manifolds*, Mem. AMS **497** (1993).
93. J. Roe, *An index theorem on open manifolds I,II* , J. Differential Geometry **27** (1988) 87-113, 115-136.
94. J. Roe, *Elliptic operators, Topology, and Asymptotic Methods*, Res. Notes in Math series **23** (1990).
95. J. Rosenberg, *C^*-algebras, positive scalar curvature and the Novikov conjecture*, Publ. Math d'IHES **58** (1983) 409-424.
96. J. Rosenberg, *The KO assembly map and positive scalar curvature*, LNM **1474** (1991) 170-182.
97. H. Short (ed.), *Notes on word hyperbolic groups, in Group Theory from a geometric viewpoint, (Short ed.)*, World Scientific.
98. L. Siebenmann and D. Sullivan, *On complexes that are Lipschitz manifolds*, Geometric Topology, 503-525 (Cantrell ed.) Academic Press, 1978.
99. M. Spivah, *A classification of theorem for fiber spaces*, Topology **6** (1969) 77-102.
100. D. Sullivan, *Hyperbolic geometry and homeomorphisms*, Geometric Topology, (Cantrell ed.) 543-555 Academic Press, 1978.
101. N. Telemann, *Index of signature operators on Lipschitz manifolds*, Publ. Math IHES No 58, 1983.
102. W. Thurston, *The topology and geometry of 3-manifolds*, Princeton notes.
103. C. T. C. Wall, *Surgery on compact manifolds*, Academic Press, 1969.
104. J. Walsh, *Dimension, cohomological dimension, and cell-like mappings* , LNM **870** 105-118.
105. S. Weinberger, *The topological classification of stratified spaces*, University of Chicago Press, 1994.

106. S. Weinberger, *The Lipschitz rationality of linear representations*, CPAM **40** (1987) 609-610.
107. S. Weinberger, *Fake G-tori,* 1992, preprint.
108. S. Weinberger, *Noneuclidean manifolds and orbifolds,* (in preparation).

DEPARTMENT OF MATHEMATICS UNIVERSITY OF PENNSYLVANIA

AMS/IP Studies in Advanced Mathematics
Volume **2**, 1997 (Part 1)

MÖBIUS ENERGIES FOR KNOTS AND LINKS, SURFACES AND SUBMANIFOLDS

ROBERT B. KUSNER AND JOHN M. SULLIVAN

ABSTRACT. There has been recent interest in knot energies, especially those which are invariant under Möbius transformations of space. We describe computer experiments with such energies, and discuss ways of extending these to energies for higher-dimensional submanifolds. The Appendix gives a table of computed energy-minimizing knots and links through eight crossings.

1. INTRODUCTION

Is there an optimal way to tie a knot in space, or to embed a more general submanifold? And is there a natural way to evolve any embedding isotopically to an optimal one, so that we could detect whether two embeddings are isotopic?

An approach to such questions is to associate to any submanifold an energy, and look for minimizers or critical points of this energy. If the energy is infinite for immersions which are not embeddings, then presumably its gradient flow will prevent self-crossings and preserve isotopy type. One way to get an energy with an infinite barrier against self-crossings is to think of spreading charge along the submanifold and then consider the electrostatic potential. Such an energy for knots was introduced by Ohara [21] and studied by Freedman, He and Wang [12]. We define an analogous energy for k-dimensional submanifolds in $\mathbb{R}^n$ (or S^n). Our energy is again a repulsive potential between points on the submanifold, depending only on first-order data. It is given by a regularized inverse power law, with the power chosen to make the

---

1991 *Mathematics Subject Classification.* 57M25, also 57Q45, 57-04, 49Q15, 53C40, 53A07.

Key words and phrases. Knot, energy, submanifold.

This research was supported in part by the NSF (Kusner) and by the Geometry Center (Sullivan), and at MSRI by NSF grant DMS-9022140.

energy scale-invariant and the regularization to make it invariant under conformal (Möbius) transformations of the ambient space.

The gradient flow of this energy appears to lead to optimal embeddings, both theoretically and computationally [20]. In particular, we have done extensive computer experiments with knots and links, which seem to indicate that the energy will untangle many complicated curves to the minimum representative for their knot type. However, we also find certain links with several distinct local minima at different energy values. Our Appendix gives a table of computed energy minimizers for knots and links up through eight crossings.

We would like to thank Ken Brakke for his help in adding features to the `evolver`, and John Conway, Peter Doyle, and Bill Thurston for useful discussions. We dedicate this paper to our grandmothers Esther Paley Kusner (1895-1994) and Elizabeth Gummere Peplow (1915-1993).

2. Defining Möbius Energies

Recall Coulomb's Law which asserts that the potential energy between a pair of unit point charges at points x and y in $\mathbb{R}^3$ is given by the reciprocal $1/|x-y|$ of their distance in space. If we imagine charge uniformly spread over a k-dimensional oriented submanifold M of $\mathbb{R}^n$, the total energy would be given by a double integral over all pairs of points on M of some inverse power of distance. Although for physical charges in $\mathbb{R}^n$ we might think of using the power $n-2$, we prefer to choose the power $2k$, which makes the integrand scale-invariant. (Without scale invariance, we would need a constraint on the size of M to get nontrivial energy minima.)

Of course, the description we have given so far ignores the fact that such an integrand will blow up as x approaches y, in such a way that the integral will be infinite for any M. So we include a regularizing factor f and define

$$E_f(M) = \iint_{M\times M} \frac{f(x,y)}{|x-y|^{2k}}\, d\,\mathrm{vol}_M(x)\, d\,\mathrm{vol}_M(y).$$

If we did not have the factor of f, this integrand clearly would be scale-invariant. In fact, it is also easy to show (see Figure 1) that it would be invariant under inversion ($x \mapsto x/|x|^2 =: \tilde{x}$), and hence under the full conformal group of Möbius transformations of $\mathbb{R}^n \cup \{\infty\}$. Note that when computing the energy, we can view M as a submanifold of $S^n \subset \mathbb{R}^{n+1}$ via stereographic projection, instead of $\mathbb{R}^n$. This follows because stereographic projection from S^n to $\mathbb{R}^n \cup \{\infty\}$ extends to a

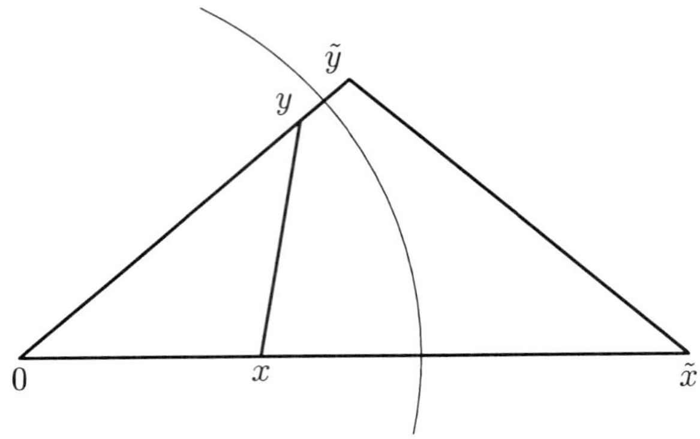

FIGURE 1. The similar triangles show $|\tilde{x}||\tilde{y}| \,/\, |\tilde{x} - \tilde{y}|^2 = |x||y| \,/\, |x - y|^2$. Since the conformal expansion factor for inversion from M to $\tilde{M}$ at x is $|\tilde{x}| \,/\, |x|$, the volume element changes by the kth power $d\,\mathrm{vol}_{\tilde{M}}(\tilde{x}) = (|\tilde{x}| \,/\, |x|)^k d\,\mathrm{vol}_M(x)$. Combining these facts shows the integrand of our energy is Möbius-invariant.

Möbius transformation of $\mathbb{R}^{n+1} \cup \{\infty\}$, and the formula for energy is independent of the ambient dimension.

Thus we would like to choose our regularizer f (which is supposed to vanish as x approaches y) to be independent of scale and also to be Möbius-invariant. We will allow this function f to depend on first-order information—the tangent planes to M at the points x and y—although this was suppressed in the notation used above.

Given a point $x \in M$ and any other point p in space, there is a unique round k-sphere $S_x(p)$ tangent to M at x and passing through p. Thus given two points x and y of M, we have two oriented k-spheres $S_x(y)$ and $S_y(x)$ which meet at equal angles at x and y. These spheres, and in particular the angle at which they meet, are defined in a Möbius-invariant manner.

By the angle between these k-spheres, we mean the angle between their tangent k-planes at points of intersection. In fact, a configuration of two oriented k-planes in $\mathbb{R}^n$ is described by k principal angles $\alpha_1, \dots \alpha_k$, but perhaps most useful is the combined angle α whose cosine is the inner product of two simple unit k-vectors $u_1 \wedge \cdots \wedge u_k$ and $v_1 \wedge \cdots \wedge v_k$ representing the two planes:

$$\cos\alpha = \prod \cos\alpha_i = \det[u_i \cdot v_j] = (u_1 \wedge \cdots \wedge u_k) \cdot (v_1 \wedge \cdots \wedge v_k).$$

We propose taking $f(x, y)$ to be some function of these angles α_i between the spheres $S_x(y)$ and $S_y(x)$. It should be nonnegative, to keep the energy well-behaved, and should vanish when the angles are zero, in order to cancel the singularity in the integrand. We would like the energy E_f to have the following basic properties, which qualify it as a "knot energy":

- $E_f(M)$ is nonnegative, and zero only for $M = S^k$, the round k-sphere;
- $E_f(M)$ is infinite for immersions which are not embeddings, creating a barrier against M "crossing itself";
- $E_f(M)$ is finite for all compact k-dimensional embedded smooth submanifolds $M \subset \mathbb{R}^n$.

The first two properties will be true for essentially any f which is a nonnegative function of the angle α, vanishing only at 0. The third property follows if f vanishes sufficiently fast at 0 to regularize the integral.

Note that our Möbius energies $E_f(M)$ are somewhat like "quadratic forms" on the space of oriented submanifolds. We can also examine the associated "bilinear form" $E_f(M, N)$, given by the double integral over $M \times N$. If we interpret M and N as chains or integral currents, then indeed this cross-energy will be linear in each argument, but only for positive multiples. We can think of the first surface M defining a potential P^f_M at all points of space. Then $E_f(M, N)$ is just the integral of this potential over the points of N. Since f depends on the tangent planes of M and N, the potential P^f_M is a function not merely of points in space, but of k-vectors at those points. In the language of geometric measure theory, P^f_M is a parametric integrand, and in fact we might think of E_f as given by a bilinear parametric integrand.

3. The Excess-Length Picture and a Standard Choice of Regularization

One good choice for the regularization in the energy E_f is $f_0 := (1 - \cos\alpha)^k$. For the remainder of this paper, we will study mostly this particular energy; we will write simply E for the energy E_{f_0} with this choice of f. This energy E generalizes the energy $\tilde{E}$ for knots $K \subset \mathbb{R}^3$ studied in [21] and [12]:

$$\tilde{E}(K) = \iint_{K\times K} \left(\frac{1}{|x-y|^2} - \frac{1}{d_K(x,y)^2} \right) ds_K(x)\, ds_K(y),$$

where $d_K(x, y)$ is the shorter arclength distance within K from x to y. In fact, Peter Doyle and Oded Schramm [10] introduced, in the one dimensional case, the idea of a regularization by a multiplicative factor

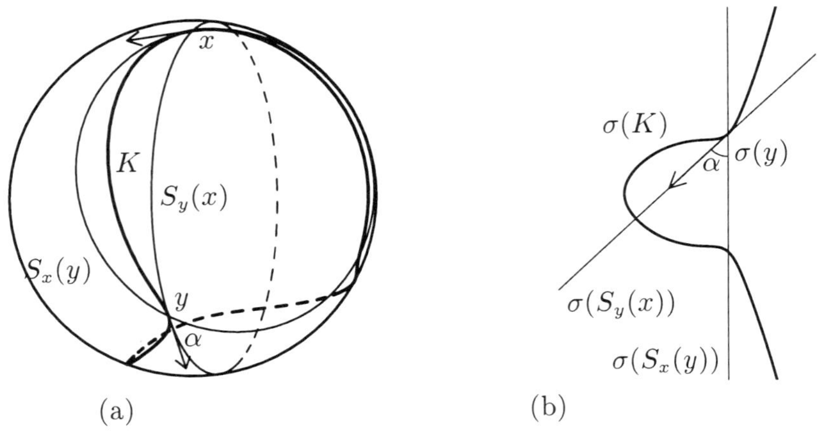

FIGURE 2. The excess-length picture shows how to calculate the potential $P_K(x)$ for a curve K in the sphere. When we apply the stereographic projection σ (from x as north pole) the spheres $S_x(y)$ and $S_y(x)$ become straight lines at angle α. The integral of $1 - \cos\alpha$ is the excess length of $\sigma(K)$ over that of the straight line.

depending on angle, and observed that $E(K) = \tilde{E}(K) - 4$ when the factor used is $(1 - \cos\alpha)$.

One way they explained this equivalence of E and $\tilde{E}$ is through a picture which interprets the potential $P_K(x) := P_K^{f_0}(x)$ as an excess length. Given a curve K in S^n, we want to evaluate P_K on a tangent direction at some point $x \in S^n$. To compute this, rotate the sphere so that x is at the north pole, and then stereographically project to $\mathbb{R}^n$ (sending x to infinity). Rotate this euclidean space so that the given tangent direction at x becomes the vertical direction. (Figure 2 shows this in the case $n = 2$, when we are projecting a curve from the two-sphere to the plane; although here all embedded curves are unknotted, the energy still makes sense.) If $\sigma(K)$ is the stereographic projection of the curve K, then we can check that $P_K(x)$ equals the integral $\int_{\sigma(K)}(1 - \cos\alpha)\, ds_{\sigma(K)}$, where in this picture the angle α is simply the angle that the tangent to $\sigma(K)$ makes with the vertical. (Our stereographic projection is scaled nonstandardly, to take the equatorial sphere to a sphere of half the size; the extra factor of $1/2$ is needed because we want to use an inversion in a unit sphere in $\mathbb{R}^{n+1}$.)

If K is a closed curve, and x is not on K, this integral is simply the length of $\sigma(K)$, because the $\cos\alpha$ term integrates to zero. Thus the potential at points not on K does not depend on tangent directions, and in fact is the same as it would be if we used no regularization, setting

$f = 1$. Integrating along a second, disjoint curve L, we see that the cross energy $E(K, L)$ equals the unregularized energy $E_1(K, L)$. (This is what Dave Auckly and Lorenzo Sadun [3] refer to as the additive link property for the energy E.)

But of course for x on K, we need the regularization, as $\sigma(K)$ is an infinite curve in $\mathbb{R}^n$, asymptotic at either end to a vertical line. Notice that the difference between the length of the curve $\sigma(K)$ and its vertical progression is given by the integral of $1 - \cos\alpha$, which gave the potential P_K. Each term is infinite, but their difference is the excess length of $\sigma(K)$ compared with the vertical straight line, which is finite if K is smooth enough at x.

One might hope to use the same picture to define energies for k-dimensional submanifolds. If a submanifold M is stereographically projected from a point $x \in M$, the image is asymptotic at infinity to a flat k-plane. However, this image submanifold has infinite excess area when compared to this k-plane, since its distance from the k-plane varies as we approach infinity in different directions (unless x is an umbilic point of M). Thus a naïve definition of energy in terms of excess area will not work; in other words the energy $E_{(1-\cos\alpha)}$ is not finite for dimensions $k > 1$. Our choice of $f_0 = (1 - \cos\alpha)^k$ overcomes this difficulty, although we do not have an equally nice geometric picture for the resulting energy E.

4. Other Möbius-Invariant Knot Energies

The Möbius energy E has been useful—theoretically, and from the perspective of computer experiments—but other Möbius-invariant knot energies are possible. First we might consider another regularizer f in our general energy E_f. Recent discussions with Doyle, Schramm, and Bill Thurston have focused our attention on $f = |\sin\alpha|$ as giving an interesting energy for curves. One problem here is that f is not differentiable, so E_f does not have a well-behaved gradient flow, and it is hard to model numerically. Higher powers of this function give regularizers for k-submanifolds; perhaps here there are also good choices for f other than powers of a curve regularizer.

Our E_f is defined in terms of first-order information at the two points x and y in the double integral. If we allow the use of higher-order information, there are other possibilities for the regularization. For two-dimensional surfaces, Auckly and Sadun [3] have suggested a regularization using the squared-mean-curvature integral (second-order data), but this is difficult to bound below.

We have also proposed a "holomorphic" energy for embedded Riemann surfaces by considering a relative energy within conformal classes. We choose a reference embedding of a surface in space; then the energy of any conformally equivalent embedding is given by comparing the straight-line distance between a pair of points on the surface with the corresponding distance on the reference surface. We have yet to find a proper regularization for this energy, but for spheres there is of course a unique conformal class, and the round sphere serves as a natural reference surface. This idea of relative energy should extend to rather general subsets of $\mathbb{R}^n$, and in fact we have already succeeded in regularizing it for embedded 1-complexes or "knotted graphs" [20].

5. Prime Decomposition

We are interested in minimizing E_f within isotopy classes. Given a submanifold $M \subset \mathbb{R}^n$, we write $[M] = \{N\colon\ N \sim M\}$ for its isotopy class, and

$$E_f([M]) = \inf_{N \sim M} E_f(N)$$

for the infimum energy. One basic result of [12] is that prime knots have E-minimizing representatives—this infimum is achieved. On the other hand, it seems that under energy minimization, composite knots decompose into their summands in a natural way. More generally, given a pair of k-dimensional submanifolds M and N, there are two ways to naturally combine or add their isotopy classes $[M]$ and $[N]$, in such a manner that minimizing the energy seems to separate the two pieces again.

Most trivial is the disjoint union $[M] \sqcup [N]$, obtained by embedding M and N in disjoint balls in $\mathbb{R}^n$. It is clear that by placing M and N far apart from each other (or equivalently scaling each one down) we can make their cross-energy $E_f(M, N)$ arbitrarily small. Since the energy of the union is the sum of the self-energies and the cross-energy, it is thus clear that

$$E_f([M] \sqcup [N]) = E_f([M]) + E_f([N]).$$

Even a submanifold of several topological components may not be decomposable in this way as a disjoint union; in this case we say it is essentially linked.

There is also a natural notion of the connected sum $[M] \# [N]$, which is well defined when M and N are both connected. (Of course, if M or N has more than one component, we simply must specify which components are to be connected.) We say a submanifold is prime if the only way it can be decomposed as a connected sum is when one

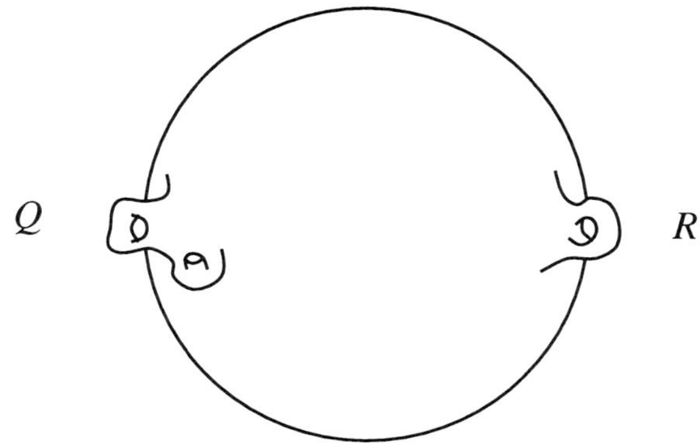

FIGURE 3. After applying Möbius transformations to Q in $[M]$ and R in $[N]$ to get large round pieces, we can weld them together to make this representative for $[M]\#[N]$, with energy not much more than $E_f(Q) + E_f(R)$.

summand is isotopic to a trivial S^k. Using Möbius invariance, we find that

$$E_f([M]\#[N]) \leq E_f([M]) + E_f([N]).$$

To check this, consider $Q \sim M$ and $R \sim N$ each having E_f within any given $\varepsilon > 0$ of the respective infimum values. Deform a small neighborhood of some point on Q to be flat, changing the energy only by ε, and then apply a Möbius transformation mapping this small piece of a k-plane to almost all of a round S^k. Apply the same procedure to some point on R. Welding the resulting submanifolds together in the obvious way, we get a submanifold P in $[M]\#[N]$ (resembling a round k-spherical "head" with little copies of Q and R attached as small "ears"—see Figure 3) with

$$E_f(P) \leq E_f(Q) + E_f(R) + 3\varepsilon,$$

where the last ε includes the interaction terms between the two "ears". Infimizing yields the desired inequality.

In fact, we conjecture equality holds for infima of the energy E, and moreover, that minimization of E leads to a natural "conformal connected sum decomposition" of a submanifold into E-minimizing, essentially linked, prime submanifolds. This phenomenon has been observed in our computational investigations of the energy E for knots, links and surfaces in $\mathbb{R}^3$, described in the next section. Figure 4 shows what we expect is a minimizing sequence for the energy of a connected sum of two trefoil knots. (Note that we consider left- and right-handed trefoils

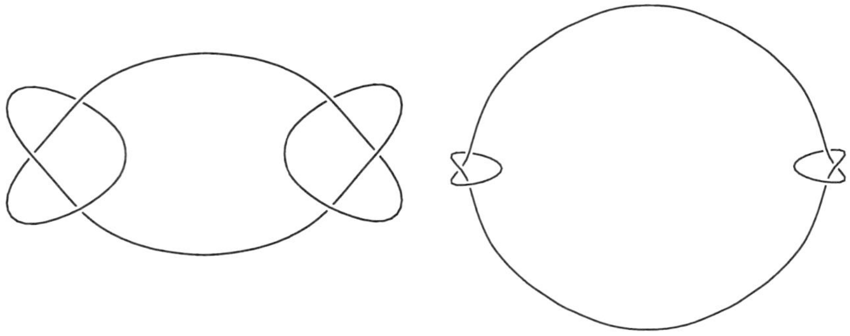

FIGURE 4. We expect that an energy-minimizing sequence for this knot type, the connected sum of a left-handed and a right-handed trefoil knot, will begin as above. The knot (a) has energy approximately 150, while (b) has energy approximately 142. The limit energy of the sequence is approximately 140.824, or exactly twice the energy of the critical trefoil of Section 9.

to be distinct isotopy classes, and that this matters when taking connected sum. Of course they have the same energy, so are not considered separately in our later knot tables.)

6. DISCRETIZATION AND COMPUTER EXPERIMENTS

In order to gain some intuition into the behavior of the Möbius energy, especially for knots and links, we have implemented various discretizations of E and its gradient flow, and carried out computer experiments using Brakke's `evolver` [4]. All our discrete models work with polyhedral surfaces or polygonal knots and links. These have infinite energy E, since they have sharp corners, so we must work with some discretization of the energy, which is supposed to model the energy of a nearby smooth curve or surface.

One discretization for curves, the *cosine energy*, places point charges at the midpoints of all edges of the polygon. The charge at x equals the length l_x of the edge. For this energy, we sum $(1 - \cos\alpha)l_x l_y/|x - y|^2$ over all pairs $x \neq y$; here α is the angle between two circles passing through the midpoints x and y. In fact, α can be more easily computed as the angle between the edge at x, and the edge at y reflected in the perpendicular bisector of $\overline{xy}$ (so that it also passes through x).

If we approximate some smooth curve by polygonal segments, this

Number of edges	80	160	320	640
Cosine energy	70.85	70.52	70.44	70.42
Edge energy	72.24	71.21	70.79	70.59
Vertex energy	69.16	69.84	70.14	70.28

TABLE 1. These discrete energies were computed for minimal polygonal trefoil knots with equal edge lengths in S^3, approximations to the symmetric critical trefoil whose true energy $E \approx 70.41204$ is computed in Section 9.

discretization seems to give energy values quite close to the true energy of the smooth curve, even when the polygonal approximation is relatively coarse. However, its gradient flow is problematic, since some edges tend to get very short, and then sometimes fail to line up with their neighbors. We can avoid this problem by adding a "Hooke" term to the energy, which tends to keep all the edges in the polygon at some fixed length by pretending the edges are stiff springs with that equilibrium length.

The *edge energy* discretization is the same, but without the factor of $(1 - \cos\alpha)$ in the summand. This models the unregularized energy E_1, but of course the discrete sum is finite. The idea is that if we wrote down a discretization for the formula for $\tilde{E}$, the regularization term would depend only on the edge lengths of the polygon, and not on its position in space. This subtracted term in fact is smallest when the edge lengths are equal, so leaving it out of the energy we minimize just helps keep edge lengths equal (as the Hooke energy did). Note also that the exact integral of $1/|x-y|^2$ over a pair of disjoint line segments in space is not an elementary function of their endpoints; this is why we concentrate the charges at the edge midpoints in the edge and cosine energies.

Finally, the *vertex energy* places a charge at each vertex v of the polygonal knot. The charge l_v equals the average of the adjacent edge lengths. Again we merely sum $l_v l_w / |v-w|^2$ without any regularization. This vertex energy is the one we have used for most of our gradient flows. For informational purposes, we also compute the sum of $l_v l_w / d^2(v,w)$, which is subtracted from the vertex or edge energies as a regularization. We report the values of all three discretized energies for our final knots. All three seem to converge as the number of segments used increases, as shown in Table 1, though the cosine energy is by far the most accurate. This table reports energies computed for polygonal approximations to a critical trefoil knot whose energy can be computed exactly (see Section 9).

There have been some previous experiments with the energy $\tilde{E}$ by other researchers: Kazushi Ahara [2] has used a method like our vertex energy with Hooke terms to compute several simple examples, and a program of Steve Bryson [5] was able to approximate the minimal trefoil. Others, including Buck, Fukuhara, Gunn, Orloff, Simon [6, 7, 13, 14] have done experiments with other repulsive energies for explicitly polyhedral knots, not all of which model any energy of smooth knots.

Of course, the vertex-based discretization (by which we flow) does not have an infinite barrier to changing knot type. Two segments can cross each other in space if their endpoints stay far away. But in practice, this does not happen as long as we keep edge lengths short enough near tight crossings. We can do this either with the Hooke energy, forcing all edges to be short, or by selectively refining edges whose contribution to the knot energy has become large during the evolution. The latter method (implemented only very recently) seems preferable, as it concentrates the vertices where they are needed, and in the case of links does not constrain the relative lengths of the components.

As always with the `evolver`, formulas for the exact gradients of the discrete energies are compiled with the computer code, and these are used in the conjugate gradient method to flow towards a critical point.

We have also implemented a discretization of the energy for two-dimensional surfaces, though not for submanifolds of arbitrary dimension. Here, one discretization is like the vertex energy, ignoring regularization and placing a charge at each vertex equal to one third of the area of the surrounding triangles. This energy seems to work nicely for surfaces in $\mathbb{R}^4$, but for surfaces in $\mathbb{R}^3$ it is too rigid: the high power in the $1/r^4$ replusive energy (needed for surfaces) means that vertices are influenced mostly just by their nearest neighbors. Thus the discrete surface seems to get locked into a particular, nearly equilateral triangulation, without much freedom to move. It is also not clear if this energy models any E_f.

For surfaces in space, we have had more success with a discretized $(1 - \cos\alpha)^2$ energy, which places a charge at the center of each face, equal to its area, and computes the angle between the tangent planes at pairs of faces. We have computed, for instance, a tube around a trefoil knot (see Figure 5) with energy about 638, but we have yet to do comprehensive experiments with this energy. As with the similar cosine energy for links, we must pay special attention to keep the triangulation from degenerating during the evolution.

FIGURE 5. This torus, in the isotopy class of a tubed trefoil knot, seems to minimize energy, with $E \approx 638$. In S^3, it would evidently be the orbit of a small circle under a rigid rotation.

7. UNTANGLING UNKNOTS

For any prime knot type, the existence of an E-minimizer is guaranteed by the result of [12] already mentioned. But this leaves open the interesting question of whether the same knot type might have other critical points for E, and in particular further local minima. Of course, the E-gradient flow will be most useful for classifying knots if no other minima exist.

Of special interest are tangled unknots. There is no efficient algorithm known for untangling an unknotted curve, though according to Hatcher's solution [15] of the Smale Conjecture, there is no obstruction to finding a flow which evolves any unknotted curve to a round circle. We would be surprised if the E-flow accomplished this, but we were equally amazed to see that the (discrete) flow did untangle the example in Figure 6 (suggested by [12]). Although initially the evolution seems to lead to a large loop caught in a tight slip-knot, perhaps it is the Möbius invariance which lets different parts of the curve grow or shrink as necessary to untangle the unknot to a round circle. The full process is shown in the six-minute video [23], along with other examples of evolutions towards E-minimizing knots and links.

Of course, our experiments showing how this one curve untangles leave open the basic question: are there any E-critical unknotted curves

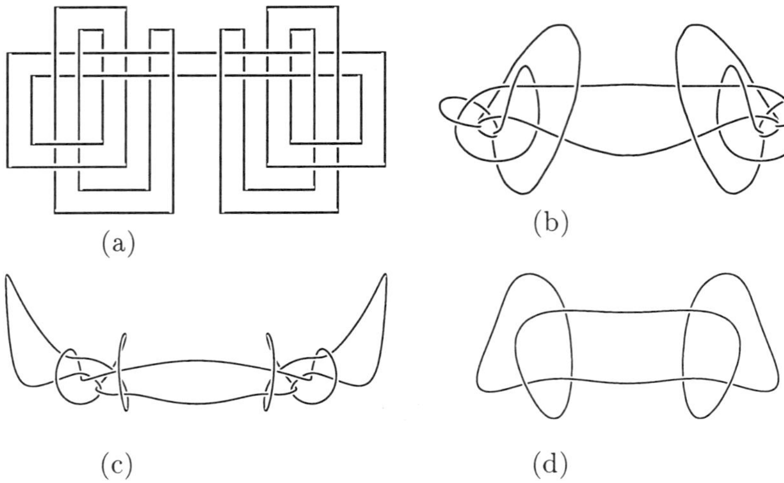

FIGURE 6. The tangled unknot (a) quickly rounds off to the smooth curve (b). Eventually loops grow as needed to swing out around the knobs, giving the curve (c), which can then shrink these loops to pull them through. The curve (d) clearly has no further obstacles to becoming a round circle, and indeed quickly evolves there.

besides the round circle? A negative answer would give an elegant analytic proof of Hatcher's theorem.

8. Crossing Numbers

Knots and links are often studied by means of planar projections with marked crossings. Any two projections of equivalent knots can be obtained from each other by a sequence of Reidemeister moves (see, for example, [1]). The topological crossing number $c([K])$ of the link type $[K]$ is defined as the minimum number of crossings in any planar projection.

From a three-dimensional geometric perspective, perhaps more interesting is the average crossing number $A(K)$ of the space curve K, which is the average number of crossings in planar projections of the curve, averaged over all possible orthogonal projections. This can be computed by a formula of Gauss:

$$4\pi A(K) = \iint_{K\times K} \frac{|(u\times v)\cdot(x-y)|}{|x-y|^3}\, ds_K(x)\, ds_K(y),$$

where u and v are the unit tangent vectors to K at x and y, respectively. In our computations, we discretize this for a polygonal link as a double sum over all pairs of edges, using the edge vectors for u and v, and the edge midpoints for x and y. It is clear that $A(K) \geq c([K])$ for any link.

One result of [12] is a relation between the energy E and the crossing number, namely $E(K) \geq 2\pi c([K])$. This follows from a similar relation between the energy of any curve and its average crossing number, once we apply a Möbius transformation to send one point of the curve to infinity. Also, [12] shows that $E(K) \geq \frac{6}{11} 2\pi A(K) - \frac{56}{11}$. These results are proved there only for knots, although it is easy to check they remain true for our link energy. Note, in this context, that [12] normalizes the energy for links differently than we would, counting cross terms between different components only half as much as the self-energy terms. We believe our normalization is more natural, providing a good ordering of all links by energy, independent of how many components they have.

Since the standard projection of any knot type has a minimal number of crossings, we usually like to reduce any projection of that knot to the standard one only using Reidemeister moves which decrease the number of crossings. But sometimes this is impossible. The projection of our initial curve shown in Figure 6(a) has 32 crossings, but it is easy to check that no moves are applicable except the ones which increase crossing number. It seems that at least four extra crossings must be introduced to move this diagram to the zero-crossing picture of the unknot. Therefore, before our experiments, it was reasonable to think that the corresponding three-dimensional (but nearly planar) example might not untangle under the E-flow.

9. Critical Knots and Links from Group Actions

One can also ask about the existence of further critical points for truly knotted curves. Perhaps the simplest of these are the (p, q) torus knots and links. These curves can be realized as (unions of) orbits of points under rigid rotations of S^3, which rotate one two-plane at speed p and the perpendicular one at speed q. (These can also be viewed as Möbius circle actions on $\mathbb{R}^3 \cup \infty$.) In earlier work with Denise Kim [17], explicit E-critical torus knots and links were constructed, by varying the ratio of radii of the torus containing these orbits until we reach a critical ratio, then using the "principle of symmetric criticality".

Having reduced the energy to a function on the space of orbits (which is finite dimensional, and in this case can be parametrized by the ratio of radii), this principle asserts that at a critical point of the reduced energy, the corresponding knot or link is critical for E among *all* variations. To see this, note that if the gradient of E did not vanish, we could simply average it over the orbit to get a variation through orbits which changed

(a) (b)

FIGURE 7. The E-critical orbital $(3,5)$ torus knot (a), with energy just over 265, is not stable, while the knot (b), with energy numerically computed to be slightly under 260, seems stable and presumably is the minimizer.

(to first order) the reduced energy, contradicting the fact that we are at a critical point on the orbit space.

It can be shown [20] that for large p and q (each at least 3) these orbital (p,q) torus knots and links are unstable for E (compare Figure 7), hence they are not the E-minimizers guaranteed to exist by [12]; in particular, there is more than one critical point for E within each of these isotopy types. On the other hand, the orbital $(2,q)$ torus knots do appear to be the E-minimizers, at least experimentally; as $q \to \infty$, these converge to an E-critical double-helix (see Figure 8) whose pitch, close to 1.454, is a universal constant.

The energies of these orbital critical points can be computed by an analytic formula, derived by Gil Stengle and the first author (see [17]). This is obtained by using the circle action to reduce the double integral for E to an explicit single integral of a meromorphic differential around the unit circle $S^1 \subset \mathbb{C}$, and evaluating this as a residue sum. For example, the critical trefoil (the $(2,3)$ torus knot, see Figure 9) has Möbius energy

$$E = 4\pi^2 \min_r (4r^2+9) \sum \operatorname{Res} \frac{-z^2}{r^2 z(z^2-1)^2 + (z^3-1)^2} \approx 70.41204,$$

where the sum is over poles within the unit circle, and the minimum is achieved at $r \approx 1.857$.

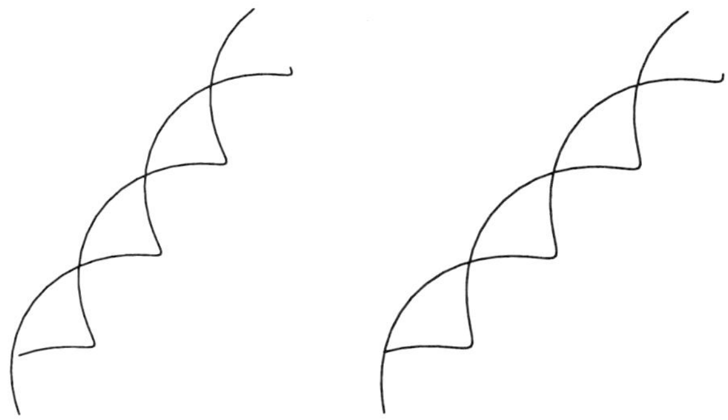

FIGURE 8. This double helix, critical for E, is a limit of minimizing $(2, q)$ torus knots for large q; many knot minimizers seem to include segments which look like a piece of this helix. We show a stereo pair of pictures; to see the stereoscopic effect, look at one figure with the left eye and the other with the right eye (either by crossing the eyes or by straightening them, perhaps with the aid of a stereo viewer).

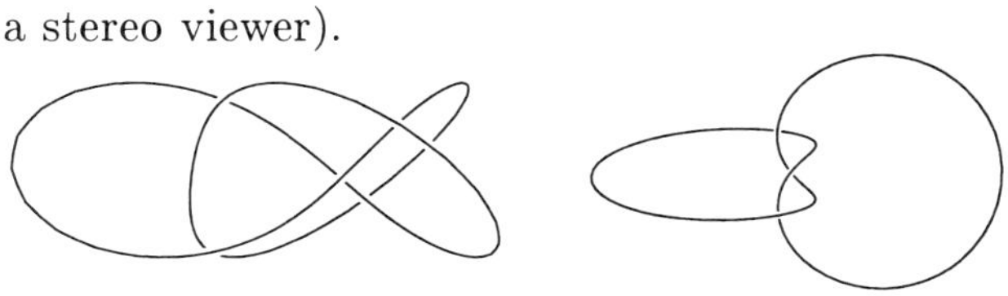

(a) (b)

(c) (d)

FIGURE 9. All these are views of the same critical trefoil, the presumed energy minimizer, differing only by Möbius transformations. The view (a), which is a different planar projection of the same space curve shown in the Appendix, is a $(2, 3)$-torus knot in space, while (b) is dually a $(3, 2)$-torus knot. The view (c) shows how any curve can be shown as a large circle with a small "ear", while (d) is a randomly chosen view.

10. Hopf Links and Electrons on S^2

We can find more examples of links with several E-critical points, and presumably several local minima, by examining the special case of Hopf links with the methods of the previous section. Recall that the Hopf fibration of S^3 is given by the orbits of the usual action of $S^1 \subset \mathbb{C}^*$ on $S^3 \subset \mathbb{C}^2$. Each fiber is a great circle, and each pair of fibers has linking number 1. We call the union of any p fibers a p-component geometric Hopf link; these are all isotopic for a given p. We can find E-critical points for this link type among such geometric Hopf links by lifting a finite set of points from S^2, the orbit space for the Hopf action.

Explicitly, suppose that in our p-component Hopf link $(\Gamma_1, ..., \Gamma_p)$, each component Γ_j corresponds to the point x_j on S^2. Then we compute

$$E(\Gamma_1, ..., \Gamma_p) = 4\pi^2 \sum_{i \neq j} \frac{1}{d_{ij}},$$

where $d_{ij} = |x_i - x_j|$ is the straight line distance in $\mathbb{R}^3$ between the points x_i and x_j. In other words, the Möbius energy of a geometric Hopf link is simply (a multiple of) the ordinary Coulomb energy of the corresponding point charges in $\mathbb{R}^3$ constrained to lie on the round sphere. Thus, in this case, the residue sum for E has a simple geometric interpretation; it would be interesting to know whether this interpretation could be extended beyond Hopf links to torus knots and links.

When the number of components p is less than 4, it is easy to see that there is only one critical configuration of charges on S^2 (or thus of Hopf circles in S^3). This is the global minimum for the Coulomb energy, with the points spaced equally around the equator, corresponding to (p, p) torus links. But when $p = 4$ there are two distinct critical configurations: the equatorial configuration (or geometric $(4, 4)$ torus link) is an unstable equilibrium, while the global minimum has points at the vertices of a regular tetrahedron (the link having fallen off the torus in S^3).

In the early part of this century, just before the discovery of quantum mechanics, there was interest in this problem, because it was thought that stable configurations of electrons on a sphere might explain the periodic table of elements. With this in mind, configurations of $p \leq 8$ points were analyzed in detail by Föppl [11].

To examine the structure of critical geometric Hopf links in general, consider a variation which moves each of the corresponding points x_i by Δx_i, and set $u_{ij} = (x_i - x_j)/d_{ij}$ and $v_{ij} = (\Delta x_i - \Delta x_j)/d_{ij}$. Then to

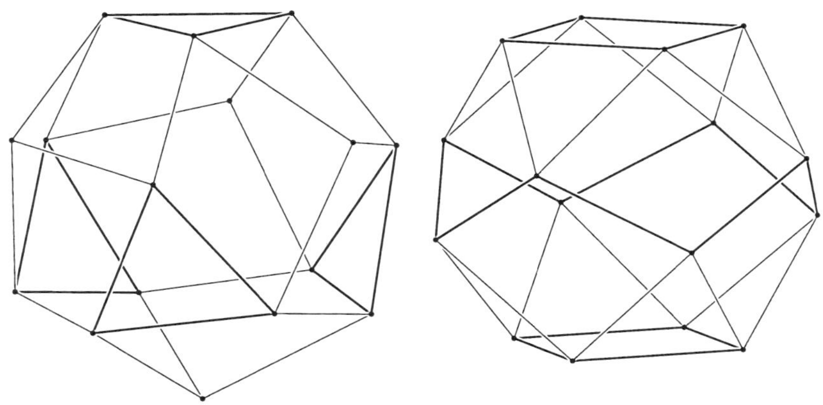

(a) (b)

FIGURE 10. Two stable configurations of sixteen point charges on the sphere, shown with the closest pairs joined with edges. Configuration (a) has tetrahedral symmetry and lower energy; configuration (b) has amphichiral D_8 symmetry and slightly higher energy.

second order, the change in energy is

$$\Delta E = 2\pi^2 \sum_{i \neq j} \frac{1}{d_{ij}} \left(2v_{ij} \cdot u_{ij} - v_{ij} \cdot v_{ij} + 6(v_{ij} \cdot u_{ij})^2\right).$$

Of course, if the points x_i are to remain on S^2, we must also impose on the variation the constraint $\Delta x_i \cdot x_i = 0$ for $i = 1, \ldots, p$.

We have done extensive numerical experiments to find such stable configurations of p point charges on S^2. These indicate that for $p < 16$ there is a single minimum. When $p = 16$ there seem to be two stable configurations, shown in Figure 10. One has tetrahedral A_4 symmetry of order 12 and $E \approx 7336.010$; the other has amphichiral dihedral D_8 symmetry of order 16 and $E \approx 7336.697$; the first local minimum seems to attract about three-quarters of the random configurations we start with. We have also found two distinct stable configurations of 22 points (with amphichiral D_4 and D_{10} symmetry). Our experiments indicate that for $p > 36$ there are always several local minima, while for $p \leq 36$ there is a unique minimum, except in the three cases $p = 16$, $p = 22$, and $p = 35$.

Our experiments also suggest that the corresponding geometric Hopf links are Möbius-energy stable. In general, we conjecture that the Morse index for an equilibrium configuration of point charges equals that of the corresponding geometric Hopf link. The configuration space of distinct points on S^2 has plenty of cohomology, so we expect to find many unstable extrema by viewing the reduced energy as an equivariant Morse function [19]. (In this context, Kawazumi has investigated critical points of a logarithmic repulsive potential [16].)

11. Surfaces and Submanifolds

In the case of surfaces and higher dimensional submanifolds, we know much less about the existence of E-stationary examples or E-minimizers. Presumably, for instance, the round S^2 is the unique critical point for spheres in $\mathbb{R}^3$ (again consistent with the Smale Conjecture), and prime knotted spheres in $\mathbb{R}^4$ have E-minimizing representatives.

Using a simple scaling argument we are able to prove some partial regularity results in general dimensions. For example, tangent cones (if they exist) to k-dimensional E-finite submanifolds must be flat k-planes [20]. Presumably E-minimizers have tangent cones everywhere, from which it follows that they are at least C^1 submanifolds; in fact, we expect all E-minimizers to be real analytic submanifolds.

To explicitly compute the energy E of a submanifold it is helpful to rewrite the formulas for the angle α. If two points x and y on $M \subset \mathbb{R}^n$ are separated by the vector $r = x - y$, and the tangent spaces to M at these points are spanned by $s_1, s_2, \ldots, s_k$ and by $t_1, t_2, \ldots, t_k$, then we do not have to find the spheres $S_x(y)$ and $S_y(x)$ in order to determine how they meet. Instead, note that $S_x(y)$ is tangent to M at x, while the tangent space to $S_y(x)$ there is obtained by reflecting the tangent space $T_y(M)$ in the vector r, and is thus spanned by the vectors $\bar{s}_i := s_i - 2(r \cdot s_i)r/r^2$. Observe that the $\bar{s}_i$ have the same inner products with each other as the s_i do; however, this basis for the tangent space to $S_y(x)$ has the wrong orientation, because of the reflection.

Define a $(k+1)\times(k+1)$ matrix A by setting $A_{00} := r^2/2$, $A_{0j} := r \cdot t_j$, $A_{i0} := s_i \cdot r$, $A_{ij} := s_i \cdot t_j$. Thus we have

$$A = \begin{pmatrix} r^2/2 & \ldots & r \cdot t_j & \ldots \\ \vdots & \ddots & \vdots & \\ s_i \cdot r & \ldots & s_i \cdot t_j & \ldots \\ \vdots & & \vdots & \ddots \end{pmatrix}.$$

By row reduction applied to the leading row and column of A, we find $\det A = \frac{r^2}{2} \det[\bar{s}_i \cdot t_j]$. This latter determinant is the one used to find $\cos\alpha$, so we get

$$\frac{-r^2}{2} \cos\alpha \, |s_1 \wedge \cdots \wedge s_k| \, |t_1 \wedge \cdots \wedge t_k| = \det A,$$

where the minus sign comes from the reversal of orientation by the reflection.

Note that by choosing principal angles we can arrange the s_i and t_j to be orthogonal sets of vectors with the property that $s_i \cdot t_j = 0$ unless $i = j$. Then the formula above simplifies to

$$(1) \qquad \frac{-r^2}{2} \cos\alpha \prod_{j=1}^{k} |s_j||t_j| = \left(\frac{r^2}{2} - \sum_{j=1}^{k} \frac{(s_j \cdot r)(r \cdot t_j)}{s_j \cdot t_j} \right) \prod_{j=1}^{k} s_j \cdot t_j .$$

As an example, let us compute the energy of a k-dimensional Clifford torus of radii $r_1, \ldots, r_k$. We have

$$T^k = S^1(r_1) \times \cdots S^1(r_k) \subset S^{2k-1} \subset \mathbb{C}^k,$$

parameterized by the map $f(\theta_1, \ldots, \theta_k) = (r_1 e^{i\theta_1}, \ldots, r_k e^{i\theta_k})$. This embedding is homogeneous, so the energy density is constant; thus we can compute $E(T)$ with a single integral over T:

$$(2) \ E(T) = \iint_{T \times T} \left(\frac{1 - \cos\alpha}{r^2} \right)^k dx\, dy = \operatorname{vol}(T) \int_T \left(\frac{1 - \cos\alpha}{r^2} \right)^k dx$$

for any fixed y, say $y = f(0, \ldots, 0)$. We find it easier, though, to compute this integral not by fixing this y and letting $x = f(\phi_1, \ldots, \phi_k)$ vary, but by rotating so that $x = f(\theta_1, \ldots, \theta_k)$ and $y = f(-\theta_1, \ldots, -\theta_k)$, where $2\theta_j = \phi_j$. The tangent vectors at these points x and y are $s_j = (0, \ldots, 0, i r_j e^{i\theta_j}, 0, \ldots, 0)$ and $t_j = (0, \ldots, 0, i r_j e^{-i\theta_j}, 0, \ldots, 0)$. The difference vector is $r = 2i(r_1 \sin\theta_1, \ldots, r_k \sin\theta_k)$.

It is a simple matter to compute the inner products:

$$s_i \cdot s_j = t_i \cdot t_j = \delta_{ij} r_j^2, \qquad s_i \cdot t_j = \delta_{ij} r_j^2 \cos\phi_j,$$

$$s_j \cdot r = t_j \cdot r = r_j^2 \sin^2\theta_j,$$

$$r \cdot r = 4 \sum r_j^2 \sin^2\theta_j = 2 \sum r_j^2 (1 - \cos\phi_j).$$

Define an angle ϕ by $\cos\phi = \prod \cos\phi_j$. Then using the formula (1) for $\cos\alpha$, we find that

$$\cos\alpha = \cos\phi \, \frac{\sum r_j^2 (1/\cos\phi_j - 1)}{\sum r_j^2 (1 - \cos\phi_j)},$$

which gives

$$\frac{1 - \cos\alpha}{r \cdot r} = \frac{1}{2} \, \frac{\sum r_j^2 (1 - \cos\phi_j)(1 - \cos\phi/\cos\phi_j)}{\left(\sum r_j^2 (1 - \cos\phi_j)\right)^2}.$$

FIGURE 11. The minimal Clifford torus in S^3, shown stereographically projected to $\mathbb{R}^3$, has energy $E = 5\pi^3/3$, presumably the lowest of any nonspherical surface.

For $k = 2$ we can now explicitly evaluate the integral (2). We find

$$\begin{aligned} E(T) &= \operatorname{area}(T) \int_0^{2\pi}\int_0^{2\pi} \left(\frac{1-\cos\alpha}{r\cdot r}\right)^2 r_1 d\phi_1\, r_2 d\phi_2 \\ &= 4\pi^2 r_1^2 r_2^2 \int_0^{\pi}\int_0^{\pi} \left(\frac{(r_1^2+r_2^2)(1-\cos\phi_1)(1-\cos\phi_2)}{\left(r_1^2(1-\cos\phi_1)+r_2^2(1-\cos\phi_2)\right)^2}\right)^2 d\phi_1\, d\phi_2 \\ &= \frac{2\pi^2 r_1^2(r_1^2+r_2^2)^2}{r_2} \int_0^{\pi} \frac{\sqrt{1-\cos\phi_2}\,(r_1^2+3(1-\cos\phi_2)r_2^2)}{(2r_1^2+(1-\cos\phi_2)r_2^2)^{\frac{7}{2}}}\, d\phi_2 \\ &= \frac{\pi^3}{6}\frac{3r_1^4+14r_1^2r_2^2+3r_2^4}{r_1 r_2(r_1^2+r_2^2)}. \end{aligned}$$

This is a rational function of the r_j, homogeneous of degree 0. It has a global minimum at $r_1 = r_2$, corresponding to the minimal Clifford torus in S^3, the lift of the equator under the Hopf map. We conjecture that this surface (see Figure 11) has the minimum Möbius energy for any unknotted torus in S^3; its energy is $5\pi^3/3$. In fact, this should be the absolute minimum for E among all nonspherical embedded surfaces in S^3 (or $\mathbb{R}^3$).

By the technique of symmetric criticality mentioned earlier, we do know that this surface is a critical point for E, and for any k, in fact, the k-torus of equal radii in S^{2k-1} is a critical point for the k-dimensional energy E.

12. A Table of Knots and Links Minimizing Möbius Energy

We have computed experimentally (with help from a group of undergraduate students at the 1993 Five Colleges Geometry Institute [18]) what seem to be E-minimizers for all the prime, essential knots and links with less than nine crossings; these are shown in the Appendix. Most of these knots have a two-bridge or rational tangle decomposition into segments where two particular strands twist around each other a certain number of half-turns [8]. This decomposition seems to be reflected in the shapes of the energy minimizers: each twisted piece resembles several half-turns of the E-critical double-helix (see Figure 8), up to Möbius transformations (which can send a half-turn of the helix into a pair of large arcs near infinity). It would be nice to prove that this is the shape of a minimizer, but at present it is not even known whether, for instance, a minimizer is a real analytic curve.

We computed an approximation to each knot or link by evolving at least 9000 steps with the conjugate gradient method, refining as necessary when edges had high energy. We included all the knots and links through eight crossings as well as a few nine crossing knots and the $(2,15)$ torus knot for comparison.

In Table 2, the first column lists the name of the knot or link from the standard tables [22] and its Conway notation [9]. The next three columns list the energy of our approximation, computed with each of the three discretizations. Finally, we list the average crossing number, and the number of edges, for this polygonal link. Note that the average crossing number would change somewhat if we applied a Möbius transformation to our link, though the other values should stay constant. If we had started with different initial configurations, we might get to the same minimizer in a different picture, and the value shown for A would be different.

We believe that the cosine energy is significantly more accurate than the other discretizations, and have ordered the table by its values. Note that the edge energy tends to be a bit higher, and the vertex energy a bit lower.

Remember from [12] that the minimum energy E for a knot type is at least $2\pi c$, where c is the topological crossing number. Our experiments suggest this inequality is far from sharp: the minimum of the ratio E/c seems closer to $2\pi^2$, achieved by the Hopf link 2_1^2. However, we still expect our ordering of knots by Möbius energy to list knots of small crossing number first. In fact, this is reflected in our table, though it seems that nonalternating links have significantly less energy than their crossing number would suggest—evidently two over-crossings in a row

Knot or Link		E_{cos}	E_{edge}	E_{vert}	A	N_{e}
0_1		0.0	0.1	−0.1	0.0	256
2_1^2	2	39.5	39.9	39.2	2.3	192
3_1	3	70.4	70.8	70.1	4.1	264
4_1^2	4	99.1	99.5	98.7	5.5	320
4_1	22	104.9	105.2	104.6	5.5	602
5_1	5	126.8	127.3	126.3	6.9	360
5_2	32	134.6	135.0	134.1	7.0	574
6_3^3	2,2,2−	136.8	137.1	136.5	6.9	534
5_1^2	212	138.9	139.3	138.4	7.1	674
6_1^2	6	154.1	154.6	153.6	8.2	584
6_1	42	162.8	163.4	162.1	8.4	640
6_2^2	33	164.5	164.9	163.8	8.4	684
7_7^2	3,2,2−	166.0	166.4	165.7	8.5	560
6_2	312	168.5	169.1	167.9	8.6	746
6_3^2	222	170.0	169.8	169.4	8.5	722
6_3	2112	172.9	173.2	172.4	8.7	914
7_8^2	21,2,2−	173.9	174.3	173.5	8.9	592
6_1^3	2,2,2	174.6	175.2	174.2	8.6	756
7_1	7	181.0	181.7	180.4	9.7	504
7_2	52	190.3	191.0	189.6	9.8	728
7_3	43	192.7	193.2	191.9	9.7	756
8_7^3	4,2,2−	194.2	194.7	193.8	9.8	612
7_1^2	412	196.6	197.2	195.9	9.9	820
8_{19}	3,3,2−	197.0	197.3	196.7	10.2	680
7_4	313	197.7	198.1	197.1	9.9	940
6_2^3	6*	197.9	198.3	197.4	9.2	492
7_3^2	232	198.7	199.1	198.0	9.8	796
7_5	322	199.7	200.0	199.0	9.9	902
8_{15}^2	22,2,2−	203.3	203.8	202.9	10.2	826
7_2^2	3112	203.6	203.8	203.2	10.1	1224
7_6	2212	203.7	204.0	203.1	10.2	970
8_8^3	31,2,2−	203.7	204.3	203.2	10.3	700
8_{20}	3,21,2−	203.9	204.4	203.4	10.4	720
7_4^2	3,2,2	204.4	205.1	203.7	10.0	808
7_1^3	2,2,2+	205.4	206.0	204.7	10.3	890
8_3^4	2,2,2,2−−	206.0	206.5	205.5	9.8	672
7_7	21112	207.1	207.0	206.5	10.2	1080
8_1^2	8	207.9	208.7	207.1	11.0	512
8_{16}^2	211,2,2−	208.2	208.5	207.6	10.3	912
7_5^2	21,2,2	208.9	209.4	208.2	10.1	916
8_{21}	21,21,2−	209.5	210.0	209.0	10.5	852
8_9^3	(2,2)(2,2−)	216.3	216.5	215.8	10.5	758
8_2^4	2,2,2,2−	217.2	217.7	216.6	10.9	752
8_{10}^3	(2,2)−(2,2)	217.4	217.7	216.9	10.6	720
8_1	62	217.4	218.3	216.7	11.1	798
8_2^2	53	220.3	220.9	219.5	11.1	776
8_3	44	220.9	221.1	220.1	11.1	836
9_{43}^2	5,2,2−	221.9	224.1	223.7	11.3	1568
8_2	512	224.1	224.8	223.3	11.3	898
8_6^2	242	226.2	226.8	225.4	11.4	934
8_4	413	226.2	226.7	225.4	11.3	878
8_3^2	422	227.9	227.4	227.2	11.2	998
8_6	332	228.6	229.2	227.8	11.3	914
8_4^2	323	229.6	229.7	228.9	11.2	974
7_6^2	6*2	231.1	231.6	230.6	10.7	760
8_7	4112	231.3	231.2	230.5	11.4	880
8_8	2312	232.4	232.9	231.6	11.5	1002
8_1^3	4,2,2	232.7	233.5	231.8	11.3	862
8_9	3113	233.2	233.3	232.4	11.4	906
8_{11}	3212	233.7	233.8	232.9	11.4	1034
8_5	3,3,2	234.0	234.6	233.2	11.3	954
8_3^3	2,2,2++	234.1	234.0	233.4	11.5	1220
9_1	9	234.6	235.6	233.7	12.3	576
8_{12}	2222	235.1	234.6	234.4	11.4	1036
8_5^2	3122	235.2	235.3	234.8	11.5	1406
8_7^2	21212	237.8	237.5	237.1	11.6	1176
8_{13}	31112	237.9	238.1	237.1	11.5	982
8_{10}	3,21,2	238.7	239.3	237.9	11.5	1006
8_{14}	22112	238.7	238.7	237.9	11.6	956
8_2^3	31,2,2	238.9	239.4	238.1	11.5	900
8_{12}^2	21,2,2+	239.3	239.1	238.6	11.7	1180
8_9^2	22,2,2	239.6	240.0	238.9	11.6	1044
8_{11}^2	3,2,2+	241.6	241.3	241.2	11.6	1700
8_{10}^2	211,2,2	243.0	243.2	242.3	11.6	1114
8_1^4	2,2,2,2	243.1	243.8	242.3	11.5	844
8_{15}	21,21,2	243.1	243.0	242.5	11.7	1154
8_4^3	(2,2)(2,2)	245.8	246.5	245.1	11.5	906
8_8^2	211112	246.0	248.1	247.7	11.7	1868
8_5^3	6*3	260.5	261.2	259.8	12.0	828
8_6^3	6*2:.20	264.4	264.8	263.7	12.1	916
8_{16}	6*2.20	264.5	264.8	263.9	12.4	862
8_{14}^2	6*2:.2	264.8	265.1	264.1	12.1	894
8_{17}	6*2.2	265.1	265.7	264.4	12.3	832
8_{13}^2	6*21	266.0	266.2	265.3	12.2	998
8_{18}	8*	283.9	284.3	283.4	12.6	844
9_{31}	2111112	284.1	284.2	283.9	13.3	2616
9_{40}	9*	329.6	329.9	329.4	14.3	1640
15_1	15	394.2	395.6	392.8	20.2	1022

TABLE 2. Approximate Möbius energies of links through eight crossings (see text).

require less twisting from a three-dimensional perspective. The ordering of rational (or two-bridge) knots by our energy seems quite predictable from their Conway names (their structure as rational tangles). Indeed, for each k, the lowest energy alternating k-crossing link in the table is the $(2, k)$ torus link with notation k. The highest energy two-bridge link is $21 \cdots 12$.

Many of the non-alternating links in the table have the notation $p, q, 2-$ and in each case, this link is very close in energy to $p1q$, a link with one less crossing. Note that the link $p, 2, 2-$ consists of a $(2, p)$ torus link together with a core circle of the torus, linking it twice.

Suppose we look at the highest energy knots and links for a fixed number of crossings. These in general seem to be knots based (in Conway's nomenclature) on the planar diagrams 6*, 8*, etc., in which all regions have at least three sides. In fact, 6* (the Borromean rings, 6_2^3) and 8* (8_{18}), which each have significantly more energy than all others with the same number of crossings, each fit into the class of so-called "Turk's-head" knots, with very symmetric planar diagrams. The energy minimizer stays close to this plane, with the strands weaving up and down only slightly, and presumably this accounts for the high energy. The highest-energy link of seven crossings, 6*2 (7_6^2) is in fact the Borromean rings with one of the six crossings replaced by a double half-twist. It thus also has a symmetric planar picture, although we would have to apply Möbius transformations to the picture in the Appendix to see this.

We have included in the table four knots of nine crossings, which we believe to be extreme for energy. We expect that 5,2,2− (9_{43}^2) has the lowest energy of any nine crossing link. The $(2, 9)$ torus knot 9 (9_1) presumably has the lowest energy among alternating knots and links. The knot 2111112 (9_{31}) is presumably highest-energy among arithmetic nine-crossing knots. And 9* (9_{40}) should have the highest energy overall. We have also included the $(2, 15)$ torus knot to indicate the limiting helical behavior.

The Appendix shows the approximate minimizer for each link, in a stereo pair. We made no particular effort to choose an optimal projection or Möbius representative for the minimizers; often, as for 7_3^2, there is some conformal symmetry that fails to be Euclidean for our representative. To see the stereoscopic effect, look at the left figure with the left eye and the right one with the right eye (by straightening the eyes, perhaps with the aid of a stereo viewer or just a card held between the images). Upon request, we can supply a version of these pictures printed instead for cross-eyed viewing.

Appendix

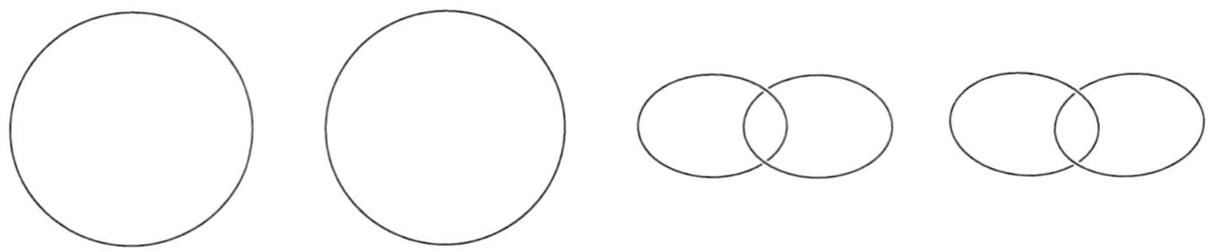

0_1 2^2_1 2

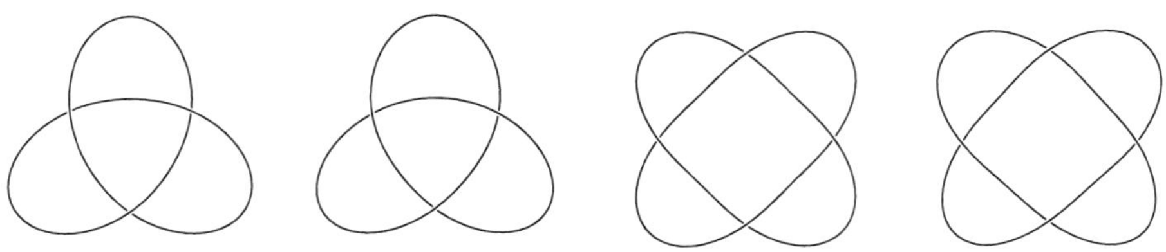

3_1 3 4^2_1 4

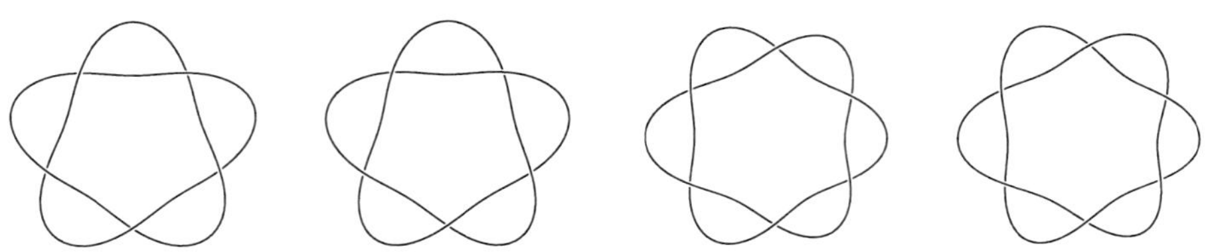

5_1 5 6^2_1 6

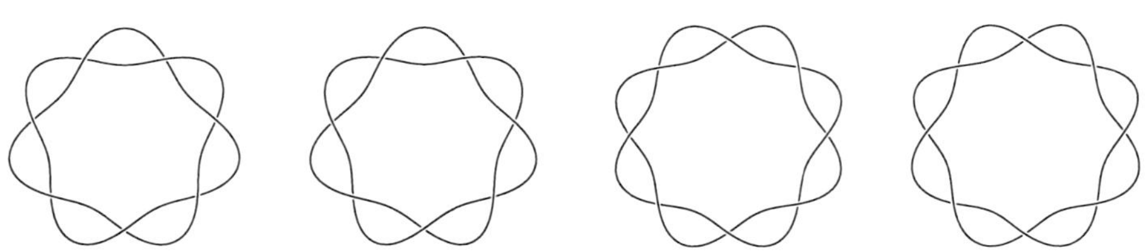

7_1 7 8^2_1 8

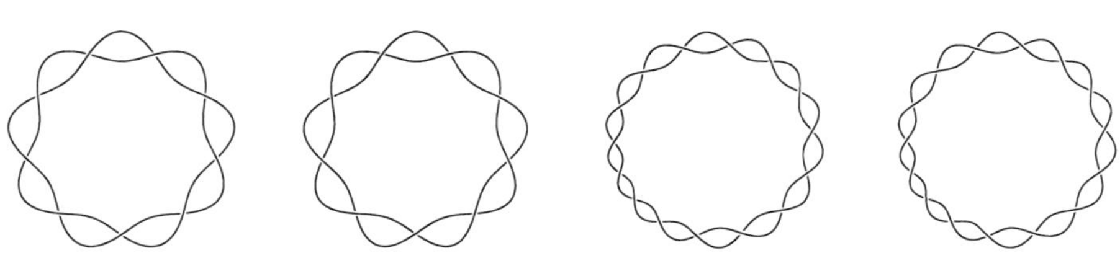

9_1 9 15_1 15

$(2, q)$ torus knots and links

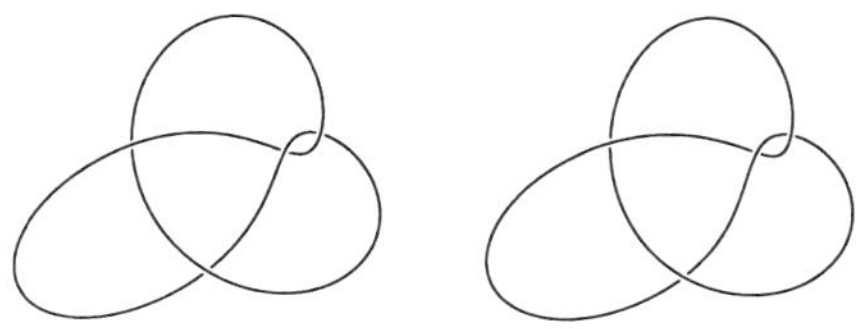

4_1 22

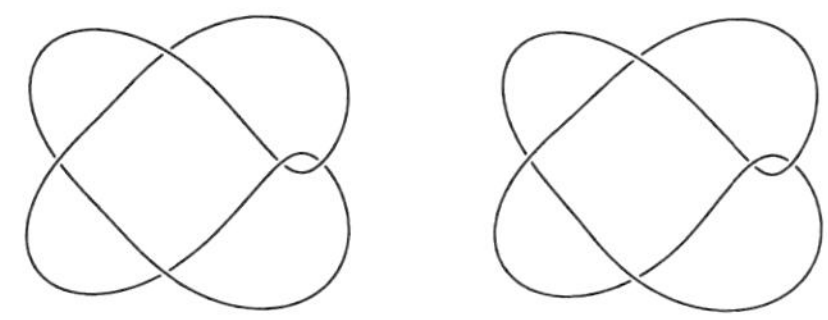

5_2 32

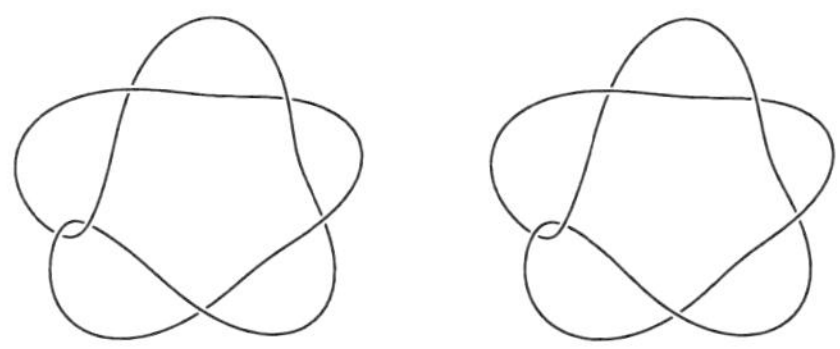

6_1 42

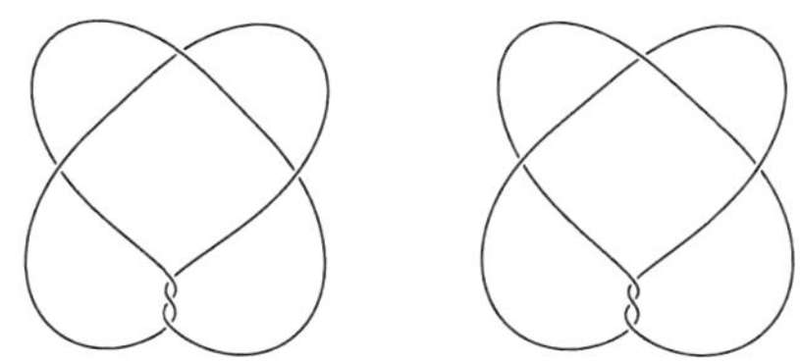

6_2^2 33

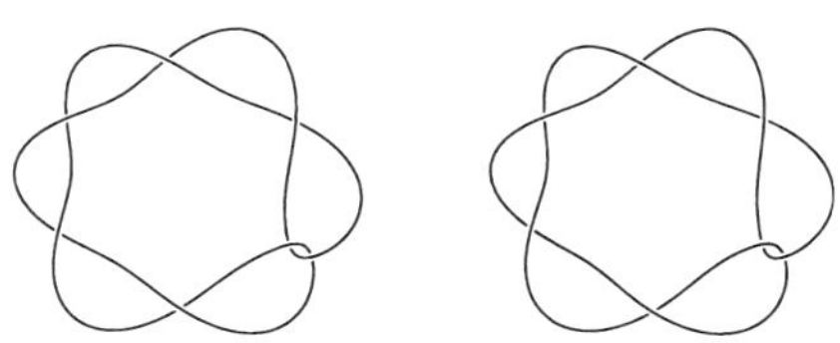

7_2 52

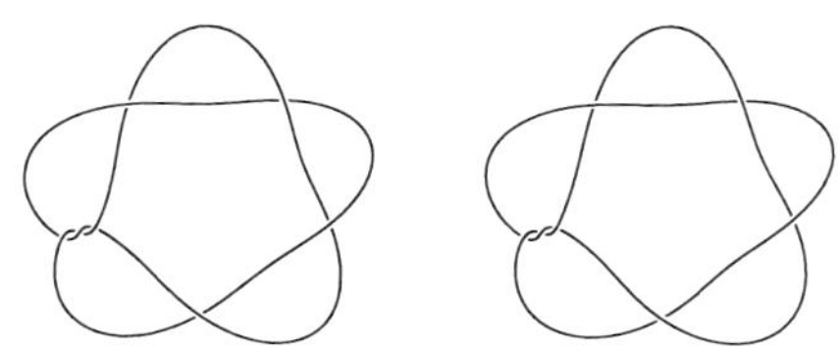

7_3 43

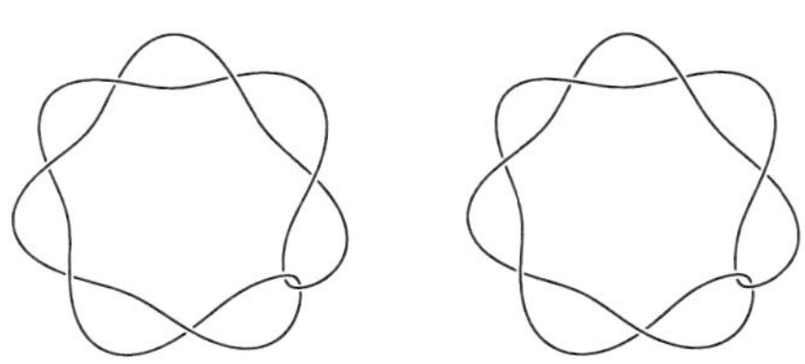

8_1 62

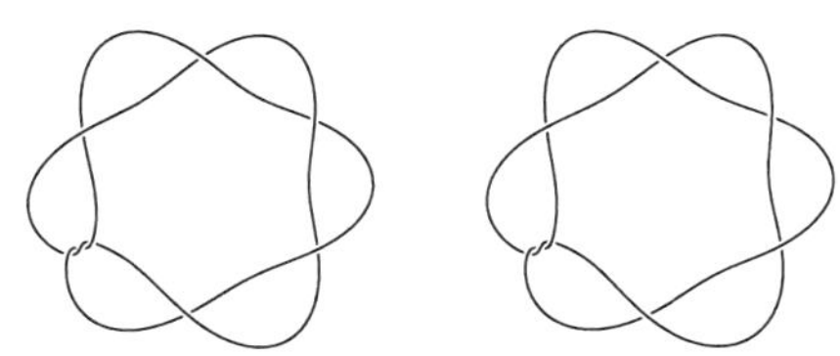

8_2^2 53

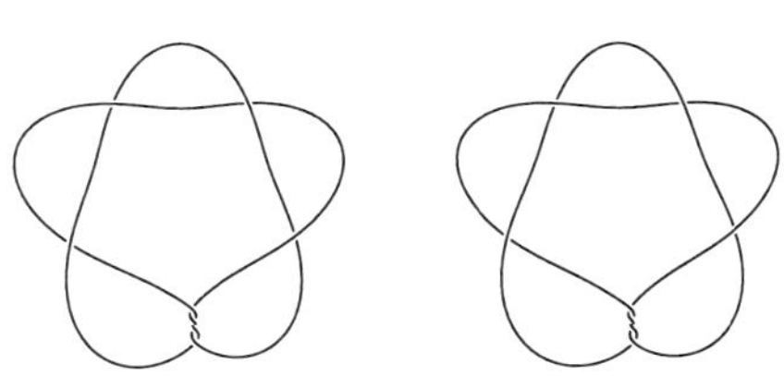

8_3 44

Rational links with two-term continued fractions

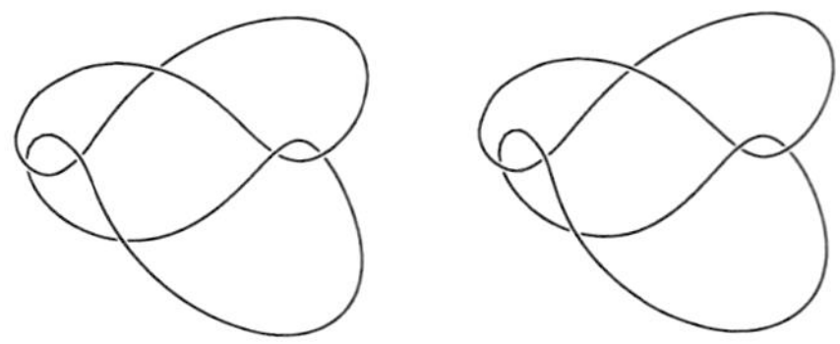

5^2_1 212

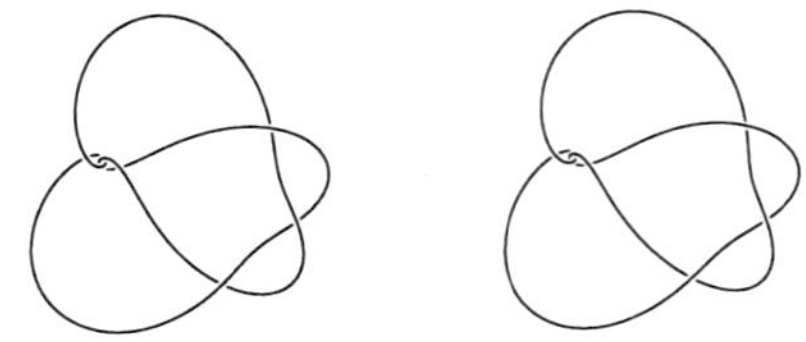

6_2 312

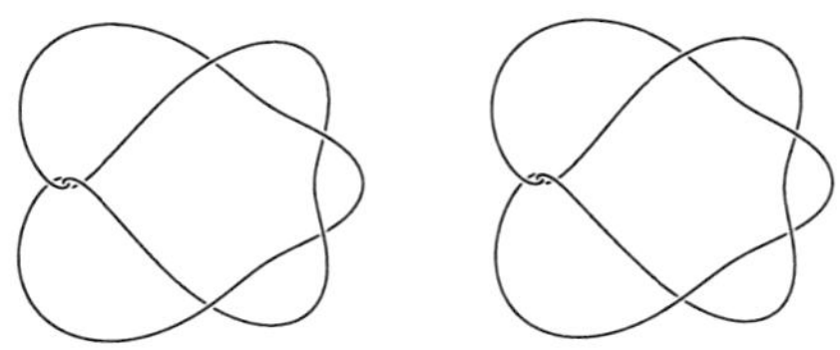

7^2_1 412

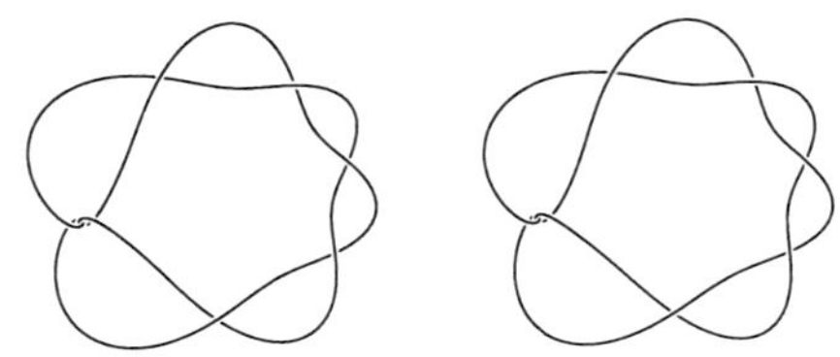

8_2 512

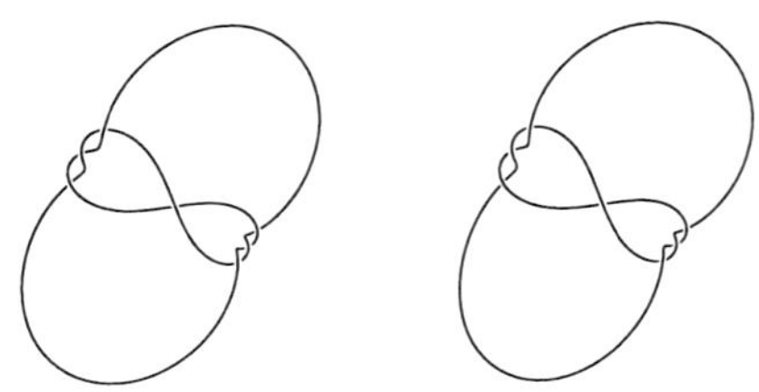

7_4 313

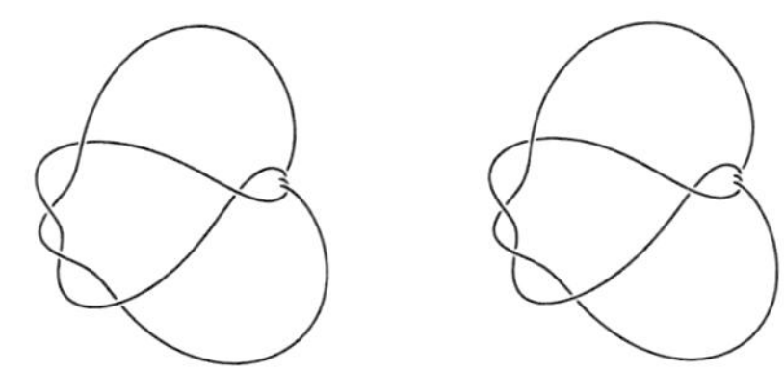

8_4 413

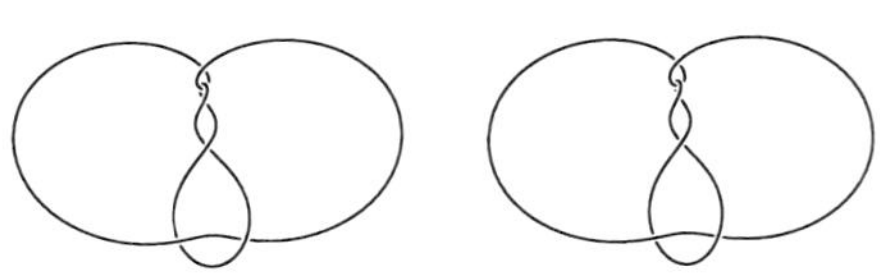

6^2_3 222

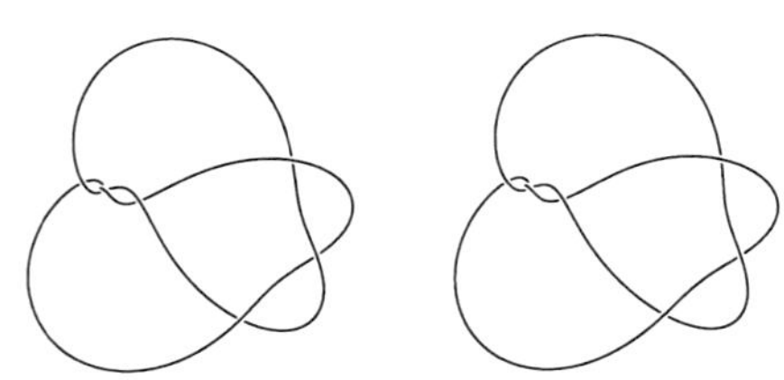

7_5 322

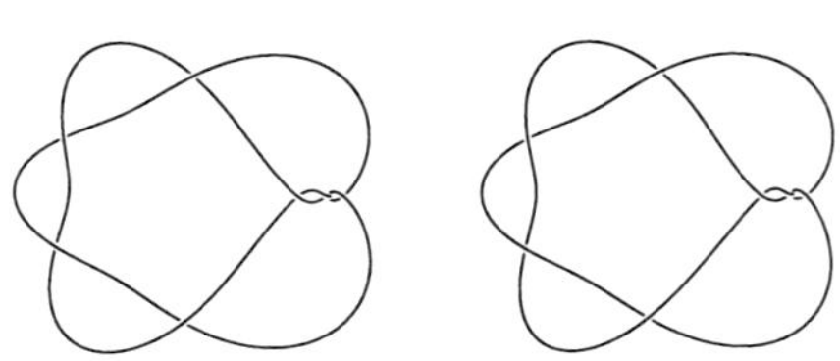

8^2_3 422

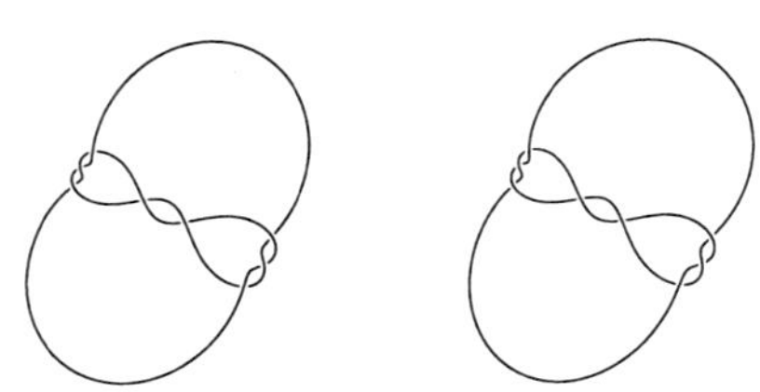

8^2_4 323

Most rational links with three-term continued fractions

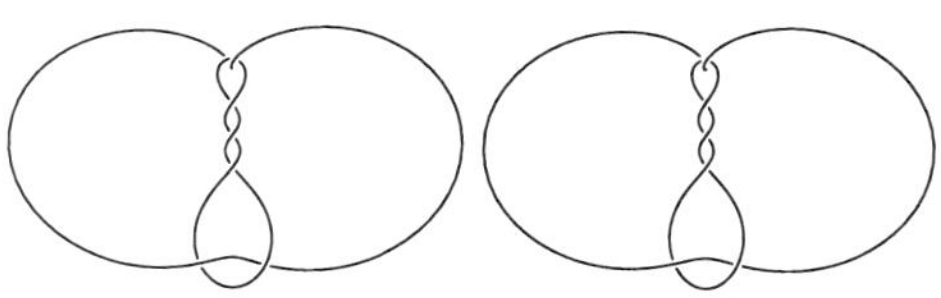

7_3^2 232

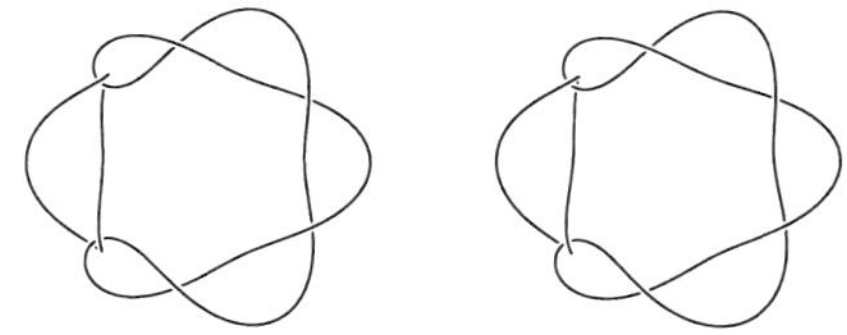

8_6^2 242

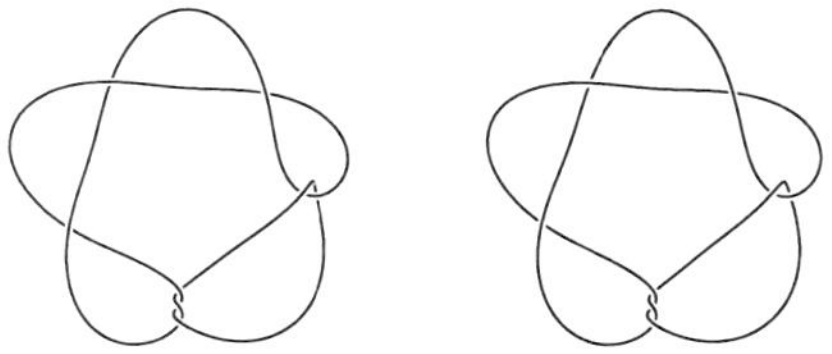

8_6 332

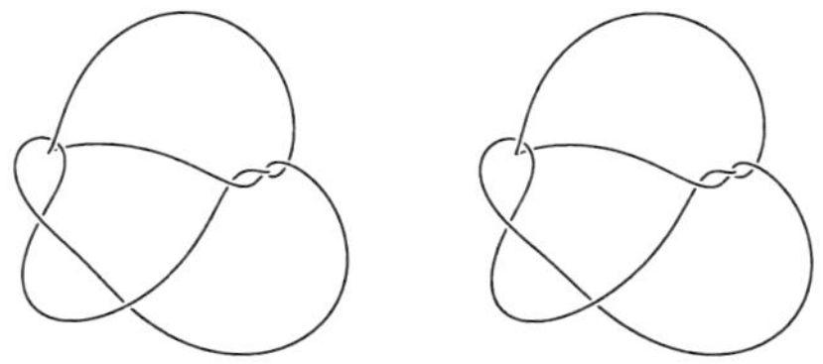

8_{12} 2222

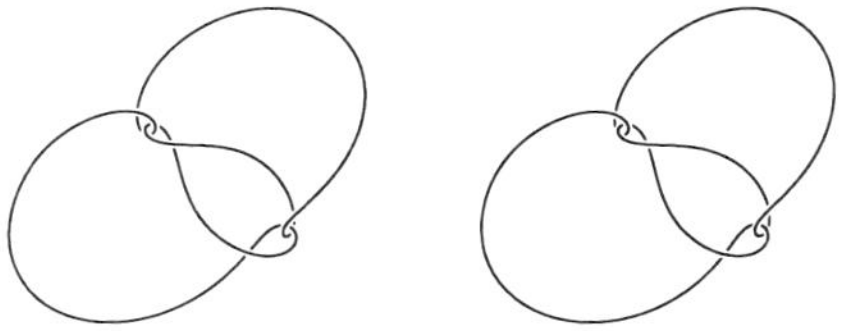

6_3 2112

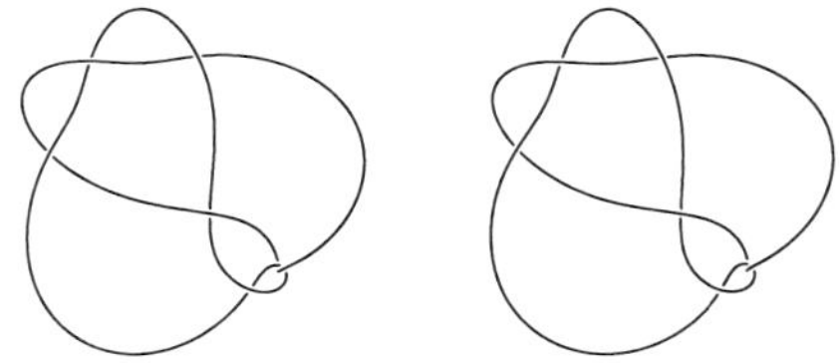

7_2^2 3112

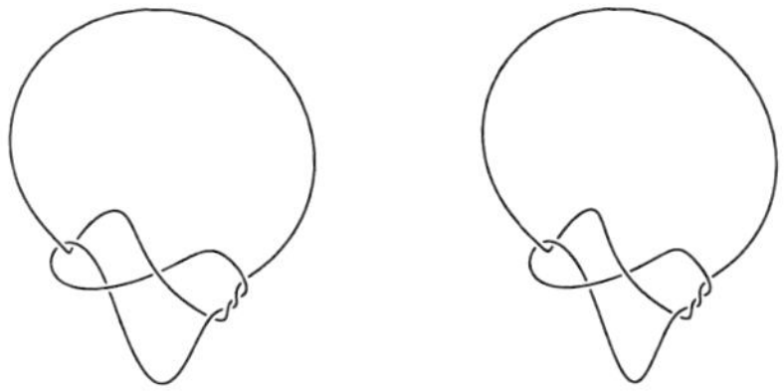

8_7 4112

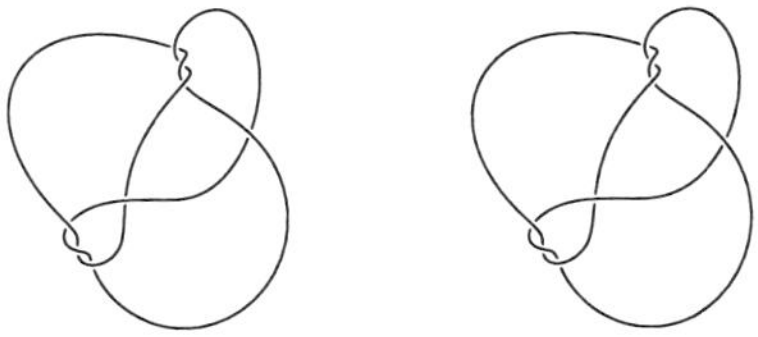

8_9 3113

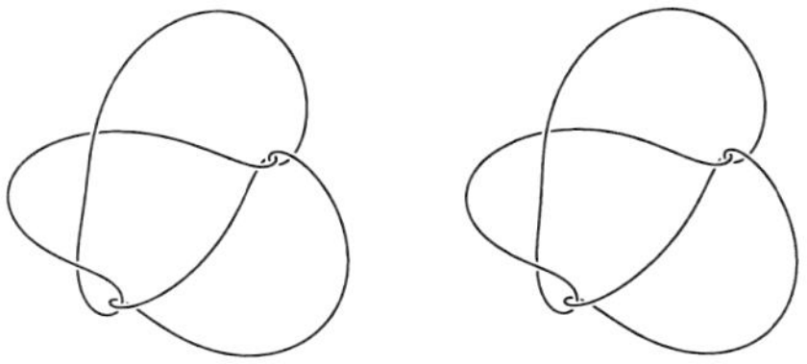

7_6 2212

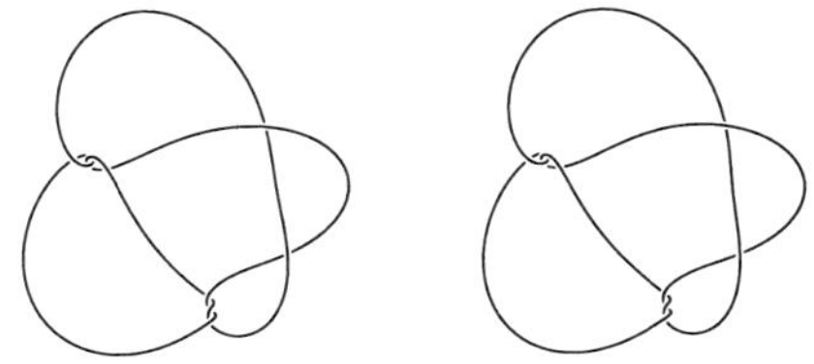

8_{11} 3212

More rational links with three- or four-term continued fractions

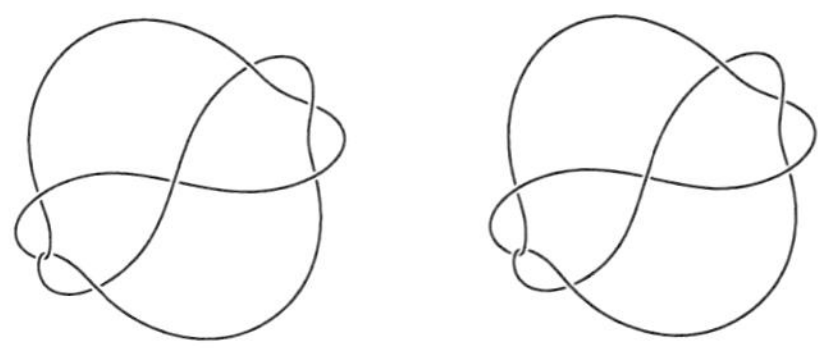

8^2_5 3122

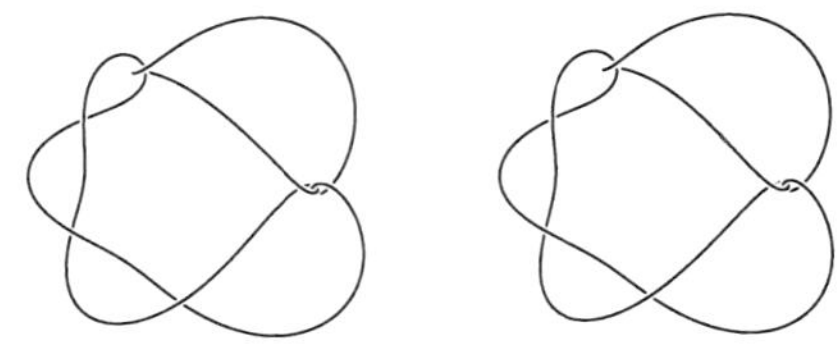

8_8 2312

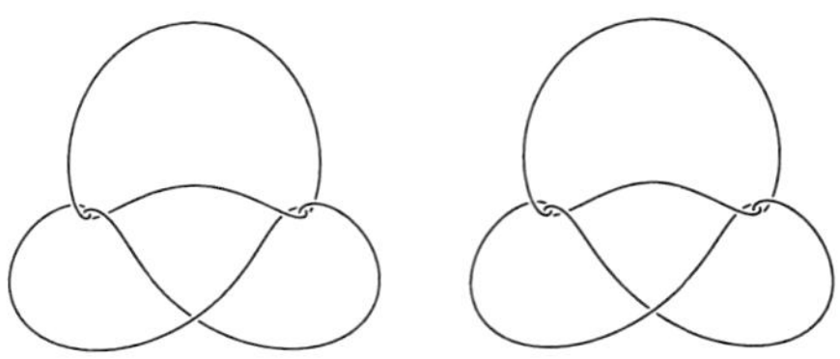

7_7 21112

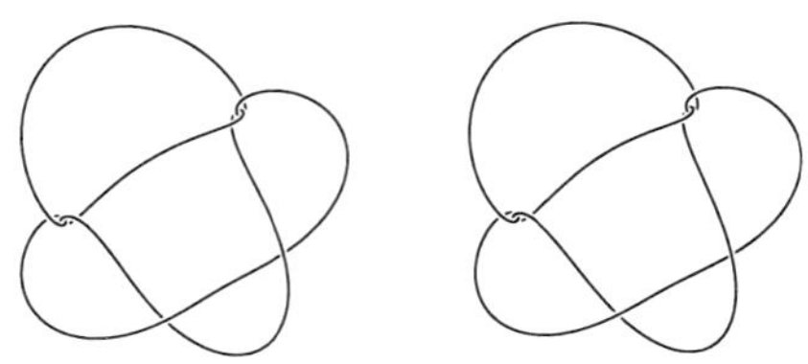

8^2_7 21212

8_{13} 31112

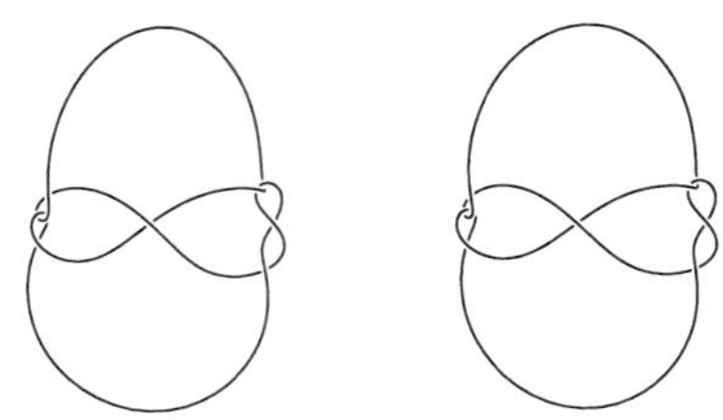

8_{14} 22112

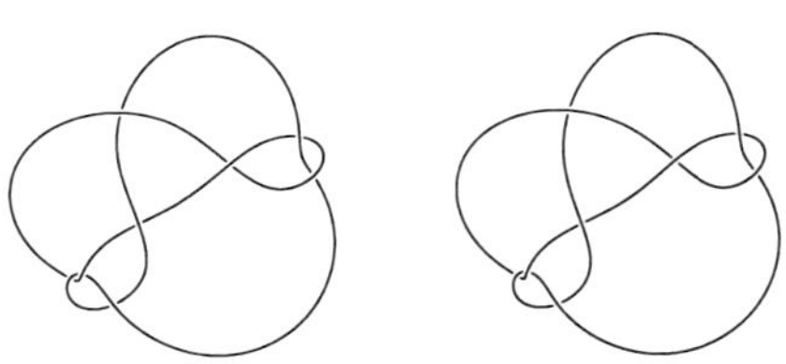

8^2_8 211112

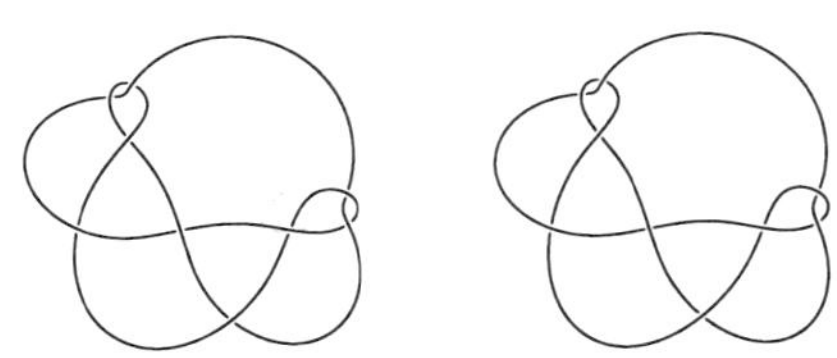

9_{31} 2111112

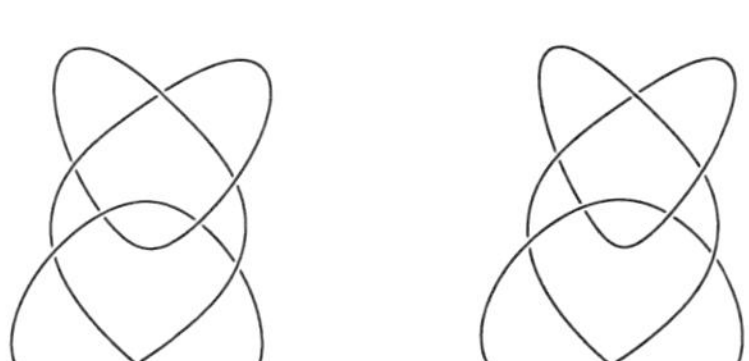

8_{19} 3,3,2-

Rational links with long continued fractions, and the $(3,4)$ torus knot

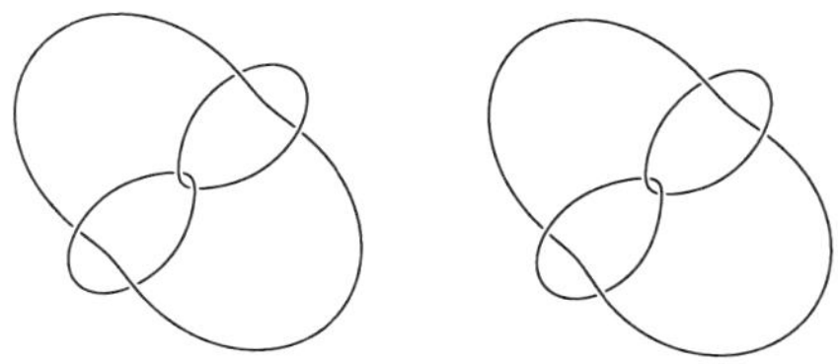

6^3_1 2,2,2

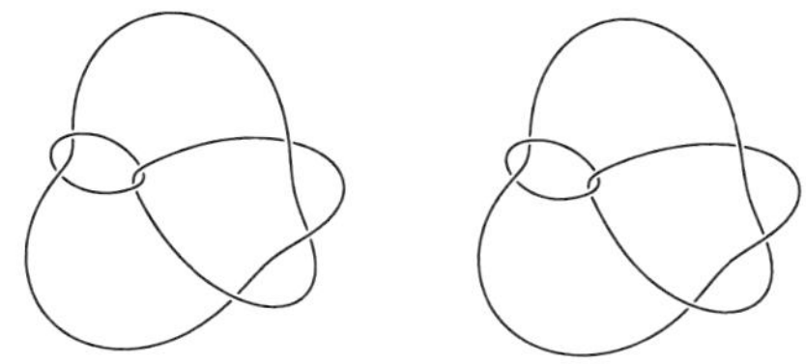

7^2_4 3,2,2

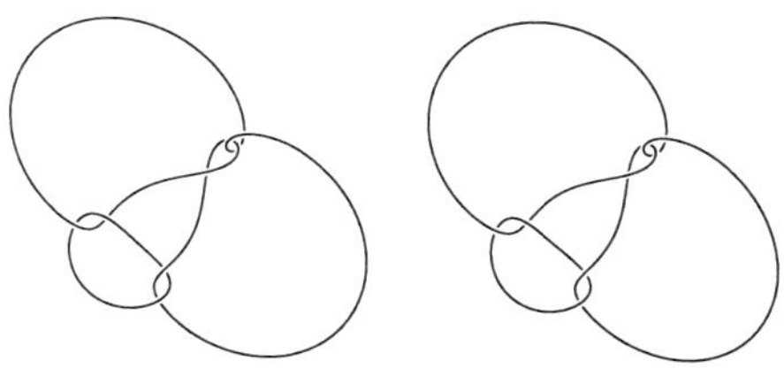

7^2_5 21,2,2

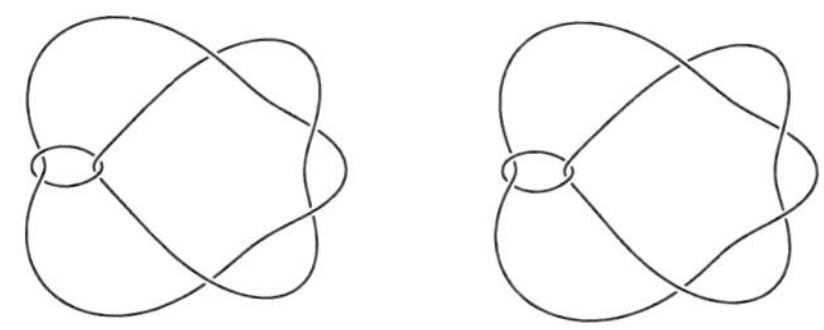

8^3_1 4,2,2

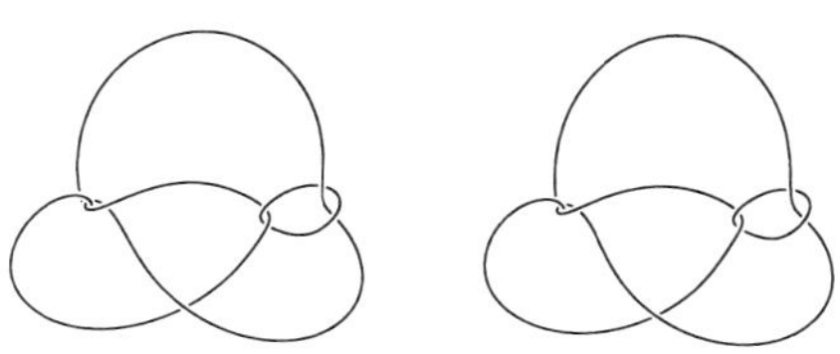

8^2_{10} 211,2,2

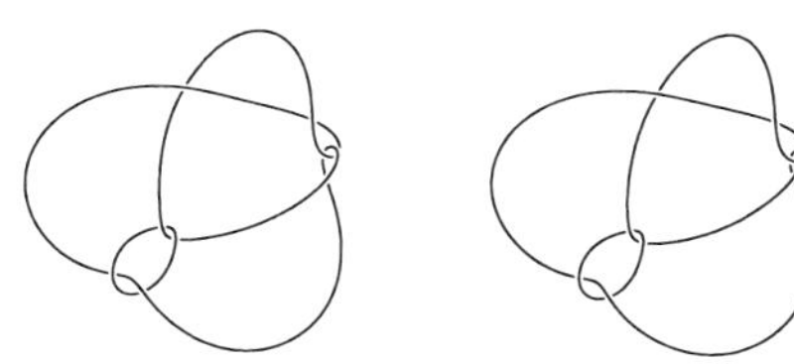

8^2_9 22,2,2

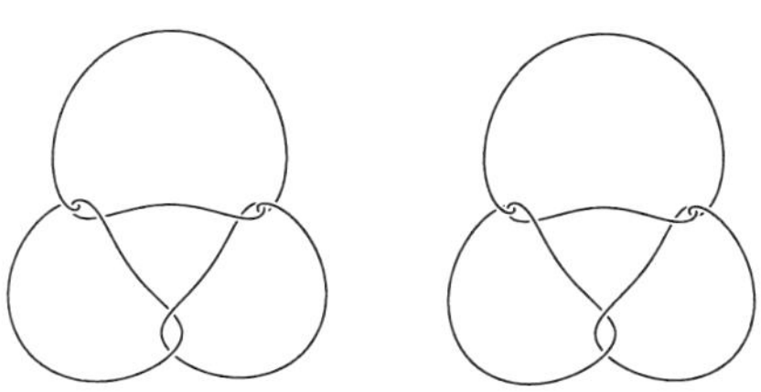

8_{15} 21,21,2

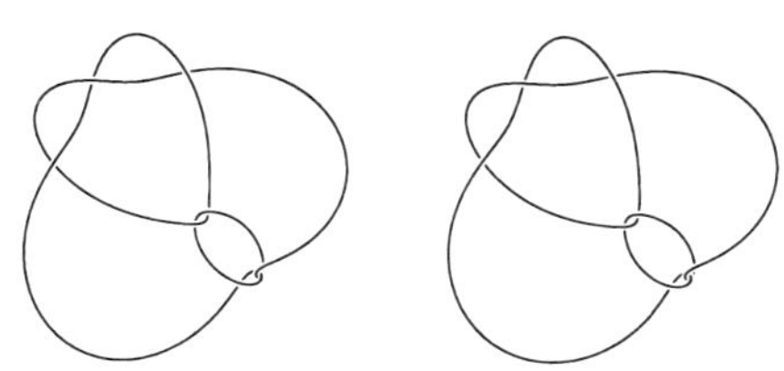

8_{10} 3,21,2

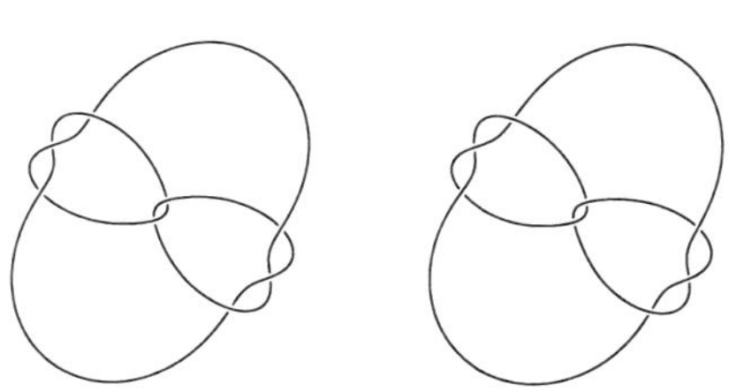

8_5 3,3,2

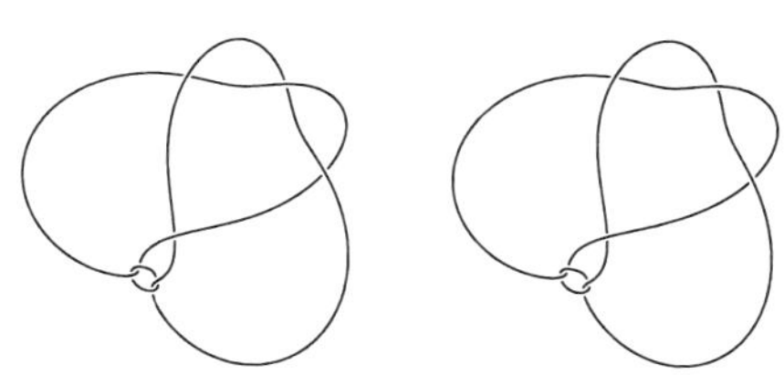

8^3_2 31,2,2

Links with Conway notation a, b, c

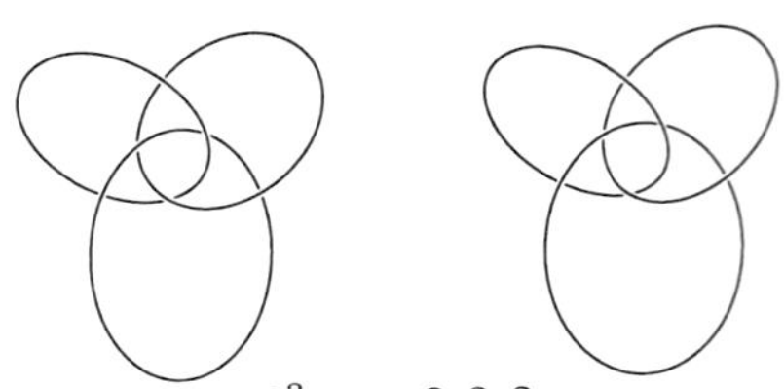

6^3_3 2,2,2−

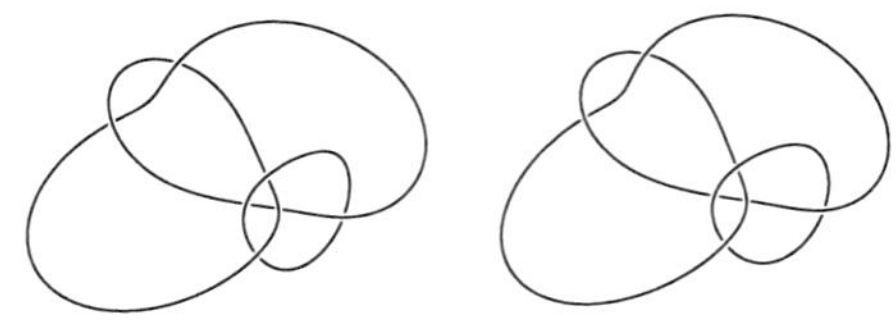

7^2_8 21,2,2−

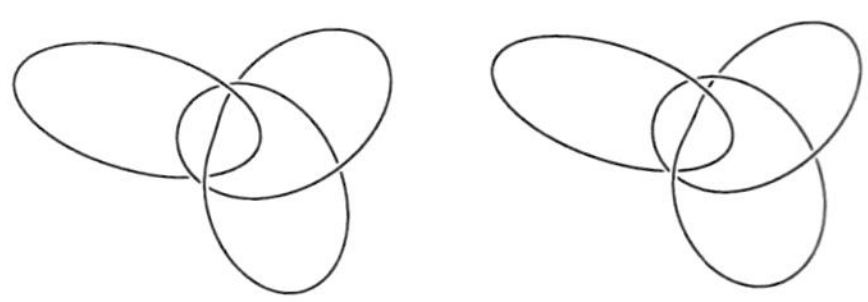

7^2_7 3,2,2−

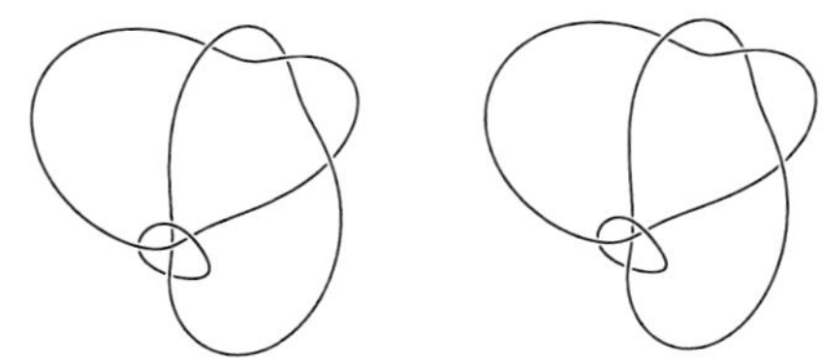

8^3_8 31,2,2−

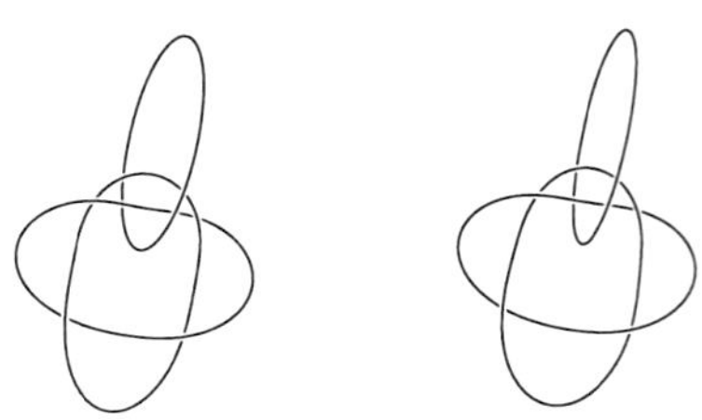

8^3_7 4,2,2−

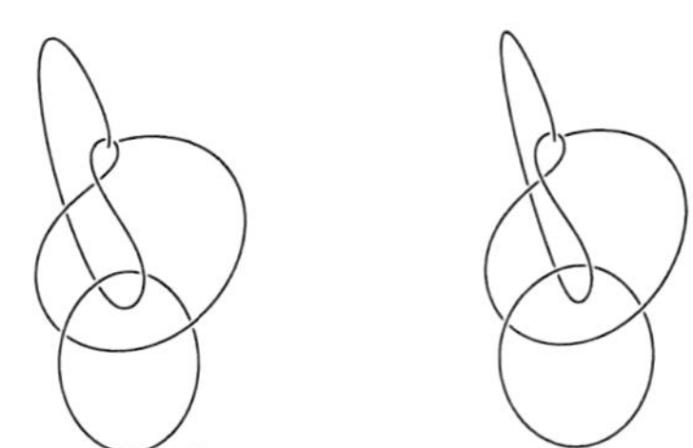

8^2_{16} 211,2,2−

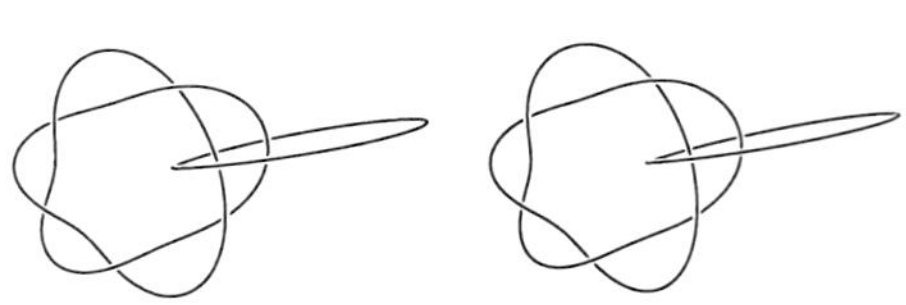

9^2_{43} 5,2,2−

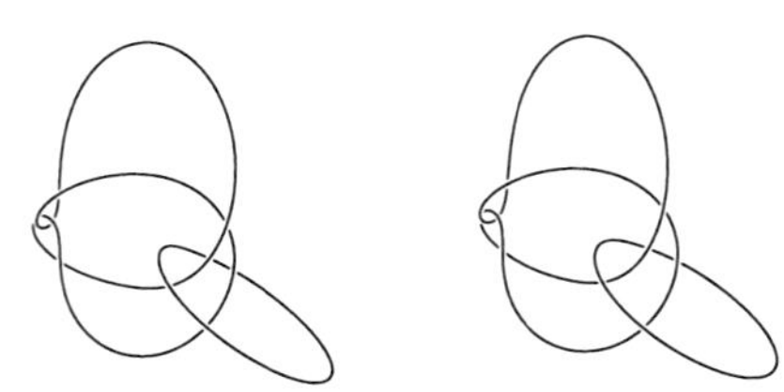

8^2_{15} 22,2,2−

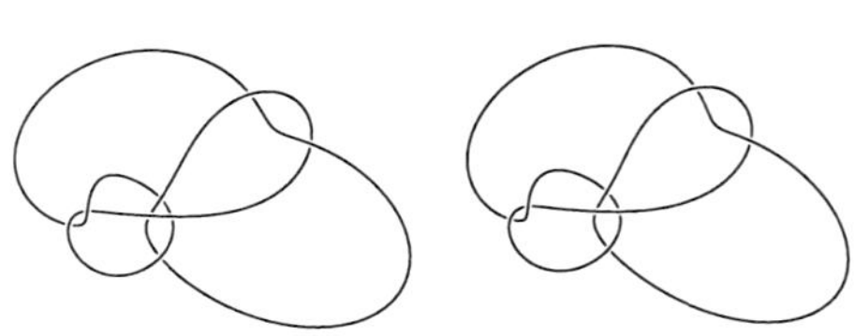

8_{21} 21,21,2−

8_{20} 3,21,2−

Most nonalternating links

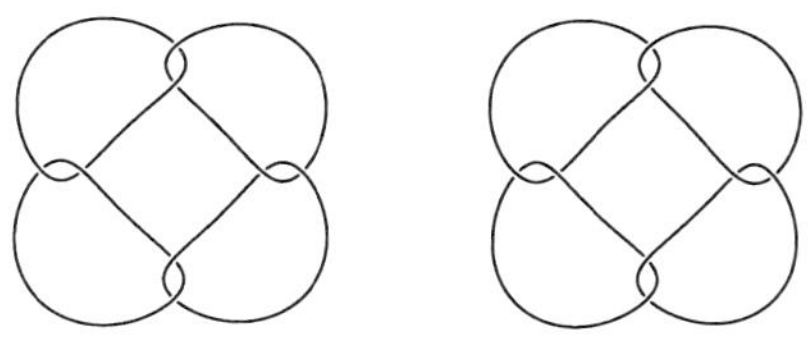

8^4_1 2,2,2,2

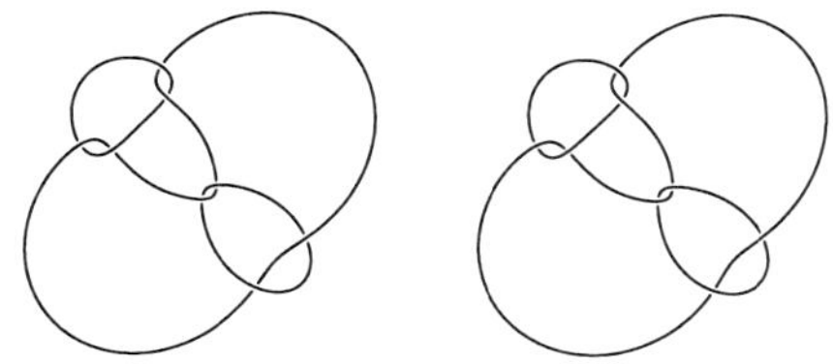

8^3_4 (2,2)(2,2)

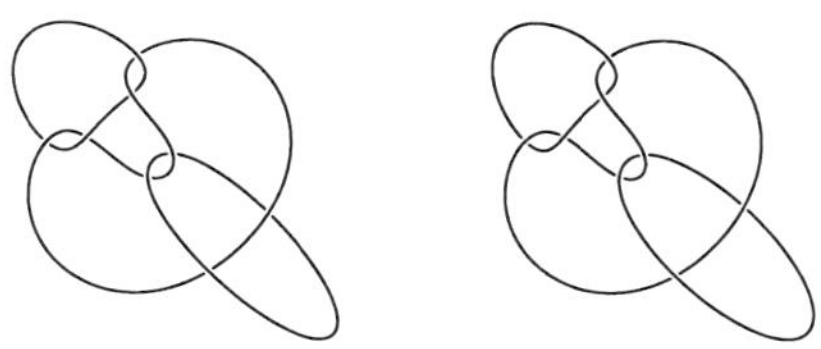

8^3_9 (2,2)(2,2−)

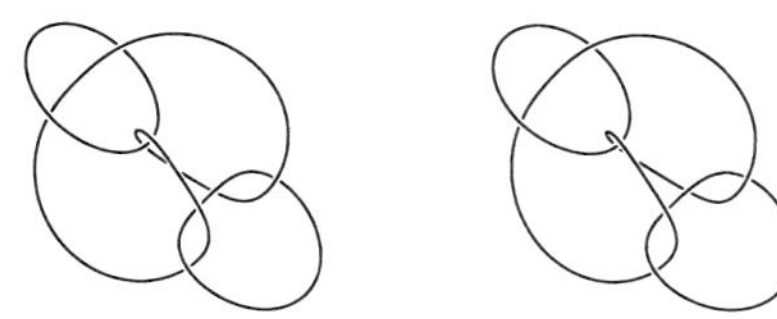

8^3_{10} (2,2)−(2,2)

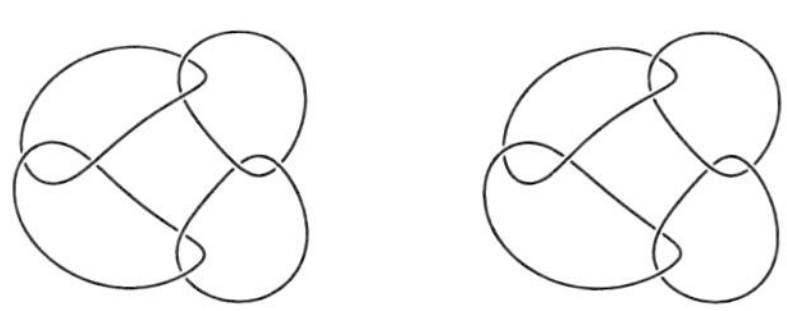

8^4_2 2,2,2,2−

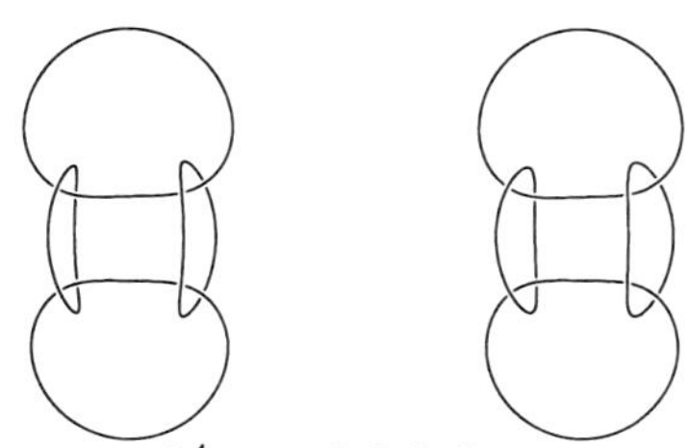

8^4_3 2,2,2,2−−

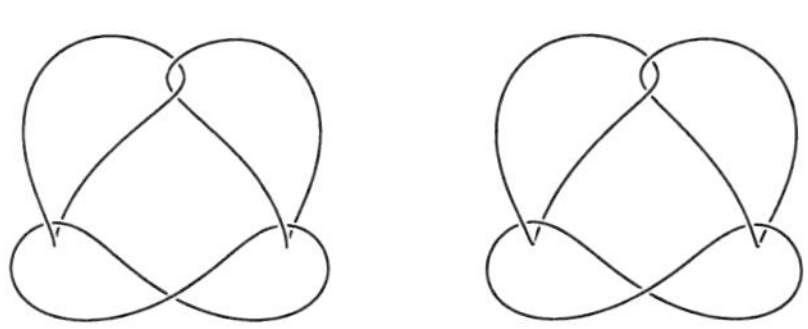

7^3_1 2,2,2+

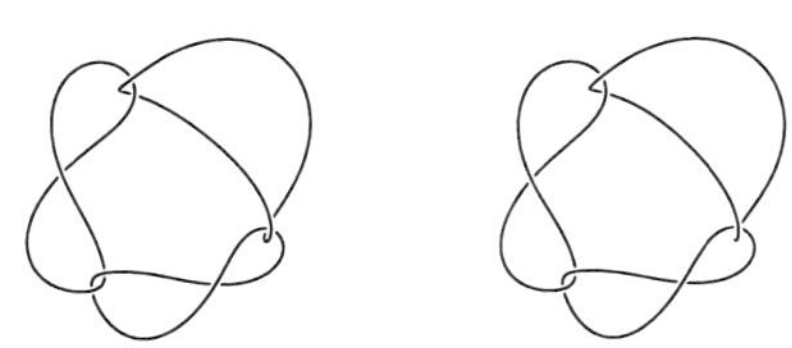

8^3_3 2,2,2++

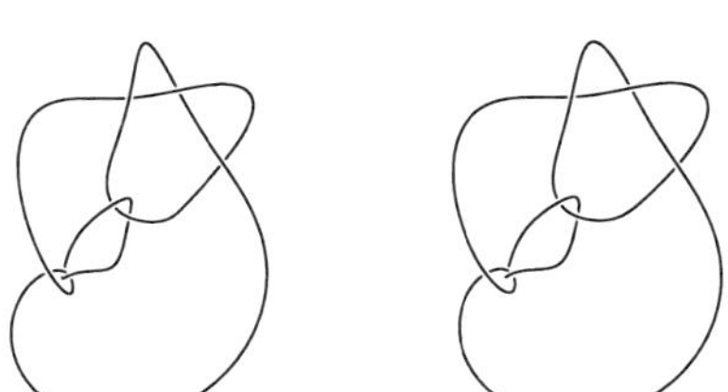

8^2_{11} 3,2,2+

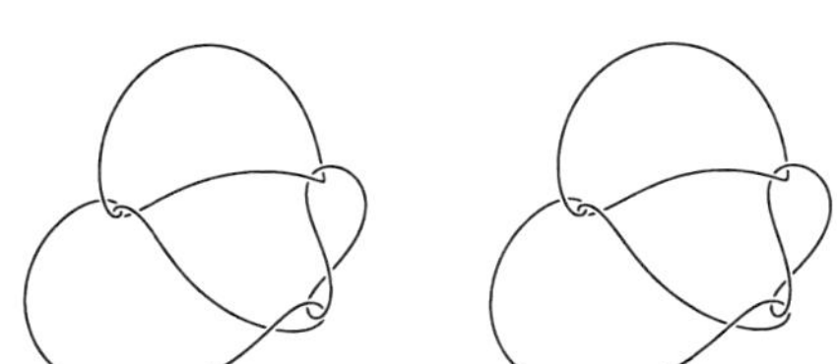

8^2_{12} 21,2,2+

The remaining links based on 1*

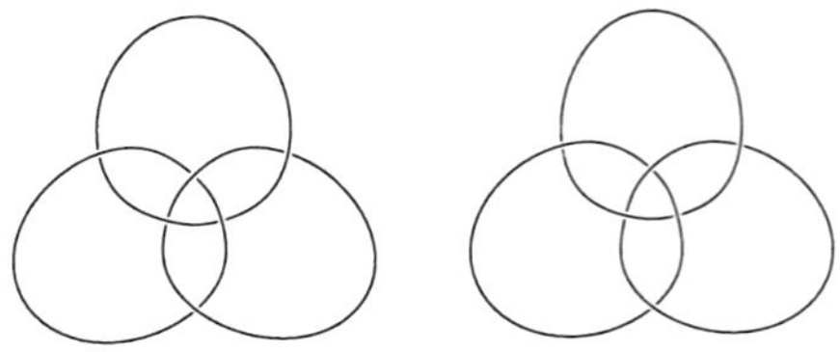

6^3_2 6*

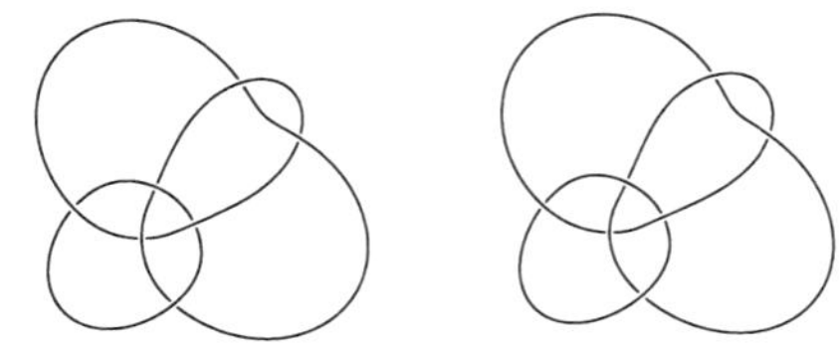

7^2_6 6*2

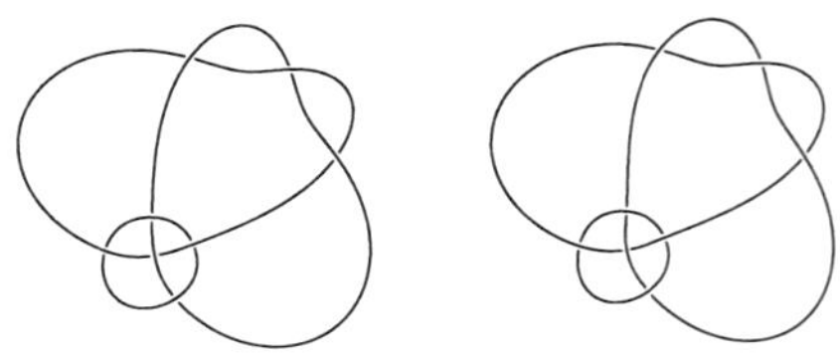

8^3_5 6*3

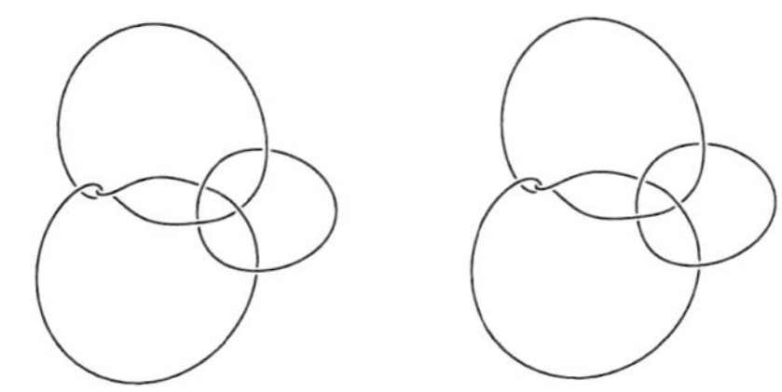

8^2_{13} 6*21

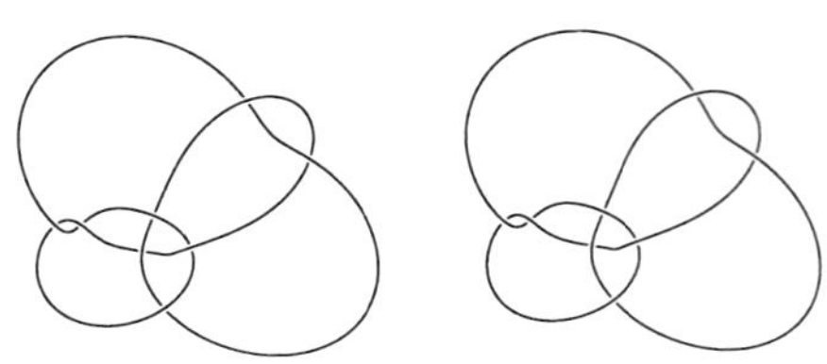

8_{17} 6*2.2

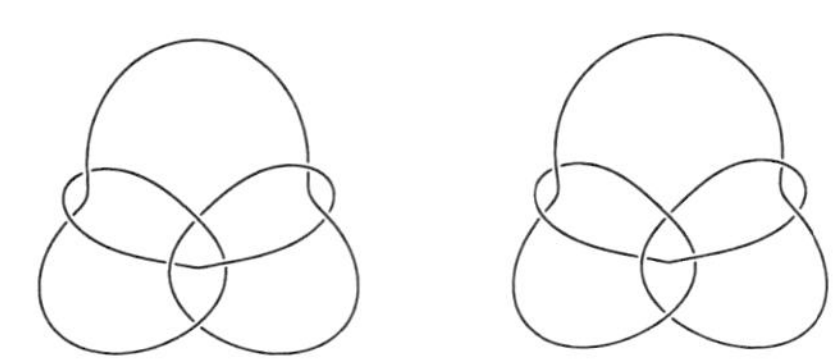

8_{16} 6*2.20

8^2_{14} 6*2:.2

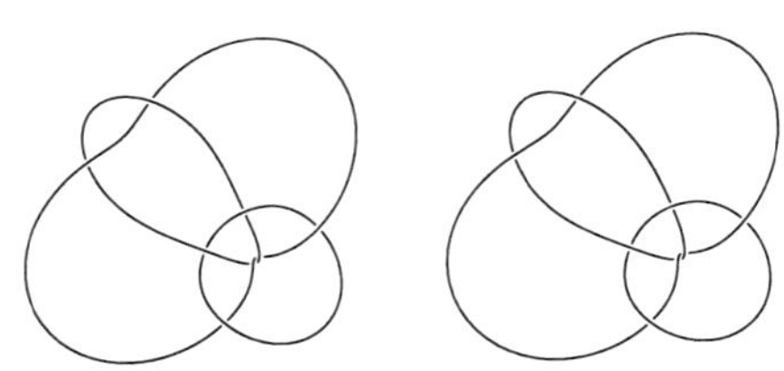

8^3_6 6*2:.20

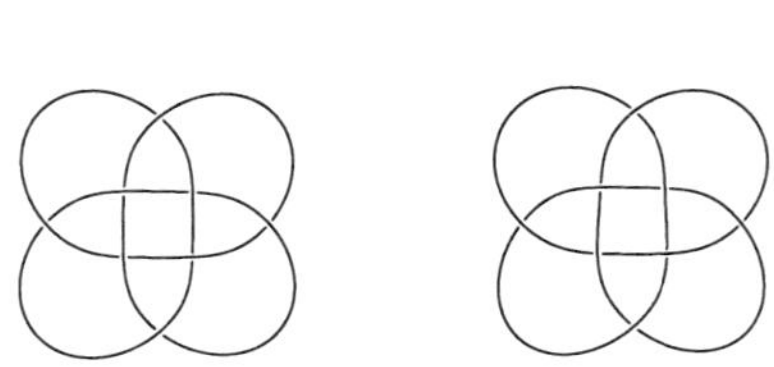

8_{18} 8*

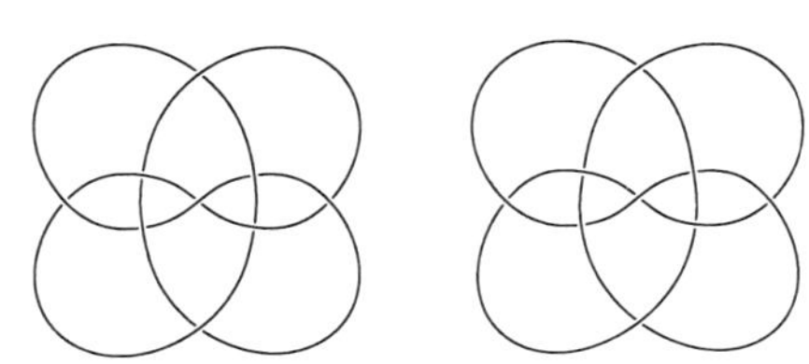

9_{40} 9*

The links based on 6*, 8* or 9*

References

1. Colin C. Adams, *The Knot Book*, W. H. Freeman, 1994.
2. Kazushi Ahara, *The Numerical Analysis of an Energy of a Knot*, preprint with video, Univ. of Tokyo.
3. David Auckly and Lorenzo Sadun, *A Family of Möbius Invariant 2-Knot Energies*, in these proceedings.
4. Kenneth A. Brakke, *The Surface Evolver*, Exper. Math. **1** (1992), 141–165.
5. Steve Bryson, Michael H. Freedman, Zheng-Xu He, and Zhenghan Wang, *Möbius Invariance of Knot Energy*, Bull. AMS **28** (1993), 99–103.
6. Greg Buck and J. Orloff, *Computing Canonical Conformations for Knots*, Topology Appl. **51** (1993), 247–253.
7. Greg Buck and Jonathan Simon, *Knots as Dynamical Systems*, Topology Appl. **51** (1993), 229–246.
8. Gerhard Burde and Heiner Zieschang, *Knots*, Studies in Math., de Gruyter, Berlin, 1985.
9. John H. Conway, *An Enumeration of Knots and Links and Some of Their Related Properties*, Computational Problems in Abstract Algebra (J. Leech, ed.), Pergamon, 1970, Proc. Conf. Oxford, 1967, pp. 329–358.
10. Peter Doyle and Oded Schramm, personal communications.
11. L. Föppl, *Stabile Anordnungen von Elektronen im Atom*, J. reine angew. Math. **141** (1912), 251–302.
12. Michael H. Freedman, Zheng-Xu He, and Zhenghan Wang, *On the Möbius Energy of Knots and Unknots*, Annals of Math. **139** (1994), no. 1, 1–50.
13. Shinji Fukuhara, *Energy of a Knot*, A Fête of Topology (Y. Matsumoto, T. Mizutani, and S. Morita, eds.), Academic Press, 1988, pp. 443–451.
14. Charlie Gunn, personal communication.
15. Allen E. Hatcher, *A Proof of the Smale Conjecture* $\mathit{Diff}(S^3) \simeq O(4)$, Annals of Math. **117** (1983), 553–607.
16. Nariya Kawazumi, *On the Homotopy Type of the Moduli Space of n-Point Sets of P^1*, J. Fac. Sci., Univ. Tokyo, IA **37** (1990), 263–287.
17. Denise Kim and Rob Kusner, *Torus Knots Extremizing the Möbius Energy*, Experimental Math. **2** (1993), 1–9.
18. Rob Kusner, Peter Norman, and John Sullivan, *Möbius Energy Minimizing Knot Catalog*, Preprint III.18, GANG/UMass Amherst, 1993.
19. Robert B. Kusner and John M. Sullivan, *Möbius Energy of Hopf Links, and Electrons on the Sphere*, in preparation.
20. Robert B. Kusner and John M. Sullivan, *A Nonlocal, Möbius-Invariant Energy for Embedded Submanifolds*, in preparation.
21. Jun O'Hara, *Energy of a Knot*, Topology **30** (1991), 241–247.
22. Dale Rolfsen, *Knots and Links*, Publish or Perish, Wilmington, 1976.
23. John M. Sullivan, *Knots Minimizing a Möbius-Invariant Energy*, VHS video tape (Nov. 1993) available from the Geometry Center, Minneapolis.

University of Massachusetts, Amherst, MA 01003
E-mail address: Kusner@GANG.UMass.Edu

University of Minnesota, Minneapolis, MN 55455
E-mail address: Sullivan@Geom.UMn.Edu